U0229389

NONLINEAR PHYSICAL SCIENCE
非线性物理科学

NONLINEAR PHYSICAL SCIENCE

Nonlinear Physical Science focuses on recent advances of fundamental theories and principles, analytical and symbolic approaches, as well as computational techniques in nonlinear physical science and nonlinear mathematics with engineering applications.

Topics of interest in *Nonlinear Physical Science* include but are not limited to:

- New findings and discoveries in nonlinear physics and mathematics
- Nonlinearity, complexity and mathematical structures in nonlinear physics
- Nonlinear phenomena and observations in nature and engineering
- Computational methods and theories in complex systems
- Lie group analysis, new theories and principles in mathematical modeling
- Stability, bifurcation, chaos and fractals in physical science and engineering
- Discontinuity, synchronization and natural complexity in physical sciences
- Nonlinear chemical and biological physics

Albert C. J. Luo

Bifurcation Dynamics in Polynomial Discrete Systems

多项式离散系统的分岔动力学

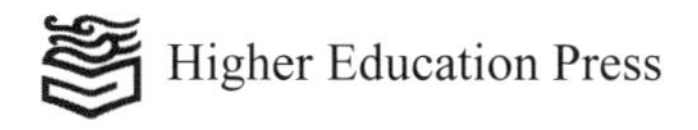
Higher Education Press

Author
Albert C. J. Luo
Department of Mechanical and Industrial Engineering
Southern Illinois University Edwardsville
Edwardsville, IL
USA

图书在版编目（CIP）数据

多项式离散系统的分岔动力学 = Bifurcation Dynamics in Polynomial Discrete Systems : 英文 / (美) 罗朝俊 (Albert C. J. Luo) 著 . -- 北京 : 高等教育出版社 , 2021.5
（非线性物理科学）
ISBN 978-7-04-055783-1

Ⅰ. ①多… Ⅱ. ①罗… Ⅲ. ①离散系统 – 英文 Ⅳ. ①O231

中国版本图书馆 CIP 数据核字（2021）第 043112 号

多项式离散系统的分岔动力学
DUOXIANGSHI LISANXITONG DE FENCHA DONGLIXUE

策划编辑 吴晓丽　　责任编辑 吴晓丽　　封面设计 杨立新
责任印制 赵义民

出版发行 高等教育出版社
社　　址 北京市西城区德外大街4号
邮政编码 100120
印　　刷 北京中科印刷有限公司
开　　本 787 mm× 1092 mm　1/16
印　　张 28
字　　数 720 千字
购书热线 010-58581118
咨询电话 400-810-0598
网　　址 http://www.hep.edu.cn
http://www.hep.com.cn
网上订购 http://www.hepmall.com.cn
http://www.hepmall.com
http://www.hepmall.cn
版　　次 2021 年 5 月第 1 版
印　　次 2021 年 5 月第 1 次印刷
定　　价 199.00 元

本书如有缺页、倒页、脱页等质量问题，请到所购图书销售部门联系调换

物 料 号 55783-00

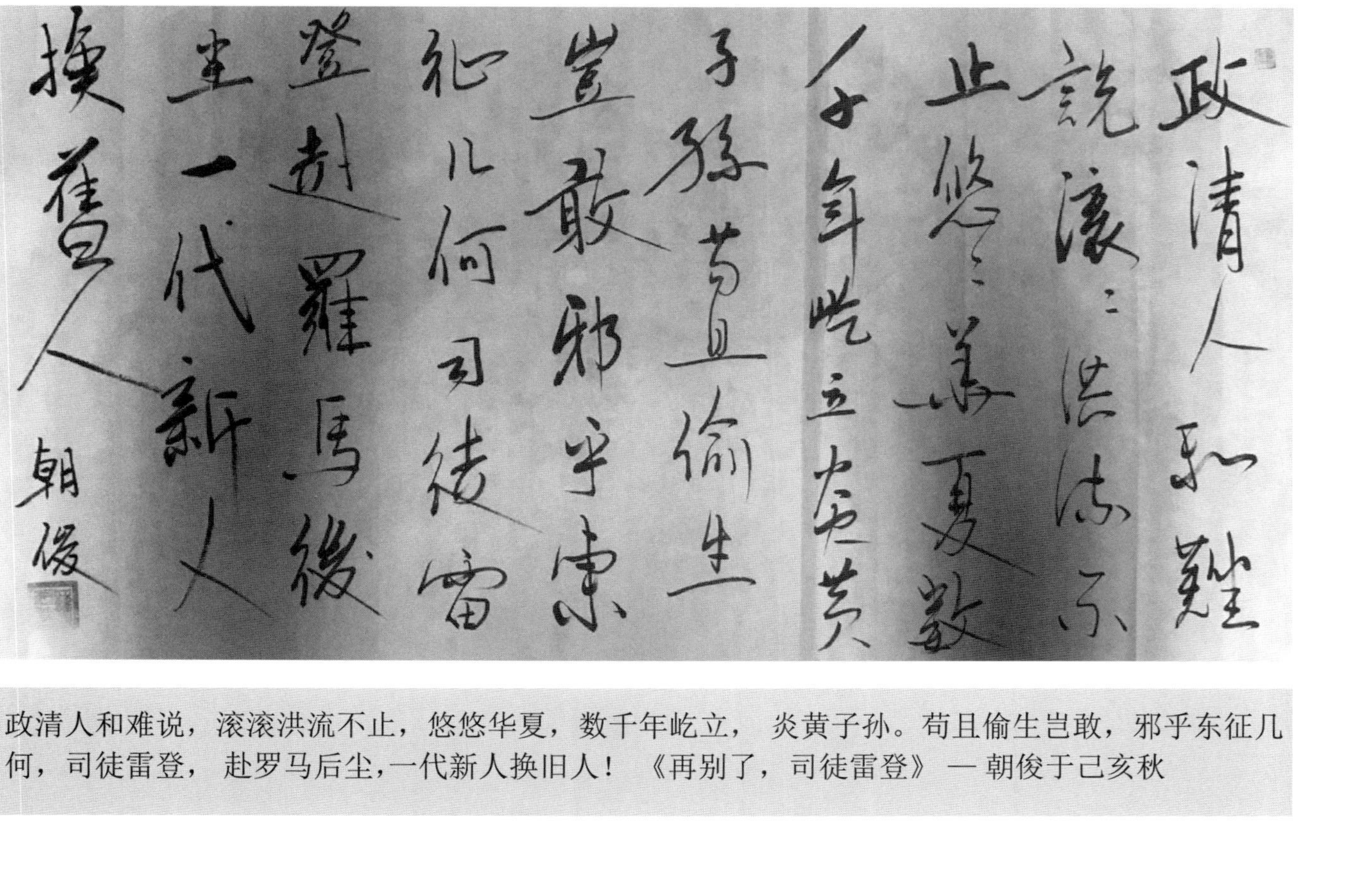

政清人和难说，滚滚洪流不止，悠悠华夏，数千年屹立， 炎黄子孙。苟且偷生岂敢，邪乎东征几何，司徒雷登， 赴罗马后尘，一代新人换旧人！ 《再别了，司徒雷登》 一 朝俊于己亥秋

Preface

This book is the second part for bifurcation and stability of nonlinear discrete systems. The first part mainly presents a local theory for monotonic and oscillatory stability and bifurcations of nonlinear discrete systems, and such monotonic and oscillatory stability and bifurcations on specific eigenvectors of the corresponding linearized discrete systems are discussed. In this book, the bifurcation dynamics of one-dimensional polynomial nonlinear discrete systems is presented and bifurcation trees caused by period-doubling and monotonic saddle-node bifurcations are discussed for forward and backward polynomial discrete systems. The mechanism of bifurcation trees caused by monotonic saddle-node bifurcations is determined. The appearing and switching bifurcations of simple and higher-order period-1 fixed-points are discussed in this book. From this book, one will find more interesting research results in nonlinear discrete systems.

This book consists of five chapters. In Chap. 1, a global bifurcation theory for quartic polynomial discrete systems is presented, and the bifurcation trees through period-doubling and monotonic saddle-node bifurcations are discussed for forward and backward quadratic discrete systems. In Chap. 2, a global bifurcation theory for cubic polynomial discrete systems is discussed, and the bifurcation trees through the period-doubling and monotonic saddle-node bifurcation are also presented, which is different from the quadratic discrete system. In Chap. 3, a global bifurcation theory for quartic polynomial discrete systems is presented for extension to the $(2m)$th degree polynomial discrete systems. The bifurcation and stability of the $(2m)$th and $(2m+1)$th degree polynomial discrete systems are presented in Chaps. 4 and 5 as a general theory of stability and bifurcations for polynomial nonlinear discrete systems.

Finally, the author hopes the materials presented herein can last long for science and engineering. Some typos and errors may exist in the book, which can be corrected by readers during reading. Herein, the author would like to thank all supporting people during the difficult time period.

Edwardsville, IL, USA Albert C. J. Luo

Contents

1 Quadratic Nonlinear Discrete Systems 1
1.1 Linear Discrete Systems 1
1.2 Forward Quadratic Discrete Systems 7
1.2.1 Period-1 Appearing Bifurcations 7
1.2.2 Period-1 Switching Bifurcations 21
1.3 Backward Quadratic Discrete Systems 28
1.3.1 Backward Period-1 Appearing Bifurcations 28
1.3.2 Backward Period-1 Switching Bifurcations 39
1.4 Forward Bifurcation Trees 44
1.4.1 Period-2 Appearing Bifurcations 44
1.4.2 Period-Doubling Renormalization 53
1.4.3 Period-*n* Appearing and Period-Doublization 62
1.4.4 Period-*n* Bifurcation Trees 71
1.5 Backward Bifurcation Trees 75
1.5.1 Backward Period-2 Quadratic Discrete Systems 75
1.5.2 Backward Period-Doubling Renormalization 79
1.5.3 Backward Period-*n* Appearing and Period-Doublization . . . 82
1.5.4 Backward Period-*n* Bifurcation Trees 92
References 92

2 Cubic Nonlinear Discrete Systems 93
2.1 Period-1 Cubic Discrete Systems 93
2.2 Period-1 to Period-2 Bifurcation Trees 104
2.3 Higher-Order Period-1 Switching Bifurcations 114
2.4 Direct Cubic Polynomial Discrete Systems 117
2.5 Forward Cubic Discrete Systems 121
2.5.1 Period-Doubled Cubic Discrete Systems 121
2.5.2 Period-Doubling Renormalization 128
2.5.3 Period-*n* Appearing and Period-Doublization 138

2.5.4 Sampled Period-n Appearing Bifurcations . . . 147
2.6 Backward Cubic Nonlinear Discrete Systems . . . 150
2.6.1 Backward Period-2 Cubic Discrete Systems . . . 150
2.6.2 Backward Period-Doubling Renormalization . . . 153
2.6.3 Backward Period-n Appearing and Period-Doublization . . . 157
Reference . . . 166

3 Quartic Nonlinear Discrete Systems . . . 167
3.1 Period-1 Appearing Bifurcations . . . 167
3.2 Period-1 to Period-2 Bifurcation Trees . . . 179
3.3 Higher-Order Period-1 Quartic Discrete Systems . . . 188
3.4 Period-1 Switching Bifurcations . . . 200
3.4.1 Simple Period-1 Switching Bifurcations . . . 201
3.4.2 Higher-Order Period-1 Switching Bifurcations . . . 215
3.5 Forward Quartic Discrete Systems . . . 223
3.5.1 Period-2 Quartic Discrete Systems . . . 224
3.5.2 Period-Doubling Renormalization . . . 227
3.5.3 Period-n Appearing and Period-Doublization . . . 231
3.6 Backward Quartic Discrete Systems . . . 239
3.6.1 Backward Period-2 Quartic Discrete Systems . . . 239
3.6.2 Backward Period-Doubling Renormalization . . . 243
3.6.3 Backward Period-n Appearing and Period-Doublization . . . 247
Reference . . . 256

4 $(2m)^{th}$-Degree Polynomial Discrete Systems . . . 257
4.1 Global Stability and Bifurcations . . . 257
4.2 Simple Fixed-Point Bifurcations . . . 276
4.2.1 Appearing Bifurcations . . . 276
4.2.2 Switching Bifurcations . . . 282
4.2.3 Switching-Appearing Bifurcations . . . 289
4.3 Higher-Order Fixed-Point Bifurcations . . . 294
4.3.1 Appearing Bifurcations . . . 294
4.3.2 Switching Bifurcations . . . 305
4.3.3 Switching-Appearing Bifurcations . . . 311
4.4 Forward Bifurcation Trees . . . 318
4.4.1 Period-Doubled $(2m)^{th}$-Degree Polynomial Discrete Systems . . . 318
4.4.2 Renormalization and Period-Doubling . . . 321
4.4.3 Period-n Appearing and Period-Doublization . . . 326
References . . . 334

5 $(2m+1)^{th}$-Degree Polynomial Discrete Systems . . . 335
5.1 Global Stability and Bifurcations . . . 335
5.2 Simple Fixed-Point Bifurcations . . . 354
5.2.1 Appearing Bifurcations . . . 354

5.2.2 Switching Bifurcations 369
5.2.3 Switching-Appearing Bifurcations 374
5.3 Higher-Order Fixed-Point Bifurcations 379
5.3.1 Higher-Order Fixed-Point Bifurcations 379
5.3.2 Switching Bifurcations 398
5.3.3 Switching-Appearing Bifurcations 407
5.4 Forward Bifurcation Trees 410
5.4.1 Period-Doubled $(2m+1)^{\text{th}}$-Degree Polynomial Systems 410
5.4.2 Renormalization and Period-Doubling 415
5.4.3 Period-n Appearing and Period-Doublization 419
References 428

Index 429

Chapter 1
Quadratic Nonlinear Discrete Systems

In this Chapter, the global stability and bifurcation of quadratic nonlinear discrete systems are discussed. Appearing and switching bifurcations of simple period-1 fixed-points are discussed. Period-1 and period-2 bifurcation trees with global stability are presented for forward and backward quadratic discrete systems. The period-2 appearing and period-doubling renormalizations of quadratic discrete systems are discussed, and the period-n appearing and period-doublization in quadratic discrete systems are presented. Similarly, backward period-2 appearing, backward period-doubling renormalization, and backward period-n appearing and period-doublization are also discussed. The forward and backward period-n bifurcations are presented, and the corresponding bifurcation dynamics can be determined as well.

1.1 Linear Discrete Systems

In this section, the stability and stability switching of fixed-points in linear discrete systems are discussed. The monotonic and oscillatory sink and source fixed-points are discussed.

Definition 1.1 Consider a 1-dimensional linear discrete system

$$x_{k+1} = x_k + A(\mathbf{p})x_k + B(\mathbf{p}) \tag{1.1}$$

where two scalar constants $A(\mathbf{p})$ and $B(\mathbf{p})$ are determined by a vector parameter

$$\mathbf{p} = (p_1, p_2, \ldots, p_m)^{\mathrm{T}}. \tag{1.2}$$

(i) If $A(\mathbf{p}) \neq 0$, there is a fixed-point of

A. C. J Luo, *Bifurcation Dynamics in Polynomial Discrete Systems*,
Nonlinear Physical Science, https://doi.org/10.1007/978-981-15-5208-3_1

$$x_k^* = a_1(\mathbf{p}) = -\frac{B(\mathbf{p})}{A(\mathbf{p})}, \text{ with } a_0(\mathbf{p}) = A(\mathbf{p}), \tag{1.3}$$

and the corresponding discrete system becomes

$$x_{k+1} = x_k + a_0(x_k - a_1). \tag{1.4}$$

(ii) If $A(\mathbf{p}) = 0$, Eq. (1.1) becomes

$$x_{k+1} = x_k + B(\mathbf{p}). \tag{1.5}$$

(ii$_1$) For $B(\mathbf{p}) \neq 0$, the 1-dimensional linear discrete system is called a constant adding discrete system.
(ii$_2$) For $B(\mathbf{p}) = 0$, the 1-dimensional linear discrete system is called a permanent invariant discrete system.

For $\|\mathbf{p}\| \to \|\mathbf{p}_0\| = \beta$, if the following relations hold

$$A(\mathbf{p}) = a_0 = \varepsilon \to 0, B(\mathbf{p}) = \varepsilon a_1(\mathbf{p}) \to 0, \tag{1.6}$$

then there is an instant fixed-point to the vector parameter $\mathbf{p}$

$$x_k^* = a_1(\mathbf{p}). \tag{1.7}$$

Theorem 1.1 *Under assumption in Eq.* (1.6), *a standard form of the 1-dimensional discrete system in Eq.* (1.1) *is*

$$x_{k+1} = x_k + f(x_k) = x_k + a_0(x_k - a_1) \tag{1.8}$$

(i) *If* $|1 + a_0(\mathbf{p})| < 1$ *(or* $|1 + df/dx_k|_{x_k^* = a_1}| < 1$*), then the fixed-point of* $x_k^* = a_1(\mathbf{p})$ *is stable. Such a stable fixed-point is called a sink or a stable node.*

(i$_1$) *If* $a_0(\mathbf{p}) \in (-1, 0)$ *(or* $df/dx_k|_{x_k^* = a_1} \in (-1, 0)$*), then the fixed-point of* $x_k^* = a_1(\mathbf{p})$ *is monotonically stable. Such a stable fixed-point is called a monotonic sink or a monotonically stable node.*
(i$_2$) *If* $a_0(\mathbf{p}) \in (-2, -1)$ *(or* $df/dx_k|_{x_k^* = a_1} \in (-2, -1)$*), then the fixed-point* $x_k^* = a_1(\mathbf{p})$ *is oscillatorilly stable. Such a stable fixed-point is called an oscillatory sink or an oscillatorilly stable node.*
(i$_3$) *If* $a_0(\mathbf{p}) = -1$ *(or* $df/dx_k|_{x_k^* = a_1} = -1$*), then the fixed-point of* $x_k^* = a_1(\mathbf{p})$ *is invariantly stable. Such a stable fixed-point is called an invariant sink.*

(ii) *If* $|1 + a_0(\mathbf{p})| > 1$ *(or* $|1 + df/dx_k|_{x_k^* = a_1}| > 1$*), then the fixed-point of* $x_k^* = a_1(\mathbf{p})$ *is unstable. Such an unstable fixed-point is called a source or an unstable node.*

(ii$_1$) *If $a_0(\mathbf{p}) \in (0, \infty)$ (or $1 + df/dx_k|_{x_k^*=a_1} \in (1, \infty)$), then the fixed-point of $x_k^* = a_1(\mathbf{p})$ is monotonically unstable. Such a stable fixed-point is called a monotonic source or a monotonically unstable node.*

(ii$_2$) *If $a_0(\mathbf{p}) \in (\infty, -2)$ (or $1 + df/dx_k|_{x_k^*=a_1} \in (-\infty, -1)$), then fixed-point of $x_k^* = a_1(\mathbf{p})$ is oscillatorilly unstable. Such a stable fixed-point is called an oscillatory source or an oscillatorilly unstable node.*

(iii) *If $a_0(\mathbf{p}) = 0$ (or $1 + df/dx_k|_{x_k^*=a_1} = 1$), then the flow in the neighborhood of fixed-point $x_k^* = a_1(\mathbf{p})$ is invariant. Such an invariant point is called a monotonic saddle switching.*

(iv) *If $a_0(\mathbf{p}) = -2$ (or $1 + df/dx_k|_{x_k^*=a_1} = -1$), then the flow in the neighborhood of fixed-point $x_k^* = a_1(\mathbf{p})$ is flipped. Such an invariant point is called an oscillatory saddle switching.*

Proof Let $y_k = x_k - a_1$ and $y_{k+1} = x_{k+1} - x_k^*$. Thus, Eq. (1.8) becomes

$$y_{k+1} = (1 + a_0)y_k.$$

The corresponding solution is

$$y_k = (1 + a_0)^k y_0$$

where $y_0 = x_0 - a$ is an initial condition.

(i) If $|a_0(\mathbf{p}) + 1| < 1$, we have

$$\lim_{k\to\infty}(x_k - a_1) = \lim_{k\to\infty} y_k = \lim_{k\to\infty}(1 + a_0)^k y_0.$$

(i$_1$) If $a_0(\mathbf{p}) \in (-1, 0)$, we have $0 < 1 + a_0 < 1$ and

$$\lim_{k\to\infty}(x_k - a_1) = \lim_{k\to\infty} y_k = \lim_{k\to\infty}(1 + a_0)^k y_0 = 0$$
$$\Rightarrow \lim_{k\to\infty} x_k = a_1.$$

Thus, the fixed-point of $x_k^* = a_1(\mathbf{p})$ is monotonically stable. Such a fixed-point is also called a monotonic sink.

(i$_2$) If $a_0(\mathbf{p}) \in (-2, -1)$, we have $-1 < 1 + a_0 < 0$ and

$$\lim_{k\to\infty}(x_k - a_1) = \lim_{k\to\infty} y_k = \lim_{k\to\infty}|1 + a_0|^k (-1)^k y_0 = 0$$
$$\Rightarrow \lim_{k\to\infty} x_k = a_1.$$

Thus, the fixed-point of $x_k^* = a_1(\mathbf{p})$ is oscillatorilly stable. Such a fixed-point is also called an oscillatory sink.

(i$_3$) If $a_0(\mathbf{p}) = -1$, we have $1 + a_0 = 0$ and

$$
\begin{aligned}
&(x_k - a_1) = y_k = (1 + a_0)^k y_0 = 0 \\
&\Rightarrow x_k = a_1 (k = 1, 2, \ldots).
\end{aligned}
$$

Thus, the fixed-point of $x_k^* = a_1(\mathbf{p})$ is invariant. Such a fixed-point is also called an invariant sink.

So the fixed-point of $x_k^* = a_1(\mathbf{p})$ is stable. The fixed-point is called a sink or stable node.

(ii) If $|a_0(\mathbf{p}) + 1| < 1$, we have

(ii$_1$) If $a_0(\mathbf{p}) \in (0, \infty)$, we have $1 + a_0 > 1$ and

$$
\begin{aligned}
&\lim_{k\to\infty} (x_k - a_1) = \lim_{k\to\infty} y_k = \lim_{k\to\infty} (1 + a_0)^k y_0 = \infty \\
&\Rightarrow \lim_{k\to\infty} x_k = \infty.
\end{aligned}
$$

Thus, the fixed-point of $x_k^* = a_1(\mathbf{p})$ is monotonically unstable. Such a fixed-point is also called a monotonic source.

(ii$_2$) If $a_0(\mathbf{p}) \in (-\infty, -2)$, we have $1 + a_0 < -1$ and

$$
\begin{aligned}
\lim_{k\to\infty} (x_k - a_1) &= \lim_{k\to\infty} y_k = \lim_{k\to\infty} |1 + a_0|^k (-1)^k y_0 \\
&= \begin{cases} \infty, k = 2l \to \infty, \\ -\infty, k = 2l + 1 \to \infty; \end{cases} \\
\Rightarrow \lim_{k\to\infty} x_k &= \begin{cases} \infty, k = 2l \to \infty, \\ -\infty, k = 2l + 1 \to \infty. \end{cases}
\end{aligned}
$$

Thus, the fixed-point of $x_k^* = a_1(\mathbf{p})$ is oscillatorilly unstable. Such a fixed-point is also called an oscillatory source.

So the fixed-point of $x_k^* = a_1(\mathbf{p})$ is unstable. The fixed-point is called an oscillatory source or an oscillatorilly unstable node.

(iii) If $a_0(\mathbf{p}) = 0$, we have $1 + a_0 = 1$ and

$$
\lim_{k\to\infty} (x_k - a_1) = \lim_{k\to\infty} y_k = \lim_{k\to\infty} (1 + a_0)^k y_0 = y_0.
$$

So the fixed-point of $x_k^* = a_1(\mathbf{p})$ is invariant. The fixed-point is called a monotonic saddle switching.

(iv) If $a_0(\mathbf{p}) = -2$, we have $1 + a_0 = -1$ and

$$
\lim_{k\to\infty} (x_k - a_1) = \lim_{k\to\infty} y_k = \lim_{k\to\infty} (-1)^k y_0 = \begin{cases} y_0, k = 2l, \\ -y_0, k = 2l + 1. \end{cases}
$$

So the fixed-point of $x_k^* = a_1(\mathbf{p})$ is flipped. The fixed-point is called an oscillatory saddle switching.

The theorem is proved. ■

As in Luo (2010, 2012), the theory for the positive and negative mappings in discrete systems are used herein. If the discrete system in Eq. (1.1) is a positive mapping for x_k $(k = 1, 2, \ldots)$ via x_0, then the corresponding negative mapping is from

$$x_{k+1} = x_k + A(\mathbf{p})x_k + B(\mathbf{p}) \tag{1.9}$$

for x_k $(k = -1, -2, -3, \ldots)$ via x_0 with the corresponding stability determined by $dx_k/dx_{k+1}|_{x_{k+1}^*}$. Such a negative mapping is equivalent to the following mapping

$$x_k = x_{k+1} + A(\mathbf{p})x_{k+1} + B(\mathbf{p}) \tag{1.10}$$

for x_k $(k = 1, 2, 3, \ldots)$ via x_0 with the corresponding stability determined by $dx_{k+1}/dx_k|_{x_k^*}$. Such a linear discrete system with the negative mapping is called a linear backward discrete system. The linear discrete system with a positive mapping is called a linear forward discrete system.

Definition 1.2 Consider a 1-dimensional, linear, backward discrete system

$$x_k = x_{k+1} + A(\mathbf{p})x_{k+1} + B(\mathbf{p}) \tag{1.11}$$

where two scalar constants $A(\mathbf{p})$ and $B(\mathbf{p})$ are determined by a vector parameter

$$\mathbf{p} = (p_1, p_2, \ldots, p_m)^{\mathrm{T}}. \tag{1.12}$$

(i) If $A(\mathbf{p}) \neq 0$, there is a fixed-point of

$$x_k^* = a_1(\mathbf{p}) = -\frac{B(\mathbf{p})}{A(\mathbf{p})}, \text{ with } a_0(\mathbf{p}) = A(\mathbf{p}) \tag{1.13}$$

and the corresponding backward discrete system becomes

$$x_k = x_{k+1} + a_0(x_{k+1} - a_1). \tag{1.14}$$

(ii) If $A(\mathbf{p}) = 0$, Eq. (1.1) becomes

$$x_k = x_{k+1} + B(\mathbf{p}). \tag{1.15}$$

For $B(\mathbf{p}) \neq 0$, the 1-dimensional backward discrete system is called a constant adding discrete system.
For $B(\mathbf{p}) = 0$, the 1-dimensional backward discrete system is called a permanent invariant discrete system.

(iii) For $\|\mathbf{p}\| \to \|\mathbf{p}_0\| = \beta$, if the following relations hold

$$A(\mathbf{p}) = a_0 = \varepsilon \to 0, B(\mathbf{p}) = \varepsilon a_1(\mathbf{p}) \to 0, \tag{1.16}$$

then there is an instant fixed-point to the vector parameter $\mathbf{p}$

$$x_k^* = a_1(\mathbf{p}). \tag{1.17}$$

Theorem 1.2 *Under assumption in Eq.* (1.16), *a standard form of the 1-dimensional, linear, backward discrete system in Eq.* (1.11) is

$$x_k = x_{k+1} + f(x_{k+1}) = x_{k+1} + a_0(x_{k+1} - a_1). \tag{1.18}$$

(i) *If* $|(1 + a_0(\mathbf{p}))^{-1}| < 1$ *(or* $|(1 + df/dx_{k+1}|_{x_{k+1}^* = a_1})^{-1}| < 1$*), then the fixed-point of* $x_k^* = a_1(\mathbf{p})$ *is stable. Such a stable fixed-point is called a sink or a stable node.*

  (i$_1$) *If* $a_0(\mathbf{p}) \in (-\infty, 0)$ *(or* $(1 + df/dx_{k+1}|_{x_{k+1}^* = a_1})^{-1} \in (-1, 0)$*), then the fixed-point of* $x_k^* = a_1(\mathbf{p})$ *is oscillatorilly stable. Such a stable fixed-point is called an oscillatory sink or an oscillatorilly stable node.*

  (i$_2$) *If* $a_0(\mathbf{p}) \in (0, \infty)$ *(or* $(1 + df/dx_{k+1}|_{x_{k+1}^* = a_1})^{-1} \in (0, 1)$*), then fixed-point* $x_k^* = a_1(\mathbf{p})$ *is monotonically stable. Such a stable fixed-point is called a monotonic sink or a monotonically stable node.*

(ii) *If* $|(1 + a_0(\mathbf{p}))^{-1}| > 1$ *(or* $|(1 + df/dx_{k+1}|_{x_{k+1}^* = a_1})^{-1}| > 1$*), then the fixed-point of* $x_k^* = a_1(\mathbf{p})$ *is unstable. Such an unstable fixed-point is called a source or an unstable node.*

  (ii$_1$) *If* $a_0(\mathbf{p}) \in (0, -1)$ *(or* $(1 + df/dx_{k+1}|_{x_{k+1}^* = a_1})^{-1} \in (1, \infty)$*), then the fixed-point of* $x_k^* = a_1(\mathbf{p})$ *is monotonically unstable. Such a stable fixed-point is called a monotonic source or a monotonically unstable node.*

  (ii$_2$) *If* $a_0(\mathbf{p}) \in (-1, -2)$ *(or* $(1 + df/dx_{k+1}|_{x_{k+1}^* = a_1})^{-1} \in (-\infty, -1)$*), then fixed-point* $x_k^* = a_1(\mathbf{p})$ *is oscillatory unstable. Such a stable fixed-point is called an oscillatory source or an oscillatorilly unstable node.*

(iii) *If* $a_0(\mathbf{p}) = 0$ *(or* $(1 + df/dx_{k+1}|_{x_{k+1}^* = a_1})^{-1} = 1$*), then the discrete flow in the neighborhood of fixed-point* $x_k^* = a_1(\mathbf{p})$ *is invariant. Such an invariant point is called a monotonic saddle switching.*

(iv) *If* $a_0(\mathbf{p}) = -2$ *(or* $(1 + df/dx_{k+1}|_{x_{k+1}^* = a_1})^{-1} = -1$*), then the discrete flow in the neighborhood of fixed-point* $x_k^* = a_1(\mathbf{p})$ *is flipped. Such an invariant point is called an oscillatory saddle switching.*

Proof The proof is similar to the proof of Theorem 1.1. This theorem is proved.

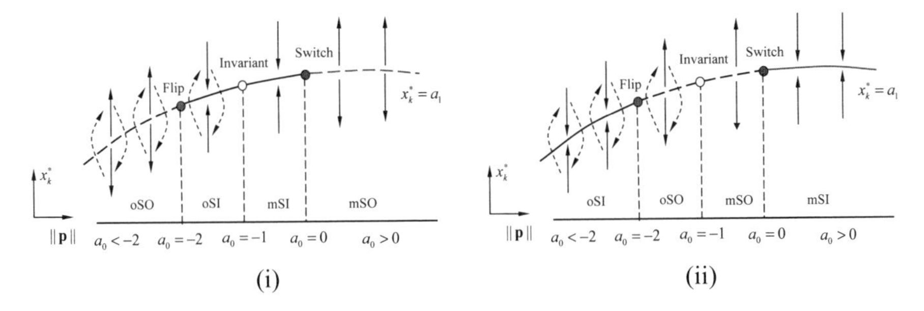

Fig. 1.1 Stability of single fixed-point in the 1-dimensional linear discrete system. (i) Positive mapping (forward), (ii) Negative mapping (backward). Stable and unstable fixed-points are represented by solid and dashed curves, respectively. The stability switches are labelled by solid circular symbols. (Flip: flipped switch for oscillatorilly stable to unstable fixed-point; Switch: for monotonically stable and unstable fixed-points; Invariant: invariant sink; mSO: monotonic source; mSI: monotonic sink; oSO: oscillatory source; oSI: oscillatory sink.)

To illustrate the stability of fixed-points, one fixed-point of $x_k^* = a_1(\mathbf{p})$ changes with a vector parameter $\mathbf{p}$. The stability of such a fixed-point is determined by the constant $a_0(\mathbf{p})$. The stability switching is at the boundary $\mathbf{p}_0 \in \partial\Omega_{12}$ with $a_0 = 0$. The stability of the fixed-points for the positive (forward) and negative (backward) maps in the 1-dimensional discrete system is presented in Fig. 1.1i and ii, respectively. The stable and unstable portions of the fixed-point are presented by the solid and dash curves, respectively.

1.2 Forward Quadratic Discrete Systems

In this section, the stability of fixed-points in 1-dimensional quadratic nonlinear discrete systems are discussed. The *upper-saddle-node* and *lower-saddle-node* appearing and switching bifurcations are presented. The nonlinear discrete systems with positive and negative mappings will be discussed. As in Luo (2010, 2012), the discrete system with a map with a positive (forward) iteration is called a positive (forward) discrete system, and the discrete system with a map with a negative (backward) iteration is called a negative (backward) discrete system.

1.2.1 Period-1 Appearing Bifurcations

For one of the simplest nonlinear discrete systems, consider a positive (forward) quadratic nonlinear discrete system first.

Definition 1.3 Consider a 1-dimensional quadratic nonlinear discrete system as

$$x_{k+1} = x_k + f(x_k, \mathbf{p}) = x_k + A(\mathbf{p})x_k^2 + B(\mathbf{p})x_k + C(\mathbf{p}) \tag{1.19}$$

where three scalar constants $A(\mathbf{p}) \neq 0$, $B(\mathbf{p})$ and $C(\mathbf{p})$ are determined by a vector parameter

$$\mathbf{p} = (p_1, p_2, \ldots, p_m)^{\mathrm{T}}. \tag{1.20}$$

(i) If

$$\Delta = B^2 - 4AC < 0 \quad \text{for} \quad \mathbf{p} \in \Omega_1 \subset \mathbf{R}^m, \tag{1.21}$$

then the quadratic discrete system does not have any fixed-points. The discrete flow without fixed-points is called a non-fixed-point discrete flow.

(i_1) If $a_0(\mathbf{p}) = A(\mathbf{p}) > 0$, the non-fixed-point discrete flow is called a positive discrete flow.
(i_2) If $a_0(\mathbf{p}) = A(\mathbf{p}) < 0$, the non-fixed-point discrete flow is called a negative discrete flow.

(ii) If

$$\Delta = B^2 - 4AC > 0 \quad \text{for} \quad \mathbf{p} \in \Omega_2 \subset \mathbf{R}^m, \tag{1.22}$$

then the quadratic discrete system has two different simple fixed-points as

$$x_k^* = a_1 \text{ and } x_k^* = a_2, \tag{1.23}$$

and the corresponding standard form is

$$x_{k+1} = x_k + a_0(x_k - a_1)(x_k - a_2), \tag{1.24}$$

where

$$a_0 = A(\mathbf{p}),\ a_{1,2} = \frac{-B(\mathbf{p}) \pm \sqrt{\Delta}}{2A(\mathbf{p})} \text{ with } a_1 < a_2. \tag{1.25}$$

(iii) If

$$\Delta = B^2 - 4AC = 0 \quad \text{for} \quad \mathbf{p} = \mathbf{p}_0 \in \partial\Omega_{12} \subset \mathbf{R}^{m-1}, \tag{1.26}$$

then the quadratic nonlinear discrete system has a double repeated fixed-point, i.e.,

$$x_k^* = a_1 \text{ and } x_k^* = a_1, \tag{1.27}$$

with the corresponding standard form of

$$x_{k+1} = x_k + a_0(x_k - a_1)^2, \tag{1.28}$$

where

$$a_0 = A(\mathbf{p}_0), \quad \text{and} \quad a_1 = a_2 = -\frac{B(\mathbf{p}_0)}{2A(\mathbf{p}_0)}. \tag{1.29}$$

Such a discrete flow with the fixed-point of $x_k^* = x_{k+1}^* = a_1(\mathbf{p})$ is called a monotonic *saddle* discrete flow of the second order.

(iii$_1$) If $a_0(\mathbf{p}) > 0$, then the fixed-point of $x_k^* = x_{k+1}^* = a_1(\mathbf{p})$ is a *monotonic upper-saddle of the second-order*.

(iii$_2$) If $a_0(\mathbf{p}) < 0$, then the fixed-point of $x_k^* = x_{k+1}^* = a_1(\mathbf{p})$ is a *monotonic lower-saddle of the second-order*.

(iv) The fixed-point of $x_k^* = a_1$ for two fixed-points vanishing or appearance is called a monotonic *saddle-node* appearing bifurcation of the second-order at a point $\mathbf{p} = \mathbf{p}_0 \in \partial\Omega_{12}$, and the bifurcation condition is

$$\Delta = B^2 - 4AC = 0. \tag{1.30}$$

(iv$_1$) If $a_0(\mathbf{p}) > 0$, the bifurcation at $x_k^* = x_{k+1}^* = a_1(\mathbf{p})$ for two fixed-points appearance or vanishing is called a monotonic *upper-saddle-node appearing* bifurcation of the second-order.

(iv$_2$) If $a_0(\mathbf{p}) < 0$, the bifurcation at $x_k^* = x_{k+1}^* = a_1(\mathbf{p})$ for two fixed-points appearance or vanishing is called a monotonic *lower-saddle-node appearing* bifurcation of the second-order.

Theorem 1.3

(i) *Under a condition of*

$$\Delta = B^2 - 4AC < 0, \tag{1.31}$$

a standard form of the quadratic nonlinear discrete system in Eq. (1.19) *is*

$$x_{k+1} = x_k + a_0[(x_k - \frac{1B}{2A})^2 + \frac{1}{4A^2}(-\Delta)] \tag{1.32}$$

with $a_0 = A(\mathbf{p})$, *which has a non-fixed-point flow.*

(i$_1$) *If* $a_0(\mathbf{p}) > 0$, *the non-fixed-point discrete flow is called a positive discrete flow.*

(i$_2$) *If* $a_0(\mathbf{p}) > 0$, *the non-fixed-point discrete flow is called a negative discrete flow.*

(ii) *Under a condition of*

$$\Delta = B^2 - 4AC > 0, \tag{1.33}$$

a standard form of the 1-dimensional discrete system in Eq. (1.19) *is*

$$x_{k+1} = x_k + f(x_k, \mathbf{p}) = x_k + a_0(x_k - a_1)(x_k - a_2). \tag{1.34}$$

(ii$_1$) *For* $a_0(\mathbf{p}) > 0$, *the following cases exist.*

(ii$_{1a}$) *The fixed-point of* $x_k^* = x_{k+1}^* = a_1(\mathbf{p})$ *is*

- *monotonically stable (a monotonic sink) if* $df/dx_k|_{x_k^*=a_1} \in (-1, 0)$;
- *invariantly stable (an invariant sink) if* $df/dx_k|_{x_k^*=a_1} = -1$;
- *oscillatorilly stable (an oscillatory sink) if* $df/dx_k|_{x_k^*=a_1} \in (-2, -1)$;
- *flipped if* $df/dx_k|_{x_k^*=a_1} = -2$ *(an oscillatory upper-saddle of the second order for* $d^2f/dx_k^2|_{x_k^*=a_2} = a_0 > 0$*);*
- *oscillatorilly unstable (an oscillatory source) if* $df/dx_k|_{x_k^*=a_1} \in (-\infty, -2)$.

(ii$_{1b}$) *The fixed-point of* $x_k^* = a_2(\mathbf{p})$ *is monotonically unstable (a monotonic source) if* $df/dx_k|_{x_k^*=a_2} \in (0, \infty)$.

(ii$_2$) *For* $a_0(\mathbf{p}) < 0$, *the following cases exist.*

(ii$_{2a}$) *The fixed-point of* $x_k^* = x_{k+1}^* = a_2(\mathbf{p})$ *is*

- *monotonically stable (a monotonic sink) if* $df/dx_k|_{x_k^*=a_2} \in (-1, 0)$;
- *invariantly stable (an invariant sink) if* $df/dx_k|_{x_k^*=a_2} = -1$;
- *oscillatorilly stable (an oscillatory sink) if* $df/dx|_{x_k^*=a_2} \in (-2, -1)$;
- *flipped if* $df/dx_k|_{x_k^*=a_2} = -2$ *(an oscillatory lower-saddle of the second-order for* $d^2f/dx_k^2|_{x_k^*=a_2} = a_0 < 0$*);*
- *oscillatorilly unstable (an oscillatory source) if* $df/dx_k|_{x_k^*=a_2} \in (-\infty, -2)$.

(ii$_{2b}$) *The fixed-point of* $x_k^* = x_{k+1}^* = a_1(\mathbf{p})$ *is monotonically unstable (a monotonic source) if* $df/dx_k|_{x_k^*=a_1} \in (0, \infty)$.

(iii) *Under a condition of*

$$\Delta = B^2 - 4AC = 0, \tag{1.35}$$

a standard form of the 1-dimensional discrete system in Eq. (1.19) *is*

$$x_{k+1} = x_k + f(x_k, \mathbf{p}) = x_k + a_0(x_k - a_1)^2. \tag{1.36}$$

(iii$_1$) *If $a_0(\mathbf{p}) > 0$, then the fixed-point of $x_k^* = x_{k+1}^* = a_1(\mathbf{p})$ is a monotonic* upper-saddle *of the second-order* if $d^2f/dx_k^2|_{x_k^*=a_1} > 0$. *The bifurcation at $x_k^* = x_{k+1}^* = a_1(\mathbf{p})$ for the appearance or vanishing of two simple fixed-points is called a monotonic* upper-*saddle-node appearing bifurcation of the second-order.*

(iii$_2$) *If $a_0(\mathbf{p}) < 0$, then the fixed-point of $x_k^* = x_{k+1}^* = a_1(\mathbf{p})$ is a monotonic* lower-saddle *of the second order with $d^2f/dx_k^2|_{x_k^*=a_1} < 0$. The bifurcation at $x_k^* = x_{k+1}^* = a_1(\mathbf{p})$ for the appearance or vanishing of two simple fixed-points is called a monotonic* lower-saddle-node appearing *bifurcation of the second-order.*

Proof

(i) Consider

$$\Delta = B^2 - 4AC < 0.$$

(i$_1$) If $a_0 > 0$, we have

$$x_{k+1} - x_k = a_0[(x_k - \frac{B}{2A})^2 + \frac{1}{4A^2}(-\Delta)] > 0.$$

Thus, such a non-fixed-point discrete flow is called a positive discrete flow.

(i$_2$) If $a_0 < 0$, we have

$$x_{k+1} - x_k = a_0[(x_k - \frac{B}{2A})^2 + \frac{1}{4A^2}(-\Delta)] < 0.$$

Thus, such a non-fixed-point discrete flow is called a negative discrete flow.

(ii) Let $\Delta x_{k(i)} = x_k - a_i \;\; (i = 1, 2)$ and $x_{k+1(i)} = \Delta x_{k+1(i)}$. Equation (1.34) becomes

$$\Delta x_{k+1(i)} = [1 + a_0(a_i - a_j)]\Delta x_{k(i)} + a_0 \Delta x_{k(i)}^2 \; (i, j \in \{1, 2\}, j \neq i).$$

Because Δx_i is arbitrary small, we have

$$\Delta x_{k+1(i)} = \lambda_i \Delta x_{k(i)} \;\text{ for }\; \lambda_i \equiv 1 + df/dx_k|_{x_k^*=a_i} = 1 + a_0(a_i - a_j).$$

The corresponding solution is

$$\Delta x_{k(i)} = (\lambda_i)^k \Delta x_{0(i)}$$

where $\Delta x_{0(i)} = x_{0(i)} - a_i$ is an initial condition.

(ii$_a$) For $\lambda_i \in (0,1)$ (or $df/dx_k|_{x_k^*=a_i} \in (-1,0)$), we have

$$\lim_{k\to\infty}(x_{k(i)} - a_i) = \lim_{k\to\infty}\Delta x_{k(i)} = \lim_{k\to\infty}(\lambda_i)^k \Delta x_{0(i)} = 0 \Rightarrow \lim_{k\to\infty} x_{k(i)} = a_i.$$

Consider

$$\lambda_i = 1 + a_0(a_i - a_j) \in (0,1) \Rightarrow a_0(a_i - a_j) \in (-1,0).$$

(ii$_{a1}$) For $a_0 > 0$, we have

$$a_i < a_j \Rightarrow x_k^* = a_1.$$

(ii$_{a2}$) For $a_0 < 0$, we have

$$a_i > a_j \Rightarrow x_k^* = a_2.$$

Thus, the fixed-point of $x_k^* = a_i$ is monotonically stable.

(ii$_b$) For $\lambda_i = 0$ (or $df/dx_k|_{x_k^*=a_i} = -1$), we have

$$(x_{k(i)} - a_i) = \Delta x_{k(i)} = (\lambda_i)^k \Delta x_{0(i)} = 0 \Rightarrow x_{k(i)} = a_i (i = 1,2).$$

Consider

$$\lambda_i = 1 + a_0(a_i - a_j) = 0 \Rightarrow a_0(a_i - a_j) = -1.$$

(ii$_{b1}$) For $a_0 > 0$, we have

$$a_i < a_j \Rightarrow x_k^* = a_1.$$

(ii$_{b2}$) For $a_0 < 0$, we have

$$a_i > a_j \Rightarrow x_k^* = a_2.$$

Thus, the fixed-point of $x_k^* = a_i$ is invariantly stable.

(ii$_c$) For $\lambda_i \in (-1,0)$ (or $df/dx_k|_{x_k^*=a_i} \in (-2,-1)$), we have

$$\lim_{k\to\infty}(x_{k(i)} - a_i) = \lim_{k\to\infty}\Delta x_{k(i)} = \lim_{k\to\infty}\lambda_i^k \Delta x_{0(i)}$$
$$= \lim_{k\to\infty}(|\lambda_i|)^k(-1)^k \Delta x_{0(i)} = \begin{cases} 0^- & \text{for } k = 2l-1, \\ 0^+ & \text{for } k = 2l. \end{cases}$$

Thus

$$\lim_{k\to\infty} x_{k(i)} = \begin{cases} a_i^- & \text{for } k = 2l-1, \\ a_i^+ & \text{for } k = 2l. \end{cases}$$

Consider

$$\lambda_i = 1 + a_0(a_i - a_j) \in (0,1) \Rightarrow a_0(a_i - a_j) \in (-1,0).$$

($\mathrm{ii_{c1}}$) For $a_0 > 0$, we have

$$a_i < a_j \Rightarrow x_k^* = a_1.$$

($\mathrm{ii_{c2}}$) For $a_0 < 0$, we have

$$a_i > a_j \Rightarrow x_k^* = a_2.$$

Thus, the fixed-point of $x_k^* = a_i$ is monotonically stable.

($\mathrm{ii_d}$) For $\lambda_i = -1$ (or $df/dx_k|_{x_k^*=a_i} = -2$), we have

$$\lim_{k\to\infty}(x_{k(i)} - a_i) = \lim_{k\to\infty}\Delta x_{k(i)} = \lim_{k\to\infty}(-1)^k \Delta x_{0(i)},$$

so

$$\lim_{k\to\infty} x_{k(i)} = \begin{cases} a_i + \Delta x_{0(i)} & \text{for } k = 2l-1, \\ a_i + \Delta x_{0(i)} a_i + \Delta x_{0(i)} & \text{for } k = 2l. \end{cases}$$

Consider

$$\lambda_i = 1 + a_0(a_i - a_j) = -1 \Rightarrow a_0(a_i - a_j) = -2.$$

($\mathrm{ii_{d1}}$) For $a_0 > 0$, we have

$$a_i < a_j \Rightarrow x_k^* = a_1.$$

For the fixed-point of $x_k^* = a_1$, we have

$$\Delta x_{k+1} = (-1 + a_0 \Delta x_k) \Delta x_k.$$

Therefore, the fixed-point of $x_k^* = a_1$ is an oscillatory lower-saddle of the second-order.

(ii_{d2}) For $a_0 < 0$, we have

$$a_i > a_j \Rightarrow x_k^* = a_2.$$

For the fixed-point of $x_k^* = a_2$, we have

$$\Delta x_{k+1} = (-1 + a_0 \Delta x_k) \Delta x_k.$$

Thus, the fixed-point of $x_k^* = a_2$ is an oscillatory upper-saddle of the second-order.

(ii_e) For $\lambda_i < -1$ (or $df/dx_k|_{x_k^*=a_i} \in (-\infty, -2)$), we have

$$\begin{aligned}\lim_{k\to\infty} (x_{k(i)} - a_i) &= \lim_{k\to\infty} \Delta x_{k(i)} = \lim_{k\to\infty} \lambda_i^k \Delta x_{0(i)} \\ &= \lim_{k\to\infty} (|\lambda_i|)^k (-1)^k \Delta x_{0(i)} = \begin{cases} -\infty & \text{for } k = 2l - 1, \\ +\infty & \text{for } k = 2l. \end{cases}\end{aligned}$$

Thus

$$\lim_{k\to\infty} x_{k(i)} = \begin{cases} a_i - \infty & \text{for } k = 2l - 1, \\ a_i + \infty & \text{for } k = 2l \end{cases}$$

and

$$\lambda_i = 1 + a_0(a_i - a_j) < -1 \Rightarrow a_0(a_i - a_j) < -2.$$

(ii_{e1}) For $a_0 > 0$, we have

$$a_i < a_j \Rightarrow x_k^* = a_1.$$

(ii_{e2}) For $a_0 < 0$, we have

$$a_i > a_j \Rightarrow x_k^* = a_2.$$

Thus, the fixed-point of $x_k^* = a_i$ is oscillatorilly stable.

(ii_f) If $\lambda_i > 1$ (or $df/dx_k|_{x_k^*=a_i} \in (0, \infty)$), we have the following cases.

$$\lim_{k\to\infty}(x_{k(i)} - a_i) = \lim_{k\to\infty}\Delta x_{k(i)} = \lim_{k\to\infty}(\lambda_i)^k \Delta x_{0(i)} = \infty \Rightarrow \lim_{k\to\infty} x_{k(i)} = \infty$$

and

$$\lambda_i = 1 + a_0(a_i - a_j) \in (1, \infty) \Rightarrow a_0(a_i - a_j) \in (0, \infty).$$

($\mathrm{ii}_{\mathrm{f1}}$) For $a_0 > 0$, we have

$$a_i > a_j \Rightarrow x_k^* = a_2.$$

($\mathrm{ii}_{\mathrm{f2}}$) For $a_0 < 0$, we have

$$a_i < a_j \Rightarrow x_k^* = a_1.$$

Thus, the fixed-point of $x_k^* = a_i$ is monotonically unstable.

(iii) If $a_1(\mathbf{p}) = a_2(\mathbf{p})$, $\lambda_i = 1$ (or $df/dx_k|_{x_k^*=a_i} = 0$) and we have

$$\Delta x_{k+1} = \Delta x_k + a_0 \Delta x_k^2 = (1 + a_0 \Delta x_k)\Delta x_k \quad \text{and} \quad \Delta x_k = x_k - x_k^*.$$

(iii_1) For $a_0 > 0$, $\Delta x_{k+1} > \Delta x_k > 0$ if $\Delta x_k > 0$ and $0 > \Delta x_{k+1} > \Delta x_k$ if $\Delta x_k < 0$. So a flow of x_k reaches to $x_k^* = a_1$ from the initial point of $x_{k0} < a_1$ and it goes to the positive infinity from $x_{k0} > a_1$. Such a fixed-point is monotonically unstable of the second order, which is called a monotonic *upper-saddle of the second-order*.

(iii_2) Similarly, for $a_0 < 0$, $0 < \Delta x_{k+1} < \Delta x_k$ if $\Delta x_k > 0$ and $\Delta x_{k+1} < \Delta x_k < 0$ if $\Delta x_k < 0$. So a flow of x_k reaches to $x_k^* = a_1$ from the initial point of $x_{k0} > a_1$ and it goes to the negative infinity from $x_{k0} < a_1$. Such a fixed-point is monotonically unstable of the second order, which is called a monotonic *lower-saddle of the second-order*.

The theorem is proved. ■

The stability and bifurcation of fixed-points for the quadratic nonlinear discrete system in Eq. (1.19) are illustrated in Fig. 1.2. The stable and unstable fixed-points varying with the vector parameter are depicted by solid and dashed curves, respectively. The bifurcation point of fixed-points occurs at the double-repeated fixed-points at $\mathbf{p}_0 \in \partial\Omega_{12}$. In Fig. 1.2i, for $a_0 > 0$, the fixed-point of $x_k^* = a_2$ for $\Delta > 0$ is monotonically unstable, and the fixed-point of $x_k^* = a_1$ in a small neighborhood of $\Delta = 0^+$ is monotonically stable. The fixed point of $x_k^* = a_1$ can be a monotonic sink, a zero-invariant sink, an oscillatory sink, a flipped invariance and an oscillatory source. The monotonic bifurcation of two simple fixed-points also occurs at $\Delta = 0$. The discrete flow of x_k is a forward *upper-branch* discrete flow for $a_0 > 0$, and the fixed-point $x_k^* = -B(\mathbf{p}_0)/2A(\mathbf{p}_0)$ at $\Delta = 0$ is termed a monotonic *upper-saddle of the second-order*. Such a bifurcation is termed a monotonic *upper-*

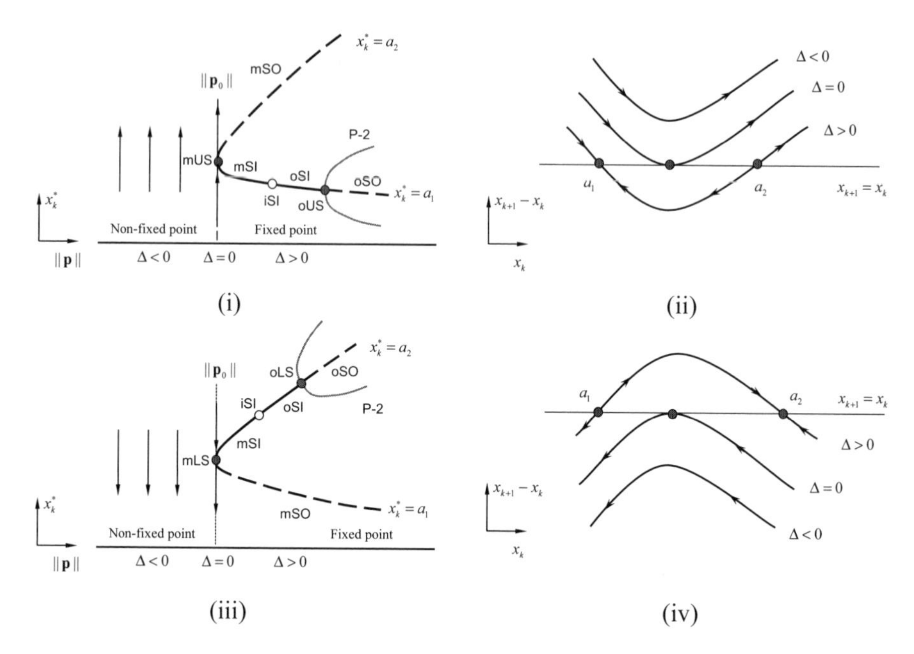

Fig. 1.2 Stability and bifurcation of two fixed-points in the quadratic forward discrete system: (i) a *monotonic upper-saddle-node* bifurcation and (ii) phase portrait ($a_0 > 0$), (iii) a *monotonic lower-saddle-node* bifurcation and (iv) phase portrait ($a_0 < 0$). Stable and unstable fixed-points are represented by solid and dashed curves, respectively. P-2 is for period-2 fixed-points for the quadratic forward discrete system. (mSO: monotonic source; mSI: monotonic sink; oSO: oscillatory source; oSI: oscillatory sink; mLS: monotonic lower-saddle; mUS: monotonic upper-saddle; oUS: oscillatory upper-saddle; oLS: oscillatory lower-saddle; iSI: invariant sink.)

saddle-node bifurcation. For $x_k^* = a_1$ with $df/dx_k|_{x_k^*=a_1} = -2$, a period-doubling bifurcation is an oscillatory upper-saddle-node bifurcation, and for $df/dx_k|_{x_k^*=a_1} \in (-\infty, -2)$, the period-2 fixed points exists. For $\Delta<0$, no any fixed-points exist. For $\Delta<0$ and $a_0 > 0$, the discrete flow of x_k is always toward the positive direction due to $x_{k+1} - x_k = a_0[(x_k + \frac{B}{2A})^2 + (-\frac{\Delta}{4A^2})] > 0$. The corresponding phase portrait is presented in Fig. 1.2ii. In Fig. 1.2iii, the fixed-point of $x_k^* = a_1$ for $a_0<0$ is monotonically unstable, and the fixed-point of $x_k^* = a_2$ in a small neighborhood of $\Delta = 0^+$ is monotonically stable. The fixed point of $x_k^* = a_2$ can be a monotonic sink, a zero-invariant sink, an oscillatory sink, a flipped invariance and an oscillatory source. The monotonic bifurcation of the two simple fixed-points also occurs at $\Delta = 0$. The discrete flow of x_k is a forward *lower-branch* discrete flow for $a_0<0$, and the fixed-point of $x_k^* = -B(\mathbf{p}_0)/2A(\mathbf{p}_0)$ at $\Delta = 0$ is termed a monotonic *lower-saddle of the second-order*. Such a bifurcation is termed a monotonic *lower-saddle-node* bifurcation. For $x_k^* = a_2$ with $df/dx_k|_{x_k^*=a_2} = -2$, a period-doubling bifurcation is an oscillatory lower-saddle-node bifurcation, and for $df/dx_k|_{x_k^*=a_2} \in (-\infty, -2)$, the period-2 fixed points exists. For $\Delta<0$ and $a_0<0$, the discrete flow of x_k is always toward the negative direction due to $x_{k+1} - x_k = a_0[(x_k + \frac{B}{2A})^2$

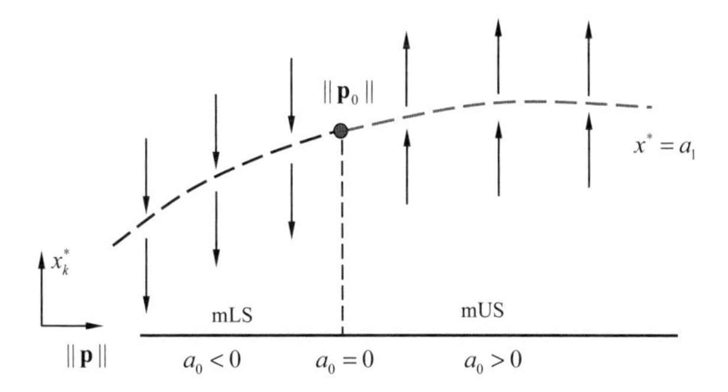

Fig. 1.3 Stability and bifurcation of a repeated fixed-point of the second order in the quadratic forward discrete system. Unstable fixed-points is represented by a dashed curve. The stability switching from the monotonic lower-saddle to monotonic upper-saddle is labelled by a circular symbol. (mLS: monotonic lower-saddle; mUS: monotonic upper-saddle.)

$+(-\frac{\Delta}{4A^2})]<0$. The corresponding phase portrait is presented in Fig. 1.2iv. The period-2 fixed-points based on the quadratic forward discrete map are also presented through red curves, labelled by P-2.

To illustrate the stability and bifurcation of fixed-points with singularity in a 1-dimensional, quadratic nonlinear discrete system, the fixed-point of $x_{k+1} - x_k = a_0(x_k - a_1)^2$ is presented in Fig. 1.3. The monotonic *upper-saddle* and *lower-saddle* fixed-points of $x_k^* = a_1$ with the second-order are unstable, which are depicted by dashed curves. At $a_0 = 0$, the monotonic *upper-saddle* and *lower-saddle* fixed-points will be switched, which is marked by a circular symbol.

Consider a symmetric case for the appearing bifurcations in forward quadratic discrete systems. Such a special case can help one further understand the appearing bifurcation and the corresponding stability.

Definition 1.4 If $B(\mathbf{p}) = 0$ in Eq. (1.19), a 1-dimensional quadratic discrete system is

$$x_{k+1} = x_k + A(\mathbf{p})x_k^2 + C(\mathbf{p}). \tag{1.37}$$

(i) For $A(\mathbf{p}) \times C(\mathbf{p}) > 0$, the discrete system does not have any fixed-points.

- (i_1) The non-fixed-point discrete flow of the discrete system is called a positive discrete flow if $A(\mathbf{p}) > 0$.
- (i_2) The non-fixed-point discrete flow of the discrete system is called a negative discrete flow if $A(\mathbf{p}) < 0$.

(ii) For $A(\mathbf{p}) \times C(\mathbf{p}) < 0$, the corresponding standard form is

$$x_{k+1} = x_k + a_0(x_k + a)(x_k - a) \tag{1.38}$$

with two symmetric fixed-points

$$x_k^* = -a \text{ and } x_k^* = a,$$
$$\text{with } a_0 = A(\mathbf{p}_0) \text{ and } a = \sqrt{-\frac{C(\mathbf{p}_0)}{A(\mathbf{p}_0)}}. \tag{1.39}$$

(iii) For $C(\mathbf{p}_0) = 0$, the corresponding standard form with $\Delta = 0$ is

$$x_{k+1} = x_k + a_0 x_k^2 \tag{1.40}$$

with two fixed-points of

$$x_k^* = -a = -0 \text{ and } x_k^* = a = +0. \tag{1.41}$$

Such a fixed-point of $x_k^* = 0$ is called a monotonic *saddle* of the second-order.

(iii$_1$) If $a_0 > 0$, the fixed-point is a monotonic *upper-saddle* of the second-order.
(iii$_2$) If $a_0 < 0$, the fixed-point is a monotonic *lower-saddle* of the second-order.

(iv) The fixed-point of $x_k^* = 0$ for two fixed-points appearance or vanishing is called a monotonic *saddle-node appearing* bifurcation of the second-order at a point $\mathbf{p} = \mathbf{p}_0 \in \partial\Omega_{12}$, and the appearing bifurcation condition is

$$C(\mathbf{p}_0) = 0. \tag{1.42}$$

Theorem 1.4

(i) *Under a condition of*

$$A(\mathbf{p}) \times C(\mathbf{p}) > 0, \tag{1.43}$$

a standard form of the 1-dimensional forward, quadratic discrete system in Eq. (1.37) *is*

$$x_{k+1} = x_k + f(x_k, \mathbf{p}) = a_0(x_k^2 + \frac{C}{A}). \tag{1.44}$$

(i$_1$) *For* $A(\mathbf{p}) > 0$, *the non-fixed-point flow of the discrete system is a positive discrete flow.*
(i$_2$) *For* $A(\mathbf{p}) < 0$, *the non-fixed-point flow of the discrete system is a negative discrete flow.*

(ii) *Under a condition of*

$$A(\mathbf{p}) \times C(\mathbf{p}) < 0, \quad (1.45)$$

a standard form of the 1-dimensional forward, quadratic nonlinear discrete system in Eq. (1.37) *is*

$$x_{k+1} = x_k + f(x_k, \mathbf{p}) = a_0(x_k + a)(x_k - a). \quad (1.46)$$

(ii_1) *For* $a_0(\mathbf{p}) > 0$, *there are two cases:*

(ii_{1a}) *The fixed-point of* $x_k^* = -a$ *is:*

- *monotonically stable (a monotonic sink) if* $df/dx_k|_{x_k^*=-a} \in (-1, 0)$;
- *invariantly stable (an invariant sink) if* $df/dx_k|_{x_k^*=-a} = -1$;
- *oscillatorilly stable (an oscillatory sink) if* $df/dx_k|_{x_k^*=-a} \in (-2, -1)$;
- *flipped if* $df/dx_k|_{x_k^*=-a} = -2$ *(an oscillatory upper-saddle of the second-order under* $d^2f/dx_k^2|_{x_k^*=-a} = a_0 > 0$*)*;
- *oscillatorilly unstable (an oscillatory source) if* $df/dx_k|_{x_k^*=-a} \in (-\infty, -2)$.

(ii_{1b}) *The fixed-point of* $x_k^* = a$ *is monotonically unstable (a monotonic source) if* $df/dx_k|_{x_k^*=a} > 0$.

(ii_2) *For* $a_0(\mathbf{p}) < 0$, *there are two cases:*

(ii_{2a}) *The fixed-point of* $x_k^* = -a$ *is monotonically unstable (a monotonic source) if* $df/dx_k|_{x_k^*=-a} > 0$.

(ii_{2b}) *The fixed-point of* $x_k^* = a$ *is:*

- *monotonically stable (a monotonic sink) if* $df/dx_k|_{x_k^*=a} \in (-1, 0)$;
- *invariantly stable (an invariant sink) if* $df/dx_k|_{x_k^*=a} = -1$;
- *oscillatorilly stable (an oscillatory source) if* $df/dx_k|_{x_k^*=a} \in (-2, -1)$;
- *flipped if* $df/dx_k|_{x_k^*=a} = -2$ *(an oscillatory lower-saddle of the second-order with* $d^2f/dx_k^2|_{x_k^*=a} = a_0 < 0$*)*;
- *oscillatorilly unstable (an oscillator source) if* $df/dx_k|_{x_k^*=a} \in (-\infty, -2)$.

(iii) *Under a condition of*

$$C(\mathbf{p}) = 0, \quad (1.47)$$

a standard form of the 1-dimensional discrete system in Eq. (1.37) *is*

$$x_{k+1} = x_k + f(x_k, \mathbf{p}) = x_k + a_0 x_k^2. \tag{1.48}$$

(iii$_1$) *If* $a_0(\mathbf{p}) > 0$, *then the fixed-point of* $x^* = 0$ *is a monotonic* upper-saddle *of the second-order for* $d^2f/dx^2|_{x^*=0} > 0$. *Such a bifurcation for two fixed-points appearance or vanishing is a monotonic* upper-saddle-node *appearing bifurcation of the second-order.*

(iii$_2$) *If* $a_0(\mathbf{p}) < 0$, *then the fixed-point* $x^* = 0$ *is a monotonic* lower-saddle *of the second-order for* $d^2f/dx^2|_{x^*=0} < 0$. *Such a bifurcation for two fixed-points appearance or vanishing is a monotonic* lower-saddle-node appearing *bifurcation of the second-order.*

Proof The proof is similar to Theorem 1.3. The theorem is proved. ■

The stability and bifurcation of fixed-points for the quadratic nonlinear system in Eq. (1.37) are illustrated in Fig. 1.4 as a special case of the discrete system in Eq. (1.19) with $B(\mathbf{p}) = 0$. The stable and unstable fixed-points varying with the vector parameter are depicted by solid and dashed curves, respectively. The bifurcation of fixed-point occurs at the double-repeated fixed-point at the boundary of $\mathbf{p}_0 \in \partial\Omega_{12}$. In Fig. 1.4i, for $\Delta = -4AC > 0$ and $a_0 = A > 0$, the fixed-point of $x_k^* = a > 0$ for $C < 0$ is unstable, and the fixed-point of $x_k^* = -a < 0$ for $C < 0$ is from monotonically stable to oscillatorilly unstable. The bifurcation of fixed-point also occurs at $C = 0$. The discrete flow of x_k is a forward *upper-branch* discrete flow for $a_0 > 0$, and the fixed-point of $x_k^* = 0$ at $C = 0$ is termed an monotonic *upper-saddle of the second-order*. Such a bifurcation is termed a monotonic *upper-saddle-node* bifurcation of the second-order. For $\Delta = -4AC < 0$ and $a_0 = A > 0$, we have $C > 0$. Thus, no any fixed-point exists because of $x_{k+1} - x_k = Ax_k^2 + C > 0$. Such a 1-dimensional discrete system is termed a non-fixed-point discrete system. For $a_0 = A > 0$ and $C > 0$, the discrete flow of x_k is always toward the positive direction. In Fig. 1.4(ii), for $\Delta = -4AC > 0$ and $a_0 = A < 0$, the fixed-point of $x_k^* = a$ for $C > 0$ is unstable, and the fixed-point of $x_k^* = -a$ for $C > 0$ is from monotonically stable to oscillatorilly unstable. The bifurcation of fixed-point also occurs at $C = 0$. The discrete flow of x_k for the bifurcation point is a forward monotonic *lower-branch* discrete flow for $a_0 = A < 0$, and the bifurcation point of the fixed-point at $x_k^* = 0$ for $C = 0$ is termed a monotonic *lower-saddle of the second-order*. Such a bifurcation is termed a monotonic *lower-saddle-node* bifurcation of the second-order. For $\Delta = -4AC < 0$ and $a_0 = A < 0$, we have $C < 0$. For $a_0 = A < 0$ and $C < 0$, the discrete flow of x_k is always toward the negative direction without any fixed-points because of $x_{k+1} - x_k = Ax_k^2 + C < 0$.

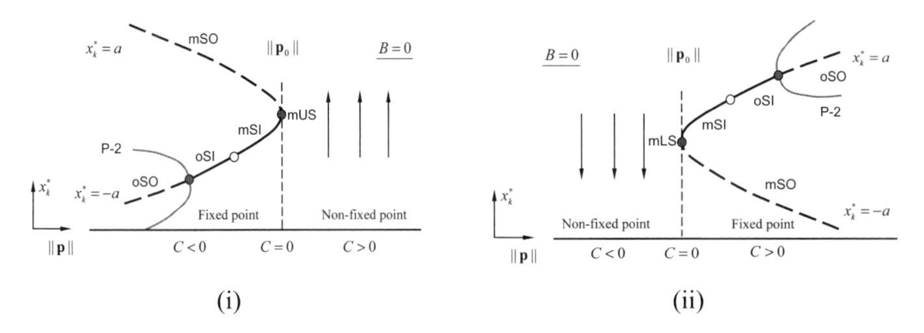

Fig. 1.4 Stability and bifurcation of two fixed-points in the quadratic forward discrete system: (i) a monotonic *upper-saddle*-node bifurcation ($a_0 > 0$), (ii) a monotonic *lower-saddle-node* bifurcation ($a_0 < 0$). Stable and unstable fixed-points are represented by solid and dashed curves, respectively. (mSO: monotonic source; mSI: monotonic sink; oSO: oscillatory source; oSI: oscillatory sink; mLS: monotonic lower-saddle; mUS: monotonic upper-saddle; oUS: oscillatory upper-saddle; oLS: oscillatory lower-saddle; iSI: invariant sink.)

1.2.2 Period-1 Switching Bifurcations

Definition 1.5 Consider a 1-dimensional discrete system in Eq. (1.19) as

$$\begin{aligned} x_{k+1} &= x_k + A(\mathbf{p})x_k^2 + B(\mathbf{p})x_k + C(\mathbf{p}) \\ &= x_k + a_0(\mathbf{p})(x_k - a(\mathbf{p}))(x_k - b(\mathbf{p})). \end{aligned} \tag{1.49}$$

(i) For $a < b$, the corresponding standard form is

$$x_{k+1} = x_k + a_0(x_k - a)(x_k - b) \tag{1.50}$$

with two fixed-points

$$\begin{aligned} &x_k^* = a_1 = a \text{ and } x_k^* = a_2 = b \\ &\text{with } \Delta = a_0^2(a-b)^2 > 0. \end{aligned} \tag{1.51}$$

(ii) For $a > b$, the corresponding standard form is

$$x_{k+1} = x_k + a_0(x_k - b)(x_k - a) \tag{1.52}$$

with two fixed-points of

$$\begin{aligned} &x_k^* = a_1 = b \text{ and } x_k^* = a_2 = a \\ &\text{with } \Delta = a_0^2(a-b)^2 > 0. \end{aligned} \tag{1.53}$$

(iii) For $a = b$, the corresponding standard form is

$$x_{k+1} = x_k + a_0(x_k - a)^2 \tag{1.54}$$

with a repeated fixed-point of $x_k^* = a$. Such a fixed-point is called a monotonic *saddle* of the second-order.

(iii$_1$) If $a_0 > 0$, the fixed-point is a *monotonic upper-saddle* of the second-order.

(iii$_2$) If $a_0 < 0$, the fixed-point is a *monotonic lower-saddle* of the second-order.

(iv) The fixed-point of $x_k^* = a$ for two fixed-points switching is called a monotonic *saddle-node* switching bifurcation of fixed-points at a point $\mathbf{p} = \mathbf{p}_0 \in \partial\Omega_{12}$, and the bifurcation condition is

$$\Delta = a_0^2(a - b)^2 = 0 \text{ or } a = b. \tag{1.55}$$

Theorem 1.5

(i) *Under a condition of*

$$a < b \text{ and } \Delta = a_0^2(a - b)^2 > 0 \tag{1.56}$$

a standard form of the 1-dimensional discrete system in Eq. (1.49) *is*

$$x_{k+1} = x_k + f(x_k, \mathbf{p}) = x_k + a_0(x_k - a)(x_k - b). \tag{1.57}$$

(i$_1$) *For $a_0(\mathbf{p}) > 0$, there are two cases:*

(i$_{1a}$) *The fixed-point of $x_k^* = a$ is:*

- *monotonically stable (a monotonic sink) if $df/dx_k|_{x_k^*=a} \in (-1, 0)$;*
- *invariantly stable (an invariant sink) if $df/dx_k|_{x_k^*=a} = -1$;*
- *oscillatorilly stable (an oscillatory source) if $df/dx_k|_{x_k^*=a} \in (-2, -1)$;*
- *flipped if $df/dx_k|_{x_k^*=a} = -2$ (an oscillatory lower-saddle of the second-order with $d^2f/dx_k^2|_{x_k^*=a} = a_0 < 0$);*
- *oscillatorilly unstable (an oscillator source) if $df/dx_k|_{x_k^*=a} \in (-\infty, -2)$.*

(i$_{1b}$) *The fixed-point of $x_k^* = b$ is monotonically unstable (a monotonic source) if $df/dx_k|_{x_k^*=b} > 0$.*

(i$_2$) *For $a_0(\mathbf{p}) < 0$, there are two cases:*

(i$_{2a}$) *The fixed-point of $x_k^* = a$ is monotonically unstable (a monotonic source) if $df/dx_k|_{x_k^*=a} > 0$.*

(i_{2b}) *The fixed-point of* $x_k^* = b$ *is:*

- *monotonically stable (a monotonic sink) if* $df/dx_k|_{x_k^*=b} \in (-1, 0)$;
- *invariantly stable (an invariant sink) if* $df/dx_k|_{x_k^*=b} = -1$;
- *oscillatorilly stable (an oscillatory source) if* $df/dx_k|_{x_k^*=b} \in (-2, -1)$;
- *flipped if* $df/dx_k|_{x_k^*=b} = -2$ *(an oscillatory lower-saddle of the second-order with* $d^2f/dx_k^2|_{x_k^*=b} = a_0 < 0$*);*
- *oscillatorilly unstable (an oscillatory source) if* $df/dx_k|_{x_k^*=b} \in (-\infty, -2)$.

(ii) *Under a condition of*

$$a > b \text{ and } \Delta = a_0^2(a-b)^2 > 0 \tag{1.58}$$

a standard form of the 1-dimensional discrete system in Eq. (1.49) *is*

$$x_{k+1} = x_k + f(x_k, \mathbf{p}) = x_k + a_0(x_k - b)(x_k - a). \tag{1.59}$$

(ii_1) *For* $a_0(\mathbf{p}) > 0$, *there are two cases:*

(ii_{1a}) The fixed-point of $x_k^* = a$ is monotonically unstable with $df/dx_k|_{x_k^*=a} > 0$.

(ii_{1b}) The fixed-point of $x_k^* = b$ is:

- *monotonically stable (a monotonic sink) if* $df/dx_k|_{x_k^*=b} \in (-1, 0)$;
- *invariantly stable (an invariant sink) if* $df/dx_k|_{x_k^*=b} = -1$;
- *oscillatorilly stable (an oscillatory sink) if* $df/dx_k|_{x_k^*=b} \in (-2, -1)$;
- *flipped if* $df/dx_k|_{x_k^*=b} = -2$ *(an oscillatory upper-saddle of the second-order with* $d^2f/dx_k^2|_{x_k^*=b} = a_0 > 0$*);*
- *oscillatorilly unstable (an oscillatory source) if* $df/dx_k|_{x_k^*=b} \in (-\infty, -2)$.

(ii_2) *For* $a_0(\mathbf{p}) < 0$, *there are two cases:*

(ii_{2a}) *The fixed-point of* $x_k^* = a$ *is:*

- *monotonically stable (a monotonic sink) if* $df/dx_k|_{x_k^*=a} \in (-1, 0)$;
- *invariantly stable (an invariant sink) if* $df/dx_k|_{x_k^*=a} = -1$;
- *oscillatorilly stable (an oscillatory sink) if* $df/dx_k|_{x_k^*=a} \in (-2, -1)$;
- *flipped if* $df/dx_k|_{x_k^*=a} = -2$ *(an oscillatory lower-saddle of the second-order with* $d^2f/dx_k^2|_{x_k^*=a} = a_0 < 0$*);*
- *oscillatorilly unstable (an oscillatory source) if* $df/dx_k|_{x_k^*=a} \in (-\infty, -2)$.

(ii_{2b}) *The fixed-point of* $x_k^* = b$ *is monotonically unstable (a monotonic source) if* $df/dx_k|_{x_k^*=b} > 0$.

(iii) *For $a = b$, the corresponding standard form with $\Delta = 0$ is*

$$x_{k+1} = x_k + f(x, \mathbf{p}) = x_k + a_0(x - a)^2. \tag{1.60}$$

(iii₁) *If $a_0(\mathbf{p}) > 0$, then the fixed-point $x^* = a$ is a monotonically* upper-saddle *of the second order with $d^2f/dx^2|_{x^*=a} > 0$. The fixed-point $x_k^* = a$ for two fixed-points switching is a monotonic* upper-saddle-node switching *bifurcation of the second order.*

(iii₂) *If $a_0(\mathbf{p}) < 0$, then the fixed-point $x_k^* = a$ is a monotonic* lower-saddle *of the second-order with $d^2f/dx^2|_{x^*=a} < 0$. The fixed-point $x^* = a$ for two fixed-points switching is a monotonic* lower-saddle-node switching *bifurcation of the second-order.*

Proof The theorem can be proved as for Theorem 1.3. ■

Definition 1.6 If $C(\mathbf{p}) = 0$ in Eq. (1.19), a 1-dimensional quadratic discrete system is

$$x_{k+1} = x_k + A(\mathbf{p})x_k^2 + B(\mathbf{p})x_k. \tag{1.61}$$

(i) For $A(\mathbf{p}) \times B(\mathbf{p}) < 0$, the corresponding standard form is

$$x_{k+1} = x_k + a_0 x_k (x_k - a) \tag{1.62}$$

with two fixed-points of

$$x_k^* = a_1 = 0 \text{ and } x_k^* = a_2 = a > 0$$
$$\text{with } a_0 = A(\mathbf{p}) \text{ and } a = -\frac{B(\mathbf{p})}{A(\mathbf{p})}. \tag{1.63}$$

(ii) For $A(\mathbf{p}) \times B(\mathbf{p}) > 0$, the corresponding standard form is

$$x_{k+1} = x_k + a_0(x_k - a)x_k \tag{1.64}$$

with two fixed-points of

$$x_k^* = a_1 = a < 0 \text{ and } x_k^* = a_2 = 0. \tag{1.65}$$

(iii) For $B(\mathbf{p}) = 0$, the corresponding standard form is

$$x_{k+1} = x_k + a_0 x_k^2 \tag{1.66}$$

with a double fixed-point of $x_k^* = 0$. Such a fixed-point is called a monotonic *saddle* of the second order.

(iii₁) If $a_0 > 0$, the fixed-point is a monotonic *upper-saddle* of the second order.

(iii$_2$) If $a_0 < 0$, the fixed-point is a monotonic *lower-saddle* of the second order.

(iv) The bifurcation of $x^* = 0$ for two fixed-points switching is called a monotonic *saddle-node switching* bifurcation at a point $\mathbf{p} = \mathbf{p}_0 \in \partial\Omega_{12}$, and the bifurcation condition is

$$B(\mathbf{p}_0) = 0. \tag{1.67}$$

Theorem 1.6

(i) *Under a condition of*

$$A(\mathbf{p}) \times B(\mathbf{p}) < 0, \tag{1.68}$$

a standard form of the 1-dimensional discrete system in Eq. (1.61) *is*

$$x_{k+1} = x_k + f(x_k, \mathbf{p}) = x_k + a_0 x_k (x_k - a). \tag{1.69}$$

(i$_1$) *For* $a_0(\mathbf{p}) > 0$ *there are two cases:*

(i$_{1a}$) *The fixed-point of* $x_k^* = 0$ *is:*

- *monotonically stable (a monotonic sink) if* $df/dx_k|_{x_k^*=0} \in (-1, 0)$;
- *invariantly stable (an invariant sink) if* $df/dx_k|_{x_k^*=0} = -1$;
- *oscillatorilly stable (an oscillatory sink) if* $df/dx_k|_{x_k^*=0} \in (-2, -1)$;
- *flipped if* $df/dx_k|_{x_k^*=0} = -2$ *(an oscillatory upper-saddle of the second-order with* $d^2f/dx_k^2|_{x_k^*=a} = a_0 > 0$*);*
- *oscillatorilly unstable (an oscillatory source) if* $df/dx_k|_{x_k^*=0} \in (-\infty, -2)$.

(i$_{1b}$) *The fixed-point of* $x_k^* = a$ *is monotonically unstable with* $df/dx_k|_{x_k^*=a} > 0$.

(i$_2$) *For* $a_0(\mathbf{p}) < 0$, *there are two cases:*

(i$_{2a}$) *The fixed-point of* $x_k^* = 0$ *is monotonically unstable with* $df/dx_k|_{x_k^*=0} > 0$.

(i$_{2b}$) *The fixed-point of* $x_k^* = a$ *is:*

- *monotonically stable (a monotonic sink) if* $df/dx_k|_{x_k^*=a} \in (-1, 0)$;
- *invariantly stable (an invariant sink) if* $df/dx_k|_{x_k^*=a} = -1$;
- *oscillatorilly stable (an oscillatory sink) if* $df/dx_k\,|_{x_k^*=a} \in (-2, -1)$;
- *flipped if* $df/dx_k|_{x_k^*=a} = -2$ *(an oscillatory lower-saddle of the second kind with* $d^2f/dx_k^2|_{x_k^*=a} = a_0 < 0$*);*
- *oscillatorilly unstable (an oscillatory source) if* $df/dx_k|_{x_k^*=a} \in (-\infty, -2)$.

(ii) *Under a condition of*

$$A(\mathbf{p}) \times B(\mathbf{p}) > 0, \tag{1.70}$$

a standard form of the 1-dimensional quadratic system in Eq. (1.61) *is*

$$x_{k+1} = x_k + a_0(x_k - a)x_k. \tag{1.71}$$

(ii_1) *For* $a_0(\mathbf{p}) > 0$, *there are two cases:*

($\mathrm{ii}_{1\mathrm{a}}$) *The fixed-point of* $x_k^* = 0$ *is monotonically unstable with* $df/dx_k|_{x_k^*=0} > 0$.

($\mathrm{ii}_{1\mathrm{b}}$) *The fixed-point of* $x_k^* = a$ *is:*

- *monotonically stable (a monotonic sink) if* $df/dx_k|_{x_k^*=a} \in (-1, 0)$;
- *invariantly stable (an invariant sink) if* $df/dx_k|_{x_k^*=a} = -1$;
- *oscillatorilly stable (an oscillatory sink) if* $df/dx_k|_{x_k^*=a} \in (-2, -1)$;
- *flipped if* $df/dx_k|_{x_k^*=a} = -2$ *(an oscillatory upper-saddle of the second-order with* $d^2f/dx_k^2|_{x_k^*=a} = a_0 > 0$);
- *oscillatorilly unstable (an oscillatory source) if* $df/dx_k|_{x_k^*=a} \in (-\infty, -2)$.

(ii_2) *For* $a_0(\mathbf{p}) < 0$, *there are two cases:*

($\mathrm{ii}_{2\mathrm{a}}$) *The fixed-point of* $x_k^* = 0$ *is:*

- *monotonically stable (a monotonic sink) if* $df/dx_k|_{x_k^*=0} \in (-1, 0)$;
- *invariantly stable (an invariant sink) if* $df/dx_k|_{x_k^*=0} = -1$;
- *oscillatorilly stable (an oscillatory sink) if* $df/dx_k|_{x_k^*=0} \in (-2, -1)$;
- *flipped if* $df/dx_k|_{x_k^*=0} = -2$ *(an oscillatory lower-saddle of the second-order for* $d^2f/dx_k^2|_{x_k^*=0} < 0$);
- *oscillatorilly unstable (an oscillatory source) if* $df/dx_k|_{x_k^*=0} \in (-\infty, -2)$.

($\mathrm{ii}_{2\mathrm{b}}$) *The fixed-point of* $x_k^* = a$ *is monotonically unstable if* $df/dx_k|_{x_k^*=a} > 0$.

(iii) *For* $B(\mathbf{p}) = 0$, *the corresponding standard form with* $\Delta = 0$ *is*

$$x_{k+1} = x_k + f(x_k, \mathbf{p}) = x_k + a_0 x_k^2. \tag{1.72}$$

(iii_1) *If* $a_0(\mathbf{p}) > 0$, *then the fixed-point of* $x_k^* = 0$ *is a monotonic* upper-saddle *of the second-order with* $d^2f/dx_k^2|_{x_k^*=0} > 0$. *The fixed-point of* $x_k^* = 0$ *for two fixed-point switching is a monotonic* upper-saddle-node *switching bifurcation of the second-order.*

(iii_2) *If* $a_0(\mathbf{p}) < 0$, *then the fixed-point of* $x_k^* = 0$ *is a monotonic* lower-saddle *of the second-order with* $d^2f/dx_k^2|_{x_k^*=0} < 0$. *The fixed-point of* $x_k^* = 0$ *for two fixed-points switching is a monotonic* lower-saddle-node switching *bifurcation of the second-order.*

Proof The theorem can be proved as for Theorem 1.3. ■

The stability and bifurcation of two fixed-points for the 1-dimensional forward discrete system in Eq. (1.49) with $\Delta = B^2 - 4AC = a_0^2(a-b)^2 \geq 0$ are presented in Fig. 1.5. The stable and unstable fixed-points varying with the vector parameter are depicted by solid and dashed curves, respectively. The bifurcation point of fixed-points occurs at the double-repeated fixed-point at the boundary of $\mathbf{p}_0 \in \partial\Omega_{12}$. With varying parameters, the two fixed-points of $x_k^* = a, b$ equal each other (i.e., $x_k^* = a = b$). Such a fixed-point is a switching bifurcation point at $x_k^* = a = b$ for $\Delta = 0$. The fixed-points of $x_k^* = a, b$ with $\Delta \geq 0$ are presented in Fig. 1.5i and ii for $a_0 > 0$ and $a_0 < 0$, respectively. The quadratic discrete system in Eq. (1.61) is as a special case of the discrete system in Eq. (1.19) with $C(\mathbf{p}) = 0$. Thus $\Delta = B^2 - 4AC = B^2 \geq 0$. The fixed-points exist in the entire domain. In Fig. 1.4iii, for $a_0 > 0$ and $B < 0$, the fixed-point of $x_k^* = 0$ is unstable, and the fixed-point of $x_k^* = a$ is from monotonically stable to oscillatorilly unstable. However, for $a_0 > 0$ and $B > 0$, the fixed-point of $x_k^* = a$ is stable, and the fixed-point of $x_k^* = 0$ is from monotonically stable to oscillatorilly unstable. The

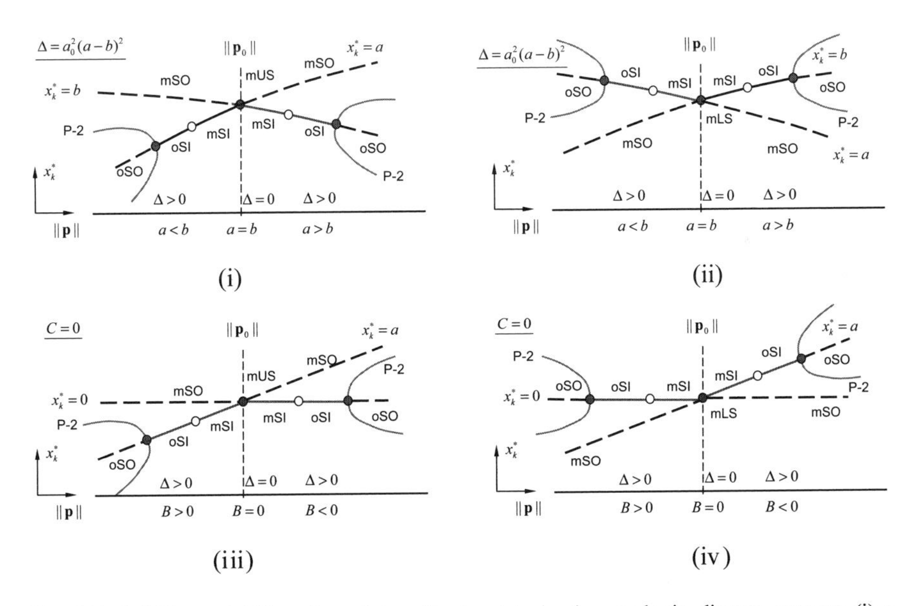

Fig. 1.5 Stability and bifurcation of two fixed-points in the quadratic discrete system: (i) a monotonic *upper-saddle*-node switching bifurcation ($a_0 > 0$), (ii) a monotonic *lower-saddle-node* switching bifurcation ($a_0 < 0$); (iii) a monotonic upper-saddle-node switching bifurcation ($a_0 > 0$), (iv) a monotonic *lower-saddle-node* switching bifurcation ($a_0 < 0$). Stable and unstable fixed-points are represented by solid and dashed curves, respectively. (mSO: monotonic source; mSI: monotonic sink; oSO: oscillatory source; oSI: oscillatory sink; mLS: monotonic lower-saddle; mUS: monotonic upper-saddle; oUS: oscillatory upper-saddle; oLS: oscillatory lower-saddle; iSI: invariant sink.)

bifurcation of fixed-points occurs at $B = 0$. The discrete flow of x_k $(k = 1, 2, 3\ldots)$ is a forward *upper-saddle* discrete flow for $a_0 > 0$, and the fixed-point of $x_k^* = 0$ at $B = 0$ is termed a monotonic *upper-saddle of the second-order*. Such a bifurcation is termed a monotonic *upper-saddle-node* bifurcation of the second-order. In Fig. 1.5iv, for $a_0 < 0$ and $B < 0$, the fixed-points of $x_k^* = a$ are unstable, and the fixed-point of $x_k^* = 0$ is from monotonically stable to oscillatorilly unstable. However, for $a_0 < 0$ and $B > 0$, the fixed-points of $x_k^* = 0$ are unstable, and the fixed-point of $x_k^* = a$ is from monotonically stable to oscillatorilly unstable. The bifurcation of fixed-points also occurs at $B = 0$. The discrete flow of $x_k (k = 1, 2, \ldots)$ is a forward *lower-saddle discrete* flow for $a_0 < 0$, and the fixed-point of $x_k^* = 0$ at $B = 0$ is termed a monotonic *lower-saddle of the second-order*. Such a bifurcation is termed a monotonic *lower-saddle-node* switching bifurcation of the second-order.

1.3 Backward Quadratic Discrete Systems

For one of the simplest nonlinear discrete systems, consider a backward discrete quadratic nonlinear system as follows.

1.3.1 Backward Period-1 Appearing Bifurcations

Definition 1.7 Consider a 1-dimensional, backward, quadratic discrete system

$$x_k = x_{k+1} + A(\mathbf{p})x_{k+1}^2 + B(\mathbf{p})x_{k+1} + C(\mathbf{p}) \tag{1.73}$$

where three scalar constants $A(\mathbf{p}) \neq 0$, $B(\mathbf{p})$ and $C(\mathbf{p})$ are determined by a vector parameter

$$\mathbf{p} = (p_1, p_2, \ldots, p_m)^{\mathrm{T}}. \tag{1.74}$$

(i) If

$$\Delta = B^2 - 4AC < 0 \quad \text{for} \quad \mathbf{p} \in \Omega_1 \subset \mathbf{R}^m, \tag{1.75}$$

then the negative quadratic discrete system does not have any fixed-points. The flow without fixed-points is called a non-fixed-point flow.

(i_1) If $a_0(\mathbf{p}) = A(\mathbf{p}) > 0$, the non-fixed-point flow in the backward discrete system is called a negative backward discrete flow.

(i_2) If $a_0(\mathbf{p}) = A(\mathbf{p}) < 0$, the non-fixed-point flow in the backward quadratic discrete systems is called a positive backward discrete flow.

(ii) If

$$\Delta = B^2 - 4AC > 0 \quad \text{for} \quad \mathbf{p} \in \Omega_2 \subset \mathbf{R}^m \tag{1.76}$$

then the backward quadratic discrete system has two simple fixed-points as

$$x_{k+1}^* = x_k^* = a_1 \text{ and } x_{k+1}^* = x_k^* = a_2, \tag{1.77}$$

and the corresponding standard form is

$$x_k = x_{k+1} + a_0(x_{k+1} - a_1)(x_{k+1} - a_2), \tag{1.78}$$

where

$$a_0 = A(\mathbf{p}),\ a_{1,2} = \frac{-B(\mathbf{p}) \pm \sqrt{\Delta}}{2A(\mathbf{p})} \text{ with } a_1 < a_2. \tag{1.79}$$

(iii) If

$$\Delta = B^2 - 4AC = 0 \quad \text{for} \quad \mathbf{p} = \mathbf{p}_0 \in \partial\Omega_{12} \subset \mathbf{R}^{m-1}, \tag{1.80}$$

then the backward quadratic discrete system has a repeated fixed-point, i.e.,

$$x_{k+1}^* = a_1 \text{ and } x_{k+1}^* = a_1, \tag{1.81}$$

with the corresponding standard form of

$$x_k = x_{k+1} + a_0(x_{k+1} - a_1)^2, \tag{1.82}$$

where

$$a_0 = A(\mathbf{p}_0), \text{ and } a_1 = a_2 = -\frac{B(\mathbf{p}_0)}{2A(\mathbf{p}_0)}. \tag{1.83}$$

Such a discrete flow with the fixed-point of $x_k^* = a_1$ is called a monotonic saddle backward discrete flow of the second order.

(iii$_1$) If $a_0(\mathbf{p}) > 0$, then the fixed-point of $x_{k+1}^* = a_1(\mathbf{p})$ is a monotonic *lower-saddle*.

(iii$_2$) If $a_0(\mathbf{p}) < 0$, then the fixed-point of $x_{k+1}^* = a_1(\mathbf{p})$ is a monotonic *upper-saddle*.

(iv) The fixed-point of $x_{k+1}^* = a_1$ for two fixed-points vanishing or appearance is called a *saddle-node* bifurcation of the second-order at a point $\mathbf{p} = \mathbf{p}_0 \in \partial\Omega_{12}$, and the backward bifurcation condition is

$$\Delta = B^2 - 4AC = 0. \tag{1.84}$$

(iv$_1$) If $a_0(\mathbf{p}) > 0$, the bifurcation at $x^*_{k+1} = a_1(\mathbf{p})$ for the appearance or vanishing of two fixed-points is called a monotonic *lower-saddle-node* appearing bifurcation of the second-order.

(iv$_2$) If $a_0(\mathbf{p}) < 0$, the bifurcation at $x^*_{k+1} = a_1(\mathbf{p})$ for the appearance or vanishing of two fixed-points is called a monotonic *upper-saddle-node* appearing bifurcation of the second-order.

Theorem 1.7

(i) *Under a condition of*

$$\Delta = B^2 - 4AC < 0, \tag{1.85}$$

a standard form of the 1-dimensional backward discrete system in Eq. (1.73) *is*

$$x_k = x_{k+1} + a_0[(x_{k+1} - \frac{1B}{2A})^2 + \frac{1}{4A^2}(-\Delta)] \tag{1.86}$$

with $a_0 = A(\mathbf{p})$, *which has a non-fixed-point flow.*

(i$_1$) *If* $a_0(\mathbf{p}) > 0$, *the non-fixed-point discrete flow is called a positive backward discrete flow.*

(i$_2$) *If* $a_0(\mathbf{p}) > 0$, *the non-fixed-point discrete flow is called a negative backward discrete flow.*

(ii) *Under a condition of*

$$\Delta = B^2 - 4AC > 0, \tag{1.87}$$

a standard form of the 1-dimensional backward discrete system in Eq. (1.73) *is*

$$x_k = x_{k+1} + f(x_{k+1}, \mathbf{p}) = x_{k+1} + a_0(x_{k+1} - a_1)(x_{k+1} - a_2). \tag{1.88}$$

(ii$_1$) *For* $a_0(\mathbf{p}) > 0$, *the following cases exist.*

(ii$_{1a}$) *The fixed-point of* $x^*_{k+1} = x^*_k = a_1(\mathbf{p})$ *is*

- *monotonically unstable (a monotonic source) if* $df/dx_{k+1}|_{x^*_{k+1}=a_1} \in (-1, 0)$;
- *infinitely unstable (or an infinite source) if* $df/dx_{k+1}|_{x^*_{k+1}=a_1} = -1$, *which is*
  - *monotonically unstable with a positive infinity eigenvalue (a monotonic positive infinity source) if* $df/dx_{k+1}|_{x^*_{k+1}=a_1} = -1^+$;

  - *oscillatorilly unstable with a negative infinity eigenvalue (an oscillatory, negative infinity source) if* $df/dx_{k+1}|_{x^*_{k+1}=a_1} = -1^-$;

- *oscillatorilly unstable (an oscillatory source) if* $df/dx_{k+1}|_{x^*_{k+1}=a_1} \in (-2,-1)$;
- *flipped if* $df/dx_{k+1}|_{x^*_{k+1}=a_1} = -2$ *(an oscillatory lower-saddle of the second order with* $d^2f/dx^2_{k+1}|_{x^*_{k+1}=a_1} = a_0 > 0$*)*;
- oscillatorilly stable (an oscillatory sink) if $df/dx_{k+1}|_{x^*_{k+1}=a_1} \in (-\infty,-2)$.

($\mathrm{ii}_{1\mathrm{b}}$) *The fixed-point of* $x^*_{k+1} = x^*_k = a_2(\mathbf{p})$ *is monotonically stable (a monotonic sink) if* $df/dx_{k+1}|_{x^*_{k+1}=a_2} \in (0,\infty)$.

(ii_2) *For* $a_0(\mathbf{p})<0$, *the following cases exist.*

($\mathrm{ii}_{2\mathrm{a}}$) *The fixed-point of* $x^*_{k+1} = x^*_k = a_2(\mathbf{p})$ *is*

- *monotonically unstable (a monotonic source) if* $df/dx_{k+1}|_{x^*_{k+1}=a_2} \in (-1,0)$;
- *infinitely unstable (an infinite source) if* $df/dx_{k+1}|_{x^*_{k+1}=a_2} = -1$, *which is*

  - *monotonically unstable with a positive infinity eigenvalue (a monotonically positive infinite source) if* $df/dx_{k+1}|_{x^*_{k+1}=a_2} = -1^+$;
  - *oscillatorilly unstable with a negative infinity eigenvalue (or an oscillatory negative infinity source) if* $df/dx_{k+1}|_{x^*_{k+1}=a_2} = -1^-$;

- *oscillatorilly unstable (an oscillatory source) if* $df/dx_{k+1}|_{x^*_{k+1}=a_2} \in (-2,-1)$;
- *flipped if* $df/dx_{k+1}|_{x^*_{k+1}=a_2} = -2$ *(an oscillatory upper-saddle of the second-order with* $d^2f/dx^2_{k+1}|_{x^*_{k+1}=a_1} = a_0<0$*)*;
- *oscillatorilly stable (an oscillatory sink) if* $df/dx_{k+1}|_{x^*_{k+1}=a_2} \in (-\infty,-2)$.

($\mathrm{ii}_{2\mathrm{b}}$) *The fixed-point of* $x^*_{k+1} = x^*_k = a_1(\mathbf{p})$ *is monotonically stable (a monotonic sink) if* $df/dx_{k+1}|_{x^*_{k+1}=a_2} \in (0,\infty)$.

(iii) *Under a condition of*

$$\Delta = B^2 - 4AC = 0, \tag{1.89}$$

a standard form of the 1-dimensional backward discrete system in Eq. (1.73) *is*

$$x_k = x_{k+1} + f(x_{k+1}, \mathbf{p}) = x_{k+1} + a_0(x_{k+1} - a_1)^2. \tag{1.90}$$

(iii$_1$) If $a_0(\mathbf{p}) > 0$, then the fixed-point of $x_{k+1}^* = x_k^* = a_1(\mathbf{p})$ is an monotonic *lower-saddle* of the second-order with $d^2f/dx_{k+1}^2|_{x_{k+1}^*=a_1} > 0$. The bifurcation at $x_k^* = a_1(\mathbf{p})$ for two-fixed-points appearance or vanishing is called a monotonic lower-saddle-node appearing bifurcation of the second-order.

(iii$_2$) If $a_0(\mathbf{p}) < 0$, then the fixed-point of $x_{k+1}^* = x_k^* = a_1(\mathbf{p})$ is a monotonic upper-*saddle* of the second order with $d^2f/dx_{k+1}^2|_{x_{k+1}^*=a_1} < 0$. The bifurcation at $x_k^* = a_1(\mathbf{p})$ for two-fixed-points appearance or vanishing is called a monotonic *upper-saddle-node* appearing bifurcation of the second-order.

Proof

(i) For

$$\Delta = B^2 - 4AC < 0,$$

(i$_1$) if $a_0 > 0$, we have

$$x_{k+1} - x_k = -a_0[(x_{k+1} - \frac{B}{2A})^2 + \frac{1}{4A^2}(-\Delta)] < 0,$$

thus, such a non-fixed-point discrete flow is called a negative backward discrete flow;

(i$_2$) if $a_0 < 0$, we have

$$x_{k+1} - x_k = -a_0[(x_{k+1} - \frac{B}{2A})^2 + \frac{1}{4A^2}(-\Delta)] > 0,$$

thus, such a non-fixed-point discrete flow is called a positive backward discrete flow.

(ii) Let $\Delta x_{k(i)} = x_k - a_i$ $(i = 1, 2)$ and $x_{k+1(i)} - a_i = \Delta x_{k+1(i)}$. Thus, Eq. (1.78) becomes

$$\Delta x_{k(i)} = [1 + a_0(a_i - a_j)]\Delta x_{k+1(i)} + a_0 \Delta x_{k+1(i)}^2 \; (i, j \in \{1, 2\}, j \neq i).$$

Because Δx_i is arbitrary small, we have

$$\Delta x_{k+1(i)} = \lambda_i \Delta x_{k(i)} \text{ for } \lambda_i \equiv (1 + df/dx_{k+1}|_{x_{k+1}^*=a_i})^{-1} = [1 + a_0(a_i - a_j)]^{-1}.$$

The corresponding solution is

$$\Delta x_{k(i)} = (\lambda_i)^k \Delta x_{0(i)}$$

where $\Delta x_{0(i)} = x_{0(i)} - a_i$ is an initial condition.

(ii$_\text{a}$) If $\lambda_i > 1$, we have

$$\lim_{k\to\infty}(x_{k(i)} - a_i) = \lim_{k\to\infty}\Delta x_{k(i)} = \lim_{k\to\infty}(\lambda_i)^k \Delta x_{0(i)} = \infty \Rightarrow \lim_{k\to\infty} x_{k(i)} = \infty$$

and

$$\lambda_i = \frac{1}{1 + a_0(a_i - a_j)} \in (1, \infty),$$
$$\Rightarrow a_0(a_i - a_j) = df/dx_{k+1}|_{x^*_{k+1}=a_i} \in (-1, 0).$$

(ii$_\text{a1}$) For $a_0 > 0$, we have

$$a_i < a_j \Rightarrow x^*_{k+1} = x^*_k = a_1.$$

(ii$_\text{a2}$) For $a_0 < 0$, we have

$$a_i > a_j \Rightarrow x^*_k = x^*_{k+1} = a_1.$$

Thus, the fixed-point of $x^*_k = x^*_{k+1} = a_i$ is monotonically unstable.

(ii$_\text{b}$) If $\lambda_i = \pm\infty$, we have

$$\lim_{k\to\infty}(x_{k(i)} - a_i) = \lim_{k\to\infty}\Delta x_{k(i)} = \lim_{k\to\infty}(\lambda_i)^k \Delta x_{0(i)} = \infty \Rightarrow \lim_{k\to\infty} x_{k(i)} = \infty$$

and

$$\lambda_i = \frac{1}{1 + a_0(a_i - a_j)} = \pm\infty$$
$$\Rightarrow a_0(a_i - a_j) = df/dx_{k+1}|_{x^*_{k+1}=a_i} = -1^{\pm}.$$

(ii$_\text{b1}$) For $a_0 > 0$, we have

$$a_i > a_j \Rightarrow x^*_k = x^*_{k+1} = a_1.$$

Thus, the fixed-point of $x^*_k = x^*_{k+1} = a_1$ is monotonically unstable with a positive infinity eigenvalue for $df/dx_{k+1}|_{x^*_{k+1}=a_1} = -1^+$, and the fixed-point of $x^*_k = x^*_{k+1} = a_1$ is oscillatorilly unstable with a negative infinity eigenvalue for $df/dx_{k+1}|_{x^*_{k+1}=a_1} = -1^-$.

(ii_{b2}) For $a_0 < 0$, we have

$$a_i < a_j \Rightarrow x_k^* = x_{k+1}^* = a_2.$$

Thus, the fixed-point of $x_k^* = x_{k+1}^* = a_2$ is oscillatorilly unstable with positive infinity for $df/dx_{k+1}|_{x_{k+1}^*=a_2} = -1^+$, and the fixed-point of $x_k^* = x_{k+1}^* = a_2$ is monotonically unstable with negative infinity for $df/dx_{k+1}|_{x_{k+1}^*=a_2} = -1^-$.

(ii_c) For $\lambda_i < -1$, we have

$$\begin{aligned}\lim_{k\to\infty}(x_{k(i)} - a_i) &= \lim_{k\to\infty}\Delta x_{k(i)} = \lim_{k\to\infty}\lambda_i^k \Delta x_{0(i)} \\ &= \lim_{k\to\infty}(|\lambda_i|)^k(-1)^k \Delta x_{0(i)} = \begin{cases} -\infty & \text{for } k = 2l-1, \\ +\infty & \text{for } k = 2l. \end{cases}\end{aligned}$$

Thus

$$\lim_{k\to\infty} x_{k(i)} = \begin{cases} a_i - \infty & \text{for } k = 2l-1, \\ a_i + \infty & \text{for } k = 2l, \end{cases}$$

and

$$\begin{aligned}&\lambda_i = \frac{1}{1 + a_0(a_i - a_j)} \in (-\infty, -1) \\ &\Rightarrow a_0(a_i - a_j) = df/dx_{k+1}|_{x_{k+1}^*=a_i} \in (-2, -1).\end{aligned}$$

(ii_{c1}) For $a_0 > 0$, we have

$$a_i < a_j \Rightarrow x_k^* = a_1.$$

(ii_{c2}) For $a_0 < 0$, we have

$$a_i > a_j \Rightarrow x_k^* = a_2.$$

Thus, the fixed-point of $x_k^* = a_i$ $(i \in \{1, 2\})$ is oscillatorilly unstable.

(ii_d) For $\lambda_i = -1$, we have

$$\lim_{k\to\infty}(x_{k(i)} - a_i) = \lim_{k\to\infty}\Delta x_{k(i)} = \lim_{k\to\infty}(-1)^k \Delta x_{0(i)},$$

so

$$\lim_{k\to\infty} x_{k(i)} = \begin{cases} a_i + \Delta x_{0(i)} & \text{for } k = 2l-1, \\ a_i + \Delta x_{0(i)} & \text{for } k = 2l. \end{cases}$$

Consider

$$\lambda_i = \frac{1}{1 + a_0(a_i - a_j)} = -1$$
$$\Rightarrow a_0(a_i - a_j) = df/dx_{k+1}|_{x^*_{k+1}=a_i} = -2.$$

($\mathrm{ii_{d1}}$) For $a_0 > 0$, we have

$$a_i < a_j \Rightarrow x^*_k = x^*_{k+1} = a_1.$$

Thus, the fixed-point of $x^*_k = a_1$ is an oscillatory lower-saddle.

($\mathrm{ii_{d2}}$) For $a_0 < 0$, we have

$$a_i > a_j \Rightarrow x^*_k = x^*_{k+1} = a_2.$$

Thus, the fixed-point of $x^*_k = a_2$ is an oscillatory upper-saddle.

($\mathrm{ii_e}$) For $\lambda_i \in (-1, 0)$, we have

$$\begin{aligned} \lim_{k\to\infty} (x_{k(i)} - a_i) &= \lim_{k\to\infty} \Delta x_{k(i)} = \lim_{k\to\infty} \lambda_i^k \Delta x_{0(i)} \\ &= \lim_{k\to\infty} (|\lambda_i|)^k (-1)^k \Delta x_{0(i)} \\ &= \begin{cases} 0^- & \text{for } k = 2l-1, \\ 0^+ & \text{for } k = 2l. \end{cases} \end{aligned}$$

Thus

$$\lim_{k\to\infty} x_{k(i)} = \begin{cases} a_i^- & \text{for } k = 2l-1, \\ a_i^+ & \text{for } k = 2l. \end{cases}$$

Consider

$$\lambda_i = \frac{1}{1 + a_0(a_i - a_j)} \in (-1, 0)$$
$$\Rightarrow a_0(a_i - a_j) = df/dx_{k+1}|_{x^*_{k+1}=a_i} \in (-\infty, -2).$$

($\mathrm{ii_{e1}}$) For $a_0 > 0$, we have

$$a_i < a_j \Rightarrow x_k^* = x_{k+1}^* = a_1.$$

($\mathrm{ii_{e2}}$) For $a_0 < 0$, we have

$$a_i > a_j \Rightarrow x_k^* = x_{k+1}^* = a_2.$$

Thus, the fixed-point of $x_k^* = a_i$ $(i \in \{1, 2\})$ is oscillatorilly stable.

($\mathrm{ii_f}$) For $\lambda_i \in (0, 1)$, we have

$$\begin{aligned}&\lim_{k\to\infty}(x_{k(i)} - a_i) = \lim_{k\to\infty}\Delta x_{k(i)} = \lim_{k\to\infty}(\lambda_i)^k \Delta x_{0(i)} = 0\\&\Rightarrow \lim_{k\to\infty} x_{k(i)} = a_i.\end{aligned}$$

Consider

$$\begin{aligned}&\lambda_i = \frac{1}{1 + a_0(a_i - a_j)} \in (0, 1)\\&\Rightarrow a_0(a_i - a_j) = df/dx_{k+1}|_{x_{k+1}^* = a_i} \in (0, \infty).\end{aligned}$$

($\mathrm{ii_{f1}}$) For $a_0 > 0$, we have

$$a_i > a_j \Rightarrow x_k^* = x_{k+1}^* = a_2.$$

($\mathrm{ii_{f2}}$) For $a_0 < 0$, we have

$$a_i < a_j \Rightarrow x_k^* = x_{k+1}^* = a_1.$$

Thus, the fixed-point of $x_k^* = a_i$ is monotonically stable.

($\mathrm{ii_g}$) For $\lambda_i = 0$, we have

$$(x_{k(i)} - a_i) = \Delta x_{k(i)} = (\lambda_i)^k \Delta x_{0(i)} = 0 \Rightarrow x_{k(i)} = a_i.$$

Consider

$$\begin{aligned}&\lambda_i = \frac{1}{1 + a_0(a_i - a_j)} = 0^{\pm}\\&\Rightarrow a_0(a_i - a_j) = df/dx_{k+1}|_{x_{k+1}^* = a_i} = \pm\infty.\end{aligned}$$

(ii_{g1}) For $a_0 > 0$, due to $a_1 < a_2$, we have

$$x_k^* = x_{k+1}^* = a_2 \to \infty \text{ with monotonic convergence,}$$
$$x_k^* = x_{k+1}^* = a_1 \to -\infty \text{ with oscillatory convergence.}$$

(ii_{g2}) For $a_0 < 0$, due to $a_1 < a_2$, we have

$$x_k^* = x_{k+1}^* = a_2 \to \infty \text{ with oscillatory convergence,}$$
$$x_k^* = x_{k+1}^* = a_1 \to -\infty \text{ with monotonic convergence.}$$

(iii) If $a_1(\mathbf{p}) = a_2(\mathbf{p})$, $\lambda_i = 1$ and we have

$$\Delta x_k = \Delta x_{k+1} + a_0 \Delta x_{k+1}^2 = (1 + a_0 \Delta x_{k+1}) \Delta x_{k+1}$$

(iii_1) For $a_0 > 0$, $0 < \Delta x_{k+1} < \Delta x_k$ if $\Delta x_{k+1} > 0$ and $\Delta x_{k+1} < \Delta x_k < 0$ if $\Delta x_{k+1} < 0$. So a backward discrete flow of x_k reaches to $x_{k+1}^* = a_1$ from the initial point of $x_{k0} > a_1$ and it goes to the negative infinity from $x_{k0} < a_1$. Such a fixed-point is unstable of the second order, which is called a monotonic *lower-saddle*.

(iii_2) Similarly, for $a_0 < 0$, $\Delta x_{k+1} > \Delta x_k > 0$ if $\Delta x_k > 0$ and $0 > \Delta x_{k+1} > \Delta x_k$ if $\Delta x_k < 0$. So a backward discrete flow of x_k reaches to $x_k^* = a_1$ from the initial point of $x_{k0} < a_1$ and it goes to the positive infinity from $x_{k0} > a_1$. Such a fixed-point is unstable of the second order, which is called a monotonic *upper-saddle*.

The theorem is proved. ■

The stability and bifurcation of fixed-points for the 1-dimensional backward discrete system in Eq. (1.73) are illustrated in Fig. 1.6. The stable and unstable fixed-points varying with the vector parameter are depicted by the solid and dashed curves, respectively. The bifurcation point of fixed-points occurs at the repeated fixed-points at the boundary of $\mathbf{p}_0 \in \partial\Omega_{12}$. In Fig. 1.6i, for $a_0 > 0$, the fixed-point of $x_{k+1}^* = a_2$ for $\Delta > 0$ are monotonically stable, and the fixed-point of $x_{k+1}^* = a_1$ for $\Delta > 0$ is from monotonically unstable to oscillatorilly stable. The corresponding period-2 fixed-points are determined through the oscillatory-lower-saddle bifurcation, which is called a period-doubling bifurcation. The period-2 fixed-points are unstable. The bifurcation of fixed-points also occurs at $\Delta = 0$, and the discrete flow of x_k is a backward *lower-branch discrete* flow for $a_0 > 0$, and the fixed-point $x_{k+1}^* = -B(\mathbf{p}_0)/2A(\mathbf{p}_0)$ at $\Delta = 0$ is termed a monotonic *lower-saddle of the second-order*. Such a bifurcation is termed a monotonic *lower-saddle-node*

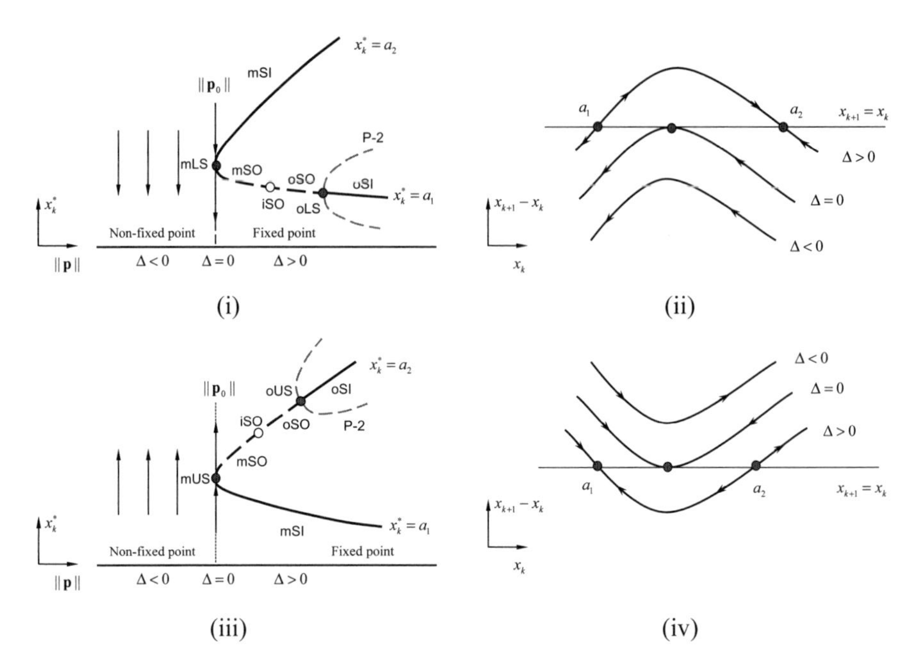

Fig. 1.6 Stability and bifurcation of two fixed-points in the quadratic backward discrete system: (i) a monotonic lower-*saddle-node* bifurcation and (ii) phase portrait ($a_0 > 0$), (iii) a monotonic *upper-saddle-node* bifurcation and (iv) phase portrait ($a_0 < 0$). Stable and unstable fixed-points are represented by solid and dashed curves, respectively. The red color curve is for period-2 fixed-point. (mSO: monotonic source; mSI: monotonic sink; oSO: oscillatory source; oSI: oscillatory sink; mLS: monotonic lower-saddle; mUS: monotonic upper-saddle; oUS: oscillatory upper-saddle; oLS: oscillatory lower-saddle; iSI: invariant sink.)

bifurcation. For $\Delta < 0$, no any fixed-points exist. Such a 1-dimensional backward discrete system is termed the non-fixed-point backward discrete system. For $\Delta < 0$ and $a_0 > 0$, the discrete flow of x_{k+1} is always toward the negative direction due to $x_{k+1} - x_k = -a_0[(x_{k+1} + \frac{B}{2A})^2 + (-\frac{\Delta}{4A^2})] < 0$. The corresponding phase portrait is presented in Fig. 1.6ii. In Fig. 1.6iii, the fixed-point of $x^*_{k+1} = a_1$ for $a_0 < 0$ are a monotonically stable, and the fixed-points of $x^*_{k+1} = a_2$ for $a_0 < 0$ are from monotonically unstable to oscillatorilly stable. The corresponding period-2 fixed-points are determined through the oscillatory-upper-saddle bifurcation, which is called a period-doubling bifurcation. The period-2 fixed-points are unstable. The bifurcation of fixed-points also occurs at $\Delta = 0$. The discrete flow of x_k is a backward *upper-branch discrete* flow for $a_0 < 0$, and the fixed-point of $x^*_{k+1} = -B(\mathbf{p}_0)/2A(\mathbf{p}_0)$ at $\Delta = 0$ is termed an monotonic *upper-saddle of the second-order*. Such a bifurcation of fixed-points is termed a monotonic *upper-saddle-node* bifurcation. For $\Delta < 0$ and $a_0 < 0$, the backward discrete flow of

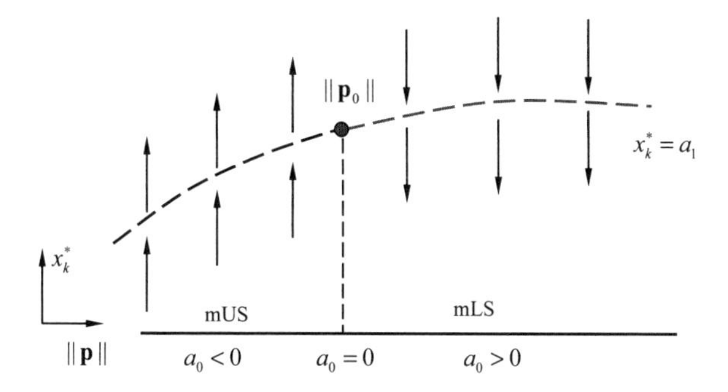

Fig. 1.7 Stability and bifurcation of a double fixed-point of the second order in the quadratic backward discrete system. Unstable fixed-points is represented by a dashed curve. The stability switching from the lower-saddle to upper-saddle is labelled by a circular symbol. (mUS: monotonic upper-saddle; mLS: monotonic lower-saddle.)

x_k is always toward the positive direction due to $x_{k+1} - x_k = -a_0[(x_{k+1} + \frac{B}{2A})^2 + (-\frac{\Delta}{4A^2})] > 0$. The corresponding phase portrait is presented in Fig. 1.6iv.

To illustrate the stability and bifurcation of fixed-points with singularity in a 1-dimensional, backward system, the fixed-point of $x_{k+1} - x_k = -a_0(x_{k+1} - a_1)^2$ is presented in Fig. 1.7. The fixed-point of $x_{k+1}^* = a_1$ for $a_0 < 0$ and $a_0 > 0$ is the monotonic *upper-saddle* and *lower-saddle* of the second-order, respectively. The monotonic *upper-saddle* and *lower-saddle* fixed-points of $x_{k+1}^* = a_1$ with the second-order are unstable, which are depicted by dashed curves. At $a_0 = 0$, the monotonic *upper-saddle* and *lower-saddle* fixed-points will be switched, which is marked by a circular symbol.

1.3.2 Backward Period-1 Switching Bifurcations

Definition 1.8 Consider a 1-dimensional, backward discrete system in Eq. (1.73) as

$$\begin{aligned} x_k &= x_{k+1} + A(\mathbf{p})x_{k+1}^2 + B(\mathbf{p})x_{k+1} + C(\mathbf{p}) \\ &= x_{k+1} + a_0(\mathbf{p})(x_{k+1} - a(\mathbf{p}))(x_{k+1} - b(\mathbf{p})). \end{aligned} \tag{1.91}$$

(i) For $a < b$, the corresponding standard form is

$$x_k = x_{k+1} + a_0(x_{k+1} - a)(x_{k+1} - b) \tag{1.92}$$

with two fixed-points

$$\begin{aligned} & x_k^* = x_{k+1}^* = a_1 = a \text{ and } x_k^* = x_{k+1}^* = a_2 = b \\ & \text{with } \Delta = a_0^2(a-b)^2 > 0. \end{aligned} \tag{1.93}$$

(ii) For $a > b$, the corresponding standard form is

$$x_k = x_{k+1} + a_0(x_{k+1} - b)(x_{k+1} - a) \tag{1.94}$$

with two fixed-points of

$$\begin{aligned} & x_k^* = x_{k+1}^* = a_1 = b \text{ and } x_k^* = x_{k+1}^* = a_2 = a \\ & \text{with } \Delta = a_0^2(a-b)^2 > 0. \end{aligned} \tag{1.95}$$

(iii) For $a = b$, the corresponding standard form is

$$x_k = x_{k+1} + a_0(x_{k+1} - a)^2 \tag{1.96}$$

with a repeated fixed-point of $x_k^* = x_{k+1}^* = a$. Such a fixed-point is called a monotonic *saddle* of the second order.

(iii_1) If $a_0 > 0$, the fixed-point is a *monotonic lower-saddle* of the second order.

(iii_2) If $a_0 < 0$, the fixed-point is a *monotonic upper-saddle* of the second order.

(iv) The fixed-point of $x_k^* = x_{k+1}^* = a$ for two fixed-points switching is called a *saddle-node switching* bifurcation point of fixed-point at a point $\mathbf{p} = \mathbf{p}_0 \in \partial\Omega_{12}$, and the bifurcation condition is

$$\Delta = a_0^2(a-b)^2 = 0 \text{ or } a = b. \tag{1.97}$$

Theorem 1.8

(i) *Under a condition of*

$$a < b \text{ and } \Delta = a_0^2(a-b)^2 > 0 \tag{1.98}$$

a standard form of the 1-dimensional, backward discrete system in Eq. (1.91) *is*

$$x_k = x_{k+1} + f(x_{k+1}, \mathbf{p}) = x_{k+1} + a_0(x_{k+1} - a)(x_{k+1} - b). \tag{1.99}$$

(i_1) *For* $a_0(\mathbf{p}) > 0$, *there are two cases:*

(i_{1a}) *The fixed-point of* $x_k^* = x_{k+1}^* = a$ *is:*

- *monotonically unstable (a monotonic source) if* $df/dx_{k+1}|_{x_{k+1}^*=a} \in (-1,0)$;
- *infinitely unstable if* $df/dx_{k+1}|_{x_{k+1}^*=a} = -1$, *which is*
  - *monotonically unstable with a positive infinity eigenvalue with* $df/dx_{k+1}|_{x_{k+1}^*=a} = -1^+$;
  - *oscillatorilly unstable with a negative infinity eigenvalue with* $df/dx_{k+1}|_{x_{k+1}^*=a} = -1^-$;
- *oscillatorilly unstable (an oscillatory source) if* $df/dx_{k+1}|_{x_{k+1}^*=a} \in (-2,-1)$;
- *flipped if* $df/dx_{k+1}|_{x_{k+1}^*=a} = -2$ *with an oscillatory lower-saddle of the second-order for* $d^2f/dx_{k+1}^2|_{x_{k+1}^*=a} = a_0 > 0$;
- *oscillatorilly stable (an oscillatory sink) if* $df/dx_{k+1}|_{x_{k+1}^*=a} \in (-\infty,-2)$.

(i_{1b}) *The fixed-point of* $x_k^* = x_{k+1}^* = b$ *is monotonically stable (a monotonic sink) if* $df/dx_{k+1}|_{x_{k+1}^*=b} \in (0,\infty)$.

(i_2) *For* $a_0(\mathbf{p}) < 0$, *there are two cases:*

(i_{2a}) *The fixed-point of* $x_k^* = x_{k+1}^* = a$ *is monotonically stable (a monotonic sink) if* $df/dx_{k+1}|_{x_{k+1}^*=a} \in (0,\infty)$.

(i_{2b}) *The fixed-point of* $x_k^* = x_{k+1}^* = b$ *is:*

- *monotonically unstable (a monotonic source) if* $df/dx_{k+1}|_{x_{k+1}^*=b} \in (-1,0)$;
- *infinitely unstable if* $df/dx_{k+1}|_{x_{k+1}^*=b} = -1$, *which is*
  - *monotonically unstable with a positive infinity eigenvalue if* $df/dx_{k+1}|_{x_{k+1}^*=b} = -1^+$;
  - *oscillatorilly unstable with a negative infinity eigenvalue if* $df/dx_{k+1}|_{x_{k+1}^*=b} = -1^-$;
- *oscillatorilly unstable (an oscillatory source) if* $df/dx_{k+1}|_{x_{k+1}^*=b} \in (-2,-1)$;
- *flipped if* $df/dx_{k+1}|_{x_{k+1}^*=b} = -2$ *with an oscillatory upper-saddle of the second-order with* $d^2f/dx_{k+1}^2|_{x_{k+1}^*=b} = a_0 < 0$;
- *oscillatorilly stable (an oscillatory sink) if* $df/dx_{k+1}|_{x_{k+1}^*=b} \in (-\infty,-2)$.

(ii) *Under a condition of*

$$a > b \text{ and } \Delta = a_0^2(a-b)^2 > 0 \tag{1.100}$$

a standard form of the 1-dimensional, backward discrete system in Eq. (1.91) is

$$x_k = x_{k+1} + f(x_{k+1}, \mathbf{p}) = x_{k+1} + a_0(x_{k+1} - b)(x_{k+1} - a). \tag{1.101}$$

(ii$_1$) *For* $a_0(\mathbf{p}) > 0$*, there are two cases:*

(ii$_{1a}$) *The fixed-point of* $x_k^* = x_{k+1}^* = a$ *is monotonically stable (a monotonic sink) if* $df/dx_{k+1}|_{x_{k+1}^*=a} > 0$.

(ii$_{1b}$) *The fixed-point of* $x_k^* = x_{k+1}^* = b$ *is:*

- *monotonically unstable (a monotonic source) if* $df/dx_{k+1}|_{x_{k+1}^*=b} \in (-1, 0)$;
- *infinitely unstable if* $df/dx_{k+1}|_{x_{k+1}^*=b} = -1$*, which is:*
  - *monotonically unstable with a positive infinity eigenvalue if* $df/dx_{k+1}|_{x_{k+1}^*=b} = -1^+$;
  - *oscillatorilly unstable with a positive infinity eigenvalue if* $df/dx_{k+1}|_{x_{k+1}^*=b} = -1^-$;
- *oscillatorilly unstable (an oscillatory source) if* $df/dx_{k+1}|_{x_{k+1}^*=b} \in (-2, -1)$;
- *flipped if* $df/dx_{k+1}|_{x_{k+1}^*=b} = -2$ *with an oscillatory lower-saddle of the second-order with* $d^2f/dx_{k+1}^2|_{x_{k+1}^*=b} = a_0 > 0$;
- *oscillatorilly stable (an oscillatory sink) if* $df/dx_{k+1}|_{x_{k+1}^*=b} \in (-\infty, -2)$.

(ii$_2$) *For* $a_0(\mathbf{p}) < 0$*, there are two cases:*

(ii$_{2a}$) *The fixed-point of* $x_k^* = x_{k+1}^* = a$ *is:*

- *monotonically unstable (a monotonic source) if* $df/dx_{k+1}|_{x_{k+1}^*=a} \in (-1, 0)$;
- *infinitely unstable if* $df/dx_{k+1}|_{x_{k+1}^*=a} = -1$;
  - *monotonically unstable with a positive infinity eigenvalue if* $df/dx_{k+1}|_{x_{k+1}^*=a} = -1^+$;
  - *oscillatorilly unstable with a negative infinity eigenvalue if* $df/dx_{k+1}|_{x_{k+1}^*=a} = -1^-$;

- *oscillatorilly stable (an oscillatory sink) if* $df/dx_{k+1}|_{x^*_{k+1}=a} \in (-2,-1)$;
- *flipped if* $df/dx_{k+1}|_{x^*_{k+1}=a} = -2$ *with an oscillatory upper-saddle of the second-order for* $d^2f/dx^2_{k+1}|\ _{x^*_{k+1}=b} = a_0 < 0$;
- *oscillatorilly stable with* $df/dx_{k+1}|_{x^*_{k+1}=a} \in (-\infty,-2)$.

(ii$_{2b}$) *The fixed-point of* $x^*_k = x^*_{k+1} = b$ *is monotonically stable (a monotonic sink) if* $df/dx_{k+1}|_{x^*_{k+1}=b} > 0$.

(iii) For $a = b$, the corresponding standard form with $\Delta = 0$ is

$$x_k = x_{k+1} + f(x_{k+1}, \mathbf{p}) = x_{k+1} + a_0(x_{k+1} - a)^2. \tag{1.102}$$

(iii$_1$) *If* $a_0(\mathbf{p}) > 0$, *then the fixed-point of* $x^*_k = x^*_{k+1} = a$ *is an monotonic lower-*saddle *of the second-order with* $d^2f/dx^2_{k+1}|_{x^*_{k+1}=a} > 0$. *The fixed-point of* $x^*_k = x^*_{k+1} = a$ *for the switching of two fixed-points is a monotonic* lower-saddle-node *switching bifurcation of the second-order.*

(iii$_2$) *If* $a_0(\mathbf{p}) < 0$, *then the fixed-point of* $x^*_k = x^*_{k+1} = a$ *is a monotonic* upper-saddle *of the second order with* $d^2f/dx^2_{k+1}|_{x^*_{k+1}=a} < 0$. *The fixed-point of* $x^*_k = a$ *for the switching of two fixed-points is a monotonic* upper-saddle-node *switching bifurcation of the second-order.*

Proof The theorem can be proved as for Theorem 1.7. ■

The stability and bifurcation of two fixed-points for the 1-dimensional backward discrete system in Eq. (1.91) with $\Delta = B^2 - 4AC = a_0^2(a-b)^2 \geq 0$ are presented in Fig. 1.8. The stable and unstable fixed-points varying with the vector parameter are depicted by solid and dashed curves, respectively. The bifurcation point of fixed-points occurs at the repeated fixed-point at the boundary of $\mathbf{p}_0 \in \partial\Omega_{12}$. With

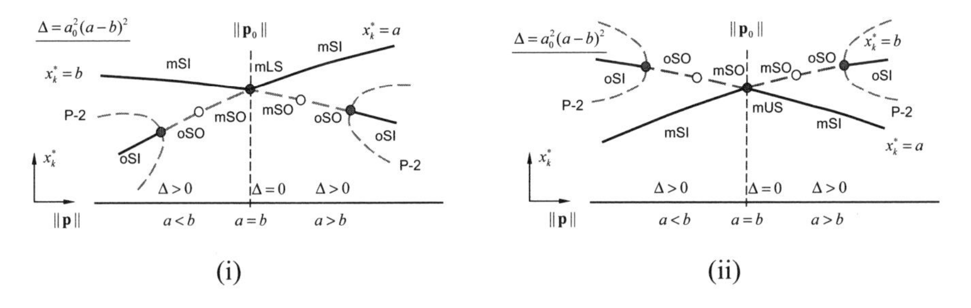

Fig. 1.8 Stability and bifurcation of two fixed-points in the quadratic backward discrete system: (i) a monotonic lower-*saddle*-node switching bifurcation ($a_0 > 0$), (ii) a monotonic *upper-saddle-node* switching bifurcation ($a_0 < 0$). Stable and unstable fixed-points are represented by solid and dashed curves, respectively. (mSO: monotonic source; mSI: monotonic sink; oSO: oscillatory source; oSI: oscillatory sink; mLS: monotonic lower-saddle; mUS: monotonic upper-saddle; oUS: oscillatory upper-saddle; oLS: oscillatory lower-saddle; iSI: invariant sink.)

varying parameters, the two fixed-points of $x_{k+1}^* = a, b$ equal each other (i.e., $x_k^* = a = b$). Such a fixed-point is a bifurcation point at $x_{k+1}^* = a = b$ for $\Delta = 0$. The fixed-points of $x_k^* = a, b$ with $\Delta \geq 0$ are presented in Fig. 1.8i and ii for $a_0 > 0$ and $a_0 < 0$, respectively. In Fig. 1.8i the switching bifurcation is a monotonic lower-saddle bifurcation. In Fig. 1.8ii the switching bifurcation is a monotonic upper-saddle bifurcation. The stable fixed-point is a monotonic sink, but the unstable fixed point is from a monotonic source, monotonic source with a positive infinity eigenvalue to oscillatory source with a negative infinity eigenvalue, flipped invariance with the oscillatory lower- or upper-saddle to the oscillatory sink. The period-2 fixed-point are unstable, which are generated through the oscillatory lower- or upper-saddle bifurcations.

1.4 Forward Bifurcation Trees

In this section, the analytical bifurcation scenario will be discussed. The period-doubling bifurcation scenario will be discussed first through nonlinear renormalization techniques, and the bifurcation scenario based on the saddle-node bifurcation will be discussed, which is independent of period-1 fixed-points.

1.4.1 Period-2 Appearing Bifurcations

After the period-doubling bifurcation of a period-1 fixed-point, the period-doubled fixed-points can be obtained. Consider the period-doubling solutions for a forward discrete quadratic nonlinear system.

Theorem 1.9 *Consider a 1-dimensional quadratic nonlinear discrete system*

$$x_{k+1} = x_k + A(\mathbf{p})x_k^2 + B(\mathbf{p})x_k + C(\mathbf{p}) \tag{1.103}$$

where three scalar constants $A(\mathbf{p}) \neq 0$, $B(\mathbf{p})$ *and* $C(\mathbf{p})$ *are determined by a vector parameter*

$$\mathbf{p} = (p_1, p_2, \ldots, p_m)^{\mathrm{T}}. \tag{1.104}$$

Under a condition of

$$\Delta = B^2 - 4AC > 0, \tag{1.105}$$

there is a standard form of the 1-dimensional discrete system in Eq. (1.103) *as*

$$x_{k+1} = x_k + f(x_k, \mathbf{p}) = x_k + a_0(x_k - a_1)(x_k - a_2) \tag{1.106}$$

where

$$a_0 = A(\mathbf{p}),\ a_{1,2} = \frac{-B(\mathbf{p}) \pm \sqrt{\Delta}}{2A(\mathbf{p})} \text{ with } a_1 < a_2, \tag{1.107}$$

If the fixed-point of $x_k^* = x_{k+1}^* = a_i(\mathbf{p})$ *with* $df/dx_k|_{x_k^*=a_i} = -2$ *is period-doubled, there is a form for period-2 fixed-points as*

$$x_{k+2} = x_k + a_0(x_k - a_1)(x_k - a_2)(A_1 x_k^2 + B_1 x_k + C_1) \tag{1.108}$$

where

$$A_1 = a_0^2, B_1 = a_0[2 - a_0(a_2 + a_1)], C_1 = 2 - a_0(a_2 + a_1) + a_0^2 a_1 a_2. \tag{1.109}$$

(i) *For* $a_0(\mathbf{p}) > 0$, *there are two cases:*

(i_1) *Under*

$$\Delta_1 = B_1^2 - 4A_1C_1 > 0, \tag{1.110}$$

there is a standard form as

$$\begin{aligned} x_{k+2} &= x_k + a_0^3(x_k - a_1)(x_k - a_2)(x_k - b_1)(x_k - b_2) \\ &= x_k + a_0^3 \textstyle\prod_{i=1}^{2\times 2^1} (x_k - a_i^{(2)}) \end{aligned} \tag{1.111}$$

where

$$\begin{aligned} &\{a_i^{(2)}, i = 1, 2, \ldots, 4\} = \text{sort}\,\{a_1, a_2, b_1, b_2\}, \\ &b_{1,2} = -\frac{1}{2a_0^2}B_1 \pm \frac{1}{2a_0^2}\sqrt{\Delta_1} \\ &\quad\ = -\frac{1}{2a_0}[2 - a_0(a_1 + a_2)] \pm \frac{1}{2a_0^2}\sqrt{\Delta_1}, \\ &\Delta_1 = -a_0^2[2 + a_0(a_1 - a_2)][2 + a_0(a_2 - a_1)]. \end{aligned} \tag{1.112}$$

(i_2) *Under an oscillatory upper-saddle-node bifurcation of*

$$\begin{aligned} &\frac{dx_{k+1}}{dx_k}|_{x_k^*=a_1} = 1 + a_0(a_1 - a_2) = -1 \\ &\Rightarrow 2 + a_0(a_1 - a_2) = 0, \\ &\frac{d^2 x_{k+1}}{dx_k^2}|_{x_k^*=a_1} = a_0 > 0, \end{aligned} \tag{1.113}$$

the second quadratics of the period-2 fixed points has

$$\begin{aligned} \Delta_1 &= B_1^2 - 4A_1C_1 \\ &= -a_0^2[2 + a_0(a_1 - a_2)][2 + a_0(a_2 - a_1)] = 0, \\ b_{1,2} &= -\frac{1}{2a_0}[2 - a_0(a_1 + a_2)] \\ &= -\frac{1}{2a_0}[2 + a_0(a_1 - a_2)] + a_1 = a_1 \end{aligned} \tag{1.114}$$

and the corresponding standard form becomes

$$x_{k+2} = x_k + a_0^3(x_k - a_1)^3(x_k - a_2). \tag{1.115}$$

(ii) *For $a_0(\mathbf{p})<0$, there are two cases:*

(ii$_1$) *Under a condition of*

$$\Delta_1 = B_1^2 - 4A_1C_1 > 0, \tag{1.116}$$

there is a standard form as

$$\begin{aligned} x_{k+2} &= x_k + a_0^3(x_k - a_1)(x_k - a_2)(x_k - b_1)(x_k - b_2) \\ &= x_k + a_0^3 \textstyle\prod_{i=1}^{2^2}(x_k - a_i^{(2)}) \end{aligned} \tag{1.117}$$

where

$$\begin{aligned} \{a_i^{(2)}, i = 1, 2, \ldots, 4\} &= \text{sort}\{a_1, a_2, b_1, b_2\}, \\ b_{1,2} &= -\frac{1}{2a_0^2}B_1 \pm \frac{1}{2a_0^2}\sqrt{\Delta_1} \\ &= -\frac{1}{2a_0}[2 - a_0(a_2 + a_1)] \pm \frac{1}{2a_0^2}\sqrt{\Delta_1}, \\ \Delta_1 &= -a_0^2[2 + a_0(a_1 - a_2)][2 + a_0(a_2 - a_1)]. \end{aligned} \tag{1.118}$$

(ii$_2$) *Under an oscillatory lower-saddle-node bifurcation of*

$$\begin{aligned} &\frac{dx_{k+1}}{dx_k}|_{x_k^*=a_2} = 1 + a_0(a_2 - a_1) = -1 \\ &\Rightarrow 2 + a_0(a_2 - a_1) = 0, \\ &\frac{d^2x_{k+1}}{dx_k^2}|_{x_k^*=a_1} = a_0 < 0, \end{aligned} \tag{1.119}$$

the second quadratics of the period-2 fixed-points has

$$\begin{aligned}\Delta_1 &= B_1^2 - 4A_1C_1 \\ &= -a_0^2[2+a_0(a_1-a_2)][2+a_0(a_2-a_1)] = 0, \\ b_{1,2} &= -\frac{1}{2a_0}[2-a_0(a_1+a_2)] \\ &= -\frac{1}{2a_0}[2+a_0(a_2-a_1)]+a_2 = a_2\end{aligned} \tag{1.120}$$

and the corresponding standard form becomes

$$x_{k+2} = x_k + a_0^3(x_k-a_1)(x_k-a_2)^3. \tag{1.121}$$

Proof The proof is straightforward through the simple algebraic manipulation. Consider

$$Ax_k^2 + Bx_k + C = 0,$$

$$\Delta = B^2 - 4AC \geq 0,$$

we have

$$a_0 = A(\mathbf{p}),\ a_{1,2} = \frac{-B(\mathbf{p}) \pm \sqrt{\Delta}}{2A(\mathbf{p})} \text{ with } a_1 < a_2.$$

Thus, we have

$$Ax_k^2 + Bx_k + C = (x_k - a_1)(x_k - a_2).$$

Therefore,

$$x_{k+1} = x_k + a_0(x_k - a_1)(x_k - a_2).$$

For $x_{k+1}^* = x_k^* = a_i$ $(i \in \{1,2\}$, if

$$\frac{dx_{k+1}}{dx_k}|_{x_k^*=a_i} = 1 + a_0(a_i - a_j) = -1,\ \frac{d^2x_{k+1}}{dx_k^2}|_{x_k^*=a_i} = a_0 \neq 0,$$

then period-2 fixed-points exists for the quadratic discrete system. Thus, consider the corresponding second iteration gives

$$x_{k+2} = x_{k+1} + a_0 \prod_{i_1=1}^2 (x_{k+1} - a_i^{(1)}).$$

The period-2 discrete system of the quadratic discrete system is

$$\begin{aligned} x_{k+2} &= x_k + [a_0 \prod_{i_1=1}^{2} (x_k - a_{i_1}^{(1)})]\{1 + \prod_{i_1=1}^{2} [1 + a_0 \prod_{i_3=1, i_2 \neq i_2}^{2} (x_k - a_{i_2}^{(1)})]\} \\ &= x_k + [a_0 \prod_{i_1=1}^{2} (x_k - a_{i_1}^{(1)})][A_1 x_k^2 + B_1 x_k + C_1)] \end{aligned}$$

where

$$A_1 = a_0^2, B_1 = a_0[2 - a_0(a_2 + a_1)], C_1 = 2 - a_0(a_2 + a_1) + a_0^2 a_1 a_2.$$

If

$$A_1 x_k^2 + B_1 x_k + C_1 = 0,$$

we have

$$\begin{aligned} \Delta_1 &= B_1^2 - 4A_1 C_1 = -a_0^2[2 + a_0(a_1 - a_2)][2 + a_0(a_2 - a_1)] \geq 0, \\ b_{1,2} &= -\frac{1}{2a_0}[2 - a_0(a_1 + a_2)] \pm \frac{1}{2a_0^2}\sqrt{\Delta_1}. \end{aligned}$$

Thus

$$A_1 x_k^2 + B_1 x_k + C_1 = a_0^2 \prod_{j_2=1}^{2} (x_k - b_{j_2}),$$

and

$$\begin{aligned} x_{k+2} &= x_k + a_0^3 (x_k - a_1)(x_k - a_2)(x_k - b_1)(x_k - b_2) \\ &= x_k + a_0^3 \prod_{i=1}^{2^2} (x_k - a_i^{(2)}) \end{aligned}$$

where

$$\{a_i^{(2)}, i = 1, 2, \ldots, 4\} = \text{sort}\{a_1, a_2, b_1, b_2\}.$$

For the period-1 discrete systems,

$$x_{k+1} = x_k + a_0 \prod_{i=1}^{2} (x_k - a_i).$$

(I) For $a_0 > 0$, the fixed-point of $x_k^* = a_2$ is monotonically unstable due to

$$\frac{dx_{k+1}}{dx_k}|_{x_k^*=a_2} = 1 + a_0(a_2 - a_1) \in (1, \infty),$$

and the fixed-point of $x_k^* = a_1$ is from monotonically stable to oscillatorilly unstable due to

$$\frac{dx_{k+1}}{dx_k}|_{x_k^*=a_1} = 1 + a_0(a_1 - a_2) \in (-\infty, 1).$$

Under

$$\begin{aligned}
&\frac{dx_{k+1}}{dx_k}|_{x_k^*=a_1} = 1 + a_0(a_1 - a_2) = -1 \\
&\Rightarrow 2 + a_0(a_1 - a_2) = 0 \text{ or } a_0(a_1 - a_2) = -2, \\
&\frac{d^2x_{k+1}}{dx_k^2}|_{x_k^*=a_1} = a_0 > 0,
\end{aligned}$$

there is a flipped discrete system of the oscillatory upper-saddle of the second order. Thus, for the period-2 discrete system,

$$\begin{aligned}
\Delta_1 &= B_1^2 - 4A_1C_1 = -a_0^2[2 + a_0(a_1 - a_2)][2 + a_0(a_2 - a_1)] = 0, \\
b_{1,2} &= -\frac{1}{2a_0}[2 - a_0(a_1 + a_2)] = -\frac{1}{2a_0}[2 + a_0(a_1 - a_2)] + a_1 = a_1
\end{aligned}$$

and the corresponding standard form of the period-2 discrete system becomes

$$x_{k+2} = x_k + a_0^3(x_k - a_1)^3(x_k - a_2)$$

with

$$\begin{aligned}
\frac{dx_{k+2}}{dx_k}|_{x_k^*=a_1} &= 1 + 3a_0^3(x_k - a_1)^2(x_k - a_2) + a_0^3(x_k - a_1)^3|_{x_k^*=a_1} = 1, \\
\frac{d^2x_{k+2}}{dx_k^2}|_{x_k^*=a_1} &= 6a_0^3(x_k - a_1)(x_k - a_2) + 6a_0^3(x_k - a_1)^2|_{x_k^*=a_1} = 0, \\
\frac{d^3x_{k+2}}{dx_k^3}|_{x_k^*=a_1} &= 6a_0^3(x_k - a_2) + 18a_0^3(x_k - a_1)|_{x_k^*=a_1} = 6a_0^3(a_1 - a_2) < 0.
\end{aligned}$$

Therefore, $x_k^* = a_1$ for the period-2 discrete system is a monotonic sink of the third-order.

(II) Similarly, for $a_0(\mathbf{p}) < 0$, the fixed-point of $x_k^* = a_1$ is monotonically unstable due to

$$\frac{dx_{k+1}}{dx_k}|_{x_k^*=a_1} = 1 + a_0(a_1 - a_2) \in (1, \infty),$$

and the fixed-point of $x_k^* = a_2$ is for monotonically stable to oscillatorilly unstable due to

$$\frac{dx_{k+1}}{dx_k}|_{x_k^*=a_2} = 1 + a_0(a_2 - a_1) \in (-\infty, 1).$$

Under

$$\begin{aligned}
&\frac{dx_{k+1}}{dx_k}|_{x_k^*=a_2} = 1 + a_0(a_2 - a_1) = -1 \\
&\Rightarrow 2 + a_0(a_2 - a_1) = 0 \text{ or } a_0(a_2 - a_1) = -2, \\
&\frac{d^2x_{k+1}}{dx_k^2}|_{x_k^*=a_2} = a_0 < 0,
\end{aligned}$$

there is a flipped discrete system of the oscillatory lower-saddle of the second order. Thus, for the period-2 discrete system,

$$\begin{aligned}
&\Delta_1 = B_1^2 - 4A_1C_1 = -a_0^2[2 + a_0(a_1 - a_2)][2 + a_0(a_2 - a_1)] = 0, \\
&b_{1,2} = -\frac{1}{2a_0}[2 - a_0(a_1 + a_2)] = -\frac{1}{2a_0}[2 + a_0(a_2 - a_1)] + a_2 = a_2
\end{aligned}$$

and the corresponding standard form of the period-2 discrete system becomes

$$x_{k+2} = x_k + a_0^3(x_k - a_1)(x_k - a_2)^3$$

with

$$\begin{aligned}
&\frac{dx_{k+2}}{dx_k}|_{x_k^*=a_2} = 1 + 3a_0^3(x_k - a_1)(x_k - a_2)^2 + a_0^3(x_k - a_2)^3|_{x_k^*=a_2} = 1, \\
&\frac{d^2x_{k+2}}{dx_k^2}|_{x_k^*=a_2} = 6a_0^3(x_k - a_1)(x_k - a_2) + 6a_0^3(x_k - a_2)^2|_{x_k^*=a_2} = 0, \\
&\frac{d^3x_{k+2}}{dx_k^3}|_{x_k^*=a_2} = 6a_0^3(x_k - a_1) + 18a_0^3(x_k - a_2)|_{x_k^*=a_2} = 6a_0^3(a_2 - a_1) < 0.
\end{aligned}$$

Thus, $x_k^* = a_2$ for the period-2 discrete system is a monotonic sink of the third-order.

This theorem is proved. ■

Based on the standardization, the above theorem can be stated as follows.

Theorem 1.10 *Consider a 1-dimensional quadratic nonlinear discrete system as*

$$x_{k+1} = x_k + A(\mathbf{p})x_k^2 + B(\mathbf{p})x_k + C(\mathbf{p}) \tag{1.122}$$

where three scalar constants $A(\mathbf{p}) \neq 0$, $B(\mathbf{p})$ *and* $C(\mathbf{p})$ *are determined by a vector parameter*

$$\mathbf{p} = (p_1, p_2, \ldots, p_m)^{\mathrm{T}}. \tag{1.123}$$

Under a condition of

$$\Delta = B^2 - 4AC > 0, \tag{1.124}$$

there is a standard form as

$$\begin{aligned} x_{k+1} &= x_k + f(x_k, \mathbf{p}) = x_k + a_0(x_k^2 + B_1^{(1)} x_k + C_1^{(1)}) \\ &= x_k + a_0(x_k - a_1)(x_k - a_2) \\ &= x_k + a_0 \prod_{i=1}^{2} (x_k - a_i^{(1)}) \end{aligned} \tag{1.125}$$

where

$$\begin{aligned} a_0 &= A(\mathbf{p}), B_1^{(1)} = \frac{B}{A}, C_1^{(1)} = \frac{C}{A}; \\ b_1^{(1)} &= -\frac{1}{2}(B_1^{(1)} + \sqrt{\Delta^{(1)}}), b_2^{(1)} = -\frac{1}{2}(B_1^{(1)} - \sqrt{\Delta^{(1)}}), \\ \Delta^{(1)} &= (B_1^{(1)})^2 - 4C_1^{(1)} \geq 0; \\ \cup_{i=1}^{2}\{a_i^{(1)}\} &= \mathrm{sort}\{\cup_{i=1}^{2}\{b_i^{(1)}\}\}, a_i^{(1)} \leq a_{i+1}^{(1)} \text{ for } i = 1, 2. \end{aligned} \tag{1.126}$$

(i) *Consider a forward period-2 discrete system of Eq.* (1.122) as

$$\begin{aligned} x_{k+2} &= x_k + [a_0 \prod_{i_1=1}^{2} (x_k - a_{i_1}^{(1)})]\{1 + \prod_{i_1=1}^{2} [1 + a_0 \prod_{i_2=1, i_2 \neq i_1}^{2} (x_k - a_{i_2}^{(1)})]\} \\ &= x_k + [a_0 \prod_{i_1=1}^{2} (x_k - a_{i_1}^{(1)})][a_0^2 (x_k^2 + B_1^{(2)} x_k + C_1^{(2)})] \\ &= x_k + [a_0 \prod_{j_1=1}^{2} (x_k - a_{i_1}^{(1)})][a_0^2 \prod_{j_2=1}^{2^2-2} (x_k - b_{j_2}^{(2)})] \\ &= x_k + a_0^{1+2} \prod_{i=1}^{4} (x_k - a_i^{(2)}) \end{aligned} \tag{1.127}$$

where

$$\begin{aligned} b_{1,2}^{(2)} &= -\frac{1}{2}(B_1^{(2)} + \sqrt{\Delta^{(2)}}), b_2^{(2)} = -\frac{1}{2}(B_1^{(2)} - \sqrt{\Delta^{(2)}}), \\ \Delta^{(2)} &= (B_1^{(2)})^2 - 4C_1^{(2)} \geq 0, \end{aligned} \tag{1.128}$$

with fixed-points

$$\begin{aligned}&x^*_{k+2}=x^*_k=a^{(2)}_i,(i=1,2,\ldots,4)\\&\cup_{i=1}^{4}\{a^{(2)}_i\}=\mathrm{sort}\{\cup_{j_1=1}^{2}\{a^{(1)}_{j_1}\},\cup_{j_2=1}^{2}\{b^{(2)}_{j_2}\}\}\\&\text{with } a^{(2)}_i<a^{(2)}_{i+1}.\end{aligned}\tag{1.129}$$

(ii) *For a fixed-point of* $x^*_{k+1}=x^*_k=a^{(1)}_{i_1}$ $(i_1\in\{1,2\})$, *if*

$$\frac{dx_{k+1}}{dx_k}\Big|_{x^*_k=a^{(1)}_{i_1}}=1+a_0(a^{(1)}_{i_1}-a^{(1)}_{i_2})=-1,\tag{1.130}$$

with

- *an oscillatory upper-saddle-node bifurcation* $(d^2x_{k+1}/dx_k^2|_{x^*_k=a_1}=a_0>0)$,
- *an oscillatory lower-saddle-node bifurcation* $(d^2x_{k+1}/dx_k^2|_{x^*_k=a_1}=a_0<0)$,

then the following relations satisfy

$$a^{(1)}_{i_1}=-\frac{1}{2}B^{(2)}_{i_1},\Delta^{(2)}_{i_1}=(B^{(2)}_1)^2-4C^{(2)}_1=0,\tag{1.131}$$

and there is a period-2 discrete system of the quadratic discrete system in Eq. (1.122) as

$$x_{k+2}=x_k+a_0^3(x_k-a^{(1)}_{i_1})^3(x_k-a^{(2)}_{i_2}).\tag{1.132}$$

For $i_1,i_2\in\{1,2\}$, $i_1\neq i_2$ *with*

$$\begin{aligned}&\frac{dx_{k+2}}{dx_k}\Big|_{x^*_k=a^{(1)}_{i_1}}=1,\ \frac{d^2x_{k+2}}{dx_k^2}\Big|_{x^*_k=a^{(1)}_{i_1}}=0;\\&\frac{d^3x_{k+2}}{dx_k^3}\Big|_{x^*_k=a^{(1)}_{i_1}}=6a_0^3(a^{(1)}_{i_1}-a^{(2)}_{i_2})=-12a_0^2<0.\end{aligned}\tag{1.133}$$

Thus, x_{k+2} *at* $x^*_k=a^{(1)}_{i_1}$ *is a monotonic sink of the third-order, and the corresponding bifurcations is a monotonic sink bifurcation for the period-2 discrete system.*

(ii$_1$) *The period-2 fixed-points are trivial and unstable if*

$$x^*_{k+2}=x^*_k=a^{(1)}_{i_1}\text{ for } i_1=1,2.\tag{1.134}$$

(ii$_2$) *The period-2 fixed-points are non-trivial and stable if*

$$x^*_{k+2}=x^*_k=b^{(2)}_{i_1}\text{ for } i_1=1,2.\tag{1.135}$$

Proof See the proof of theorem 1.9. This theorem is proved. ■

1.4.2 Period-Doubling Renormalization

The generalized cases of period-doublization of quadratic discrete systems are presented through the following theorem. The analytical period-doubling trees can be developed for quadratic discrete systems.

Theorem 1.11 *Consider a 1-dimensional quadratic nonlinear discrete system*

$$\begin{aligned} x_{k+1} &= x_k + A(\mathbf{p})x_k^2 + B(\mathbf{p})x_k + C(\mathbf{p}) \\ &= x_k + a_0 \prod_{i=1}^{2} (x_k - a_i^{(1)}). \end{aligned} \tag{1.136}$$

(i) *After l-times period-doubling bifurcations, a period-*$2^l (l = 1, 2, \ldots)$ *discrete system for the quadratic discrete system in Eq.* (1.136) *is given through the nonlinear renormalization as*

$$\begin{aligned} x_{k+2^l} &= x_k + [a_0^{(2^{l-1})} \prod_{i_1=1}^{2^{2^{l-1}}} (x_k - a_{i_1}^{(2^{l-1})})] \\ &\quad \times \{1 + \prod_{i_1=1}^{2^{2^{l-1}}} [1 + a_0^{(2^{l-1})} \prod_{i_2=1, i_2 \neq i_1}^{2^{2^{l-1}}} (x_k - a_{i_2}^{(2^{l-1})})]\} \\ &= x_k + [a_0^{(2^{l-1})} \prod_{i_1=1}^{2^{2^{l-1}}} (x_k - a_{i_1}^{(2^{l-1})})] \\ &\quad \times [(a_0^{(2^{l-1})})^{2^{2^{l-1}}} \prod_{j_1=1}^{2^{2^l-1} - 2^{2^{l-1}-1}} (x_k^2 + B_{j_2}^{(2^l)} x_k + C_{j_2}^{(2^l)})] \\ &= x_k + [a_0^{(2^{l-1})} \prod_{i_1=1,}^{2^{2^{l-1}}} (x_k - a_{i_1}^{(2^{l-1})})] \\ &\quad \times [(a_0^{(2^{l-1})})^{2^{2^{l-1}}} \prod_{i_2=1}^{2^{2^l-1} - 2^{2^{l-1}-1}} (x_k - b_{i_2,1}^{(2^l)})(x_k - b_{i_2,2}^{(2^l)})] \\ &= x_k + (a_0^{(2^{l-1})})^{1+2^{2^{l-1}}} \prod_{i=1}^{2^{2^l}} (x_k - a_i^{(2^l)}) \\ &= x_k + a_0^{(2^l)} \prod_{i=1}^{2^{2^l}} (x_k - a_i^{(2^l)}) \end{aligned} \tag{1.137}$$

with

$$\begin{aligned} \frac{dx_{k+2^l}}{dx_k} &= 1 + a_0^{(2^l)} \sum_{i_1=1}^{2^{2^l}} \prod_{i_2=1, i_2 \neq i_1}^{2^{2^l}} (x_k - a_{i_2}^{(2^l)}), \\ \frac{d^2 x_{k+2^l}}{dx_k^2} &= a_0^{(2^l)} \sum_{i_1=1}^{2^{2^l}} \sum_{i_2=1, i_2 \neq i_1}^{2^{2^l}} \prod_{i_3=1, i_3 \neq i_1, i_2}^{2^{2^l}} (x_k - a_{i_3}^{(2^l)}), \\ &\vdots \\ \frac{d^r x_{k+2^l}}{dx_k^r} &= a_0^{(2^l)} \sum_{i_1=1}^{2^{2^l}} \cdots \sum_{i_r=1, i_r \neq i_1, i_2, \ldots, i_{r-1}}^{2^{2^l}} \prod_{i_{r+1}=1, i_3 \neq i_1, i_2 \ldots, i_r}^{2^{2^l}} (x_k - a_{i_{r+1}}^{(2^l)}) \\ &\text{for } r \leq 2^{2^l}, \end{aligned} \tag{1.138}$$

where

$$
\begin{aligned}
& a_0^{(2)} = (a_0)^{1+2}, a_0^{(2^l)} = (a_0^{(2^{l-1})})^{1+2^{2^{l-1}}}, l=1,2,3,\ldots; \\
& \cup_{i=1}^{2^{2^l}}\{a_i^{(2^l)}\} = \text{sort}\{\cup_{i_1=1}^{2^{2^{l-1}}}\{a_{i_1}^{(2^l)}\}, \cup_{i_2=1}^{M_2}\{b_{i_2,1}^{(2^l)}, b_{i_2,2}^{(2^l)}\}\}, a_i^{(2^l)} \le a_{i+1}^{(2^l)}; \\
& b_{i,1}^{(2^l)} = -\frac{1}{2}(B_i^{(2^l)} + \sqrt{\Delta_i^{(2^l)}}), b_{i,2}^{(2^{l-1})} = -\frac{1}{2}(B_i^{(2^l)} - \sqrt{\Delta_i^{(2^l)}}), \\
& \Delta_i^{(2^l)} = (B_i^{(2^l)})^2 - 4C_i^{(2^l)} \ge 0 \text{ for } i \in \cup_{q_1=1}^{N_1} I_{q_1}^{(2^{l-1})} \cup \cup_{q_2=1}^{N_2} I_{q_2}^{(2^{l-1})}, \\
& I_{q_1}^{(2^{l-1})} = \{l_{(q_1-1)\times 2^{l-1}+1}, l_{(q_1-1)\times 2^{l-1}+2}, \ldots, l_{q_1\times 2^{l-1}}\} \\
& \qquad \subseteq \{1,2,\ldots,M_1\} \cup \{\varnothing\}, \\
& \text{for } q_1 \in \{1,2,\ldots,N_1\}, M_1 = N_1 \times 2^{l-1}; \\
& I_{q_2}^{(2^l)} = \{l_{(q_2-1)\times 2^l+1}, l_{(q_2-1)\times 2^l+2}, \ldots, l_{q_2\times 2^l}\} \\
& \qquad \subseteq \{M_1+1, M_1+2, \ldots, M_2\} \cup \{\varnothing\}, \\
& \text{for } q_2 \in \{1,2,\ldots,N_2\}, M_2 = 2^{2^l-1} - 2^{2^{l-1}-1}; \\
& b_{i,1}^{(2^l)} = -\frac{1}{2}(B_i^{(2^l)} + \mathbf{i}\sqrt{|\Delta_i^{(2^l)}|}), \; b_{i,2}^{(2^l)} = -\frac{1}{2}(B_i^{(2^l)} - \mathbf{i}\sqrt{|\Delta_i^{(2^l)}|}), \\
& \Delta_i^{(2^l)} = (B_i^{(2^l)})^2 - 4C_i^{(2^l)} < 0, \mathbf{i} = \sqrt{-1}, \\
& i \in J^{(2^l)} = \{l_{N_2\times 2^l}, l_{N_2\times 2^l}, \ldots, l_{M_2}\} \\
& \qquad \subset \{M_1+1, M_1+2, \ldots, M_2\} \cup \{\varnothing\}
\end{aligned}
\tag{1.139}
$$

with fixed-points

$$
\begin{aligned}
& x_{k+2^l}^* = x_k^* = a_i^{(2^l)}, (i = 1,2,\ldots,2^{2^l}) \\
& \cup_{i=1}^{2^{2^l}}\{a_i^{(2^l)}\} = \text{sort}\{\cup_{i_1=1}^{2^{2^{l-1}}}\{a_{i_1}^{(2^{l-1})}\}, \cup_{i_2=1}^{M_2}\{b_{i_2,1}^{(2^l)}, b_{i_2,2}^{(2^l)}\}\} \\
& \text{with } a_i^{(2^l)} < a_{i+1}^{(2^l)}.
\end{aligned}
\tag{1.140}
$$

(ii) *For a fixed-point of* $x_{k+2^{l-1}}^* = x_k^* = a_{i_1}^{(2^{l-1})}$ $(i_1 \in I_q^{(2^{l-1})} \subset \{1,2,\ldots,2^{(2^{l-1})}\})$, *if*

$$
\begin{aligned}
& \frac{dx_{k+2^{l-1}}}{dx_k}\Big|_{x_k^*=a_{i_1}^{(2^{l-1})}} = 1 + a_0^{(2^{l-1})} \Pi_{i_2=1, i_2\ne i_1}^{2^{2^{l-1}}} (a_{i_1}^{(2^{l-1})} - a_{i_2}^{(2^{l-1})}) = -1, \\
& \frac{d^2x_{k+2^{l-1}}}{dx_k^2}\Big|_{x_k^*=a_{i_1}^{(2^{l-1})}} > 0 \textit{ for the second-order oscillatory upper-saddle}, \\
& \frac{d^2x_{k+2^{l-1}}}{dx_k^2}\Big|_{x_k^*=a_{i_1}^{(2^{l-1})}} < 0 \textit{ for the second-order oscillatory lower-saddle},
\end{aligned}
\tag{1.141}
$$

then there is a period-2^l *fixed-point discrete system*

$$x_{k+2^l} = x_k + a_0^{(2^l)} \prod_{i_1 \in I_q^{(2^{l-1})}} (x_k - a_{i_1}^{(2^{l-1})})^3 \prod_{j_2=1}^{2^{2^l}} (x_k - a_{j_2}^{(2^l)})^{(1-\delta(i_1,j_2))} \quad (1.142)$$

where

$$\delta(i_1,j_2) = 1 \text{ if } a_{j_2}^{(2^l)} = a_{i_1}^{(2^{l-1})},\ \delta(i_1,j_2) = 0 \text{ if } a_{j_2}^{(2^l)} \neq a_{i_1}^{(2^{l-1})} \quad (1.143)$$

with

$$\begin{aligned} &\frac{dx_{k+2^l}}{dx_k}\Big|_{x_k^*=a_{i_1}^{(2^{l-1})}} = 1, \frac{d^2x_{k+2^l}}{dx_k^2}\Big|_{x_k^*=a_{i_1}^{(2^{l-1})}} = 0; \\ &\frac{d^3x_{k+2^l}}{dx_k^3}\Big|_{x_k^*=a_{i_1}^{(2^{l-1})}} = 6a_0^{(2^l)} \prod_{i_2 \in I_q^{(2^{l-1})}, i_2 \neq i_1} (a_{i_1}^{(2^{l-1})} - a_{i_2}^{(2^{l-1})})^3 \\ &\qquad \times \prod_{j_2=1}^{2^{2^l}} (a_{i_1}^{(2^{l-1})} - a_{j_2}^{(2^l)})^{(1-\delta(i_2,j_2))} < 0 \\ &(i_1 \in I_q^{(2^{l-1})}, q \in \{1,2,\ldots,N_1\}) \end{aligned} \quad (1.144)$$

Thus, x_{k+2^l} *at* $x_k^* = a_{i_1}^{(2^{l-1})}$ *is*

- *a monotonic sink of the third-order if* $d^3x_{k+2^l}/dx_k^3|_{x_k^*=a_{i_1}^{(2^{l-1})}} < 0$*;*
- *a monotonic source of the third-order if* $d^3x_{k+2^l}/dx_k^3|_{x_k^*=a_{i_1}^{(2^{l-1})}} > 0$.

(ii$_1$) *The period-*2^l *fixed-points are trivial if*

$$x_{k+2^l}^* = x_k^* = a_{i_1}^{(2^{l-1})} \text{ for } i_1 = 1,2,\ldots,2^{2^{l-1}}. \quad (1.145)$$

(ii$_2$) *The period-*2^l *fixed-points are non-trivial if*

$$\begin{aligned} &x_{k+2^l}^* = x_k^* = b_{i_1,1}^{(2^l)}, b_{i_1,2}^{(2^l)} \\ &i_1 \in \{1,2,\ldots,M_2\} \cup \{\varnothing\}. \end{aligned} \quad (1.146)$$

Such a period-2^l *fixed-point is*

- *monotonically unstable if* $dx_{k+2^l}/dx_k|_{x_k^*=a_{i_1}^{(2^l)}} \in (1,\infty)$*;*
- *monotonically invariant if* $dx_{k+2^l}/dx_k|_{x_k^*=a_{i_1}^{(2^l)}} = 1$*, which is*
  - *a monotonic upper-saddle of the* $(2l_1)^{\text{th}}$ *order for* $d^{2l_1}x_{k+2^l}/dx_k^{2l_1}|_{x_k^*} > 0$*;*
  - *a monotonic lower-saddle the* $(2l_1)^{\text{th}}$ *order for* $d^{2l_1}x_{k+2^l}/dx_k^{2l_1}|_{x_k^*} < 0$*;*
  - *a monotonic source of the* $(2l_1+1)^{\text{th}}$ *order for* $d^{2l_1+1}x_{k+2^l}/dx_k^{2l_1+1}|_{x_k^*} > 0$*;*
  - *a monotonic sink the* $(2l_1+1)^{\text{th}}$ *order for* $d^{2l_1+1}x_{k+2^l}/dx_k^{2l_1+1}|_{x_k^*} < 0$*;*

- *monotonically unstable if* $dx_{k+2^l}/dx_k|_{x_k^*=a_{i_1}^{(2^l)}} \in (0,1)$;
- *invariantly zero-stable if* $dx_{k+2^l}/dx_k|_{x_k^*=a_{i_1}^{(2^l)}} = 0$;
- *oscillatorilly stable if* $dx_{k+2^l}/dx_k|_{x_k^*=a_{i_1}^{(2^l)}} \in (-1,0)$;
- *flipped if* $dx_{k+2^l}/dx_k|_{x_k^*=a_{i_1}^{(2^{l-1})}} = -1$, *which is*
  - *an oscillatory upper-saddle of the* $(2l_1)^{\text{th}}$ *order if* $d^{2l_1}x_{k+2^l}/dx_k^{2l_1}|_{x_k^*} > 0$;
  - *an oscillatory lower-saddle the* $(2l_1)^{\text{th}}$ *order with* $d^{2l_1}x_{k+2^l}/dx_k^{2l_1}|_{x_k^*} < 0$;
  - *an oscillatory source of the* $(2l_1+1)^{\text{th}}$ *order if* $d^{2l_1+1}x_{k+2^l}/dx_k^{2l_1+1}|_{x_k^*} < 0$;
  - *an oscillatory sink the* $(2l_1+1)^{\text{th}}$ *order with* $d^{2l_1+1}x_{k+2^l}/dx_k^{2l_1+1}|_{x_k^*} > 0$;
- *oscillatorilly unstable if* $dx_{k+2^l}/dx_k|_{x_k^*=a_{i_1}^{(2^l)}} \in (-\infty,-1)$.

Proof Through the nonlinear renormalization, this theorem can be proved.

(I) For a quadratic system, if the period-1 fixed-points exists, there is a following expression.

$$x_{k+1} = x_k + a_0 \prod_{i_1=1}^{2}(x_k - a_{i_1}^{(1)}).$$

For $x_{k+1}^* = x_k^* = a_{i_1}^{(1)}$ $(i_1 \in \{1,2\}$, if

$$\frac{dx_{k+1}}{dx_k}|_{x_k^*=a_{i_1}^{(1)}} = 1 + a_0^{(1)}(a_{i_1}^{(2)} - a_{j_2}^{(2)}) = -1,$$
$$\frac{d^2x_{k+1}}{dx_k^2}|_{x_k^*=a_{i_1}^{(2)}} = a_0^{(1)} = a_0 \neq 0$$

then period-2 fixed-points exists for the quadratic discrete system. Thus, consider the corresponding second iteration gives

$$x_{k+2} = x_{k+1} + a_0 \prod_{i_1=1}^{2}(x_{k+1} - a_{i_1}^{(1)}).$$

The period-2 discrete system of the quadratic discrete system is

$$\begin{aligned} x_{k+2} &= x_k + [a_0 \prod_{i_1=1}^{2}(x_k - a_{i_1}^{(1)})]\{1 + \prod_{i_1=1}^{2}[1 + a_0 \prod_{i_2=1, i_2\neq i_1}^{2}(x_k - a_{i_2}^{(1)})]\} \\ &= x_k + [a_0 \prod_{i_1=1}^{2}(x_k - a_{i_1}^{(1)})][a_0^2(x_k^2 + B_1^{(2)}x_k + C_1^{(2)})]. \end{aligned}$$

If

$$x_k^2 + B_1^{(2)}x_k + C_1^{(2)} = 0,$$

we have

$$b_1^{(2)} = -\frac{1}{2}(B_1^{(2)} + \sqrt{\Delta^{(2)}}), b_2^{(2)} = -\frac{1}{2}(B_1^{(2)} - \sqrt{\Delta^{(2)}})$$
$$\Delta^{(2)} = (B_1^{(2)})^2 - 4C_1^{(2)} \geq 0.$$

Thus

$$x_k^2 + B_1^{(2)} x_k + C_1^{(2)} = \prod_{j_2=1}^{2}(x_k - b_{j_2}^{(2)}),$$

and

$$\begin{aligned} x_{k+2} &= x_k + [a_0 \prod_{j_1=1}^{2}(x_k - a_{i_1}^{(1)})][a_0^2 \prod_{j_2=1}^{2^2-2}(x_k - b_{j_2}^{(2)})] \\ &= x_k + a_0^{(2)} \prod_{i=1}^{4}(x_k - a_i^{(2)}), \end{aligned}$$

where $a_0^{(2)} = a_0^{1+2}$. For a fixed-point of $x_{k+2}^* = x_k^* = a_{i_1}^{(2)}$ $(i_1 \in \{1, 2, \ldots, 4\})$, if

$$\begin{aligned} \frac{dx_{k+2}}{dx_k}\Big|_{x_k^* = a_{i_1}^{(2)}} &= 1 + a_0^{(2)} \prod_{i_2=1, i_2 \neq i_1}^{4}(a_{i_1}^{(2)} - a_{i_2}^{(2)}) = -1, \\ \frac{d^2 x_{k+2}}{dx_k^2}\Big|_{x_k^* = a_{i_1}^{(2)}} &= a_0^{(2)} \sum_{i_2=1, i_2 \neq i_1}^{4} \prod_{i_3=1, i_3 \neq i_1, i_2}^{4}(a_{i_1}^{(2)} - a_{i_3}^{(2)}) \neq 0, \end{aligned}$$

then, a period-2 discrete system of a quadratic discrete system has a period-doubling bifurcation.

(II) Such a period-2 discrete system can be renormalized nonlinearly. For $k = k_1 + 2$, the previous period-2 discrete system becomes

$$x_{k_1+2+2} = x_{k_1+2} + a_0^{(2)} \prod_{i=1}^{4}(x_{k_1+2} - a_i^{(2)}).$$

Because k_1 is an index for iteration, it can be replaced by k. Thus, an equivalent form for the foregoing equations becomes

$$x_{k+2^2} = x_{k+2^1} + a_0^{(2^1)} \prod_{i=1}^{2^{2^1}}(x_{k+2^1} - a_i^{(2^1)}),$$

with

$$x_{k+2^1} = x_k + a_0^{(2^1)} \prod_{i=1}^{2^{2^1}} (x_k - a_i^{(2^1)})$$

x_{k+2^2} can be expressed as

$$\begin{aligned} x_{k+2^2} &= x_k + a_0^{(2^1)} \prod_{i_1=1}^{2^{2^1}} (x_k - a_{i_1}^{(2^1)}) \{1 + \prod_{i_1=1}^{2^{2^1}} [1 + a_0 a_0^{(2^1)} \prod_{i_2=1, i_2 \neq i_1}^{2^{2^1}} (x_k - a_{i_2}^{(2^1)})]\} \\ &= x_k + (a_0^{(2)})^{1+2^{2^1}} \prod_{i_1=1}^{2^{2^1}} (x_k - a_{i_1}^{(2^1)}) \prod_{i_2=1}^{(2^{2^2} - 2^{2^1})/2} [(x_k^2 + B_{i_2}^{(2^2)} x_k + C_{i_2}^{(2^2)})]. \end{aligned}$$

If

$$x_k^2 + B_{i_2}^{(2^2)} x_k + C_{i_2}^{(2^2)} = 0$$

for $i_2 \in \cup_{q_1=1}^{N_1} I_{q_1}^{(2^{2-1})} \cup \cup_{q_2=1}^{N_2} I_{q_2}^{(2^2)}$ with

$$\begin{aligned} & I_q^{(2^{2-1})} = \{l_{(q-1)\times 2^1+1}, l_{(q-1)\times 2^1+2}\} \subseteq \{1, 2, \ldots, M_1\}, \\ & \text{for } q \in \{1, 2, \ldots, N_1\}, M_1 = N_1 \times 2^{21-1}, N_1 = 1 \\ & I_q^{(2^2)} = \{l_{(q-1)\times 2^2+1}, l_{(q-1)\times 2^2+2}, \ldots, l_{q\times 2^2}\} \subseteq \{M_1+1, M_1+2, \ldots, M\}, \\ & \text{for } q \in \{1, 2, \ldots, N_2\}, M_2 = (2^{2^2} - 2^{2^1})/2, \end{aligned}$$

then we have

$$\begin{aligned} b_{i_2,1}^{(2^2)} &= -\frac{1}{2}(B_1^{(2^2)} + \sqrt{\Delta^{(2^2)}}), b_{i_2,2}^{(2^2)} = -\frac{1}{2}(B_1^{(2^2)} - \sqrt{\Delta^{(2^2)}}) \\ \Delta^{(2^2)} &= (B_1^{(2^2)})^2 - 4C_1^{(2^2)} \geq 0. \end{aligned}$$

Thus

$$x_k^2 + B_{i_2}^{(2^2)} x_k + C_{i_2}^{(2^2)} = (x_k - b_{i_2,1}^{(2^2)})(x_k - b_{i_2,2}^{(2^2)}),$$

and

$$\begin{aligned} x_{k+2^2} &= x_k + (a_0^{(2^1)})^{1+2^{2^1}} \prod_{j_1=1}^{2^{2^1}} (x_k - a_{j_1}^{(2^1)}) \prod_{q=1}^{N} \prod_{j_2 \in I_q^{(2^2)}} (x_k - b_{j_2,1}^{(2^2)})(x_k - b_{j_2,2}^{(2^2)}) \\ &\quad \times \prod_{j_3 \in J^{(2^2)}} (x_k - b_{j_3,1}^{(2^2)})(x_k - b_{j_3,2}^{(2^2)}) \\ &= x_k + (a_0^{(4)}) \prod_{i=1}^{2^{2^2}} (x_k - a_i^{(2^2)}) \end{aligned}$$

where

$$\begin{aligned}
&\cup_{i=1}^{2^{2^2}}\{a_i^{(2^2)}\} = \text{sort}\{\cup_{i_1=1}^{2^{2^1}}\{a_{i_1}^{(2)}\}, \cup_{q=1}^{N}\cup_{i_2\in I_q^{(2^2)}}\{b_{i_2,1}^{(2^2)}, b_{i_2,2}^{(2^2)}\}\},\\
&a_i^{(2^2)} \le a_{i+1}^{(2^2)}; \Delta_{i_2}^{(2^2)} = (B_{i_2}^{(2^2)})^2 - 4C_{i_2}^{(2^2)} \ge 0,\\
&b_{i_2,1}^{(2^2)} = -\frac{1}{2}(B_{i_2}^{(2^2)} + \sqrt{\Delta_{i_2}^{(2^2)}}), b_{i_2,2}^{(2^2)} = -\frac{1}{2}(B_{i_2}^{(2^2)} - \sqrt{\Delta_{i_2}^{(2^2)}})\\
&i_2 \in \cup_{q_1=1}^{N_1} I_{q_1}^{(2^{2-1})} \bigcup \cup_{q_2=1}^{N_2} I_{q_2}^{(2^2)},\\
&I_{q_1}^{(2^{2-1})} = \{l_{(q_1-1)\times 2^1+1}, l_{(q_1-1)\times 2^1+2}\} \subseteq \{1, 2, \ldots, M_1\},\\
&\text{for } q \in \{1, 2, \ldots, N_1\}, M_1 = N_1 \times 2^1, N_1 = 1;\\
&I_{q_2}^{(2^2)} = \{l_{(q_2-1)\times 2^2+1}, l_{(q_2-1)\times 2^2+2}, \ldots, l_{q_2\times 2^2}\}\\
&\qquad \subseteq \{M_1+1, M_1+2, \ldots, M_2\},\\
&\text{for } q_2 \in \{1, 2, \ldots, N_2\}, M_2 = 2^{2^2-1} - 2^{2^1-1};\\
&\Delta_{i_2}^{(2^2)} = (B_{i_2}^{(2^2)})^2 - 4C_{i_2}^{(2^2)} < 0, \mathbf{i} = \sqrt{-1},\\
&b_{i_2,1}^{(2^2)} = -\frac{1}{2}(B_{i_2}^{(2^2)} + \mathbf{i}\sqrt{|\Delta_{i_2}^{(2^2)}|}), b_{i_2,2}^{(2^2)} = -\frac{1}{2}(B_{i_2}^{(2^2)} - \mathbf{i}\sqrt{|\Delta_{i_2}^{(2^2)}|}),\\
&i_2 \in J^{(2^2)} = \{l_{N_2\times 2^2+1}, l_{N_2\times 2^2+2}, \ldots, l_{M_2}\}\\
&\qquad\qquad \subset \{M_1+1, M_1+2, \ldots, M_2\}.
\end{aligned}$$

Nontrivial period-2^2 fixed-points are

$$\begin{aligned}
&x_{k+2^2}^* = x_k^* = a_i^{(4)} \in \cup_{i_2=1}^{M_2}\{b_{i_2,1}^{(2^2)}, b_{i_2,2}^{(2^2)}\},\\
&i \in \{1, 2, \ldots, 2^{2^2}\} \cup \{\varnothing\},
\end{aligned}$$

and trivial period-2^2 fixed-points are

$$\begin{aligned}
&x_{k+2^2}^* = x_k^* = a_i^{(4)} \in \cup_{i_1=1}^{2^{2^1}}\{a_{i_1}^{(2^1)}\},\\
&i \in \{1, 2, \ldots, 2^{2^2}\}.
\end{aligned}$$

Similarly, the period-2^l discrete system ($l = 1, 2, \ldots$) of the quadratic discrete system in Eq. (1.136) can be developed through the above nonlinear renormalization and the corresponding fixed-points can be obtained.

(III) Consider a period-2^{l-1} discrete system as

$$x_{k+2^{l-1}} = x_k + a_0^{(2^{l-1})} \prod_{i=1}^{2^{2^{l-1}}} (x_k - a_i^{(2^{l-1})}).$$

From the nonlinear renormalizations, let $k_1 = k + 2^{l-1}$, we have

$$x_{k_1+2^{l-1}+2^{l-1}} = x_{k_1+2^{l-1}} + a_0^{(2^{l-1})} \prod_{i=1}^{2^{2^{l-1}}} (x_{k_1+2^{l-1}} - a_i^{(2^{l-1})}).$$

Because k_1 is an index for iteration, it can be replaced by k. Thus, an equivalent form for the foregoing equations becomes

$$x_{k+2^l} = x_{k+2^{l-1}} + a_0^{(2^{l-1})} \prod_{i=1}^{2^{2^{l-1}}} (x_{k+2^{l-1}} - a_i^{(2^{l-1})}).$$

With the period-2^{l-1} discrete system, the foregoing equations becomes

$$\begin{aligned}
x_{k+2^l} &= x_k + [a_0^{(2^{l-1})} \prod_{i_1=1}^{2^{2^{l-1}}} (x_k - a_{i_1}^{(2^{l-1})})] \\
&\quad \times \{1 + \prod_{i_1=1}^{2^{2^{l-1}}} [1 + a_0^{(2^{l-1})} \prod_{i_2=1, i_2 \neq i_1}^{2^{2^{l-1}}} (x_k - a_{i_2}^{(2^{l-1})})]\} \\
&= x_k + [a_0^{(2^{l-1})} \prod_{i_1=1}^{2^{2^{l-1}}} (x_k - a_{i_1}^{(2^{l-1})})] \\
&\quad \times [(a_0^{(2^{l-1})})^{2^{2^{l-1}}} \prod_{i_2=1}^{2^{2^l-1}-2^{2^{l-1}-1}} (x_k^2 + B_{i_2}^{(2^l)} x_k + C_{i_2}^{(2^l)})].
\end{aligned}$$

If

$$x_k^2 + B_{i_2}^{(2^l)} x_k + C_{i_2}^{(2^l)} = 0$$

for $i_2 \in \cup_{q_1=1}^{N_1} I_{q_1}^{(2^{l-1})} \cup \cup_{q_2=1}^{N_2} I_{q_2}^{(2^{l-1})}$ with

$$\begin{aligned}
I_{q_1}^{(2^{l-1})} &= \{l_{(q-1)\times 2^{l-1}+1}, l_{(q-1)\times 2^{l-1}+2}, \ldots, l_{q\times 2^{l-1}}\} \\
&\quad \subseteq \{1, 2, \ldots, M_1\} \cup \{\varnothing\}, \\
&\text{for } q_1 = 1, 2, \ldots, N_1; M_1 = N_1 \times 2^{l-1}; \\
I_{q_2}^{(2^l)} &= \{l_{(q-1)\times 2^l+1}, l_{(q-1)\times 2^l+2}, \ldots, l_{q\times 2^l}\} \\
&\quad \subseteq \{M_1+1, M_1+2, \ldots, M_2\} \cup \{\varnothing\}, \\
&\text{for } q_2 = 1, 2, \ldots, N_2; M_2 = 2^{2^l-1} - 2^{2^{l-1}-1};
\end{aligned}$$

then we have

$$\begin{aligned}
&b_{i_2,1}^{(2^l)} = -\frac{1}{2}(B_{i_2}^{(2^l)} + \sqrt{\Delta_{i_2}^{(2^l)}}), b_{i_2,2}^{(2^l)} = -\frac{1}{2}(B_{i_2}^{(2^l)} - \sqrt{\Delta_{i_2}^{(2^l)}}) \\
&\Delta_{i_2}^{(2^l)} = (B_{i_2}^{(2^l)})^2 - 4C_{i_2}^{(2^l)} \geq 0.
\end{aligned}$$

Thus

$$x_k^2 + B_{i_2}^{(2^l)} x_k + C_{i_2}^{(2^l)} = (x_k - b_{i_2,1}^{(2^l)})(x_k - b_{i_2,2}^{(2^l)}).$$

Therefore

$$\begin{aligned}
x_{k+2^l} &= x_k + [a_0^{(2^{l-1})} \prod_{i_1=1,}^{2^{2^{l-1}}} (x_k - a_{i_1}^{(2^{l-1})})] \\
&\quad \times [(a_0^{(2^{l-1})})^{2^{2^{l-1}}} \prod_{i_2=1}^{2^{2^l}-2^{2^{l-1}}} (x_k - b_{i_2,1}^{(2^l)})(x_k - b_{i_2,2}^{(2^l)})] \\
&= x_k + (a_0^{(2^{l-1})})^{1+2^{2^{l-1}}} \prod_{i=1}^{2^{2^l}} (x_k - a_i^{(2^l)}) \\
&= x_k + a_0^{(2^l)} \prod_{i=1}^{2^{2^l}} (x_k - a_i^{(2^l)})
\end{aligned}$$

where

$$\begin{aligned}
& a_0^{(2)} = (a_0)^{1+2}, a_0^{(2^l)} = (a_0^{(2^{l-1})})^{1+2^{2^{l-1}}}, l = 1, 2, 3, \ldots; \\
& \cup_{i=1}^{2^{2^l}} \{a_i^{(2^l)}\} = \text{sort}\{\cup_{i_1=1}^{2^{2^{l-1}}} \{a_{i_1}^{(2^l)}\}, \cup_{i_2=1}^{M_2} \{b_{i_2,1}^{(2^l)}, b_{i_2,2}^{(2^l)}\}\}, a_i^{(2^l)} \le a_{i+1}^{(2^l)}; \\
& b_{i,1}^{(2^l)} = -\frac{1}{2}(B_i^{(2^l)} + \sqrt{\Delta_i^{(2^l)}}), b_{i,2}^{(2^{l-1})} = -\frac{1}{2}(B_i^{(2^l)} - \sqrt{\Delta_i^{(2^l)}}), \\
& \Delta_i^{(2^l)} = (B_i^{(2^l)})^2 - 4C_i^{(2^l)} \ge 0 \text{ for } i \in \cup_{q_1=1}^{N_1} I_{q_1}^{(2^{l-1})} \cup \cup_{q_2=1}^{N_2} I_{q_2}^{(2^{l-1})}, \\
& I_{q_1}^{(2^{l-1})} = \{l_{(q_1-1)\times 2^{l-1}+1}, l_{(q_1-1)\times 2^{l-1}+2}, \ldots, l_{q_1 \times 2^{l-1}}\} \\
& \qquad \subseteq \{1, 2, \ldots, M_1\} \cup \{\varnothing\}, \\
& \text{for } q_1 \in \{1, 2, \ldots, N_1\}, M_1 = N_1 \times 2^{l-1}; \\
& I_{q_2}^{(2^l)} = \{l_{(q_2-1)\times 2^l+1}, l_{(q_2-1)\times 2^l+2}, \ldots, l_{q_2 \times 2^l}\} \\
& \qquad \subseteq \{M_1+1, M_1+2, \ldots, M_2\} \cup \{\varnothing\}, \\
& \text{for } q_2 \in \{1, 2, \ldots, N_2\}, M_2 = 2^{2^l-1} - 2^{2^{l-1}-1}; \\
& b_{i,1}^{(2^l)} = -\frac{1}{2}(B_i^{(2^l)} + \mathbf{i}\sqrt{|\Delta_i^{(2^l)}|}), \; b_{i,2}^{(2^l)} = -\frac{1}{2}(B_i^{(2^l)} - \mathbf{i}\sqrt{|\Delta_i^{(2^l)}|}), \\
& \Delta_i^{(2^l)} = (B_i^{(2^l)})^2 - 4C_i^{(2^l)} < 0, \mathbf{i} = \sqrt{-1}, \\
& i \in J^{(2^l)} = \{l_{N_2\times 2^l+1}, l_{N_2\times 2^l+2}, \ldots, l_{M_2}\} \\
& \qquad \subset \{M_1+1, M_1+2, \ldots, M_2\} \cup \{\varnothing\}.
\end{aligned}$$

For the period-2^l discrete system, we have

$$x_{k+2^l} = x_k + a_0^{(2^l)} \prod_{i=1}^{2^{2^l}} (x_k - a_i^{(2^l)})$$

and the local stability and bifurcation at $x_k^* = a_i^{(2^l)}$ $(i \in \{1, 2, \ldots, 2^{2^l}\})$ can be determined by

$$\frac{dx_{k+2^l}}{dx_k} = 1 + a_0^{(2^l)} \sum_{i_1=1}^{2^{2^l}} \prod_{i_2=1, i_2 \neq i_1}^{2^{2^l}} (x_k - a_{i_2}^{(2^l)}),$$
$$\frac{d^2 x_{k+2^l}}{dx_k^2} = a_0^{(2^l)} \sum_{i_1=1}^{2^{2^l}} \sum_{i_2=1, i_2 \neq i_1}^{2^{2^l}} \prod_{i_3=1, i_3 \neq i_1, i_2}^{2^{2^l}} (x_k - a_{i_3}^{(2^l)}),$$
$$\vdots$$
$$\frac{d^r x_{k+2^l}}{dx_k^r} = a_0^{(2^l)} \sum_{i_1=1}^{2^{2^l}} \cdots \sum_{i_r=1, i_r \neq i_1, i_2, \ldots, i_{r-1}}^{2^{2^l}} \prod_{i_{r+1}=1, i_{r+1} \neq i_1, i_2 \ldots, i_r}^{2^{2^l}} (x_k - a_{i_{r+1}}^{(2^l)})$$
$$\text{for } r \le 2^{2^l};$$

and the period-doubling bifurcations are determined by

$$\frac{dx_{k+2^l}}{dx_k}\Big|_{x_k^* = a_i^{(2^l)}} = 1 + a_0^{(2^l)} \sum_{i_1=1}^{2^{2^l}} \prod_{i_2=1, i_2 \neq i_1}^{2^{2^l}} (x_k - a_{i_2}^{(2^l)})\Big|_{x_k^* = a_i^{(2^l)}} = -1,$$
$$\frac{d^2 x_{k+2^l}}{dx_k^2}\Big|_{x_k^* = a_i^{(2^l)}} = a_0^{(2^l)} \sum_{i_1=1}^{2^{2^l}} \sum_{i_2=1, i_2 \neq i_1}^{2^{2^l}} \prod_{i_3=1, i_3 \neq i_1, i_2}^{2^{2^l}} (x_k - a_{i_3}^{(2^l)})\Big|_{x_k^* = a_i^{(2^l)}} \neq 0.$$

Non-trivial period-2^l fixed-points are

$$x_{k+2^2}^* = x_k^* = a_i^{(2^l)} \in \cup_{i_2=1}^{M_2} \{ b_{i_2,1}^{(2^l)}, b_{i_2,2}^{(2^l)} \},$$
$$i \in \{1, 2, \ldots, 2^{2^l}\},$$

and trivial period-2^l fixed-points are

$$x_{k+2^2}^* = x_k^* = a_i^{(2^l)} \in \cup_{i_1=1}^{2^{2^{l-1}}} \{ a_{i_1}^{(2^{l-1})} \},$$
$$i \in \{1, 2, \ldots, 2^{2^l}\}.$$

This theorem can be easily proved. ■

1.4.3 Period-n Appearing and Period-Doublization

The period-n discrete system for the quadratic nonlinear discrete systems will be discussed, and the period-doublization of the period-n quadratic discrete system is discussed through the nonlinear renormalization.

Theorem 1.12 *Consider a 1-dimensional quadratic nonlinear discrete system*

$$\begin{aligned} x_{k+1} &= x_k + A(\mathbf{p})x_k^2 + B(\mathbf{p})x_k + C(\mathbf{p}) \\ &= x_k + a_0 \prod_{i=1}^{2}(x_k - a_i^{(1)}). \end{aligned} \tag{1.147}$$

(i) *After* n-*times iterations, a period*-n *discrete system for the quadratic discrete system in Eq.* (1.147) *is*

$$\begin{aligned} x_{k+n} &= x_k + a_0 \prod_{i_1=1}^{2}(x_k - a_{i_2})\{1 + \sum_{j=1}^{n} Q_j\} \\ &= x_k + (a_0)^{2^n-1} \prod_{i_1=1}^{2}(x_k - a_{i_1})[\prod_{j_2=1}^{2^{n-1}-1}(x_k^2 + B_{j_2}^{(n)} x_k + C_{j_2}^{(n)})] \\ &= x_k + a_0^{(n)} \prod_{i=1}^{2^n}(x_k - a_i^{(n)}) \end{aligned} \tag{1.148}$$

with

$$\begin{aligned} &\frac{dx_{k+n}}{dx_k} = 1 + a_0^{(n)} \sum_{i_1=1}^{2^n} \prod_{i_2=1, i_2 \neq i_1}^{2^n} (x_k - a_{i_2}^{(n)}), \\ &\frac{d^2 x_{k+n}}{dx_k^2} = a_0^{(n)} \sum_{i_1=1}^{2^n} \sum_{i_2=1, i_2 \neq i_1}^{2^n} \prod_{i_3=1, i_3 \neq i_1, i_2}^{2^n} (x_k - a_{i_3}^{(n)}), \\ &\vdots \\ &\frac{d^r x_{k+n}}{dx_k^r} = a_0^{(n)} \sum_{i_1=1}^{2^n} \cdots \sum_{i_r=1, i_r \neq i_1, i_2 \ldots, i_{r-1}}^{2^n} \prod_{i_{r+1}=1, i_{r+1} \neq i_1, i_2 \ldots, i_r}^{2^n} (x_k - a_{i_{r+1}}^{(n)}) \\ &\text{for } r \leq 2^n; \end{aligned} \tag{1.149}$$

where

$$\begin{aligned} &a_0^{(n)} = (a_0)^{2^n-1}, Q_1 = 0, Q_2 = \prod_{i_2=1}^{2}[1 + a_0 \prod_{i_1=1, i_1 \neq i_2}^{2}(x_k - a_{i_1}^{(1)})], \\ &Q_n = \prod_{i_n=1}^{2}[1 + a_0(1 + Q_{n-1}) \prod_{i_{n-1}=1, i_{n-1} \neq i_n}^{2}(x_k - a_{i_{n-1}}^{(1)})],\ n = 3, 4, \ldots; \\ &\cup_{i=1}^{2^n}\{a_i^{(n)}\} = \text{sort}\{\cup_{i_1=1}^{2}\{a_{i_1}^{(1)}\}, \cup_{i_2=1}^{M}\{b_{i_2,1}^{(n)}, b_{i_2,2}^{(n)}\}\}; \\ &b_{i_2,1}^{(n)} = -\frac{1}{2}(B_{i_2}^{(n)} + \sqrt{\Delta_{i_2}^{(n)}}), b_{i_2,2}^{(n)} = -\frac{1}{2}(B_{i_2}^{(n)} - \sqrt{\Delta_{i_2}^{(n)}}), \\ &\Delta_{i_2}^{(n)} = (B_{i_2}^{(n)})^2 - 4C_{i_2}^{(n)} \geq 0 \text{ for } i_2 \in \cup_{q=1}^{N} I_q^{(n)}, \\ &I_q^{(n)} = \{l_{(q-1)\times n+1}, l_{(q-1)\times n+2}, \ldots, l_{q\times n}\} \subseteq \{1, 2, \ldots, M\} \cup \{\varnothing\}, \\ &\text{for } q \in \{1, 2, \ldots, N\}, M = 2^{n-1} - 1; \\ &b_{i,1}^{(n)} = -\frac{1}{2}(B_i^{(n)} + \mathbf{i}\sqrt{|\Delta_i^{(n)}|}),\ b_{i,2}^{(n)} = -\frac{1}{2}(B_i^{(n)} - \mathbf{i}\sqrt{|\Delta_i^{(n)}|}), \\ &\Delta_i^{(n)} = (B_i^{(n)})^2 - 4C_i^{(n)} < 0,\ \mathbf{i} = \sqrt{-1} \\ &i \in \{l_{N\times n+1}, l_{N\times n+2}, \ldots, l_M\} \subset \{1, 2, \ldots, M\} \cup \{\varnothing\}; \end{aligned} \tag{1.150}$$

with fixed-points

$$
\begin{aligned}
&x_{k+n}^{*} = x_k^{*} = a_i^{(n)}, (i = 1, 2, \ldots, 2^n) \\
&\cup_{i=1}^{2^n}\{a_i^{(n)}\} = \text{sort}\{\cup_{i_1=1}^{2}\{a_{i_1}^{(1)}\}, \cup_{i_2=1}^{M}\{b_{i_2,1}^{(n)}, b_{i_2,2}^{(n)}\}\} \\
&\text{with } a_i^{(n)} < a_{i+1}^{(n)}.
\end{aligned}
\tag{1.151}
$$

(ii) *For a fixed-point of* $x_{k+n}^{*} = x_k^{*} = a_{i_1}^{(n)}$ $(i_1 \in I_q^{(n)},\ q \in \{1, 2, \ldots, N\})$, *if*

$$
\frac{dx_{k+n}}{dx_k}\Big|_{x_k^{*}=a_{i_1}^{(n)}} = 1 + a_0^{(n)} \prod_{i_2=1, i_2 \neq i_1}^{2^n} (a_{i_1}^{(n)} - a_{i_2}^{(n)}) = 1,
\tag{1.152}
$$

then there is a new discrete system for onset of the qth*-set of period-*n *fixed-points based on the second-order monotonic saddle-node bifurcation as*

$$
x_{k+n} = x_k + a_0^{(n)} \prod_{i_1 \in I_q^{(n)}} (x_k - a_{i_1}^{(n)})^2 \prod_{i_2=1}^{2^n} (x_k - a_{i_2}^{(n)})^{(1-\delta(i_1, j_2))}
\tag{1.153}
$$

where

$$
\delta(i_1, j_2) = 1 \text{ if } a_{j_2}^{(n)} = a_{i_1}^{(n)},\ \delta(i_1, j_2) = 0 \text{ if } a_{j_2}^{(n)} \neq a_{i_1}^{(n)}.
\tag{1.154}
$$

(ii_1) *If*

$$
\begin{aligned}
&\frac{dx_{k+n}}{dx_k}\Big|_{x_k^{*}=a_{i_1}^{(n)}} = 1 \quad (i_1 \in I_q^{(n)}), \\
&\frac{d^2 x_{k+n}}{dx_k^2}\Big|_{x_k^{*}=a_{i_1}^{(n)}} = 2a_0^{(n)} \prod_{i_1 \in I_q^{(n)}, i_2 \neq i_1} (a_{i_1}^{(n)} - a_{i_2}^{(n)})^2 \\
&\qquad \times \prod_{i_3=1}^{2^n} (a_{i_1}^{(n)} - a_{i_3}^{(n)})^{(1-\delta(i_2, j_2))} \neq 0
\end{aligned}
\tag{1.155}
$$

x_{k+n} at $x_k^{*} = a_{i_1}^{(n)}$ is

- *a monotonic lower-saddle of the second-order for* $d^2 x_{k+n}/dx_k^2|_{x_k^{*}=a_{i_1}^{(n)}} < 0$;
- *a monotonic upper-saddle of the second-order for* $d^2 x_{k+n}/dx_k^2|_{x_k^{*}=a_{i_1}^{(n)}} > 0$.

(ii_2) *The period-*n *fixed-points* $(n = 2^{n_1} \times m)$ *are trivial if*

$$
\left.\begin{array}{l}
x_{k+n}^{*}=x_{k}^{*}=a_{j_1}^{(n)}\in\{\cup_{i_i=1}^{2}\{a_{i_1}^{(1)}\},\cup_{i_2=1}^{2^{2^{n_1-1}m}}\{a_{i_2}^{(2^{n_1-1}m)}\}\} \\
\text{for } n_1=1,2,\ldots;m=2l_1+1;j_1\in\{1,2,\ldots,2^n\}\cup\{\varnothing\}
\end{array}\right\}
$$
$$
\text{for } n\neq 2^{n_2}, \tag{1.156}
$$
$$
\left.\begin{array}{l}
x_{k+n}^{*}=x_{k}^{*}=a_{j_1}^{(n)}\in\{\cup_{i_2=1}^{2^{2^{n_1-1}m}}\{a_{i_2}^{(2^{n_1-1}m)}\}\} \\
\text{for } n_1=1,2,\ldots;m=1;j_1\in\{1,2,\ldots,2^n\}\cup\{\varnothing\}
\end{array}\right\}
$$
$$
\text{for } n=2^{n_2}.
$$

(ii_3) *The period-*n *fixed-points* $(n=2^{n_1}\times m)$ *are non-trivial if*

$$
\left.\begin{array}{l}
x_{k+n}^{*}=x_{k}^{*}=a_{j_1}^{(n)}\notin\{\cup_{i_i=1}^{2}\{a_{i_1}^{(1)}\},\cup_{i_2=1}^{2^{2^{n_1-1}m}}\{a_{i_2}^{(2^{n_1-1}m)}\}\} \\
\text{for } n_1=1,2,\ldots;m=2l_1+1;j_1\in\{1,2,\ldots,2^n\}\cup\{\varnothing\}
\end{array}\right\}
$$
$$
\text{for } n\neq 2^{n_2}, \tag{1.157}
$$
$$
\left.\begin{array}{l}
x_{k+n}^{*}=x_{k}^{*}=a_{j_1}^{(n)}\notin\{\cup_{i_2=1}^{2^{2^{n_1-1}m}}\{a_{i_2}^{(2^{n_1-1}m)}\}\} \\
\text{for } n_1=1,2,\ldots;m=1;j_1\in\{1,2,\ldots,2^n\}\cup\{\varnothing\}
\end{array}\right\}
$$
$$
\text{for } n=2^{n_2}.
$$

*Such a period-*n *fixed-point is*

- *monotonically unstable if* $dx_{k+n}/dx_k|_{x_k^*=a_{i_1}^{(n)}}\in(1,\infty)$;
- *monotonically invariant if* $dx_{k+n}/dx_k|_{x_k^*=a_{i_1}^{(n)}}=1$, *which is*
  - *a monotonic upper-saddle of the* $(2l_1)^{\text{th}}$ *order for* $d^{2l_1}x_{k+n}/dx_k^{2l_1}|_{x_k^*}>0$;
  - *a monotonic lower-saddle the* $(2l_1)^{\text{th}}$ *order for* $d^{2l_1}x_{k+n}/dx_k^{2l_1}|_{x_k^*}<0$;
  - *a monotonic source of the* $(2l_1+1)^{\text{th}}$ *order for* $d^{2l_1+1}x_{k+n}/dx_k^{2l_1+1}|_{x_k^*}>0$;
  - *a monotonic sink the* $(2l_1+1)^{\text{th}}$ *order for* $d^{2l_1+1}x_{k+n}/dx_k^{2l_1+1}|_{x_k^*}<0$;
- *monotonically stable if* $dx_{k+n}/dx_k|_{x_k^*=a_{i_1}^{(n)}}\in(0,1)$;
- *invariantly zero-stable if* $dx_{k+n}/dx_k|_{x_k^*=a_{i_1}^{(n)}}=0$;
- *oscillatorilly stable if* $dx_{k+n}/dx_k|_{x_k^*=a_{i_1}^{(n)}}\in(-1,0)$;
- *flipped if* $dx_{k+n}/dx_k|_{x_k^*=a_{i_1}^{(n)}}=-1$, *which is*
  - *an oscillatory upper-saddle of the* $(2l_1)^{\text{th}}$ *order for* $d^{2l_1}x_{k+n}/dx_k^{2l_1}|_{x_k^*}>0$;
  - *an oscillatory lower-saddle the* $(2l_1)^{\text{th}}$ *order for* $d^{2l_1}x_{k+n}/dx_k^{2l_1}|_{x_k^*}<0$;

- *an oscillatory source of the* $(2l_1+1)^{\text{th}}$ *order for* $d^{2l_1+1}x_{k+n}/dx_k^{2l_1+1}|_{x_k^*}<0$;
- *an oscillatory sink the* $(2l_1+1)^{\text{th}}$ *order for* $d^{2l_1+1}x_{k+n}/dx_k^{2l_1+1}|_{x_k^*}>0$;

- *oscillatorilly unstable if* $dx_{k+n}/dx_k|_{x_k^*=a_{i_1}^{(n)}}\in(-\infty,-1)$.

(iii) *For a fixed-point of* $x_{k+n}^*=x_k^*=a_{i_1}^{(n)}$ $(i_1\in I_q^{(n)},\ q\in\{1,2,\ldots,N\})$, *there is a period-doubling of the* q^{th}*-set of period-*n *fixed-points if*

$$\frac{dx_{k+n}}{dx_k}|_{x_k^*=a_{i_1}^{(n)}}=1+a_0^{(n)}\prod\nolimits_{j_2=1,j_2\neq i_1}^{2^n}(a_{i_1}^{(n)}-a_{j_2}^{(n)})=-1 \tag{1.158}$$

with

- *an oscillatory upper-saddle for* $d^2x_{k+n}/dx_k^2|_{x_k^*=a_{i_1}^{(n)}}>0$;
- *an oscillatory lower saddle for* $d^2x_{k+n}/dx_k^2|_{x_k^*=a_{i_1}^{(n)}}<0$.

The corresponding period-$2\times n$ *discrete system of the quadratic discrete system in Eq. (1.147) is*

$$x_{k+2\times n}=x_k+a_0^{(2\times n)}\prod\nolimits_{i_1\in I_q^{(n)}}(x_k-a_{i_1}^{(n)})^3\prod\nolimits_{i_2=1}^{2^{2\times n}}(x_k-a_{i_2}^{(2\times n)})^{(1-\delta(i_1,i_2))} \tag{1.159}$$

with

$$\begin{aligned}
&\frac{dx_{k+2\times n}}{dx_k}|_{x_k^*=a_{i_1}^{(n)}}=1,\ \frac{d^2x_{k+2\times n}}{dx_k^2}|_{x_k^*=a_{i_1}^{(n)}}=0;\\
&\frac{d^3x_{k+2\times n}}{dx_k^3}|_{x_k^*=a_{i_1}^{(n)}}=6a_0^{(2\times n)}\prod\nolimits_{i_2\in I_q^{(n)},i_2\neq i_1}(a_{i_1}^{(n)}-a_{i_2}^{(n)})^3\\
&\qquad\times\prod\nolimits_{i_3=1}^{2^{2\times n}}(a_{i_1}^{(n)}-a_{i_3}^{(2\times n)})^{(1-\delta(i_1,i_3))}\neq 0.
\end{aligned} \tag{1.160}$$

Thus, $x_{k+2\times n}$ *at* $x_k^*=a_{i_1}^{(n)}$ *for* $i_1\in I_q^{(n)}$, $q\in\{1,2,\ldots,N\}$ *is*

- *a monotonic sink of the third-order if* $d^3x_{k+2\times n}/dx_k^3|_{x_k^*=a_{i_1}^{(n)}}<0$,
- *a monotonic source of the third-order if* $d^3x_{k+2\times n}/dx_k^3|_{x_k^*=a_{i_1}^{(n)}}>0$.

(iv) *After* l*-times period-doubling bifurcations of period-*n *fixed points, a period-*$2^l\times n$ *discrete system of the quadratic discrete system in Eq. (1.147) is*

$$
\begin{aligned}
x_{k+2^l\times n} &= x_k + [a_0^{(2^{l-1}\times n)} \prod_{i_1=1}^{2^{2^{l-1}\times n}} (x_k - a_{i_1}^{(2^{l-1}\times n)})] \\
&\quad \times \{1 + \prod_{i_1=1}^{2^{2^{l-1}\times n}} [1 + a_0^{(2^{l-1}\times n)} \prod_{i_2=1, i_2\neq i_1}^{2^{2^{l-1}\times n}} (x_k - a_{i_2}^{(2^{l-1}\times n)})]\} \\
&= x_k + [a_0^{(2^{l-1}\times n)} \prod_{i_1=1}^{2^{2^{l-1}\times n}} (x_k - a_{i_1}^{(2^{l-1}\times n)})] \\
&\quad \times [(a_0^{(2^{l-1}\times n)})^{2^{2^{l-1}\times n}} \prod_{j_1=1}^{2^{2^l\times n-1}-2^{2^{l-1}\times n-1}} (x_k^2 + B_{j_2}^{(2^l\times n)} x_k + C_{j_2}^{(2^l\times n)})] \\
&= x_k + [a_0^{(2^{l-1}\times n)} \prod_{i_1=1}^{2^{2^{l-1}\times n}} (x_k - a_{i_1}^{(2^{l-1}\times n)})] \\
&\quad \times [(a_0^{(2^{l-1}\times n)})^{2^{2^{l-1}\times n}} \prod_{j_2=1}^{2^{2^l\times n-1}-2^{2^{l-1}\times n-1}} (x_k - b_{j_2}^{(2^l\times n)})] \\
&= x_k + (a_0^{(2^{l-1}\times n)})^{2^{2^{l-1}\times n}} \prod_{i=1}^{2^{2^l\times n}} (x_k - a_i^{(2^l\times n)}) \\
&= x_k + a_0^{(2^l\times n)} \prod_{i=1}^{2^{2^l\times n}} (x_k - a_i^{(2^l\times n)})
\end{aligned} \tag{1.161}
$$

with

$$
\begin{aligned}
\frac{dx_{k+2^l\times n}}{dx_k} &= 1 + a_0^{(2^l\times n)} \sum_{i_1=1}^{2^{2^l\times n}} \prod_{i_2=1, i_2\neq i_1}^{2^{2^l\times n}} (x_k - a_{i_2}^{(2^l\times n)}), \\
\frac{d^2 x_{k+2^l\times n}}{dx_k^2} &= a_0^{(2^l\times n)} \sum_{i_1=1}^{2^{2^l\times n}} \sum_{i_2=1, i_2\neq i_1}^{2^{2^l\times n}} \prod_{i_3=1, i_3\neq i_1, i_2}^{2^{2^l\times n}} (x_k - a_{i_3}^{(2^l\times n)}), \\
&\vdots \\
\frac{d^r x_{k+2^l\times n}}{dx_k^r} &= a_0^{(2^l\times n)} \sum_{i_1=1}^{2^{2^l\times n}} \cdots \sum_{i_r=1, i_r\neq i_1, i_2\ldots, i_{r-1}}^{2^{2^l\times n}} \prod_{i_{r+1}=1, i_{r+1}\neq i_1, i_2\ldots, i_r}^{2^{2^l\times n}} (x_k - a_{i_{r+1}}^{(2^l\times n)})
\end{aligned} \tag{1.162}
$$

for $r \le 2^{2^l\times n}$,

where

$$
\begin{aligned}
a_0^{(2\times n)} &= (a_0^{(n)})^{1+2^n}, a_0^{(2^l\times n)} = (a_0^{(2^{l-1}\times n)})^{1+2^{2^{l-1}\times n}}, l = 1, 2, 3, \ldots; \\
\cup_{i=1}^{2^{(2^l\times n)}} \{a_i^{(2^l\times n)}\} &= \text{sort}\{\cup_{i_1=1}^{2^{2^{l-1}\times n}} \{a_{i_1}^{(2^{l-1}\times n)}\}, \cup_{i_2=1}^{M_2} \{b_{i_2,1}^{(2^l\times n)}, b_{i_2,2}^{(2^l\times n)}\}\}; \\
b_{i,1}^{(2^l\times n)} &= -\frac{1}{2}(B_i^{(2^l\times n)} + \sqrt{\Delta_i^{(2^l\times n)}}), \\
b_{i,2}^{(2^l\times n)} &= -\frac{1}{2}(B_i^{(2^l\times n)} - \sqrt{\Delta_i^{(2^l\times n)}}), \\
\Delta_i^{(2^l\times n)} &= (B_i^{(2^l\times n)})^2 - 4C_i^{(2^l\times n)} \ge 0
\end{aligned}
$$

$$
\begin{aligned}
&\text{for } i \in \cup_{q_1=1}^{N_1} I_{q_1}^{(2^{l-1}\times n)} \bigcup \cup_{q_2=1}^{N_2} I_{q_2}^{(2^l\times n)} \\
&I_{q_1}^{(2^{l-1}\times n)} = \{l_{(q_1-1)\times(2^{l-1}\times n)+1}, l_{(q_1-1)\times(2^{l-1}\times n)+2}, \ldots, l_{q_1\times(2^{l-1}\times n)}\} \\
&\qquad \subseteq \{1,2,\ldots,M_1\} \cup \{\varnothing\}, \\
&\text{for } q_1 \in \{1,2,\ldots,N_1\}, M_1 = N_1 \times (2^{l-1} \times n); \\
&I_{q_2}^{(2^l\times n)} = \{l_{(q_2-1)\times(2^l\times n)+1}, l_{(q_2-1)\times(2^l\times n)+2}, \ldots, l_{q_2\times(2^l\times n)}\} \\
&\qquad \subseteq \{M_1+1, M_1+2,\ldots,M_2\} \cup \{\varnothing\}, \\
&\text{for } q_2 \in \{1,2,\ldots,N_2\}, M_2 = (2^{2^l\times n-1} - 2^{2^{l-1}\times n-1}); \\
&b_{i,1}^{(2^l\times n)} = -\frac{1}{2}(B_i^{(2^l\times n)} + \mathbf{i}\sqrt{|\Delta_i^{(2^l\times n)}|}), \\
&b_{i,2}^{(2^l\times n)} = -\frac{1}{2}(B_i^{(2^l\times n)} - \mathbf{i}\sqrt{|\Delta_i^{(2^l\times n)}|}), \\
&\Delta_i^{(2^l\times n)} = (B_i^{(2^l\times n)})^2 - 4C_i^{(2^l\times n)} < 0, \mathbf{i} = \sqrt{-1}, \\
&i \in \{l_{N_2\times(2^l\times n)+1}, l_{N_2\times(2^l\times n)+2}, \ldots, l_{M_2}\} \\
&\quad \subset \{M_1+1, M_1+2,\ldots,M_2\} \cup \{\varnothing\}
\end{aligned} \tag{1.163}
$$

with fixed-points

$$
\begin{aligned}
&x_{k+2^l\times n}^* = x_k^* = a_i^{(2^l\times n)}, (i = 1,2,\ldots,2^{2^l\times n}) \\
&\cup_{i=1}^{2^{2^l\times n}} \{a_i^{(2^l\times n)}\} = \text{sort}\{\cup_{i=1}^{2^{2^{l-1}\times n}} \{a_i^{(2^{l-1}\times n)}\}, \cup_{i_2=1}^{M_2}\{b_{i_2,1}^{(2^l\times n)}, b_{i_2,2}^{(2^l\times n)}\}\} \\
&\text{with } a_i^{(2^l\times n)} < a_{i+1}^{(2^l\times n)}.
\end{aligned} \tag{1.164}
$$

(v) *For a fixed-point of* $x_{k+(2^l\times n)}^* = x_k^* = a_{i_1}^{(2^{l-1}\times n)}$ $(i_1 \in I_q^{(2^{l-1}\times n)} \subseteq \{1,2,\ldots,M_1\})$, *there is a period-*$(2^l \times n)$ *discrete system if*

$$
\frac{dx_{k+2^{l-1}\times n}}{dx_k}\Big|_{x_k^*=a_{i_1}^{(2^{l-1}\times n)}} = 1 + a_0^{(2^{l-1}\times n)} \prod_{i_2=1, i_2\neq i_1}^{2^{2^{l-1}\times n}} (a_{i_1}^{(2^{l-1}\times n)} - a_{i_2}^{(2^{l-1}\times n)}) = -1 \tag{1.165}
$$

with

- *an oscillatory upper-saddle for* $d^2x_{k+2^{l-1}\times n}/dx_k^2|_{x_k^*=a_{i_1}^{(2^{l-1}\times n)}} > 0$;
- *an oscillatory lower-saddle for* $d^2x_{k+2^{l-1}\times n}/dx_k^2|_{x_k^*=a_{i_1}^{(2^{l-1}\times n)}} < 0$.

The corresponding period-$(2^l \times n)$ *discrete system is*

$$\begin{aligned} x_{k+2^l\times n} &= x_k + a_0^{(2^l\times n)} \prod\nolimits_{i_1\in I_{q_1}^{(2^{l-1}\times n)}} (x_k - a_{i_1}^{(2^{l-1}\times n)})^3 \\ &\times \prod\nolimits_{j_2=1}^{2^{2^l\times n}} (x_k - a_{j_2}^{(2^l\times n)})^{(1-\delta(i_1,j_2))}, \end{aligned} \tag{1.166}$$

where

$$\delta(i_1,j_2) = 1 \text{ if } a_{j_2}^{(2^l\times n)} = a_{i_1}^{(2^{l-1}\times n)},\ \delta(i_1,j_2) = 0 \text{ if } a_{j_2}^{(2^l\times n)} \neq a_{i_1}^{(2^{l-1}\times n)} \tag{1.167}$$

with

$$\begin{aligned} &\frac{dx_{k+2^l\times n}}{dx_k}\Big|_{x_k^*=a_{i_1}^{(2^{l-1}\times n)}} = 1,\ \frac{d^2x_{k+2^l\times n}}{dx_k^2}\Big|_{x_k^*=a_{i_1}^{(2^{l-1}\times n)}} = 0; \\ &\frac{d^3x_{k+2^l\times n}}{dx_k^3}\Big|_{x_k^*=a_{i_1}^{(2^{l-1})}} = 6a_0^{(2^l\times n)} \prod\nolimits_{i_2\in I_{q_1}^{(2^{l-1}\times n)},i_2\neq i_1} (a_{i_1}^{(2^{l-1}\times n)} - a_{i_2}^{(2^{l-1}\times n)})^3 \\ &\qquad \times \prod\nolimits_{j_2=1}^{2^{2^l\times n}} (a_{i_1}^{(2^{l-1}\times n)} - a_{j_2}^{(2^l\times n)})^{(1-\delta(i_2,j_2))} \neq 0 \\ &(i_1 \in I_q^{(2^{l-1}\times n)}, q \in \{1,2,\ldots,N_1\}). \end{aligned} \tag{1.168}$$

Thus, $x_{k+2^l\times n}$ *at* $x_k^* = a_{i_1}^{(2^{l-1}\times n)}$ *is*

- *a monotonic sink of the third-order if* $d^3x_{k+2^l\times n}/dx_k^3|_{x_k^*=a_{i_1}^{(2^{l-1})}} < 0$*;*
- *a monotonic source of the third-order if* $d^3x_{k+2^l\times n}/dx_k^3|_{x_k^*=a_{i_1}^{(2^{l-1})}} > 0$.

(v$_1$) *The period-*$2^l \times n$ *fixed-points are trivial if*

$$\left.\begin{aligned} &x_{k+2^l\times n}^* = x_k^* = a_j^{(2^l\times n)} \in \{\cup_{i_1=1}^{2}\{a_{i_1}^{(1)}\}, \cup_{i_2=1}^{2^{2^{l-1}\times n}}\{a_{i_2}^{(2^{l-1}\times n)}\}\} \\ &\text{for } j = 1,2,\cdots,2^{2^l\times n} \end{aligned}\right\}$$
$$\text{for } n \neq 2^{n_1},$$
$$\left.\begin{aligned} &x_{k+2^l\times n}^* = x_k^* = a_j^{(2^l\times n)} \in \cup_{i_2=1}^{2^{2^{l-1}\times n}}\{a_{i_2}^{(2^{l-1}\times n)}\} \\ &\text{for } j = 1,2,\cdots,2^{2^l\times n} \end{aligned}\right\} \tag{1.169}$$
$$\text{for } n = 2^{n_1}.$$

(v$_2$) *The period-*$2^l \times n$ *fixed-points are non-trivial if*

$$
\left.\begin{array}{l}
x^*_{k+2^l\times n} = x^*_k = a_j^{(2^l\times n)} \notin \{\cup_{i_i=1}^{2}\{a_{i_1}^{(1)}\}, \cup_{i_2=1}^{2^{2^{l-1}\times n}}\{a_{i_2}^{(2^{l-1}\times n)}\}\} \\
\text{for } j=1,2,\cdots,2^{(2^l\times n)}
\end{array}\right\}
$$
$$
\text{for } n\neq 2^{n_1},
$$
$$
\left.\begin{array}{l}
x^*_{k+2^l\times n} = x^*_k = a_j^{(2^l\times n)} \notin \cup_{i_2=1}^{2^{2^{l-1}\times n}}\{a_{i_2}^{(2^{l-1}\times n)}\} \\
\text{for } j=1,2,\cdots,2^{2^l\times n}
\end{array}\right\}
$$
$$
\text{for } n = 2^{n_1}. \tag{1.170}
$$

Such a period-$2^l\times n$ *fixed-point is*

- *monotonically unstable if* $dx_{k+2^l\times n}/dx_k|_{x_k^*=a_{i_1}^{(2^l\times n)}}\in(1,\infty)$;
- *monotonically invariant if* $dx_{k+2^l\times n}/dx_k|_{x_k^*=a_{i_1}^{(2^l\times n)}}=1$, *which is*

  - *a monotonic upper-saddle of the* $(2l_1)^{\text{th}}$ *order for* $d^{2l_1}x_{k+2^l\times n}/dx_k^{2l_1}|_{x_k^*}>0$ *(independent* $(2l_1)$*-branch appearance);*
  - *a monotonic lower-saddle the* $(2l_1)^{\text{th}}$ *order for* $d^{2l_1}x_{k+2^l\times n}/dx_k^{2l_1}|_{x_k^*}<0$ *(independent* $(2l_1)$*-branch appearance);*
  - *a monotonic source of the* $(2l_1+1)^{\text{th}}$ *order for* $d^{2l_1+1}x_{k+2^l\times n}/dx_k^{2l_1+1}|_{x_k^*}>0$ *(dependent* $(2l_1+1)$*-branch appearance from one branch);*
  - *a monotonic sink the* $(2l_1+1)^{\text{th}}$ *order for* $d^{2l_1+1}x_{k+2^l\times n}/dx_k^{2l_1+1}|_{x_k^*}<0$ *(dependent* $(2l_1+1)$*-branch appearance from one branch);*

- *monotonically stable if* $dx_{k+2^l\times n}/dx_k|_{x_k^*=a_{i_1}^{(2^l\times n)}}\in(0,1)$;
- *invariantly zero-stable if* $dx_{k+2^l\times n}/dx_k|_{x_k^*=a_{i_1}^{(2^l\times n)}}=0$;
- *oscillatorilly stable if* $dx_{k+2^l\times n}/dx_k|_{x_k^*=a_{i_1}^{(2^l\times n)}}\in(-1,0)$;
- *flipped if* $dx_{k+2^l\times n}/dx_k|_{x_k^*=a_{i_1}^{(2^l\times n)}}=-1$, *which is*

  - *an oscillatory upper-saddle of the* $(2l_1)^{\text{th}}$ *order if* $d^{2l_1}x_{k+2^l\times n}/dx_k^{2l_1}|_{x_k^*}>0$;
  - *an oscillatory lower-saddle the* $(2l_1)^{\text{th}}$ *order with* $d^{2l_1}x_{k+2^l\times n}/dx_k^{2l_1}|_{x_k^*}<0$;
  - *an oscillatory source of the* $(2l_1+1)^{\text{th}}$ *order if* $d^{2l_1+1}x_{k+2^l\times n}/dx_k^{2l_1+1}|_{x_k^*}<0$;
  - *an oscillatory sink the* $(2l_1+1)^{\text{th}}$ *order with* $d^{2l_1+1}x_{k+2^l\times n}/dx_k^{2l_1+1}|_{x_k^*}>0$;

- *oscillatorilly unstable if* $dx_{k+2^l\times n}/dx_k|_{x_k^*=a_{i_1}^{(2^l\times n)}}\in(-\infty,-1)$.

Proof Through the nonlinear renormalization, the proof of this theorem is similar to the proof of Theorem 1.11. This theorem can be easily proved. ■

1.4.4 Period-n Bifurcation Trees

Consider a period-n discrete system of the quadratic system as

$$x_{k+n} = x_k + a_0^{(n)} \prod_{i=1}^{2^n} (x_k - a_i^{(n)}) \quad (1.171)$$

where $a_0^{(n)} = (a_0)^{2^n - 1}$.

For $n = 1$, Eq. (1.171) gives a period-1 discrete system of the quadratic system as

$$x_{k+1} = x_k + a_0^{(1)} \prod_{i=1}^{2} (x_k - a_i^{(1)}). \quad (1.172)$$

- If $a_i^{(1)} (i = 1, 2)$ are complex, none of fixed-points exists in such a quadratic discrete system.
- If $a_i^{(1)} (i = 1, 2)$ are real, two fixed-points exist in such a quadratic discrete system.

For $n = 2$, Eq. (1.171) gives a period-2 discrete system of the quadratic system as

$$x_{k+2} = x_k + a_0^{(2)} \prod_{i=1}^{2^2} (x_k - a_i^{(2)}). \quad (1.173)$$

- If $a_i^{(2)} (i = 1, 2, \ldots, 2^2)$ are complex, the period-2 discrete system does not have any fixed-points.
- If two of $a_i^{(2)} (i = 1, 2, \ldots, 2^2)$ are real, the period-2 discrete system possesses two fixed-points, which are trivial. The two fixed-points are the same as the period-1 fixed-points.
- If all of $a_i^{(2)} (i = 1, 2, \ldots, 2^2)$ are real, the period-2 discrete system possesses four fixed-points, including two trivial fixed-points for period-1 and two non-trivial fixed-points for period-2. Such two non-trivial fixed points are generated through period-doubling bifurcation, and both of fixed-points are stable for period-2.

Thus, the period-2 discrete system has one set of period-2 fixed-points on the period-1 to period-2 period-doubling bifurcation tree. Without any independent period-2 fixed-points exists.

For $n = 3$, Eq. (1.171) gives a period-3 discrete system of the quadratic system as

$$x_{k+3} = x_k + a_0^{(3)} \prod_{i=1}^{2^3} (x_k - a_i^{(3)}). \tag{1.174}$$

- If $a_i^{(3)} (i = 1, 2, \ldots, 2^3)$ are complex, the period-3 discrete system does not have any fixed-points.
- If two of $a_i^{(3)} (i = 1, 2, \ldots, 2^3)$ are real, the period-3 discrete system possesses two trivial fixed-points which are the same as the period-1 fixed-points.
- If all of $a_i^{(3)} (i = 1, 2, \ldots, 2^3)$ are real, the period-3 discrete system possesses eight fixed-points, including two trivial fixed-points for period-1 and six non-trivial fixed-points for period-3. Such non-trivial fixed points are generated through the monotonic upper-saddle or monotonic lower-saddle bifurcations. The period-3 fixed-points are independent of the trivial fixed-points for period-1.

Thus, the period-3 discrete system has at most one set of period-3 fixed-points, which is independent of the period-1 fixed-points.

For $n = 4$, Eq. (1.171) gives a period-4 discrete system of the quadratic system as

$$x_{k+4} = x_k + a_0^{(4)} \prod_{i=1}^{2^4} (x_k - a_i^{(4)}). \tag{1.175}$$

- If $a_i^{(4)} (i = 1, 2, \ldots, 2^4)$ are complex, the period-4 discrete system does not have any fixed-points.
- If two of $a_i^{(4)} (i = 1, 2, \ldots, 2^4)$ are real, the period-4 discrete system possesses two trivial fixed-points which are the same as the period-1 fixed-points.
- If four of $a_i^{(4)} (i = 1, 2, \ldots, 2^4)$ are real, the period-4 discrete system possesses four trivial fixed-points which are the same as the period-1 and period-2 fixed-points.
- If eight of $a_i^{(4)} (i = 1, 2, \ldots, 2^4)$ are real, the period-4 discrete system possesses eight fixed-points, including two trivial fixed-points for period-1, two trivial fixed-points for period-2, and four non-trivial fixed-points for period-4. Such non-trivial fixed points are stable, which are generated through the period-doubling bifurcations. All trivial fixed-points for period-4 are unstable.
- If all of $a_i^{(4)} (i = 1, 2, \ldots, 2^4)$ are real, in addition to the period-4 fixed-points by the period-doubling bifurcation, the period-4 discrete system possesses eight non-trivial fixed-points for period-4, which are generated by the monotonic upper-saddle or lower-saddle bifurcations. The period-4 fixed-points are independent of the trivial fixed-points.

Thus, the period-4 discrete system has at most two sets of period-4 fixed-points, one is dependent on the period-1 to period-4 period-doubling tree, and one set of period-4 fixed-points is independent of the period-1 to period-4 period-doubling bifurcation tree.

For $n = 5$, Eq. (1.171) gives a period-5 discrete system of the quadratic system as

$$x_{k+5} = x_k + a_0^{(5)} \prod_{i=1}^{2^5} (x_k - a_i^{(5)}). \tag{1.176}$$

- If $a_i^{(5)} (i = 1, 2, \ldots, 2^5)$ are complex, the period-5 discrete system does not have any fixed-points.
- If two of $a_i^{(5)} (i = 1, 2, \ldots, 2^5)$ are real, the period-5 discrete system possesses two trivial fixed-points which are the same as the period-1 fixed-points.
- If twelve (12) of $a_i^{(5)} (i = 1, 2, \ldots, 2^5)$ are real, the period-5 discrete system possesses 12 fixed-points, including two trivial fixed-points for period-1 and ten (10) non-trivial fixed-points for one set of period-5. Such non-trivial fixed points are generated through the monotonic upper-saddle or monotonic lower-saddle bifurcations. The period-5 fixed-points are independent of the trivial fixed-points for period-5.
- If twenty-two (22) of $a_i^{(5)} (i = 1, 2, \ldots, 2^5)$ are real, the period-5 discrete system possesses 20 fixed-points, including two (2) trivial fixed-points for period-1 and 20 non-trivial fixed-points for two sets of period-5.
- If thirty-two (32) of $a_i^{(5)} (i = 1, 2, \ldots, 2^5)$ are real, the period-5 discrete system possesses 32 fixed-points, including two (2) trivial fixed-points for period-1 and 30 non-trivial fixed-points for three sets of period-5.

Thus, the period-5 discrete system has at most three (3) sets of period-5 fixed-points independent of the trivial period-5 fixed-points from the period-1 fixed-points.

For $n = 6$, Eq. (1.171) gives a period-6 discrete system of the quadratic system as

$$x_{k+6} = x_k + a_0^{(6)} \prod_{i=1}^{2^6} (x_k - a_i^{(6)}). \tag{1.177}$$

- If $a_i^{(6)} (i = 1, 2, \ldots, 2^6)$ are complex, the period-6 discrete system does not have any fixed-points.
- If two of $a_i^{(6)} (i = 1, 2, \ldots, 2^6)$ are real, the period-6 discrete system possesses two trivial fixed-points which are the same as the period-1 fixed-points.
- If twelve (14) of $a_i^{(6)} (i = 1, 2, \ldots, 2^6)$ are real, the period-6 discrete system possesses 14 fixed-points, including two (2) trivial fixed-points for period-1, six (6) trivial fixed-points for period-3, and six (6) non-trivial fixed-points for

period-6. Such non-trivial fixed points are generated through the period-3 period-doubling bifurcation. The six trivial fixed-points for period-3 are unstable. The six non-trivial fixed-points for period-6 are stable.

- If twenty-two (26) of $a_i^{(6)}(i = 1, 2, \ldots, 2^6)$ are real, the period-6 discrete system possesses 26 fixed-points, including two (2) trivial fixed-points for period-1, 12 non-trivial fixed-points on the period-doubling bifurcation trees of period-3, and 12 non-trivial fixed-points for period-6 caused by monotonic upper- and lower-saddle-nodes bifurcations.
- If thirty-two (62) of $a_i^{(6)}(i = 1, 2, \ldots, 2^6)$ are real, in addition to 14 fixed-points for period-1 and period-3 bifurcation tree, there are 4 sets of period-6 fixed-points, which are generated through the monotonic upper- and lower-saddle-node bifurcations.

Thus, the period-6 discrete system has at most six sets of period-6 fixed-points including five sets of independent period-6 fixed-point, one set of period-6 fixed-points on the period-3 period-doubling bifurcation tree, and the period-1 fixed-pionts. For such a period-6 discrete system, there exist two complex fixed-points.

Similarly, other period-n discrete systems can be discussed. From the previous discussion, the period-n fixed-points for a quadratic discrete system are tabulated in Table 1.1. The dependent sets of period-n fixed-points are on the period-doubling bifurcation trees. The independent sets of period-n fixed-points are generated through monotonic saddle-node bifurcations. From analytical expressions, the maximum sets of period-n fixed-points includes dependent and independent sets of period-n fixed-points. In addition to the period-1 trivial fixed-points, other fixed-points on the bifurcation trees relative to period-n fixed points are also trivial. From the period-n discrete systems of a quadratic discrete system, period-1 to period-4 bifurcation trees are sketched through period-n discrete systems, as shown in Fig. 1.9. The solid and dashed curves are for stable and unstable period-n fixed-points, respectively. The red and dark red colors are for period-n fixed points dependent on and independent of period-1 fixed-points on the bifurcation trees. The period-n fixed-points on the other period-doubling bifurcations are said to be dependent. The dependent period-n fixed-points are obtained from period-doubling bifurcation. However, the onsets of period-n fixed-points are not based on period-doubling bifurcations, which are said to be independent. The onsets of such independent period-n fixed-points are based on the monotonic saddle-node bifurcations. The numerical examples can be found from Luo and Guo (2013). Such dependent and independent period-n fixed-points for quadratic systems are presented. Through such a way, one can find all possible period-n fixed points.

Table 1.1 Period-n fixed-points for a quadratic discrete system

	Dependent sets	Independent sets	Maximum sets	Trivial fixed-points
P-1	N/A	1	1	N/A
P-2	(1)P-1 to P-2	N/A	1	(1)P-1
P-3	N/A	1	1	(1)P-1
P-4	(1)P-1 to P-4	1	2	(1)P-1 to P-2
P-5	N/A	3	3	(1)P-1
P-6	(1)P-3 to P-6	4	5	(1)P-1, (1)P-3
P-7	N/A	9	9	(1)P-1
P-8	(1)P-1 to P-8 (1)P-4 to P-8	14	16	(1)P-1 to P-4 (1)P-4
P-9	N/A	18	18	(1)P-1
P-10	(3)P-5 to P-10	48	51	(1)P-1 (1)P-5
P-11	N/A	93	93	(1)P-1
P-12	(1)P-3 to P-12 (4)P-6 to P-12	165	170	(1)P-1 (1)P-3 to P-6 (4)P-6

1.5 Backward Bifurcation Trees

In this section, the analytical bifurcation scenario for backward quadratic discrete systems will be discussed as in a similar fashion through nonlinear renormalization techniques, and the backward bifurcation scenario based on the monotonic saddle-node bifurcations will be discussed, which is independent of period-1 fixed-points.

1.5.1 Backward Period-2 Quadratic Discrete Systems

After the backward period-doubling bifurcation of a period-1 fixed-point, the backward period-doubled fixed-points can be obtained and the corresponding stability is determined through dx_{k+1}/dx_k.

Theorem 1.13 *Consider a 1-dimensional backward quadratic discrete system as*

$$x_k = x_{k+1} + A(\mathbf{p})x_{k+1}^2 + B(\mathbf{p})x_{k+1} + C(\mathbf{p}) \tag{1.178}$$

where three scalar constants $A(\mathbf{p}) \neq 0$, $B(\mathbf{p})$ and $C(\mathbf{p})$ are determined by a vector parameter

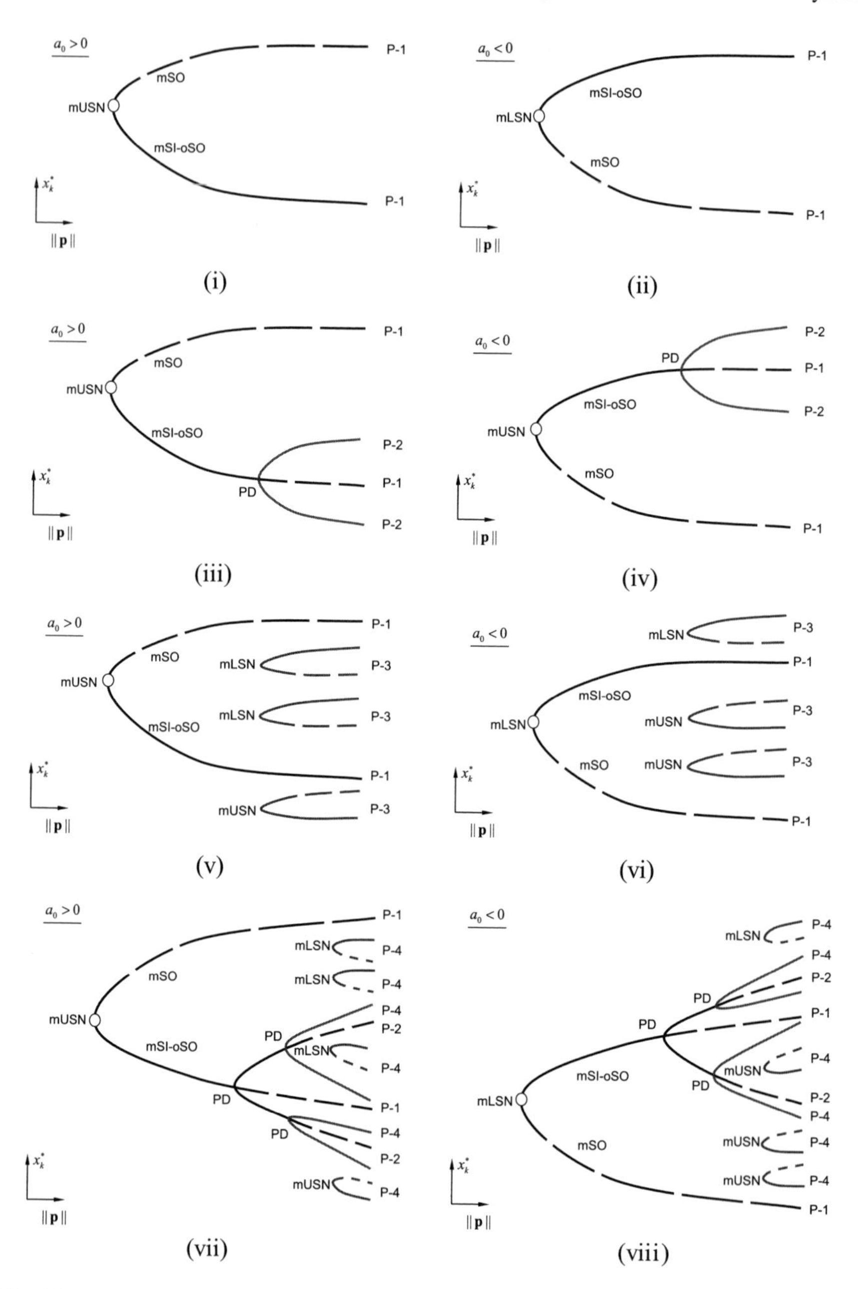

Fig. 1.9 Sketched bifurcation trees based on period-doubling and monotonic saddle-node bifurcations. (i)-(viii) period-1 to period-4 bifurcation trees, based on period-n discrete systems. mUSN: monotonic upper-saddle-node, mLSN: monotonic lower-saddle-node. PD: period-doubling bifurcation. The solid and dashed curves are for stable and unstable fixed-points. The red and dark red colors are for dependent and independent period-n fixed points. mSO: monotonic source, mSI-oSO: monotonic sink to oscillatory source.

$$\mathbf{p} = (p_1, p_2, \ldots, p_m)^{\mathrm{T}}. \tag{1.179}$$

Under a condition of

$$\Delta = B^2 - 4AC > 0, \tag{1.180}$$

there is a standard form for the backward discrete system as

$$\begin{aligned} x_k &= x_{k+1} + f(x_{k+1}, \mathbf{p}) = x_{k+1} + a_0(x_{k+1}^2 + B_1^{(1)} x_{k+1} + C_1^{(1)}) \\ &= x_{k+1} + a_0(x_{k+1} - a_1)(x_{k+1} - a_2) \\ &= x_{k+1} + a_0 \prod_{i=1}^{2} (x_{k+1} - a_i^{(1)}) \end{aligned} \tag{1.181}$$

where

$$\begin{aligned} &a_0 = A(\mathbf{p}), B_1^{(1)} = \frac{B}{A}, C_1^{(1)} = \frac{C}{A}; \\ &b_1^{(1)} = -\frac{1}{2}(B_1^{(1)} + \sqrt{\Delta^{(1)}}), b_2^{(1)} = -\frac{1}{2}(B_1^{(1)} - \sqrt{\Delta^{(1)}}), \\ &\Delta^{(1)} = (B_1^{(1)})^2 - 4C_1^{(1)} \geq 0; \\ &\cup_{i=1}^{2}\{a_i^{(1)}\} = \mathrm{sort}\{\cup_{i=1}^{2}\{b_i^{(1)}\}\},\ a_i^{(1)} \leq a_{i+1}^{(1)} \text{ for } i = 1, 2. \end{aligned} \tag{1.182}$$

(i) *Consider a backward period-2 discrete system of Eq.* (1.178) *as*

$$\begin{aligned} x_k &= x_{k+2} + [a_0 \prod_{i_1=1}^{2} (x_{k+2} - a_{i_1}^{(1)})]\{1 + \prod_{i_1=1}^{2} [1 + a_0 \prod_{i_2=1, i_2 \neq i_1}^{2} (x_{k+2} - a_{i_2}^{(1)})]\} \\ &= x_{k+2} + [a_0 \prod_{i_1=1}^{2} (x_{k+2} - a_{i_1}^{(1)})][a_0^2 (x_{k+2}^2 + B_1^{(2)} x_{k+2} + C_1^{(2)})] \\ &= x_{k+2} + [a_0 \prod_{j_1=1}^{2} (x_{k+2} - a_{i_1}^{(1)})][a_0^2 \prod_{j_2=1}^{2^2-2} (x_{k+2} - b_{j_2}^{(2)})] \\ &= x_{k+2} + a_0^{1+2} \prod_{i=1}^{4} (x_{k+2} - a_i^{(2)}) \end{aligned} \tag{1.183}$$

where

$$\begin{aligned} &b_{1,2}^{(2)} = -\frac{1}{2}(B_1^{(2)} + \sqrt{\Delta^{(2)}}), b_2^{(2)} = -\frac{1}{2}(B_1^{(2)} - \sqrt{\Delta^{(2)}}), \\ &\Delta^{(2)} = (B_1^{(2)})^2 - 4C_1^{(2)} \geq 0, \end{aligned} \tag{1.184}$$

with fixed-points

$$\begin{aligned}&x_k^* = x_{k+2}^* = a_i^{(2)}, (i = 1, 2, \ldots, 4)\\&\cup_{i=1}^{4}\{a_i^{(2)}\} = \text{sort}\{\cup_{j_1=1}^{2}\{a_{j_1}^{(1)}\}, \cup_{j_2=1}^{2}\{b_{j_2}^{(2)}\}\}\\&\text{with } a_i^{(2)} < a_{i+1}^{(2)}.\end{aligned} \tag{1.185}$$

(ii) *For a fixed-point of* $x_k^* = x_{k+1}^* = a_{i_1}^{(1)}$ $(i_1 \in \{1, 2\})$, *if*

$$\frac{dx_k}{dx_{k+1}}|_{x_{k+1}^* = a_{i_1}^{(1)}} = 1 + a_0(a_{i_1}^{(1)} - a_{i_2}^{(1)}) = -1, \tag{1.186}$$

with

- *an oscillatory lower-saddle-node bifurcation* $(d^2x_k/dx_k^2|_{x_{k+1}^*=a_1} = a_0 > 0)$,
- *an oscillatory upper-saddle-node bifurcation* $(d^2x_k/dx_k^2|_{x_{k+1}^*=a_1} = a_0 < 0)$,

then the following relations satisfy

$$a_{i_1}^{(1)} = -\frac{1}{2}B_{i_1}^{(2)}, \Delta_{i_1}^{(2)} = (B_1^{(2)})^2 - 4C_1^{(2)} = 0, \tag{1.187}$$

and there is a backward period-2 discrete system of the quadratic discrete system in Eq. (1.178) *as*

$$x_k = x_{k+2} + a_0^3(x_{k+2} - a_{i_1}^{(1)})^3(x_{k+2} - a_{i_2}^{(2)}) \tag{1.188}$$

for $i_1, i_2 \in \{1, 2\}$, $i_1 \neq i_2$ *with*

$$\begin{aligned}&\frac{dx_k}{dx_{k+2}}|_{x_{k+2}^*=a_{i_1}^{(1)}} = 1, \frac{d^2x_k}{dx_{k+2}^2}|_{x_{k+2}^*=a_{i_1}^{(1)}} = 0;\\&\frac{d^3x_k}{dx_{k+2}^3}|_{x_{k+2}^*=a_{i_1}^{(1)}} = 6a_0^3(a_{i_1}^{(1)} - a_{i_2}^{(2)}) = -12a_0^2 < 0.\end{aligned} \tag{1.189}$$

Thus, x_k *at* $x_{k+2}^* = a_{i_1}^{(1)}$ *is a monotonic source of the third-order, and the corresponding bifurcations is a monotonic source bifurcation for the period-2 discrete system.*

(ii$_1$) *The backward period-2 fixed-points are trivial and unstable if*

$$x_k^* = x_{k+2}^* = a_{i_1}^{(1)} \text{ for } i_1 = 1, 2. \tag{1.190}$$

(ii$_2$) *The backward period-2 fixed-points are non-trivial and stable if*

$$x_k^* = x_{k+2}^* = b_{i_1}^{(2)} \text{ for } i_1 = 1, 2. \tag{1.191}$$

Proof Following the corresponding proof for the forward quadratic discrete system. This theorem can be proved. ■

1.5.2 Backward Period-Doubling Renormalization

The generalized cases of period-doublization of backward quadratic discrete systems are presented through the following theorem. The analytical period-doubling trees can be developed for backward quadratic discrete systems.

Theorem 1.14 *Consider a 1-dimensional backward quadratic discrete system as*

$$\begin{aligned} x_k &= x_{k+1} + A(\mathbf{p})x_{k+1}^2 + B(\mathbf{p})x_{k+1} + C(\mathbf{p}) \\ &= x_{k+1} + a_0 \prod\nolimits_{i=1}^{2} (x_{k+1} - a_i^{(1)}). \end{aligned} \tag{1.192}$$

(i) *After* l-*times period-doubling bifurcations, a period-*$2^l (l = 1, 2, \ldots)$ *discrete system for the quadratic discrete system in Eq.* (1.192) *is produced through the nonlinear renormalization as*

$$\begin{aligned} x_k &= x_{k+2^l} + [a_0^{(2^{l-1})} \prod\nolimits_{i_1=1}^{2^{2^{l-1}}} (x_{k+2^l} - a_{i_1}^{(2^{l-1})})] \\ &\quad \times \{1 + \prod\nolimits_{i_1=1}^{2^{2^{l-1}}} [1 + a_0^{(2^{l-1})} \prod\nolimits_{i_2=1, i_2 \neq i_1}^{2^{2^{l-1}}} (x_{k+2^l} - a_{i_2}^{(2^{l-1})})]\} \\ &= x_{k+2^l} + [a_0^{(2^{l-1})} \prod\nolimits_{i_1=1}^{2^{2^{l-1}}} (x_{k+2^l} - a_{i_1}^{(2^{l-1})})] \\ &\quad \times [(a_0^{(2^{l-1})})^{2^{2^{l-1}}} \prod\nolimits_{j_1=1}^{2^{2^{l-1}} - 2^{2^{l-1}-1}} (x_{k+2^l}^2 + B_{j_2}^{(2^l)} x_{k+2^l} + C_{j_2}^{(2^l)})] \\ &= x_{k+2^l} + [a_0^{(2^{l-1})} \prod\nolimits_{i_1=1,}^{2^{2^{l-1}}} (x_{k+2^l} - a_{i_1}^{(2^{l-1})})] \\ &\quad \times [(a_0^{(2^{l-1})})^{2^{2^{l-1}}} \prod\nolimits_{i_2=1}^{2^{2^{l-1}} - 2^{2^{l-1}-2}} (x_{k+2^l} - b_{i_2,1}^{(2^l)})(x_{k+2^l} - b_{i_2,2}^{(2^l)})] \\ &= x_{k+2^l} + (a_0^{(2^{l-1})})^{1+2^{2^{l-1}}} \prod\nolimits_{i=1}^{2^{2^l}} (x_{k+2^l} - a_i^{(2^l)}) \\ &= x_{k+2^l} + a_0^{(2^l)} \prod\nolimits_{i=1}^{2^{2^l}} (x_{k+2^l} - a_i^{(2^l)}) \end{aligned} \tag{1.193}$$

with

$$
\begin{aligned}
&\frac{dx_k}{dx_{k+2^l}} = 1 + a_0^{(2^l)} \sum_{i_1=1}^{2^{2^l}} \prod_{i_2=1, i_2 \neq i_1}^{2^{2^l}} (x_{k+2^l} - a_{i_2}^{(2^l)}),\\
&\frac{d^2 x_k}{dx_{k+2^l}^2} = a_0^{(2^l)} \sum_{i_1=1}^{2^{2^l}} \sum_{i_2=1, i_2 \neq i_1}^{2^{2^l}} \prod_{i_3=1, i_3 \neq i_1, i_2}^{2^{2^l}} (x_{k+2^l} - a_{i_3}^{(2^l)}),\\
&\vdots\\
&\frac{d^r x_k}{dx_{k+2^l}^r} = a_0^{(2^l)} \sum_{i_1=1}^{2^{2^l}} \cdots \sum_{i_r=1, i_{r+1} \neq i_1, i_2 \ldots, i_{r-1}}^{2^{2^l}} \prod_{i_{r+1}=1, i_{r+1} \neq i_1, i_2 \ldots, i_r}^{2^{2^l}} (x_{k+2^l} - a_{i_{r+1}}^{(2^l)})\\
&\text{for } r \leq 2^{2^l}
\end{aligned}
\tag{1.194}
$$

where

$$
\begin{aligned}
&a_0^{(2)} = (a_0)^{1+2}, a_0^{(2^l)} = (a_0^{(2^{l-1})})^{1+(2^{l-1})^2}, l = 2, 3, \ldots;\\
&\cup_{i=1}^{2^{2^l}} \{a_i^{(2^l)}\} = \text{sort}\{\cup_{i_1=1}^{2^{2^{l-1}}} \{a_{i_1}^{(2^l)}\}, \cup_{i_2=1}^{M_2} \{b_{i_2,1}^{(2^l)}, b_{i_2,2}^{(2^l)}\}\}, a_i^{(2^l)} \leq a_{i+1}^{(2^l)};\\
&b_{i,1}^{(2^l)} = -\frac{1}{2}(B_i^{(2^l)} + \sqrt{\Delta_i^{(2^l)}}), b_{i,2}^{(2^{l-1})} = -\frac{1}{2}(B_i^{(2^l)} - \sqrt{\Delta_i^{(2^l)}}),\\
&\Delta_i^{(2^l)} = (B_i^{(2^l)})^2 - 4C_i^{(2^l)} \geq 0 \,\text{for}\, i \in \cup_{q_1=1}^{N_1} I_{q_1}^{(2^{l-1})} \cup \cup_{q_2=1}^{N_2} I_{q_2}^{(2^l)},\\
&I_{q_1}^{(2^{l-1})} = \{l_{(q_1-1)\times 2^{l-1}+1}, l_{(q_1-1)\times 2^{l-1}+2}, \ldots, l_{q_1 \times 2^{l-1}}\}\\
&\qquad \subseteq \{1, 2, \ldots, M_1\} \cup \{\varnothing\},\\
&\text{for}\, q_1 \in \{1, 2, \ldots, N_1\}, M_1 = N_1 \times 2^{l-1};\\
&I_{q_2}^{(2^l)} = \{l_{(q_2-1)\times 2^l+1}, l_{(q_2-1)\times 2^l+2}, \ldots, l_{q_2 \times 2^{l-1}}\}\\
&\qquad \subseteq \{M_1+1, M_1+2, \ldots, M_2\} \cup \{\varnothing\},\\
&\text{for}\, q_2 \in \{1, 2, \ldots, N_2\}, M_2 = 2^{2^l-1} - 2^{2^{l-1}-1};\\
&b_{i,1}^{(2^l)} = -\frac{1}{2}(B_i^{(2^l)} + \mathbf{i}\sqrt{|\Delta_i^{(2^l)}|}), \, b_{i,2}^{(2^l)} = -\frac{1}{2}(B_i^{(2^l)} - \mathbf{i}\sqrt{|\Delta_i^{(2^l)}|}),\\
&\Delta_i^{(2^l)} = (B_i^{(2^l)})^2 - 4C_i^{(2^l)} < 0, \, \mathbf{i} = \sqrt{-1},\\
&i \in J^{(2^l)} = \{l_{N_2\times 2^l+1}, l_{N_2\times 2^l+2}, \ldots, l_{M_2}\}\\
&\qquad \subset \{M_1+1, M_1+2, \ldots, M_2\} \cup \{\varnothing\}
\end{aligned}
\tag{1.195}
$$

with fixed-points

$$
\begin{aligned}
&x_k^* = x_{k+2^l}^* = a_i^{(2^l)}, (i = 1, 2, \ldots, 2^{2l})\\
&\cup_{i=1}^{2^{2l}} \{a_i^{(2^l)}\} = \text{sort}\{\cup_{i_1=1}^{2^{2^{l-1}}} \{a_{i_1}^{(2^{l-1})}\}, \cup_{i_2=1}^{M_2} \{b_{i_2,1}^{(2^{l-1})}, b_{i_2,2}^{(2^{l-1})}\}\}\\
&\text{with}\, a_i^{(2^l)} < a_{i+1}^{(2^l)}.
\end{aligned}
\tag{1.196}
$$

(ii) *For a fixed-point of* $x_k^* = x_{k+2^{l-1}}^* = a_{i_1}^{(2^{l-1})}$ $(i_1 \in I_q^{(2^l)} \subset \{1,2,\ldots,2^{2^l}\})$, *if*

$$\frac{dx_k}{dx_{k+2^{l-1}}}\Big|_{x^*_{k+2^{l-1}}=a_{i_1}^{(2^{l-1})}} = 1 + a_0^{(2^{l-1})} \prod_{i_2=1,i_2\neq i_1}^{2^{2^{l-1}}} (a_{i_1}^{(2^{l-1})} - a_{i_2}^{(2^{l-1})}) = -1 \quad (1.197)$$

then there is a backward period-2^l *fixed-point discrete system*

$$x_k = x_{k+2^l} + a_0^{(2^l)} \prod_{i_1\in I_q^{(2^{l-1})}} (x_{k+2^l} - a_{i_1}^{(2^{l-1})})^3 \prod_{j_2=1}^{2^{2^l}} (x_{k+2^l} - a_{j_2}^{(2^l)})^{(1-\delta(i_1,j_2))} \quad (1.198)$$

where

$$\delta(i_1,j_2) = 1 \text{ if } a_{j_2}^{(2^l)} = a_{i_1}^{(2^{l-1})},\ \delta(i_1,j_2) = 0 \text{ if } a_{j_2}^{(2^l)} \neq a_{i_1}^{(2^{l-1})} \quad (1.199)$$

with

$$\begin{aligned} &\frac{dx_k}{dx_{k+2^l}}\Big|_{x_k^*=a_{i_1}^{(2^{l-1})}} = 1,\ \frac{d^2x_k}{dx_{k+2^l}^2}\Big|_{x_k^*=a_{i_1}^{(2^{l-1})}} = 0;\\ &\frac{d^3x_k}{dx_{k+2^l}^3}\Big|_{x_k^*=a_{i_1}^{(2^{l-1})}} = 6a_0^{(2^l)} \prod_{i_2\in I_q^{(2^{l-1})},i_2\neq i_1} (a_{i_1}^{(2^{l-1})} - a_{i_2}^{(2^{l-1})})^3\\ &\qquad\qquad \times \prod_{j_2=1}^{2^{2^l}} (a_{i_1}^{(2^{l-1})} - a_{j_2}^{(2^l)})^{(1-\delta(i_2,j_2))} < 0\\ &(i_1 \in I_q^{(2^{l-1})}, q \in \{1,2,\ldots,N_1\}). \end{aligned} \quad (1.200)$$

Thus, x_k *at* $x_{k+2^l}^* = a_{i_1}^{(2^{l-1})}$ *is*

- *a monotonic source of the third-order if* $d^3x_k/dx_{k+2^l}^3\big|_{x^*_{k+2^l}=a_{i_1}^{(2^{l-1})}} < 0$;
- *a monotonic sink of the third-order if* $d^3x_k/dx_{k+2^l}^3\big|_{x^*_{k+2^l}=a_{i_1}^{(2^{l-1})}} > 0$.

(ii_1) *The backward period-*2^l *fixed-points are trivial if*

$$x_k^* = x_{k+2^l}^* = a_{i_1}^{(2^{l-1})} \text{ for } i_1 = 1,2,\ldots,2^{2^{l-1}}. \quad (1.201)$$

(ii_2) *The backward period-*2^l *fixed-points are non-trivial if*

$$\begin{aligned} &x_k^* = x_{k+2^l}^* = b_{j_1,1}^{(2^l)}, b_{j_1,2}^{(2^l)}\\ &j_1 \in \cup_{q=2}^{N} I_q^{(2^l)} \subseteq \{1,2,\ldots,2^{2^l}\} \cup \{\varnothing\}. \end{aligned} \quad (1.202)$$

Such a backward period-2^l *fixed-point is*

- *monotonically stable if* $dx_k/dx_{k+2^l}|_{x^*_{k+2^l}=a^{(2^l)}_{i_1}} \in (1,\infty)$;
- *monotonically invariant if* $dx_k/dx_{k+2^l}|_{x^*_{k+2^l}=a^{(2^l)}_{i_1}} = 1$, *which is*
  - *a monotonic lower-saddle of the* $(2l_1)^{\text{th}}$ *order for* $d^{2l_1}x_k/dx^{2l_1}_{k+2^l}|_{x^*_{k+2^l}} > 0$;
  - *a monotonic upper-saddle the* $(2l_1)^{\text{th}}$ *order for* $d^{2l_1}x_k/dx^{2l_1}_{k+2^l}|_{x^*_{k+2^l}} < 0$;
  - *a monotonic sink of the* $(2l_1+1)^{\text{th}}$ *order for* $d^{2l_1+1}x_k/dx^{2l_1+1}_{k+2^l}|_{x^*_{k+2^l}} > 0$;
  - *a monotonic source the* $(2l_1+1)^{\text{th}}$ *order for* $d^{2l_1+1}x_k/dx^{2l_1+1}_{k+2^l}|_{x^*_{k+2^l}} < 0$;
- *monotonically unstable if* $dx_k/dx_{k+2^l}|_{x^*_{k+2^l}=a^{(2^l)}_{i_1}} \in (0,1)$;
- *monotonically unstable with infinite eigenvalue if* $dx_k/dx_{k+2^l}|_{x^*_{k+2^l}=a^{(2^l)}_{i_1}} = 0^+$;
- *oscillatorilly unstable with infinite eigenvalue if* $dx_k/dx_{k+2^l}|_{x^*_{k+2^l}=a^{(2^l)}_{i_1}} = 0^-$;
- *oscillatorilly unstable if* $dx_k/dx_{k+2^l}|_{x^*_{k+2^l}=a^{(2^l)}_{i_1}} \in (-1,0)$;
- *flipped if* $dx_k/dx_{k+2^l}|_{x^*_{k+2^l}=a^{(2^l-1)}_{i_1}} = -1$, *which is*
  - *an oscillatory lower-saddle of the* $(2l_1)^{\text{th}}$ *order if* $d^{2l_1}x_k/dx^{2l_1}_{k+2^l}|_{x^*_{k+2^l}} > 0$;
  - *an oscillatory upper-saddle* the $(2l_1)^{\text{th}}$ order with $d^{2l_1}x_k/dx^{2l_1}_{k+2^l}|_{x^*_{k+2^l}} < 0$;
  - *an oscillatory source of the* $(2l_1+1)^{\text{th}}$ *order if* $d^{2l_1+1}x_k/dx^{2l_1+1}_{k+2^l}|_{x^*_{k+2^l}} > 0$;
  - *an oscillatory sink the* $(2l_1+1)^{\text{th}}$ *order with* $d^{2l_1+1}x_k/dx^{2l_1+1}_{k+2^l}|_{x^*_{k+2^l}} < 0$;
- *oscillatorilly stable if* $dx_k/dx_{k+2^l}|_{x^*_{k+2^l}=a^{(2^l)}_{i_1}} \in (-\infty,-1)$.

Proof Through the nonlinear renormalization, following the forward case, this theorem can be proved. ■

1.5.3 *Backward Period-*n *Appearing and Period-Doublization*

The period-n discrete system for backward quadratic nonlinear discrete systems will be discussed, and the period-doublization of a backward period-n discrete system is discussed through the nonlinear renormalization.

Theorem 1.15 *Consider a 1-dimensional backward quadratic discrete system as*

$$
\begin{aligned}
x_k &= x_{k+1} + A(\mathbf{p})x_{k+1}^2 + B(\mathbf{p})x_{k+1} + C(\mathbf{p}) \\
&= x_{k+1} + a_0 \prod_{i=1}^{2}(x_{k+1} - a_i^{(1)}).
\end{aligned} \tag{1.203}
$$

(i) *After* n-*times iterations, a period*-n *discrete system for the quadratic discrete system in Eq.* (1.203) *is*

$$
\begin{aligned}
x_k &= x_{k+n} + a_0 \prod_{i_1=1}^{2}(x_{k+n} - a_{i_1}^{(1)})\{1 + \sum_{i=1}^{n} Q_j\} \\
&= x_k + a_0^{2^n-1} \prod_{i_1=1}^{2}(x_{k+n} - a_{i_1}^{(1)})[\prod_{j_2=1}^{2^{n-1}-1}(x_{k+n}^2 + B_{j_2}^{(n)} x_{k+n} + C_{j_2}^{(n)})] \\
&= x_k + a_0^{(n)} \prod_{i=1}^{2^n}(x_{k+n} - a_i^{(n)})
\end{aligned} \tag{1.204}
$$

with

$$
\begin{aligned}
&\frac{dx_k}{dx_{k+n}} = 1 + a_0^{(n)} \sum_{i_1=1}^{2^n} \prod_{i_2=1, i_2 \neq i_1}^{2^n}(x_{k+n} - a_{i_2}^{(n)}), \\
&\frac{d^2 x_k}{dx_{k+n}^2} = a_0^{(n)} \sum_{i=1}^{2^n} \sum_{i_2=1, i_2 \neq i_1}^{2^n} \prod_{i_3=1, i_3 \neq i_1, i_2}^{2^n}(x_{k+n} - a_{i_3}^{(n)}), \\
&\vdots \\
&\frac{d^r x_k}{dx_{k+n}^r} = a_0^{(n)} \sum_{i_1=1}^{2^n} \cdots \sum_{i_r=1, i_r \neq 1, i_2, \ldots, i_{r-1}}^{2^n} \prod_{i_{r+1}=1, i_{r+1} \neq i_1, i_2 \ldots, i_r}^{2^n}(x_{k+n} - a_{i_{r+1}}^{(n)}) \\
&\text{for } r \leq 2^n;
\end{aligned} \tag{1.205}
$$

where

$$
\begin{aligned}
&a_0^{(n)} = (a_0)^{2^n-1}, Q_1 = 0, Q_2 = \prod_{i_2=1}^{2}[1 + a_0 \prod_{i_1=1, i_1 \neq i_2}^{2}(x_{k+n} - a_{i_1}^{(1)})], \\
&Q_n = \prod_{i_n=1}^{2}[1 + a_0(1 + Q_{n-1} \prod_{i_{n-1}=1, i_{n-1} \neq i_n}^{2}(x_{k+n} - a_{i_{n-1}}^{(1)}))],\ n = 3, 4, \ldots; \\
&\cup_{i=1}^{2^n}\{a_i^{(n)}\} = \text{sort}\{\cup_{i_1=1}^{2}\{a_{i_1}^{(1)}\}, \cup_{i_2 \in =1}^{M}\{b_{i_2,1}^{(n)}, b_{i_2,2}^{(n)}\}\}; \\
&b_{i_2,1}^{(n)} = -\frac{1}{2}(B_{i_2}^{(n)} + \sqrt{\Delta_{i_2}^{(n)}}), b_{i_2,2}^{(n)} = -\frac{1}{2}(B_{i_2}^{(n)} - \sqrt{\Delta_{i_2}^{(n)}}), \\
&\Delta_{i_2}^{(n)} = (B_{i_2}^{(n)})^2 - 4C_{i_2}^{(n)} \geq 0 \text{ for } i_2 \in \cup_{q=1}^{N} I_q^{(n)}, \\
&I_q^{(n)} = \{l_{(q-1)\times n+1}, l_{(q-1)\times n+2}, \ldots, l_{q\times n}\} \subseteq \{1, 2, \ldots, M\} \cup \{\varnothing\}, \\
&\text{for } q \in \{1, 2, \ldots, N\}, M = 2^{n-1} - 1; \\
&b_{i,1}^{(n)} = -\frac{1}{2}(B_i^{(n)} + \mathbf{i}\sqrt{|\Delta_i^{(n)}|}),\ b_{i,2}^{(n)} = -\frac{1}{2}(B_i^{(n)} - \mathbf{i}\sqrt{|\Delta_i^{(n)}|}), \\
&\Delta_i^{(n)} = (B_i^{(n)})^2 - 4C_i^{(n)} < 0,\ \mathbf{i} = \sqrt{-1} \\
&i \in \{l_{N\times n+1}, l_{N\times n+2}, \ldots, l_M\} \subset \{1, 2, \ldots, M\} \cup \{\varnothing\};
\end{aligned} \tag{1.206}
$$

with fixed-points

$$\begin{aligned}
&x_k^* = x_{k+n}^* = a_i^{(n)}, (i = 1, 2, \ldots, 2^n) \\
&\cup_{i=1}^{2^n}\{a_i^{(n)}\} = \text{sort}\{\cup_{i_1=1}^{2}\{a_{i_1}^{(1)}\}, \cup_{i_1=1}^{M}\{b_{i_2,1}^{(n)}, b_{i_2,2}^{(n)}\}\} \\
&\text{with } a_i^{(n)} < a_{i+1}^{(n)}.
\end{aligned} \tag{1.207}$$

(ii) *For a backward fixed-point of* $x_k^* = x_{k+n}^* = a_{i_1}^{(n)}$ $(i_1 \in I_q^{(n)},\ q \in \{1, 2, \ldots, N\})$, *if*

$$\frac{dx_k}{dx_{k+n}}\Big|_{x_{k+n}^* = a_{i_1}^{(n)}} = 1 + a_0^{(n)} \prod_{i_2=1, i_2 \neq i_1}^{2^n} (a_{i_1}^{(n)} - a_{i_2}^{(n)}) = 1, \tag{1.208}$$

then there is a new discrete system for onset of the qth*- set of period-*n *fixed-points based on the second-order monotonic saddle-node bifurcation as*

$$x_k = x_{k+n} + a_0^{(n)} \prod_{i_1 \in I_q^{(n)}} (x_{k+n} - a_{i_1}^{(n)})^2 \prod_{j_2=1}^{2^n} (x_{k+n} - a_{j_2}^{(n)})^{(1-\delta(i_1, j_2))} \tag{1.209}$$

where

$$\delta(i_1, j_2) = 1 \text{ if } a_{j_2}^{(n)} = a_{i_1}^{(n)},\ \delta(i_1, j_2) = 0 \text{ if } a_{j_2}^{(n)} \neq a_{i_1}^{(n)}. \tag{1.210}$$

(ii$_1$) *If*

$$\begin{aligned}
&\frac{dx_k}{dx_{k+n}}\Big|_{x_{k+n}^* = a_{i_1}^{(n)}} = 1\ (i_1 \in I_q^{(n)}), \\
&\frac{d^2x_k}{dx_{k+n}^2}\Big|_{x_{k+n}^* = a_{i_1}^{(n)}} = 2a_0^{(n)} \prod_{i_2 \in I_q^{(n)}, i_2 \neq i_1} (a_{i_1}^{(n)} - a_{i_2}^{(n)})^2 \\
&\qquad \times \prod_{j_2=1}^{2^n} (a_{i_1}^{(n)} - a_{j_2}^{(n)})^{(1-\delta(i_2, j_2))} \neq 0
\end{aligned} \tag{1.211}$$

x_k at $x_{k+n}^* = a_{i_1}^{(n)}$ is

- *a monotonic upper-saddle of the second-order for* $d^2x_k/dx_{k+n}^2|_{x_{k+n}^*} = a_{i_1}^{(n)} < 0$;
- *a monotonic lower-saddle of the second-order for* $d^2x_k/dx_{k+n}^2|_{x_{k+n}^*} = a_{i_1}^{(n)} > 0$.

(ii$_2$) *The backward period-*n *fixed-points* $(n = 2^{n_1} \times m)$ *are trivial*

$$\left.\begin{array}{l} x_k^* = x_{k+n}^* = a_{j_1}^{(n)} \in \{\cup_{i_l=1}^{2}\{a_{i_1}^{(1)}\}, \cup_{i_2=1}^{2^{2^{n_1-1}m}}\{a_{i_2}^{(2^{n_1-1}m)}\}\} \\ \text{for } n_1 = 1,2,\ldots; m = 2l_1+1; j_1 \in \{1,2,\ldots,2^n\} \cup \{\varnothing\} \end{array}\right\}$$
$$\text{for } n \neq 2^{n_2},$$
$$\left.\begin{array}{l} x_k^* = x_{k+n}^* = a_{j_1}^{(n)} \in \{\cup_{i_2=1}^{2^{2^{n_1-1}m}}\{a_{i_2}^{(2^{n_1-1}m)}\}\} \\ \text{for } n_1 = 1,2,\ldots; m = 1; j_1 \in \{1,2,\ldots,2^n\} \cup \{\varnothing\} \end{array}\right\}$$
$$\text{for } n = 2^{n_2}. \tag{1.212}$$

(ii$_3$) *The period*-n *fixed-points* $(n = 2^{n_1} \times m)$ *are non-trivial if*

$$\left.\begin{array}{l} x_k^* = x_{k+n}^* = a_{j_1}^{(n)} \notin \{\cup_{i_l=1}^{2}\{a_{i_1}^{(1)}\}, \cup_{i_2=1}^{2^{2^{n_1-1}m}}\{a_{i_2}^{(2^{n_1-1}m)}\}\} \\ \text{for } n_1 = 1,2,\ldots; m = 2l_1+1; j_1 \in \{1,2,\ldots,2^n\} \cup \{\varnothing\} \end{array}\right\}$$
$$\text{for } n \neq 2^{n_2},$$
$$\left.\begin{array}{l} x_k^* = x_{k+n}^* = a_{j_1}^{(n)} \notin \{\cup_{i_2=1}^{2^{2^{n_1-1}m}}\{a_{i_2}^{(2^{n_1-1}m)}\}\} \\ \text{for } n_1 = 1,2,\ldots; m = 1; j_1 \in \{1,2,\ldots,2^n\} \cup \{\varnothing\} \end{array}\right\}$$
$$\text{for } n = 2^{n_2}. \tag{1.213}$$

Such a backward period-n *fixed-point is*

- *monotonically stable if* $dx_k/dx_{k+n}|_{x_{k+n}^*=a_{i_1}^{(n)}} \in (1,\infty)$;
- *monotonically invariant if* $dx_k/dx_{k+n}|_{x_{k+n}^*=a_{i_1}^{(n)}} = 1$, *which is*
  - *a monotonic lower-saddle of the* $(2l_1)^{\text{th}}$ *order for* $d^{2l_1}x_k/dx_{k+n}^{2l_1}|_{x_{k+n}^*} > 0$;
  - *a monotonic upper-saddle the* $(2l_1)^{\text{th}}$ *order for* $d^{2l_1}x_k/dx_{k+n}^{2l_1}|_{x_{k+n}^*} < 0$;
  - *a monotonic source of the* $(2l_1+1)^{\text{th}}$ *order for* $d^{2l_1+1}x_k/dx_{k+n}^{2l_1+1}|_{x_{k+n}^*} < 0$;
  - *a monotonic sink the* $(2l_1+1)^{\text{th}}$ *order for* $d^{2l_1+1}x_k/dx_{k+n}^{2l_1+1}|_{x_{k+n}^*} > 0$;
- *monotonically unstable if* $dx_k/dx_{k+n}|_{x_{k+n}^*=a_{i_1}^{(n)}} \in (0,1)$;
- *monotonically infinity-unstable if* $dx_k/dx_{k+n}|_{x_{k+n}^*=a_{i_1}^{(n)}} = 0^+$;
- *oscillatorilly infinity-unstable if* $dx_k/dx_{k+n}|_{x_{k+n}^*=a_{i_1}^{(n)}} = 0^-$;
- *oscillatorilly unstable if* $dx_k/dx_{k+n}|_{x_{k+n}^*=a_{i_1}^{(n)}} \in (-1,0)$;
- *flipped if* $dx_k/dx_{k+n}|_{x_{k+n}^*=a_{i_1}^{(n)}} = -1$, *which is*
  - *an oscillatory lower-saddle of the* $(2l_1)^{\text{th}}$ *order for* $d^{2l_1}x_k/dx_{k+n}^{2l_1}|_{x_{k+n}^*} > 0$;
  - *an oscillatory upper-saddle the* $(2l_1)^{\text{th}}$ *order for* $d^{2l_1}x_k/dx_{k+n}^{2l_1}|_{x_{k+n}^*} < 0$;
  - *an oscillatory source of the* $(2l_1+1)^{\text{th}}$ *order for* $d^{2l_1+1}x_k/dx_{k+n}^{2l_1+1}|_{x_{k+n}^*} > 0$;
  - *an oscillatory sink the* $(2l_1+1)^{\text{th}}$ *order for* $d^{2l_1+1}x_k/dx_{k+n}^{2l_1+1}|_{x_{k+n}^*} < 0$;

- *oscillatorilly stable if* $dx_k/dx_{k+n}|_{x_{k+n}^*=a_{i_1}^{(n)}} \in (-\infty, -1)$.

(iii) *For a fixed-point of* $x_k^* = x_{k+n}^* = a_{i_1}^{(n)}$ $(i_1 \in I_q^{(n)}, q \in \{1, 2, \ldots, N\})$, *there is a backward period-doubling of the* q^{th}*-set of period-*n *fixed-points if*

$$\frac{dx_k}{dx_{k+n}}|_{x_{k+n}^*=a_{i_1}^{(n)}} = 1 + a_0^{(n)} \prod_{j_2=1, j_2\neq i_1}^{2^n} (a_{i_1}^{(n)} - a_{j_2}^{(n)}) = -1 \tag{1.214}$$

with

- *an oscillatory lower-saddle for* $d^2x_k/dx_{k+n}^2|_{x_k^*=a_{i_1}^{(n)}} > 0$;
- *an oscillatory upper-saddle for* $d^2x_k/dx_{k+n}^2|_{x_{k+n}^*=a_{i_1}^{(n)}} < 0$.

The corresponding period-$2 \times n$ *discrete system of the quadratic discrete system in Eq.* (1.203) *is*

$$\begin{aligned} x_k = x_{k+2\times n} + a_0^{(2\times n)} &\prod_{i_1 \in I_q^n} (x_{k+2\times n} - a_{i_1}^{(n)})^3 \\ &\times \prod_{i_2=1}^{2^{2\times n}} (x_{k+2\times n} - a_{i_2}^{(2\times n)})^{(1-\delta(i_1, i_2))} \end{aligned} \tag{1.215}$$

with

$$\begin{aligned} &\frac{dx_k}{dx_{k+2\times n}}|_{x_{k+2\times n}^*=a_{i_1}^{(n)}} = 1, \frac{d^2x_k}{dx_{k+2\times n}^2}|_{x_{k+2\times n}^*=a_{i_1}^{(n)}} = 0; \\ &\frac{d^3x_k}{dx_{k+2\times n}^3}|_{x_{k+2\times n}^*=a_{i_1}^{(n)}} = 6a_0^{(2\times n)} \prod_{i_2 \in I_q^n, i_2\neq i_1} (a_{i_1}^{(n)} - a_{i_2}^{(n)})^3 \\ &\qquad\qquad \times \prod_{i_3=1}^{2^{2\times n}} (a_{i_1}^{(n)} - a_{i_3}^{(2\times n)})^{(1-\delta(i_1, i_3))} \neq 0. \end{aligned} \tag{1.216}$$

Thus, x_k *at* $x_{k+n}^* = a_{i_1}^{(n)}$ *for* $i_1 \in I_q^{(n)}$, $q \in \{1, 2, \ldots, N\}$ *is*

- *a monotonic source of the third-order if* $d^3x_k/dx_{k+2\times n}^3|_{x_{k+2\times n}^*=a_{i_1}^{(n)}} < 0$,
- *a monotonic sink of the third-order if* $d^3x_k/dx_{k+2\times n}^3|_{x_{k+2\times n}^*=a_{i_1}^{(n)}} > 0$.

(iv) *After* l*-times period-doubling bifurcations of period-*n *fixed points, a backward period-*$2^l \times n$ *discrete system of the backward quadratic discrete system in Eq.* (1.203) *is*

$$
\begin{aligned}
x_k &= x_{k+2^l\times n} + [a_0^{(2^{l-1}\times n)} \prod_{i_1=1}^{2^{2^{l-1}\times n}} (x_{k+2^l\times n} - a_{i_1}^{(2^{l-1}\times n)})] \\
&\quad \times \{1 + \prod_{i_1=1}^{2^{2^{l-1}\times n}} [1 + a_0^{(2^{l-1}\times n)} \prod_{i_2=1, i_2\neq i_1}^{2^{2^{l-1}\times n}} (x_{k+2^l\times n} - a_{i_2}^{(2^{l-1}\times n)})]\} \\
&= x_{k+2^l\times n} + [a_0^{(2^{l-1}\times n)} \prod_{i_1=1}^{2^{2^{l-1}\times n}} (x_{k+2^l\times n} - a_{i_1}^{(2^{l-1}\times n)})] \\
&\quad \times [(a_0^{(2^{l-1}\times n)})^{2^{2^{l-1}\times n}} \prod_{j_1=1}^{2^{2^l\times n-1}-2^{2^{l-1}\times n-1}} (x_{k+2^l\times n}^2 + B_{j_2}^{(2^l\times n)} x_{k+2^l\times n} + C_{j_2}^{(2^l\times n)})] \\
&= x_{k+2^l\times n} + [a_0^{(2^{l-1}\times n)} \prod_{i_1=1}^{2^{2^{l-1}\times n}} (x_{k2^l\times n} - a_{i_1}^{(2^{l-1}\times n)})] \\
&\quad \times [(a_0^{(2^{l-1}\times n)})^{2^{(2^{l-1}\times n)}} \prod_{j_2=1}^{2^{2^l\times n-1}-2^{2^{l-1}\times n)-1}} (x_{k+2^l\times n} - b_{j_2}^{(2^l\times n)})] \\
&= x_{k+2^l\times n} + (a_0^{(2^{l-1}\times n)})^{2^{2^{l-1}\times n}} \prod_{i=1}^{2^{2^l\times n}} (x_{k+2^l\times n} - a_i^{(2^l\times n)}) \\
&= x_{k+2^l\times n} + a_0^{(2^l\times n)} \prod_{i=1}^{2^{(2^l\times n)}} (x_{k+2^l\times n} - a_i^{(2^l\times n)})
\end{aligned}
\tag{1.217}
$$

with

$$
\begin{aligned}
&\frac{dx_k}{dx_{k+2^l\times n}} = 1 + a_0^{(2^l\times n)} \sum_{i_1=1}^{2^{2^l\times n}} \prod_{i_2=1, i_2\neq i_1}^{2^{2^l\times n}} (x_{k+2^l\times n} - a_{i_2}^{(2^l\times n)}), \\
&\frac{d^2x_k}{dx_{k+2^l\times n}^2} = a_0^{(2^l\times n)} \sum_{i_1=1}^{2^{2^l\times n}} \sum_{i_2=1, i_2\neq i_1}^{2^{2^l\times n}} \prod_{i_3=1, i_3\neq i_1, i_2}^{2^{2^l\times n}} (x_{k+2^l\times n} - a_{i_3}^{(2^l\times n)}), \\
&\vdots \\
&\frac{d^rx_k}{dx_{k+2^l\times n}^r} = a_0^{(2^l\times n)} \sum_{i_1=1}^{2^{2^l\times n}} \cdots \sum_{i_r=1, i_r\neq i_1, i_2\ldots, i_{r-1}}^{2^{2^l\times n}} \prod_{i_{r+1}=1, i_{r+1}\neq i_1, i_2\ldots, i_r}^{2^{2^l\times n}} (x_{k+2^l\times n} - a_{i_{r+1}}^{(2^l\times n)}) \\
&\text{for } r \le 2^{2^l\times n};
\end{aligned}
\tag{1.218}
$$

where

$$
\begin{aligned}
&a_0^{(2\times n)} = (a_0^{(n)})^{1+2^{2\times n}}, a_0^{(2^l\times n)} = (a_0^{(2^{l-1}\times n)})^{1+2^{2^{l-1}\times n}}, l = 1, 2, 3, \ldots; \\
&\cup_{i=1}^{2^{2^l\times n}} \{a_i^{(2^l\times n)}\} = \text{sort}\{\cup_{i_1=1}^{2^{(2^{l-1}\times n)}} \{a_{i_1}^{(2^{l-1}\times n)}\}, \cup_{i_2=1}^{M_2} \{b_{i_2,1}^{(2^l\times n)}, b_{i_2,2}^{(2^l\times n)}\}\}; \\
&b_{i,1}^{(2^l\times n)} = -\tfrac{1}{2}(B_i^{(2^l\times n)} + \sqrt{\Delta_i^{(2^l\times n)}}), \\
&b_{i,2}^{(2^l\times n)} = -\tfrac{1}{2}(B_i^{(2^l\times n)} - \sqrt{\Delta_i^{(2^l\times n)}}), \\
&\Delta_i^{(2^l\times n)} = (B_i^{(2^l\times n)})^2 - 4C_i^{(2^l\times n)} \ge 0
\end{aligned}
$$

$$
\begin{aligned}
&\text{for } i \in \cup_{q_1=1}^{N_1} I_{q_1}^{(2^{l-1}\times n)} \bigcup \cup_{q_2=1}^{N_2} I_{q_2}^{(2^l\times n)},\\
&I_{q_1}^{(2^{l-1}\times n)} = \{l_{(q_1-1)\times(2^{l-1}\times n)+1}, l_{(q_1-1)\times(2^{l-1}\times n)+2}, \ldots, l_{q_1\times(2^{l-1}\times n)}\}\\
&\qquad \subseteq \{1,2,\ldots,M_1\} \cup \{\varnothing\},\\
&\text{for } q_1 \in \{1,2,\ldots,N_1\}, M_1 = N_1 \times (2^{l-1}\times n);\\
&I_{q_2}^{(2^l\times n)} = \{l_{(q_2-1)\times(2^l\times n)+1}, l_{(q_2-1)\times(2^l\times n)+2}, \ldots, l_{q_2\times(2^l\times n)}\}\\
&\qquad \subseteq \{M_1+1, M_1+2, \ldots, M_2\} \cup \{\varnothing\},\\
&\text{for } q_2 \in \{1,2,\ldots,N_2\}, M_2 = 2^{2^l-1} - 2^{2^{l-1}-1};\\
&b_{i,1}^{(2^l\times n)} = -\frac{1}{2}(B_i^{(2^l\times n)} + \mathbf{i}\sqrt{|\Delta_i^{(2^l\times n)}|}),\\
&b_{i,2}^{(2^l\times n)} = -\frac{1}{2}(B_i^{(2^l\times n)} - \mathbf{i}\sqrt{|\Delta_i^{(2^l\times n)}|}),\\
&\Delta_i^{(2^l\times n)} = (B_i^{(2^l\times n)})^2 - 4C_i^{(2^l\times n)} < 0, \ \mathbf{i} = \sqrt{-1},\\
&i \in \{l_{N\times(2^l\times n)+1}, l_{N\times(2^l\times n)+2}, \ldots, l_{M_2}\}\\
&\quad \subset \{M_1+1, M_1+2, \ldots, M_2\} \cup \{\varnothing\}
\end{aligned}
\tag{1.219}
$$

with fixed-points

$$
\begin{aligned}
&x_k^* = x_{k+2^l\times n}^* = a_i^{(2^l\times n)}, (i = 1,2,\ldots,2^{2^l\times n})\\
&\cup_{i=1}^{2^{2^l\times n}} \{a_i^{(2^l\times n)}\} = \text{sort}\{\cup_{i_1=1}^{2^{2^{l-1}\times n}} \{a_{i_1}^{(2^{l-1}\times n)}\}, \cup_{i_2=1}^{M_2}\{b_{i_2,1}^{(2^l\times n)}, b_{i_2,2}^{(2^l\times n)}\}\}\\
&\text{with } a_i^{(2^l\times n)} < a_{i+1}^{(2^l\times n)}.
\end{aligned}
\tag{1.220}
$$

(v) *For a fixed-point of* $x_k^* = x_{k+2^l\times n}^* = a_{i_1}^{(2^{l-1}\times n)}$ $(i_1 \in I_q^{(2^{l-1}\times n)} \subseteq \{1,2,\ldots,M_1\})$, *there is a period-*$2^l \times n$ *discrete system if*

$$
\frac{dx_k}{dx_{k+2^{l-1}\times n}}\Big|_{x_{k+2^{l-1}\times n}^* = a_{i_1}^{(2^{l-1}\times n)}} = 1 + a_0^{(2^{l-1}\times n)} \prod_{i_2=1, i_2\neq i_1}^{2^{2^{l-1}\times n}} (a_{i_1}^{(2^{l-1}\times n)} - a_{i_2}^{(2^{l-1}\times n)}) = -1
\tag{1.221}
$$

with

- *an oscillatory lower-saddle for* $d^2x_k/dx_{k+2^{l-1}\times n}^2|_{x_{k+2^{l-1}\times n}^* = a_{i_1}^{(n)}} > 0$;
- *an oscillatory upper-saddle for* $d^2x_k/dx_{k+2^{l-1}\times n}^2|_{x_{k+2^{l-1}\times n}^* = a_{i_1}^{(n)}} < 0$.

The corresponding period-$(2^l \times n)$ *discrete system is*

$$\begin{aligned} x_k &= x_{k+2^l\times n} + a_0^{(2^l\times n)} \prod_{i_1\in I_q^{(2^{l-1}\times n)}} (x_{k+2^l\times n} - a_{i_1}^{(2^{l-1}\times n)})^3 \\ &\quad \times \prod_{i_2=1}^{2^{2^l\times n}} (x_{k+2^l\times n} - a_{i_2}^{(2^l\times n)})^{(1-\delta(i_1,i_2))} \end{aligned} \tag{1.222}$$

where

$$\delta(i_1,j_2) = 1 \text{ if } a_{j_2}^{(2^l\times n)} = a_{i_1}^{(2^{l-1}\times n)},\ \delta(i_1,j_2) = 0 \text{ if } a_{j_2}^{(2^l\times n)} \neq a_{i_1}^{(2^{l-1}\times n)} \tag{1.223}$$

with

$$\begin{aligned} &\frac{dx_k}{dx_{k+2^l\times n}}\Big|_{x^*_{k+2^l\times n}=a_{i_1}^{(2^{l-1}\times n)}} = 1,\ \frac{d^2x_{k+2^l\times n}}{dx_k^2}\Big|_{x^*_{k+2^l\times n}=a_{i_1}^{(2^{l-1}\times n)}} = 0; \\ &\frac{d^3x_k}{dx^3_{k+2^l\times n}}\Big|_{x^*_{k+2^l\times n}=a_{i_1}^{(2^{l-1}\times n)}} = 6a_0^{(2^l\times n)} \prod_{i_2\in I_q^{(2^{l-1}\times n)}, i_2\neq i_1} (a_{i_1}^{(2^{l-1}\times n)} - a_{i_2}^{(2^{l-1}\times n)})^3 \\ &\qquad\qquad \times \prod_{i_3=1}^{2^{2^l\times n}} (a_{i_1}^{(2^{l-1}\times n)} - a_{i_3}^{(2^l\times n)})^{(1-\delta(i_2,i_3))} \neq 0 \\ &\left(i_1 \in I_q^{(2^{l-1}\times n)}, q \in \{1,2,\ldots,N_1\}\right). \end{aligned} \tag{1.224}$$

Thus, x_k at $x^*_{k+2^l\times n} = a_{i_1}^{(2^{l-1}\times n)}$ is

- *a monotonic source of the third-order if $d^3x_k/dx^3_{k+2^l\times n}|_{x^*_{k+2^l\times n}=a_{i_1}^{(2^{l-1}\times n)}} < 0$;*
- *a monotonic sink of the third-order if $d^3x_k/dx^3_{k+2^l\times n}|_{x^*_{k+2^l\times n}=a_{i_1}^{(2^{l-1}\times n)}} > 0$.*

(v_1) *The backward period-$2^l \times n$ fixed-points are trivial if*

$$\left.\begin{aligned} &\left.\begin{aligned} &x_k^* = x^*_{k+2^l\times n} = a_j^{(2^l\times n)} \in \{\cup_{i_1=1}^{2}\{a_{i_1}^{(1)}\}, \cup_{i_2=1}^{2^{2^{l-1}\times n}}\{a_{i_2}^{(2^{l-1}\times n)}\}\} \\ &\text{for } j = 1,2,\ldots,2^{(2^l\times n)} \end{aligned}\right\} \\ &\text{for } n \neq 2^{n_1}, \\ &\left.\begin{aligned} &x_k^* = x^*_{k+2^l\times n} = a_j^{(2^l\times n)} \in \cup_{i_2=1}^{2^{2^{l-1}\times n}}\{a_{i_2}^{(2^{l-1}\times n)}\} \\ &\text{for } j = 1,2,\ldots,2^{2^l\times n} \end{aligned}\right\} \\ &\text{for } n = 2^{n_1}. \end{aligned}\right. \tag{1.225}$$

(v_2) The backward period-$2^l \times n$ fixed-points are non-trivial if

$$\left.\begin{array}{l} x_k^* = x_{k+2^l\times n}^* = a_j^{(2^l\times n)} \notin \{\cup_{i_i=1}^{2}\{a_{i_1}^{(1)}\}, \cup_{i_2=1}^{2^{2^{l-1}\times n}}\{a_{i_2}^{(2^{l-1}\times n)}\}\} \\ \text{for } j=1,2,\ldots,2^{(2^l\times n)} \end{array}\right\}$$
$$\text{for } n \neq 2^{n_1},$$
$$\left.\begin{array}{l} x_k^* = x_{k+2^l\times n}^* = a_j^{(2^l\times n)} \notin \{\cup_{i_2=1}^{2^{2^{l-1}\times n}}\{a_{i_2}^{(2^{l-1}\times n)}\} \\ \text{for } j=1,2,\ldots,2^{2^l\times n} \end{array}\right\}$$
$$\text{for } n = 2^{n_1}. \tag{1.226}$$

Such a backward period-$2^l \times n$ *fixed-point is*

- *monotonically stable if* $dx_k/dx_{k+2^l\times n}|_{x_{k+2^l\times n}^*=a_{i_1}^{(2^l\times n)}} \in (1,\infty)$*;*
- *monotonically invariant if* $dx_k/dx_{k+2^l\times n}|_{x_{k+2^l\times n}^*=a_{i_1}^{(2^l\times n)}} = 1$*, which is*
  - *a monotonic lower-saddle of the* $(2l_1)^{\text{th}}$ *order for* $dx_k/dx_{k+2^l\times n}|_{x_{k+2^l\times n}^*} > 0$ *(independent* $(2l_1)$*-branch appearance);*
  - *a monotonic upper-saddle the* $(2l_1)^{\text{th}}$ *order for* $dx_k/dx_{k+2^l\times n}|_{x_{k+2^l\times n}^*} < 0$ *(independent* $(2l_1)$*-branch appearance)*
  - *a monotonic sink of the* $(2l_1+1)^{\text{th}}$ *order for* $dx_k/dx_{k+2^l\times n}|_{x_{k+2^l\times n}^*} > 0$ *(dependent* $(2l_1+1)$*-branch appearance from one branch);*
  - *a monotonic source the* $(2l_1+1)^{\text{th}}$ order for $dx_k/dx_{k+2^l\times n}|_{x_{k+2^l\times n}^*} < 0$ *(dependent* $(2l_1+1)$*-branch appearance from one branch);*
- *monotonically unstable if* $dx_k/dx_{k+2^l\times n}|_{x_{k+2^l\times n}^*=a_{i_1}^{(2^l\times n)}} \in (0,1)$*;*
- *monotonically infinity-unstable if* $dx_k/dx_{k+2^l\times n}|_{x_{k+2^l\times n}^*=a_{i_1}^{(2^l\times n)}} = 0^+$*;*
- *monotonically infinity-unstable if* $dx_k/dx_{k+2^l\times n}|_{x_{k+2^l\times n}^*=a_{i_1}^{(2^l\times n)}} = 0^+$*;*
- *oscillatorilly unstable if* $dx_k/dx_{k+2^l\times n}|_{x_{k+2^l\times n}^*=a_{i_1}^{(2^l\times n)}} \in (-1,0)$*;*
- *flipped if* $dx_k/dx_{k+2^l\times n}|_{x_{k+2^l\times n}^*=a_{i_1}^{(2^l\times n)}} = -1$*, which is*
  - *an oscillatory lower-saddle of the* $(2l_1)^{\text{th}}$ *order if* $d^{2l_1}x_k/dx_{k+2^l\times n}^{2l_1}|_{x_{k+2^l\times n}^*} > 0$*;*
  - *an oscillatory upper-saddle the* $(2l_1)^{\text{th}}$ *order with* $d^{2l_1}x_k/dx_{k+2^l\times n}^{2l_1}|_{x_{k+2^l\times n}^*} < 0$*;*
  - *an oscillatory source of the* $(2l_1+1)^{\text{th}}$ *order if* $d^{2l_1+1}x_k/dx_{k+2^l\times n}^{2l_1+1}|_{x_{k+2^l\times n}^*} > 0$*;*
  - *an oscillatory sink the* $(2l_1+1)^{\text{th}}$ *order with* $d^{2l_1+1}x_k/dx_{k+2^l\times n}^{2l_1+1}|_{x_{k+2^l\times n}^*} < 0$*;*
- *oscillatorilly stable if* $dx_k/dx_{k+2^l\times n}|_{x_{k+2^l\times n}^*=a_{i_1}^{(2^l\times n)}} \in (-\infty,-1)$*.*

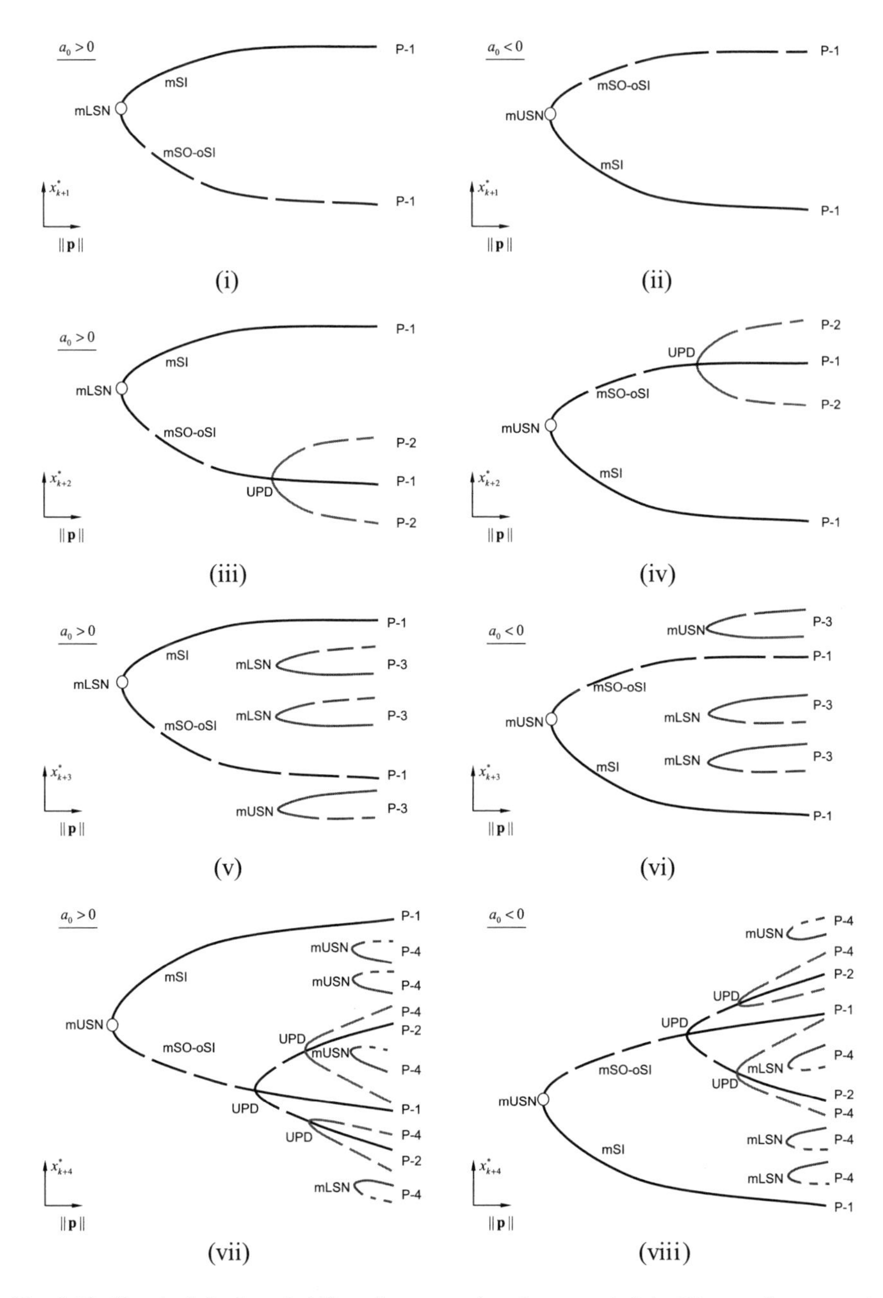

Fig. 1.10 Sketched backward bifurcation trees based on period-doubling and monotonic saddle-node bifurcations for backward period-n discrete systems of quadratic discrete systems. (i)–(viii) period-1 to period-4 bifurcation trees, based on period-n discrete systems. mUSN: monotonic upper-saddle-node, mLSN: monotonic lower-saddle-node. UPD: unstable period-doubling bifurcation. The solid and dashed curves are for stable and unstable fixed-points. The red and dark red colors are for dependent and independent period-n fixed points. mSI: monotonic sink, mSO-oSI: monotonic source to oscillatory sink.

Proof Through the nonlinear renormalization, the proof of this theorem can follow the proof for the forward discrete system. This theorem can be easily proved. ■

1.5.4 *Backward Period-*n *Bifurcation Trees*

Similarly, from the period-n discrete systems of a backward quadratic discrete system, period-1 to period-4 bifurcation trees are sketched through period-n discrete systems, as shown in Fig. 1.10. As for the forward quadratic discrete system, solid and dashed curves are for stable and unstable period-n fixed-points. The red and dark red colors are for period-n fixed points dependent on and independent of period-1 fixed-points on the bifurcation trees. The backward period-n fixed-points on the other period-doubling bifurcations are also said to be dependent. The dependent period-n fixed-points are obtained from unstable period-doubling bifurcation. However, the onsets of period-n fixed-points are based on unstable period-doubling bifurcations, which are also said to be independent as well. The onsets of such independent backward period-n fixed-points are based on the monotonic saddle-node bifurcations.

References

Luo ACJ (2010) A Ying-Yang theory for nonlinear discrete dynamical systems. International Journal of Bifurcation and Chaos 20(4):1085–1098

Luo ACJ (2012) Regularity and Complexity in Dynamical Systems. Springer, New York

Luo ACJ (2019) The stability and bifurcation of fixed-points in low-degree polynomial systems. Journal of Vibration Testing and System Dynamics 3(4):403–451

Luo ACJ, Guo Y (2013) Vibro-Impact Dynamics. Wiley, New York

Chapter 2
Cubic Nonlinear Discrete Systems

In this Chapter, the stability and stability switching of fixed-points in cubic polynomial discrete systems are discussed. As in Luo (2019), the monotonic *upper-saddle-node* and *lower-saddle-node* appearing and switching bifurcations are discussed and the third-order monotonic sink and source switching bifurcations are discussed as well. The third-order monotonic sink and source flower-bundle switching bifurcations for simple fixed-points are presented. The third-order monotonic sink and source switching bifurcations for monotonic *saddle* and nodes are discovered. Graphical illustrations of global stability and bifurcations are presented. The bifurcation trees for cubic nonlinear discrete systems are discussed through period-doublization and monotonic saddle-node bifurcations.

2.1 Period-1 Cubic Discrete Systems

In this section, period-1 fixed-points in cubic nonlinear discrete systems will be discussed, and the stability and bifurcation conditions will be developed.

Definition 2.1 Consider a cubic nonlinear discrete system

$$\begin{aligned} x_{k+1} &= x_k + A(\mathbf{p})x_k^3 + B(\mathbf{p})x_k^2 + C(\mathbf{p})x_k + D(\mathbf{p}) \\ &\equiv x_k + a_0(\mathbf{p})(x_k - a(\mathbf{p}))[x_k^2 + B_1(\mathbf{p})x_k + C_1(\mathbf{p})] \end{aligned} \tag{2.1}$$

where four scalar constants $A(\mathbf{p}) \neq 0$, $B(\mathbf{p})$, $C(\mathbf{p})$ and $D(\mathbf{p})$ are determined by

$$\begin{aligned} &A = a_0, B = (-a + B_1)a_0, C = (-aB_1 + C_1)a_0, D = -aa_0C_1, \\ &\mathbf{p} = (p_1, p_2, \ldots, p_m)^{\mathrm{T}}. \end{aligned} \tag{2.2}$$

A. C. J Luo, *Bifurcation Dynamics in Polynomial Discrete Systems*,
Nonlinear Physical Science, https://doi.org/10.1007/978-981-15-5208-3_2

(i) If

$$\Delta_1 = B_1^2 - 4C_1 < 0 \text{ for } \mathbf{p} \in \Omega_1 \subset \mathbf{R}^m \tag{2.3}$$

then the cubic nonlinear discrete system has a simple fixed-point only as

$$x_k^* = a \text{ for } \mathbf{p} \in \Omega_1 \subset \mathbf{R}^m \tag{2.4}$$

and the standard form of such a 1-dimensional system is

$$x_{k+1} = x_k + a_0(x_k - a)(x_k^2 + B_1 x_k + C_1). \tag{2.5}$$

(ii) If

$$\Delta_1 = B_1^2 - 4C_1 > 0 \text{ for } \mathbf{p} \in \Omega_2 \subset \mathbf{R}^m \tag{2.6}$$

then there are three fixed-points with

$$\begin{aligned}
&a_0 = A(\mathbf{p}),\ b_{1,2} = -\tfrac{1}{2}(B_1(\mathbf{p}) \mp \sqrt{\Delta_1}) \text{ with } b_1 > b_2;\\
&a_1 = \min\{a, b_1, b_2\}, a_3 = \max\{a, b_1, b_2\}, a_2 \in \{a, b_1, b_2\} \neq \{a_1, a_3\},\\
&\Delta_{ij} = (a_i - a_j)^2 > 0 \text{ for } i, j \in \{1, 2, 3\} \text{ but } i \neq j.
\end{aligned} \tag{2.7}$$

(ii_1) If

$$a_i \neq a_j \text{ with } \Delta_{ij} = (a_i - a_j)^2 > 0 \text{ for } i, j \in \{1, 2, 3\} \text{ but } i \neq j. \tag{2.8}$$

the cubic forward discrete system has three different, simple fixed-points as

$$x_k^* = a_1,\ x_k^* = a_2,\ x_k^* = a_3 \tag{2.9}$$

and the corresponding standard form is

$$x_{k+1} = x_k + a_0(x_k - a_1)(x_k - a_2)(x_k - a_3). \tag{2.10}$$

(ii_2) If at $\mathbf{p} = \mathbf{p}_1$

$$\begin{aligned}
&a_1 = b_2,\ a_2 = a,\ a_3 = b_1;\\
&\Delta_{12} = (a_1 - a_2) = (a - b_2)^2 = 0,
\end{aligned} \tag{2.11}$$

the cubic nonlinear discrete system has a double-repeated fixed-point and a simple fixed-point as

$$x_k^* = a_1,\ x_k^* = a_1,\ x_k^* = a_2 \tag{2.12}$$

and the corresponding standard form is

$$x_{k+1} = x_k + a_0(x_k - a_1)^2(x_k - a_2). \tag{2.13}$$

Such a discrete flow at the fixed-point of $x_k^* = a_1$ is called a monotonic-*saddle* discrete flow of the second-order.

The fixed-point of $x_k^* = a_1$ for two different fixed-points switching is called a switching bifurcation point of fixed-point at $\mathbf{p} = \mathbf{p}_1$ with the second-order multiplicity, and the switching bifurcation condition is

$$a = b_1 = \min\{-\tfrac{1}{2}(B_1(\mathbf{p}) + \sqrt{\Delta_1}), -\tfrac{1}{2}(B_1(\mathbf{p}) - \sqrt{\Delta_1})\} \tag{2.14}$$

(ii$_3$) If at $\mathbf{p} = \mathbf{p}_2$,

$$\begin{aligned} &a_2 = b_1,\ a_3 = a,\ a_1 = b_2, \\ &\Delta_{23} = (a_2 - a_3) = (a - b_1)^2 = 0, \end{aligned} \tag{2.15}$$

the cubic forward discrete system has three fixed-points as

$$x_k^* = a_1,\ x_k^* = a_2,\ \ x_k^* = a_2 \tag{2.16}$$

and the corresponding standard form is

$$x_{k+1} = x_k + a_0(x_k - a_1)(x_k - a_2)^2. \tag{2.17}$$

Such a discrete flow at the fixed-point of $x_k^* = a_2$ is called a monotonic *saddle* discrete flow of the second-order.

The fixed-point of $x_k^* = a_2$ for two different fixed-points switching is called a switching bifurcation point of fixed-point at a point $\mathbf{p} = \mathbf{p}_1$ with the second-order multiplicity, and the switching bifurcation condition is

$$a = b_2 = \max\{-\tfrac{1}{2}(B_1(\mathbf{p}) + \sqrt{\Delta_1}), -\tfrac{1}{2}(B_1(\mathbf{p}) - \sqrt{\Delta_1})\} \tag{2.18}$$

(ii$_3$) If at $\mathbf{p} = \mathbf{p}_3$,

$$\begin{aligned} &a_1 = b_2,\ a_2 = a,\ a_3 = b_1, \\ &\Delta_{12} = (a_1 - a_2)^2 = (a - b_2)^2 = 0, \\ &\Delta_{23} = (a_2 - a_3)^2 = (a - b_1)^2 = 0, \\ &\Delta_{13} = (a_1 - a_3)^2 = (b_2 - b_1)^2 = 0, \end{aligned} \tag{2.19}$$

the cubic nonlinear system has three repeated fixed-point as

$$x_k^* = a_1,\ x_k^* = a_2 \text{ and } x_k^* = a_3 \tag{2.20}$$

and the corresponding standard form is

$$x_{k+1} = x_k + a_0(x_k - a)^3. \tag{2.21}$$

Such a discrete flow at the fixed-point of $x_k^* = a_1$ is called a monotonic sink or source discrete flow of the third-order.

The fixed-point of $x_k^* = a_1$ at a point $\mathbf{p} = \mathbf{p}_3$ for three different fixed-points switching is called a switching bifurcation point of fixed-point with the third-order multiplicity, and the switching bifurcation condition is

$$a = b = -\frac{1}{2}B_1(\mathbf{p}). \tag{2.22}$$

(iii) If

$$\Delta_1 = B_1^2 - 4A_1C_1 = 0 \text{ for } \mathbf{p} = \mathbf{p}_0 \in \partial\Omega_{12} \subset \mathbf{R}^{m-1}, \tag{2.23}$$

then there exist

$$a_0 = A(\mathbf{p}_0), \text{ and } b_1 = b_2 = b = -\frac{1}{2}B_1(\mathbf{p}_0). \tag{2.24}$$

(iii$_1$) For $a<b$, the cubic nonlinear system has a double-repeated fixed-point plus a monotonic *lower-branch* simple fixed-point

$$x_k^* = a_1 = a,\ x_k^* = a_2 = b \text{ and } x_k^* = a_2 = b \tag{2.25}$$

with the corresponding standard form of

$$x_{k+1} = x_k + a_0(x_k - a_1)(x_k - a_2)^2. \tag{2.26}$$

Such a discrete flow at the fixed-point of $x^* = a_2$ is called a monotonic *saddle* discrete flow of the second-order.

The fixed-point of $x_k^* = a_2$ for two different fixed-points appearing is called an appearing bifurcation point of fixed-points at a point $\mathbf{p} = \mathbf{p}_0 \in \partial\Omega_{12}$ with the second-order multiplicity, and the appearing bifurcation condition is

$$\Delta_1 = B_1^2 - 4C_1 = 0 \text{ with } a<b. \tag{2.27}$$

(iii$_2$) For $a > b$, the cubic nonlinear system has a double-repeated fixed-point plus a simple fixed-point

$$x_k^* = a_1 = b \text{ and } x_k^* = a_1 = b, x_k^* = a_2 = a \tag{2.28}$$

with the corresponding standard form of

$$x_{k+1} = x_k + a_0(x_k - a_1)^2(x_k - a_2). \tag{2.29}$$

Such a discrete flow at the fixed-point of $x_k^* = a_1$ is called a monotonic *saddle* discrete flow of the second order.

The fixed-point of $x_k^* = a_1 = b$ for two different fixed-point appearing is called a bifurcation point of fixed-point at a point $\mathbf{p} = \mathbf{p}_0 \in \partial\Omega_{12}$ with the *lower-branch* second-order multiplicity, and the bifurcation appearing condition is also

$$\Delta_1 = B_1^2 - 4C_1 = 0 \text{ with } a > b. \tag{2.30}$$

(iii$_3$) For $a = b$, the cubic forward discrete system has a triple-repeated fixed-point as

$$x_k^* = a_1 = a \text{ and } x_k^* = a_1 = a, x_k^* = a_2 = a \tag{2.31}$$

with the corresponding standard form of

$$x_{k+1} = x_k + a_0(x_k - a_1)^3. \tag{2.32}$$

Such a discrete flow at the fixed-point of $x^* = a_1$ is called a monotonic source or sink discrete flow of the third-order.

The fixed-point of $x^* = a_1 = a$ for three fixed-points switching or two fixed-points switching is called a switching bifurcation of fixed-point at a point $\mathbf{p} = \mathbf{p}_0 \in \partial\Omega_{12}$ with the third-order multiplicity, and the switching bifurcation condition is

$$\Delta_1 = B_1^2 - 4C_1 = 0 \text{ with } a = b. \tag{2.33}$$

From the previous definitions, the conditions of stability and bifurcation in forward cubic nonlinear discrete systems are stated in the following theorem.

Theorem 2.1

(i) *Under a condition of*

$$\Delta_1 = B_1^2 - 4C_1 < 0 \tag{2.34}$$

a standard form of the 1-dimensional discrete system in Eq. (2.1) is

$$x_{k+1} = x_k + f(x_{k+1}, \mathbf{p}) = x_k + a_0(x_k - a_1)[(x_k + \tfrac{1}{2}B_1)^2 + \tfrac{1}{4}(-\Delta_1)]. \tag{2.35}$$

(i_1) *If* $a_0(\mathbf{p}) > 0$, then the fixed-point of $x_k^* = a_1$ is monotonically unstable (a monotonic source) with $df/dx_k|_{x_k^*=a_1} \in (0, \infty)$.

(i_2) *If* $a_0(\mathbf{p}) < 0$, *then the fixed-point of* $x_k^* = a_1$ *is*

- *monotonically stable with* $df/dx_k|_{x_k^*=a_1} \in (-1, 0)$ *(a monotonic sink);*
- *invariantly stable with* $df/dx_k|_{x_k^*=a_1} = -1$ *(an invariant sink);*
- *oscillatorilly stable with* $df/dx_k|_{x_k^*=a_1} \in (-2, -1)$ *(an oscillatory sink);*
- *flipped with* $df/dx_k|_{x_k^*=a_1} = -2$, *which is*
  - *an oscillatory upper-saddle of the second-order for* $d^2f/dx_k^2|_{x_k^*=a_1} > 0$;
  - *an oscillatory lower-saddle of the second-order for* $d^2f/dx_k^2|_{x_k^*=a_1} < 0$;
  - *an oscillatory source of the third-order for* $d^2f/dx_k^2|_{x_k^*=a_1} = 0$ *and* $d^3f/dx_k^3|_{x_k^*=a_1} < 0$;
- *oscillatorilly unstable with* $df/dx_k|_{x_k^*=a_1} \in (-\infty, -2)$ *(an oscillatory source).*

(i_3) If $a_0(\mathbf{p}) = 0$, *then the fixed-point of* $x_k^* = a_1$ is *stability switching.*

(ii) *Under the conditions of*

$$\begin{aligned} &\Delta_1 = B_1^2 - 4C_1 > 0, \\ &a_1, a_2, a_3 = \text{sort}\{b_2, a, b_1\}, a_i \neq a_j, a_i < a_{i+1}; \\ &\Delta_{ij} = (a_i - a_j)^2 \neq 0 \text{ for } i, j \in \{1, 2, 3\}, \end{aligned} \tag{2.36}$$

a standard form of the 1-dimensional forward discrete system in Eq. (2.31) is

$$x_{k+1} = x_k + f(x_k, \mathbf{p}) = x_k + a_0(x_k - a_1)(x_k - a_2)(x_k - a_3). \tag{2.37}$$

(ii_{1a}) *if* $a_0(\mathbf{p}) > 0$, *then the fixed-points of* $x_k^* = a_1$ *is monotonically unstable with* $df/dx_k|_{x_k^*=a_1} \in (0, \infty)$ *(a monotonic source).*

(ii_{1b}) *If* $a_0(\mathbf{p}) > 0$, *then the fixed-points of* $x_k^* = a_2$ *is*

- *monotonically stable with* $df/dx_k|_{x_k^*=a_2} \in (-1, 0)$ *(a monotonic sink);*
- *invariantly stable with* $df/dx_k|_{x_k^*=a_2} = -1$ *(an invariant sink);*
- *oscillatorilly stable with* $df/dx_k|_{x_k^*=a_2} \in (-2, -1)$ *(an oscillatory sink).*
- *flipped with* $df/dx_k|_{x_k^*=a_2} = -2$, *which is*
  - *an oscillatory upper-saddle of the second-order for* $d^2f/dx_k^2|_{x_k^*=a_2} > 0$;
  - *an oscillatory lower-saddle of the second-order for* $d^2f/dx_k^2|_{x_k^*=a_2} < 0$;
  - *an oscillatory sink of the third-order for* $d^2f/dx_k^2|_{x_k^*=a_2} = 0$ and $d^3f/dx_k^3|_{x_k^*=a_2} > 0$;
- *oscillatorilly unstable with* $df/dx_k|_{x_k^*=a_2} \in (-\infty, -2)$ *(an oscillatory source).*

(ii_{1c}) *If* $a_0(\mathbf{p}) > 0$, *then the fixed-points of* $x_k^* = a_3$, *is monotonically unstable with* $df/dx_k|_{x_k^*=a_3} \in (0, \infty)$ *(a monotonic source).*

(ii_{2a}) If $a_0(\mathbf{p}) < 0$, *then the fixed-points of* $x_k^* = a_1$ *is*

- *monotonically stable with* $df/dx_k|_{x_k^*=a_1} \in (-1, 0)$ *(a monotonic sink);*
- invariantly stable with $df/dx_k|_{x_k^*=a_1} = -1$ *(an invariant sink);*
- *oscillatorilly stable with* $df/dx_k|_{x_k^*=a_1} \in (-2, -1)$ *(an oscillatory sink);*
- *flipped with* $df/dx_k|_{x_k^*=a_1} = -2$ *(an oscillatory upper-saddle of the second-order if* $d^2f/dx_k^2|_{x_k^*=a_1} > 0$*);*
- *oscillatorilly unstable with* $df/dx_k|_{x_k^*=a_2} \in (-\infty, -2)$ *(an oscillatory source).*

(ii_{2b}) *If* $a_0(\mathbf{p}) < 0$, *then the fixed-points of* $x_k^* = a_2$ is *monotonically unstable with* $df/dx_k|_{x_k^*=a_2} \in (0, \infty)$ *(a monotonic source).*

(ii_{2c}) *If* $a_0(\mathbf{p}) < 0$, *then the fixed-points of* $x_k^* = a_3$ *is*

- *monotonically stable with* $df/dx_k|_{x_k^*=a_3} \in (-1, 0)$ *(a monotonic sink);*
- *invariantly stable with* $df/dx_k|_{x_k^*=a_3} = -1$ *(an invariant sink);*

- *oscillatorilly stable with* $df/dx_k|_{x_k^*=a_3} \in (-2,-1)$ *(an oscillatory source);*
- *flipped with* $df/dx_k|_{x_k^*=a_3} = -2$ *(an oscillatory lower-saddle of the second-order for* $d^2f/dx_k^2|_{x_k^*=a_3} < 0$*);*
- *oscillatorilly unstable with* $df/dx_k|_{x_k^*=a_3} \in (-\infty,-2)$ *(an oscillatory source).*

(iii) *Under a condition of*

$$\begin{array}{l}\Delta_1 = B_1^2 - 4C_1 > 0,\\ a_1, a_2, a_3 = \text{sort}\{b_2, \text{a}, b_1\}, a_i \neq a_j, a_i \leq a_{i+1}\\ \Delta_{12} = (a_1 - a_2)^2 = 0, \text{ for } i,j \in \{1,2,3\}\end{array} \tag{2.38}$$

a standard form of the 1-dimensional forward discrete system in Eq. (2.1) is

$$x_{k+1} = x_k + f(x_k, \mathbf{p}) = x_k + a_0(x_k - a_1)^2(x_k - a_2). \tag{2.39}$$

(iii$_{1a}$) *If* $a_0(\mathbf{p}) > 0$, *then the fixed-points of* $x_k^* = a_1$ *is monotonically unstable* with $d^2f/dx_k^2|_{x_k^*=a_1} < 0$ *(a monotonic* lower-saddle of the second-order).

(iii$_{1b}$) *If* $a_0(\mathbf{p}) > 0$, *then the fixed-point of* $x_k^* = a_2$ *is monotonically unstable with* $df/dx_k|_{x_k^*=a_2} \in (0,\infty)$ *(a monotonic* source).

(iii$_{1c}$) *The bifurcation of fixed-point at* $x_k^* = a_1$ *for the two different fixed-points switching is a monotonic* lower-saddle-node switching *bifurcation of the second-order at a point* $\mathbf{p} = \mathbf{p}_1$.

(iii$_{2a}$) *If* $a_0(\mathbf{p}) < 0$, *then the fixed-point of* $x_k^* = a_1$ *is monotonically unstable* with $d^2f/dx_k^2|_{x_k^*=a_1} > 0$ *(a monotonic* upper-saddle of the second-order).

(iii$_{2b}$) *If* $a_0(\mathbf{p}) < 0$, then the fixed-point of $x_k^* = a_2$ is

- *monotonically stable with* $df/dx_k|_{x_k^*=a_2} \in (-1,0)$ *(a monotonic sink);*
- *invariantly stable with* $df/dx_k|_{x_k^*=a_2} = -1$ *(an invariant sink);*
- *oscillatorilly stable with* $df/dx_k|_{x_k^*=a_2} \in (-2,-1)$ *(an oscillatory sink);*
- *flipped with* $df/dx_k|_{x_k^*=a_2} = -2$ *(an oscillatory lower-saddle of the second-order if* $d^2f/dx_k^2|_{x_k^*=a_2} < 0$*);*
- *oscillatorilly unstable with* $df/dx_k|_{x_k^*=a_2} \in (-\infty,-2)$ *(an oscillatory source).*

(iii$_{2c}$) *The bifurcation of fixed-point at* $x_k^* = a_1$ *for the two different fixed-point switching is a monotonic* upper-saddle-node *switching bifurcation of the second order at a point* $\mathbf{p} = \mathbf{p}_1$.

(iv) *For*

$$\begin{aligned} &\Delta_1 = B_1^2 - 4C_1 > 0, \\ &a_1, a_2, a_3 = sort\{b_2, \mathrm{a}, b_1\}, a_i \neq a_j, a_i \leq a_{i+1} \\ &\Delta_{23} = (a_2 - a_3)^2 = 0, \ \text{for } i, j \in \{1, 2, 3\} \end{aligned} \tag{2.40}$$

a standard form of the 1-dimensional discrete system in Eq. (2.1) is

$$x_{k+1} = x_k + f(x_k, \mathbf{p}) = x_k + a_0(x_k - a_1)(x_k - a_2)^2. \tag{2.41}$$

(iv$_{1a}$) *If* $a_0(\mathbf{p}) > 0$, *then the fixed-points of* $x_k^* = a_2$ *is monotonically unstable* with $d^2f/dx_k^2|_{x_k^*=a_2} > 0$ *(a monotonic* upper-saddle of the second-order).

(iv$_{1b}$) *If* $a_0(\mathbf{p}) > 0$, *then the fixed-point of* $x_k^* = a_1$ *is monotonically unstable with* $df/dx_k|_{x_k^*=a_1} \in (0, \infty)$ *(a monotonic* source).

(iv$_{1c}$) *The bifurcation of fixed-point at* $x_k^* = a_2$ *for the two different fixed-points switching is a* monotonic upper-saddle-node *switching bifurcation of the second order at a point* $\mathbf{p} = \mathbf{p}_1$.

(iv$_{2a}$) *If* $a_0(\mathbf{p}) < 0$, *then the fixed-point of* $x_k^* = a_2$ *is monotonically unstable* with $d^2f/dx_k^2|_{x_k^*=a_2} < 0$ *(a monotonic* lower-saddle of the second-order).

(iv$_{2b}$) *If* $a_0(\mathbf{p}) < 0$, *then the fixed-point of* $x_k^* = a_1$ is

- *monotonically stable with* $df/dx_k|_{x_k^*=a_1} \in (-1, 0)$ *(a monotonic sink);*
- *invariantly stable with* $df/dx_k|_{x^*=a_1} = -1$ (an invariant sink);
- *oscillatorilly stable with* $df/dx_k|_{x^*=a_1} \in (-2, -1)$ *(an oscillatory sink);*
- *flipped with* $df/dx_k|_{x_k^*=a_1} = -2$ *(an oscillatory upper-saddle of the second*-order for $d^2f/dx_k^2|_{x_k^*=a_1} > 0$);
- *oscillatorilly unstable with* $df/dx_k|_{x_k^*=a_1} \in (-\infty, -2)$ *(an oscillatory source).*

(iv$_{2c}$) *The bifurcation of fixed-point at* $x_k^* = a_2$ *for the two different fixed-point switching is a monotonic* lower-saddle-node *bifurcation of the second-order at a point* $\mathbf{p} = \mathbf{p}_1$.

(v) *For*

$$\begin{aligned} &\Delta_1 = B_1^2 - 4C_1 \geq 0, \\ &a_1, a_2, a_3 = \text{sort}\{b_2, \text{a}, b_1\}, a_i \neq a_j, a_i \leq a_{i+1}, \\ &\Delta_{ij} = (a_i - a_j)^2 = 0 \text{ for } i,j = 1,2,3 \text{ but } i \neq j, \end{aligned} \tag{2.42}$$

a standard form of the 1-dimensional forward discrete system in Eq. (2.1) is

$$x_{k+1} = x_k + f(x_k, \mathbf{p}) = x_k + a_0(x_k - a_1)^3. \tag{2.43}$$

($\text{v}_{1\text{a}}$) *If* $a_0(\mathbf{p}) > 0$, *then the fixed-point of* $x_k^* = a_1$ *is monotonically unstable (a third-order monotonic source,* $d^3f/dx_k^3|_{x_k^*=a_1} > 0$).

($\text{v}_{1\text{b}}$) *The bifurcation of fixed-point at* $x_k^* = a_1$ *for three different fixed-points switching is a monotonic source switching bifurcation of the third-order at a point* $\mathbf{p} = \mathbf{p}_1$.

($\text{v}_{2\text{a}}$) *If* $a_0(\mathbf{p}) < 0$, *then the fixed-point of* $x^* = a_1$ *is monotonically stable (a third-order monotonic sink,* $d^3f/dx_k^3|_{x_k^*=a_1} < 0$).

($\text{v}_{2\text{b}}$) *The bifurcation of fixed-point at* $x^* = a_2$ *for three fixed-points switching is a monotonic* sink *switching bifurcation of the third-order at a point* $\mathbf{p} = \mathbf{p}_1$.

(vi) *For*

$$\begin{aligned} &\Delta_1 = B_1^2 - 4A_1C_1 = 0,\ a < b \\ &a_1 = a,\ a_2 = b, \Delta_{12} = (a_1 - a_2)^2 \neq 0 \end{aligned} \tag{2.44}$$

at $\mathbf{p} = \mathbf{p}_0 \in \partial\Omega_{12} \subset \mathbf{R}^{m-1}$, *a standard form of the 1-dimensional discrete system is*

$$x_{k+1} = x_k + f(x_k, \mathbf{p}) = x_k + a_0(x_k - a_1)(x_k - a_2)^2. \tag{2.45}$$

($\text{vi}_{1\text{a}}$) *If* $a_0(\mathbf{p}) > 0$, *then the fixed-point of* $x_k^* = a_1$ *is monotonically unstable (a monotonic source,* $df/dx_k|_{x_k^*=a_1} > 0$).

($\text{vi}_{1\text{b}}$) *If* $a_0(\mathbf{p}) > 0$, *then the fixed-point of* $x_k^* = a_2$ *is monotonically unstable (a monotonic upper-saddle of the second-order,* $d^2f/dx_k^2|_{x_k^*=a_2} > 0$).

($\text{vi}_{1\text{c}}$) *The bifurcation of fixed-point at* $x^* = a_2$ *for two different fixed-point vanishing or appearance is a monotonic* upper-saddle-node appearing *bifurcation of the second-order at a point* $\mathbf{p} = \mathbf{p}_0 \in \partial\Omega_{12}$.

($\text{vi}_{2\text{a}}$) *If* $a_0(\mathbf{p}) < 0$, *then the fixed-point of* $x_k^* = a_1$ is

- *monotonically stable with* $df/dx_k|_{x_k^*=a_1} \in (-1,0)$ *(a monotonic sink);*
- *invariantly stable with* $df/dx_k|_{x^*=a_1} = -1$ *(an invariant sink);*
- *oscillatorilly stable with* $df/dx_k|_{x^*=a_1} \in (-2,-1)$ *(an oscillatory sink);*
- *flipped with* $df/dx_k|_{x_k^*=a_1} = -2$ *(an oscillatory upper-saddle of the second-order for* $d^2f/dx_k^2|_{x_k^*=a_1} > 0$*);*
- *oscillatorilly unstable with* $df/dx_k|_{x_k^*=a_1} \in (-\infty,-2)$ *(an oscillatory source).*

(vi_{2b}) *If* $a_0(\mathbf{p}) < 0$, *then the fixed-point of* $x_k^* = a_2$ *is monotonically unstable (a monotonic* lower-saddle of the second-order, $d^2f/dx_k^2|_{x_k^*=a_2} < 0$).

(vi_{2c}) *The bifurcation of fixed-point at* $x^* = a_2$ *for two different fixed-points vanishing or appearance is a monotonic* lower-saddle-node *appearing bifurcation of the second order at a point* $\mathbf{p} = \mathbf{p}_0 \in \partial\Omega_{12}$.

(vii) For

$$\begin{aligned} &\Delta_1 = B_1^2 - 4A_1C_1 = 0,\ a > b \\ &a_1 = b,\ a_2 = a,\ \Delta_{12} = (a_1 - a_2)^2 \neq 0 \end{aligned} \tag{2.46}$$

at $\mathbf{p} = \mathbf{p}_0 \in \partial\Omega_{12} \subset \mathbf{R}^{m-1}$, *a standard form of the 1-dimensional, forward discrete system is*

$$x_{k+1} = x_k + f(x_k,\mathbf{p}) = x_k + a_0(x_k - a_1)^2(x_k - a_2). \tag{2.47}$$

(vii_{1a}) *If* $a_0(\mathbf{p}) > 0$, *then the fixed-point of* $x_k^* = a_1$ *is monotonically unstable (a monotonic* lower-saddle of the second-order, $d^2f/dx_k^2|_{x_k^*=a_1} < 0$).

(vii_{1b}) *If* $a_0(\mathbf{p}) > 0$, *then the fixed-point of* $x_k^* = a_2$ *is monotonically unstable (a monotonic source,* $df/dx_k|_{x_k^*=a_2} > 0$*).*

(vii_{1c}) *The bifurcation of fixed-point at* $x^* = a_1$ *for two different fixed-points vanishing or appearance is a monotonic* lower-saddle-node *appearing bifurcation of the second order at a point* $\mathbf{p} = \mathbf{p}_0 \in \partial\Omega_{12}$.

(vii_{2a}) *If* $a_0(\mathbf{p}) < 0$, *then the fixed-point of* $x_k^* = a_1$ *is monotonically unstable (a monotonically* upper-saddle of the second-order, $d^2f/dx_k^2|_{x_k^*=a_1} > 0$).

(vii_{2b}) *If* $a_0(\mathbf{p}) < 0$, *then the fixed-point of* $x_k^* = a_2$ is

- *monotonically stable with* $df/dx_k|_{x_k^*=a_2} \in (-1,0)$ *(a monotonic sink);*
- *invariantly stable with* $df/dx_k|_{x^*=a_2} = -1$ *(an invariant sink);*
- *oscillatorilly stable with* $df/dx_k|_{x^*=a_2} \in (-2,-1)$ *(an oscillatory sink);*

- *flipped with* $df/dx_k|_{x_k^*=a_2} = -2$ *(an oscillatory lower-saddle of the second-order if* $d^2f/dx_k^2|_{x_k^*=a_2} < 0$*);*
- *oscillatorilly unstable with* $df/dx_k|_{x_k^*=a_2} \in (-\infty, -2)$ *(an oscillatory source).*

(vii$_{2c}$) *The bifurcation of fixed-point at* $x^* = a_1$ *for two different fixed-points vanishing or appearance is a monotonically* upper-saddle-node *appearing bifurcation of the second-order at a point* $\mathbf{p} = \mathbf{p}_0 \in \partial\Omega_{12}$.

(viii) *For*

$$\begin{aligned} &\Delta_1 = B_1^2 - 4A_1C_1 = 0, a = b \\ &a_2 = a, a_2 = a_3 = b, \\ &\Delta_{12} = (a_1 - a_2)^2 = 0 \end{aligned} \tag{2.48}$$

at $\mathbf{p} = \mathbf{p}_0 \in \partial\Omega_{12} \subset \mathbf{R}^{m-1}$, *a standard form of the 1-dimensional, forward discrete system is*

$$x_{k+1} = x_{k+1} + f(x_k, \mathbf{p}) = a_0(x_k - a_1)^3. \tag{2.49}$$

(viii$_{1a}$) *If* $a_0(\mathbf{p}) > 0$, *then the fixed-point of* $x_k^* = a_1$ *is monotonically unstable (a third-order monotonic source,* $d^3f/dx_k^3|_{x_k^*=a_1} > 0$*).*

(viii$_{1b}$) *The bifurcation of fixed-point at* $x_k^* = a_1$ *for one fixed-point to three different three fixed-point switching is a monotonic source switching bifurcation of the third order at a point* $\mathbf{p} = \mathbf{p}_0 \in \partial\Omega_{12}$.

(viii$_{2a}$) *If* $a_0(\mathbf{p}) < 0$, *then the fixed-point of* $x_k^* = a_1$ *is monotonically stable (a third-order monotonic sink,* $d^3f/dx_k^3|_{x_k^*=a_1} < 0$*).*

(viii$_{2b}$) *The bifurcation of fixed-point at* $x_k^* = a_1$ *for one fixed-point to three different three fixed-point switching is a monotonic sink switching bifurcation of the third order at a point* $\mathbf{p} = \mathbf{p}_0 \in \partial\Omega_{12}$.

Proof The proof is similar to Theorem 1.2. ■

2.2 Period-1 to Period-2 Bifurcation Trees

In this section, period-1 stability and bifurcation of cubic nonlinear discrete systems are discussed graphically and period-2 fixed-points are also presented for a better understanding of complex bifurcations.

The 1-dimensional cubic nonlinear discrete system can be expressed by a factor of $(x_k - a)$ and a quadratic form of $a_0(x_k^2 + B_1x_k + C_1)$ as in Eq. (2.1). Three period-1 fixed-points do not have any intersections. Thus, only one bifurcation occurs at $\Delta_1 = B_1^2 - 4C_1 = 0$. The bifurcation of fixed-points occurs at the double or triple repeated fixed-point at the boundary of $\mathbf{p}_0 \in \partial\Omega_{12}$. For $\Delta_1 = B_1^2 - 4C_1 > 0$, $x_k^2 + B_1 x_k + C_1 = 0$ gives two fixed-points of $x_k^* = b_1, b_2$. For $a_0 > 0$, if $a > \max\{b_1, b_2\}$, then the fixed-point of $x_k^* = a_3 = a$ is monotonically unstable, and the fixed-point of $x_k^* = a_2 = \max\{b_1, b_2\}$ is from monotonically stable to oscillatorilly unstable, and the fixed-points of $x_k^* = a_1 = \min\{b_1, b_2\}$ is monotonically unstable.

For $\Delta_1 = B_1^2 - 4C_1 < 0$, $x_k^2 + B_1 x_k + C_1 = 0$ does not have any real solutions. For $\Delta_1 = B_1^2 - 4C_1 = 0$, $x_k^2 + B_1 x_k + C_1 = 0$ has a double repeated fixed-point of $x^* = b = -\frac{1}{2}B_1$. The condition of $\Delta_1 = B_1^2 - 4C_1 = 0$ gives

$$B_1^2 = 4C_1. \tag{2.50}$$

From Eq. (2.2), one obtains

$$B_1 = a + \frac{B}{A} \text{ and } C_1 = \frac{C}{A} + a(a + \frac{B}{A}). \tag{2.51}$$

Thus, equation (2.50) gives

$$a = -\frac{B}{3A} \pm \frac{2}{3A}\sqrt{B^2 - 3AC}. \tag{2.52}$$

Further, the double repeated fixed-point of $x_k^* = b = -\frac{1}{2}B_1$ is given by

$$x_k^* = b = -\frac{B}{3A} \mp \frac{1}{3A}\sqrt{B^2 - 3AC}. \tag{2.53}$$

If $B^2 > 3AC$, such a double repeated fixed-point exists. If $B^2 < 3AC$, such a double repeated fixed-point does not exist. From Eq. (2.51), another fixed-point is $x_k^* = a$, which is different from $x_k^* = b$. If $B^2 = 3AC$, such a double repeated fixed-point with fixed-point of $x_k^* = a$ has an intersected point at $x_k^* = -\frac{B}{3A}$.

The bifurcation diagram for $a > \max\{b_1, b_2\}$ and $a_0 > 0$ is presented in Fig. 2.1 (i). The stable and unstable fixed-points varying with the vector parameter are presented by solid and dashed curves, respectively. Such a fixed-point of $x_k^* = b$ is a monotonic *lower-saddle-node* (mLSN) appearing or vanishing bifurcation.

- The fixed-point of $x_k^* = a$ is a monotonic source (mSO), which is monotonically unstable.
- The fixed-point of $x_k^* = \max\{b_1, b_2\}$ is
  - a monotonic sink (mSI),
  - an invariant sink (iSI),
  - an oscillatory sink (oSI),
  - an oscillatory saddle bifurcation (oUS or oLS),
  - an oscillatory source (oSO).

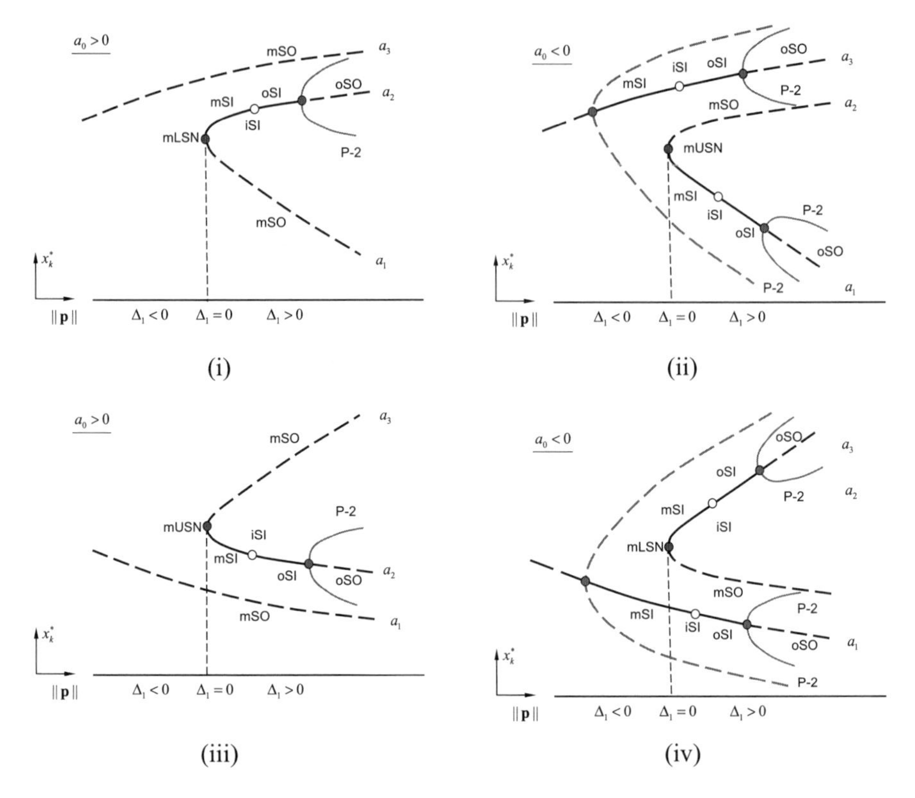

Fig. 2.1 Stability and bifurcation of three independent fixed-points in the 1-dimensional, cubic nonlinear discrete system: For $a > \{b_1, b_2\}$: (i) a mLSN bifurcation ($a_0 > 0$), (ii) a mUSN bifurcation ($a_0 < 0$). For $a < \{b_1, b_2\}$: (iii) a mUSN bifurcation ($a_0 > 0$), (iv) a mLSN bifurcation ($a_0 < 0$). mLSN: monotonic *lower-saddle-node*, mUSN: monotonic *upper-saddle-node*. Stable and unstable fixed-points are represented by solid and dashed curves, respectively. The bifurcation points are marked by circular symbols. (mSO: monotonic source; mSI: monotonic sink; oSO: oscillatory source; oSI: oscillatory sink; mLS: monotonic lower-saddle; mUS: monotonic upper-saddle; oUS: oscillatory upper-saddle; oLS: oscillatory lower-saddle; iSI: invariant sink). The period-2 fixed-points are presented on the period-1 bifurcation trees through red curves.

- The fixed-point of $x_k^* = \min\{b_1, b_2\}$ is a monotonic source (mSO), which is monotonically unstable.

However, the bifurcation diagram for $a > \max\{b_1, b_2\}$ and $a_0 < 0$ is presented in Fig. 2.1(ii). The fixed-point of $x_k^* = b$ is a monotonic *upper-saddle-node* (mUSN) appearing or vanishing bifurcation.

- The fixed-point of $x_k^* = a$ is
  - a monotonic sink (mSI) first,
  - an invariant sink (iSI),
  - an oscillatory sink (oSI),

- an oscillatory saddle bifurcation (oUS or oLS),
- an oscillatory source (oSO).

- The fixed-point of $x_k^* = \max\{b_1, b_2\}$ is a monotonic source (mSO).
- The fixed-point of $x_k^* = \min\{b_1, b_2\}$ is
  - a monotonic sink (mSI) first,
  - an invariant sink (iSI),
  - an oscillatory sink (oSI),
  - an oscillatory saddle bifurcation (oUS, oLS),
  - an oscillatory source (oSO).

The bifurcation diagram for $a < \min\{b_1, b_2\}$ and $a_0 > 0$ is presented in Fig. 2.1(iii). The fixed-point of $x_k^* = b$ is a monotonic *upper-saddle-node* (mUSN) appearing or vanishing bifurcation.

- The fixed-point of $x_k^* = \max\{b_1, b_2\}$ is a monotonic source (mSO).
- The fixed-point of $x_k^* = \min\{b_1, b_2\}$ is
  - a monotonic sink (mSI) first,
  - an invariant sink (iSI),
  - an oscillatory sink (oSO),
  - an oscillatory saddle bifurcation (oUS, oLS),
  - an oscillatory source (oSO).
- The fixed-point of $x_k^* = a$ is a monotonic source (mSO).

The bifurcation diagram for $a < \min\{b_1, b_2\}$ and $a_0 < 0$ is presented in Fig. 2.1(iv). The fixed-point of $x_k^* = b$ is a monotonic *lower-saddle-node* (mLSN) appearing or vanishing bifurcation.

- The fixed-point of $x_k^* = \max\{b_1, b_2\}$ is
  - a monotonic sink (mSI) first,
  - an invariant sink (iSI),
  - an oscillatory sink (oSI),
  - an oscillatory saddle bifurcation (oUS, oLS),
  - an oscillatory source (oSO).
- The fixed-point of $x_k^* = \min\{b_1, b_2\}$ is a monotonic source (mSO).
- The fixed-point of $x_k^* = a$ is
  - a monotonic sink (mSI) first,
  - an invariant sink (iSI),
  - an oscillatory sink (oSI),
  - an oscillatory saddle bifurcation (oUS, oLS),
  - an oscillatory source (oSO).

Table 2.1 Stability and bifurcation of a 1-dimensional cubic nonlinear discrete system ($x_{k+1} = x_k + a_0(x_k - a)[x_k^2 + B_1(\mathbf{p})x_k + C_1(\mathbf{p})]$)

Figure 2.1		a_1	a_2	a_3	Bifurcation
(i) $a_0 > 0$	$a > \max\{b_1, b_2\}$	mSO	mSI-oSO	mSO	2nd mLSN
(ii) $a_0 < 0$	$a > \max\{b_1, b_2\}$	mSI-oSO	mSO	mSI-oSO	2nd mUSN
(iii) $a_0 < 0$	$a < \min\{b_1, b_2\}$	mSO	mSI-oSO	mSO	2nd mUSN
(iv) $a_0 < 0$	$a < \min\{b_1, b_2\}$	mSI-oSO	mSO	mSI-oSO	2nd mLSN

Notice that $b_{1,2} = -\frac{1}{2}(B_1 \pm \sqrt{\Delta_1})$, $\Delta_1 = B_1^2 - 4C_1$. Bifurcation condition: $\Delta_1 = 0$. (mSO: monotonic source; mSI: monotonic sink; oSO: oscillatory source; oSI: oscillatory sink; mLS: monotonic lower-saddle; mUS: monotonic upper-saddle; oUS: oscillatory upper-saddle; oLS: oscillatory lower-saddle; mSI-oSO: for monotonic sink to oscillatory source via the oscillatory sink.)

The stability and bifurcations of fixed-points of the 1-dimensional cubic nonlinear, forward discrete system are summarized in Table 2.1. The period-2 fixed-points are also presented as well through the period-doubling.

For $\Delta_1 = B_1^2 - 4C_1 \geq 0$, the 1-dimensional cubic nonlinear, forward discrete system in Eq. (2.1) have three fixed-points. Three fixed-points are $x_k^* = a, b_1, b_2$. Assume $a_i \leq a_{i+1}$ for $i = 1, 2$ with $a_{1,2,3} = \text{sort}(a, b_1, b_2)$. With varying parameters, two of three fixed-points (i.e., $a_i = a_j$ for $i, j \in \{1, 2, 3\}$ but $i \neq j$) will be intersected each other with the corresponding discriminant of $\Delta_{ij} = (a_i - a_j)^2 = 0$, and in the vicinity of the intersection point, $\Delta_{ij} = (a_i - a_j)^2 > 0$. The two intersected points of $a = b_{1,2}$ gives

$$a = -\frac{B_1}{2} \pm \frac{1}{2}\sqrt{B_1^2 - 4C_1}, \tag{2.54}$$

or

$$a^2 + aB_1 = -C_1. \tag{2.55}$$

With Eqs. (2.2) or (2.51), the foregoing equation gives

$$a = b_{1,2} = -\frac{B}{3A} \pm \frac{1}{3A}\sqrt{B^2 - 3AC}. \tag{2.56}$$

If $B^2 > 3AC$, such a intersected point of $x^* = a$ and $x^* = b_1$ or $x^* = b_2$ exists. If $B^2 < 3AC$, such an intersected point does not exist. Such the intersection point is for the two fixed-point switching, which is called the monotonic *saddle-node* bifurcation. The stability and bifurcation diagrams for $a_0 > 0$ and $a_0 < 0$ are presented in Fig. 2.2(i) and (ii), respectively. Three fixed-points are intersected at a point with $\Delta_{ij} = (a_i - a_j)^2 = 0$ and $a_1 = a_2 = a_3 = -\frac{B}{3A}$, and in the vicinity of the intersection point, $\Delta_{ij} = (a_i - a_j)^2 > 0$ for $i, j = 1, 2, 3$ but $i \neq j$. The intersection points for $a_0 > 0$ and $a_0 < 0$ are called the monotonic source and sink bifurcations of the third

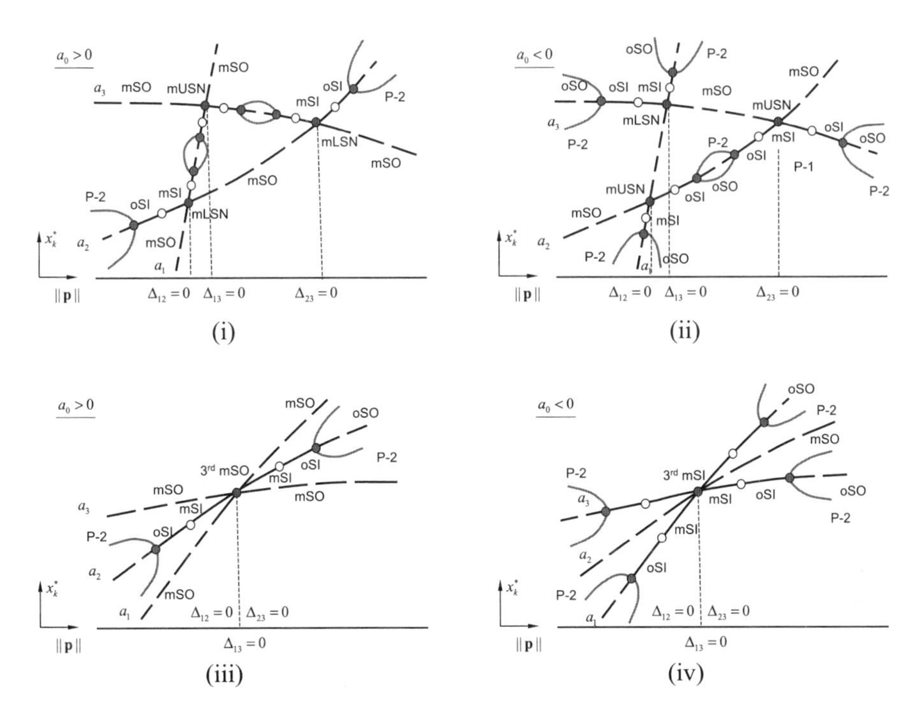

Fig. 2.2 Stability and bifurcation of fixed-points switching in the 1-dimensional, cubic nonlinear discrete system. For two fixed-points switching: (i) $a_0 > 0$, (ii) $a_0 < 0$. For three fixed-point switching: (iii) 3rd order monotonic source bifurcation ($a_0 > 0$), (iv) 3rd order monotonic sink bifurcation ($a_0 < 0$). mLSN: mono *lower-saddle-node*, mUSN: monotonic *upper-saddle-node*. Stable and unstable fixed-points are represented by solid and dashed curves, respectively. The bifurcation points are marked by circular symbols. (mSO: monotonic source; mSI: monotonic sink; oSO: oscillatory source; oSI: oscillatory sink; mLS: monotonic lower-saddle; mUS: monotonic upper-saddle; oUS: oscillatory upper-saddle; oLS: oscillatory lower-saddle; iSI: invariant sink.) The period-2 fixed-points are presented on the period-1 bifurcation trees through red curves.

order, respectively. The corresponding stability and bifurcation diagrams for three fixed-points switching are presented in Fig. 2.2(iii) and (iv). The period-2 fixed-points are also presented as well through the period-doubling.

In the 1-dimensional cubic nonlinear, forward discrete system of Eq. (2.1), $x_k^2 + B_1 x_k + C_1 = 0$ gives two fixed-points of $x_k^* = b_1, b_2$ for $\Delta_1 = B_1^2 - 4C_1 > 0$. One of the two fixed-points has one intersection with $x_k^* = a$ and there are three different fixed-points for $a = a_2 \in (\min\{b_1, b_2\}, \max\{b_1, b_2\})$. For this case, the intersection point occurs at $a = \min\{b_1, b_2\}$ for $\mathbf{p}_1 \in \partial\Omega_{23}$ or $a = \max\{b_1, b_2\}$ for $\mathbf{p}_2 \in \partial\Omega_{23}$. The bifurcation of fixed-point occurs at the double repeated fixed-point at $\Delta_1 = B_1^2 - 4C_1 = 0$ for $\mathbf{p}_0 \in \partial\Omega_{12}$. Such a bifurcation is a monotonic *lower-* or *upper-saddle-node* bifurcation. For $a = -\frac{1}{2}B_1$ with $\Delta_1 = B_1^2 - 4C_1 = 0$, three fixed-points are repeated with three multiplicity. The intersected point of $a = -\frac{1}{2}B_1$ with Eq. (2.51) gives

$$a = -\frac{1}{2}(a + \frac{B}{A}). \tag{2.57}$$

Thus

$$a = -\frac{B}{3A}. \tag{2.58}$$

Such a bifurcation at the intersection point is also a third-order monotonic *source* or *sink* bifurcation. The bifurcation diagrams for six cases of three fixed-points with one intersection are presented in Fig. 2.3(i)–(vi) and the stability and bifurcations are listed in Table 2.2. The corresponding period-2 fixed points are sketched as well.

The 1-dimensional cubic nonlinear system is expressed by a factor of $(x_k - a)$ and a quadratic form of $a_0(x_k^2 + B_1 x_k + C_1)$ as in Eq. (2.1). For $\Delta_1 = B_1^2 - 4C_1 > 0$, $x_k^2 + B_1 x_k + C_1 = 0$ gives two fixed-points of $x_k^* = b_1, b_2$. The two fixed-points *do not* have any intersections with $x_k^* = a$. For $\Delta_1 = B_1^2 - 4C_1 = 0$, there are two parameters of $\mathbf{p}_1 \in \partial\Omega_{12}$ and $\mathbf{p}_2 \in \partial\Omega_{12}$, and the two double repeated fixed-points are at $x_k^*(\mathbf{p}_i) = -\frac{1}{2}B_1(\mathbf{p}_i)$ $(i = 1, 2)$. With the two repeated fixed-points, the two fixed-points of $x_k^* = b_1, b_2$ formed a closed path in the bifurcation diagram. The bifurcation points of fixed-point occur at the two double repeated fixed-points of $\Delta_1 = B_1^2 - 4C_1 = 0$ for $\mathbf{p}_i \in \partial\Omega_{12}$ $(x_k^* = b_1, b_2)$. Such a bifurcation at the intersection point is also a monotonic *lower or upper-saddle-node* bifurcation. The stable and unstable fixed-points varying with the vector parameter are also represented by solid and dashed curves, respectively. The bifurcation diagrams for four cases of three fixed-points are presented in Fig. 2.4(i)–(vi), and the stability and bifurcations are summarized in Table 2.3.

If the two repeated fixed-points have two intersections with $x_k^* = a(\mathbf{p}_i)$ $(i = 1, 2)$, i.e., $a(\mathbf{p}_i) = -\frac{1}{2}B_1(\mathbf{p}_i)$. The two triple repeated fixed-points at $x_k^* = a(\mathbf{p}_i)$ $(i = 1, 2)$ are the third-order, monotonic sink or source bifurcations. The stability and bifurcation diagrams of fixed-points are formed by the fixed-point of $x_k^* = a(\mathbf{p})$ and the closed loop of fixed-points of $x_k^* = b_1, b_2$, as shown in Fig. 2.4(v) and (vi) for $a_0 > 0$ and $a_0 < 0$, respectively. The stability and bifurcations are also summarized in Table 2.3, and the corresponding period-2 fixed-points are sketched as well.

In the 1-dimensional cubic nonlinear system in Eq. (2.1), $x_k^2 + B_1 x_k + C_1 = 0$ for $\Delta_1 = B_1^2 - 4C_1 > 0$ gives two fixed-points of $x_k^* = b_1, b_2$, which have an intersection with $x_k^* = a$. The intersected point are at $a = b_1$ or $a = b_2$ with Eq. (2.55). The double repeated fixed-point requires $\Delta_1 = B_1^2 - 4C_1 = 0$ and the two fixed-points of $x_k^* = a, b_1$ under $\Delta_1 = B_1^2 - 4C_1 > 0$ and $x_k^* = b_2$ for $\Delta_1 = B_1^2 - 4C_1 < 0$. Similarly, the two fixed-points of $x_k^* = a, b_2$ under $\Delta_1 = B_1^2 - 4C_1 > 0$ and $x_k^* = b_2$ for $\Delta_1 = B_1^2 - 4C_1 < 0$. Such a bifurcation for two fixed-points appearance and vanishing is called a monotonic *lower* or *upper-saddle*-node appearing bifurcation. The stable and unstable fixed-points varying with the vector

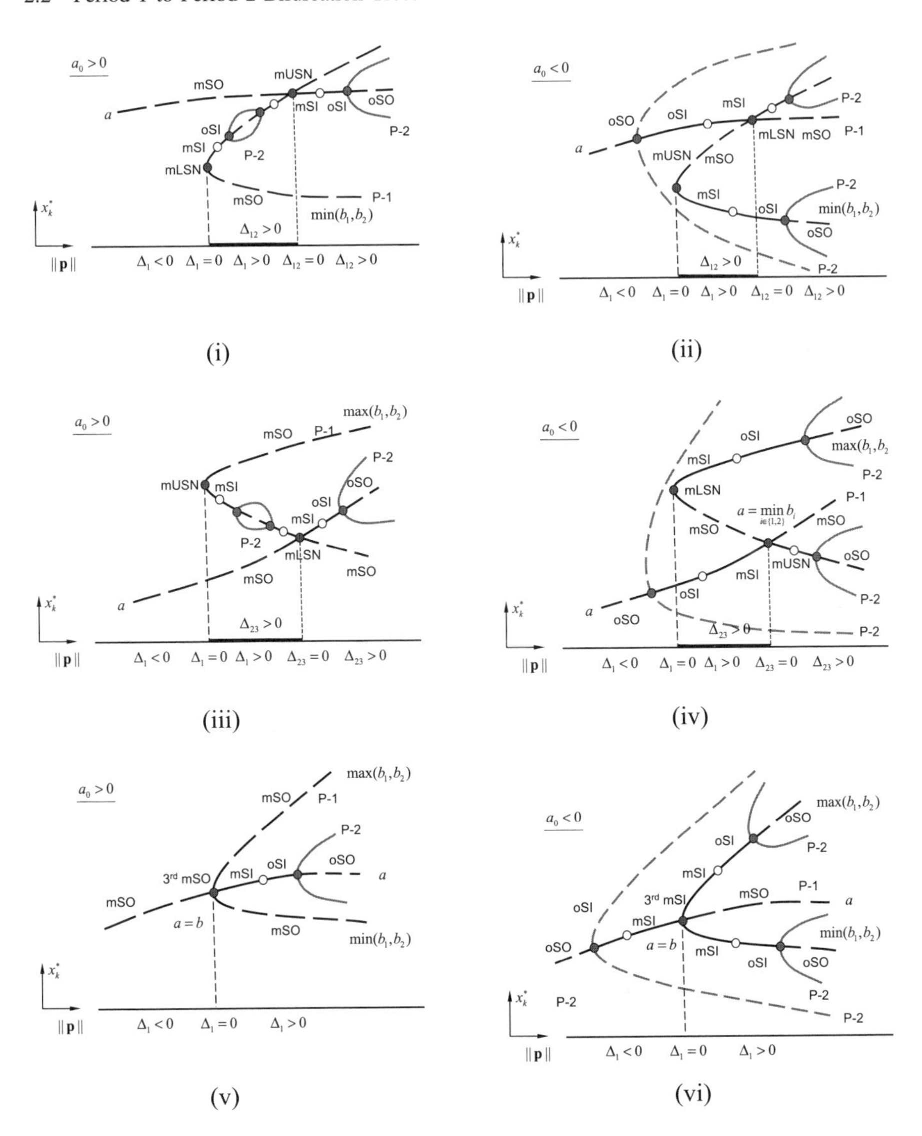

Fig. 2.3 Stability and bifurcation of fixed-points in the 1-dimensional, cubic nonlinear discrete system before the oscillatory saddle bifurcation for sink branches: (i) the mLSN ($\Delta_1 = 0$) and mUSN ($a = \max\{b_1, b_2\}$) bifurcations ($a_0 > 0$), (ii) the mUSN ($\Delta_1 = 0$) and mLSN ($a = \max\{b_1, b_2\}$) bifurcations ($a_0 < 0$). (iii) the mUSN ($\Delta_1 = 0$) and mLSN ($a = \min\{b_1, b_2\}$) bifurcations ($a_0 > 0$), (iv) the mLSN ($\Delta_1 = 0$) and mUSN ($a = \min\{b_1, b_2\}$) bifurcations ($a_0 < 0$), (v) the third order mSO bifurcation ($\Delta_1 = 0$ and $a = b$) ($a_0 > 0$), (vi) the third-order mSI bifurcation ($\Delta_1 = 0$ and $a = b$) ($a_0 < 0$). The bifurcation points are marked by circular symbols. (mSO: monotonic source; mSI: monotonic sink; oSO: oscillatory source; oSI: oscillatory sink; mLS: monotonic lower-saddle; mUS: monotonic upper-saddle; oUS: oscillatory upper-saddle; oLS: oscillatory lower-saddle; iSI: invariant sink.) The period-2 fixed-points are presented on the period-1 bifurcation trees through the red curves.

Table 2.2 Stability and bifurcation of a 1-dimensional cubic nonlinear system ($\dot{x} = a_0(x-a)[x^2 + B_1(\mathbf{p})x + C_1(\mathbf{p})]$, $a \in (\min\{b_1, b_2\}, \max\{b_1, b_2\})$)

Figure 2.3	a_1	a_2	a_3	B-I	B-II	B-III
(i) $a_0 > 0$	mSO	mSI-oSO-mSI	mSO	2nd mLSN	2nd mUSN	$a = \max\{b_1, b_2\}$
(ii) $a_0 < 0$	mSI-oSO-mSI	mSO	mSI-oSO-mSI	2nd mUSN	2nd mLSN	$a = \max\{b_1, b_2\}$
(iii) $a_0 > 0$	mSO	mSI-oSO-mSI	mSO	2nd mUSN	2nd mUSN	$a = \min\{b_1, b_2\}$
(iv) $a_0 < 0$	mSI-oSO-mSI	mSO	mSI-oSO-mSI	2nd mLSN	2nd mUSN	$a = \min\{b_1, b_2\}$
(v) $a_0 > 0$	mSO	mSI-oSO-mSI	mSO	$\Delta_1 = 0$	$a = -\frac{1}{2}B_1$	3rd order SO
(vi) $a_0 < 0$	mSI-oSO-mSI	mSO	mSI-oSO-mSI	$\Delta_1 = 0$	$a = -\frac{1}{2}B_1$	3rd order SI

Notice that $b_{1,2} = -\frac{1}{2}(B_1 \pm \sqrt{\Delta_1})$, $\Delta_1 = B_1^2 - 4C_1$. Bifurcation-I (B-I): $\Delta_1 = 0$. Bifurcation-II (B-II): $a = \max\{b_1, b_2\}$ or $a = \min\{b_1, b_2\}$. Bifurcation-III (B-III): $\Delta_1 = 0$ and $a = -\frac{B_1}{2A_1}$. mLSN: *monotonic lower-saddle-node*, mUSN: *monotonic upper-saddle-node*. mSO: monotonic source, mSI: monotonic sink. mSI-oSO-mSI: from monotonic sink to oscillatory source then to monotonic sink.

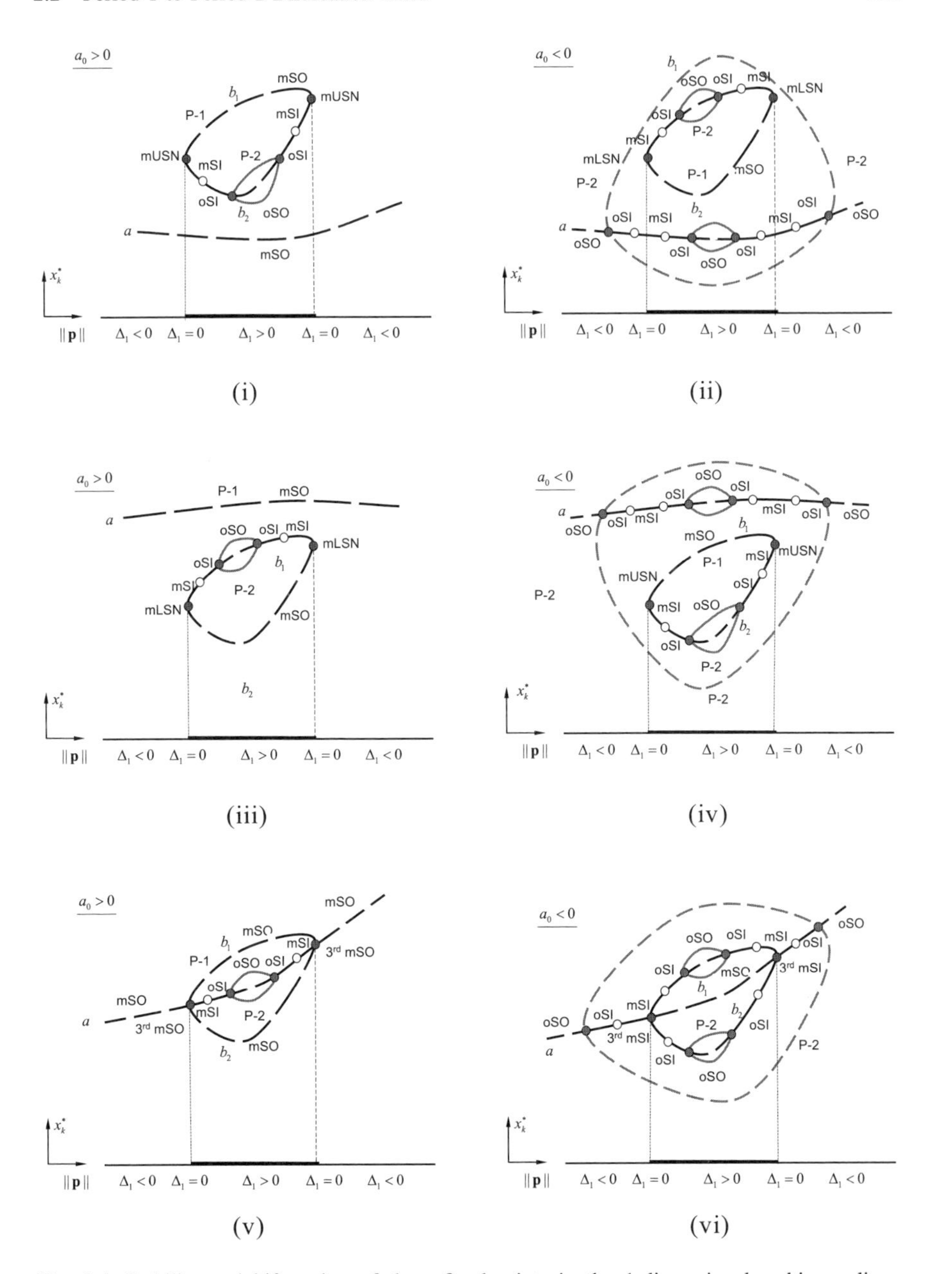

Fig. 2.4 Stability and bifurcation of three fixed-points in the 1-dimensional, cubic nonlinear discrete system: For $a < \{b_1, b_2\}$: (i) two mUSN bifurcations ($a_0 > 0$), (ii) two mLSN bifurcations ($a_0 < 0$). For $a > \{b_1, b_2\}$: (iii) two mLSN bifurcations ($a_0 > 0$), (iv) two mUSN bifurcations ($a_0 < 0$). (v) two 3rd order mSO bifurcations ($a_0 > 0$), (vi) two 3rd mSI bifurcations ($a_0 < 0$).Stable and unstable fixed-points are represented by solid and dashed curves, respectively. The bifurcation points are marked by circular symbols. (mLSN: monotonic *lower-saddle-node;* mUSN: monotonic *upper-saddle-node*; (mSO: monotonic source; mSI: monotonic sink; oSO: oscillatory source; oSI: oscillatory sink; mLS: monotonic lower-saddle; mUS: monotonic upper-saddle.) The period-2 fixed-points are presented on the period-1 bifurcation trees through red curves.

Table 2.3 Stability and bifurcation of a 1-dimensional cubic nonlinear system ($x_{k+1} = x_k + a_0 (x_k - a)[x_k^2 + B_1(\mathbf{p})x_k + C_1(\mathbf{p})]$, $a \in (\min\{b_1, b_2\}, \max\{b_1, b_2\})$)

Figure 2.4	a	b_1	b_2	B-I	B-II	B-III
(i) $a_0 > 0$	mSO	mSI-oSO-mSI	mSO	mUSN	mUSN	–
(ii) $a_0 < 0$	mSI-oSO	mSO	mSI-oSO	mLSN	mLSN	–
(iii) $a_0 > 0$	mSO	mSI-oSO-mSI	mSO	mLSN	mLSN	–
(iv) $a_0 < 0$	mSI-oSO	mSO	mSI-oSO	mUSN	mUSN	–
(v) $a_0 < 0$	mSO	mSI-oSO-mSI	mSO	–	–	3rd mSO
(iv) $a_0 < 0$	mSI-oSO-mSI	mSO	mSI-oSO-mSI	–	–	3rd mSI

Notice that $b_{1,2} = -\frac{1}{2}(B_1 \pm \sqrt{\Delta_1})$, $\Delta_1 = B_1^2 - 4C_1$. Bifurcation-I (B-I): $\Delta_1 = 0$. Bifurcation-II (B-II): $a = \max\{b_1, b_2\}$. Bifurcation-III(B-III): $a = \min\{b_1, b_2\}$. (mLSN: monotonic *lower-saddle-node;* mUSN: monotonic *upper-saddle-node*; mSI-oSO: monotonic sink to oscillatory source via oscillatory sink.)

parameter are also represented by solid and dashed curves, respectively. The bifurcation diagrams for four cases of three fixed-points are presented in Fig. 2.5 (i)–(vi). If the double repeated fixed-point has an intersection with $x_k^* = a(\mathbf{p}_0) = -\frac{1}{2}B_1 = -\frac{B}{3A}$. The two triple repeated fixed-points of $x_k^* = a(\mathbf{p}_0)$ for $a_0 > 0$ and $a_0 < 0$ are the third-order monotonic sink and source bifurcations, respectively. The stability and bifurcation diagrams of fixed-points are shown in Fig. 2.5(v) and (vi). The period-2 fixed-points are sketched through the red curves.

2.3 Higher-Order Period-1 Switching Bifurcations

Consider a 1-dimensional, cubic nonlinear, forward discrete system with a double repeated fixed-point and one simple fixed-point.

(i) For $b < a$, the discrete system is

$$x_{k+1} = x_k + a_0(\mathbf{p})(x_k - b(\mathbf{p}))^2(x_k - a(\mathbf{p})), \tag{2.59}$$

For such a system, if $a_0 > 0$, the double repeated fixed-point of $x_k^* = b$ is a monotonic *lower-saddle*, which is unstable, and the simple fixed-point of $x_k^* = a$ is a monotonic source, which is monotonically unstable. If $a_0 < 0$, the double repeated fixed-point of $x_k^* = b$ is a monotonic *upper-saddle*, which is monotonically unstable, and the simple fixed-point of $x_k^* = a$ is from a monotonic sink to the oscillatory source. Such a fixed-point is from monotonically stable to oscillatorilly unstable.

(ii) For $b > a$, the 1-dimensional cubic nonlinear, forward discrete system is

$$x_{k+1} = x_k + a_0(\mathbf{p})(x_k - a(\mathbf{p}))(x_k - b(\mathbf{p}))^2. \tag{2.60}$$

For such a system, if $a_0 > 0$, the double-repeated fixed-point of $x_k^* = b$ is a monotonic *upper-saddle*, which is monotonically unstable, and the simple

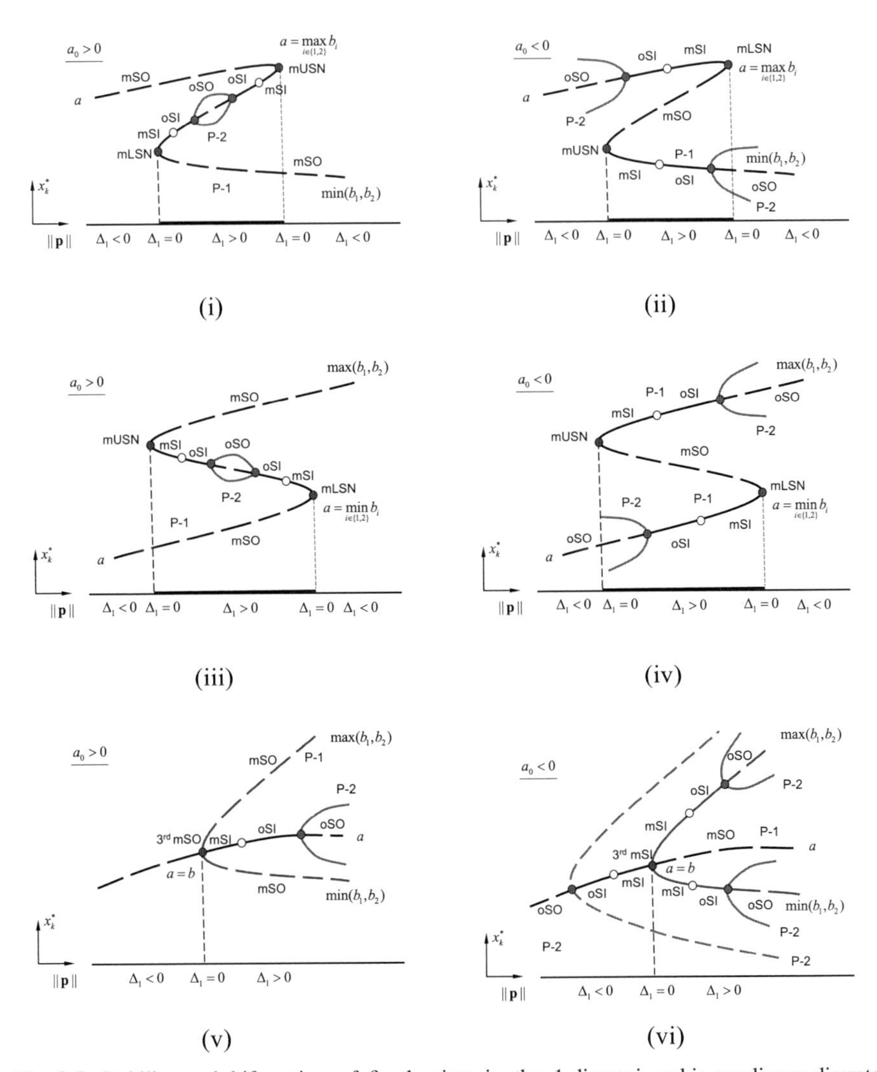

Fig. 2.5 Stability and bifurcation of fixed-points in the 1-dimensi, cubic nonlinear discrete system: (i) the LSN ($\Delta_1 = 0$) and mUSN ($a = \max\{b_1, b_2\}$) bifurcations ($a_0 > 0$), (ii) the mUSN ($\Delta_1 = 0$) and mLSN ($a = \max\{b_1, b_2\}$) bifurcations ($a_0 < 0$). (iii) the USN ($\Delta_1 = 0$) and mLSN ($a = \min\{b_1, b_2\}$) bifurcations ($a_0 > 0$), (iv) the mLSN ($\Delta_1 = 0$) and USN ($a = \min\{b_1, b_2\}$) bifurcations ($a_0 < 0$). (v) the third order mSO bifurcation ($\Delta_1 = 0$ and $a = b$) ($a_0 > 0$), (vi) the third order mSI bifurcation ($\Delta_1 = 0$ and $a = b$) ($a_0 < 0$). Stable and unstable fixed-points are represented by solid and dashed curves, respectively. The bifurcation points are marked by circular symbols. (mLSN: monotonic lower-saddle-node; mUSN: monotonic upper-saddle-node; mSO: monotonic source; mSI: monotonic sink; oSO: oscillatory source; oSI: oscillatory sink; mLS: monotonic lower-saddle; mUS: monotonic upper-saddle.) The period-2 fixed-points are presented on the period-1 bifurcation trees through red curves.

fixed-point of $x_k^* = a$ is a monotonic source, which is monotonically unstable. If $a_0 < 0$, the double fixed-point of $x_k^* = b$ is a monotonic *lower-saddle*, which is monotonically unstable, and the simple fixed-point of $x_k^* = a$ is from a monotonic sink to oscillatory source. Such a fixed-point is from monotonically stable to oscillatorilly unstable.

(iii) For $b = a$, the discrete system on the boundary is

$$x_{k+1} = x_k + a_0(\mathbf{p})(x_k - b(\mathbf{p}))^3. \tag{2.61}$$

For such a system, if $a_0 > 0$, the triple fixed-point of $x_k^* = b$ with the third multiplicity is a source switching bifurcation of the third-order for the (mUS: mSO) to (mSO:mLS) fixed-point. If $a_0 < 0$, the triple fixed-point of $x_k^* = b$ with the third multiplicity is a sink switching bifurcation of the third- order for the (mLS:mSI-oSO) to (mSO:mUS) fixed-point. With parameter changes, the bifurcation diagram for the cubic nonlinear system is presented in Fig. 2.6. The acronyms mLSN, mUSN, mSI-oSO, and mSO are for monotonic *lower-saddle-node*, monotonic *upper-saddle-node*, monotonic sink to oscillatory source, and monotonic source, respectively. Stable and unstable fixed-points are represented by solid and dashed curves, respectively. The bifurcation point is marked by a circular symbol.

To illustrate the stability and bifurcation of fixed-point with singularity in a 1-dimensional, cubic nonlinear system, the fixed-point of $x_{k+1} = x_k + a_0(x_k - a_1)^3$ is presented in Fig. 2.7. The third order monotonic sink and source of fixed-points of $x^* = a_1$ with the third order multiplicity are stable and unstable, respectively. The stable and unstable fixed-points are depicted by solid and dashed curves, respectively. At $a_0 = 0$, the fixed-points with the 3rd order monotonic sink and source are switched, which is marked by a circular symbol.

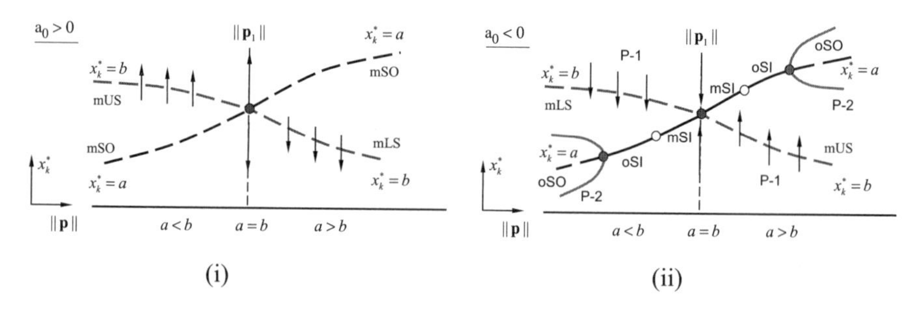

Fig. 2.6 Stability and bifurcation of a triple fixed-point with a simple fixed-point in a 1-dimensional, cubic nonlinear discrete system: (i) a 3rd order source switching bifurcation for (mUS:mSO) to (mSO:mLS) switching ($a_0 > 0$), (ii) a 3rd order monotonic sink switching bifurcation ($a_0 < 0$) for (mLS:mSI-oSO) to (mSI-oSO:mUS) switching. Stable and unstable fixed-points are represented by solid and dashed curves, respectively. The period-2 fixed-points are depicted through red curves.

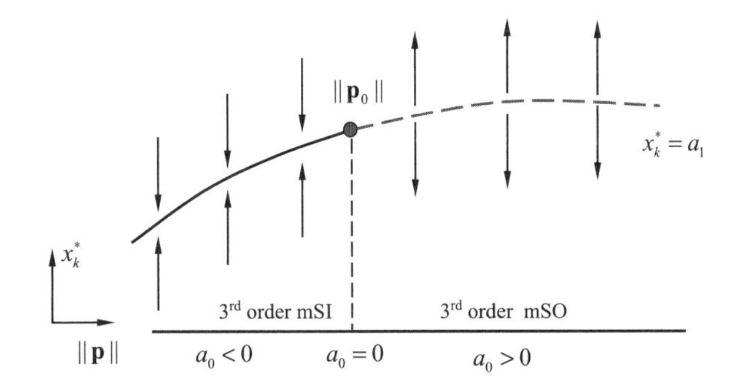

Fig. 2.7 Stability of a triple fixed-point in the 1-dimensional, cubic nonlinear, forward discrete system: Stable and unstable fixed-points are represented by solid and dashed curves, respectively. The stability switching is labelled by a circular symbol.

2.4 Direct Cubic Polynomial Discrete Systems

For the 1-dimensional, cubic nonlinear, forward discrete systems, the stability and bifurcation of fixed-points can be described through an alternative way as follows.

Definition 2.2 Consider a 1-dimensional, cubic nonlinear discrete system

$$\begin{aligned} x_{k+1} &= x_k + A(\mathbf{p})x_k^3 + B(\mathbf{p})x_k^2 + C(\mathbf{p})x_k + D(\mathbf{p}) \\ &\equiv a_0(\mathbf{p})[(x_k + \frac{B}{3A})^3 + p(\mathbf{p})(x_k + \frac{B}{3A}) + q(\mathbf{p})] \end{aligned} \tag{2.62}$$

where four scalar constants $A(\mathbf{p}) \neq 0, B(\mathbf{p}), C(\mathbf{p})$ and $D(\mathbf{p})$ satisfy

$$\begin{aligned} & A = a_0, p = \frac{C}{A} - \frac{B^2}{3A^2}, q = \frac{D}{A} - \frac{BC}{3A^2} + \frac{2B^3}{27A^3} \\ & \mathbf{p} = (p_1, p_2, \ldots, p_m)^{\mathrm{T}}. \end{aligned} \tag{2.63}$$

(i) If

$$\Delta = \frac{q^2}{4} + \frac{p^3}{27} > 0, \tag{2.64}$$

the cubic nonlinear discrete system has one fixed-point as

$$x_k^* = a \equiv (-\frac{q}{2} + \sqrt{\Delta})^{1/3} + (-\frac{q}{2} - \sqrt{\Delta})^{1/3} - \frac{B}{3A} \tag{2.65}$$

and the corresponding standard form is

$$\begin{aligned} x_{k+1} &= x_k + a_0(x_k - a)[(x_k + \frac{1}{2}B_1)^2 + \frac{1}{4}(-\Delta_1)] \\ &= a_0(x_k - a)[x_k^2 + B_1 x_k + C_1)] \end{aligned} \tag{2.66}$$

where

$$A = a_0, B = (-a + B_1)a_0, C = (-aB_1 + C_1)a_0, D = -aa_0C_1. \quad (2.67)$$

(ii) If

$$\Delta = \frac{q^2}{4} + \frac{p^3}{27} < 0 \quad (2.68)$$

the cubic nonlinear discrete system has three fixed-points as

$$\begin{aligned}
&x_k^* = a = (-\tfrac{q}{2} + \sqrt{\Delta})^{1/3} + (-\tfrac{q}{2} - \sqrt{\Delta})^{1/3} - \frac{B}{3A}, \\
&x_k^* = b_1 = \omega(-\tfrac{q}{2} + \sqrt{\Delta})^{1/3} + \omega^2(-\tfrac{q}{2} - \sqrt{\Delta})^{1/3} - \frac{B}{3A}, \\
&x_k^* = b_2 = \omega^2(-\tfrac{q}{2} + \sqrt{\Delta})^{1/3} + \omega(-\tfrac{q}{2} - \sqrt{\Delta})^{1/3} - \frac{B}{3A}; \\
&\omega = \frac{-1 + \mathbf{i}\sqrt{3}}{2}, \omega^2 = \frac{-1 - \mathbf{i}\sqrt{3}}{2}, \mathbf{i} = \sqrt{-1}.
\end{aligned} \quad (2.69)$$

The corresponding standard form is

$$x_{k+1} = x_k + a_0(x_k - a_1)(x_k - a_2)(x_k - a_3) \quad (2.70)$$

where

$$\begin{aligned}
&a_1 = \min(b_1, b_2, a),\ a_3 = \max(b_1, b_2, a), \\
&a_2 \in \{b_1, b_2, a\},\ a_2 \neq \{a_1, a_3\}
\end{aligned} \quad (2.71)$$

(iii) If

$$\Delta = \frac{q^2}{4} + \frac{p^3}{27} = 0, \frac{q^2}{4} = -\frac{p^3}{27} \neq 0 \quad (2.72)$$

the 1-dimensional, cubic nonlinear discrete system has the a double repeated fixed-point with the second multiplicity plus a simple fixed-point as

$$\begin{aligned}
&x_k^* = a = 2(-\tfrac{q}{2})^{1/3} - \frac{B}{3A}, \\
&x_k^* = b_1 = b,\ x_k^* = b_2 = b; \\
&b = \omega(-\tfrac{q}{2})^{1/3} + \omega^2(-\tfrac{q}{2})^{1/3} - \frac{B}{3A}, \\
&\omega = \frac{-1 + \mathbf{i}\sqrt{3}}{2}, \omega^2 = \frac{-1 - \mathbf{i}\sqrt{3}}{2}, \mathbf{i} = \sqrt{-1}.
\end{aligned} \quad (2.73)$$

(iii$_1$) The corresponding standard form for $a<b$ is

$$x_{k+1} = x_k + a_0(x_k - a)(x_k - b)^2. \tag{2.74}$$

Such a discrete flow with the fixed-point of $x_k^* = b$ is called

- a monotonic *upper-saddle* discrete flow of the second-order at a point $\mathbf{p} = \mathbf{p}_1 \in \partial\Omega_{12}$ for $a_0 > 0$;
- a monotonic *lower*-saddle discrete flow of the second-order at a point $\mathbf{p} = \mathbf{p}_1 \in \partial\Omega_{12}$ for $a_0 < 0$.

The bifurcation of fixed-point at $x_k^* = b$ for two different fixed-points appearance or vanishing is called

- a monotonic *upper-saddle-node* appearing bifurcation of the second-order at a point $\mathbf{p} = \mathbf{p}_1 \in \partial\Omega_{12}$ for $a_0 > 0$;
- a monotonic *lower-saddle-node* appearing bifurcation of the second-order at a point $\mathbf{p} = \mathbf{p}_1 \in \partial\Omega_{12}$ for $a_0 < 0$.

The corresponding upper (or lower)-saddle-node appearing bifurcation condition is

$$\Delta = \frac{q^2}{4} + \frac{p^3}{27} = 0, \frac{q^2}{4} = -\frac{p^3}{27} \neq 0, a<b. \tag{2.75}$$

(iii$_2$) The corresponding standard form for $a > b$ is

$$x_{k+1} = x_k + a_0(x_k - b)^2(x_k - a). \tag{2.76}$$

Such a discrete flow with the fixed-point of $x_k^* = b$ is called

- a monotonic *lower-saddle* discrete flow of the second-order at a point $\mathbf{p} = \mathbf{p}_1 \in \partial\Omega_{12}$ for $a_0 > 0$;
- a monotonic *upper-saddle* discrete flow of the second-order at a point $\mathbf{p} = \mathbf{p}_1 \in \partial\Omega_{12}$ for $a_0 < 0$.

The bifurcation of fixed-point at $x^* = b$ for two fixed-points appearing or vanishing is called

- a monotonic *lower-saddle-node* appearing bifurcation of the second-order at a point $\mathbf{p} = \mathbf{p}_1 \in \partial\Omega_{12}$ for $a_0 > 0$;
- a monotonic *upper-saddle-node* appearing bifurcation of the second order at a point $\mathbf{p} = \mathbf{p}_1 \in \partial\Omega_{12}$ for $a_0 < 0$.

and the corresponding lower (or upper)-saddle-node appearing bifurcation condition is

$$\Delta = \tfrac{1}{4}q^2 + \tfrac{1}{27}p^3 = 0, \tfrac{1}{4}q^2 = -\tfrac{1}{27}p^3 \neq 0, a > b. \tag{2.77}$$

(iv) If

$$\Delta = \frac{1}{4}q^2 + \frac{1}{27}p^3 = 0,\ \frac{1}{4}q^2 = -\frac{1}{27}p^3 = 0, \tag{2.78}$$

the 1-dimensional discrete system has a triple fixed-point as

$$x_k^* = a = -\frac{B}{3A}, x_k^* = b_1 = -\frac{B}{3A},\ x_k^* = b_2 = -\frac{B}{3A}. \tag{2.79}$$

The corresponding standard form is

$$x_{k+1} = x_k + a_0(x_k - a)^3. \tag{2.80}$$

Such a discrete flow at the fixed-point of $x_k^* = a$ is called

- a monotonic source discrete flow of the third-order for $a_0 > 0$;
- a monotonic sink discrete flow of the third-order for $a_0 < 0$.

The bifurcation of fixed-point at $x_k^* = -\frac{B}{3A}$ for one fixed-point to three fixed-points is called

- a monotonic source switching bifurcation of the third-order at $\mathbf{p} = \mathbf{p}_1 \in \partial\Omega_{12}$ for $a_0 > 0$;
- a monotonic sink switching bifurcation of the third-order at $\mathbf{p} = \mathbf{p}_1 \in \partial\Omega_{12}$ for $a_0 > 0$.

The corresponding switching bifurcation condition of the third-order source and sink is

$$\Delta = \frac{1}{4}q^2 + \frac{1}{27}p^3 = 0,\ \frac{1}{4}q^2 = -\frac{1}{27}p^3 = 0. \tag{2.81}$$

From the afore-described stability and bifurcation of the 1-dimensional, cubic nonlinear forward discrete systems, the stability and bifurcations of fixed-point in Eq. (2.62) are similar to Theorem 2.1.

The 1-dimensional cubic nonlinear, forward discrete system has the following four cases:

(i) One real solution of simple fixed-point of $x_k^* = a$ requires $\Delta = (\frac{q}{2})^2 + (\frac{p}{3})^3 > 0$ for Eq. (2.62), equivalent to $\Delta_1 = B_1^2 - 4C_1 < 0$ for Eq. (2.1).

$$Aa^3 + Ba^2 + Ca + D = 0. \tag{2.82}$$

(ii) Three different solutions of simple fixed-points of $x_k^* = a, b_1, b_2$ require $\Delta < 0$ for Eq. (2.62), equivalent to $\Delta_1 = B_1^2 - 4C_1 > 0$ for Eq. (2.1).

$$Aa^3 + Ba^2 + Ca + D = 0 \text{ and } b_{1,2} = \frac{1}{2}(B_1 \pm \sqrt{B_1^2 - 4C_1}) \tag{2.83}$$

(iii) The double repeated fixed-point requires $\Delta = 0$ and $(\frac{q}{2})^2 = -(\frac{p}{3})^3 \neq 0$ for Eq. (2.62), equivalent to $\Delta_1 = B_1^2 - 4C_1 = 0$ or $a = b_{1,2}$ for Eq. (2.1).

$$a = b_{1,2} = -\frac{B}{3A} \pm \frac{1}{3A}\sqrt{B^2 - 3AC} \text{ with } B^2 > 3AC. \tag{2.84}$$

(iv) The triple repeated fixed-point requires $\Delta = 0$ and $(\frac{q}{2})^2 = -(\frac{p}{3})^3 = 0$ for Eq. (2.62), equivalent to $\Delta_1 = B_1^2 - 4C_1 = 0$ and $a = b_{1,2}$ for Eq. (2.1).

$$a = b_{1,2} = -\frac{B}{3A} \text{ with } B^2 = 3AC. \tag{2.85}$$

2.5 Forward Cubic Discrete Systems

In this section, the analytical bifurcation scenario will be discussed. The period-doubling bifurcation scenario will be discussed first through nonlinear renormalization techniques, and the bifurcation scenario based on the saddle-node bifurcation will be discussed, which is independent of period-1 fixed-points.

2.5.1 Period-Doubled Cubic Discrete Systems

After the period-doubling bifurcation of a period-1 fixed-point, the period-doubled fixed-points in the cubic discrete system can be obtained. Consider the period-doubling solutions for a forward cubic nonlinear discrete system first.

Theorem 2.2 *Consider a 1-dimensional cubic nonlinear discrete system as*

$$\begin{aligned} x_{k+1} &= x_k + A(\mathbf{p})x_k^3 + B(\mathbf{p})x_k^2 + C(\mathbf{p})x_k + D(\mathbf{p}) \\ &= x_k + a_0(\mathbf{p})(x_k - a(\mathbf{p}))[x_k^2 + B_1(\mathbf{p})x_k + C_1(\mathbf{p})] \end{aligned} \tag{2.86}$$

where four scalar constants $A(\mathbf{p}) \neq 0, B(\mathbf{p}), C(\mathbf{p})$ and $D(\mathbf{p})$ *are determined by*

$$\begin{aligned} &A = a_0, B = (-a + B_1)a_0, C = (-aB_1 + C_1)a_0, D = -aa_0C_1, \\ &\mathbf{p} = (p_1, p_2, \ldots, p_m)^{\mathrm{T}}. \end{aligned} \tag{2.87}$$

Under

$$\Delta_1 = B_1^2 - 4C_1 < 0, \tag{2.88}$$

the standard form of such a 1-dimensional forward discrete system is

$$x_{k+1} = x_k + a_0(x_k - a)(x_k^2 + B_1 x_k + C_1). \tag{2.89}$$

Under

$$\Delta_1 = B_1^2 - 4C_1 > 0, \tag{2.90}$$

the standard form of such a 1-dimensional forward discrete system is

$$x_{k+1} = x_k + a_0(x_k - a_1)(x_k - a_2)(x_k - a_3). \tag{2.91}$$

Thus, a general standard form of such a 1-dimensional cubic discrete system is

$$\begin{aligned} x_{k+1} &= x_k + f(x_k, \mathbf{p}) = x_k + Ax_k^3 + Bx_k^2 + Cx_k + D \\ &\equiv x_k + a_0(x_k - a)[x_k^2 + B_1 x_k + C_1] \\ &= x_k + a_0 \textstyle\prod_{i=1}^{3}(x_k - a_i^{(1)}) \end{aligned} \tag{2.92}$$

where

$$\begin{aligned} &a_0 = A(\mathbf{p}),\ b_{1,2}^{(1)} = -\tfrac{1}{2}(B_1(\mathbf{p}) \mp \sqrt{\Delta^{(1)}}) \text{ for } \Delta^{(1)} > 0; \\ &a_1^{(1)} = \min\{a, b_1, b_2\}, a_3^{(1)} = \max\{a, b_1, b_2\}, a_1^{(2)} \in \{a, b_1, b_2\} \neq \{a_1, a_3\}, \\ &b_{1,2}^{(1)} = -\tfrac{1}{2}(B_1(\mathbf{p}) \mp \mathbf{i}\sqrt{|\Delta^{(1)}|}),\ \mathbf{i} = \sqrt{-1} \text{ for } \Delta^{(1)} < 0 \\ &a_1^{(1)} = a, a_2^{(1)} = b_1^{(1)}, a_3^{(1)} = b_2^{(1)}. \end{aligned} \tag{2.93}$$

(i) *Consider a forward period-2 discrete system of Eq. (2.86) as*

$$\begin{aligned} x_{k+2} &= x_k + [a_0 \textstyle\prod_{i_1=1}^{3}(x_k - a_{i_1}^{(1)})]\{1 + \prod_{i_1=1}^{3}[1 + a_0 \prod_{i_2=1, i_2 \neq i_1}^{3}(x_k - a_{i_2}^{(1)})]\} \\ &= x_k + [a_0 \textstyle\prod_{i_1=1}^{3}(x_k - a_{i_1}^{(1)})][a_0^3 \prod_{i_2=1}^{(3^2-3)/2}(x_k^2 + B_{i_2}^{(2)} x_k + C_{i_2}^{(2)})] \\ &= x_k + [a_0 \textstyle\prod_{j_1=1}^{3}(x_k - a_{i_1}^{(1)})][a_0^3 \prod_{j_2=1}^{(3^2-3)/2}(x_k - b_{j_2,1}^{(2)})(x_k - b_{j_2,2}^{(2)})] \\ &= x_k + a_0^{1+3} \textstyle\prod_{i=1}^{3^2}(x_k - a_i^{(2)}) \end{aligned} \tag{2.94}$$

where

$$
\begin{aligned}
&b_{i,1}^{(2)} = -\tfrac{1}{2}(B_i^{(2)} + \sqrt{\Delta_i^{(2)}}), b_{i,2}^{(2)} = -\tfrac{1}{2}(B_i^{(2)} - \sqrt{\Delta_i^{(2)}}),\\
&\Delta_i^{(2)} = (B_i^{(2)})^2 - 4C_i^{(2)} \geq 0,\ i \in I_q^{(2^0)};\\
&I_q^{(2^0)} = \{l_{(q-1)\times 2^0 \times m_1 + 1}, l_{(q-1)\times 2^0 \times m_1 + 2}, \cdots, l_{q\times 2^0 \times m_1}\},\\
&m_1 \in \{1,2\}, q \in \{1,2\};\\
&b_{i,1}^{(2)} = -\tfrac{1}{2}(B_i^{(2)} + \mathbf{i}\sqrt{|\Delta^{(2)}|}), b_{i,2}^{(2)} = -\tfrac{1}{2}(B_i^{(2)} - \mathbf{i}\sqrt{|\Delta_i^{(2)}|}),\\
&\mathbf{i} = \sqrt{-1}, \Delta_i^{(2)} = (B_i^{(2)})^2 - 4C_i^{(2)} < 0,
\end{aligned}
\tag{2.95}
$$

with fixed-points

$$
\begin{aligned}
&x_{k+2}^* = x_k^* = a_i^{(2)}, (i = 1, 2, \ldots, 3^2)\\
&\cup_{i=1}^{3^2}\{a_i^{(2)}\} = \text{sort}\{\cup_{j_1=1}^{3}\{a_{j_1}^{(1)}\}, \cup_{j_2=1}^{3}\{b_{j_2,1}^{(2)}, b_{j_2,2}^{(2)}\}\}\\
&\text{with } a_i^{(2)} < a_{i+1}^{(2)}.
\end{aligned}
\tag{2.96}
$$

(ii) *For a fixed-point of* $x_{k+1}^* = x_k^* = a_{i_1}^{(1)}$ $(i_1 \in \{1, 2, 3\})$, if

$$
\frac{dx_{k+1}}{dx_k}\Big|_{x_k^* = a_{i_1}^{(1)}} = 1 + a_0 \textstyle\prod_{i_2=1, i_2 \neq i_1}^{3}(a_{i_1}^{(1)} - a_{i_2}^{(1)}) = -1,
\tag{2.97}
$$

with

- *an oscillatory upper-saddle-node bifurcation* $(d^2x_{k+1}/dx_k^2|_{x_k^*=a_{i_1}^{(1)}} > 0)$,
- *an oscillatory lower-saddle-node bifurcation* $(d^2x_{k+1}/dx_k^2|_{x_k^*=a_{i_1}^{(1)}} < 0)$,
- *a third-order oscillatory sink bifurcation* $(d^3x_{k+1}/dx_k^3|_{x_k^*=a_{i_1}^{(1)}} > 0)$,
- *a third-order oscillatory source bifurcation* $(d^3x_{k+1}/dx_k^3|_{x_k^*=a_{i_1}^{(1)}} < 0)$,

then the following relations satisfy

$$
a_{i_1}^{(1)} = -\tfrac{1}{2}B_{i_1}^{(2)}, \Delta_{i_1}^{(2)} = (B_{i_1}^{(2)})^2 - 4C_{i_1}^{(2)} = 0,
\tag{2.98}
$$

and there is a period-2 discrete system of the cubic discrete system in Eq. (2.86), as

$$
x_{k+2} = x_k + a_0^4 \textstyle\prod_{i_1 \in I_q^{(2^0)}}(x_k - a_{i_1}^{(1)})^3 \prod_{i_2=1}^{3^2}(x_k - a_{i_2}^{(2)})^{(1-\delta(i_1,i_2))}
\tag{2.99}
$$

for $i_1 \in \{1,2,3\}$, $i_1 \neq i_2$ *with*

$$\frac{dx_{k+2}}{dx_k}|_{x_k^*=a_{i_1}^{(1)}} = 1, \frac{d^2x_{k+2}}{dx_k^2}|_{x_k^*=a_{i_1}^{(1)}} = 0; \tag{2.100}$$

- x_{k+2} *at* $x_k^* = a_{i_1}^{(1)}$ *is a monotonic sink of the third-orderif*

$$\begin{aligned}\frac{d^3x_{k+2}}{dx_k^3}|_{x_k^*=a_{i_1}^{(1)}} &= 6a_0^4 \Pi_{i_2 \in I_q^{(2^0)}, i_2 \neq i_1} (x_k - a_{i_1}^{(1)})^3 \\ &\times \Pi_{i_3=1}^{3^2} (a_{i_1}^{(1)} - a_{i_3}^{(2)})^{(1-\delta(i_2,i_3))} < 0,\end{aligned} \tag{2.101}$$

  and the corresponding bifurcations is a third-order monotonic sink bifurcation for the period-2 discrete system;

- x_{k+2} *at* $x_k^* = a_{i_1}^{(1)}$ *is a monotonic source of the third-order if*

$$\begin{aligned}\frac{d^3x_{k+2}}{dx_k^3}|_{x_k^*=a_{i_1}^{(1)}} &= 6a_0^4 \prod_{i_2 \in I_q^{(2^0)}, i_2 \neq i_1} (x_k - a_{i_1}^{(1)})^3 \\ &\times \prod_{i_3=1}^{3^2} (a_{i_1}^{(1)} - a_{i_3}^{(2)})^{(1-\delta(i_2,i_3))} > 0,\end{aligned} \tag{2.102}$$

  and the corresponding bifurcations is a third-order monotonic source bifurcation for the period-2 discrete system.

(ii$_1$) The period-2 fixed-points are trivial and unstable if

$$x_{k+2}^* = x_k^* = a_{i_1}^{(1)} \text{ for } i_1 = 1,2,3. \tag{2.103}$$

(ii$_2$) The period-2 fixed-points are non-trivial and stable if

$$x_{k+2}^* = x_k^* = b_{i_1,1}^{(2)}, b_{i_1,2}^{(2)} \text{ for } i_1 = 1,2,3. \tag{2.104}$$

Proof The proof is straightforward through the simple algebraic manipulation. Consider

$$Ax_k^2 + Bx_k + C = 0.$$

Under

$$\Delta = B^2 - 4AC \geq 0,$$

we have

$$a_0 = A(\mathbf{p}),\ b_{1,2} = \frac{-B(\mathbf{p}) \pm \sqrt{\Delta}}{2A(\mathbf{p})} \text{ with } b_1 < b_2.$$

Under

$$\Delta = B^2 - 4AC < 0,$$

we have

$$a_0 = A(\mathbf{p}),\ b_{1,2} = \frac{-B(\mathbf{p}) \pm \mathbf{i}\sqrt{|\Delta|}}{2A(\mathbf{p})},\ \mathbf{i} = \sqrt{-1}.$$

Thus, we have

$$Ax_k^2 + Bx_k + C = (x_k - a_{i_2})(x_k - a_{i_3}).$$

Therefore,

$$x_{k+1} = x_k + a_0(x_k - a_1)(x_k - a_2)(x_k - a_3).$$

where

$$\begin{aligned} &\{a_1, a_2, a_3\} = \mathrm{sort}\{a, b_1, b_2\} \text{ for real } b_{1,2}, \\ &a_1 = a,\ a_{2,3} = b_{1,2}; \end{aligned}$$

and

$$\begin{aligned} \frac{dx_{k+1}}{dx_k}|_{x_k^*=a_i} &= 1 + a_0 \textstyle\prod_{i_1=1, i_1 \neq i}^{3} (a_i - a_{i_1}), \\ \frac{d^2x_{k+1}}{dx_k^2}|_{x_k^*=a_i} &= a_0 \textstyle\sum_{i_1=1, i_1 \neq i}^{3} \prod_{i_2=1, i_2 \neq i_1}^{3} (a_i - a_{i_2}), \\ \frac{d^3x_{k+1}}{dx_k^3}|_{x_k^*=a_i} &= a_0. \end{aligned}$$

For real $x_{k+1}^* = x_k^* = a_i$ ($i \in \{1, 2, 3\}$, if

$$\frac{dx_{k+1}}{dx_k}|_{x_k^*=a_i} = 1 + a_0 \textstyle\prod_{i_1=1, i_1 \neq i}^{3} (a_i - a_{i_1}) = -1,$$

with

- an oscillatory upper-saddle-saddle bifurcation ($d^2x_{k+1}/dx_k^2|_{x_k^*=a_i} > 0$),
- an oscillatory lower-saddle-node bifurcation ($d^2x_{k+1}/dx_k^2|_{x_k^*=a_i} < 0$),
- a third-order oscillatory sink bifurcation ($d^2x_{k+1}/dx_k^2|_{x_k^*=a_i} = 0$ and $d^3x_{k+1}/\ dx_k^3|_{x_k^*=a_i} > 0$),
- a third-order oscillatory source bifurcation ($d^2x_{k+1}/dx_k^2|_{x_k^*=a_i} = 0$ and $d^3x_{k+1}/\ dx_k^3|_{x_k^*=a_i} < 0$),

then period-2 fixed-points exists for the cubic discrete system. The period-2 discrete system of the cubic discrete system is

$$\begin{aligned}x_{k+2} &= x_k + [a_0 \prod_{i_1=1}^{3}(x_k - a_{i_1}^{(1)})]\{1 + \prod_{i_1=1}^{3}[1 + a_0 \prod_{i_2=1, i_2\neq i_1}^{3}(x_k - a_{i_2}^{(1)})]\}\\ &= x_k + [a_0 \prod_{i_1=1}^{3}(x_k - a_{i_1}^{(1)})][(a_0)^3 \prod_{i_2=1}^{(3^2-3)/2}(x_k^2 + B_{i_2}^{(2)}x_k + C_{i_2}^{(2)})]\end{aligned}$$

If

$$x_k^2 + B_i^{(2)}x_k + C_i^{(2)} = 0,$$

we have

$$\begin{aligned}&b_{i,1}^{(2)} = -\frac{1}{2}(B_i^{(2)} + \sqrt{\Delta_i^{(2)}}), b_{i,2}^{(2)} = -\frac{1}{2}(B_i^{(2)} - \sqrt{\Delta_i^{(2)}}),\\ &\Delta_i^{(2)} = (B_i^{(2)})^2 - 4C_i^{(2)} \geq 0,\ i \in I_q^{(2^0)};\\ &I_q^{(2^0)} = \{l_{(q-1)\times 2^0\times m_1+1}, l_{(q-1)\times 2^0\times m_1+2}, \cdots, l_{q\times 2^0\times m_1}\},\\ &m_1 \in \{1,2\}, q \in \{1,2\};\\ &b_{i,1}^{(2)} = -\frac{1}{2}(B_i^{(2)} + \mathbf{i}\sqrt{|\Delta^{(2)}|}), b_{i,2}^{(2)} = -\frac{1}{2}(B_i^{(2)} - \mathbf{i}\sqrt{|\Delta_i^{(2)}|}),\\ &\mathbf{i} = \sqrt{-1}, \Delta_i^{(2)} = (B_i^{(2)})^2 - 4C_i^{(2)} < 0,\end{aligned}$$

Thus

$$x_k^2 + B_i^{(2)}x_k + C_i^{(2)} = \prod_{j=1}^{2}(x_k - b_{i,j}^{(2)}),$$

and

$$\begin{aligned}x_{k+2} &= x_k + a_0^4 \prod_{i_1=1}^{3}(x_k - a_{i_1}^{(1)}) \prod_{i_2=1}^{(3^2-3)/2}(x_k - b_{i_2,1}^{(2)})(x_k - b_{i_2,2}^{(2)})\\ &= x_k + a_0^4 \prod_{i=1}^{3^{2^1}}(x_k - a_i^{(2)})\end{aligned}$$

where

$$\{a_i^{(2)}, i = 1, 2, \ldots, 3^2\} = \text{sort}\{\cup_{i_1=1}^{3}\{a_{i_1}^{(1)}\}, \cup_{i_2=1}^{3}\{b_{i_2,1}^{(2)}, b_{i_2,2}^{(2)}\}\}.$$

For the period-1 cubic discrete systems,

$$x_{k+1} = x_k + a_0 \prod_{i=1}^{3}(x_k - a_i^{(1)}).$$

The fixed-point of $x_k^* = a_{i_1}^{(1)}$ ($i_1 \in \{1, 2, 3\}$) is monotonically unstable due to

$$\frac{dx_{k+1}}{dx_k}\Big|_{x_k^*=a_{i_1}^{(1)}} = 1 + a_0 \prod_{i_2=1, i_2\neq i_1}^{3}(a_{i_1}^{(1)} - a_{i_2}^{(1)}) \in (1, \infty),$$

and the fixed-point of $x_k^* = a_{i_1}^{(1)}$ ($i_1 \in \{1,2,3\}$) is from monotonically stable to oscillatorilly unstable due to

$$\frac{dx_{k+1}}{dx_k}|_{x_k^*=a_{i_1}^{(1)}} = 1 + a_0 \prod_{i_2=1, i_2\neq i_1}^{3} (a_{i_1}^{(1)} - a_{i_2}^{(1)}) \in (-\infty, 1).$$

Under

$$\frac{dx_{k+1}}{dx_k}|_{x_k^*=a_{i_1}^{(1)}} = 1 + a_0 \prod_{i_2=1, i_2\neq i_1}^{3} (a_{i_1}^{(1)} - a_{i_2}^{(1)}) = -1$$
$$\Rightarrow 2 + a_0 \prod_{i_2=1, i_2\neq i_1}^{3} (a_{i_1}^{(1)} - a_{i_2}^{(1)}) = 0,$$

- there is a flipped discrete system of the oscillatory upper-saddle of the second order if

$$\frac{d^2x_{k+1}}{dx_k^2}|_{x_k^*=a_{i_1}^{(1)}} = a_0 \sum_{i_2=1, i_2\neq i_1}^{3} \prod_{i_3=1, i_3\neq i_1, i_2}^{3} (a_{i_1}^{(1)} - a_{i_3}^{(1)}) > 0;$$

- there is a flipped discrete system of the oscillatory lower-saddle of the second order if

$$\frac{d^2x_{k+1}}{dx_k^2}|_{x_k^*=a_{i_1}^{(1)}} = a_0 \sum_{i_2=1, i_2\neq i_1}^{3} \prod_{i_3=1, i_3\neq i_1, i_2}^{3} (a_{i_1}^{(1)} - a_{i_3}^{(1)}) < 0;$$

- there is a flipped discrete system of the oscillatory sink of the third order if

$$\frac{d^2x_{k+1}}{dx_k^2}|_{x_k^*=a_{i_1}^{(1)}} = 0, \frac{d^3x_{k+1}}{dx_k^3}|_{x_k^*=a_{i_1}^{(1)}} = a_0 > 0;$$

- there is a flipped discrete system of the oscillatory source of the third order if

$$\frac{d^2x_{k+1}}{dx_k^2}|_{x_k^*=a_{i_1}^{(1)}} = 0, \frac{d^3x_{k+1}}{dx_k^3}|_{x_k^*=a_{i_1}^{(1)}} = a_0 < 0.$$

The corresponding standard form of the period-2 discrete system becomes

$$x_{k+2} = x_k + a_0^4 \prod_{i_1 \in I_q^{(2^0)}} (x_k - a_{i_1}^{(1)})^3 \prod_{i_2=1}^{3^2} (x_k - a_{i_2}^{(2)})^{(1-\delta(i_1,i_2))}$$

with

$$\frac{dx_{k+2}}{dx_k}|_{x_k^*=a_i^{(1)}} = 1, \frac{d^2x_{k+2}}{dx_k^2}|_{x_k^*=a_i^{(1)}} = 0,$$

- $x_k^* = a_i^{(1)}$ for the period-2 discrete system is a monotonic sink of the third-order if

$$\frac{d^3 x_{k+2}}{dx_k^3}\Big|_{x_k^*=a_{i_1}^{(1)}} = 6a_0^4 \prod_{i_2\in I_q^{(2^0)},i_2\neq i_1}(x_k - a_{i_1}^{(1)})^3 \prod_{i_3=1}^{3^2}(a_{i_1}^{(1)} - a_{i_3}^{(2)})^{(1-\delta(i_2,i_3))} < 0,$$

- $x_k^* = a_i^{(1)}$ for the period-2 discrete system is a monotonic source of the third-order if

$$\frac{d^3 x_{k+2}}{dx_k^3}\Big|_{x_k^*=a_{i_1}^{(1)}} = 6a_0^4 \prod_{i_2\in I_q^{(2^0)},i_2\neq i_1}(x_k - a_{i_1}^{(1)})^3 \prod_{i_3=1}^{3^2}(a_{i_1}^{(1)} - a_{i_3}^{(2)})^{(1-\delta(i_2,i_3))} > 0.$$

This theorem is proved. ■

2.5.2 *Period-Doubling Renormalization*

The generalized cases of period-doublization of cubic discrete systems are presented through the following theorem. The analytical period-doubling trees can be developed for cubic discrete systems.

Theorem 2.3 *Consider a 1-dimensional cubic nonlinear discrete system as*

$$\begin{aligned} x_{k+1} &= x_k + Ax_k^3 + Bx_k^2 + Cx_k + D \\ &= x_k + a_0 \prod_{i=1}^{3}(x_k - a_i^{(1)}). \end{aligned} \tag{2.105}$$

(i) *After l-times period-doubling bifurcations, a period- 2^l discrete system ($l = 1, 2, \ldots$) for the cubic discrete system in Eq. (2.105) is given through the nonlinear renormalization as*

$$\begin{aligned} x_{k+2^l} &= x_k + [a_0^{(2^{l-1})} \prod_{i_1=1}^{3^{2^{l-1}}}(x_k - a_{i_1}^{(2^{l-1})})] \\ &\quad \times \{1 + \prod_{i_1=1}^{3^{2^{l-1}}}[1 + a_0^{(2^{l-1})} \prod_{i_2=1,i_2\neq i_1}^{3^{2^{l-1}}}(x_k - a_{i_2}^{(2^{l-1})})]\} \\ &= x_k + [a_0^{(2^{l-1})} \prod_{i_1=1}^{3^{2^{l-1}}}(x_k - a_{i_1}^{(2^{l-1})})] \\ &\quad \times [(a_0^{(2^{l-1})})^{3^{2^{l-1}}} \prod_{j_1=1}^{(3^{2^l}-3^{2^{l-1}})/2}(x_k^2 + B_{j_2}^{(2^l)}x_k + C_{j_2}^{(2^l)})] \\ &= x_k + [a_0^{(2^{l-1})} \prod_{i_1=1,}^{3^{2^{l-1}}}(x_k - a_{i_1}^{(2^{l-1})})] \\ &\quad \times [(a_0^{(2^{l-1})})^{3^{2^{l-1}}} \prod_{i_2=1}^{(3^{2^l}-3^{2^{l-1}})/2}(x_k - b_{i_2,1}^{(2^l)})(x_k - b_{i_2,2}^{(2^l)})] \\ &= x_k + (a_0^{(2^{l-1})})^{1+3^{2^{l-1}}} \prod_{i=1}^{3^{2^l}}(x_k - a_i^{(2^l)}) \\ &= x_k + a_0^{(2^l)} \prod_{i=1}^{3^{2^l}}(x_k - a_i^{(2^l)}) \end{aligned} \tag{2.106}$$

with

$$\begin{aligned}
&\frac{dx_{k+2^l}}{dx_k} = 1 + a_0^{(2^l)} \sum_{i_1=1}^{3^{2^l}} \prod_{i_2=1, i_2 \neq i_1}^{3^{2^l}} (x_k - a_{i_2}^{(2^l)}),\\
&\frac{d^2 x_{k+2^l}}{dx_k^2} = a_0^{(2^l)} \sum_{i_1=1}^{3^{2^l}} \sum_{i_2=1, i_2 \neq i_1}^{3^{2^l}} \prod_{i_3=1, i_3 \neq i_1, i_2}^{3^{2^l}} (x_k - a_{i_3}^{(2^l)}),\\
&\vdots\\
&\frac{d^r x_{k+2^l}}{dx_k^r} = a_0^{(2^l)} \sum_{i_1=1}^{3^{2^l}} \cdots \sum_{i_r=1, i_3 \neq i_1, i_2 \ldots i_{r-1}}^{3^{2^l}} \prod_{i_{r+1}=1, i_3 \neq i_1, i_2 \ldots, i_r}^{3^{2^l}} (x_k - a_{i_{r+1}}^{(2^l)})\\
&\text{for } r \leq 3^{2^l}.
\end{aligned} \tag{2.107}$$

where

$$\begin{aligned}
&a_0^{(2)} = (a_0)^{1+3}, a_0^{(2^l)} = (a_0^{(2^{l-1})})^{1+3^{2^{l-1}}}, l = 1, 2, 3, \cdots;\\
&\cup_{i=1}^{3^{2^l}} \{a_i^{(2^l)}\} = \text{sort}\{\cup_{i_1=1}^{3^{2^{l-1}}} \{a_{i_1}^{(2^l)}\}, \cup_{i_2=1}^{M_2} \{b_{i_2,1}^{(2^l)}, b_{i_2,2}^{(2^l)}\}\}\ , a_i^{(2^l)} \leq a_{i+1}^{(2^l)};\\
&b_{i,1}^{(2^l)} = -\tfrac{1}{2}(B_i^{(2^l)} + \sqrt{\Delta_i^{(2^l)}}), b_{i,2}^{(2^l)} = -\tfrac{1}{2}(B_i^{(2^l)} - \sqrt{\Delta_i^{(2^l)}}),\\
&\Delta_i^{(2^l)} = (B_i^{(2^l)})^2 - 4C_i^{(2^l)} \geq 0 \text{ for } i \in \cup_{q_1=1}^{N_1} I_{q_1}^{(2^{l-1})} \cup \cup_{q_2=1}^{N_2} I_{q_2}^{(2^l)}\\
&I_{q_1}^{(2^{l-1})} = \{l_{(q_1-1)\times 2^{l-1}\times m_1+1}, l_{(q_1-1)\times 2^{l-1}\times m_1+2}, \cdots, l_{q_1\times 2^{l-1}\times m_1}\}\\
&\qquad \subseteq \{1, 2, \cdots, M_1\} \cup \{\varnothing\},\\
&\text{for } q_1 \in \{1, 2, \cdots, N_1\},\ M_1 = N_1 \times 2^{l-1} \text{ with } m_1 \in \{1, 2\};\\
&I_{q_2}^{(2^l)} = \{l_{(q_2-1)\times 2^l\times m_1+1}, l_{(q_2-1)\times 2^l\times m_1+2}, \cdots, l_{q_2\times 2^l\times m_1}\}\\
&\qquad \subseteq \{M_1+1, M_1+2, \cdots, M_2\} \cup \{\varnothing\},\\
&\text{for } q_2 \in \{1, 2, \cdots, N_2\},\ M_2 = (3^{2^l} - 3^{2^{l-1}})/2;\\
&b_{i,1}^{(2^l)} = -\tfrac{1}{2}(B_i^{(2^l)} + \mathbf{i}\sqrt{|\Delta_i^{(2^l)}|}),\ b_{i,2}^{(2^l)} = -\tfrac{1}{2}(B_i^{(2^l)} - \mathbf{i}\sqrt{|\Delta_i^{(2^l)}|}),\\
&\Delta_i^{(2^l)} = (B_i^{(2^l)})^2 - 4C_i^{(2^l)} < 0,\ \mathbf{i} = \sqrt{-1},\\
&i \in J^{(2^l)} = \{l_{N_2\times 2^l\times m_1+1}, l_{N_2\times 2^l\times m_1+2}, \cdots, l_{M_2}\}\\
&\qquad \subset \{M_1+1, M_1+2, \cdots, M_2\} \cup \{\varnothing\}
\end{aligned} \tag{2.108}$$

with fixed-points

$$\begin{aligned}
&x_{k+2^l}^* = x_k^* = a_i^{(2^l)}, (i = 1, 2, \ldots, 3^{2^l})\\
&\cup_{i=1}^{3^{2^l}} \{a_i^{(2^l)}\} = \text{sort}\{\cup_{i_1=1}^{3^{2^{l-1}}} \{a_{i_1}^{(2^{l-1})}\}, \cup_{i_2=1}^{M_2} \{b_{i_2,1}^{(2^l)}, b_{i_2,2}^{(2^l)}\}\}\\
&\text{with } a_i^{(2^l)} < a_{i+1}^{(2^l)}.
\end{aligned} \tag{2.109}$$

(ii) *For a fixed-point of* $x^*_{k+2^{l-1}} = x^*_k = a^{(2^{l-1})}_{i_1}$ ($i_1 \in I^{(2^{l-1})}_q \subset \{1,2,\ldots,3^{(2^{l-1})}\}$), *if*

$$\begin{aligned}
&\frac{dx_{k+2^{l-1}}}{dx_k}\Big|_{x^*_k=a^{(2^{l-1})}_{i_1}} = 1 + a^{(2^{l-1})}_0 \prod\nolimits_{i_2=1, i_2\neq j_1}^{3^{2^{l-1}}} (a^{(2^{l-1})}_{i_1} - a^{(2^{l-1})}_{i_1}) = -1,\\
&\frac{d^s x_{k+2^{l-1}}}{dx^s_k}\Big|_{x^*_k=a^{(2^{l-1})}_{i_1}} = 0,\ \text{for}\ s = 2,\ldots, r-1;\\
&\frac{d^r x_{k+2^{l-1}}}{dx^r_k}\Big|_{x^*_k=a^{(2^{l-1})}_{i_1}} \neq 0\ \text{for}\ 1<r\leq 3^{2^{l-1}},
\end{aligned} \tag{2.110}$$

with

- *a* r^{th}*-order oscillatory sink for* $d^r x_{k+2^{l-1}}/dx^r_k|_{x^*_k=a^{(2^{l-1})}_{i_1}} > 0$ *and* $r = 2l_1 + 1$;
- *a* r^{th}*-order oscillatory source for* $d^r x_{k+2^{l-1}}/dx^r_k|_{x^*_k=a^{(2^{l-1})}_{i_1}} < 0$ and $r = 2l_1 + 1$;
- *a* r^{th}*-order oscillatory upper-saddle for* $d^r x_{k+2^{l-1}}/dx^r_k|_{x^*_k=a^{(2^{l-1})}_{i_1}} > 0$ *and* $r = 2l_1$;
- *a* r^{th}*-order oscillatory lower-saddle for* $d^r x_{k+2^{l-1}}/dx^r_k|_{x^*_k=a^{(2^{l-1})}_{i_1}} < 0$ *and* $r = 2l_1$;

then there is a period- 2^l *fixed-point discrete system*

$$x_{k+2^l} = x_k + a^{(2^l)}_0 \prod\nolimits_{i_1\in I^{(2^{l-1})}_q} (x_k - a^{(2^{l-1})}_{i_1})^3 \prod\nolimits_{j_2=1}^{3^{2^l}} (x_k - a^{(2^l)}_{j_2})^{(1-\delta(i_1,j_2))} \tag{2.111}$$

where

$$\delta(i_1,j_2) = 1\ \text{if}\ a^{(2^l)}_{j_2} = a^{(2^{l-1})}_{i_1},\ \delta(i_1,j_2) = 0\ \text{if}\ a^{(2^l)}_{j_2} \neq a^{(2^{l-1})}_{i_1} \tag{2.112}$$

and

$$\frac{dx_{k+2^l}}{dx_k}\Big|_{x^*_k=a^{(2^{l-1})}_{i_1}} = 1,\ \frac{d^2 x_{k+2^l}}{dx^2_k}\Big|_{x^*_k=a^{(2^{l-1})}_{i_1}} = 0. \tag{2.113}$$

- x_{k+2^l} *at* $x^*_k = a^{(2^{l-1})}_{i_1}$ *is a monotonic sink of the third-order if*

$$\begin{aligned}
\frac{d^3 x_{k+2^l}}{dx^3_k}\Big|_{x^*_k=a^{(2^{l-1})}_{i_1}} &= 6a^{(2^l)}_0 \prod\nolimits_{i_2\in I^{(2^{l-1})}_q, i_2\neq i_1} (a^{(2^{l-1})}_{i_1} - a^{(2^{l-1})}_{i_2})^3\\
&\quad\times \prod\nolimits_{j_2=1}^{3^{2^l}} (a^{(2^{l-1})}_{i_1} - a^{(2^l)}_{j_2})^{(1-\delta(i_2,j_2))} < 0
\end{aligned} \tag{2.114}$$

$(i_1 \in I^{(2^{l-1})}_q, q \in \{1,2,\ldots,N_1\})$,

and such a bifurcation at $x_k^* = a_{i_1}^{(2^{l-1})}$ *is a third-order monotonic sink bifurcation.*

- x_{k+2^l} *at* $x_k^* = a_{i_1}^{(2^{l-1})}$ *is a monotonic source of the third-order if*

$$\frac{d^3 x_{k+2^l}}{dx_k^3}\Big|_{x_k^*=a_{i_1}^{(2^{l-1})}} = 6a_0^{(2^l)} \prod\nolimits_{i_2 \in I_q^{(2^{l-1})}, i_2 \neq i_1} (a_{i_1}^{(2^{l-1})} - a_{i_2}^{(2^{l-1})})^3 \\ \times \prod\nolimits_{j_2=1}^{3^{2^l}} (a_{i_1}^{(2^{l-1})} - a_{j_2}^{(2^l)})^{(1-\delta(i_2,j_2))} > 0 \tag{2.115}$$

$(i_1 \in I_q^{(2^{l-1})}, q \in \{1, 2, \ldots, N_1\})$

and such a bifurcation at $x_k^* = a_{i_1}^{(2^{l-1})}$ *is a third-order monotonic source bifurcation.*

(ii$_1$) The period- 2^l *fixed-points are trivial if*

$$x_{k+2^l}^* = x_k^* = a_{i_1}^{(2^{l-1})} \text{ for } i_1 = 1, 2, \ldots, 3^{2^{l-1}}, \tag{2.116}$$

(ii$_2$) The period- 2^l *fixed-points are non-trivial if*

$$x_{k+2^l}^* = x_k^* = \cup_{j_1=1}^{M_2} \{b_{j_1,1}^{(2^l)}, b_{j_1,2}^{(2^l)}\}. \tag{2.117}$$

Such a period- 2^l *fixed-point is*

- *monotonically unstable if* $dx_{k+2^l}/dx_k|_{x_k^*=a_{i_1}^{(2^l)}} \in (1, \infty)$;
- *monotonically invariant if* $dx_{k+2^l}/dx_k|_{x_k^*=a_{i_1}^{(2^l)}} = 1$, *which is*
  - *a monotonic upper-saddle of the* $(2l_1)^{\text{th}}$ *order for* $d^{2l_1} x_{k+2^l}/dx_k^{2l_1}|_{x_k^*} > 0$;
  - *a monotonic lower-saddle the* $(2l_1)^{\text{th}}$ order for $d^{2l_1} x_{k+2^l}/dx_k^{2l_1}|_{x_k^*} < 0$;
  - *a monotonic source of the* $(2l_1+1)^{\text{th}}$ *order for* $d^{2l_1+1} x_{k+2^l}/dx_k^{2l_1+1}|_{x_k^*} > 0$;
  - *a monotonic sink the* $(2l_1+1)^{\text{th}}$ *order for* $d^{2l_1+1} x_{k+2^l}/dx_k^{2l_1+1}|_{x_k^*} < 0$;
- *monotonically stable if* $dx_{k+2^l}/dx_k|_{x_k^*=a_{i_1}^{(2^l)}} \in (0, 1)$;
- *invariantly zero-stable if* $dx_{k+2^l}/dx_k|_{x_k^*=a_{i_1}^{(2^l)}} = 0$;
- *oscillatorilly stable if* $dx_{k+2^l}/dx_k|_{x_k^*=a_{i_1}^{(2^l)}} \in (-1, 0)$;
- *flipped if* $dx_{k+2^l}/dx_k|_{x_k^*=a_{i_1}^{(2^{l-1})}} = -1$, *which is*
  - *an oscillatory upper-saddle of the* $(2l_1)^{\text{th}}$ *order for* $d^{2l_1} x_{k+2^l}/dx_k^{2l_1}|_{x_k^*} > 0$;
  - *an oscillatory lower-saddle of the* $(2l_1)^{\text{th}}$ *order for* $d^{2l_1} x_{k+2^l}/dx_k^{2l_1}|_{x_k^*} < 0$;

- *an oscillatory source of the* $(2l_1+1)^{\text{th}}$ *order for* $d^{2l_1+1}x_{k+2^l}/dx_k^{2l_1+1}|_{x_k^*}<0$;
- *an oscillatory sink the* $(2l_1+1)^{\text{th}}$ *order for* $d^{2l_1+1}x_{k+2^l}/dx_k^{2l_1+1}|_{x_k^*}>0$;

- *oscillatorilly unstable if* $dx_{k+2^l}/dx_k|_{x_k^*=a_{i_1}^{(2^l)}}\in(-\infty,-1)$.

Proof Through the nonlinear renormalization, this theorem can be proved.

(I) For a cubic discrete system, if the period-1 fixed-points exists, there is a following expression.

$$x_{k+1}=x_k+a_0\prod_{i_1=1}^{3}(x_k-a_{i_1}^{(1)})$$

For $x_{k+1}^*=x_k^*=a_{i_1}^{(1)}$ ($i_1\in\{1,2,3\}$, if

$$\frac{dx_{k+1}}{dx_k}|_{x_{k+1}^*=a_{i_1}^{(1)}}=1+a_0^{(1)}\prod_{i_2=1,i_2\neq i_1}^{3}(a_{i_1}^{(1)}-a_{i_2}^{(1)})=-1,$$
$$\frac{d^2x_{k+1}}{dx_k^2}|_{x_{k+1}^*=a_{i_1}^{(2)}}=a_0^{(1)}\sum_{i_2=1,i_2\neq i_1}^{3}\prod_{i_3=1,i_3\neq i_1,i_2}^{3}(a_{i_1}^{(1)}-a_{i_4}^{(1)})\neq 0$$

with

$$\frac{d^sx_{k+1}}{dx_k^s}|_{x_{k+1}^*=a_{i_1}^{(2)}}=0,\ \text{for } s=2,\ldots,\ r-1,$$
$$\frac{d^rx_{k+1}}{dx_k^r}|_{x_{k+1}^*=a_{i_1}^{(2)}}\neq 0 \text{ for } 1<r\leq 3,$$

then period-2 fixed-points exists for the cubic discrete system. Thus, consider the corresponding second iteration gives

$$x_{k+2}=x_{k+1}+a_0\prod_{i_1=1}^{3}(x_{k+1}-a_{i_1}^{(1)}).$$

The forward period-2 discrete system of the cubic discrete system is

$$\begin{aligned}x_{k+2}&=x_k+[a_0\prod_{i_1=1}^{3}(x_k-a_{i_1}^{(1)})]\{1+\prod_{i_1=1}^{3}[1+a_0\prod_{i_2=1,i_2\neq i_1}^{3}(x_k-a_{i_2}^{(1)})]\}\\&=x_k+[a_0\prod_{i_1=1}^{3}(x_k-a_{i_1}^{(1)})][a_0^3\prod_{i_2=1}^{3^2-3^1}(x_k^2+B_{i_2}^{(2)}x_k+C_{i_2}^{(2)})].\end{aligned}$$

If

$$x_k^2+B_{i_2}^{(2)}x_k+C_{i_2}^{(2)}=0,$$

we have

$$
\begin{aligned}
&b_{i,1}^{(2)} = -\frac{1}{2}(B_i^{(2)} + \sqrt{\Delta_i^{(2)}}), b_{i,2}^{(2)} = -\frac{1}{2}(B_i^{(2)} - \sqrt{\Delta_i^{(2)}}) \\
&\Delta^{(2)} = (B_i^{(2)})^2 - 4C_i^{(2)} \geq 0, i \in I_q^{(2^0)}; \\
&I_q^{(2^0)} = \{l_{(q-1)\times 2^0 \times m_1 + 1}, l_{(q-1)\times 2^0 \times m_1 + 2}, \cdots, l_{q\times 2^0 \times m_1}\}, \\
&m_1 \in \{1, 2\}, q \in \{1, 2\}.
\end{aligned}
$$

Thus

$$
x_k^2 + B_{i_2}^{(2)} x_k + C_{i_2}^{(2)} = (x_k - b_{i_2,1}^{(2)})(x_k - b_{i_2,2}^{(2)}),
$$

and

$$
\begin{aligned}
x_{k+2} &= x_k + [a_0 \textstyle\prod_{i_1=1}^{3}(x_k - a_{i_1}^{(1)})][a_0^3 \prod_{i_2=1}^{(3^2-3^1)/2}(x_k - b_{i,1}^{(2)})(x_k - b_{i,2}^{(2)})] \\
&= x_k + a_0^{1+3} \textstyle\prod_{i=1}^{3^2}(x_k - a_i^{(2)}).
\end{aligned}
$$

For a fixed-point of $x_{k+2}^* = x_k^* = a_{i_1}^{(2)}$ $(i_1 \in \{1, 2, \ldots, 3^2\})$, if

$$
\frac{dx_{k+2}}{dx_k}\Big|_{x_k^* = a_{i_1}^{(2)}} = 1 + a_0^{(2)} \textstyle\prod_{i_2=1, i_2 \neq i_1}^{3^2} (a_{i_1}^{(2)} - a_{i_2}^{(2)}) = -1,
$$

with

$$
\begin{aligned}
&\frac{d^s x_{k+2}}{dx_k^s}\Big|_{x_k^* = a_{i_1}^{(2)}} = 0, \text{ for } s = 2, \ldots, r-1, \\
&\frac{d^r x_{k+2}}{dx_k^r}\Big|_{x_k^* = a_{i_1}^{(2)}} \neq 0 \text{ for } 1 < r \leq 3^2,
\end{aligned}
$$

then, the forward period-2 discrete system of a cubic discrete system has a period-doubling bifurcation.

(II) Such a period-2 discrete system can be renormalized nonlinearly. For $k = k_1 + 2$, the previous period-2 discrete system becomes

$$
x_{k_1+2+2} = x_{k_1+2} + a_0^{(2^1)} \textstyle\prod_{i=1}^{3^2}(x_{k_1+2} - a_i^{(2^1)})
$$

Because k_1 is index for iteration, it can be replaced by k. Thus, an equivalent form for the foregoing equations becomes

$$
x_{k+2^2} = x_{k+2^1} + a_0^{(2^1)} \textstyle\prod_{i=1}^{3^{2^1}}(x_{k+2^1} - a_i^{(2^1)}).
$$

with

$$x_{k+2^1} = x_k + a_0^{(2^1)} \prod_{i=1}^{3^2} (x_k - a_i^{(2^1)})$$

x_{k+2^2} can be expressed as

$$\begin{aligned} x_{k+2^2} &= x_k + a_0^{(2^1)} \prod_{i_1=1}^{3^{2^1}} (x_k - a_{i_1}^{(2^1)}) \{1 + \prod_{i_1=1}^{3^{2^1}} [1 + a_0^{(2^1)} \prod_{i_2=1, i_2 \neq i_1}^{3^{2^1}} (x_k - a_{i_2}^{(2^1)})]\} \\ &= x_k + (a_0^{(2^1)})^{1+3^{2^1}} \prod_{i_1=1}^{3^{2^1}} (x_k - a_{i_1}^{(2^1)}) \prod_{i_2=1}^{(3^{2^2}-3^{2^1})/2} [(x_k^2 + B_{i_2}^{(2^2)} x_k + C_{i_2}^{(2^2)})]. \end{aligned}$$

If

$$x_k^2 + B_{i_2}^{(2^2)} x_k + C_{i_2}^{(2^2)} = 0$$

for $i_2 \in \cup_{q_1=1}^{N_1} I_{q_1}^{(2^{2-1})} \bigcup \cup_{q_2=1}^{N_2} I_{q_2}^{(2^2)}$ with

$$\begin{aligned} I_{q_1}^{(2^{2-1})} &= \{l_{(q_1-1)\times 2^{2-1}\times m_1+1}, l_{(q_1-1)\times 2^{2-1}\times m_1+2}, \cdots, l_{q_1\times 2^{2-1}\times m_1}\} \\ &\subseteq \{1, 2, \cdots, M_1\} \cup \{\varnothing\}, \\ &\text{for } q_1 \in \{1, 2, \cdots, N_1\},\ M_1 = N_1 \times 2^{2-1} \times m_1, m_1 \in \{1, 2\}; \\ I_{q_2}^{(2^2)} &= \{l_{(q_2-1)\times 2^2\times m_1+1}, l_{(q_2-1)\times 2^2\times m_1+2}, \cdots, l_{q_2\times 2^2\times m_1}\} \\ &\subseteq \{M_1+1, M_1+2, \cdots, M_2\} \cup \{\varnothing\}, \\ &\text{for } q_2 \in \{1, 2, \cdots, N_2\},\ M_2 = (3^{2^2} - 3^{2^{2-1}})/2; \end{aligned}$$

then we have

$$\begin{aligned} b_{i_2,1}^{(2^2)} &= -\frac{1}{2}(B_{i_2}^{(2^2)} + \sqrt{\Delta_{i_2}^{(2^2)}}), b_{i_2,2}^{(2^2)} = -\frac{1}{2}(B_{i_2}^{(2^2)} - \sqrt{\Delta_{i_2}^{(2^2)}}) \\ \Delta_{i_2}^{(2^2)} &= (B_{i_2}^{(2^2)})^2 - 4C_{i_2}^{(2^2)} \geq 0 \end{aligned}$$

Thus

$$x_k^2 + B_{i_2}^{(2^2)} x_k + C_{i_2}^{(2^2)} = (x_k - b_{i_2,1}^{(2^2)})(x_k - b_{i_2,2}^{(2^2)}).$$

and

$$\begin{aligned} x_{k+2^2} &= x_k + (a_0^{(2)})^{1+3^{2^1}} \prod_{j_1=1}^{3^{2^1}} (x_k - a_{j_1}^{(2^1)}) \\ &\quad \times \prod_{q_1=1}^{N_1} \prod_{j_2 \in I_{q_1}^{(2^{2-1})}} (x_k - b_{j_2,1}^{(2^2)})(x_k - b_{j_2,2}^{(2^2)}) \\ &\quad \times \prod_{q_2=1}^{N_2} \prod_{j_3 \in I_{q_2}^{(2^2)}} (x_k - b_{j_3,1}^{(2^2)})(x_k - b_{j_3,2}^{(2^2)}) \\ &\quad \times \prod_{j_4 \in J^{(2^2)}} (x_k - b_{j_4,1}^{(2^2)})(x_k - b_{j_4,2}^{(2^2)}) \\ &= x_k + (a_0^{(2^2)}) \prod_{i=1}^{3^{2^2}} (x_k - a_i^{(2^2)}) \end{aligned}$$

where

$$a_0^{(2)} = (a_0)^{1+3}, a_0^{(2^2)} = (a_0^{(2^{2-1})})^{1+3^{2^{2-1}}};$$
$$\cup_{i=1}^{3^{2^2}}\{a_i^{(2^2)}\} = \text{sort}\{\cup_{i_1=1}^{3^{2^2-1}}\{a_{i_1}^{(2^2)}\}, \cup_{i_2=1}^{M_2}\{b_{i_2,1}^{(2^2)}, b_{i_2,2}^{(2^2)}\}\}, a_i^{(2^2)} \le a_{i+1}^{(2^2)};$$
$$b_{i,1}^{(2^2)} = -\frac{1}{2}(B_i^{(2^2)} + \sqrt{\Delta_i^{(2^2)}}), b_{i,2}^{(2^2)} = -\frac{1}{2}(B_i^{(2^2)} - \sqrt{\Delta_i^{(2^2)}}),$$
$$\Delta_i^{(2^2)} = (B_i^{(2^2)})^2 - 4C_i^{(2^2)} \ge 0 \text{ for } i \in \cup_{q_1=1}^{N_1} I_{q_1}^{(2^1)} \cup \cup_{q_2=1}^{N_2} I_{q_2}^{(2^2)}$$
$$I_{q_1}^{(2^1)} = \{l_{(q_1-1)\times 2^1\times m_1+1}, l_{(q_1-1)\times 2^1\times m_1+2}, \cdots, l_{q_1\times 2^1\times m_1}\}$$
$$\subseteq\{1,2,\cdots,M_1\}\cup\{\varnothing\},$$
$$\text{for } q_1 \in \{1,2,\cdots,N_1\}, M_1 = N_1\times 2^1\times m_1, m_1\in\{1,2\};$$
$$I_{q_2}^{(2^2)} = \{l_{(q_2-1)\times 2^2\times m_1+1}, l_{(q_2-1)\times 2^2\times m_1+2}, \cdots, l_{q_2\times 2^2\times m_1}\}$$
$$\subseteq\{M_1+1, M_1+2,\cdots,M_2\}\cup\{\varnothing\},$$
$$\text{for } q_2 \in \{1,2,\cdots,N_2\}, M_2 = (3^{2^2} - 3^{2^{2-1}})/2;$$
$$b_{i,1}^{(2^2)} = -\frac{1}{2}(B_i^{(2^2)} + \mathbf{i}\sqrt{|\Delta_i^{(2^2)}|}), b_{i,2}^{(2^2)} = -\frac{1}{2}(B_i^{(2^2)} - \mathbf{i}\sqrt{|\Delta_i^{(2^2)}|}),$$
$$\Delta_i^{(2^2)} = (B_i^{(2^2)})^2 - 4C_i^{(2^2)} < 0, \mathbf{i} = \sqrt{-1},$$
$$i \in J^{(2^2)} = \{l_{N_2\times 2^2\times m_1+1}, l_{N_2\times 2^2\times m_1+2}, \cdots, l_{M_2}\}$$
$$\subset \{M_1+1, M_1+2,\cdots,M_2\}\cup\{\varnothing\}.$$

Non-trivial period-2^2 fixed-points are

$$x_{k+2^2}^* = x_k^* = a_i^{(2^2)} \in \cup_{i_2=1}^{M_2}\{b_{i_2,1}^{(2^1)}, b_{i_2,2}^{(2^1)}\},$$
$$i \in \{1,2,\ldots,3^{2^1}\}.$$

and trivial period-2^2 fixed-points are

$$x_{k+2^2}^* = x_k^* = a_i^{(2^2)} \in \cup_{i_1=1}^{3^{2^{2-1}}}\{a_{i_1}^{(2^{2-1})}\},$$
$$i \in \{1,2,\ldots,3^{2^1}\}.$$

Similarly, the period-2^l discrete systems ($l = 1,2,\ldots$) of the cubic discrete system in Eq. (2.105) can be developed through the above nonlinear renormalization and the corresponding fixed-points can be obtained.

(III) Consider a period-2^{l-1} discrete system as

$$x_{k+2^{l-1}} = x_k + a_0^{(2^{l-1})}\prod_{i=1}^{3^{2^{l-1}}}(x_k - a_i^{(2^{l-1})}).$$

From the nonlinear renormalizations, let $k_1 = k + 2^{l-1}$, we have

$$x_{k_1+2^{l-1}+2^{l-1}} = x_{k_1+2^{l-1}} + a_0^{(2^{l-1})} \prod_{i=1}^{3^{2^{l-1}}} (x_{k_1+2^{l-1}} - a_i^{(2^{l-1})}).$$

Because k_1 is index for iteration, it can be replaced by k. Thus, an equivalent form for the foregoing equations becomes

$$x_{k+2^l} = x_{k+2^{l-1}} + a_0^{(2^{l-1})} \prod_{i=1}^{3^{2^{l-1}}} (x_{k+2^{l-1}} - a_i^{(2^{l-1})}).$$

With the period-2^{l-1} discrete system, the foregoing equations becomes

$$\begin{aligned} x_{k+2^l} &= x_k + [a_0^{(2^{l-1})} \prod_{i_1=1}^{3^{2^{l-1}}} (x_k - a_{i_1}^{(2^{l-1})})] \\ &\quad \times \{1 + \prod_{i_1=1}^{3^{2^{l-1}}} [1 + a_0^{(2^{l-1})} \prod_{i_2=1, i_2 \neq i_1}^{3^{2^{l-1}}} (x_k - a_{i_2}^{(2^{l-1})})]\} \\ &= x_k + [a_0^{(2^{l-1})} \prod_{i_1=1}^{3^{2^{l-1}}} (x_k - a_{i_1}^{(2^{l-1})})] \\ &\quad \times [(a_0^{(2^{l-1})})^{3^{2^{l-1}}} \prod_{i_2=1}^{(3^{2^l} - 3^{2^{l-1}})/2} (x_k^2 + B_{i_2}^{(2^l)} x_k + C_{i_2}^{(2^l)})]. \end{aligned}$$

If

$$x_k^2 + B_{i_2}^{(2^l)} x_k + C_{i_2}^{(2^l)} = 0$$

for $i_2 \in \cup_{q_1=1}^{N_1} I_{q_1}^{(2^{l-1})} \cup \cup_{q_2=1}^{N_2} I_{q_2}^{(2^l)}$ with

$$\begin{aligned} I_{q_1}^{(2^{l-1})} &= \{l_{(q_1-1)\times 2^{l-1}\times m_1+1}, l_{(q_1-1)\times 2^{l-1}\times m_1+2}, \cdots, l_{q_1\times 2^{l-1}\times m_1}\} \\ &\quad \subseteq \{1, 2, \cdots, M_1\} \cup \{\varnothing\}, \\ &\text{for } q_1 \in \{1, 2, \cdots, N_1\},\ M_1 = N_1 \times 2^{l-1} \times m_1, m_1 \in \{1, 2\}; \\ I_{q_2}^{(2^l)} &= \{l_{(q_2-1)\times 2^l\times m_1+1}, l_{(q_2-1)\times 2^l\times m_1+2}, \cdots, l_{q_2\times 2^l\times m_1}\} \\ &\quad \subseteq \{M_1+1, M_1+2, \cdots, M_2\} \cup \{\varnothing\}, \\ &\text{for } q_2 \in \{1, 2, \cdots, N_2\},\ M_2 = (3^{2^l} - 3^{2^{l-1}})/2; \end{aligned}$$

then we have

$$\begin{aligned} b_{i_2,1}^{(2^l)} &= -\frac{1}{2}(B_{i_2}^{(2^l)} + \sqrt{\Delta_{i_2}^{(2^l)}}), b_{i_2,2}^{(2^l)} = -\frac{1}{2}(B_{i_2}^{(2^l)} - \sqrt{\Delta_{i_2}^{(2^l)}}) \\ \Delta_{i_2}^{(2^l)} &= (B_{i_2}^{(2^l)})^2 - 4C_{i_2}^{(2^l)} \geq 0. \end{aligned}$$

Thus

$$x_k^2 + B_{i_2}^{(2^l)} x_k + C_{i_2}^{(2^l)} = (x_k - b_{i_2,1}^{(2^l)})(x_k - b_{i_2,2}^{(2^l)}).$$

Therefore

$$
\begin{aligned}
x_{k+2^l} &= x_k + [a_0^{(2^{l-1})} \prod_{i_1=1,}^{3^{2^{l-1}}} (x_k - a_{i_1}^{(2^{l-1})})] \\
&\quad \times [(a_0^{(2^{l-1})})^{3^{2^{l-1}}} \prod_{i_2=1}^{(3^{2^l}-3^{2^{l-1}})/2} (x_k - b_{i_2,1}^{(2^l)})(x_k - b_{i_2,2}^{(2^l)})] \\
&= x_k + (a_0^{(2^{l-1})})^{1+3^{2^{l-1}}} \prod_{i=1}^{3^{2^l}} (x_k - a_i^{(2^l)}) \\
&= x_k + a_0^{(2^l)} \prod_{i=1}^{3^{2^l}} (x_k - a_i^{(2^l)})
\end{aligned}
$$

where

$$
\begin{aligned}
& a_0^{(2)} = (a_0)^{1+3}, a_0^{(2^l)} = (a_0^{(2^{l-1})})^{1+3^{2^{l-1}}}, l = 1, 2, 3, \cdots; \\
& \cup_{i=1}^{3^{2^l}} \{a_i^{(2^l)}\} = \text{sort}\{\cup_{i_1=1}^{3^{2^{l-1}}} \{a_{i_1}^{(2^l)}\}, \cup_{i_2=1}^{M_2} \{b_{i_2,1}^{(2^l)}, b_{i_2,2}^{(2^l)}\}\} \ , a_i^{(2^l)} \le a_{i+1}^{(2^l)}; \\
& b_{i,1}^{(2^l)} = -\tfrac{1}{2}(B_i^{(2^l)} + \sqrt{\Delta_i^{(2^l)}}), b_{i,2}^{(2^l)} = -\tfrac{1}{2}(B_i^{(2^l)} - \sqrt{\Delta_i^{(2^l)}}), \\
& \Delta_i^{(2^l)} = (B_i^{(2^l)})^2 - 4C_i^{(2^l)} \ge 0 \ \text{ for } \ i \in \cup_{q_1=1}^{N_1} I_{q_1}^{(2^{l-1})} \cup \cup_{q_2=1}^{N_2} I_{q_2}^{(2^l)}, \\
& I_{q_1}^{(2^{l-1})} = \{l_{(q_1-1)\times 2^{l-1}\times m_1+1}, l_{(q_1-1)\times 2^{l-1}\times m_1+2}, \cdots, l_{q_1\times 2^{l-1}\times m_1}\} \\
& \qquad \subseteq \{1, 2, \cdots, M_1\} \cup \{\varnothing\}, \\
& \text{for } q_1 \in \{1, 2, \cdots, N_1\},\ M_1 = N_1 \times 2^{l-1} \times m_1, m_1 \in \{1, 2\}; \\
& I_{q_2}^{(2^l)} = \{l_{(q_2-1)\times 2^l\times m_1+1}, l_{(q_2-1)\times 2^l\times m_1+2}, \cdots, l_{q_2\times 2^l\times m_1}\} \\
& \qquad \subseteq \{M_1+1, M_1+2, \cdots, M_2\} \cup \{\varnothing\}, \\
& \text{for } q_2 \in \{1, 2, \cdots, N_2\},\ M_2 = (3^{2^l} - 3^{2^{l-1}})/2; \\
& b_{i,1}^{(2^l)} = -\tfrac{1}{2}(B_i^{(2^l)} + \mathbf{i}\sqrt{|\Delta_i^{(2^l)}|}),\ b_{i,2}^{(2^l)} = -\tfrac{1}{2}(B_i^{(2^l)} - \mathbf{i}\sqrt{|\Delta_i^{(2^l)}|}), \\
& \Delta_i^{(2^l)} = (B_i^{(2^l)})^2 - 4C_i^{(2^l)} < 0,\ \mathbf{i} = \sqrt{-1}, \\
& i \in J^{(2^l)} = \{l_{N_2\times 2^l\times m_1+1}, l_{N_2\times 2^l\times m_1+2}, \cdots, l_{M_2}\} \\
& \qquad \subset \{M_1+1, M_1+2, \cdots, M_2\} \cup \{\varnothing\}.
\end{aligned}
$$

For the period-2^l discrete system, we have

$$
x_{k+2^l} = x_k + a_0^{(2^l)} \prod_{i=1}^{3^{2^l}} (x_k - a_i^{(2^l)})
$$

and the local stability and bifurcation at $x_k^* = a_i^{(2^l)}$ ($i \in \{1, 2, \ldots, 3^{2^l}\}$) can be determined by

$$\frac{dx_{k+2^l}}{dx_k} = 1 + a_0^{(2^l)} \sum_{i_1=1}^{3^{2^l}} \prod_{i_2=1, i_2 \neq i_1}^{3^{2^l}} (x_k - a_{i_2}^{(2^l)}),$$

$$\frac{d^2 x_{k+2^l}}{dx_k^2} = a_0^{(2^l)} \sum_{i_1=1}^{3^{2^l}} \sum_{i_2=1, i_2 \neq i_1}^{3^{2^l}} \prod_{i_3=1, i_3 \neq i_1, i_2}^{3^{2^l}} (x_k - a_{i_3}^{(2^l)}),$$

$$\vdots$$

$$\frac{d^r x_{k+2^l}}{dx_k^r} = a_0^{(2^l)} \sum_{i_1=1}^{3^{2^l}} \cdots \sum_{i_r=1, i_r \neq i_1, i_2, \ldots i_{r-1}}^{3^{2^l}} \prod_{i_{r+1}=1, i_3 \neq i_1, i_2 \ldots, i_r}^{3^{2^l}} (x_k - a_{i_{r+1}}^{(2^l)})$$

for $r \leq 3^{2^l}$;

and the period-doubling bifurcations are determined by

$$\begin{aligned}
&\frac{dx_{k+2^l}}{dx_k}\Big|_{x_k^* = a_i^{(2^l)}} = -1, \\
&\frac{d^s x_{k+2^l}}{dx_k^s}\Big|_{x_k^* = a_i^{(2^l)}} = 0, \quad \text{for } s = 2, 3, \cdots, r-1; \\
&\frac{d^r x_{k+2^l}}{dx_k^r}\Big|_{x_k^* = a_i^{(2^l)}} \neq 0, \quad \text{for } r \leq 3^{2^l}.
\end{aligned}$$

Nontrivial period-2^l fixed-points are

$$\begin{aligned}
&x_{k+2^2}^* = x_k^* = a_i^{(2^l)} \in \cup_{i_2=1}^{M_2} \{b_{i_2,1}^{(2^l)}, b_{i_2,2}^{(2^l)}\}, \\
&i \in \{1, 2, \ldots, 3^{2^l}\};
\end{aligned}$$

and trivial period-2^l fixed-points are

$$\begin{aligned}
&x_{k+2^2}^* = x_k^* = a_i^{(2^l)} \in \cup_{i_1=1}^{3^{2^{l-1}}} \{a_{i_1}^{(2^{l-1})}\}, \\
&i \in \{1, 2, \ldots, 3^{2^l}\}.
\end{aligned}$$

This theorem is proved. ■

2.5.3 *Period*-n *Appearing and Period-Doublization*

A period-n discrete system for a cubic nonlinear discrete system will be discussed, and the period-doublization of period-n discrete systems is discussed through the nonlinear renormalization.

Theorem 2.4 *Consider a 1-dimensional, forward cubic discrete system as*

$$\begin{aligned} x_{k+1} &= x_k + Ax_k^3 + Bx_k^2 + Cx_k + D \\ &= x_k + a_0 \prod_{i=1}^{3}(x_k - a_i^{(1)}). \end{aligned} \tag{2.118}$$

(i) *After n-times iterations, a period-n discrete system for the cubic discrete system in Eq. (2.118) is*

$$\begin{aligned} x_{k+n} &= x_k + a_0 \prod_{i_1=1}^{3}(x_k - a_{i_2})\{1 + \sum_{j=1}^{n} Q_j\} \\ &= x_k + a_0^{(3^n-1)/2} \prod_{i_1=1}^{3}(x_k - a_{i_1})[\prod_{j_2=1}^{(3^n-3)/2}(x_k^2 + B_{j_2}^{(2^l)} x_k + C_{j_2}^{(2^l)})] \\ &= x_k + a_0^{(n)} \prod_{i=1}^{3^n}(x_k - a_i^{(n)}) \end{aligned} \tag{2.119}$$

with

$$\begin{aligned} \frac{dx_{k+n}}{dx_k} &= 1 + a_0^{(n)} \sum_{i_1=1}^{3^n} \prod_{i_2=1, i_2 \neq i_1}^{3^n}(x_k - a_{i_2}^{(n)}), \\ \frac{d^2 x_{k+n}}{dx_k^2} &= a_0^{(n)} \sum_{i_1=1}^{3^n} \sum_{i_2=1, i_2 \neq i_1}^{3^n} \prod_{i_3=1, i_3 \neq i_1, i_2}^{3^n}(x_k - a_{i_3}^{(n)}), \\ &\vdots \\ \frac{d^r x_{k+n}}{dx_k^r} &= a_0^{(n)} \sum_{i_1=1}^{3^n} \cdots \sum_{i_r=1, i_r \neq i_1, i_2 \ldots, i_{r-1}}^{3^n} \prod_{i_{r+1}=1, i_{r+1} \neq i_1, i_2 \ldots, i_r}^{3^n}(x_k - a_{i_{r+1}}^{(n)}) \\ &\text{for } r \leq 3^n; \end{aligned} \tag{2.120}$$

where

$$\begin{aligned} & a_0^{(n)} = (a_0)^{(3^n-1)/2}, Q_1 = 0, Q_2 = \prod_{i_2=1}^{3}[1 + a_0 \prod_{i_1=1, i_1 \neq i_2}^{3}(x_k - a_{i_1}^{(1)})], \\ & Q_n = \prod_{i_n=1}^{3}[1 + a_0(1 + Q_{n-1}) \prod_{i_{n-1}=1, i_{n-1} \neq i_n}^{3}(x_k - a_{i_{n-1}}^{(1)})],\ n = 3, 4, \cdots; \\ & \cup_{i=1}^{3^n}\{a_i^{(n)}\} = \text{sort}\{\cup_{i_1=1}^{3}\{a_{i_1}^{(1)}\}, \cup_{i_2=1}^{M}\{b_{i_2,1}^{(n)}, b_{i_2,2}^{(n)}\}\}; \\ & b_{i_2,1}^{(n)} = -\tfrac{1}{2}(B_{i_2}^{(n)} + \sqrt{\Delta_{i_2}^{(n)}}), b_{i_2,2}^{(n)} = -\tfrac{1}{2}(B_{i_2}^{(n)} - \sqrt{\Delta_{i_2}^{(n)}}), \\ & \Delta_{i_2}^{(n)} = (B_{i_2}^{(n)})^2 - 4C_{i_2}^{(n)} \geq 0 \text{ for } i_2 \in \cup_{q=1}^{N} I_q^{(n)}, \\ & I_q^{(n)} = \{l_{(q-1)\times n+1}, l_{(q-1)\times n+2}, \cdots, l_{q\times n}\} \subseteq \{1, 2, \cdots, M\} \cup \{\varnothing\}, \\ & \text{for } q \in \{1, 2, \cdots, N\}, M = (3^n - 3)/2; \\ & b_{i,1}^{(n)} = -\tfrac{1}{2}(B_i^{(n)} + \mathbf{i}\sqrt{|\Delta_i^{(n)}|}),\ b_{i,2}^{(n)} = -\tfrac{1}{2}(B_i^{(n)} - \mathbf{i}\sqrt{|\Delta_i^{(n)}|}), \\ & \Delta_i^{(n)} = (B_i^{(n)})^2 - 4C_i^{(n)} < 0,\ \mathbf{i} = \sqrt{-1} \\ & i \in \{l_{N\times n+1}, l_{N\times n+2}, \cdots, l_M\} \subset \{1, 2, \cdots, M\} \cup \{\varnothing\}; \end{aligned} \tag{2.121}$$

with fixed-points

$$\begin{aligned}&x_{k+n}^{*}=x_{k}^{*}=a_{i}^{(n)},(i=1,2,\ldots,3^{n})\\&\cup_{i=1}^{3^{n}}\{a_{i}^{(n)}\}=\mathrm{sort}\{\cup_{i_1=1}^{3}\{a_{i_1}^{(1)}\},\cup_{i_2=1}^{M_2}\{b_{i_2,1}^{(n)},b_{i_2,2}^{(n)}\}\}\\&\text{with } a_{i}^{(n)}<a_{i+1}^{(n)}.\end{aligned}\tag{2.122}$$

(ii) *For a fixed-point of* $x_{k+n}^{*}=x_{k}^{*}=a_{i_1}^{(n)}$ $(i_1\in I_q^{(n)},\ q\in\{1,2,\ldots,N\})$, *if*

$$\frac{dx_{k+n}}{dx_k}|_{x_k^*=a_{i_1}^{(n)}}=1+a_0^{(n)}\prod_{i_2=1,i_2\neq i_1}^{3^n}(a_{i_1}^{(n)}-a_{i_2}^{(2^l)})=1,\tag{2.123}$$

with

$$\frac{d^2x_{k+n}}{dx_k^2}|_{x_k^*=a_{i_1}^{(n)}}=a_0^{(n)}\sum_{i_2=1,i_2\neq i_1}^{3^n}\prod_{i_3=1,i_3\neq i_1,i_2}^{3^n}(a_{i_1}^{(n)}-a_{i_3}^{(2^l)})\neq 0,\tag{2.124}$$

then there is a new discrete system for onset of the qth*-set of period-n fixed-points based on the second-order monotonic saddle-node bifurcation as*

$$x_{k+n}=x_k+a_0^{(n)}\prod_{i_1\in I_q^{(n)}}(x_k-a_{i_1}^{(n)})^2\prod_{j_2=1}^{3^n}(x_k-a_{j_2}^{(n)})^{(1-\delta(i_1,j_2))}\tag{2.125}$$

where

$$\delta(i_1,j_2)=1\ \text{ if }\ a_{j_2}^{(n)}=a_{i_1}^{(n)},\ \delta(i_1,j_2)=0\ \text{ if }\ a_{j_2}^{(n)}\neq a_{i_1}^{(n)}.\tag{2.126}$$

(ii_1) *If*

$$\begin{aligned}&\frac{dx_{k+n}}{dx_k}|_{x_k^*=a_{i_1}^{(n)}}=1\ (i_1\in I_q^{(n)}),\\&\frac{d^2x_{k+n}}{dx_k^2}|_{x_k^*=a_{i_1}^{(n)}}=2a_0^{(n)}\prod_{i_1\in I_q^{(n)},i_2\neq i_1}(a_{i_1}^{(n)}-a_{i_1}^{(n)})^2\\&\qquad\times\prod_{j_2=1}^{3^n}(a_{i_1}^{(n)}-a_{j_2}^{(n)})^{(1-\delta(i_2,j_2))}\neq 0\end{aligned}\tag{2.127}$$

x_{k+n} *at* $x_k^*=a_{i_1}^{(n)}$ *is*

- *a monotonic lower-saddle of the second-order for* $d^2x_{k+n}/dx_k^2\ |_{x_k^*=a_{i_1}^{(n)}}<0$;
- *a monotonic upper-saddle of the second-order for* $d^2x_{k+n}/dx_k^2\ |_{x_k^*=a_{i_1}^{(n)}}>0$.

(ii_2) *The period-n fixed-points* ($n=2^{n_1}\times m$) *are trivial if*

$$\left.\begin{array}{l} x_k^* = x_{k+n}^* = a_{j_1}^{(n)} \in \{\cup_{i_1=1}^{3}\{a_{i_1}^{(1)}\}, \cup_{i_2=1}^{3^{2^{n_1-1}m}}\{a_{i_2}^{(2^{n_1-1}m)}\}\} \\ \text{for } n_1 = 1,2,\ldots; m = 2l_1+1; j_1 \in \{1,2,\ldots,3^n\}\cup\{\varnothing\} \end{array}\right\}$$
$$\text{for } n \neq 2^{n_2},$$
$$\left.\begin{array}{l} x_k^* = x_{k+n}^* = a_{j_1}^{(n)} \in \{\cup_{i_2=1}^{3^{2^{n_1-1}m}}\{a_{i_2}^{(2^{n_1-1}m)}\}\} \\ \text{for } n_1 = 1,2,\ldots; m = 1; j_1 \in \{1,2,\ldots,3^n\}\cup\{\varnothing\} \end{array}\right\} \tag{2.128}$$
$$\text{for } n = 2^{n_2}.$$

(ii$_3$) *The period-n fixed-points* ($n = 2^{n_1} \times m$) *are non-trivial if*

$$\left.\begin{array}{l} x_k^* = x_{k+n}^* = a_{j_1}^{(n)} \notin \{\cup_{i_1=1}^{3}\{a_{i_1}^{(1)}\}, \cup_{i_2=1}^{3^{2^{n_1-1}m}}\{a_{i_2}^{(2^{n_1-1}m)}\}\} \\ \text{for } n_1 = 1,2,\ldots; m = 2l_1+1; j_1 \in \{1,2,\ldots,3^n\}\cup\{\varnothing\} \end{array}\right\}$$
$$\text{for } n \neq 2^{n_2},$$
$$\left.\begin{array}{l} x_k^* = x_{k+n}^* = a_{j_1}^{(n)} \notin \{\cup_{i_2=1}^{3^{2^{n_1-1}m}}\{a_{i_2}^{(2^{n_1-1}m)}\}\} \\ \text{for } n_1 = 1,2,\ldots; m = 1; j_1 \in \{1,2,\ldots,3^n\}\cup\{\varnothing\} \end{array}\right\} \tag{2.129}$$
$$\text{for } n = 2^{n_2}.$$

Such a period-n fixed-point is

- *monotonically unstable if* $dx_{k+n}/dx_k|_{x_k^*=a_{i_1}^{(n)}} \in (1,\infty)$;
- *monotonically invariant if* $dx_{k+n}/dx_k|_{x_k^*=a_{i_1}^{(n)}} = 1$, *which is*
  - *a monotonic upper-saddle of the* $(2l_1)^{\text{th}}$ *order for* $d^{2l_1}x_{k+n}/dx_k^{2l_1}|_{x_k^*} > 0$;
  - *a monotonic lower-saddle the* $(2l_1)^{\text{th}}$ *order for* $d^{2l_1}x_{k+n}/dx_k^{2l_1}|_{x_k^*} < 0$;
  - *a monotonic source of the* $(2l_1+1)^{\text{th}}$ *order for* $d^{2l_1+1}x_{k+n}/dx_k^{2l_1+1}|_{x_k^*} > 0$;
  - *a monotonic sink the* $(2l_1+1)^{\text{th}}$ *order for* $d^{2l_1+1}x_{k+n}/dx_k^{2l_1+1}|_{x_k^*} < 0$;
- *monotonically stable if* $dx_{k+n}/dx_k|_{x_k^*=a_{i_1}^{(n)}} \in (0,1)$;
- *invariantly zero-stable if* $dx_{k+n}/dx_k|_{x_k^*=a_{i_1}^{(n)}} = 0$;
- *oscillatorilly stable if* $dx_{k+n}/dx_k|_{x_k^*=a_{i_1}^{(n)}} \in (-1,0)$;
- *flipped if* $dx_{k+n}/dx_k|_{x_k^*=a_{i_1}^{(n)}} = -1$, *which is*
  - *an oscillatory upper-saddle of the* $(2l_1)^{\text{th}}$ *order for* $d^{2l_1}x_{k+n}/dx_k^{2l_1}|_{x_k^*} > 0$;
  - *an oscillatory lower-saddle the* $(2l_1)^{\text{th}}$ *order for* $d^{2l_1}x_{k+n}/dx_k^{2l_1}|_{x_k^*} < 0$;

- *an oscillatory source of the* $(2l_1+1)^{\text{th}}$ *order for* $d^{2l_1+1}x_{k+n}/\,dx_k^{2l_1+1}|_{x_k^*}<0$;
- *an oscillatory sink the* $(2l_1+1)^{\text{th}}$ *order for* $d^{2l_1+1}x_{k+n}/dx_k^{2l_1+1}|_{x_k^*}>0$;

• *oscillatorilly unstable if* $dx_{k+n}/dx_k|_{x_k^*=a_{i_1}^{(n)}}\in(-\infty,-1)$.

(iii) *For a fixed-point of* $x_{k+n}^*=x_k^*=a_{i_1}^{(n)}$ $(i_1\in I_q^{(n)},q\in\{1,2,\ldots,N\})$, *there is a period-doubling of the* q^{th}*-set of period-n fixed-points if*

$$\begin{aligned}&\frac{dx_{k+n}}{dx_k}|_{x_k^*=a_{i_1}^{(n)}}=1+a_0^{(n)}\prod_{j_2=1,j_2\neq i_1}^{3^n}(a_{i_1}^{(n)}-a_{j_2}^{(n)})=-1,\\&\frac{d^sx_{k+n}}{dx_k^s}|_{x_k^*=a_{i_1}^{(n)}}=0,\ \text{for } s=2,\ldots,\ r-1;\\&\frac{d^rx_{k+n}}{dx_k^r}|_{x_k^*=a_{i_1}^{(n)}}\neq 0\text{ for }1<r\leq 3^n\end{aligned}\tag{2.130}$$

with

- a rth*-order oscillatory source for* $d^rx_{k+n}/dx_k^r|_{x_k^*=a_{i_1}^{(n)}}<0$ *and* $r=2l_1+1$;
- a rth*-order oscillatory sink for* $d^rx_{k+n}/dx_k^r|_{x_k^*=a_{i_1}^{(n)}}>0$ *and* $r=2l_1+1$;
- a rth*-order oscillatory upper-saddle for* $d^rx_{k+n}/dx_k^r|_{x_k^*=a_{i_1}^{(n)}}>0$ *and* $r=2l_1$;
- a rth*-order oscillatory lower-saddle for* $d^rx_{k+n}/dx_k^r|_{x_k^*=a_{i_1}^{(n)}}<0$ and $r=2l_1$.

The corresponding period- $2\times n$ *discrete system of the cubic discrete system in Eq. (2.118) is*

$$x_{k+2\times n}=x_k+a_0^{(2\times n)}\prod_{i_1=I_q^{(n)}}(x_k-a_{i_1}^{(n)})^3\prod_{j_2=1}^{3^{2\times n}}(x_k-a_{j_2}^{(2\times n)})^{(1-\delta(i_1,j_2))}\tag{2.131}$$

with

$$\begin{aligned}&\frac{dx_{k+2\times n}}{dx_k}|_{x_k^*=a_{i_1}^{(n)}}=1,\ \frac{d^2x_{k+2\times n}}{dx_k^2}|_{x_k^*=a_{i_1}^{(n)}}=0;\\&\frac{d^3x_{k+2\times n}}{dx_k^3}|_{x_k^*=a_{i_1}^{(n)}}=6a_0^{(2\times n)}\prod_{i_2\in I_q^{(n)},i_2\neq i_1}(a_{i_1}^{(n)}-a_{i_2}^{(n)})^3\\&\qquad\times\prod_{j_2=1}^{3^{2\times n}}(a_{i_1}^{(n)}-a_{j_2}^{(2\times n)})^{(1-\delta(i_1,j_2))}.\end{aligned}\tag{2.132}$$

Thus, $x_{k+2\times n}$ *at* $x_k^* = a_{i_1}^{(n)}$ *for* $i_1 \in I_q^{(n)}$, $q \in \{1, 2, \ldots, N\}$ is

- *a monotonic sink of the third-order if* $d^3 x_{k+2\times n}/dx_k^3|_{x_k^*=a_{i_1}^{(n)}} < 0$,
- *a monotonic source of the third-order if* $d^3 x_{k+2\times n}/dx_k^3|_{x_k^*=a_{i_1}^{(n)}} > 0$.

(iv) *After l-times period-doubling bifurcations of period-n fixed points, a period-*$2^l \times n$ *discrete system of the cubic discrete system in Eq.* (1.118) *is*

$$
\begin{aligned}
x_{k+2^l\times n} &= x_k + [a_0^{(2^{l-1}\times n)} \prod_{i_1=1}^{3^{2^{l-1}\times n}} (x_k - a_{i_1}^{(2^{l-1}\times n)})] \\
&\quad \times \{1 + \prod_{i_1=1}^{3^{2^{l-1}\times n}} [1 + a_0^{(2^{l-1})} \prod_{i_2=1, i_2\neq i_1}^{3^{2^{l-1}\times n}} (x_k - a_{i_2}^{(2^{l-1}\times n)})]\} \\
&= x_k + [a_0^{(2^{l-1}\times n)} \prod_{i_1=1}^{3^{2^{l-1}\times n}} (x_k - a_{i_1}^{(2^{l-1}\times n)})] \\
&\quad \times [(a_0^{(2^{l-1}\times n)})^{3^{2^{l-1}\times n}} \prod_{j_1=1}^{(3^{2^l\times n} - 3^{2^{l-1}\times n})/2} (x_k^2 + B_{j_2}^{(2^l\times n)} x_k + C_{j_2}^{(2^l\times n)})] \\
&= x_k + [a_0^{(2^{l-1}\times n)} \prod_{i_1=1}^{3^{2^{l-1}\times n}} (x_k - a_{i_1}^{(2^{l-1}\times n)})] \\
&\quad \times [(a_0^{(2^{l-1}\times n)})^{3^{2^{l-1}\times n}} \prod_{j_2=1}^{(3^{2^l\times n} - 3^{2^{l-1}\times n})/2} (x_k - b_{j_2,1}^{(2^l\times n)})(x_k - b_{j_2,2}^{(2^l\times n)})] \\
&= x_k + (a_0^{(2^{l-1}\times n)})^{1+3^{2^{l-1}\times n}} \prod_{i=1}^{3^{2^l\times n}} (x_k - a_i^{(2^l\times n)}) \\
&= x_k + a_0^{(2^l\times n)} \prod_{i=1}^{3^{2^l\times n}} (x_k - a_i^{(2^l\times n)})
\end{aligned}
\tag{2.133}
$$

with

$$
\begin{aligned}
\frac{dx_{k+2^l\times n}}{dx_k} &= 1 + a_0^{(2^l\times n)} \sum_{i_1=1}^{3^{2^l\times n}} \prod_{i_2=1, i_2\neq i_1}^{3^{2^l\times n}} (x_k - a_{i_2}^{(2^l\times n)}), \\
\frac{d^2 x_{k+2^l\times n}}{dx_k^2} &= a_0^{(2^l\times n)} \sum_{i_1=1}^{3^{2^l\times n}} \sum_{i_2=1, i_2\neq i_1}^{3^{2^l\times n}} \prod_{i_3=1, i_3\neq i_1, i_2}^{3^{2^l\times n}} (x_k - a_{i_3}^{(2^l\times n)}), \\
&\vdots \\
\frac{d^r x_{k+2^l\times n}}{dx_k^r} &= a_0^{(2^l\times n)} \sum_{i_1=1}^{3^{2^l\times n}} \cdots \sum_{i_r=1, i_r\neq i_1, i_2\ldots, i_{r-1}}^{3^{2^l\times n}} \prod_{i_{r+1}=1, i_{r+1}\neq i_1, i_2\ldots, i_r}^{3^{2^l\times n}} (x_k - a_{i_{r+1}}^{(2^l\times n)}) \\
&\text{for } r \leq 3^{2^l\times n};
\end{aligned}
\tag{2.134}
$$

where

$$
\begin{aligned}
&a_0^{(2\times n)} = (a_0^{(n)})^{1+3^{2\times n}}, a_0^{(2^l\times n)} = (a_0^{(2^{l-1}\times n)})^{1+3^{2^{l-1}\times n}}, l = 1,2,3,\ldots;\\
&\cup_{i=1}^{3^{2^l\times n}}\{a_i^{(2^l\times n)}\} = \mathrm{sort}\{\cup_{i_1=1}^{3^{2^{l-1}\times n}}\{a_{i_1}^{(2^{l-1}\times n)}\}, \cup_{i_2=1}^{M_2}\{b_{i_2,1}^{(2^l\times n)}, b_{i_2,2}^{(2^l\times n)}\}\};\\
&b_{i,1}^{(2^l\times n)} = -\frac{1}{2}(B_i^{(2^l\times n)} + \sqrt{\Delta_i^{(2^l\times n)}}),\\
&b_{i,2}^{(2^l\times n)} = -\frac{1}{2}(B_i^{(2^l\times n)} - \sqrt{\Delta_i^{(2^l\times n)}}),\\
&\Delta_i^{(2^l\times n)} = (B_i^{(2^l\times n)})^2 - 4C_i^{(2^l\times n)} \geq 0\\
&\text{for } i \in \cup_{q_1=1}^{N_1} I_{q_1}^{(2^{l-1}\times n)} \bigcup \cup_{q_2=1}^{N_2} I_{q_2}^{(2^l\times n)},\\
&I_{q_1}^{(2^{l-1}\times n)} = \{l_{(q_1-1)\times(2^{l-1}\times n)+1}, l_{(q_1-1)\times(2^{l-1}\times n)+2}, \cdots, l_{q_1\times(2^{l-1}\times n)}\}\\
&\qquad \subseteq \{1,2,\ldots,M_1\} \cup \{\varnothing\},\\
&\text{for } q_1 \in \{1,2,\ldots,N_1\}, M_1 = N_1 \times n;\\
&I_{q_2}^{(2^l\times n)} = \{l_{(q_2-1)\times(2^l\times n)+1}, l_{(q_2-1)\times(2^l\times n)+2}, \cdots, l_{q_2\times(2^l\times n)}\}\\
&\qquad \subseteq \{M_1+1, M_1+2,\ldots,M_2\} \cup \{\varnothing\},\\
&\text{for } q_2 \in \{1,2,\ldots,N_2\}, M_2 = (3^{2^l\times n} - 3^{2^{l-1}\times n})/2;\\
&b_{i,1}^{(2^l\times n)} = -\frac{1}{2}(B_i^{(2^l\times n)} + \mathbf{i}\sqrt{|\Delta_i^{(2^l\times n)}|}),\\
&b_{i,2}^{(2^l\times n)} = -\frac{1}{2}(B_i^{(2^l\times n)} - \mathbf{i}\sqrt{|\Delta_i^{(2^l\times n)}|}),\\
&\Delta_i^{(2^l\times n)} = (B_i^{(2^l\times n)})^2 - 4C_i^{(2^l\times n)} < 0, \ \mathbf{i} = \sqrt{-1},\\
&i \in \{l_{N\times(2^{l-1}\times n)+1}, l_{N\times(2^{l-1}\times n)+2}, \cdots, l_{M_2}\}\\
&\subset \{M_1+1, M_1+2,\ldots,M_2\} \cup \{\varnothing\}
\end{aligned}
\tag{2.135}
$$

with fixed-points

$$
\begin{aligned}
&x_{k+2^l\times n}^* = x_k^* = a_i^{(2^l\times n)}, (i = 1,2,\ldots,3^{2^l\times n})\\
&\cup_{i=1}^{3^{2^l\times n}}\{a_i^{(2^l\times n)}\} = \mathrm{sort}\{\cup_{i=1}^{3^{2^{l-1}\times n}}\{a_i^{(2^{l-1}\times n)}\}, \cup_{i=1}^{M_2}\{b_{i,1}^{(2^{l-1}\times n)}, b_{i,2}^{(2^{l-1}\times n)}\}\}\\
&\text{with } a_i^{(2^l\times n)} < a_{i+1}^{(2^l\times n)}.
\end{aligned}
\tag{2.136}
$$

(v) *For a fixed-point of* $x_{k+(2^l\times n)}^* = x_k^* = a_{i_1}^{(2^{l-1}\times n)}$ $(i_1 \in I_q^{(2^{l-1}\times n)}, q \in \{1,2,\ldots,N_1\})$, *there is a period-* $(2^l \times n)$ *discrete system if*

$$\begin{aligned}
&\frac{dx_{k+2^{l-1}\times n}}{dx_k}\Big|_{x_k^*=a_{i_1}^{(2^{l-1}\times n)}} = 1 + a_0^{(2^{l-1}\times n)} \prod_{i_2=1, i_2\neq i_1}^{3^{2^{l-1}\times n}} (a_{i_1}^{(2^{l-1}\times n)} - a_{i_2}^{(2^{l-1}\times n)}) = -1,\\
&\frac{d^s x_{k+2^{l-1}\times n}}{dx_k^s}\Big|_{x_k^*=a_{i_1}^{(2^{l-1}\times n)}} = 0,\ \text{for } s=2,\ldots,r-1;\\
&\frac{d^r x_{k+2^{l-1}\times n}}{dx_k^r}\Big|_{x_k^*=a_{i_1}^{(2^{l-1}\times n)}} \neq 0 \text{ for } 1<r\leq 3^{2^{l-1}\times n}
\end{aligned} \tag{2.137}$$

with

- *a rth-order oscillatory sink for* $d^r x_{k+2^l\times n}/dx_k^r|_{x_k^*=a_{i_1}^{(2^l\times n)}} > 0$ *and* $r = 2l_1+1$;
- *a rth-order oscillatory source for* $d^r x_{k+2^l\times n}/dx_k^r|_{x_k^*=a_{i_1}^{(2^l\times n)}} < 0$ *and* $r = 2l_1+1$;
- *a rth-order oscillatory upper-saddle for* $d^r x_{k+2^l\times n}/dx_k^r|_{x_k^*=a_{i_1}^{(2^l\times n)}} > 0$ *and* $r = 2l_1$;
- *a rth-order oscillatory lower-saddle for* $d^r x_{k+2^l\times n}/dx_k^r|_{x_k^*=a_{i_1}^{(2^l\times n)}} < 0$ and $r = 2l_1$;

The corresponding period- $2^l\times n$ *discrete system is*

$$\begin{aligned}
x_{k+2^l\times n} = x_k + a_0^{(2^l\times n)} &\prod_{i_1\in I_q^{(2^{l-1}\times n)}} (x_k - a_{i_1}^{(2^{l-1}\times n)})^3\\
&\times \prod_{j_2=1}^{3^{2^l\times n}} (x_k - a_{j_2}^{(2^l\times n)})^{(1-\delta(i_1,j_2))}
\end{aligned} \tag{2.138}$$

where

$$\delta(i_1,j_2) = 1 \text{ if } a_{j_2}^{(2^l\times n)} = a_{i_1}^{(2^{l-1}\times n)},\ \delta(i_1,j_2) = 0 \text{ if } a_{j_2}^{(2^l\times n)} \neq a_{i_1}^{(2^{l-1}\times n)} \tag{2.139}$$

with

$$\begin{aligned}
&\frac{dx_{k+2^l\times n}}{dx_k}\Big|_{x_k^*=a_{i_1}^{(2^{l-1}\times n)}} = 1,\ \frac{d^2 x_{k+2^l\times n}}{dx_k^2}\Big|_{x_k^*=a_{i_1}^{(2^{l-1}\times n)}} = 0,\\
&\frac{d^3 x_{k+2^l\times n}}{dx_k^3}\Big|_{x_k^*=a_{i_1}^{(2^{l-1})}} = 6a_0^{(2^l\times n)} \prod_{i_2\in I_q^{(2^{l-1}\times n)}, i_2\neq i_1} (a_{i_1}^{(2^{l-1}\times n)} - a_{i_2}^{(2^{l-1}\times n)})^3\\
&\qquad\qquad \times \prod_{j_2=1}^{3^{2^l\times n}} (a_{i_1}^{(2^{l-1}\times n)} - a_{j_2}^{(2^l\times n)})^{(1-\delta(i_2,j_2))} \neq 0\\
&(i_1\in I_q^{(2^{l-1}\times n)}, q\in\{1,2,\ldots,N_1\}).
\end{aligned} \tag{2.140}$$

$x_{k+2^l\times n}$ at $x_k^* = a_{i_1}^{(2^{l-1}\times n)}$ is

- *a monotonic sink of the third-order for* $d^3x_{k+2^l\times n}/dx_k^3|_{x_k^*=a_{i_1}^{(2^{l-1})}} < 0$;
- *a monotonic source of the third-order for* $d^3x_{k+2^l\times n}/dx_k^3|_{x_k^*=a_{i_1}^{(2^{l-1})}} > 0$.

(v$_1$) *The period-* $2^l\times n$ *fixed-points are trivial if*

$$\left.\begin{array}{l} x_{k+2^l\times n}^* = x_k^* = a_j^{(2^l\times n)} \in \{\cup_{i_i=1}^{3}\{a_{i_1}^{(1)}\}, \cup_{i_2=1}^{3^{2^{l-1}\times n}}\{a_{i_2}^{(2^{l-1}\times n)}\}\} \\ \text{for } j=1,2,\ldots,2^{(2^l\times n)} \end{array}\right\}$$
$$\text{for } n\neq 2^{n_1}$$
$$\left.\begin{array}{l} x_{k+2^l\times n}^* = x_k^* = a_j^{(2^l\times n)} \in \cup_{i_2=1}^{3^{2^{l-1}\times n}}\{a_{i_2}^{(2^{l-1}\times n)}\} \\ \text{for } j=1,2,\ldots,3^{2^l\times n} \end{array}\right\}$$
$$\text{for } n=2^{n_1}. \tag{2.141}$$

(v$_2$) *The period-* $2^l\times n$ *fixed-points are non-trivial if*

$$\left.\begin{array}{l} x_{k+2^l\times n}^* = x_k^* = a_j^{(2^l\times n)} \notin \{\cup_{i_i=1}^{3}\{a_{i_1}^{(1)}\}, \cup_{i_2=1}^{3^{2^{l-1}\times n}}\{a_{i_2}^{(2^{l-1}\times n)}\}\} \\ \text{for } j=1,2,\ldots,3^{2^l\times n} \end{array}\right\}$$
$$\text{for } n\neq 2^{n_1}$$
$$\left.\begin{array}{l} x_{k+2^l\times n}^* = x_k^* = a_j^{(2^l\times n)} \notin \cup_{i_2=1}^{3^{2^{l-1}\times n}}\{a_{i_2}^{(2^{l-1}\times n)}\} \\ \text{for } j=1,2,\ldots,3^{2^l\times n} \end{array}\right\}$$
$$\text{for } n=2^{n_1}. \tag{2.142}$$

Such a period- $2^l\times n$ *fixed-point is*

- *monotonically unstable if* $dx_{k+2^l\times n}/dx_k|_{x_k^*=a_{i_1}^{(2^l\times n)}} \in (0,\infty)$;
- *monotonically invariant if* $dx_{k+2^l\times n}/dx_k|_{x_k^*=a_{i_1}^{(2^l\times n)}} = 1$, *which is*
  - *a monotonic upper-saddle of the* $(2l_1)^{\text{th}}$ *order for* $d^{2l_1}x_{k+2^l\times n}/dx_k^{2l_1}|_{x_k^*} > 0$ *(independent* $(2l_1)$*-branch appearance);*
  - *a monotonic lower-saddle the* $(2l_1)^{\text{th}}$ order for $d^{2l_1}x_{k+2^l\times n}/dx_k^{2l_1}|_{x_k^*} < 0$ *(independent* $(2l_1)$*-branch appearance)*
  - *a monotonic source of the* $(2l_1+1)^{\text{th}}$ *order for* $d^{2l_1+1}x_{k+2^l\times n}/dx_k^{2l_1+1}|_{x_k^*} > 0$ *(dependent* $(2l_1+1)$*-branch appearance from one branch);*
  - *a monotonic sink the* $(2l_1+1)^{\text{th}}$ *order for* $d^{2l_1+1}x_{k+2^l\times n}/dx_k^{2l_1+1}|_{x_k^*} < 0$ *(dependent* $(2l_1+1)$*-branch appearance from one branch);*

- *monotonically stable if* $dx_{k+2^l\times n}/dx_k|_{x_k^*=a_{i_1}^{(2^l\times n)}} \in (0,1)$;
- *invariantly zero-stable if* $dx_{k+2^l\times n}/dx_k|_{x_k^*=a_{i_1}^{(2^{l-1}\times n)}} = 0$;
- *oscillatorilly stable if* $dx_{k+2^l\times n}/dx_k|_{x_k^*=a_{i_1}^{(2^l\times n)}} \in (-1,0)$;
- *flipped if* $dx_{k+2^l\times n}/dx_k|_{x_k^*=a_{i_1}^{(2^l\times n)}} = -1$, *which is*
  - *an oscillatory upper-saddle of the* $(2l_1)^{\text{th}}$ *order for* $d^{2l_1}x_{k+2^l\times n}/dx_k^{2l_1}|_{x_k^*} > 0$;
  - *an oscillatory lower-saddle the* $(2l_1)^{\text{th}}$ *order for* $d^{2l_1}x_{k+2^l\times n}/dx_k^{2l_1}|_{x_k^*} < 0$;
  - *an oscillatory source of the* $(2l_1+1)^{\text{th}}$ *order for* $d^{2l_1+1}x_{k+2^l\times n}/\, dx_k^{2l_1+1}|_{x_k^*} < 0$;
  - *an oscillatory sink the* $(2l_1+1)^{\text{th}}$ *order for* $d^{2l_1+1}x_{k+2^l\times n}/dx_k^{2l_1+1}|_{x_k^*} > 0$;
- *oscillatorilly unstable if* $dx_{k+2^l\times n}/dx_k|_{x_k^*=a_{i_1}^{(2^l\times n)}} \in (-\infty,-1)$.

Proof Through the nonlinear renormalization, the proof of this theorem is similar to the proof of Theorem 2.3. This theorem can be easily proved. ■

2.5.4 *Sampled Period-*n *Appearing Bifurcations*

Consider a period-n discrete system of the cubic discrete system as

$$x_{k+n} = x_k + a_0^{(n)} \prod_{i=1}^{3^n} (x_k - a_i^{(n)}) \tag{2.143}$$

where $a_0^{(n)} = (a_0)^{(3^n-1)/2}$.

For $n = 1$, Eq. (2.143) gives a period-1 discrete system of the cubic system as

$$x_{k+1} = x_k + a_0^{(1)} \prod_{i=1}^{3} (x_k - a_i^{(1)}). \tag{2.144}$$

- If two of $a_i^{(1)}$ $(i = 1,2,3)$ are complex, only one fixed-point exists in such a cubic discrete system.
- If $a_i^{(1)}$ $(i = 1,2,3)$ are real, three fixed-points exist in such a cubic discrete system.

For $n = 2$, equation (2.143) gives a period-2 discrete system of the cubic system as

$$x_{k+n} = x_k + a_0^{(2)} \prod_{i=1}^{3^2} (x_k - a_i^{(2)}). \tag{2.145}$$

- If eight of $a_i^{(2)}$ $(i = 1, 2, \ldots, 3^2)$ are complex, the period-2 discrete system has only one trivial fixed-point.
- If three of $a_i^{(2)}$ $(i = 1, 2, \ldots, 3^2)$ are real, the period-2 discrete system possesses three fixed-points. One fixed-point is trivial from period-1 and two fixed-points are for period-2, or the three fixed-points are the same as the period-1 fixed-points. Such two non-trivial fixed points are generated through period-doubling bifurcation.
- If five of $a_i^{(2)}$ $(i = 1, 2, \ldots, 3^2)$ are real, the period-2 discrete system possesses five fixed-points, including three trivial fixed-points for period-1 and two non-trivial fixed-points for period-2. Such two non-trivial fixed points are generated through period-doubling bifurcation, and both of fixed-points are stable for period-2.
- If nine of $a_i^{(2)}$ $(i = 1, 2, \ldots, 3^2)$ are real, the period-2 discrete system possesses nine fixed-points, including three trivial fixed-points for period-1 and six non-trivial fixed-points for period-2. Such six non-trivial fixed points are generated through different period-doubling bifurcation. Two sets of period-2 fixed-points are stable, and one set of period-2 fixed-points is unstable. With three unstable trivial period-2 fixed-points, we have five unstable fixed points.

Thus, the period-2 discrete system of the cubic nonlinear discrete system has three sets of period-2 fixed-points on the period-1 to period-2 period-doubling bifurcation tree. No any independent period-2 fixed-points exists. For numerical simulations, one set of stable asymmetric period-2 fixed points can be obtained.

For $n = 3$, Eq. (2.143) gives a period-3 discrete system of the cubic nonlinear discrete system as

$$x_{k+n} = x_k + a_0^{(3)} \prod_{i=1}^{3^3} (x_k - a_i^{(3)}). \tag{2.146}$$

- If one of $a_i^{(3)}$ $(i = 1, 2, \ldots, 3^3)$ is real, the period-3 discrete system does have one trivial fixed-point from period-1.
- If three of $a_i^{(3)}$ $(i = 1, 2, \ldots, 3^3)$ is real, the period-3 discrete system does have three trivial fixed-points from period-1.
- If nine of $a_i^{(3)}$ $(i = 1, 2, \ldots, 3^3)$ are real, the period-3 discrete system possesses three trivial fixed-points and one set of six period-3 fixed points.
- If all of $a_i^{(3)}$ $(i = 1, 2, \ldots, 3^3)$ are real, the period-3 discrete system possesses 27 fixed-points, including three trivial fixed-points for period-1 and 4 sets of non-trivial fixed-points. Such non-trivial fixed points are generated through the

monotonic upper-saddle or monotonic lower-saddle bifurcations. The period-3 fixed-points are independent of the trivial fixed-points for period-1.

Thus, the period-3 discrete system has at most four sets of period-3 fixed-points, which are independent of the period-1fixed-points.

For $n = 4$, Eq. (2.143) gives a period-4 discrete system of the cubic system as

$$x_{k+4} = x_k + a_0^{(4)} \prod_{i=1}^{3^4} (x_k - a_i^{(4)}). \tag{2.147}$$

- If one of $a_i^{(4)}$ $(i = 1, 2, \ldots, 3^4)$ are real, the period-4 discrete system does have one trivial fixed-point from period-1.
- If three of $a_i^{(4)}$ $(i = 1, 2, \ldots, 3^4)$ are real, the period-4 discrete system possesses three trivial fixed-points from period-1 or period-2.
- If nine of $a_i^{(4)}$ $(i = 1, 2, \ldots, 3^4)$ are real, the period-4 discrete system possesses nine trivial fixed-points which are the same as the period-1 and period-2 fixed-points.
- If 17 of $a_i^{(4)}$ $(i = 1, 2, \ldots, 3^4)$ are real, the period-4 discrete system possesses eight fixed-points, including three trivial fixed-points for period-1, six trivial fixed-points for period-2, and eight non-trivial fixed-points for period-4. Such non-trivial fixed points are stable, which are generated through the period-doubling bifurcations from 4 period-2 branches. All trivial fixed-points for period-4 are unstable.
- If all of $a_i^{(4)}$ $(i = 1, 2, \ldots, 3^4)$ are real, in addition to the period-4 fixed-points by the period-doubling bifurcation, the period-4 discrete system possesses eight sets of non-trivial period-4 fixed-points, which are generated by the monotonic upper-saddle or lower-saddle bifurcations. The period-4 fixed-points are independent of the trivial fixed-points.

Thus, the period-4 discrete system has at most nine sets of period-4 fixed-points, two sets are dependent on the period-1 to period-4 period-doubling trees, and eights set of period-4 fixed-points are independent of the period-1 to period-4 period-doubling bifurcation trees.

Similarly, other period-n discrete systems can be discussed. From the previous discussion, the period-n fixed-points for a cubic discrete system are tabulated in Table 2.4. The dependent sets of period-n fixed-points are on the period-doubling bifurcation trees. The independent sets of period-n fixed-points are generated through monotonic saddle-node bifurcations. From analytical expressions, the maximum sets of period-n fixed-points includes dependent and independent sets of period-n fixed-points. In addition to the period-1 trivial fixed-points, other fixed-points on the bifurcation trees relative to period-n fixed points are also trivial.

Table 2.4 Period-n fixed-points for a cubic discrete system

	Dependent sets	Independent sets	Maximum sets	Trivial fixed-points
P-1	N/A	3	3	N/A
P-2	(1)P-1 to P-2	1-3	3	(3)P-1
P-3	N/A	4	4	(3)P-1
P-4	(2)P-1 to P-4	8	10	(2)P-1 to P-2
P-5	N/A	24	24	(3)P-1
P-6	(4)P-3 to P-6	57	61	(3)P-1, (4)P-3
P-7	N/A	156	156	(3)P-1
P-8	(2)P-1 to P-8 (8)P-4 to P-8	401	411	(2)P-1 to P-4 (8)P-4
P-9	N/A	1093	1093	(3)P-1
P-10	(24)P-5 to P-10	2928	2952	(3)P-1 (24)P-5
P-11	N/A	8052	8052	(1)P-1
P-12	(4)P-3 to P-12 (57)P-6 to P-12	22082	22143	(3)P-1 (4)P-3 to P-6 (57)P-6

2.6 Backward Cubic Nonlinear Discrete Systems

In this section, the analytical bifurcation trees for backward cubic nonlinear discrete systems will be discussed, as in a similar fashion, through nonlinear renormalization techniques, and the bifurcation scenario based on the monotonic saddle-node bifurcations will be discussed, which is independent of period-1 fixed-points.

2.6.1 Backward Period-2 Cubic Discrete Systems

After the period-doubling bifurcation of a period-1 fixed-point in the backward cubic nonlinear discrete system, the period-doubled fixed-points can be obtained. Consider the period-doubling solutions for a backward cubic nonlinear discrete system as follows.

Theorem 2.5 *Consider a 1-dimensional backward cubic nonlinear discrete system as*

$$\begin{aligned} x_k &= x_{k+1} + A(\mathbf{p})x_{k+1}^3 + B(\mathbf{p})x_{k+1}^2 + C(\mathbf{p})x_{k+1} + D(\mathbf{p}) \\ &= x_{k+1} + a_0(\mathbf{p})(x_{k+1} - a(\mathbf{p}))[x_{k+1}^2 + B_1(\mathbf{p})x_{k+1} + C_1(\mathbf{p})] \end{aligned} \tag{2.148}$$

where four scalar constants $A(\mathbf{p}) \neq 0$,$B(\mathbf{p})$,$C(\mathbf{p})$ and $D(\mathbf{p})$ are determined by

$$\begin{aligned} &A = a_0, B = (-a + B_1)a_0, C = (-aB_1 + C_1)a_0, D = -aa_0C_1, \\ &\mathbf{p} = (p_1, p_2, \ldots, p_m)^{\mathrm{T}}. \end{aligned} \tag{2.149}$$

Under

$$\Delta_1 = B_1^2 - 4C_1 < 0, \tag{2.150}$$

the standard form of such a 1-dimensional backward cubic discrete system is

$$x_k = x_{k+1} + a_0(x_{k+1} - a)(x_{k+1}^2 + B_1 x_{k+1} + C_1). \tag{2.151}$$

Under

$$\Delta_1 = B_1^2 - 4C_1 > 0, \tag{2.152}$$

the standard form of such a backward cubic discrete system is

$$x_k = x_{k+1} + a_0(x_{k+1} - a_1)(x_{k+1} - a_2)(x_{k+1} - a_3). \tag{2.153}$$

Thus, a general standard form of such a backward cubic discrete system is

$$\begin{aligned} x_k &= x_{k+1} + f(x_{k+1}, \mathbf{p}) = x_{k+1} + Ax_{k+1}^3 + Bx_{k+1}^2 + Cx_{k+1} + D \\ &\equiv x_{k+1} + a_0(x_{k+1} - a)[x_{k+1}^2 + B_1 x_{k+1} + C_1]) \\ &= x_{k+1} + a_0 \prod_{i=1}^{3} (x_{k+1} - a_i^{(1)}) \end{aligned} \tag{2.154}$$

where

$$\begin{aligned} &a_0 = A(\mathbf{p}),\ b_{1,2}^{(1)} = -\tfrac{1}{2}(B_1(\mathbf{p}) \mp \sqrt{\Delta^{(1)}}) \ \text{ for } \ \Delta^{(1)} > 0; \\ &a_1^{(1)} = \min\{a, b_1, b_2\}, a_3^{(1)} = \max\{a, b_1, b_2\}, a_1^{(2)} \in \{a, b_1, b_2\} \neq \{a_1, a_3\}, \\ &b_{1,2}^{(1)} = -\tfrac{1}{2}(B_1(\mathbf{p}) \mp \mathbf{i}\sqrt{\Delta^{(1)}}),\ \mathbf{i} = \sqrt{-1} \text{ for } \Delta^{(1)} < 0 \\ &a_1^{(1)} = a, a_2^{(1)} = b_1^{(1)}, a_3^{(1)} = b_2^{(1)} \end{aligned} \tag{2.155}$$

(i) *Consider a backward period-2 discrete system of Eq. (2.148) as*

$$\begin{aligned} x_k &= x_{k+2} + [a_0 \prod_{i_1=1}^{3} (x_{k+2} - a_{i_1}^{(1)})] \\ &\quad \times \{1 + \prod_{i_1=1}^{3} [1 + a_0 \prod_{i_2=1, i_2 \neq i_1}^{3} (x_{k+2} - a_{i_2}^{(1)})]\} \\ &= x_{k+2} + [a_0 \prod_{i_1=1}^{3} (x_{k+2} - a_{i_1}^{(1)})][a_0^3 \prod_{i_2=1}^{3} (x_{k+2}^2 + B_{i_2}^{(2)} x_{k+2} + C_{i_2}^{(2)})] \\ &= x_{k+2} + [a_0 \prod_{j_1=1}^{3} (x_{k+2} - a_{i_1}^{(1)})][a_0^3 \prod_{j_2=1}^{3^2-3} (x_{k+2} - b_{j_2}^{(2)})] \\ &= x_{k+2} + a_0^{1+3} \prod_{i=1}^{3^2} (x_{k+2} - a_i^{(2)}) \end{aligned} \tag{2.156}$$

where

$$\begin{aligned}
&b_{i,1}^{(2)} = -\frac{1}{2}(B_i^{(2)} + \sqrt{\Delta^{(2)}}), b_{i,2}^{(2)} = -\frac{1}{2}(B_i^{(2)} - \sqrt{\Delta_i^{(2)}}),\\
&\Delta_i^{(2)} = (B_i^{(2)})^2 - 4C_i^{(2)} \geq 0, i \in I_q^{(2^0)};\\
&I_q^{(2^0)} = \{l_{(q-1)\times 2^0\times m_1+1}, l_{(q-1)\times 2^0\times m_1+2}, \cdots, l_{q\times 2^0\times m_1}\},\\
&m_1 \in \{1,2\}, q \in \{1,2\};\\
&b_{i,1}^{(2)} = -\frac{1}{2}(B_i^{(2)} + \mathbf{i}\sqrt{\Delta^{(2)}}), b_{i,2}^{(2)} = -\frac{1}{2}(B_i^{(2)} - \mathbf{i}\sqrt{\Delta_i^{(2)}}),\\
&\mathbf{i} = \sqrt{-1}, \Delta_i^{(2)} = (B_i^{(2)})^2 - 4C_i^{(2)} < 0,
\end{aligned} \tag{2.157}$$

with backward fixed-points

$$\begin{aligned}
&x_k^* = x_{k+2}^* = a_i^{(2)}, (i = 1, 2, \ldots, 3^2)\\
&\cup_{i=1}^{3^2}\{a_i^{(2)}\} = \mathrm{sort}\{\cup_{j_1=1}^{3}\{a_{j_1}^{(1)}\}, \cup_{j_2=1}^{3}\{b_{j_2,1}^{(2)}, b_{j_2,2}^{(2)}\}\}\\
&\text{with } a_i^{(2)} < a_{i+1}^{(2)}.
\end{aligned} \tag{2.158}$$

(ii) *For a backward fixed-point of* $x_{k+1}^* = x_k^* = a_{i_1}^{(1)}$ $(i_1 \in \{1,2,3\})$, *if*

$$\frac{dx_k}{dx_{k+1}}\Big|_{x_{k+1}^* = a_{i_1}^{(1)}} = 1 + a_0 \prod_{i_2=1, i_2\neq i_1}^{3} (a_{i_1}^{(1)} - a_{i_2}^{(1)}) = -1, \tag{2.159}$$

with

- *an oscillatory lower-saddle-saddle bifurcation* $(d^2x_k/dx_{k+1}^2|_{x_{k+1}^*=a_{i_1}^{(1)}} > 0)$,
- *an oscillatory upper-saddle-node bifurcation* $(d^2x_k/dx_{k+1}^2|_{x_{k+1}^*=a_{i_1}^{(1)}} < 0)$,
- *a third-order oscillatory source bifurcation* $(d^3x_k/dx_{k+1}^3|_{x_{k+1}^*=a_{i_1}^{(1)}} > 0)$,
- *a third-order oscillatory sink bifurcation* $(d^3x_k/dx_{k+1}^3|_{x_{k+1}^*=a_{i_1}^{(1)}} < 0)$,

then the following relations satisfy

$$a_{i_1}^{(1)} = -\frac{1}{2}B_{i_1}^{(2)}, \Delta_{i_1}^{(2)} = (B_{i_1}^{(2)})^2 - 4C_{i_1}^{(2)} = 0, \tag{2.160}$$

and there is a backward period-2 discrete system of the cubic discrete system in Eq. (2.148) *as*

$$x_k = x_{k+2} + a_0^4 \prod_{i_1 \in I_q^{(2^0)}} (x_{k+2} - a_{i_1}^{(1)})^3 \prod_{i_2=1}^{3^3} (x_{k+2} - a_{i_2}^{(2)})^{(1-\delta(i_1,i_2))} \tag{2.161}$$

for $i_1 \in \{1,2,3\}$, $i_1 \neq i_2$ *with*

$$\frac{dx_k}{dx_{k+2}}|_{x^*_{k+2}=a^{(1)}_{i_1}} = 1, \ \frac{d^2x_k}{dx^2_{k+2}}|_{x^*_{k+2}=a^{(1)}_{i_1}} = 0; \tag{2.162}$$

- *x_k at $x^*_{k+2} = a^{(1)}_{i_1}$ is a monotonic source of the third-order if*

$$\begin{aligned} \frac{d^3x_k}{dx^3_{k+2}}|_{x^*_k=a^{(1)}_{i_1}} &= 6a_0^4 \prod\nolimits_{i_1 \in I_q^{(20)}, i_2 \neq i_1} (a^{(1)}_{i_1} - a^{(1)}_{i_2})^3 \\ &\times \prod\nolimits_{i_3=1}^{3^3} (a^{(1)}_{i_1} - a^{(2)}_{i_3})^{(1-\delta(i_2,i_3))} < 0, \end{aligned} \tag{2.163}$$

and the corresponding bifurcations is a third-order monotonic source bifurcation for the backward period-2 discrete system;

- *x_{k+2} at $x^*_k = a^{(1)}_{i_1}$ is a monotonic sink of the third-order if*

$$\begin{aligned} \frac{d^3x_k}{dx^3_{k+2}}|_{x^*_k=a^{(1)}_{i_1}} &= 6a_0^4 \prod\nolimits_{i_1 \in I_q^{(20)}, i_2 \neq i_1} (a^{(1)}_{i_1} - a^{(1)}_{i_2})^3 \\ &\times \prod\nolimits_{i_3=1}^{3^3} (a^{(1)}_{i_1} - a^{(2)}_{i_3})^{(1-\delta(i_2,i_3))} > 0, \end{aligned} \tag{2.164}$$

and the corresponding bifurcations is a third-order monotonic sink bifurcation for the backward period-2 discrete system.

(ii_1) *The backward period-2 fixed-points are trivial and unstable if*

$$x^*_k = x^*_{k+2} = a^{(1)}_{i_1} \text{ for } i_1 = 1, 2, 3. \tag{2.165}$$

(ii_2) *The backward period-2 fixed-points are non-trivial and stable if*

$$x^*_k = x^*_{k+2} = b^{(2)}_{i_1,1}, b^{(2)}_{i_1,2} \text{ for } i_1 = 1, 2, 3. \tag{2.166}$$

Proof Following the corresponding proof for the forward cubic discrete system. This theorem can be proved. ■

2.6.2 Backward Period-Doubling Renormalization

The generalized cases of period-doublization of backward cubic discrete systems are presented through the following theorem. The analytical backward period-doubling trees can be developed for backward cubic discrete systems.

Theorem 2.6 *Consider a 1-dimensional backward cubic discrete system as*

$$\begin{aligned} x_k &= x_{k+1} + A(\mathbf{p})x_{k+1}^3 + B(\mathbf{p})x_{k+1}^2 + C(\mathbf{p})x_{k+1} + D(\mathbf{p}) \\ &= x_{k+1} + a_0 \prod_{i=1}^{3} (x_{k+1} - a_i^{(1)}). \end{aligned} \tag{2.167}$$

(i) *After l-times backward period-doubling bifurcations, a backward period-* 2^l *discrete system* $(l = 1, 2, \ldots)$ *for the backward cubic discrete system in Eq. (2.167) is given through the nonlinear renormalization as*

$$\begin{aligned} x_k &= x_{k+2^l} + [a_0^{(2^{l-1})} \prod_{i_1=1}^{3^{2^{l-1}}} (x_{k+2^l} - a_{i_1}^{(2^{l-1})})] \\ &\quad \times \{1 + \prod_{i_1=1}^{3^{2^{l-1}}} [1 + a_0^{(2^{l-1})} \prod_{i_2=1, i_2 \neq i_1}^{3^{2^{l-1}}} (x_{k+2^l} - a_{i_2}^{(2^{l-1})})]\} \\ &= x_{k+2^l} + [a_0^{(2^{l-1})} \prod_{i_1=1}^{3^{2^{l-1}}} (x_{k+2^l} - a_{i_1}^{(2^{l-1})})] \\ &\quad \times [(a_0^{(2^{l-1})})^{2^{2^{l-1}}} \prod_{j_1=1}^{(3^{2^l} - 3^{2^{l-1}})/2} (x_{k+2^l}^2 + B_{j_2}^{(2^l)} x_{k+2^l} + C_{j_2}^{(2^l)})] \\ &= x_{k+2^l} + [a_0^{(2^{l-1})} \prod_{i_1=1,}^{3^{2^{l-1}}} (x_{k+2^l} - a_{i_1}^{(2^{l-1})})] \\ &\quad \times [(a_0^{(2^{l-1})})^{2^{2^{l-1}}} \prod_{i_2=1}^{(3^{2^l} - 3^{2^{l-1}})/2} (x_{k+2^l} - b_{i_2,1}^{(2^l)})(x_{k+2^l} - b_{i_2,2}^{(2^l)})] \\ &= x_{k+2^l} + (a_0^{(2^{l-1})})^{1+3^{2^{l-1}}} \prod_{i=1}^{3^{2^l}} (x_{k+2^l} - a_i^{(2^l)}) \\ &= x_{k+2^l} + a_0^{(2^l)} \prod_{i=1}^{3^{2^l}} (x_{k+2^l} - a_i^{(2^l)}) \end{aligned} \tag{2.168}$$

with for $r \leq 2^{2^l}$

$$\begin{aligned} \frac{dx_k}{dx_{k+2^l}} &= 1 + a_0^{(2^l)} \sum_{i_1=1}^{3^{2^l}} \prod_{i_2=1, i_2 \neq i_1}^{3^{2^l}} (x_{k+2^l} - a_{i_2}^{(2^l)}), \\ \frac{d^2 x_k}{dx_{k+2^l}^2} &= a_0^{(2^l)} \sum_{i_1=1}^{3^{2^l}} \sum_{i_2=1, i_2 \neq i_1}^{3^{2^l}} \prod_{i_3=1, i_3 \neq i_1, i_2}^{3^{2^l}} (x_{k+2^l} - a_{i_3}^{(2^l)}), \\ &\vdots \\ \frac{d^r x_k}{dx_{k+2^l}^r} &= a_0^{(2^l)} \sum_{i_1=1}^{3^{2^l}} \cdots \sum_{i_r=1, i_r \neq i_1, i_2 \cdots, i_{r-1}}^{3^{2^l}} \prod_{i_{r+1}=1, i_{r+1} \neq i_1, i_2 \cdots, i_r}^{3^{2^l}} (x_{k+2^l} - a_{i_{r+1}}^{(2^l)}) \end{aligned} \tag{2.169}$$

where

$$
\begin{aligned}
&a_0^{(2)} = (a_0)^{1+3}, a_0^{(2^l)} = (a_0^{(2^{l-1})})^{1+3^{2^{l-1}}}, l = 2, 3, \cdots ;\\
&\cup_{i=1}^{3^{2^l}}\{a_i^{(2^l)}\} = \text{sort}\{\cup_{i_1=1}^{3^{2^{l-1}}}\{a_{i_1}^{(2^l)}\}, \cup_{i_2 \in=1}^{M_2}\{b_{i_2,1}^{(2^l)}, b_{i_2,2}^{(2^l)}\}\}\ , a_i^{(2^l)} \le a_{i+1}^{(2^l)};\\
&b_{i,1}^{(2^l)} = -\tfrac{1}{2}(B_i^{(2^l)} + \sqrt{\Delta_i^{(2^l)}}), b_{i,2}^{(2^{l-1})} = -\tfrac{1}{2}(B_i^{(2^l)} - \sqrt{\Delta_i^{(2^l)}}),\\
&\Delta_i^{(2^l)} = (B_i^{(2^l)})^2 - 4C_i^{(2^l)} \ge 0 \ \text{ for } \ i \in \cup_{q_1=1}^{N_1} I_{q_1}^{(2^{l-1})} \cup \cup_{q_2=1}^{N_2} I_{q_2}^{(2^l)},\\
&I_{q_1}^{(2^{l-1})} = \{l_{(q_1-1)\times 2^{l-1}\times m_1+1}, l_{(q_1-1)\times 2^{l-1}\times m_1+2}, \cdots, l_{q_1\times 2^{l-1}\times m_1}\}\\
&\qquad \subseteq \{1, 2, \cdots, M_1\} \cup \{\varnothing\},\\
&\text{for } q_1 \in \{1, 2, \cdots, N_1\}, M_1 = N_1 \times 2^{l-1} \times m_1, m_1 \in \{1, 2\};\\
&I_{q_2}^{(2^l)} = \{l_{(q_2-1)\times 2^l\times m_1+1}, l_{(q_2-1)\times 2^l\times m_1+2}, \cdots, l_{q_2\times m_1\times 2^l}\}\\
&\qquad \subseteq \{M_1+1, M_1+2, \cdots, M_2\} \cup \{\varnothing\},\\
&\text{for } q_2 \in \{1, 2, \cdots, N_2\}, M_2 = (3^{2^l} - 3^{2^{l-1}})/2;\\
&b_{i,1}^{(2^l)} = -\tfrac{1}{2}(B_i^{(2^l)} + \mathbf{i}\sqrt{|\Delta_i^{(2^l)}|}),\ b_{i,2}^{(2^l)} = -\tfrac{1}{2}(B_i^{(2^l)} - \mathbf{i}\sqrt{|\Delta_i^{(2^l)}|}),\\
&\Delta_i^{(2^l)} = (B_i^{(2^l)})^2 - 4C_i^{(2^l)} < 0,\ \mathbf{i} = \sqrt{-1},\\
&i \in J^{(2^l)} = \{l_{N_2\times 2^l\times m_1+1}, l_{N_2\times 2^l\times m_1+2}, \cdots, l_{M_2}\}\\
&\qquad \subset \{1, 2, \cdots, M_2\} \cup \{\varnothing\}
\end{aligned} \tag{2.170}
$$

with fixed-points

$$
\begin{aligned}
&x_k^* = x_{k+2^l}^* = a_i^{(2^l)}, (i = 1, 2, \ldots, 3^{2^l})\\
&\cup_{i=1}^{3^{2^l}}\{a_i^{(2^l)}\} = \text{sort}\{\cup_{i=1}^{2^{2^{l-1}}}\{a_i^{(2^{l-1})}\}, \cup_{i=1}^{M_2}\{b_{i,1}^{(2^{l-1})}, b_{i,2}^{(2^{l-1})}\}\}\\
&\text{with } a_i^{(2^l)} < a_{i+1}^{(2^l)}.
\end{aligned} \tag{2.171}
$$

(ii) *For a backward fixed-point of* $x_k^* = x_{k+2^{l-1}}^* = a_{i_1}^{(2^{l-1})} (i_1 \in I_q^{(2^{l-1})}, q \in \{1, 2, \ldots, N_1\})$*, if*

$$
\frac{dx_k}{dx_{k+2^{l-1}}}\Big|_{x_{k+2^{l-1}}^* = a_{i_1}^{(2^{l-1})}} = 1 + a_0^{(2^{l-1})} \prod_{i_2=1, i_2 \ne i_1}^{3^{2^{l-1}}} (a_{i_1}^{(2^{l-1})} - a_{i_2}^{(2^{l-1})}) = -1 \tag{2.172}
$$

then there is a backward period- 2^l *fixed-point discrete system*

$$
x_{k^l} = x_{k+2^l} + a_0^{(2^l)} \prod_{i_1 \in I_q^{(2^{l-1})}} (x_{k+2^l} - a_{i_1}^{(2^{l-1})})^3 \prod_{j_2=1}^{3^{2^l}} (x_{k+2^l} - a_{j_2}^{(2^l)})^{(1-\delta(i_1, j_2))} \tag{2.173}
$$

where

$$\delta(i_1,j_2)=1 \text{ if } a_{j_2}^{(2^l)}=a_{i_1}^{(2^{l-1})},\ \delta(i_1,j_2)=0 \text{ if } a_{j_2}^{(2^l)}\neq a_{i_1}^{(2^{l-1})} \tag{2.174}$$

with

$$\begin{aligned}
&\frac{dx_k}{dx_{k+2^l}}\Big|_{x^*_{k+2^l}=a_{i_1}^{(2^{l-1})}}=1,\ \frac{d^2x_k}{dx^2_{k+2^l}}\Big|_{x^*_{k+2^l}=a_{i_1}^{(2^{l-1})}}=0;\\
&\frac{d^3x_k}{dx^3_{k+2^l}}\Big|_{x^*_{k+2^l}=a_{i_1}^{(2^{l-1})}}=6a_0^{(2^l)}\prod\nolimits_{i_2\in I_q^{(2^{l-1})},i_2\neq i_1}(a_{i_1}^{(2^{l-1})}-a_{i_2}^{(2^{l-1})})^3\\
&\qquad\qquad\times\prod\nolimits_{j_2=1}^{3^{2^l}}(a_{i_1}^{(2^{l-1})}-a_{j_2}^{(2^l)})^{(1-\delta(i_2,j_2))}\neq 0\\
&(i_1\in I_q^{(2^{l-1})},q\in\{1,2,\ldots,N_1\})
\end{aligned} \tag{2.175}$$

x_k at $x^*_{k+2^l}=a_{i_1}^{(2^{l-1})}$ *is*

- *a monotonic sink of the third-order if* $d^3x_k/dx^3_{k+2^l}|_{x^*_{k+2^l}=a_{i_1}^{(2^{l-1})}}>0$;
- *a monotonic source of the third-order if* $d^3x_k/dx^3_{k+2^l}|_{x^*_{k+2^l}=a_{i_1}^{(2^{l-1})}}<0$.

(ii$_1$) *The backward period-* 2^l *fixed-points are trivial if*

$$x_k^*=x^*_{k+2^l}=a_{i_1}^{(2^{l-1})} \text{ for } i_1=1,2,\cdots,3^{2^{l-1}}, \tag{2.176}$$

(ii$_2$) *The backward period-* 2^l *fixed-points are non-trivial if*

$$\begin{aligned}
&x_k^*=x^*_{k+2^l}=b_{j_1,1}^{(2^l)},b_{j_1,2}^{(2^l)}\\
&j_1\in 1,2,\ldots,M_2\}\cup\{\varnothing\}
\end{aligned}. \tag{2.177}$$

Such a period- 2^l *fixed-point is*

- *monotonically stable if* $dx_k/dx_{k+2^l}|_{x^*_{k+2^l}=a_{i_1}^{(2^l)}}\in(0,\infty)$;
- *monotonically invariant if* $dx_{k+2^l}/dx_k|_{x_k^*=a_{i_1}^{(2^l)}}=1$, *which is*
  - *a monotonic lower-saddle of the* $(2l_1)^{\text{th}}$ *order for* $d^{2l_1}x_k/dx^{2l_1}_{k+2^l}|_{x^*_{k+2^l}}>0$;
  - *a monotonic upper-saddle the* $(2l_1)^{\text{th}}$ *order for* $d^{2l_1}x_k/dx^{2l_1}_{k+2^l}|_{x^*_{k+2^l}}<0$;
  - *a monotonic sink of the* $(2l_1+1)^{\text{th}}$ *order for* $d^{2l_1+1}x_k/dx^{2l_1+1}_{k+2^l}|_{x^*_{k+2^l}}>0$;
  - *a monotonic source the* $(2l_1+1)^{\text{th}}$ *order for* $d^{2l_1+1}x_k/dx^{2l_1+1}_{k+2^l}|_{x^*_{k+2^l}}<0$;
- *monotonically unstable if* $dx_k/dx_{k+2^l}|_{x^*_{k+2^l}=a_{i_1}^{(2^l)}}\in(0,1)$;
- *monotonically infinity-unstable if* $dx_k/dx_{k+2^l}|_{x^*_{k+2^l}=a_{i_1}^{(2^l)}}=0^+$;

- *oscillatorilly infinity-unstable if* $dx_k/dx_{k+2^l}|_{x^*_{k+2^l}=a^{(2^l)}_{i_1}} = 0^-$;
- *oscillatorilly unstable if* $dx_k/dx_{k+2^l}|_{x^*_{k+2^l}=a^{(2^l)}_{i_1}} \in (-1,0)$;
- *flipped if* $dx_k/dx_{k+2^l}|_{x^*_{k+2^l}=a^{(2^{l-1})}_{i_1}} = -1$, *which is*
  - *an oscillatory lower-saddle of the* $(2l_1)^{\text{th}}$ *order if* $d^{2l_1}x_k/dx^{2l_1}_{k+2^l}|_{x^*_{k+2^l}} > 0$;
  - *an oscillatory upper-saddle the* $(2l_1)^{\text{th}}$ *order with* $d^{2l_1}x_k/dx^{2l_1}_{k+2^l}|_{x^*_{k+2^l}} < 0$;
  - *an oscillatory sink of the* $(2l_1+1)^{\text{th}}$ *order if* $d^{2l_1+1}x_k/dx^{2l_1+1}_{k+2^l}|_{x^*_{k+2^l}} < 0$;
  - *an oscillatory source the* $(2l_1+1)^{\text{th}}$ *order with* $d^{2l_1+1}x_k/dx^{2l_1+1}_{k+2^l}|_{x^*_{k+2^l}} > 0$;
- *oscillatorilly stable if* $dx_k/dx_{k+2^l}|_{x^*_{k+2^l}=a^{(2^l)}_{i_1}} \in (-\infty,-1)$.

Proof Through the nonlinear renormalization, following the forward case, this theorem can be proved. ■

2.6.3 *Backward Period-*n *Appearing and Period-Doublization*

The period-n discrete system for backward cubic nonlinear discrete systems will be discussed, and the backward period-doublization of period-n discrete systems is discussed through the nonlinear renormalization.

Theorem 2.7 *Consider a 1-dimensional backward cubic nonlinear discrete system*

$$\begin{aligned} x_k &= x_{k+1} + Ax^3_{k+1} + Bx^2_{k+1} + Cx_{k+1} + D \\ &= x_{k+1} + a_0 \textstyle\prod_{i=1}^3 (x_{k+1} - a^{(1)}_i). \end{aligned} \tag{2.178}$$

(i) *After n-times iterations, a backward period-n discrete system for the cubic discrete system in Eq. (2.178) is*

$$\begin{aligned} x_k &= x_{k+n} + a_0 \textstyle\prod_{i_1=1}^3 (x_{k+n} - a_{i_2})\{1 + \sum_{j=1}^n Q_j\} \\ &= x_{k+n} + a_0^{3^n-1} \textstyle\prod_{i_1=1}^3 (x_{k+n} - a_{i_1}) \\ &\quad \times [\textstyle\prod_{j_2=1}^{(3^n-3)/2} (x^2_{k+n} + B^{(2^l)}_{j_2} x_{k+n} + C^{(2^l)}_{j_2})] \\ &= x_{k+n} + a_0^{(n)} \textstyle\prod_{i=1}^{3^n} (x_{k+n} - a^{(n)}_i) \end{aligned} \tag{2.179}$$

with for $r \leq 3^n$

$$\begin{aligned}
\frac{dx_k}{dx_{k+n}} &= 1 + a_0^{(n)} \sum_{i_1=1}^{3^n} \prod_{i_2=1, i_2 \neq i_1}^{3^n} (x_{k+n} - a_{i_2}^{(n)}), \\
\frac{d^2 x_k}{dx_{k+n}^2} &= a_0^{(n)} \sum_{i_1=1}^{3^n} \sum_{i_2=1, i_2 \neq i_1}^{3^n} \prod_{i_3=1, i_3 \neq i_1, i_2}^{3^n} (x_{k+n} - a_{i_3}^{(n)}), \\
&\vdots \\
\frac{d^r x_k}{dx_{k+n}^r} &= a_0^{(n)} \sum_{i_1=1}^{3^n} \cdots \sum_{i_r=1, i_r \neq i_1, i_2 \ldots, i_{r-1}}^{3^n} \prod_{i_{r+1}=1, i_{r+1} \neq i_1, i_2 \ldots, i_r}^{3^n} (x_{k+n} - a_{i_{r+1}}^{(n)})
\end{aligned} \tag{2.180}$$

where

$$\begin{aligned}
& a_0^{(n)} = (a_0)^{(3^n-1)/2}, Q_1 = 0, Q_2 = \prod_{i_2=1}^{3} [1 + a_0 \prod_{i_1=1, i_1 \neq i_2}^{3} (x_{k+n} - a_{i_1}^{(1)})], \\
& Q_n = \prod_{i_n=1}^{3} [1 + a_0 (1 + Q_{n-1}) \prod_{i_{n-1}=1, i_{n-1} \neq i_n}^{3} (x_{k+n} - a_{i_{n-1}}^{(1)})], \ n = 3, 4, \ldots; \\
& \cup_{i=1}^{3^n} \{a_i^{(n)}\} = \text{sort}\{\cup_{i_1=1}^{3} \{a_{i_1}^{(1)}\}, \cup_{i_1=1}^{M} \{b_{i_2,1}^{(n)}, b_{i_2,2}^{(n)}\}\}; \\
& b_{i_2,1}^{(n)} = -\frac{1}{2}(B_{i_2}^{(n)} + \sqrt{\Delta_{i_2}^{(n)}}), b_{i_2,2}^{(n)} = -\frac{1}{2}(B_{i_2}^{(n)} - \sqrt{\Delta_{i_2}^{(n)}}), \\
& \Delta_{i_2}^{(n)} = (B_{i_2}^{(n)})^2 - 4C_{i_2}^{(n)} \geq 0 \text{ for } i_2 \in \cup_{q=1}^{N} I_q^{(n)}, \\
& I_q^{(n)} = \{l_{(q-1)\times n+1}, l_{(q-1)\times n+2}, \ldots, l_{q\times n}\} \subseteq \{1, 2, \ldots, M\} \cup \{\varnothing\}, \\
& \text{for } q \in \{1, 2, \ldots, N\}, M = (3^n - 3)/2; \\
& b_{i,1}^{(n)} = -\frac{1}{2}(B_i^{(n)} + \mathbf{i}\sqrt{|\Delta_i^{(n)}|}), \ b_{i,2}^{(n)} = -\frac{1}{2}(B_i^{(n)} - \mathbf{i}\sqrt{|\Delta_i^{(n)}|}), \\
& \Delta_i^{(n)} = (B_i^{(n)})^2 - 4C_i^{(n)} < 0, \ \mathbf{i} = \sqrt{-1} \\
& i \in \{l_{N\times n+1}, l_{N\times n+2}, \ldots, l_M\} \subset \{1, 2, \ldots, M\} \cup \{\varnothing\};
\end{aligned} \tag{2.181}$$

with backward fixed-points

$$\begin{aligned}
& x_{k+n}^* = x_k^* = a_i^{(n)}, (i = 1, 2, \ldots, 3^n) \\
& \cup_{i=1}^{3^n} \{a_i^{(n)}\} = \text{sort}\{\cup_{i=1}^{3} \{a_i^{(1)}\}, \cup_{i=1}^{M_2} \{b_{i,1}^{(n)}, b_{i,2}^{(n)}\}\} \\
& \text{with } a_i^{(n)} < a_{i+1}^{(n)}.
\end{aligned} \tag{2.182}$$

(ii) *For a backward fixed-point of* $x_k^* = x_{k+n}^* = a_{i_1}^{(n)}$ $(i_1 \in I_q^{(n)}, q \in \{1, 2, \ldots, N\})$, *if*

$$\frac{dx_k}{dx_{k+n}}\Big|_{x_k^* = a_{i_1}^{(n)}} = 1 + a_0^{(n)} \prod_{i_2=1, i_2 \neq i_1}^{3^n} (a_{i_1}^{(n)} - a_{i_2}^{(2^l)}) = 1, \tag{2.183}$$

with

$$\frac{d^2x_k}{dx_{k+n}^2}|_{x_{k+n}^*=a_{i_1}^{(n)}} = a_0^{(n)}\sum_{i_2=1,i_2\neq i_1}^{3^n}\prod_{i_3=1,i_3\neq i_1,i_2}^{3^n}(a_{i_1}^{(n)} - a_{i_3}^{(2^l)}) \neq 0, \quad (2.184)$$

then there is a new discrete system for onset of the qth-set of period-n fixed-points based on the second-order monotonic saddle-node bifurcation as

$$x_k = x_{k+n} + a_0^{(n)}\prod_{i_1\in I_q^{(n)}}(x_{k+n} - a_{i_1}^{(n)})^2\prod_{j_2=1}^{3^n}(x_{k+n} - a_{j_2}^{(n)})^{(1-\delta(i_1,j_2))} \quad (2.185)$$

where

$$\delta(i_1,j_2) = 1 \text{ if } a_{j_2}^{(n)} = a_{i_1}^{(n)},\ \delta(i_1,j_2) = 0 \text{ if } a_{j_2}^{(n)} \neq a_{i_1}^{(n)}. \quad (2.186)$$

(ii$_1$) *If*

$$\begin{aligned} &\frac{dx_k}{dx_{k+n}}|_{x_{k+n}^*=a_{i_1}^{(n)}} = 1 \quad (i_1 \in I_q^{(n)}),\\ &\frac{d^2x_k}{dx_{k+n}^2}|_{x_{k+n}^*=a_{i_1}^{(n)}} = 2a_0^{(n)}\prod_{i_2\in I_q^{(n)},i_2\neq i_1}(a_{i_1}^{(n)} - a_{i_2}^{(n)})^2\\ &\qquad\qquad \times\prod_{j_2=1}^{3^n}(a_{i_1}^{(n)} - a_{j_2}^{(n)})^{(1-\delta(i_2,j_2))} \neq 0 \end{aligned} \quad (2.187)$$

x_k at $x_{k+n}^* = a_{i_1}^{(n)}$ is

- *a monotonic upper-saddle of the second-order for* $d^2x_k/dx_{k+n}^2|_{x_{k+n}^*=a_{i_1}^{(n)}} < 0$;
- *a monotonic lower-saddle of the second-order for* $d^2x_k/dx_{k+n}^2|_{x_{k+n}^*=a_{i_1}^{(n)}} > 0$.

(ii$_2$) *The backward period-n fixed-points ($n = 2^{n_1} \times m$) are trivial*

$$\left.\begin{aligned} &x_{k+n}^* = x_k^* = a_{j_1}^{(n)} \in \{\cup_{i_1=1}^{3}\{a_{i_1}^{(1)}\},\cup_{i_2=1}^{3^{2^{n_1-1}m}}\{a_{i_2}^{(2^{n_1-1}m)}\}\}\\ &\text{for } n_1 = 1,2,\ldots; m = 2l_1+1; j_1 \in \{1,2,\ldots,3^n\}\cup\{\varnothing\} \end{aligned}\right\}$$
$$\text{for } n \neq 2^{n_2},$$
$$\left.\begin{aligned} &x_{k+n}^* = x_k^* = a_{j_1}^{(n)} \in \cup_{i_2=1}^{3^{2^{n_1-1}m}}\{a_{i_2}^{(2^{n_1-1}m)}\}\\ &\text{for } n_1 = 1,2,\ldots; m = 1; j_1 \in \{1,2,\ldots,3^n\}\cup\{\varnothing\} \end{aligned}\right\}$$
$$\text{for } n = 2^{n_2}. \quad (2.188)$$

(ii$_3$) *The backward period-n fixed-points ($n = 2^{n_1} \times m$) are non-trivial if*

$$\left.\begin{array}{l} x_k^* = x_{k+n}^* = a_{j_1}^{(n)} \notin \{\cup_{i_1=1}^{3}\{a_{i_1}^{(1)}\}, \cup_{i_2=1}^{3^{2^{n_1-1}m}}\{a_{i_2}^{(2^{n_1-1}m)}\}\} \\ \text{for } n_1 = 1,2,\ldots; m = 2l_1+1; j_1 \in \{1,2,\ldots,3^n\}\cup\{\varnothing\} \end{array}\right\}$$
$$\text{for } n \neq 2^{n_2},$$
$$\left.\begin{array}{l} x_k^* = x_{k+n}^* = a_{j_1}^{(n)} \notin \cup_{i_2=1}^{3^{2^{n_1-1}m}}\{a_{i_2}^{(2^{n_1-1}m)}\} \\ \text{for } n_1 = 1,2,\ldots; m = 1; j_1 \in \{1,2,\ldots,3^n\}\cup\{\varnothing\} \end{array}\right\}$$
$$\text{for } n = 2^{n_2}. \tag{2.189}$$

Such a backward period-n fixed-point is

- *monotonically stable if* $dx_{k+n}/dx_k|_{x_k^*=a_{i_1}^{(n)}} \in (1,\infty)$;
- *monotonically invariant if* $dx_k/dx_{k+n}|_{x_{k+n}^*=a_{i_1}^{(n)}} = 1$, *which is*
  - *a monotonic lower-saddle of the* $(2l_1)^{\text{th}}$ order for $d^{2l_1}x_k/dx_{k+n}^{2l_1}|_{x_{k+n}^*} > 0$;
  - *a monotonic upper-saddle the* $(2l_1)^{\text{th}}$ *order for* $d^{2l_1}x_k/dx_{k+n}^{2l_1}|_{x_{k+n}^*} < 0$;
  - *a monotonic sink of the* $(2l_1+1)^{\text{th}}$ *order for* $d^{2l_1+1}x_k/dx_{k+n}^{2l_1+1}|_{x_{k+n}^*} > 0$;
  - *a monotonic source the* $(2l_1+1)^{\text{th}}$ *order for* $d^{2l_1+1}x_k/dx_{k+n}^{2l_1+1}|_{x_{k+n}^*} < 0$;

- *monotonically stable if* $dx_{k+n}/dx_k|_{x_k^*=a_{i_1}^{(n)}} \in (1,\infty)$;
- *monotonic infinity-unstable if* $dx_k/dx_{k+n}|_{x_{k+n}^*=a_{i_1}^{(n)}} = 0^+$;
- *oscillatory infinity-unstable if* $dx_k/dx_{k+n}|_{x_{k+n}^*=a_{i_1}^{(n)}} = 0^-$;
- *oscillatorilly unstable if* $dx_k/dx_{k+n}|_{x_{k+n}^*=a_{i_1}^{(n)}} \in (-1,0)$;
- *flipped if* $dx_k/dx_{k+n}|_{x_{k+n}^*=a_{i_1}^{(n)}} = -1$, *which is*
  - *an oscillatory lower-saddle of the* $(2l_1)^{\text{th}}$ order for $d^{2l_1}x_k/dx_{k+n}^{2l_1}|_{x_{k+n}^*} > 0$;
  - *an oscillatory upper-saddle the* $(2l_1)^{\text{th}}$ *order for* $d^{2l_1}x_k/dx_{k+n}^{2l_1}|_{x_{k+n}^*} < 0$;
  - *an oscillatory sink of the* $(2l_1+1)^{\text{th}}$ *order for* $d^{2l_1+1}x_k/dx_{k+n}^{2l_1+1}|_{x_{k+n}^*} < 0$;

– *an oscillatory source the* $(2l_1+1)^{\text{th}}$ *order for* $d^{2l_1+1}x_k/dx_{k+n}^{2l_1+1}|_{x_{k+n}^*} > 0$;

- *oscillatorilly stable if* $dx_k/dx_{k+n}|_{x_{k+n}^*=a_{i_1}^{(n)}} \in (-\infty,-1)$.

(iii) *For a backward fixed-point of* $x_k^* = x_{k+n}^* = a_{i_1}^{(n)}$ $(i_1 \in I_q^{(n)}, q \in \{1,2,\ldots,N\})$, *there is a backward period-doubling of the* q^{th}*-set of period-n fixed-points if*

$$
\begin{aligned}
&\frac{dx_k}{dx_{k+n}}|_{x_{k+n}^*=a_{i_1}^{(n)}} = 1 + a_0^{(n)} \prod\nolimits_{j_2=1,j_2\neq i_1}^{3^n} (a_{i_1}^{(n)} - a_{j_2}^{(n)}) = -1,\\
&\frac{d^s x_k}{dx_{k+n}^s}|_{x_{k+n}^*=a_{i_1}^{(n)}} = 0, \text{ for } s = 2,\ldots, r-1;\\
&\frac{d^r x_k}{dx_{k+n}^r}|_{x_{k+n}^*=a_{i_1}^{(n)}} \neq 0 \text{ for } 1<r\leq 3^n
\end{aligned}
\tag{2.190}
$$

with

- *a* r^{th}*-order oscillatory sink for* $d^r x_k/dx_{k+n}^r|_{x_{k+n}^*=a_{i_1}^{(n)}} < 0$ *and* $r = 2l_1+1$;
- a rth*-order oscillatory source for* $d^r x_k/dx_{k+n}^r|_{x_{k+n}^*=a_{i_1}^{(n)}} > 0$ *and* $r = 2l_1+1$;
- *a* r^{th}*-order oscillatory lower-saddle for* $d^r x_k/dx_{k+n}^r|_{x_{k+n}^*=a_{i_1}^{(n)}} > 0$ *and* $r = 2l_1$;
- a rth*-order oscillatory upper-saddle for* $d^r x_k/dx_{k+n}^r|_{x_{k+n}^*=a_{i_1}^{(n)}} < 0$ and $r = 2l_1$;

The corresponding period- $2\times n$ *discrete system of the backward cubic discrete system in Eq. (2.178) is*

$$
\begin{aligned}
x_k = x_{k+2\times n} &+ a_0^{(2\times n)} \prod\nolimits_{i_1\in I_q^{(n)}} (x_{k+2\times n} - a_{i_1}^{(n)})^3\\
&\times \prod\nolimits_{j_2=1}^{3^{2\times n}} (x_{k+2\times n} - a_{j_2}^{(2\times n)})^{(1-\delta(i_1,j_2))}
\end{aligned}
\tag{2.191}
$$

with

$$
\begin{aligned}
&\frac{dx_k}{dx_{k+2\times n}}|_{x_{k+2\times n}^*=a_{i_1}^{(n)}} = 1, \ \frac{d^2 x_{k+2\times n}}{dx_{k+2\times n}^2}|_{x_{k+2\times n}^*=a_{i_1}^{(n)}} = 0;\\
&\frac{d^3 x_k}{dx_{k+2\times n}^3}|_{x_{k+2\times n}^*=a_{i_1}^{(n)}} = 6a_0^{(2\times n)} \prod\nolimits_{i_2\in I_q^{(n)},i_2\neq i_1} (a_{i_1}^{(n)} - a_{i_2}^{(n)})^3\\
&\qquad\qquad \times \prod\nolimits_{j_2=1}^{3^{2\times n}} (a_{i_1}^{(n)} - a_{j_2}^{(2\times n)})^{(1-\delta(i_1,j_2))} \neq 0.
\end{aligned}
\tag{2.192}
$$

Thus, x_k *at* $x^*_{k+2\times n}=a^{(n)}_{i_1}$ *for* $i_1\in I^{(n)}_q$, $q\in\{1,2,\ldots,N\}$ *is*

- *a monotonic source of the third-order if* $d^3x_k/dx^3_{k+2\times n}|_{x^*_{k+2\times n}=a^{(n)}_{i_1}}<0$,
- *a monotonic sink of the third-order if* $d^3x_k/dx^3_{k+2\times n}|_{x^*_{k+2\times n}=a^{(n)}_{i_1}}>0$.

(iv) *After l-times backward period-doubling bifurcations of period-n* fixed points, a period-$2^l\times n$ *discrete system of the backward cubic discrete system in Eq. (2.178) is*

$$\begin{aligned}
x_k &= x_{k+2^l\times n}+[a_0^{(2^{l-1}\times n)}\prod_{i_1=1}^{3^{2^{l-1}\times n}}(x_{k+2^l\times n}-a_{i_1}^{(2^{l-1}\times n)})]\\
&\quad\times\{1+\prod_{i_1=1}^{3^{(2^{l-1}\times n)}}[1+a_0^{(2^{l-1}\times n)}\prod_{i_2=1,i_2\neq i_1}^{3^{2^{l-1}\times n}}(x_{k+2^l\times n}-a_{i_2}^{(2^{l-1}\times n)})]\}\\
&= x_{k+2^l\times n}+[a_0^{(2^{l-1}\times n)}\prod_{i_1=1}^{3^{2^{l-1}\times n}}(x_{k+2^l\times n}-a_{i_1}^{(2^{l-1}\times n)})]\\
&\quad\times[(a_0^{(2^{l-1}\times n)})^{3^{(2^{l-1}\times n)}}\prod_{j_1=1}^{(3^{2^l\times n}-3^{2^{l-1}\times n})/2}(x^2_{k+2^l\times n}+B_{j_2}^{(2^l\times n)}x_{k+2^l\times n}+C_{j_2}^{(2^l\times n)})]\\
&= x_{k+2^l\times n}+[a_0^{(2^{l-1}\times n)}\prod_{i_1=1}^{3^{2^{l-1}\times n}}(x_{k+2^l\times n}-a_{i_1}^{(2^{l-1}\times n)})]\\
&\quad\times[(a_0^{(2^{l-1}\times n)})^{3^{(2^{l-1}\times n)}}\prod_{j_2=1}^{(3^{2^l\times n}-3^{2^{l-1}\times n})/2}(x_{k+2^l\times n}-b_{j_2,1}^{(2^l\times n)})(x_{k+2^l\times n}-b_{j_2,2}^{(2^l\times n)})]\\
&= x_{k+2^l\times n}+(a_0^{(2^{l-1}\times n)})^{3^{2^{l-1}\times n}}\prod_{i=1}^{3^{2^l\times n}}(x_{k+2^l\times n}-a_i^{(2^l\times n)})\\
&= x_{k+2^l\times n}+a_0^{(2^l\times n)}\prod_{i=1}^{3^{2^l\times n}}(x_{k+2^l\times n}-a_i^{(2^l\times n)})
\end{aligned}\tag{2.193}$$

with for $r\leq 3^{2^l\times n}$

$$\begin{aligned}
\frac{dx_k}{dx_{k+2^l\times n}} &= 1+a_0^{(2^l\times n)}\sum_{i_1=1}^{3^{2^l\times n}}\prod_{i_2=1,i_2\neq i_1}^{3^{2^l\times n}}(x_{k+2^l\times n}-a_{i_2}^{(2^l\times n)}),\\
\frac{d^2x_k}{dx^2_{k+2^l\times n}} &= a_0^{(2^l\times n)}\sum_{i_1=1}^{3^{2^l\times n}}\sum_{i_2=1,i_2\neq i_1}^{3^{2^l\times n}}\prod_{i_3=1,i_3\neq i_1,i_2}^{3^{2^l\times n}}(x_{k+2^l\times n}-a_{i_3}^{(2^l\times n)}),\\
&\vdots\\
\frac{d^rx_k}{dx^r_{k+2^l\times n}} &= a_0^{(2^l\times n)}\sum_{i_1=1}^{3^{2^l\times n}}\cdots\sum_{i_r=1,i_r\neq i_1,i_2\ldots,i_{r-1}}^{3^{2^l\times n}}\prod_{i_{r+1}=1,i_{r+1}\neq i_1,i_2\ldots,i_r}^{3^{2^l\times n}}(x_{k+2^l\times n}-a_{i_{r+1}}^{(2^l\times n)})
\end{aligned}\tag{2.194}$$

where

$$
\begin{aligned}
& a_0^{(2\times n)} = (a_0^{(n)})^{1+3^{2\times n}}, a_0^{(2^l\times n)} = (a_0^{(2^{l-1}\times n)})^{1+3^{2^{l-1}\times n}}, l = 1,2,3,\ldots;\\
& \cup_{i=1}^{3^{(2^l\times n)}}\{a_i^{(2^l\times n)}\} = \mathrm{sort}\{\cup_{i_1=1}^{3^{2^{l-1}\times n}}\{a_{i_1}^{(2^{l-1}\times n)}\}, \cup_{i_2=1}^{M_2}\{b_{i_2,1}^{(2^l\times n)}, b_{i_2,2}^{(2^l\times n)}\}\};\\
& b_{i,1}^{(2^l\times n)} = -\frac{1}{2}(B_i^{(2^l\times n)} + \sqrt{\Delta_i^{(2^l\times n)}}),\\
& b_{i,2}^{(2^l\times n)} = -\frac{1}{2}(B_i^{(2^l\times n)} - \sqrt{\Delta_i^{(2^l\times n)}}),\\
& \Delta_i^{(2^l\times n)} = (B_i^{(2^l\times n)})^2 - 4C_i^{(2^l\times n)} \geq 0\\
& \text{for } i \in \cup_{q_1=1}^{N_1} I_{q_1}^{(2^l\times n)} \cup \cup_{q_2=1}^{N_2} I_{q_2}^{(2^l\times n)},\\
& I_{q_1}^{(2^{l-1}\times n)} = \{l_{(q_1-1)\times(2^{l-1}\times n)+1}, l_{(q_1-1)\times(2^{l-1}\times n)+2}, \ldots, l_{q_1\times(2^{l-1}\times n)}\}\\
& \qquad \subseteq\{1,2,\ldots,M_1\}\cup\{\varnothing\},\\
& \text{for } q_1 \in \{1,2,\ldots,N_1\}, M_1 = N_1\times(2^{l-1}\times n);\\
& I_{q_2}^{(2^l\times n)} = \{l_{(q_2-1)\times(2^l\times n)+1}, l_{(q_2-1)\times(2^l\times n)+2}, \ldots, l_{q_1\times(2^l\times n)}\}\\
& \qquad \subseteq\{M_1+1, M_1+2,\ldots,M_2\}\cup\{\varnothing\},\\
& \text{for } q_2 \in \{1,2,\ldots,N_2\}, M_2 = (3^{2^l\times n} - 3^{2^{l-1}\times n})/2;\\
& b_{i,1}^{(2^l\times n)} = -\frac{1}{2}(B_i^{(2^l\times n)} + \mathbf{i}\sqrt{|\Delta_i^{(2^l\times n)}|}),\\
& b_{i,2}^{(2^l\times n)} = -\frac{1}{2}(B_i^{(2^l\times n)} - \mathbf{i}\sqrt{|\Delta_i^{(2^l\times n)}|}),\\
& \Delta_i^{(2^l\times n)} = (B_i^{(2^l\times n)})^2 - 4C_i^{(2^l\times n)} < 0, \mathbf{i} = \sqrt{-1},\\
& i \in \{l_{N\times(2^{l-1}\times n)+1}, l_{N\times(2^{l-1}\times n)+2}, \ldots, l_{M_2}\}\\
& \quad \subset \{M_1+1, M_1+2,\ldots,M_2\}\cup\{\varnothing\}
\end{aligned}
\tag{2.195}
$$

with fixed-points

$$
\begin{aligned}
& x_k^* = x_{k+2^l\times n}^* = a_i^{(2^l\times n)}, (i = 1,2,\ldots,3^{2^l\times n})\\
& \cup_{i=1}^{3^{2^l\times n}}\{a_i^{(2^l\times n)}\} = \mathrm{sort}\{\cup_{i=1}^{3^{2^{l-1}\times n}}\{a_i^{(2^{l-1}\times n)}\}, \cup_{i=1}^{M_2}\{b_{i,1}^{(2^{l-1}\times n)}, b_{i,2}^{(2^{l-1}\times n)}\}\}\\
& \text{with } a_i^{(2^l\times n)} < a_{i+1}^{(2^l\times n)}.
\end{aligned}
\tag{2.196}
$$

(v) *For a fixed-point of* $x_k^* = x_{k+2^l\times n}^* = a_{i_1}^{(2^{l-1}\times n)}(i_1 \in I_q^{(2^{l-1}\times n)}, \ q \in \{1,2,\ldots, N_1\})$*, there is a backward period-*$(2^l\times n)$ *discrete system if*

$$
\begin{aligned}
& \frac{dx_k}{dx_{k+2^{l-1}\times n}}\Big|_{x_{k+2^{l-1}\times n}^* = a_{i_1}^{(2^{l-1}\times n)}} = -1,\\
& \frac{d^s x_k}{dx_{k+2^{l-1}\times n}^s}\Big|_{x_{k+2^{l-1}\times n}^* = a_{i_1}^{(2^{l-1}\times n)}} = 0, \text{ for } s = 2,\ldots, r-1;\\
& \frac{d^r x_k}{dx_{k+2^{l-1}\times n}^r}\Big|_{x_{k+2^{l-1}\times n}^* = a_{i_1}^{(2^{l-1}\times n)}} \neq 0 \text{ for } 1 < r \leq 3^{2^{l-1}\times n}
\end{aligned}
\tag{2.197}
$$

with

- *a* r^{th}*-order oscillatory source for* $d^r x_k/dx^r_{k+2^l\times n}|_{x^*_{k+2^{l-1}\times n}=a^{(2^l\times n)}_{i_1}} > 0$ *and* $r = 2l_1 + 1$;
- *a* r^{th}*-order oscillatory sink for* $d^r x_k/dx^r_{k+2^l\times n}|_{x^*_{k+2^{l-1}\times n}=a^{(2^l\times n)}_{i_1}} < 0$ *and* $r = 2l_1 + 1$;
- a rth*-order oscillatory upper-saddle for* $d^r x_k/dx^r_{k+2^l\times n}|_{x^*_{k+2^{l-1}\times n}=a^{(2^l\times n)}_{i_1}} < 0$ *and* $r = 2l_1$;
- a rth*-order oscillatory lower-saddle for* $d^r x_k/dx^r_{k+2^l\times n}|_{x^*_{k+2^{l-1}\times n}=a^{(2^l\times n)}_{i_1}} > 0$ *and* $r = 2l_1$.

The corresponding backward period- $(2^l \times n)$ *discrete system is*

$$\begin{aligned} x_k = x_{k+2^l\times n} + a_0^{(2^l\times n)} &\prod\nolimits_{i_1\in I_q^{(2^{l-1}\times n)}} (x_{k+2^l\times n} - a_{i_1}^{(2^{l-1}\times n)})^3 \\ &\times \prod\nolimits_{j_2=1}^{3^{2^l\times n}} (x_{k+2^l\times n} - a_{j_2}^{(2^l\times n)})^{(1-\delta(i_1,j_2))} \end{aligned} \tag{2.198}$$

where

$$\delta(i_1,j_2) = 1 \text{ if } a_{j_2}^{(2^l\times n)} = a_{i_1}^{(2^{l-1}\times n)},\ \delta(i_1,j_2) = 0 \text{ if } a_{j_2}^{(2^l\times n)} \neq a_{i_1}^{(2^{l-1}\times n)} \tag{2.199}$$

with

$$\begin{aligned} &\frac{dx_k}{dx_{k+2^l\times n}}\Big|_{x^*_{k+2^l\times n}=a_{i_1}^{(2^{l-1}\times n)}} = 1,\ \frac{d^2x_k}{dx^2_{k+2^l\times n}}\Big|_{x^*_{k+2^l\times n}=a_{i_1}^{(2^{l-1}\times n)}} = 0; \\ &\frac{d^3x_k}{dx^3_{k+2^l\times n}}\Big|_{x^*_{k+2^l\times n}=a_{i_1}^{(2^{l-1}\times n)}} = 6a_0^{(2^l\times n)} \prod\nolimits_{i_2\in I_q^{(2^{l-1}\times n)},i_2\neq i_1} (a_{i_1}^{(2^{l-1}\times n)} - a_{i_2}^{(2^{l-1}\times n)})^3 \\ &\qquad\qquad \times \prod\nolimits_{j_2=1}^{3^{2^l\times n}} (a_{i_1}^{(2^{l-1}\times n)} - a_{j_2}^{(2^l\times n)})^{(1-\delta(i_2,j_2))} \neq 0 \\ &(i_1 \in I_q^{(2^{l-1}\times n)}, q \in \{1,2,\ldots,N_1\}) \end{aligned} \tag{2.200}$$

x_k *at* $x^*_{k+2^l\times n} = a_{i_1}^{(2^{l-1}\times n)}$ is

- *a monotonic source of the third-order if* $d^3x_k/dx^3_{k+2^l\times n}|_{x^*_{k+2^l\times n}=a_{i_1}^{(2^{l-1}\times n)}} < 0$;
- *a monotonic sink of the third-order if* $d^3x_k/dx^3_{k+2^l\times n}|_{x^*_{k+2^l\times n}=a_{i_1}^{(2^{l-1}\times n)}} > 0$.

(v_1) *The backward period-* $2^l \times n$ *fixed-points are trivial if*

$$
\left.\begin{array}{l}
\left.\begin{array}{l}
x^*_{k+2^l\times n} = x^*_k = a_j^{(2^l\times n)} \in \{\cup_{i_i=1}^{3}\{a_{i_1}^{(1)}\}, \cup_{i_2=1}^{3^{2^{l-1}\times n}}\{a_{i_2}^{(2^{l-1}\times n)}\}\} \\
\text{for } j = 1,2,\ldots,3^{2^l\times n}
\end{array}\right\} \\
\text{for } n \neq 2^{n_1} \\
\left.\begin{array}{l}
x^*_{k+2^l\times n} = x^*_k = a_j^{(2^l\times n)} \in \cup_{i_2=1}^{3^{2^{l-1}\times n}}\{a_{i_2}^{(2^{l-1}\times n)}\} \\
\text{for } j = 1,2,\ldots,3^{2^l\times n}
\end{array}\right\} \\
\text{for } n = 2^{n_1}.
\end{array}\right. \tag{2.201}
$$

(v_2) *The backward period-* $2^l \times n$ *fixed-points are non-trivial if*

$$
\left.\begin{array}{l}
\left.\begin{array}{l}
x^*_{k+2^l\times n} = x^*_k = a_j^{(2^l\times n)} \notin \{\cup_{i_i=1}^{3}\{a_{i_1}^{(1)}\}, \cup_{i_2=1}^{3^{2^{l-1}\times n}}\{a_{i_2}^{(2^{l-1}\times n)}\}\} \\
\text{for } j = 1,2,\ldots,3^{2^l\times n}
\end{array}\right\} \\
\text{for } n \neq 2^{n_1} \\
\left.\begin{array}{l}
x^*_{k+2^l\times n} = x^*_k = a_j^{(2^l\times n)} \notin \cup_{i_2=1}^{3^{2^{l-1}\times n}}\{a_{i_2}^{(2^{l-1}\times n)}\} \\
\text{for } j = 1,2,\ldots,3^{2^l\times n}
\end{array}\right\} \\
\text{for } n = 2^{n_1}.
\end{array}\right. \tag{2.202}
$$

Such a backward period- $2^l \times n$ *fixed-point is*

- *monotonically stable if* $dx_k/dx_{k+2^l\times n}|_{x^*_{k+2^l\times n}=a_{i_1}^{(2^l\times n)}} \in (1,\infty)$;
- *monotonically invariant if* $dx_k/dx_{k+2^l\times n}|_{x^*_{k+2^l\times n}=a_{i_1}^{(2^l\times n)}} = 1$, *which is*
  - *a monotonic lower-saddle of the* $(2l_1)^{\text{th}}$ *order for* $d^{2l_1}x_k/dx^{2l_1}_{k+2^l\times n}|_{x^*_{k+2^l\times n}} > 0$ *(independent* $(2l_1)$*-branch appearance);*
  - *a monotonic upper-saddle the* $(2l_1)^{\text{th}}$ *order for* $d^{2l_1}x_k/dx^{2l_1}_{k+2^l\times n}|_{x^*_{k+2^l\times n}} < 0$ (independent $(2l_1)$*-branch appearance)*
  - *a monotonic sink of the* $(2l_1+1)^{\text{th}}$ *order for* $d^{2l_1+1}x_k/dx^{2l_1+1}_{k+2^l\times n}|_{x^*_{k+2^l\times n}} > 0$ *(dependent* $(2l_1+1)$*-branch appearance from one branch);*
  - *a monotonic source the* $(2l_1+1)^{\text{th}}$ *order for* $d^{2l_1+1}x_k/dx^{2l_1+1}_{k+2^l\times n}|_{x^*_{k+2^l\times n}} < 0$ *(dependent* $(2l_1+1)$*-branch appearance from one branch);*
- *monotonically unstable if* $dx_k/dx_{k+2^l\times n}|_{x^*_{k+2^l\times n}=a_{i_1}^{(2^l\times n)}} \in (0,1)$;
- *monotonically infinity-unstable if* $dx_k/dx_{k+2^l\times n}|_{x^*_{k+2^l\times n}=a_{i_1}^{(2^l\times n)}} = 0^+$;
- *oscillatorilly infinity-unstable if* $dx_k/dx_{k+2^l\times n}|_{x^*_{k+2^l\times n}=a_{i_1}^{(2^l\times n)}} = 0^-$;

- *oscillatorilly unstable if* $dx_k/dx_{k+2^l\times n}|_{x^*_{k+2^l\times n}=a^{(2^l\times n)}_{i_1}} \in (-1,0)$;
- *flipped if* $dx_k/dx_{k+2^l\times n}|_{x^*_{k+2^l\times n}=a^{(2^l\times n)}_{i_1}} = -1$, *which is*
  - *an oscillatory lower-saddle of the* $(2l_1)^{\text{th}}$ *order for* $d^{2l_1}x_k/dx^{2l_1}_{k+2^l\times n}|_{x^*_{k+2^l\times n}} > 0$;
  - *an oscillatory upper-saddle of the* $(2l_1)^{\text{th}}$ *order for* $d^{2l_1}x_k/dx^{2l_1}_{k+2^l\times n}|_{x^*_{k+2^l\times n}} < 0$;
  - *an oscillatory source of the* $(2l_1+1)^{\text{th}}$ *order for* $d^{2l_1+1}x_k/dx^{2l_1+1}_{k+2^l\times n}|_{x^*_{k+2^l\times n}} > 0$;
  - *an oscillatory sink the* $(2l_1+1)^{\text{th}}$ *order for* $d^{2l_1+1}x_k/dx^{2l_1+1}_{k+2^l\times n}|_{x^*_{k+2^l\times n}} < 0$;
- *oscillatorilly stable if* $dx_k/dx_{k+2^l\times n}|_{x^*_{k+2^l\times n}=a^{(2^l\times n)}_{i_1}} \in (-\infty,-1)$.

Proof Through the nonlinear renormalization, the proof of this theorem can follow the proof for the forward cubic discrete system. This theorem can be easily proved. ■

Reference

Luo ACJ (2019) The stability and bifurcation of fixed-points in low-degree polynomial systems. J Vib Test Syst Dyn 3(4):403–451

Chapter 3
Quartic Nonlinear Discrete Systems

In this Chapter, the stability and bifurcation of the quartic nonlinear discrete systems will be presented, which is similar to Luo (2019). The fourth-order monotonic upper-saddle and monotonic lower-saddle appearing bifurcations of two second-order monotonic upper-saddles or monotonic lower-saddles will be presented. The 3rd order monotonic sink and source switching bifurcations of monotonic lower-saddle with monotonic sink and monotonic upper-saddle with monotonic source will be discussed. Graphical illustrations of global stability and bifurcations are presented. The bifurcation trees for quartic nonlinear discrete systems are discussed through period-doublization and monotonic saddle-node bifurcations.

3.1 Period-1 Appearing Bifurcations

In this section, period-1 fixed-points in quartic nonlinear discrete systems will be discussed, and the stability and appearing and switching bifurcation conditions for period-1 fixed points will be developed.

Definition 3.1 Consider a 1-dimensional, forward, quartic nonlinear discrete system

$$
\begin{aligned}
x_{k+1} &= x_k + f(x_k, \mathbf{p}) \\
&= x_k + A(\mathbf{p})x_k^4 + B(\mathbf{p})x_k^3 + C(\mathbf{p})x_k^2 + D(\mathbf{p})x_k + E(\mathbf{p}) \\
&= x_k + a_0(\mathbf{p})[x_k^2 + B_1(\mathbf{p})x_k + C_1(\mathbf{p})][x_k^2 + B_2(\mathbf{p})x_k + C_2(\mathbf{p})]
\end{aligned}
\tag{3.1}
$$

where $A(\mathbf{p}) \neq 0$, and

$$
\mathbf{p} = (p_1, p_2, \ldots, p_m)^{\mathrm{T}}. \tag{3.2}
$$

A. C. J Luo, *Bifurcation Dynamics in Polynomial Discrete Systems*, Nonlinear Physical Science, https://doi.org/10.1007/978-981-15-5208-3_3

(i) If

$$\Delta_i = B_i^2 - 4C_i < 0 \text{ for } i = 1, 2 \tag{3.3}$$

the quartic nonlinear discrete system does not have any fixed-point, and the corresponding standard form is

$$x_{k+1} = x_k + a_0[(x_k + \frac{1}{2}B_1)^2 + \frac{1}{4}(-\Delta_1)][(x_k + \frac{1}{2}B_2)^2 + \frac{1}{4}(-\Delta_2)]. \tag{3.4}$$

The discrete flow of such a discrete system without fixed-points is called a non-fixed-points discrete flow.

(i_1) If $a_0 > 0$, the non-fixed-point discrete flow is called the positive discrete flow.

(i_2) If $a_0 < 0$, the non-fixed-point discrete flow is called the negative discrete flow.

(ii) If

$$\Delta_i = B_i^2 - 4C_i > 0 \text{ and } \Delta_j = B_j^2 - 4C_j < 0 \text{ for } i, j \in \{1, 2\}, i \neq j \tag{3.5}$$

the quartic polynomial discrete system has two simple fixed-points, i.e.,

$$x_k^* = b_1^{(i)} = -\frac{1}{2}(B_i + \sqrt{\Delta_i}), x_k^* = b_2^{(i)} = -\frac{1}{2}(B_i - \sqrt{\Delta_i}). \tag{3.6}$$

The corresponding standard form is

$$x_{k+1} = x_k + a_0(x_k - a_1)(x_k - a_2)[(x_k + \frac{1}{2}B_j)^2 + \frac{1}{4}(-\Delta_j)] \tag{3.7}$$

where

$$a_1 = \min(b_1^{(i)}, b_2^{(i)}) \text{ and } a_2 = \max(b_1^{(i)}, b_2^{(i)}) \tag{3.8}$$

Such a discrete flow of fixed-points is called a discrete flow of two simple fixed-points.

(iii) If

$$\Delta_i = B_i^2 - 4C_i = 0 \text{ and } \Delta_j = B_j^2 - 4C_j < 0 \text{ for } i, j \in \{1, 2\}, i \neq j \tag{3.9}$$

the quartic polynomial, forward discrete system has a double repeated fixed-point, i.e.,

$$x_k^* = b_1^{(i)} = -\frac{1}{2}B_i, x_k^* = b_2^{(i)} = -\frac{1}{2}B_i. \tag{3.10}$$

The corresponding standard form is

$$x_{k+1} = x_k + a_0(x_k - a_1)^2[(x_k + \tfrac{1}{2}B_j)^2 + \tfrac{1}{4}(-\Delta_j)] \tag{3.11}$$

where

$$a_1 = b_1^{(i)} = b_2^{(i)}. \tag{3.12}$$

Such a discrete flow of the fixed-point of $x_k^* = a_1$ is called

- a monotonic *upper-saddle* discrete flow of the second-order for $a_0 > 0$;
- a monotonic *lower-saddle* discrete flow of the second-order for $a_0 < 0$.

The fixed-point of $x^* = a_1$ for two fixed-points appearance or vanishing is called

- a monotonic upper-saddle appearing bifurcation of the second-order for two fixed-points at $\mathbf{p} = \mathbf{p}_1 \in \partial\Omega_{12}$ for $a_0 > 0$;
- a monotonic lower-saddle appearing bifurcation of the second-order for two fixed-points at $\mathbf{p} = \mathbf{p}_1 \in \partial\Omega_{12}$ for $a_0 < 0$.

The appearing bifurcation condition of the upper or lower-saddle is

$$\Delta_i = B_i^2 - 4C_i = 0 (i \in \{1,2\}) \text{ and } a_1 = -\tfrac{1}{2}B_i. \tag{3.13}$$

(iv) If

$$\Delta_i = B_i^2 - 4C_i \geq 0 \text{ for } i = 1,2, \tag{3.14}$$

the quartic nonlinear discrete system has four fixed-points, i.e.,

$$x_k^* = b_1^{(i)} = -\tfrac{1}{2}(B_i + \sqrt{\Delta_i}), x_k^* = b_2^{(i)} = -\tfrac{1}{2}(B_i - \sqrt{\Delta_i}) \text{ for } i = 1,2. \tag{3.15}$$

(iv_1) A standard form is

$$x_{k+1} = x_k + a_0(x_k - a_1)(x_k - a_2)(x_k - a_3)(x_k - a_4) \tag{3.16}$$

where

$$\begin{aligned} &\Delta_i = B_i^2 - 4C_i > 0, \ i = 1,2; \\ &b_k^{(1)} \neq b_l^{(2)} \text{ for } k,l \in \{1,2\}; \\ &a_{1,2,3,4} \in \{b_1^{(1)}, b_2^{(1)}, b_1^{(2)} b_2^{(2)}\} \text{ with } a_m < a_{m+1}. \end{aligned} \tag{3.17}$$

Such a discrete flow of fixed-points is called a discrete flow of four simple fixed-points.

(iv$_2$) The corresponding standard form is

$$x_{k+1} = x_k + a_0(x_k - a_{i_1})^2(x_k - a_{i_2})(x_k - a_{i_3}) \tag{3.18}$$

where

$$\begin{aligned}
&\Delta_i = B_i^2 - 4C_i > 0, \ \ \Delta_j = B_j^2 - 4C_j > 0 \text{ for } i,j = 1,2; \\
&a_{i_1} = b_k^{(i)} = b_l^{(j)}, (i,k) \neq (j,l); i,j,k,l \in \{1,2\}, \\
&a_{i_1} \notin \{a_{i_2}, a_{i_3}\} \text{ for } i_\alpha \in \{1,2,3,4\} \text{ and } \alpha \in \{1,2,3,4\}.
\end{aligned} \tag{3.19}$$

(iv$_{2a}$) Such a discrete flow of fixed-point $x_k^* = a_{i_1}$ ($a_{i_1} < \min\{a_{i_2}, a_{i_3}\}$ or $a_{i_1} > \max\{a_{i_2}, a_{i_3}\}$) is called

- a monotonic *upper-saddle* discrete flow of the second-order for $a_0 > 0$;
- a monotonic *lower-saddle* discrete flow of the second-order for $a_0 < 0$.

The fixed-point of $x_k^* = a_{i_1}$ ($a_{i_1} < \min\{a_{i_2}, a_{i_3}\}$ or $a_{i_1} > \max\{a_{i_2}, a_{i_3}\}$) for two fixed-points switching is called

- a monotonic *upper-saddle* switching bifurcation of the second-order for fixed-points at a point $\mathbf{p} = \mathbf{p}_1 \in \partial\Omega_{12}$ for $a_0 > 0$;
- a monotonic *lower-saddle* switching bifurcation of the second-order for fixed-points at a point $\mathbf{p} = \mathbf{p}_1 \in \partial\Omega_{12}$ for $a_0 < 0$.

(iv$_{2b}$) Such a discrete flow of fixed-point $x_k^* = a_{i_1}$ ($\min\{a_{i_2}, a_{i_3}\} < a_{i_1} < \max\{a_{i_2}, a_{i_3}\}$) is called

- a monotonic *lower-saddle* discrete flow of the second order for $a_0 > 0$;
- a monotonic *upper-saddle* discrete flow of the second order for $a_0 < 0$.

The fixed-point of $x_k^* = a_{i_1}$ ($\min\{a_{i_2}, a_{i_3}\} < a_{i_1} < \max\{a_{i_2}, a_{i_3}\}$) for two fixed-points switching is called

- a monotonic *lower-saddle* switching bifurcation of the second-order for fixed-points at a point $\mathbf{p} = \mathbf{p}_1 \in \partial\Omega_{12}$ for $a_0 > 0$;
- a monotonic *upper-saddle* switching bifurcation of the second-order for fixed-points at a point $\mathbf{p} = \mathbf{p}_1 \in \partial\Omega_{12}$ for $a_0 < 0$.

(iv$_{2c}$) The corresponding monotonic supper- or lower-saddle switching bifurcation condition for switching of two fixed-points is

$$\begin{aligned}
&\Delta_i = B_i^2 - 4C_i > 0 \ (i \in \{1,2\}) \text{ and} \\
&\Delta_j = B_j^2 - 4C_j > 0 \ (j \in \{1,2\}), \\
&b_k^{(i)} = b_l^{(j)}, (i,k) \neq (j,l), (i,j,k,l \in \{1,2\}).
\end{aligned} \tag{3.20}$$

(iv_{2d}) The corresponding monotonic upper-or lower-saddle appearing bifurcation condition for appearance or vanishing of two fixed-points is

$$\begin{aligned}
&\Delta_i = B_i^2 - 4C_i = 0 \ (i \in \{1,2\}) \text{ and} \\
&\Delta_j = B_j^2 - 4C_j > 0 \ (j \in \{1,2\}, j \neq i), \\
&b_k^{(i)} = b_l^{(i)}, (i,k) \neq (j,l), (i,j,k,l \in \{1,2\}).
\end{aligned} \tag{3.21}$$

(iv_3) The corresponding standard form is

$$x_{k+1} = x_k + a_0(x_k - a_{i_1})^3(x_k - a_{i_2}) \tag{3.22}$$

where

$$\begin{aligned}
&\Delta_i = B_i^2 - 4C_i > 0, \ \ \Delta_j = B_j^2 - 4C_j = 0 \text{ for } i,j = 1,2; \\
&a_{i_1} = -\tfrac{1}{2}B_j = b_l^{(i)}, a_{i_2} = b_k^{(i)}, \ k \neq l; k,l \in \{1,2\}, \\
&a_{i_1} = a_{i_3} \text{ for } i_\alpha \in \{1,2,3,4\} \text{ and } \alpha \in \{1,2,3,4\}.
\end{aligned} \tag{3.23}$$

(iv_{3a}) Such a discrete flow of the fixed-point of $x_k^* = a_{i_1}$ $(a_{i_1} < a_{i_2})$ is called

- a monotonic sink discrete flow of the third-order for $a_0 > 0$;
- a monotonic source discrete flow of the third-order for $a_0 < 0$.

The fixed-point of $x^* = a_{i_1}$ for one fixed-point to three fixed-points is called

- a monotonic sink switching bifurcation of the third-order for fixed-point at point $\mathbf{p} = \mathbf{p}_1 \in \partial\Omega_{12}$ for $a_0 > 0$;
- a monotonic source switching bifurcation of the third-order for fixed-points at point $\mathbf{p} = \mathbf{p}_1 \in \partial\Omega_{12}$ for $a_0 < 0$.

(iv_{3b}) Such a discrete flow of the fixed-point of $x_k^* = a_{i_1}$ $(a_{i_1} > a_{i_2})$ is called

- a monotonic source discrete flow of the third-order for $a_0 > 0$;
- a monotonic sink discrete flow of the third-order for $a_0 < 0$.

The fixed-point of $x^* = a_{i_1}$ for one fixed-point to three fixed-points is called

- a monotonic source switching bifurcation of the third-order at point $\mathbf{p} = \mathbf{p}_1 \in \partial\Omega_{12}$ for $a_0 > 0$;
- a monotonic sink switching bifurcation of the third-order at point $\mathbf{p} = \mathbf{p}_1 \in \partial\Omega_{12}$ for $a_0 < 0$.

(iv_{3c}) The corresponding monotonic sink or source switching bifurcation condition of the third-order is

$$\begin{aligned}&\Delta_i = B_i^2 - 4C_i > 0\ (i \in \{1,2\})\ \text{and}\\&\Delta_j = B_j^2 - 4C_j = 0\ (j \in \{1,2\});\\&b_k^{(i)} = -\tfrac{1}{2}B_j, b_k^{(i)} \neq b_l^{(i)}, (k \neq l,\ k, l \in \{1,2\}).\end{aligned} \tag{3.24}$$

(iv_4) The corresponding standard form is

$$x_{k+1} = x_k + a_0(x_k - a_1)^2(x_k - a_2)^2 \tag{3.25}$$

where

$$\begin{aligned}&\Delta_i = B_i^2 - 4C_i = 0, i = 1,2;\\&b_1^{(1)} = b_2^{(1)} = -\tfrac{1}{2}B_1, b_1^{(2)} = b_2^{(2)} = -\tfrac{1}{2}B_2, B_1 \neq B_2;\\&a_1 = \min\{-\tfrac{1}{2}B_1, -\tfrac{1}{2}B_2\}, a_2 = \max\{-\tfrac{1}{2}B_1, -\tfrac{1}{2}B_2\}.\end{aligned} \tag{3.26}$$

Such a discrete flow with the two fixed-points of $x_k^* = a_1$ and $x_k^* = a_2$ is called

- a (2nd mUS:2nd mUS) discrete flow for $a_0 > 0$;
- a (2nd mLS:2nd mLS) discrete flow for $a_0 < 0$.

The fixed-points of $x_k^* = a_1$ and $x_k^* = a_2$ for two sets of two fixed-points switching or appearing are called two upper- or lower-saddle switching or appearing bifurcations of the second-order at a point $\mathbf{p} = \mathbf{p}_1 \in \partial\Omega_{12}$, and the bifurcation condition is

$$\begin{aligned}&\Delta_i = B_i^2 - 4C_i = 0,\ i = 1,2;\\&b_1^{(1)} = b_2^{(1)} = -\tfrac{1}{2}B_1,\ b_1^{(2)} = b_2^{(2)} = -\tfrac{1}{2}B_2.\end{aligned} \tag{3.27}$$

(iv_5) The corresponding standard form is

$$x_{k+1} = x_k + a_0(x_k - a_1)^4 \tag{3.28}$$

where

$$\begin{aligned}&\Delta_i = B_i^2 - 4C_i = 0,\ i = 1,2;\\&b_1^{(i)} = b_2^{(i)} = -\tfrac{1}{2}B_i,\ B_1 = B_2.\end{aligned} \tag{3.29}$$

Such a discrete flow at the fixed-point of $x_k^* = a_1$ is called

- a monotonic *upper-saddle* discrete flow of the fourth-order for $a_0 > 0$;
- a monotonic *lower-saddle* discrete flow of the fourth-order for $a_0 < 0$.

The fixed-point of $x_k^* = a_1$ for two double repeated fixed-points switching or four simple fixed-points appearance is called

- a monotonic *upper-saddle* switching or appearing bifurcation of the fourth-order at point $\mathbf{p} = \mathbf{p}_1 \in \partial\Omega_{12}$ for $a_0 > 0$;
- a monotonic *lower-saddle* switching or appearing bifurcation of the fourth-order at point $\mathbf{p} = \mathbf{p}_1 \in \partial\Omega_{12}$ for $a_0 < 0$.

The corresponding upper- or lower-saddle bifurcation condition is

$$\Delta_i = B_i^2 - 4C_i = 0,\ a_1 = b_1^{(i)} = b_2^{(i)},\ i = 1, 2. \tag{3.30}$$

Theorem 3.1

(i) *Under conditions of*

$$\Delta_i = B_i^2 - 4C_i < 0 \quad \text{for } i = 1, 2 \tag{3.31}$$

a standard form of Eq. (3.1) is

$$\begin{aligned} x_{k+1} &= x_k + f(x_k, \mathbf{p}) \\ &= x_k + a_0[(x_k + \tfrac{1}{2}B_1)^2 + \tfrac{1}{4}(-\Delta_1)][(x_k + \tfrac{1}{2}B_2)^2 + \tfrac{1}{4}(-\Delta_2)] \end{aligned} \tag{3.32}$$

with $a_0 = A(\mathbf{p})$, which has a non-fixed-point discrete flow.

(i_1) *If $a_0(\mathbf{p}) > 0$, the non-fixed-point discrete flow is called a positive discrete flow.*
(i_2) *If $a_0(\mathbf{p}) > 0$, the non-fixed-point discrete flow is called a negative discrete flow.*

(ii) *Under a condition of*

$$\Delta_i = B_i^2 - 4C_i > 0 \text{ and } \Delta_j = B_j^2 - 4C_j < 0 \quad \text{for } i, j \in \{1, 2\}, i \neq j \tag{3.33}$$

a standard form of Eq. (3.1) *is*

$$\begin{aligned} x_{k+1} &= x_k + f(x_k, \mathbf{p}) \\ &= x_k + a_0(x_k - a_1)(x_k - a_2)[(x_k + \tfrac{1}{2}B_j)^2 + \tfrac{1}{4}(-\Delta_j)] \end{aligned} \tag{3.34}$$

where

$$
\begin{aligned}
&a_1 = \min(b_1^{(i)}, b_2^{(i)}) \text{ and } a_2 = \max(b_1^{(i)}, b_2^{(i)}),\\
&b_1^{(i)} = -\tfrac{1}{2}(B_i + \sqrt{\Delta_i}), b_2^{(i)} = -\tfrac{1}{2}(B_i - \sqrt{\Delta_i}).
\end{aligned}
\tag{3.35}
$$

(ii_{1a}) *For* $a_0(\mathbf{p}) > 0$, *the fixed-point of* $x_k^* = a_1$ *is*

- *monotonically stable (a monotonic sink) if* $df/dx_k|_{x_k^*=a_1} \in (-1, 0)$;
- *invariantly stable (an invariant sink) if* $df/dx_k|_{x_k^*=a_1} = -1$;
- *oscillatorilly stable (an oscillatory sink) if* $df/dx_k|_{x_k^*=a_1} \in (-2, -1)$;
- *slipped if* $df/dx_k|_{x_k^*=a_1} = -2$, *which is*
  - *an oscillatory upper-saddle of the second-order for* $d^2f/dx_k^2|_{x_k^*=a_1} > 0$;
  - *an oscillatory lower-saddle of the second-order for* $d^2f/dx_k^2|_{x_k^*=a_1} < 0$;
- *oscillatorilly unstable (an oscillatory source) if* $df/dx_k|_{x_k^*=a_1} \in (-\infty, -2)$.

(ii_{1b}) *For* $a_0(\mathbf{p}) > 0$, *the fixed-point of* $x_k^* = a_2$ *is monotonically unstable (a monotonic source) if* $df/dx_k|_{x_k^*=a_2} \in (0, \infty))$.

(ii_{2a}) *For* $a_0(\mathbf{p}) < 0$, *the fixed-point of* $x_k^* = a_1$ *is monotonically unstable (a monotonic source) if* $df/dx_k|_{x_k^*=a_1} \in (0, \infty)$.

(ii_{2b}) *For* $a_0(\mathbf{p}) < 0$, *the fixed-point of* $x_k^* = a_2$ *is*

- *monotonically stable (a monotonic sink) if* $df/dx_k|_{x_k^*=a_2} \in (-1, 0))$;
- *invariantly stable (an invariant sink) if* $df/dx_k|_{x_k^*=a_2} = -1$;
- *oscillatorilly stable (an oscillatory sink) if* $df/dx_k|_{x_k^*=a_2} \in (-2, -1))$;
- *slipped if* $df/dx_k|_{x_k^*=a_2} = -2$, *which is*
  - *an oscillatory upper-saddle of the second-order for* $d^2f/dx_k^2|_{x_k^*=a_2} > 0$;
  - *an oscillatory lower-saddle of the second-order for* $d^2f/dx_k^2|_{x_k^*=a_2} < 0$;
- *oscillatorilly unstable (an oscillatory source) if* $df/dx_k|_{x_k^*=a_2} \in (-\infty, -2)$.

(iii) *Under conditions of*

$$\Delta_i = B_i^2 - 4C_i = 0 \text{ and } \Delta_j = B_j^2 - 4C_j < 0 \quad \text{for } i,j \in \{1,2\}, i \neq j \tag{3.36}$$

a standard form of Eq. (3.1) *is*

$$\begin{aligned} x_{k+1} &= x_k + f(x_k, \mathbf{p}) \\ &= a_0(x_k - a_1)^2[(x_k + \tfrac{1}{2}B_j)^2 + \tfrac{1}{4}(-\Delta_j)] \end{aligned} \tag{3.37}$$

where

$$a_1 = b_1^{(i)} = b_2^{(i)} = -\tfrac{1}{2}B_i. \tag{3.38}$$

(iii$_1$) *For* $a_0(\mathbf{p}) > 0$*, the fixed-point of* $x_k^* = a_1$ *is monotonically unstable (a monotonic* upper-saddle, $d^2f/dx_k^2|_{x_k^*=a_1} > 0$).

- *Such a discrete flow at the fixed-point of* $x_k^* = a_1$ *is called a monotonic upper-saddle discrete flow of the second-order.*
- *The bifurcation of fixed-point of at* $x_k^* = a_1$ *for two fixed-points appearance or vanishing is called a monotonic* upper-saddle-node *appearing bifurcation of the second-order at a point* $\mathbf{p} = \mathbf{p}_1 \in \partial\Omega_{12}$.

(iii$_2$) *For* $a_0(\mathbf{p}) < 0$*, the fixed-point of* $x_k^* = a_1$ *is monotonically unstable (a monotonic* lower-saddle, $d^2f/dx_k^2|_{x_k^*=a_1} < 0$).

- *Such a discrete flow at the fixed-point of* $x_k^* = a_1$ *is called a monotonic lower-saddle discrete flow of the second-order.*
- *The bifurcation of fixed-point of at* $x_k^* = a_1$ *for two fixed-points appearance or vanishing is called a monotonic lower-saddle-node bifurcation of the second-order at a point* $\mathbf{p} = \mathbf{p}_1 \in \partial\Omega_{12}$.

(iv) *Under conditions of*

$$\begin{aligned} &\Delta_i = B_i^2 - 4C_i > 0, \ i = 1,2 \\ &b_k^{(1)} \neq b_l^{(2)} \text{ for } k,l \in \{1,2\}; \\ &b_1^{(i)} = -\tfrac{1}{2}(B_i + \sqrt{\Delta_i}), b_2^{(i)} = -\tfrac{1}{2}(B_i - \sqrt{\Delta_i}) \text{ for } i = 1,2 \end{aligned} \tag{3.39}$$

a standard form is

$$\begin{aligned} x_{k+1} &= x_k + f(x_k, \mathbf{p}) \\ &= x_k + a_0(x_k - a_1)(x_k - a_2)(x_k - a_3)(x_k - a_4) \end{aligned} \tag{3.40}$$

where

$$a_{1,2,3,4} \in \cup_{i=1}^2 \{b_1^{(i)}, b_2^{(i)}\} \text{ with } a_m < a_{m+1}. \tag{3.41}$$

(iv_1) *For* $a_0(\mathbf{p}) > 0$, *the fixed-points of* $x_k^* = a_1, a_2, a_3, a_4$ *are monotonically stable to oscillatorilly unstable, monotonically unstable, monotonically stable to oscillatorilly unstable, and monotonically unstable, respectively. The discrete flow is called a (mSI-oSO:mSO:mSI-oSO:mSO) discrete flow.*

(iv_2) *For* $a_0(\mathbf{p}) < 0$, *the fixed-points of* $x_k^* = a_1, a_2, a_3, a_4$ *are monotonically unstable, monotonically stable to oscillatorilly unstable, monotonically unstable, and monotonically stable to oscillatorilly unstable, respectively. The discrete flow is called a (mSO:mSI-oSO:mSO:mSI-oSO) discrete flow.*

(iv_3) *The fixed-point of* $x_k^* = a_i$ $(i = 1, 2, 3, 4)$ *is*

- *monotonically unstable (a monotonic source) if* $df/dx_k|_{x_k^*=a_i} \in (0, \infty)$;
- *monotonically stable (a monotonic sink) if* $df/dx_k|_{x_k^*=a_i} \in (-1, 0)$;
- *invariantly stable (an invariant sink) if* $df/dx_k|_{x_k^*=a_i} = -1$;
- *oscillatorilly stable (an oscillatory sink) if* $df/dx_k|_{x_k^*=a_i} \in (-2, -1)$;
- *flipped if* $df/dx|_{x^*=a_i} = -2$, *which is*
  - *an oscillatory upper-saddle of the second-order for* $d^2f/dx_k^2|_{x_k^*=a_i} > 0$;
  - *an oscillatory lower-saddle of the second-order for* $d^2f/dx_k^2|_{x_k^*=a_i} < 0$;
- *oscillatorilly unstable (an oscillatory source) if* $df/dx|_{x_k^*=a_i} \in (-\infty, -2)$.

(v) *Under conditions of*

$$\begin{array}{l}\Delta_i = B_i^2 - 4C_i > 0\ (i \in \{1,2\}) \text{ and} \\ \Delta_j = B_j^2 - 4C_j > 0 (j \in \{1,2\}), \\ b_1^{(\alpha)} = -\frac{1}{2}(B_\alpha + \sqrt{\Delta_\alpha}), b_2^{(\alpha)} = -\frac{1}{2}(B_\alpha - \sqrt{\Delta_\alpha}) \text{ for } \alpha = i,j \\ b_k^{(i)} = b_l^{(j)},\ (i,k) \neq (j,l), (i,j,k,l \in \{1,2\}),\end{array} \tag{3.42}$$

a standard form of Eq. (3.1) *is*

$$x_{k+1} = x_k + f(x_k, \mathbf{p}) = x_k + a_0(x_k - a_{i_1})^2 (x_k - a_{i_2})(x_k - a_{i_3}) \tag{3.43}$$

where

$$\begin{array}{l}a_{i_1} = b_k^{(i)} = b_l^{(j)} \in \cup_{i=1}^2 \{b_1^{(i)}, b_1^{(i)}\},\ (i,k) \neq (j,l); i,j,k,l \in \{1,2\}, \\ a_{i_1} \notin \{a_{i_2}, a_{i_3}\} \subset \cup_{i=1}^2 \{b_1^{(i)}, b_1^{(i)}\} \text{ for } i_\alpha \in \{1,2,3\} \text{ and } \alpha \in \{1,2,3\}.\end{array} \tag{3.44}$$

(v_1) *The fixed-points of* $x_k^* = a_{i_2}, a_{i_3}$ *are*

- *monotonically unstable (a monotonic source) if* $df/dx_k|_{x_k^*=a_{i_2},a_{i_3}} \in (0, \infty)$;
- *monotonically stable (a monotonic sink) if* $df/dx_k|_{x_k^*=a_{i_2},a_{i_3}} \in (-1, 0)$;
- *invariantly stable (an invariant sink) if* $df/dx_k|_{x_k^*=a_{i_2},a_{i_3}} = -1$;
- *oscillatorilly stable (an oscillatory sink) if* $df/dx_k|_{x_k^*=a_{i_2},a_{i_3}} \in (-2, -1)$;

- *flipped if* $df/dx_k|_{x_k^*=a_{i_2},a_{i_3}} = -2$, *which is*
  - *an oscillatory upper-saddle of the second-order for* $d^2f/dx_k^2|_{x_k^*=a_{i_2},a_{i_3}} > 0$;
  - *an oscillatory lower-saddle of the second-order for* $d^2f/dx_k^2|_{x_k^*=a_{i_2},a_{i_3}} < 0$;
- oscillatorilly unstable (an oscillatory source, $df/dx_k|_{x_k^*=a_{i_2},a_{i_3}} \in (-\infty, -2)$).

(v_2) *The fixed-point of* $x_k^* = a_{i_1}$ *is*

- *monotonically unstable (a monotonic upper-saddle,* $d^2f/dx_k^2|_{x_k^*=a_{i_1}} > 0$),
- *monotonically unstable (a monotonic lower-saddle,* $d^2f/dx_k^2|_{x_k^*=a_{i_1}} < 0$).

(v_3) *The bifurcation of fixed-point at* $x^* = a_{i_1}$ *for two fixed-points switching or appearing is called*

- *a monotonic upper-saddle-node* ($d^2f/dx_k^2|_{x_k^*=a_{i_1}} > 0$) *switching or appearing bifurcation of the second-order at a point* $\mathbf{p} = \mathbf{p}_1 \in \partial\Omega_{12}$;
- *a monotonic lower-saddle-node* ($d^2f/dx_k^2|_{x_k^*=a_{i_1}} < 0$) *switching or appearing bifurcation of the second-order at a point* $\mathbf{p} = \mathbf{p}_1 \in \partial\Omega_{12}$.

(vi) *Under conditions of*

$$\begin{aligned}
&\Delta_i = B_i^2 - 4C_i > 0\ (i \in \{1,2\}) \text{ and} \\
&\Delta_j = B_j^2 - 4C_j = 0\ (j \in \{1,2\}), \\
&b_1^{(i)} = -\frac{1}{2}(B_i + \sqrt{\Delta_i}), b_2^{(i)} = -\frac{1}{2}(B_i - \sqrt{\Delta_i}), \\
&b_{1,2}^{(j)} = -\frac{1}{2}B_j, b_k^{(i)} = b_l^{(i)}, (k \neq l, k, l \in \{1,2\})
\end{aligned} \tag{3.45}$$

a standard form of Eq. (3.1) *is*

$$x_{k+1} = x_k + f(x_k, \mathbf{p}) = x_k + a_0(x_k - a_{i_1})^3(x_k - a_{i_2}) \tag{3.46}$$

where

$$\begin{aligned}
&a_{i_1} = b_{1,2}^{(j)} = b_l^{(i)},\ a_{i_2} = b_k^{(i)}, a_1 < a_2; \\
&\text{for } i, j, l \in \{1,2\},\ i_\alpha \in \{1,2\} \text{ and } \alpha \in \{1,2\}.
\end{aligned} \tag{3.47}$$

(vi_1) *The fixed-point of* $x_k^* = a_{i_2}$ *is*

- *monotonically unstable (a monotonic source) if* $df/dx_k|_{x_k^*=a_{i_2}} \in (0, \infty)$;
- *monotonically stable (a monotonic sink) if* $df/dx_k|_{x_k^*=a_{i_2}} \in (-1, 0)$;
- *invariantly stable (an invariant sink) if* $df/dx_k|_{x_k^*=a_{i_2}} = -1$;

- *oscillatorilly stable (an oscillatory sink if* $df/dx_k|_{x_k^*=a_{i_2}} \in (-2,-1)$*;*
- *flipped if* $df/dx_k|_{x_k^*=a_{i_2}} = -2$*, which is*
  - *an oscillatory upper-saddle of the second-order for* $d^2f/dx_k^2|_{x_k^*=a_{i_2}} > 0$*;*
  - *an oscillatory lower-saddle of the second-order for* $d^2f/dx_k^2|_{x_k^*=a_{i_2}} < 0$*;*
- *oscillatorilly unstable (an oscillatory source) if* $df/dx_k|_{x_k^*=a_{i_2}} \in (-\infty,-2)$.

(vi$_2$) *The fixed-point of* $x_k^* = a_{i_1}$ *with* $df/dx_k|_{x_k^*=a_{i_1}} = 0$ and $d^2f/dx_k^2|_{x_k^*=a_{i_1}} = 0$ *is*

- *unstable of the third-order monotonic source for* $d^3f/dx_k^3|_{x_k^*=a_{i_1}} > 0$*;*
- *stable of the third-order monotonic sink for* $d^3f/dx_k^3|_{x_k^*=a_{i_1}} < 0$.

(vi$_3$) *The bifurcation of fixed-point at* $x_k^* = a_{i_1}$ *with* $df/dx_k|_{x_k^*=a_{i_1}} = 0$ *and* $d^2f/dx_k^2|_{x_k^*=a_{i_1}} = 0$ *for one fixed-point to three fixed-points is called*

- *a monotonic source switching bifurcation of the third-order at a point* $\mathbf{p} = \mathbf{p}_1 \in \partial\Omega_{12}$ $(d^3f/dx_k^3|_{x_k^*=a_{i_1}} > 0)$,
- *a monotonic sink switching bifurcation of the third-order at a point* $\mathbf{p} = \mathbf{p}_1 \in \partial\Omega_{12}$ $(d^3f/dx_k^3|_{x^*=a_{i_1}} < 0)$.

(vii) *Under conditions of*

$$\begin{aligned}&\Delta_i = B_i^2 - 4C_i = 0 (i \in \{1,2\}) \text{ and}\\&\Delta_j = B_j^2 - 4C_j = 0 (j \in \{1,2\}),\\&b_1^{(\alpha)} = b_2^{(\alpha)} = -\tfrac{1}{2}B_\alpha \text{ for } \alpha = i,j,\\&B_1 \neq B_2,\end{aligned} \tag{3.48}$$

a standard form of Eq. (3.1) *is*

$$x_{k+1} = x_k + f(x_k,\mathbf{p}) = x_k + a_0(x_k - a_1)^2(x_k - a_2)^2 \tag{3.49}$$

where

$$a_1 = \min\{-\tfrac{1}{2}B_1, -\tfrac{1}{2}B_2\}, a_2 = \max\{-\tfrac{1}{2}B_1, -\tfrac{1}{2}B_2\}. \tag{3.50}$$

(vii$_{1a}$) *For* $a_0(\mathbf{p}) > 0$*, the fixed-points of* $x_k^* = a_i$ $(i = 1,2)$ *are unstable of a monotonic upper-saddle of the second-order if* $d^2f/dx_k^2|_{x_k^*=a_i} > 0$.

(vii$_{1b}$) *The fixed-points of* $x_k^* = a_i$ $(i = 1,2)$ *for two fixed-points vanishing and appearance are called a monotonic upper-saddle-node appearing bifurcation of the second-order at a point* $\mathbf{p} = \mathbf{p}_1 \in \partial\Omega_{12}$.

(vii$_{2a}$) *For* $a_0(\mathbf{p})<0$, *the fixed-points of* $x_k^* = a_i$ $(i = 1,2)$ *are unstable of a monotonic lower-saddle of the second-order if* $d^2f/dx_k^2|_{x_k^*=a_i}<0$.

(vii$_{2b}$) *The fixed-point of* $x_k^* = a_i$ $(i = 1,2)$ *for two fixed-points vanishing and appearance are called a monotonic lower-saddle-node appearing bifurcation of the second-order at a point* $\mathbf{p} = \mathbf{p}_1 \in \partial\Omega_{12}$.

(viii) *Under conditions of*

$$\begin{aligned}
&\Delta_i = B_i^2 - 4C_i = 0\ (i \in \{1,2\}) \text{ and}\\
&\Delta_j = B_j^2 - 4C_j = 0\ (j \in \{1,2\}),\\
&b_1^{(\alpha)} = b_2^{(\alpha)} = -\frac{1}{2}B_\alpha \text{ for } \alpha = i,j;\\
&B_1 \neq B_2,
\end{aligned} \tag{3.51}$$

the corresponding standard form is

$$x_{k+1} = x_k + f(x_k, \mathbf{p}) = x_k + a_0(x_k - a_1)^4 \tag{3.52}$$

where

$$a_1 = -\frac{1}{2}B_1 = -\frac{1}{2}B_2. \tag{3.53}$$

(viii$_{1a}$) *For* $a_0(\mathbf{p}) > 0$, *the fixed-point of* $x_k^* = a_1$ *is unstable of a monotonic upper-saddle of the fourth-order if* $d^4f/dx_k^4|_{x_k^*=a_1} > 0$.

(viii$_{1b}$) *The fixed-point of* $x_k^* = a_1$ *for four fixed-points vanishing and appearance are called a upper-saddle-node appearing bifurcation of the fourth order at a point* $\mathbf{p} = \mathbf{p}_1 \in \partial\Omega_{12}$.

(viii$_{2a}$) *For* $a_0(\mathbf{p})<0$, *the fixed-point of* $x_k^* = a_1$ *is unstable of a monotonic lower-saddle of the fourth-order if* $d^4f/dx_k^4|_{x_k^*=a_1}<0$.

(viii$_{2b}$) *The fixed-point of* $x_k^* = a_1$ *for four fixed-points vanishing and appearance are called a monotonic lower-saddle-node appearing bifurcation of the fourth-order at a point* $\mathbf{p} = \mathbf{p}_1 \in \partial\Omega_{12}$.

Proof As for quadratic discrete systems, the proof is completed. ■

3.2 Period-1 to Period-2 Bifurcation Trees

In this section, period-1 stability and bifurcation of quartic nonlinear discrete systems are discussed graphically and period-2 fixed-points on the period-1 to period-2 bifurcation trees are also presented for a better understanding of complex bifurcations.

As discussed before, a quartic nonlinear discrete system is expressed by the product of two quadratic polynomials, i.e.,

$$\begin{aligned} x_{k+1} &= x_k + f(x_k, \mathbf{p}) \\ &= x_k + a_0(\mathbf{p})[x_k^2 + B_1(\mathbf{p})x_k + C_1(\mathbf{p})][x_k^2 + B_2(\mathbf{p})x_k + C_2(\mathbf{p})]. \end{aligned} \tag{3.54}$$

Thus, for $x_{k+1}^* = x_k^*$, the period-1 fixed-points are determined by the roots of two quadratic polynomial equations, i.e.,

$$\begin{aligned} x_k^{*2} + B_1(\mathbf{p})x_k^* + C_1(\mathbf{p}) = 0, \\ x_k^{*2} + B_2(\mathbf{p})x_k^* + C_2(\mathbf{p}) = 0. \end{aligned} \tag{3.55}$$

- If $x_k^{*2} + B_i(\mathbf{p})x_k^* + C_i(\mathbf{p}) \neq 0$ for $i = 1, 2$, such a quartic discrete system does not have any period-1 fixed-points.
- If $x_k^{*2} + B_i(\mathbf{p})x_k^* + C_i(\mathbf{p}) = 0$ for $i = 1, 2$, such a quartic discrete system has four period-1 fixed-points.
- If $x_k^{*2} + B_i(\mathbf{p})x_k^* + C_i(\mathbf{p}) = 0$ and $x_k^{*2} + B_j(\mathbf{p})x_k^* + C_j(\mathbf{p}) \neq 0$ for $i, j \in \{1, 2\}$ and $i \neq j$, such a quartic discrete system has two period-1 fixed-points.

The roots of such quadratic equations are determined by the corresponding discriminant of the quadratic equations, i.e.,

$$\Delta_i = B_i^2 - 4C_i \text{ for } i = 1, 2. \tag{3.56}$$

If $\Delta_i < 0$, the quadratic equation of $x_k^2 + B_i(\mathbf{p})x_k + C_i(\mathbf{p}) = 0$ does not have any roots. If $\Delta_i > 0$, the quadratic equation of $x_k^2 + B_i(\mathbf{p})x_k + C_i(\mathbf{p}) = 0$ has two roots. If $\Delta_i = 0$, the quadratic equation of $x_k^2 + B_i(\mathbf{p})x_k + C_i(\mathbf{p}) = 0$ has a repeated root. With parameter variation, suppose one of two quadratic polynomial equations has one root intersected with the roots of the other quadratic polynomial equation. There are six cases for $a_0 > 0$: (i) $b_2^{(i)} = b_1^{(j)}$, (ii) $b_1^{(j)} = b_1^{(i)} = b_2^{(i)} = -\frac{1}{2}B_i$. (iii) $b_1^{(i)} = b_1^{(j)}$, (iv) $b_2^{(i)} = b_2^{(j)}$ (v) $b_2^{(j)} = b_1^{(i)} = b_2^{(i)} = -\frac{1}{2}B_i$, (vi) $b_1^{(i)} = b_2^{(j)}$, as presented in Fig. 3.1. The intersected points for simple fixed-roots is a monotonic saddle-node bifurcation of the second-order for the subscritical case. The monotonic lower-saddle-node and monotonic upper-saddle-node bifurcations are shown in Fig. 3.1(i, ii) and (iv, vi), respectively. P-2 is for period-2 fixed-point. Open curves of P-2 are for mSI-oSO. Closed loops of P-2 is for mSI-oSO-mSI. The bifurcation dynamics for the 1-dimensional quartic nonlinear, forward, discrete system is determined by

$$x_{k+1} = x_k + a_0(x_k - a_{i_1})^2(x_k - a_{i_2})(x_k - a_{i_3}) \tag{3.57}$$

with $i_\alpha, \alpha \in \{1, 2, 3\}$ for four fixed-points or

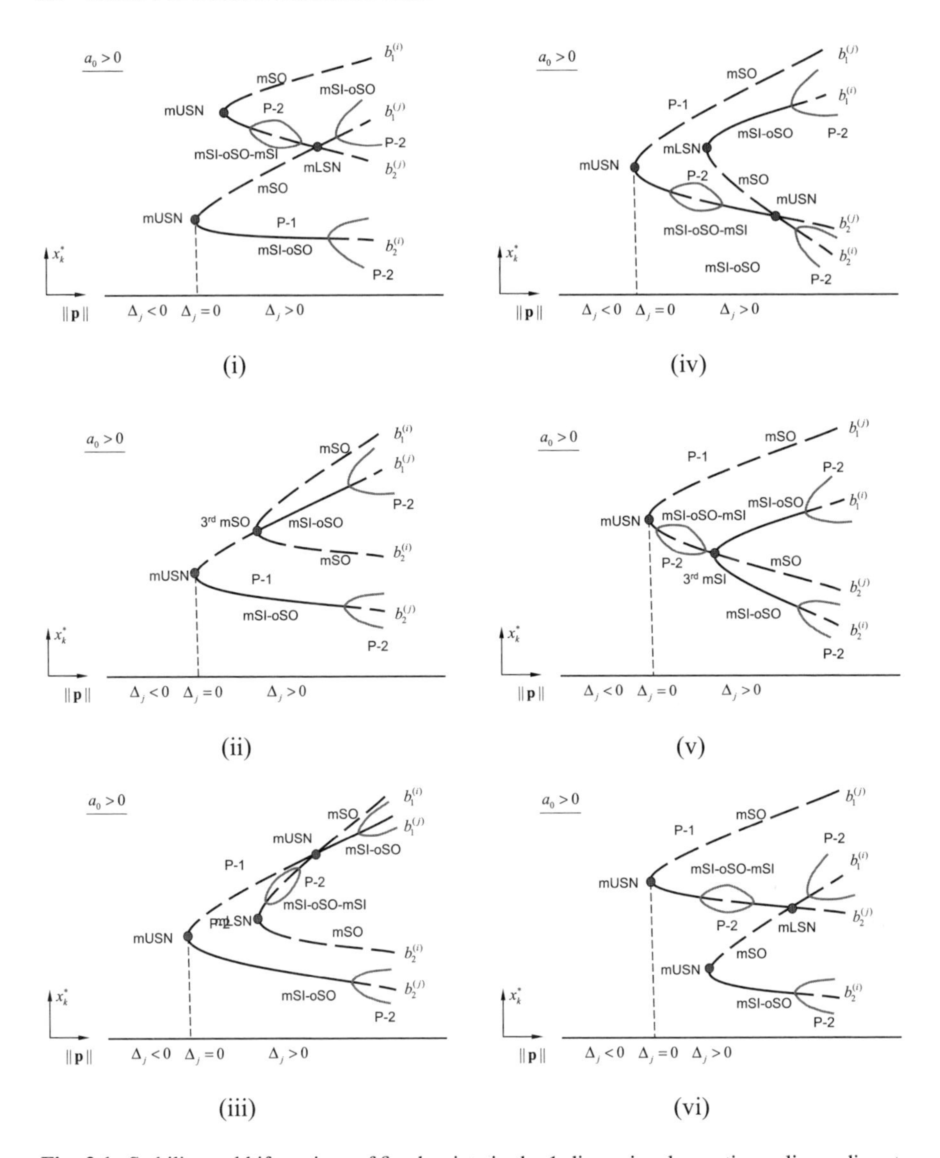

Fig. 3.1 Stability and bifurcations of fixed-points in the 1-dimensional, quartic nonlinear discrete system ($a_0 > 0$): (i) $b_2^{(i)} = b_1^{(j)}$, (ii) $b_1^{(j)} = b_1^{(i)} = b_2^{(i)} = -\frac{1}{2}B_i$. (iii) $b_1^{(i)} = b_1^{(j)}$, (iv) $b_2^{(i)} = b_2^{(j)}$ (v) $b_2^{(j)} = b_1^{(i)} = b_2^{(i)} = -\frac{1}{2}B_i$, (vi) $b_1^{(i)} = b_2^{(j)}$. mLSN: monotonic lower-saddle-node, mUSN: monotonic upper-saddle-node, mSI-oSO: monotonic sink to oscillatory source, mSI-oSO-mSI: monotonic sink to oscillatory source to monotonic sink, mSO: monotonic source. Stable and unstable fixed-points are represented by solid and dashed curves, respectively. The bifurcation points are marked by circular symbols. P-2: Period-2 fixed-point. Open curve for P-2 is for mSI-oSO. Closed loop for P-2 is for mSI-oSO-mSI.

$$x_{k+1} = x_k + a_0(x_k - a_i)^2[(x_k + \tfrac{1}{2}B_j)^2 - \tfrac{1}{4}\Delta_j] \tag{3.58}$$

with $i,j \in \{1,2\}$ for two fixed-points. If the intersected point occurs at the repeated root, the third-order monotonic source and monotonic sink switching bifurcations are presented in Fig. 3.1(ii) and (iv), respectively. The corresponding bifurcation dynamics for the 1-dimensional quartic, forward discrete system is determined by

$$x_{k+1} = x_k + a_0(x_k - a_{i_1})^3(x_k - a_{i_2}) \tag{3.59}$$

with $i_\alpha, \alpha \in \{1,2\}$. The stable and unstable fixed pints are presented by solid and dashed curves, respectively. The intersected points are marked by circular symbols, which are for bifurcation points. Without losing generality, suppose the two roots of the quadratic polynomial equation have a relation of $b_1^{(i)} > b_2^{(i)}$ for $i = 1,2$. The repeated roots of the two quadratic polynomial equations are also the monotonic upper or lower-saddle-node bifurcations for two fixed-points appearance and vanishing. The period-2 fixed-points are sketched as well. Similarly, the six cases of stability and bifurcation diagrams varying with parameter for $a_0 < 0$ are presented in Fig. 3.2. The stability and bifurcation conditions for $a_0 < 0$ are opposite to $a_0 > 0$.

If the roots of two quadratic equations do not have any intersections, the open loops for stability and bifurcation diagrams of fixed-points for $a_0 > 0$ and $a_0 < 0$ are presented in Fig. 3.3. There are four cases of open loops for $a_0 > 0$: (i) $B_i < B_j$, (ii) $B_i > B_j$, (iii) $b_2^{(j)} < -\frac{1}{2}B_i < b_1^{(j)}$, (vi) $\Delta_i = \Delta_j, B_i \neq B_j$ and four cases of open loops for $a_0 < 0$: (v) $B_i < B_j$, (vi) $B_i > B_j$, (vii) $b_2^{(j)} < -\frac{1}{2}B_i < b_1^{(j)}$, (viii) $\Delta_i = \Delta_j, B_i \neq B_j$. The two bifurcations occur at the same time because the quadratic equations have $\Delta_i = \Delta_j, B_i \neq B_j$. The bifurcation points are only for two fixed-points appearance or vanishing from the discriminants of the quadratic equations. The bifurcation dynamics for the 1-dimensional quartic discrete system is from

$$x_{k+1} = x_k + a_0(x_k - a_{i_1})^2(x_k - a_{i_2})(x_k - a_{i_3}) \tag{3.60}$$

with $i_\alpha, \alpha \in \{1,2,3\}$.

With varying vector parameter, the open loops of stability and bifurcation diagrams will become closed loops. Thus, the closed loops of stability and bifurcation diagrams of fixed-points for $a_0 > 0$ and $a_0 < 0$ are presented in Fig. 3.4. There are six cases of closed loops: (i) $B_i < B_j$, (ii) $B_i > B_j$, (iii) $b_2^{(j)} < -\frac{1}{2}B_i < b_1^{(j)}$ for $a_0 > 0$; (iv) $B_i < B_j$, (v) $B_i > B_j$, (vi) $b_2^{(j)} < -\frac{1}{2}B_i < b_1^{(j)}$ for $a_0 < 0$. For such a closed loop, the bifurcation points are the upper and lower-saddle bifurcations of the second order at both ends. The bifurcation points are determined from the discriminants of the quadratic equations. The corresponding period-2 fixed-points are sketched as well.

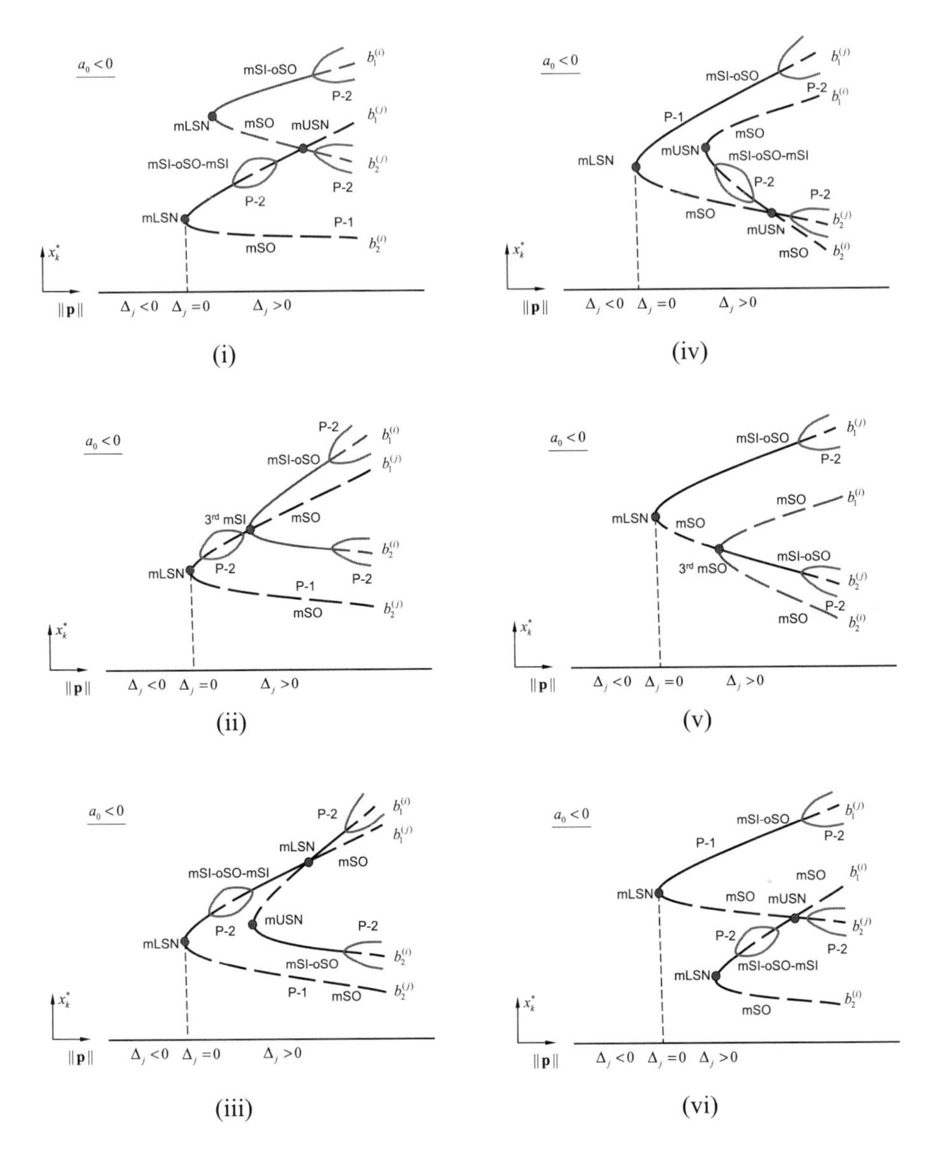

Fig. 3.2 Open loops for stability and bifurcation of fixed-points in the 1-dimeisonal, quartic nonlinear discrete system ($a_0<0$): (i) $b_2^{(i)}=b_1^{(j)}$, (ii) $b_1^{(j)}=b_1^{(i)}=b_2^{(i)}=-\frac{1}{2}B_i$. (iii) $b_1^{(i)}=b_1^{(j)}$, (iv) $b_2^{(i)}=b_2^{(j)}$ (v) $b_2^{(j)}=b_1^{(i)}=b_2^{(i)}=-\frac{1}{2}B_i$, (vi) $b_1^{(i)}=b_2^{(j)}$. mLSN: monotonic lower-saddle-node, mUSN: monotonic upper-saddle-node, mSI-oSO: monotonic sink to oscillatory source, mSO: monotonic source, mSI-oSO-mSI: monotonic sink to oscillatory source to monotonic sink. Stable and unstable fixed-points are represented by solid and dashed curves, respectively. The bifurcation points are marked by circular symbols. P-2: Period-2 fixed-point. Open curves of P-2 are for mSI-oSO. Closed loop of P-2 are for mSI-oSO-mSI.

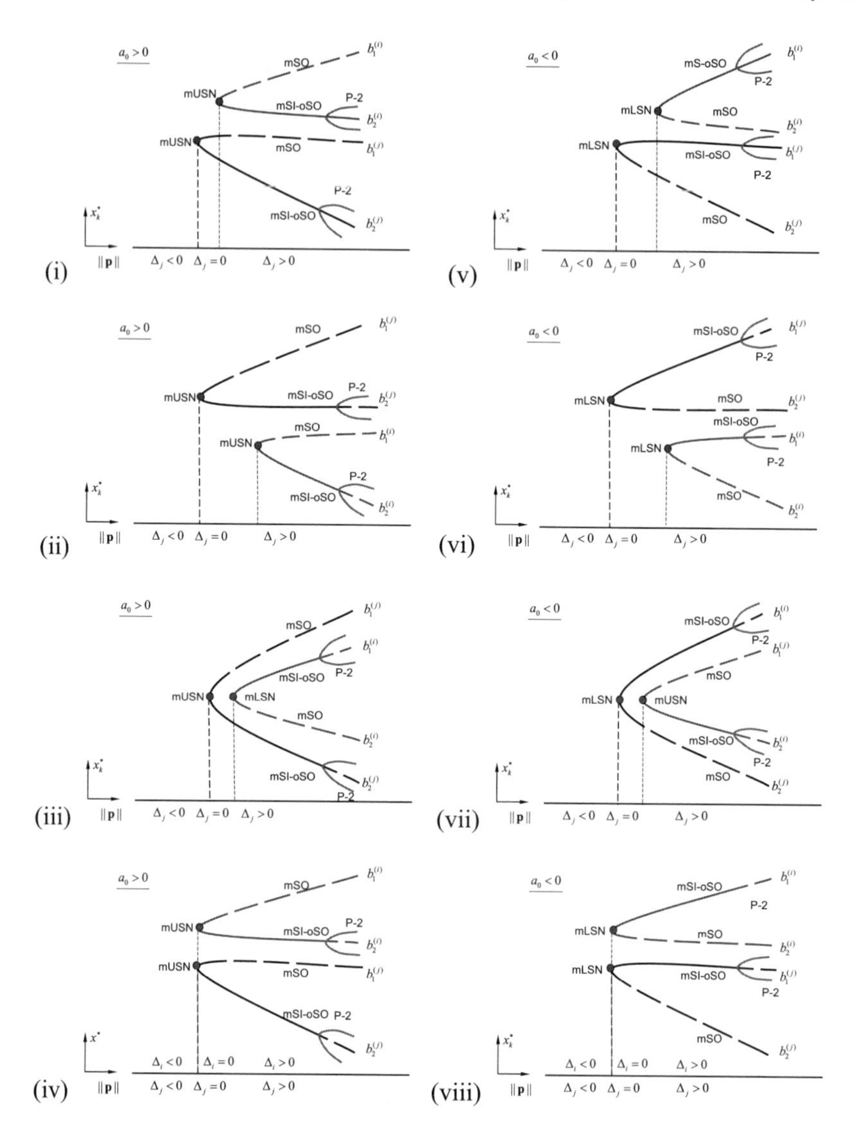

Fig. 3.3 Open loops of stability and bifurcation of fixed-points in the 1-dimeisonal, quartic nonlinear forward discrete system. ($a_0 > 0$): (i) $B_i < B_j$, (ii) $B_i > B_j$, (iii) $b_2^{(j)} < -\frac{1}{2}B_i < b_1^{(j)}$ (vi) $\Delta_i = \Delta_j, B_i \neq B_j$. ($a_0 < 0$): (v) $B_i < B_j$, (vi) $B_i > B_j$, (vii) $b_2^{(j)} < -\frac{1}{2}B_i < b_1^{(j)}$, (viii) $\Delta_i = \Delta_j, B_i \neq B_j$. mLSN: monotonic lower-saddle-node, mUSN: monotonic upper-saddle-node, mSI-oSO: monotonic sink to oscillatory source, mSO-monotonic source. Stable and unstable fixed-points are represented by solid and dashed curves, respectively. The bifurcation points are marked by circular symbols. P-2: Period-2 fixed-point. Open curves of P-2 are for oSI-oSO.

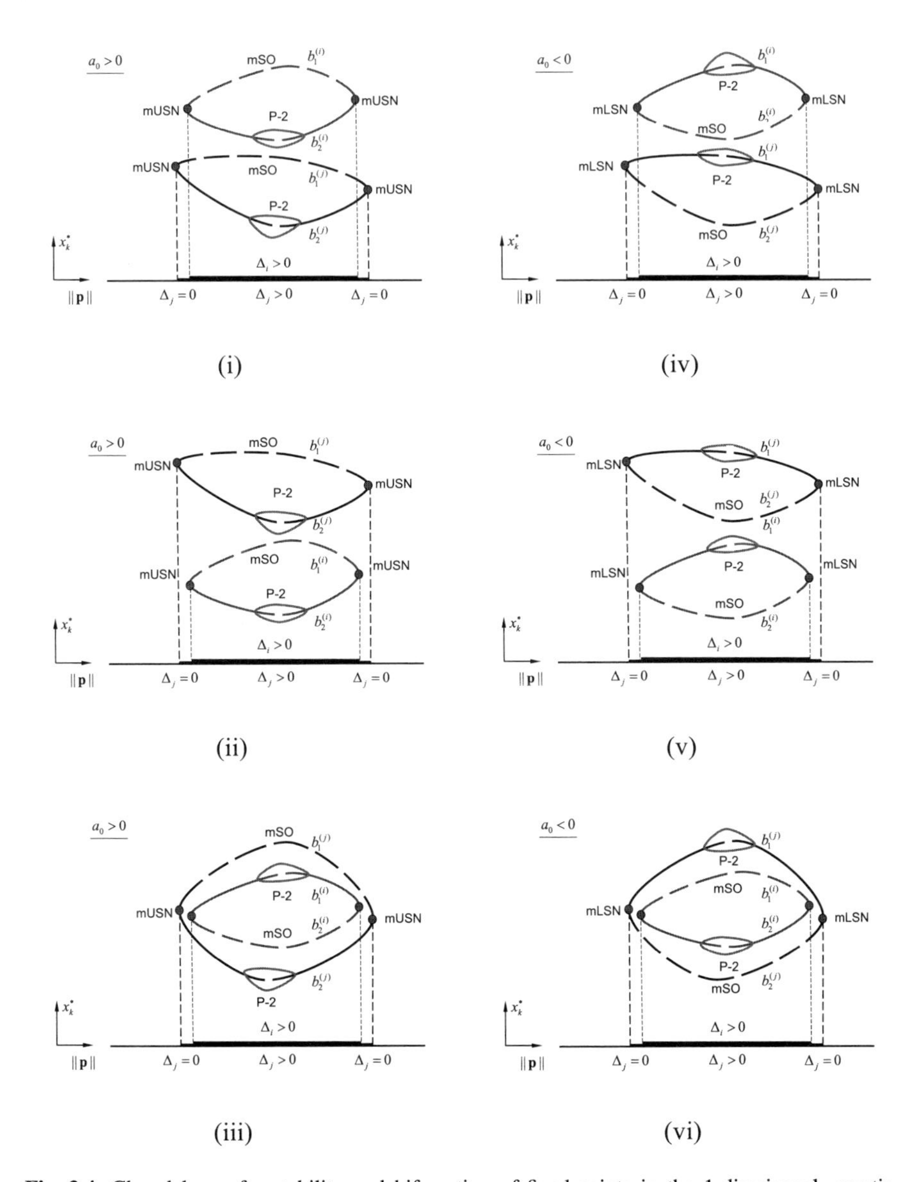

Fig. 3.4 Closed loops for stability and bifurcation of fixed-points in the 1-dimeisonal, quartic nonlinear forward discrete system: ($a_0 > 0$): (i) $B_i < B_j$, (ii) $B_i > B_j$, (iii) $b_2^{(j)} < -\frac{1}{2}B_i < b_1^{(j)}$. ($a_0 < 0$): (iv) $B_i < B_j$, (v)$B_i > B_j$, (vi) $b_2^{(j)} < -\frac{1}{2}B_i < b_1^{(j)}$. mLSN: monotonic lower-saddle-node, mUSN: monotonic upper-saddle-node, mSI-oSO-mSI: monotonic sink to oscillatory source to monotonic sink, mSO: monotonic source. Stable and unstable fixed-points are represented by solid and dashed curves, respectively. The bifurcation points are marked by circular symbols. P-2: Period-2 fixed-point. Closed loop of P-2 are for mSI-oSO-mSI.

If $\Delta_i = \Delta_j = 0$ occur at the same parameter, the bifurcation dynamics for the quartic discrete system is determined by $x_{k+1} = x_k + a_0(x_k - a_1)^4$, as shown in Fig. 3.5. There are six cases: two closed loops $(\Delta_i > \Delta_j, B_i = B_j)$: (i) $a_0 > 0$, (ii) $a_0 > 0$; two open loops $(\Delta_i > \Delta_j, B_i = B_j)$: (iii) $a_0 > 0$, (iv) $a_0 > 0$; two open loops $(B_i \leq B_j)$: (v) $a_0 > 0$, (vi) $a_0 > 0$. The bifurcation points are the monotonic upper- and lower-saddle-node bifurcations of the fourth order. In Fig. 3.5(i) and (ii), the monotonic upper and lower-saddle-node bifurcations of the fourth-order in the closed loop of bifurcation diagrams for $a_0 > 0$ and $a_0 > 0$ are presented, respectively. However, for the open loop of bifurcation diagrams, the monotonic upper- and lower-saddle-node bifurcations of the fourth-order for $a_0 > 0$ and $a_0 > 0$ are presented in Fig. 3.5(iii) and (iv), respectively. The fourth order saddle-node bifurcation possesses four branches of simple fixed-points rather than two branches of simple fixed-point for the harmonic saddle-node bifurcation of second-order. For $B_i \leq B_j$, the 4th order upper and lower-saddle node bifurcations are presented in Fig. 3.5(v) and (vi). The quadratic equation gives the two fixed-points, which are the same stability. That is, both of fixed-points for the same quadratic equation are monotonically stable or unstable.

If $\Delta_i = 0$ $(i \in \{1, 2\})$ occurs for new fixed-point appearance or vanishing only, the open loop of the bifurcation diagrams for $a_0 > 0$ is presented in Fig. 3.6. There are four cases for $a_0 > 0$: (i) $b_2^{(i)} = b_1^{(j)}$, (ii) $b_1^{(i)} = b_1^{(j)}$, (iii) $b_2^{(i)} = b_2^{(j)}$, (iv) $b_1^{(i)} = b_2^{(j)}$. The bifurcation points are the monotonic upper- and lower-saddle-node bifurcations of the second-order. In Fig. 3.7(i)–(iv), the monotonic upper and monotonic lower-saddle-node bifurcations of the second-order in the open loop of bifurcation diagrams for $a_0 < 0$ are presented. Such a diagram of stability and bifurcation possesses three monotonic saddle-node bifurcations of the second-order.

If $\Delta_i = 0$ and $\Delta_j > 0$ $(i, j \in \{1, 2\}, i \neq j)$ exist for fixed-point appearance or vanishing with the monotonic source and sink bifurcations of the third-order, one closed loop of the bifurcation diagrams for $a_0 > 0$ is presented in Fig. 3.8. Four cases for $a_0 > 0$ exist with the four monotonic saddle-node bifurcations in one closed loop: (i) $b_2^{(i)} = b_1^{(j)}$ and $b_1^{(i)} = b_2^{(j)}$, (iii) $b_1^{(i)} = b_1^{(j)}$ and $b_2^{(i)} = b_2^{(j)}$, (iv) $b_2^{(i)} = b_2^{(j)}$ and $b_1^{(i)} = b_1^{(j)}$, (vi) $b_1^{(i)} = b_2^{(j)}$ and $b_2^{(i)} = b_1^{(j)}$. Two cases for $a_0 > 0$ exist with the three monotonic saddle-node bifurcations of the second-order plus the monotonic source and monotonic sink bifurcations of the third-order in two closed loops: (ii) $b_1^{(j)} = b_1^{(i)} = b_2^{(i)} = -\frac{1}{2}B_i$ with $b_1^{(i)} = b_1^{(j)}$ and $b_2^{(i)} = b_2^{(j)}$, and (v) $b_2^{(j)} = b_1^{(i)} = b_2^{(i)} = -\frac{1}{2}B_i$ with $b_1^{(i)} = b_1^{(j)}$ and $b_2^{(i)} = b_2^{(j)}$. However, in Fig. 3.9(i)–(vi), the monotonic upper- and lower-saddle-node bifurcations of the second-order plus the monotonic source and monotonic sink bifurcations of the third-order in the closed loop of bifurcation diagrams for $a_0 < 0$ are presented.

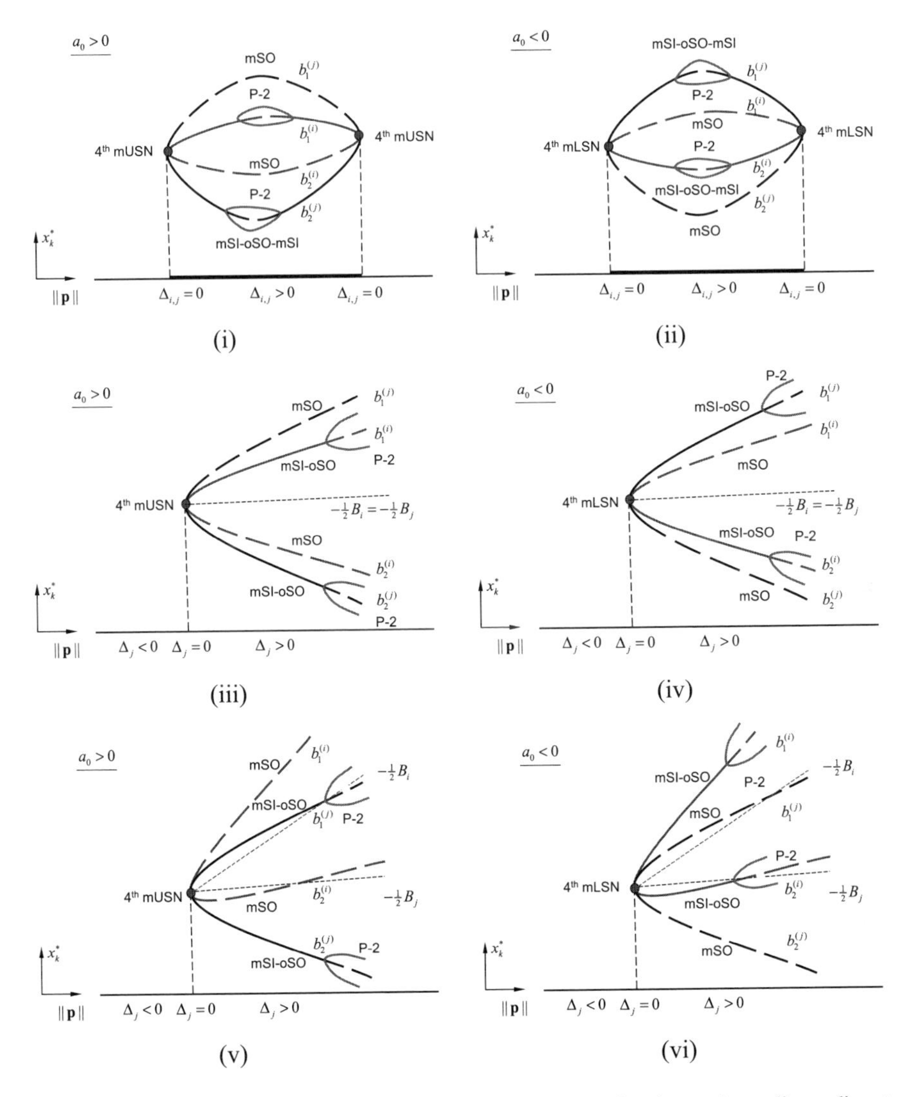

Fig. 3.5 Stability and bifurcation of fixed-points in the 1-dimensional, quartic nonlinear discrete system. Closed loop ($\Delta_i > \Delta_j, B_i = B_j$) for: (i) $a_0 > 0$, (ii) $a_0 < 0$; Open loop ($\Delta_i > \Delta_j, B_i = B_j$) for: (iii) $a_0 > 0$, (iv) $a_0 < 0$; Open loop ($B_i < B_j$) for: (v) $a_0 > 0$, (vi) $a_0 < 0$. mLSN: monotonic lower-saddle-node, mUSN: monotonic upper-saddle-node, mSI-oSO: monotonic sink to oscillatory source, mSI-oSO-mSI: monotonic sink to oscillatory source to monotonic sink, mSO: monotonic source. Stable and unstable fixed-points are represented by solid and dashed curves, respectively. The bifurcation points are marked by circular symbols. P-2: Period-2 fixed-point. Open curves of P-2 are for mSI-oSO. Closed loop of P-2 are for mSI-oSO-mSI.

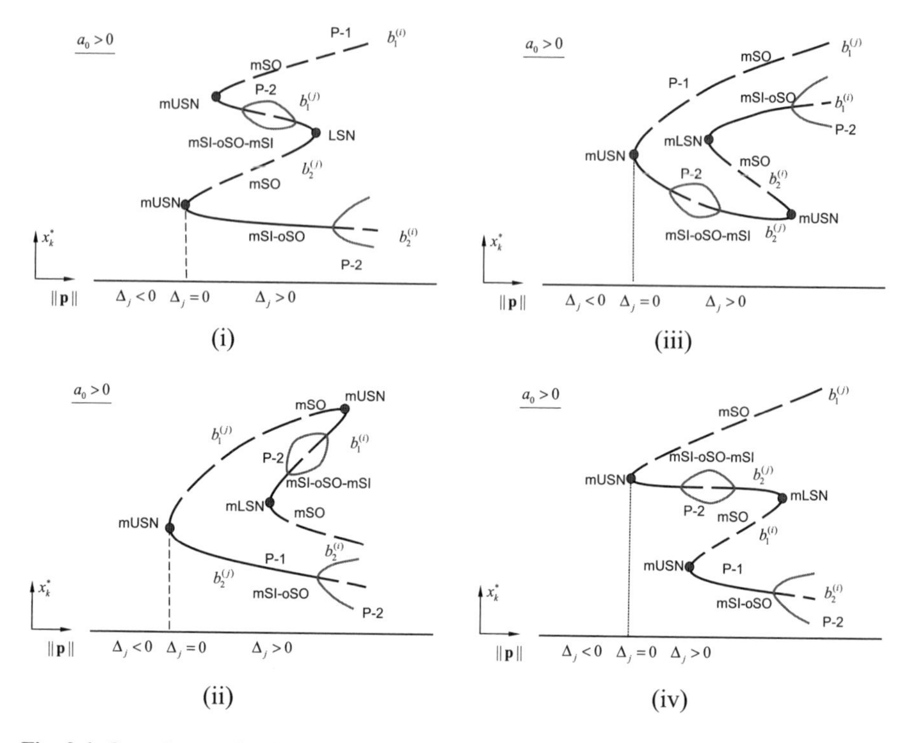

Fig. 3.6 Open loops of stability and bifurcation of fixed-points in the 1-dimensional, quartic nonlinear discrete system ($a_0 > 0$): (i) $b_2^{(i)} = b_1^{(j)}$, (ii) $b_1^{(i)} = b_1^{(j)}$, (iii) $b_2^{(i)} = b_2^{(j)}$, (iv) $b_1^{(i)} = b_2^{(j)}$. mLSN: monotonic lower-saddle-node, mUSN: monotonic upper-saddle-node, mSI-oSO: monotonic sink to oscillatory source, mSI-oSO: monotonic sink to oscillatory source to monotonic sink, mSO: monotonic source. Stable and unstable fixed-points are represented by solid and dashed curves, respectively. The bifurcation points are marked by circular symbols. P-2: Period-2 fixed-point. Open curves of P-2 are for mSI-oSO. Closed loop of P-2 are for mSI-oSO-mSI.

3.3 Higher-Order Period-1 Quartic Discrete Systems

Definition 3.2 Consider a 1-dimensional, quartic nonlinear discrete system

$$\begin{aligned} x_{k+1} &= x_k + f(x_k, \mathbf{p}) \\ &= x_k + A(\mathbf{p})x_k^4 + B(\mathbf{p})x_k^3 + C(\mathbf{p})x_k^2 + D(\mathbf{p})x_k + E(\mathbf{p}) \\ &= x_k + a_0(\mathbf{p})(x_k - b_1)^2[x_k^2 + B_2(\mathbf{p})x_k + C_2(\mathbf{p})] \end{aligned} \tag{3.61}$$

where $A(\mathbf{p}) \neq 0$, and

$$\mathbf{p} = (p_1, p_2, \ldots, p_m)^{\mathrm{T}}. \tag{3.62}$$

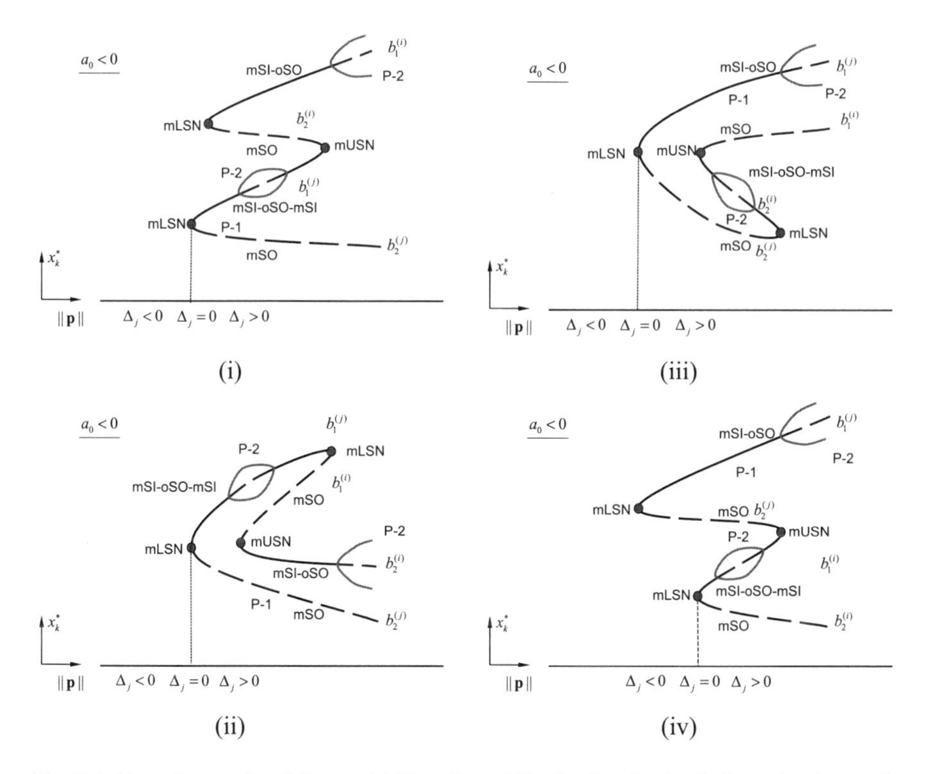

Fig. 3.7 Open loops of stability and bifurcation of fixed-points in the 1-dimensional, quartic nonlinear discrete system ($a_0 < 0$): (i) $b_2^{(i)} = b_1^{(j)}$, (ii) $b_1^{(i)} = b_1^{(j)}$, (iii) $b_2^{(i)} = b_2^{(j)}$, (iv) $b_1^{(i)} = b_2^{(j)}$. mLSN: monotonic lower-saddle-node, mUSN: monotonic upper-saddle-node, mSI-oSO: monotonic sink to oscillatory source, mSI-oSO-mSI: monotonic sink to oscillatory source to monotonic sink, mSO: monotonic source. Stable and unstable fixed-points are represented by solid and dashed curves, respectively. The bifurcation points are marked by circular symbols. P-2: Period-2 fixed-point. Open curves of P-2 are for mSI-oSO. Closed loop of P-2 are for mSI-oSO-mSI.

(i) If

$$\Delta_2 = B_2^2 - 4C_2 < 0, \tag{3.63}$$

the corresponding standard form is

$$x_{k+1} = x_k + a_0(x_k - b_1)^2[(x_k + \frac{1}{2}B_2)^2 + \frac{1}{4}(-\Delta_2)] \tag{3.64}$$

with a double fixed-point

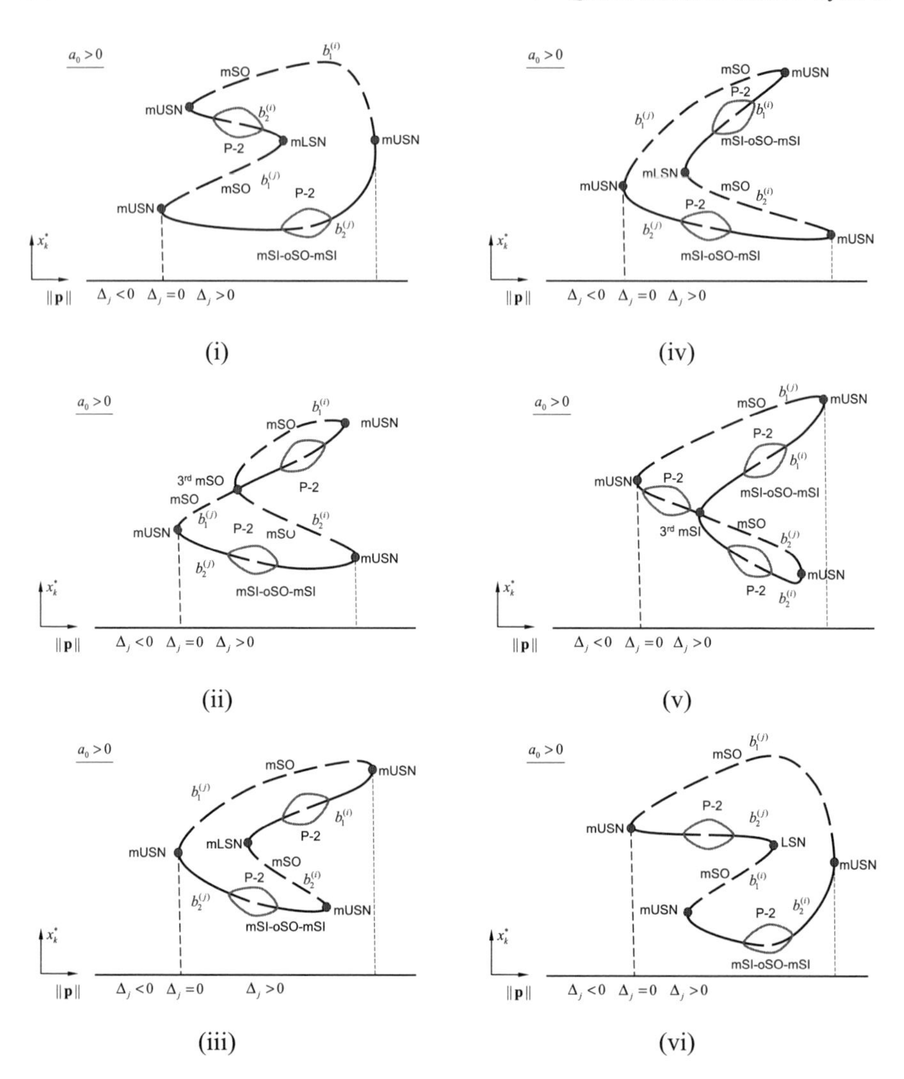

Fig. 3.8 Closed loops of stability and bifurcation of fixed-points in the 1-dimeisonal, quartic nonlinear discrete system ($a_0 > 0$): (i) $b_2^{(i)} = b_1^{(j)}$and$b_1^{(i)} = b_2^{(j)}$, (ii) $b_1^{(j)} = b_1^{(i)} = b_2^{(i)} = -\frac{1}{2}B_i$ with $b_1^{(i)} = b_1^{(j)}$ and $b_2^{(i)} = b_2^{(j)}$, (iii) $b_1^{(i)} = b_1^{(j)}$ and $b_2^{(i)} = b_2^{(j)}$, (iv) $b_2^{(i)} = b_2^{(j)}$ and $b_1^{(i)} = b_1^{(j)}$, (v) $b_2^{(j)} = b_1^{(i)} = b_2^{(i)} = -\frac{1}{2}B_i$ with $b_1^{(i)} = b_1^{(j)}$and $b_2^{(i)} = b_2^{(j)}$, (vi) $b_1^{(i)} = b_2^{(j)}$ and $b_2^{(i)} = b_1^{(j)}$. mLSN: monotonic lower-saddle-node, mUSN: monotonic upper-saddle-node, mSI-oSO-mSI: monotonic sink to oscillatory source to monotonic sink, mSO: monotonic source. Stable and unstable fixed-points are represented by solid and dashed curves, respectively. The bifurcation points are marked by circular symbols. P-2: Period-2 fixed-point. Closed loop of P-2 are for mSI-oSO-mSI.

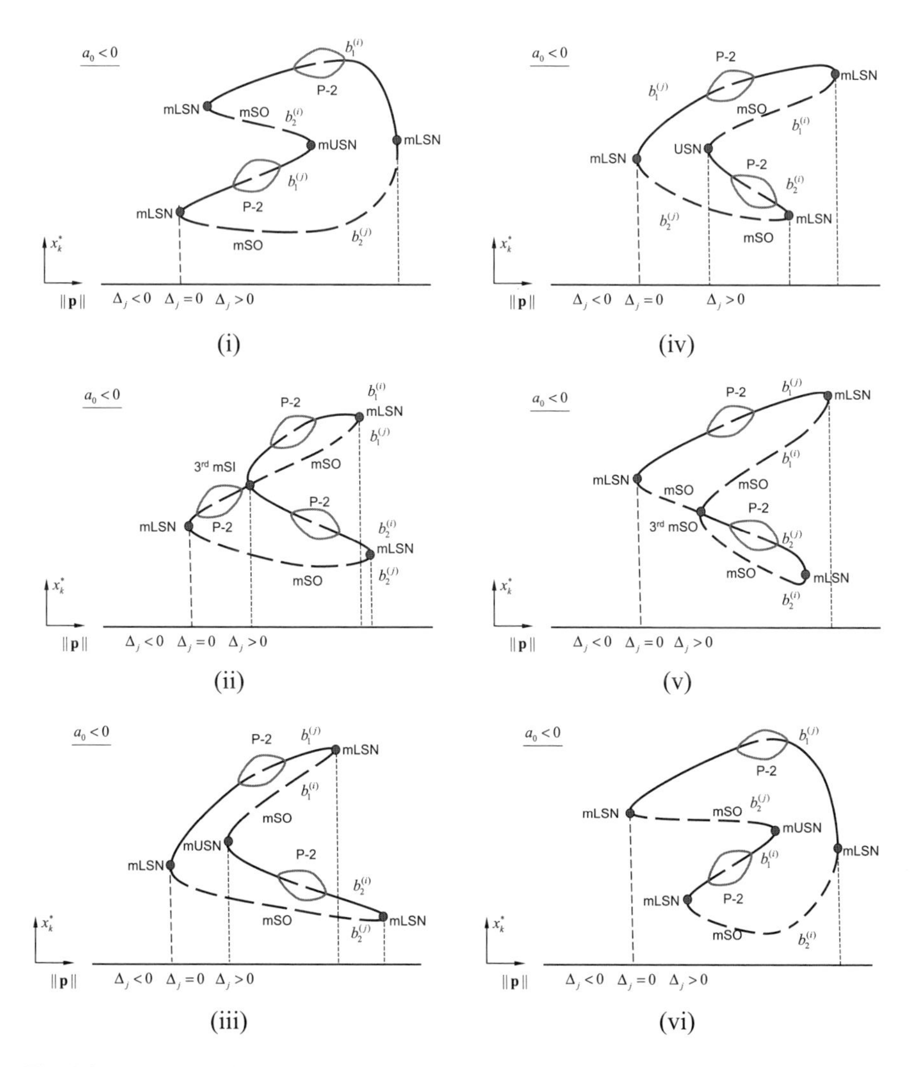

Fig. 3.9 A closed loop for stability and bifurcations of fixed-points in the 1-dimensional, quartic nonlinear forward discrete system $(a_0 < 0)$: (i) $b_2^{(i)} = b_1^{(j)}$ and $b_1^{(i)} = b_2^{(j)}$, (ii) $b_1^{(j)} = b_1^{(i)} = b_2^{(i)} = -\frac{1}{2}B_i$ with $b_1^{(i)} = b_1^{(j)}$ and $b_2^{(i)} = b_2^{(j)}$, (iii) $b_1^{(i)} = b_1^{(j)}$ and $b_2^{(i)} = b_2^{(j)}$, (iv) $b_2^{(i)} = b_2^{(j)}$ and $b_1^{(i)} = b_1^{(j)}$, (v) $b_2^{(j)} = b_1^{(i)} = b_2^{(i)} = -\frac{1}{2}B_i$ with $b_1^{(i)} = b_1^{(j)}$ and $b_2^{(i)} = b_2^{(j)}$, (vi) $b_1^{(i)} = b_2^{(j)}$ and $b_2^{(i)} = b_1^{(j)}$. mLSN: monotonic lower-saddle-node, mUSN: monotonic upper-saddle-node, mSI-oSO: monotonic sink to oscillatory source, mSO: monotonic source. Stable and unstable fixed-points are represented by solid and dashed curves, respectively. The bifurcation points are marked by circular symbols. P-2: Period-2 fixed-point. Closed loop of P-2 are for mSI-oSO-mSI.

$$x_k^* = a_1 = b_1. \tag{3.65}$$

(i_1) For $a_0 > 0$, the discrete flow is called a monotonic upper-saddle (mUS) flow of the second-order.

(i_2) For $a_0 < 0$, the discrete flow is called a monotonic lower-saddle (mLS) discrete flow of the second-order.

(ii) If

$$\Delta_2 = B_2^2 - 4C_2 > 0, \tag{3.66}$$

the quartic nonlinear forward discrete system has two fixed-points as

$$x_k^* = b_1^{(2)} = -\frac{1}{2}(B_2 + \sqrt{\Delta_2}), x_k^* = b_2^{(2)} = -\frac{1}{2}(B_2 - \sqrt{\Delta_2}). \tag{3.67}$$

(ii_1) The corresponding standard form is

$$x_{k+1} = x_k + a_0(x_k - a_1)^2(x_k - a_2)(x_k - a_3) \tag{3.68}$$

where

$$a_1 = b_1 < \min(b_1^{(2)}, b_2^{(2)}),\ a_2 = \min(b_1^{(2)}, b_2^{(2)})\ a_3 = \max(b_1^{(i)}, b_2^{(i)}). \tag{3.69}$$

($\text{ii}_{1\text{a}}$) For $a_0 > 0$, the discrete flow is called an (mUS:mSI-oSO:mSO) discrete flow.

($\text{ii}_{1\text{b}}$) For $a_0 < 0$, the discrete flow is called a (mLS:mSO:mSI-oSO) discrete flow.

(ii_2) The corresponding standard form is

$$x_{k+1} = x_k + a_0(x_k - a_1)(x_k - a_2)^2(x_k - a_3) \tag{3.70}$$

where

$$\begin{aligned} a_1 &= \min(b_1^{(2)}, b_2^{(2)}),\ a_2 = b_1 > \min(b_1^{(2)}, b_2^{(2)}), \\ a_3 &= \max(b_1^{(i)}, b_2^{(i)}) > b_1. \end{aligned} \tag{3.71}$$

($\text{ii}_{2\text{a}}$) For $a_0 > 0$, the discrete flow is called an (mSI-oSO:mLS:mSO) discrete flow.

($\text{ii}_{2\text{b}}$) For $a_0 < 0$, the discrete flow is called a (mSO:mUS:mSI-oSO) discrete flow.

(ii_3) The corresponding standard form is

$$x_{k+1} = x_k + a_0(x_k - a_1)(x_k - a_2)(x_k - a_3)^2 \tag{3.72}$$

where

$$a_1 = \min(b_1^{(2)}, b_2^{(2)}),\ a_2 = \max(b_1^{(i)}, b_2^{(i)}),\ a_3 = b_1 > \max(b_1^{(i)}, b_2^{(i)}). \tag{3.73}$$

(ii_{3a}) For $a_0 > 0$, the discrete flow is called a (mSI-oSO:mSO:mUS) discrete flow.

(ii_{3b}) For $a_0 < 0$, the discrete flow is called a (mSO:mSI-oSO:mLS) discrete flow.

(ii_4) The corresponding standard form is

$$x_{k+1} = x_k + a_0(x_k - a_1)^3(x_k - a_2) \tag{3.74}$$

where

$$a_1 = b_1 = \min(b_1^{(2)}, b_2^{(2)}),\ a_2 = \max(b_1^{(i)}, b_2^{(i)}). \tag{3.75}$$

(ii_{4a}) For $a_0 > 0$, the discrete flow is called an (3rd mSI:mSO) discrete flow. The switching bifurcation of fixed-point for (mUS:mSI-oSO:mSO)-fixed-points to (mSI-oSO:mLS:mSO)-fixed-points is called a monotonic sink switching bifurcation of the third-order.

(ii_{4b}) For $a_0 < 0$, the flow is called a (3rd mSO:mSI-oSO) discrete flow. The bifurcation of fixed-points for (mLS:mSO:mSI-oSO)-fixed-point to (mSO:mUS:mSI-oSO)-fixed-point is called a monotonic source switching bifurcation of the third-order.

(ii_5) The corresponding standard form is

$$x_{k+1} = x_k + a_0(x_k - a_1)(x_k - a_2)^3. \tag{3.76}$$

where

$$a_1 = \min(b_1^{(2)}, b_2^{(2)}),\ \ a_2 = b_1 = \max(b_1^{(i)}, b_2^{(i)}). \tag{3.77}$$

(ii_{5a}) For $a_0 > 0$, the discrete flow is called a (mSI-oSO: 3rd mSO) discrete flow. The bifurcation of fixed-point for (mSI-oSO:mSO:mUS)-fixed-points to (mSI-oSO:mUS:mSO)-fixed-point is called a monotonic source switching bifurcation of the third-order.

(ii_{5b}) For $a_0 < 0$, the flow is called a (mSO:3rd mSI) flow. The bifurcation of fixed-point for (mSO:mSI-oSO:mLS)-fixed-points to (mSO:

mUS:mSI-oSO)-fixed-points is called a monotonic sink switching bifurcation of the third order.

(iii) If

$$\Delta_2 = B_2^2 - 4C_2 = 0 \tag{3.78}$$

the quartic nonlinear discrete system has two fixed-points as

$$x^* = b_1^{(2)} = b_2^{(2)} = b_2 = -\frac{1}{2}B_2. \tag{3.79}$$

(iii_1) The corresponding standard form for $b_1 < b_2$ is

$$x_{k+1} = x_k + a_0(x_k - b_1)^2(x_k - b_2)^2. \tag{3.80}$$

(iii_{1a}) For $a_0 > 0$, the discrete flow is called a (mUS:mUS) discrete flow. The bifurcation of fixed-point for the (mUS:mSI-oSO:mSO)-fixed-points appearance is called a monotonic upper-saddle appearing bifurcation of the second-order.

(iii_{1b}) For $a_0 < 0$, the discrete flow is called a (mLS:mLS) discrete flow. The bifurcation of fixed-point for (mLS:mSO:mSI-oSO)-fixed-points appearance is called a lower-saddle appearing bifurcation of the second-order.

(iii_2) The corresponding standard form for $b_1 > b_2$ is

$$x_{k+1} = x_k + a_0(x_k - b_2)^2(x_k - b_1)^2. \tag{3.81}$$

(iii_{2a}) For $a_0 > 0$, the discrete flow is called a (mUS:mUS) discrete flow. The bifurcation of fixed-point for the (mSI-oSO:mSO:mUS)-fixed-points appearance is called a monotonic upper-saddle appearing bifurcation of the second-order.

(iii_{2b}) For $a_0 < 0$, the discrete flow is called a (mLS:mLS) discrete flow. The bifurcation of fixed-point for (mSO:mSI-oSO:mLS)-fixed-points appearance is called a monotonic lower-saddle appearing bifurcation of the second-order.

(iii_3) The corresponding standard form with $b_1 = b_2 = a_1$ is

$$x_{k+1} = x_k + a_0(x_k - a_1)^4. \tag{3.82}$$

(iii_{3a}) For $a_0 > 0$, the flow is called a monotonic upper-saddle (mUS) flow of the fourth-order. The bifurcation of fixed-point for a monotonic upper-saddle (mUS) fixed-point to (mSO:mLS:mSI-oSO)-fixed-points is called a monotonic upper-saddle-node flower-switching bifurcation of the fourth-order.

(iii_{3b}) For $a_0 < 0$, the discrete flow is called a monotonic lower-saddle (mLS) flow of the fourth-order. The bifurcation of fixed-point for a monotonic lower-saddle (mLS) fixed-point to (mSI-oSO:mUS:mSO)-fixed-points appearance is called a monotonic lower-saddle flower-switching bifurcation of the fourth-order.

Definition 3.3 Consider a 1-dimensional, quartic nonlinear discrete system

$$\begin{aligned} x_{k+1} &= x_k + f(x_k, p) \\ &= x_k + A(\mathbf{p})x_k^4 + B(\mathbf{p})x_k^3 + C(\mathbf{p})x_k^2 + D(\mathbf{p})x_k + E(\mathbf{p}) \\ &= x_k + a_0(\mathbf{p})[x_k^2 + B_1(\mathbf{p})x_k + C_1(\mathbf{p})]^2 \end{aligned} \tag{3.83}$$

where $A(\mathbf{p}) \neq 0$, and

$$\mathbf{p} = (p_1, p_2, \ldots, p_m)^{\mathrm{T}}. \tag{3.84}$$

(i) If

$$\Delta_1 = B_1^2 - 4C_1 < 0, \tag{3.85}$$

the quartic nonlinear discrete system does not have any fixed-points.

(i_1) For $a_0 > 0$, the non-fixed-point discrete flow is called a positive, monotonic discrete flow.

(i_2) For $a_0 < 0$, the non-fixed-point discrete flow is called a negative, monotonic discrete flow.

(ii) If

$$\Delta_1 = B_1^2 - 4C_1 > 0, \tag{3.86}$$

the 1-dimensional quartic nonlinear, forward discrete system has two fixed-points as

$$x_k^* = b_1^{(1)} = -\tfrac{1}{2}(B_1 + \sqrt{\Delta_1}), x_k^* = b_2^{(1)} = -\tfrac{1}{2}(B_1 - \sqrt{\Delta_1}). \tag{3.87}$$

The corresponding standard form is

$$x_{k+1} = x_k + a_0(x_k - a_1)^2(x_k - a_2)^2 \tag{3.88}$$

where

$$a_1 = \min(b_1^{(1)}, b_2^{(1)}),\ a_2 = \min(b_1^{(1)}, b_2^{(1)}). \tag{3.89}$$

(ii_1) For $a_0 > 0$, the discrete flow is called a (mUS:mUS) discrete flow.
(ii_2) For $a_0 < 0$, the discrete flow is called a (mLS:mLS) discrete flow.

(iii) If

$$\Delta_1 = B_1^2 - 4C_1 = 0, \tag{3.90}$$

the 1-dimensional quartic nonlinear forward discrete system has a repeated fixed-point as

$$x_k^* = b_1^{(1)} = b_2^{(1)} = -\tfrac{1}{2}B_1 = a_1. \tag{3.91}$$

The corresponding standard form is

$$x_{k+1} = x_k + a_0(x_k - a_1)^4. \tag{3.92}$$

(iii_1) For $a_0 > 0$, the discrete flow is called a mUS discrete flow of the fourth-order. The bifurcation of fixed-points for (mUS:mUS)-fixed-points appearance is called a monotonic upper-saddle appearing bifurcation of the fourth-order.
(iii_2) For $a_0 < 0$, the discrete flow is called a mLS discrete flow of the fourth-order. The bifurcation of fixed-point for the mLS fixed-point to (mLS:mLS)-fixed-points appearance is called a monotonic lower-saddle appearing bifurcation of the fourth-order.

From a 1-dimensional, quartic nonlinear, forward, discrete system with singularity, the monotonic upper- or lower-saddle fixed-point with and without intersection with simple fixed-points are presented in Figs. 3.10 and 3.11. In Fig. 3.10(i) and (iv), the monotonic upper-saddle fixed-point for $a_0 > 0$ does not intersect with any branch of the simple fixed-points. In Fig. 3.10(ii) and (iii), the monotonic upper-saddle fixed-point for $a_0 > 0$ intersects with one of two simple fixed-points, and the monotonic upper-saddle fixed-point switches to the monotonic lower-saddle fixed-point with monotonic source and monotonic sink fixed-points, which are called the monotonic source and monotonic sink bifurcations of the third order, accordingly. In Fig. 3.10(v), the monotonic upper-saddle fixed-point for $a_0 > 0$ intersects with a double repeated fixed-points with a monotonic upper-saddle. The intersected point is a monotonically unstable fixed-point, which is called a monotonic upper-saddle-node bifurcation of the fourth-order. In Fig. 3.10(vi), the two second-order harmonic upper-saddle fixed-points are presented. The two monotonic upper-saddle fixed-points appear at the monotonic upper-saddle bifurcation of the fourth-order.

Similarly, the monotonic lower-saddle fixed-point for $a_0 < 0$ does not intersect with any branch of the simple fixed-points, as shown in Fig. 3.11(i) and (iv). In Fig. 3.11(ii) and (iii), the monotonic lower-saddle fixed-point for $a_0 < 0$ intersects with one of two simple fixed-points, and the monotonic lower-saddle fixed-point switches to the monotonic upper-saddle fixed-point with monotonic source and monotonic sink fixed-points, which are called the monotonic source and monotonic

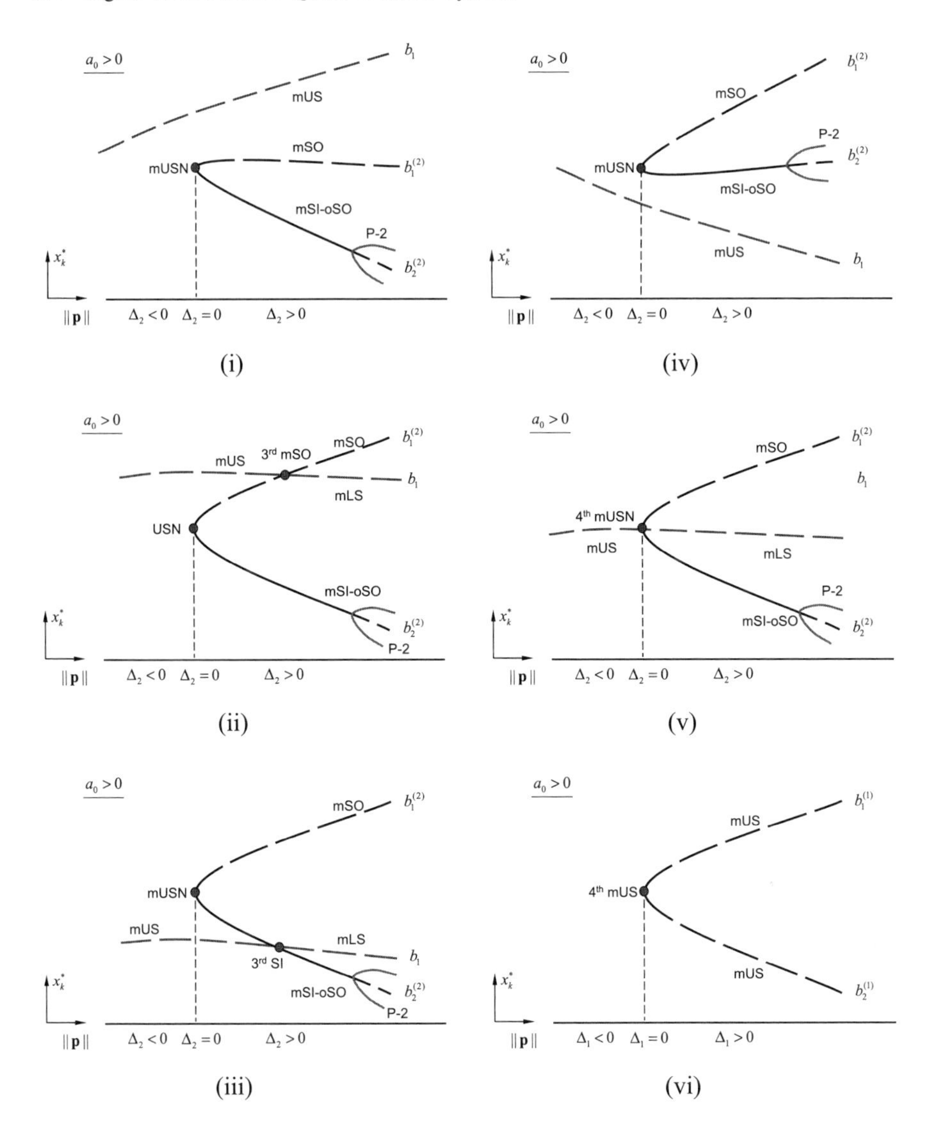

Fig. 3.10 Stability and bifurcation of three fixed-points with intersection in the 1-dimensional, quartic nonlinear forward discrete system ($a_0 > 0$): (i) without intersection $b_1 > b_1^{(2)}$, (ii) an intersection at $b_1 = b_1^{(2)}$, (iii) an intersection at $b_1 = b_2^{(2)}$, (iv) without intersection $b_1 < b_1^{(2)}$, (v) an intersection at $b_1 = -\frac{1}{2}B_1$, (vi) $\Delta_1 = 0$. mLSN: monotonic lower-saddle-node, mUSN: monotonic upper-saddle-node, mSI-oSO: monotonic sink to oscillatory source, mSO: monotonic source. Stable and unstable fixed-points are represented by solid and dashed curves, respectively. The bifurcation points are marked by circular symbols. P-2: Period-2 fixed-point. Open curves of P-2 are for mSI-oSO.

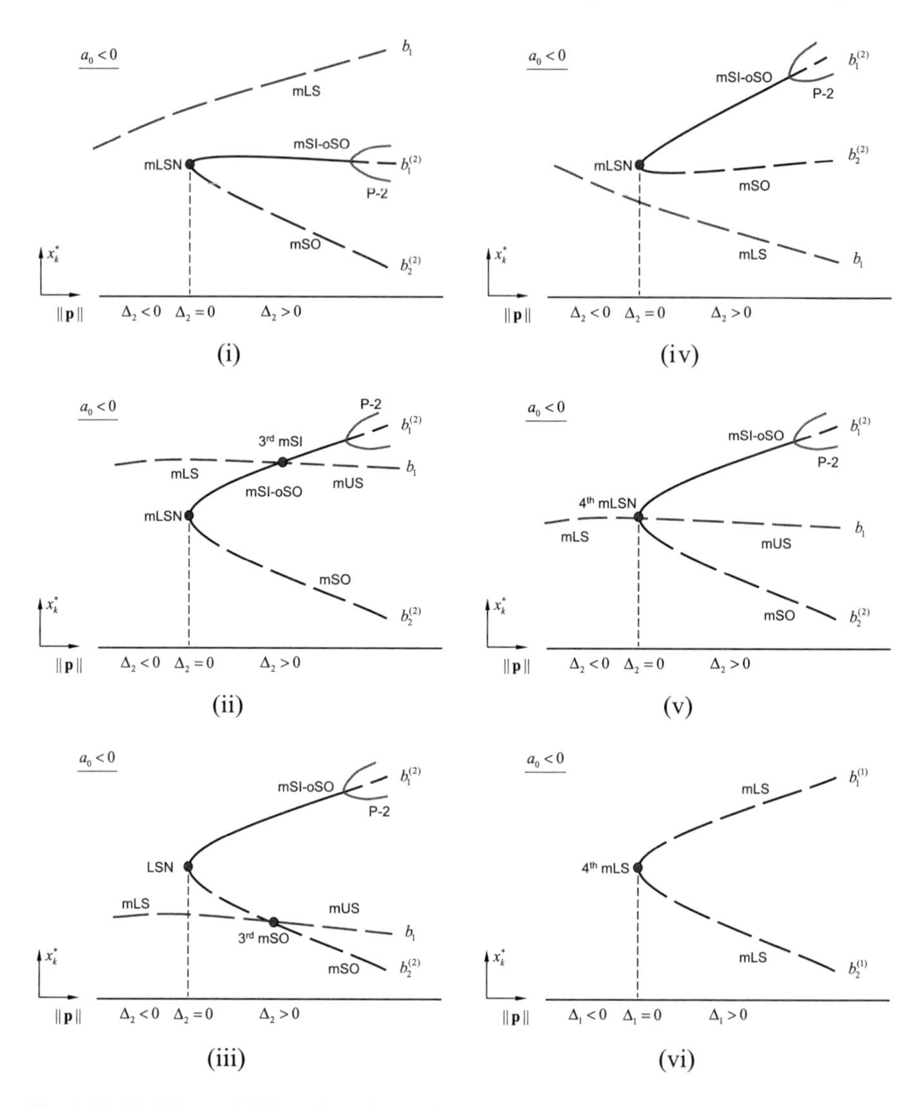

Fig. 3.11 Stability and bifurcation of three fixed-points with intersection in the 1-dimensional, quartic nonlinear forward discrete system $(a_0 < 0)$: (i) without intersection $b_1 > b_1^{(2)}$, (ii) an intersection at $b_1 = b_1^{(2)}$, (iii) an intersection at $b_1 = b_2^{(2)}$, (iv) without intersection $b_1 < b_1^{(2)}$, (v) an intersection at $b_1 = -\frac{1}{2}B_1$, (vi) $\Delta_1 = 0$. mLSN: monotonic lower-saddle-node, mUSN: monotonic upper-saddle-node, mSI: monotonic sink to oscillatory source, mSO: monotonic source. Stable and unstable fixed-points are represented by solid and dashed curves, respectively. The bifurcation points are marked by circular symbols. P-2: Period-2 fixed-point. Open curves of P-2 are for mSI-oSO.

sink bifurcations of the third order, accordingly. In Fig. 3.11(v), the monotonic lower-saddle fixed-point for $a_0 < 0$ intersects with a repeated fixed-point with a monotonic lower-saddle. The intersection point is an unstable fixed-point, which is called a 4th order monotonic lower-saddle-node bifurcation. In Fig. 3.11(vi), the two monotonic second-order lower saddle fixed-points are presented for $a_0 < 0$. The two monotonic lower-saddle fixed-points appear at the monotonic lower-saddle bifurcation of the fourth-order.

Consider a 1-dimensional, quartic nonlinear, forward, discrete system with two double fixed-points.

(i) For $b \neq a$, the forward discrete system is

$$x_{k+1} = x_k + a_0(\mathbf{p})(x_k - b(\mathbf{p}))^2(x_k - a(\mathbf{p}))^2. \tag{3.93}$$

For such a system, if $a_0 > 0$, two repeated fixed-points of $x_k^* = a, b$ are two monotonic upper-saddles, which are monotonically unstable. If $a_0 < 0$, two repeated fixed-points of $x_k^* = a, b$ are two monotonic lower-saddles, which are monotonically unstable.

(ii) For $a = b$, the discrete system on the boundary is

$$x_{k+1} = x_k + a_0(\mathbf{p})(x_k - b(\mathbf{p}))^4. \tag{3.94}$$

With parameter changes, the bifurcation diagram for the quartic nonlinear discrete system is presented in Fig. 3.12. Stable and unstable fixed-points are represented by solid and dashed curves, respectively. The bifurcation point is marked by a circular symbol. In Fig. 3.12(i), if $a_0 > 0$, two repeated fixed-points of $x_k^* = a, b$

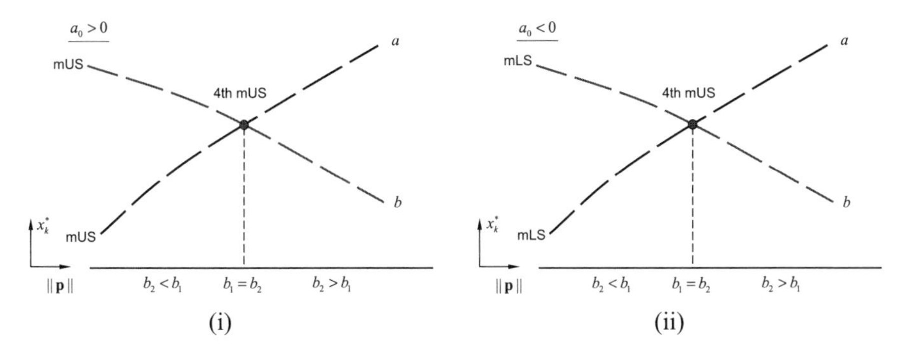

Fig. 3.12 Stability and bifurcation of two mUS or mLS fixed-points with intersection in the 1-dimensional, quartic nonlinear discrete system: (i) (mUS:mUS)-flow ($a_0 > 0$), (i) (LS:LS)-flow ($a_0 < 0$). 4th mLS: 4th order monotonic lower-saddle bifurcation, 4th mUS- 4th order monotonic upper-saddle bifurcation. Stable and unstable fixed-points are represented by solid and dashed curves, respectively. The bifurcation points are marked by circular symbols.

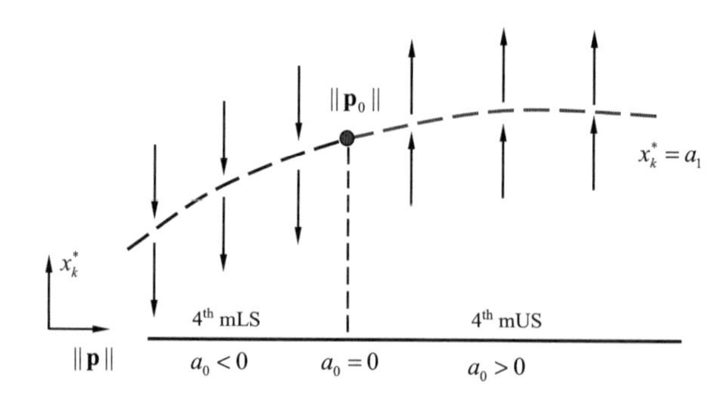

Fig. 3.13 Stability of a repeated fixed-point with the fourth multiplicity in the 1-dimensional, quartic nonlinear discrete system: Stable and unstable fixed-points are represented by solid and dashed curves, respectively. The stability switching is labelled by a circular symbol. 4th mLS: fourth-order monotonic lower-saddle bifurcation, 4th mUS: fourth-order monotonic upper-saddle bifurcation.

are the monotonic upper-saddles of the second order. The two monotonic upper-saddles intersect at a point of $x_k^* = a = b$ with the fourth multiplicity, which is a monotonic upper-saddle bifurcation of the fourth-order for the (mUS:mUS) to (mUS:mUS) fixed-points. If $a_0 < 0$, two repeated fixed-points of $x_k^* = a, b$ are the monotonic lower-saddle of the second order, which are intersected at a point of $x_k^* = a = b$, as shown in Fig. 3.12(ii). Such a quartically repeated fixed-point is called a monotonic lower-saddle bifurcation of the fourth order for the (mLS:mLS) to (mLS:mLS) fixed-point.

To illustrate the stability and bifurcation of fixed-point with singularity in a 1-dimensional, quadratic nonlinear system, the fixed-point of $x_{k+1} = x_k + a_0(x_k - a_1)^4$ is presented in Fig. 3.13. The fourth-order, monotonic upper and lower-saddles of fixed-point of $x_k^* = a_1$ with the other order multiplicity are monotonically unstable, and the monotonic upper and lower saddle fixed-points of the fourth-order are invariant. At $a_0 = 0$, the monotonic lower-saddle fixed-point switches to the monotonic upper-saddle fixed-point, which is a switching point marked by a circular symbol.

3.4 Period-1 Switching Bifurcations

For further discussion on the switching bifurcations in the quartic nonlinear system, the following definitions are presented.

3.4.1 Simple Period-1 Switching Bifurcations

Definition 3.4 Consider a 1-dimensional, quartic nonlinear discrete system

$$\begin{aligned} x_{k+1} &= x_k + f(x_k, \mathbf{p}) \\ &= x_k + A(\mathbf{p})x_k^4 + B(\mathbf{p})x_k^3 + C(\mathbf{p})x_k^2 + D(\mathbf{p})x_k + E(\mathbf{p}) \\ &= x_k + a_0(\mathbf{p})(x_k - a)(x_k - b)[x_k^2 + B_2(\mathbf{p})x_k + C_2(\mathbf{p})] \end{aligned} \tag{3.95}$$

where $A(\mathbf{p}) \neq 0$, and

$$\mathbf{p} = (p_1, p_2, \ldots, p_m)^{\mathrm{T}}. \tag{3.96}$$

(i) If

$$\begin{aligned} &\Delta_2 = B_2^2 - 4C_2 < 0; \\ &\{a_1, a_2\} = \mathrm{sort}\{a, b\}, a_1 \leq a_2, \end{aligned} \tag{3.97}$$

the quartic nonlinear discrete system has any two fixed-points. The corresponding standard form is

$$x_{k+1} = x_k + a_0(\mathbf{p})(x_k - a_1)(x_k - a_2)[(x_k + \tfrac{1}{2}B_2)^2 + \tfrac{1}{4}(-\Delta_2)]. \tag{3.98}$$

(i_1) For $a_0 > 0$, the discrete fixed-point flow is a (mSI-oSO:mSO) discrete flow.

($\mathrm{i}_{1\mathrm{a}}$) The fixed-point of $x_k^* = a_1$ is

- monotonically stable (monotonic sink) if $df/dx_k|_{x_k^*=a_1} \in (-1, 0)$;
- invariantly stable (zero-invariant sink) if $df/dx_k|_{x_k^*=a_1} = -1$;
- oscillatorilly stable (oscillatory sink) if $df/dx_k|_{x_k^*=a_1} \in (-2, -1)$;
- flipped if $df/dx_k|_{x_k^*=a_1} = -2$, where is
  - an oscillatory upper-saddle of the second-order for $d^2f/dx_k^2|_{x_k^*=a_1} > 0$;
  - an oscillatory lower-saddle of the second-order for $d^2f/dx_k^2|_{x_k^*=a_1} < 0$;
- oscillatorilly unstable (oscillatory source) if $df/dx_k|_{x_k^*=a_1} \in (-\infty, -2)$.

($\mathrm{i}_{1\mathrm{b}}$) The fixed-point of $x_k^* = a_2$ is monotonically unstable (monotonic source) if $df/dx_k|_{x_k^*=a_2} \in (0, \infty)$.

(i_2) For $a_0 < 0$, the fixed-point flow is a (mSO:mSI-oSO) flow.

(i_{2a}) The fixed-point of $x_k^* = a_1$ is monotonically unstable (monotonic source) if $df/dx_k|_{x_k^*=a_1} \in (0, \infty)$.

(i_{2b}) The fixed-point of $x_k^* = a_2$ is

- monotonically stable (monotonic sink) if $df/dx_k|_{x_k^*=a_2} \in (-1, 0)$;
- invariantly stable (zero-invariant sink) if $df/dx_k|_{x_k^*=a_2} = -1$;
- oscillatorilly stable (oscillatory sink) if $df/dx_k|_{x_k^*=a_2} \in (-2, -1)$;
- flipped if $df/dx_k|_{x_k^*=a_2} = -2$, which is
  - an oscillatory upper-saddle of the second-order for $d^2f/dx_k^2|_{x_k^*=a_2} > 0$;
  - an oscillatory lower-saddle of the second-order for $d^2f/dx_k^2|_{x_k^*=a_2} < 0$;
- oscillatorilly unstable (oscillatory source) if $df/dx_k|_{x_k^*=a_2} \in (-\infty, -2)$.

(i_3) Under

$$\Delta_{12} = (a_1 - a_2)^2 = 0 \text{ with } a_1 = a_2 \tag{3.99}$$

the quartic nonlinear discrete system has a standard form as

$$x_{k+1} = x_k + a_0(\mathbf{p})(x_k - a_1)^2[(x_k + \frac{1}{2}B_2)^2 + \frac{1}{4}(-\Delta_2)]. \tag{3.100}$$

(i_{3a}) For $a_0(\mathbf{p}) > 0$, the fixed-point of $x_k^* = a_1$ is monotonically unstable (a monotonic upper-saddle of second-order, $d^2f/dx_k^2|_{x_k^*=a_1} > 0$).

- Such a discrete flow is called a monotonic upper-saddle discrete flow of the second-order.
- The bifurcation of fixed-point at $x_k^* = a_1$ for two fixed-points switching of $x_k^* = a_1, a_2$ is called a monotonic upper-saddle-node switching bifurcation of the second-order at a point $\mathbf{p} = \mathbf{p}_1$.

(i_{3b}) For $a_0(\mathbf{p}) < 0$, the fixed-point of $x_k^* = a_1$ is monotonically unstable (an lower-saddle of second-order, $d^2f/dx_k^2|_{x^*=a_1} < 0$).

- Such a discrete flow is called a lower-saddle discrete flow of the second-order.
- The bifurcation of fixed-point at $x_k^* = a_1$ for two fixed-points switching of $x_k^* = a_1, a_2$ is called a monotonic lower-saddle-node switching bifurcation of the second-order at a point $\mathbf{p} = \mathbf{p}_1$.

(ii) If

$$\Delta_2 = B_2^2 - 4C_2 > 0, \tag{3.101}$$

the 1-dimensional quartic nonlinear discrete system has four fixed-points as

$$\begin{aligned} &x_k^* = b_1^{(2)} = -\frac{1}{2}(B_2 + \sqrt{\Delta_2}),\ x_k^* = b_2^{(2)} = -\frac{1}{2}(B_2 - \sqrt{\Delta_2}) \\ &\{a_1, a_2, a_3.a_4\} = \mathrm{sort}\{a, b, b_1^{(2)}, b_2^{(2)}\}, a_i < a_{i+1}. \end{aligned} \tag{3.102}$$

(ii_1) The corresponding standard form is

$$x_{k+1} = x_k + a_0(x_k - a_1)(x_k - a_2)(x_k - a_3)(x_k - a_4). \tag{3.103}$$

($\mathrm{ii}_{1\mathrm{a}}$) For $a_0 > 0$, the discrete flow is called an (mSI-oSO:mSO: mSI-oSO:mSO) discrete flow.
($\mathrm{ii}_{1\mathrm{b}}$) For $a_0 < 0$, the discrete flow is called an (mSO:mSI-oSOI:mSO: mSI-oSO) discrete flow.

(ii_2) The fixed-point of $x_k^* = a_{i_1}$ $(i_1 \in \{1, 2, 3, 4\})$ is

- monotonically unstable (monotonic source) if $df/dx_k|_{x_k^*=a_{i_1}} \in (0, \infty)$;
- monotonically stable (monotonic sink) if $df/dx_k|_{x_k^*=a_{i_1}} \in (-1, 0)$;
- invariantly stable (zero-invariant sink) if $df/dx_k|_{x_k^*=a_{i_1}} = -1$;
- oscillatorilly stable (oscillatory sink) if $df/dx_k|_{x_k^*=a_{i_1}} \in (-2, -1)$;
- flipped if $df/dx_k|_{x_k^*=a_{i_1}} = -2$, which is
  - an oscillatory upper-saddle of the second-order for $d^2f/dx_k^2|_{x_k^*=a_1} > 0$;
  - an oscillatory lower-saddle of the second-order for $d^2f/dx_k^2|_{x_k^*=a_1} < 0$;
- oscillatorilly unstable (oscillatory source) if $df/dx_k|_{x_k^*=a_{i_1}} \in (-\infty, -2)$.

(ii_3) Under

$$\Delta_{i_1 i_2} = (a_{i_1} - a_{i_2})^2 = 0, a_{i_1} = a_{i_2}, i_1, i_2 \in \{1, 2, 3, 4\}, i_1 \neq i_2 \tag{3.104}$$

the standard form is

$$\begin{aligned} &x_{k+1} = x_k + f(x_k, \mathbf{p}) = x_k + a_0(x_k - a_{i_1})^2(x_k - a_{i_3})(x_k - a_{i_4}) \\ &i_\alpha \in \{1, 2, 3, 4\}, \alpha = 1, 3, 4. \end{aligned} \tag{3.105}$$

($\mathrm{ii}_{3\mathrm{a}}$) The fixed-point of $x_k^* = a_{i_1}$ is monotonically unstable (a monotonic upper-saddle of second-order, $d^2f/dx_k^2|_{x_k^*=a_{i_1}} > 0$).

- Such a discrete flow is called an upper-saddle discrete flow at $x_k^* = a_{i_1}$.
- The bifurcation of fixed-point at $x_k^* = a_{i_1}$ for two fixed-points switching of $x_k^* = a_{i_1}, a_{i_2}$ is called a monotonical upper-saddle-node switching bifurcation of the second-order at a point $\mathbf{p} = \mathbf{p}_1$.

($\mathrm{ii}_{3\mathrm{b}}$) The fixed-point of $x_k^* = a_{i_1}$ is monotonically unstable (a monotonic lower-saddle of second-order, $d^2f/dx_k^2|_{x_k^*=a_{i_1}} < 0$).

- Such a discrete flow is called a lower-saddle discrete flow of the second-order at $x_k^* = a_{i_1}$.
- The bifurcation of fixed-point at $x_k^* = a_{i_1}$ for two fixed-points switching of $x_k^* = a_{i_1}, a_{i_2}$ is called a lower-saddle-node switching bifurcation of the second-order at a point $\mathbf{p} = \mathbf{p}_1$.

(ii_4) Under

$$\begin{aligned}&\Delta_{i_1 i_2} = (a_{i_1} - a_{i_2})^2 = 0, \Delta_{i_2 i_3} = (a_{i_2} - a_{i_3})^2 = 0,\\&a_{i_1} = a_{i_2} = a_3, i_1, i_2, i_3 \in \{1,2,3,4\}, i_1 \neq i_2 \neq i_3,\end{aligned} \tag{3.106}$$

the standard form is

$$\begin{aligned}&x_{k+1} = x_k + f(x_k, \mathbf{p}) = x_k + a_0(x_k - a_{i_1})^3(x_k - a_{i_4})\\&i_\alpha \in \{1,2,3,4\}, \alpha = 1, 4.\end{aligned} \tag{3.107}$$

($\mathrm{ii}_{4\mathrm{a}}$) The fixed-point of $x_k^* = a_{i_1}$ is monotonically unstable (a monotonic source of the third-order, $d^3f/dx_k^3|_{x_k^*=a_{i_1}} > 0$).

- Such a discrete flow is called a monotonic source flow of the third-order at $x_k^* = a_{i_1}$.
- The bifurcation of fixed-point at $x_k^* = a_{i_1}$ for three simple fixed-point bundle-switching of $x_k^* = a_{i_1}, a_{i_2}, a_{i_3}$ is called a source bundle-switching bifurcation of the third-order at a point $\mathbf{p} = \mathbf{p}_1$.

($\mathrm{ii}_{4\mathrm{b}}$) The fixed-point of $x_k^* = a_{i_1}$ is monotonically stable (a monotonic sink of the third-order, $d^3f/dx_k^3|_{x_k^*=a_{i_1}} < 0$).

- Such a discrete flow is called a monotonic sink discrete flow of the third-order at $x_k^* = a_{i_1}$.
- The bifurcation of fixed-point at $x_k^* = a_{i_1}$ for three simple fixed-point bundle-switching of $x_k^* = a_{i_1}, a_{i_2}, a_{i_3}$ is called a monotonic sink bundle-switching bifurcation of the third-order at a point $\mathbf{p} = \mathbf{p}_1$.

(ii_5) Under

$$\begin{aligned}&\Delta_{i_1 i_2}=(a_{i_1}-a_{i_2})^2=0, \Delta_{i_2 i_3}=(a_{i_2}-a_{i_3})^2=0,\\&\Delta_{i_3 i_4}=(a_{i_4}-a_{i_4})^2=0, a_{i_1}=a_{i_2}=a_{i_3}=a_{i_4},\\&i_1, i_2, i_3, i_4 \in\{1,2,3,4\}, i_1 \neq i_2 \neq i_3 \neq i_4\end{aligned} \tag{3.108}$$

the standard form is

$$x_{k+1}=x_k+f(x_k, \mathbf{p})=x_k+a_0(x_k-a_{i_1})^4. \tag{3.109}$$

($\mathrm{ii}_{5\mathrm{a}}$) The fixed-point of $x_k^*=a_{i_1}$ is monotonically unstable (a monotonic upper-saddle of the fourth-order, $d^4 f/dx_k^4|_{x_k^*=a_{i_1}}>0$).

- Such a discrete flow is called a monotonic upper-saddle flow of the fourth-order at $x_k^*=a_{i_1}$.
- The bifurcation of fixed-point at $x_k^*=a_{i_1}$ for four simple fixed-points bundle-switching of $x_k^*=a_{1,2,3,4}$ is called a monotonic upper-saddle-node bundle-switching bifurcation of the fourth-order at a point $\mathbf{p}=\mathbf{p}_1$.

($\mathrm{ii}_{5\mathrm{b}}$) The fixed-point of $x_k^*=a_{i_1}$ is monotonically unstable (a monotonic lower-saddle flow of the third-order, $d^4 f/dx_k^4|_{x_k^*=a_{i_1}}<0$).

- Such a discrete flow is called a monotonic lower-saddle flow of the fourth-order at $x_k^*=a_{i_1}$.
- The bifurcation of fixed-point at $x_k^*=a_{i_1}$ for four simple fixed-points bundle-switching of $x_k^*=a_{1,2,3,4}$ is called a monotonic lower-saddle-node bundle-switching bifurcation of the fourth-order at a point $\mathbf{p}=\mathbf{p}_1$.

(iii) If

$$\Delta_2=B_2^2-4C_2=0, \tag{3.110}$$

the 1-dimensional quartic nonlinear discrete system has three fixed-point as

$$\begin{aligned}&x_k^*=b_1^{(2)}=b_2^{(2)}=-\frac{1}{2}B_2;\\&\{a_1, a_2, a_3\}=\operatorname{sort}\{a, b, b_1^{(2)}=b_2^{(2)}\}, a_i<a_{i+1};\\&a_{i_1, i_2}=b_1^{(2)}=b_2^{(2)}, a_{i_3}=a, a_{i_4}=b;\\&i_\alpha \in\{1,2,3\}, \alpha \in\{1,2,3,4\}.\end{aligned} \tag{3.111}$$

The corresponding standard form is

$$x_{k+1} = x_k + f(x_k, \mathbf{p}) = x_k + a_0(x_k - a_{i_1})^2(x_k - a_{i_2})(x_k - a_{i_3}). \quad (3.112)$$

(iii_1) The fixed-point of $x_k^* = a_{i_1}$ is monotonically unstable (a monotonic upper-saddle, $d^2f/dx_k^2|_{x_k^*=a_{i_1}} > 0$).

- The discrete flow is a monotonic upper-saddle flow of the second-order at $x_k^* = a_{i_1}$.
- The bifurcation of fixed-point at $x_k^* = a_{i_1}$ for the appearing or vanishing of two simple fixed-points is called the monotonic upper-saddle-node appearing bifurcation of the second-order.

(iii_2) The fixed-point of $x_k^* = a_{i_1}$ is monotonically unstable (a monotonic lower-saddle, $d^2f/dx_k^2|_{x_k^*=a_{i_1}} < 0$).

- The discrete flow is a monotonic lower-saddle discrete flow at $x_k^* = a_{i_1}$.
- The bifurcation of fixed-point at $x_k^* = a_{i_1}$ for the appearing or vanishing of two simple fixed-points is called the monotonic lower-saddle-node appearing bifurcation of the second-order.

(iii_3) Under

$$\begin{aligned} &\Delta_{i_3 i_4} = (a_{i_3} - a_{i_4})^2 = 0, a_{i_3} = a_{i_4}, a_{i_1} \neq a_{i_3}, \\ &i_\alpha \in \{1,2,3\}, \alpha \in \{1,2,3,4\}; \end{aligned} \quad (3.113)$$

the corresponding standard form is

$$\begin{aligned} &x_{k+1} = x_k + f(x_k, \mathbf{p}) = x_k + a_0(x_k - a_{i_1})^2(x_k - a_{i_3})^2 \\ &i_\alpha \in \{1,2\}, \alpha = 1,3. \end{aligned} \quad (3.114)$$

The fixed-point of $x_k^* = a_{i_1}, a_{i_3}$ is monotonically unstable (a monotonic upper-saddle of the second-order, $d^2f/dx_k^2|_{x_k^*=a_{i_1},a_{i_3}} > 0$) and monotonically unstable (a monotonic lower-saddle of the second-order, $d^2f/dx_k^2|_{x_k^*=a_{i_1},a_{i_3}} < 0$).

- Such a discrete flow is called a (mUS:mUS) or (mLS:mLS) discrete flow.
- The bifurcation of fixed-point at $x_k^* = a_{i_1}$ for two simple fixed-point onset of $x_k^* = a_{i_1}, a_{i_2}$ and at $x_k^* = a_{i_3}$ for two fixed-point switching of $x_k^* = a_{i_3}, a_{i_4}$ is called a (mUS:mUS) or (mLS:mLS) switching bifurcation at a point $\mathbf{p} = \mathbf{p}_1$.

(iii$_4$) Under

$$\begin{aligned}&\Delta_{i_1 i_3}=(a_{i_1}-a_{i_3})^2=0, a_{i_1}=a_{i_2}, a_{i_1}=a_{i_3} a_{i_1}\neq a_{i_4},\\&i_\alpha\in\{1,2,3\}, \alpha\in\{1,2,3,4\};\end{aligned}\tag{3.115}$$

the standard form is

$$\begin{aligned}&x_{k+1}=x_k+f(x_k,\mathbf{p})=x_k+a_0(x_k-a_{i_1})^3(x_k-a_{i_4})\\&i_\alpha\in\{1,2\}, \alpha=1,4.\end{aligned}\tag{3.116}$$

(iii$_{4a}$) The fixed-point of $x_k^*=a_{i_1}$ is monotonically unstable (a third-order monotonic source, $d^3f/dx_k^3|_{x_k^*=a_{i_1}}>0$).

- Such a discrete flow is called a monotonic source pitchfork discrete flow of the third-order.
- The bifurcation of fixed-point at $x_k^*=a_{i_1}$ for one simple fixed-point of $x_k^*=a_{i_1}$ switching to three simple fixed-points of $x_k^*=a_{i_1,i_2,i_3}$ is called a monotonic upper-saddle-node pitchfork switching bifurcation of the third-order at a point $\mathbf{p}=\mathbf{p}_1$.

(iii$_{4b}$) The fixed-point of $x_k^*=a_{i_1}$ is monotonically stable (a third-order monotonic sink, $d^3f/dx_k^3|_{x_k^*=a_{i_1}}<0$).

- Such a discrete flow is called a monotonic sink discrete flow of the third-order.
- The bifurcation of fixed-point at $x_k^*=a_{i_1}$ for one simple fixed-point of $x_k^*=a_{i_1}$ switching to three simple fixed-points of $x^*=a_{i_1,i_2,i_3}$ is called a monotonic sink pitchfork switching bifurcation of the third-order at a point $\mathbf{p}=\mathbf{p}_1$.

(iii$_5$) Under

$$\begin{aligned}&\Delta_{i_1 i_3}=(a_{i_1}-a_{i_3})^2=0, \Delta_{i_3 i_4}=(a_{i_3}-a_{i_4})^2=0;\\&a_{i_1}=a_{i_2}, a_{i_1}=a_{i_3} a_{i_1}=a_{i_4},\\&i_\alpha\in\{1,2,3\}, \alpha\in\{1,2,3,4\};\end{aligned}\tag{3.117}$$

the standard form is

$$x_{k+1}=x_k+f(x_k,\mathbf{p})=x_k+a_0(x_k-a_{i_1})^4.\tag{3.118}$$

(iii$_{5a}$) For $a_0>0$, the fixed-point of $x_k^*=a_{i_1}$ is monotonically unstable (a monotonic upper-saddle of the fourth-order, $d^4f/dx_k^4|_{x_k^*=a_{i_1}}>0$).

- Such a discrete flow is called a monotonic upper-saddle discrete flow of the fourth-order.

- The bifurcation of fixed-point at $x_k^* = a_{i_1}$ for two simple fixed-points switching to four simple fixed-points is called a monotonic upper-saddle-node flower-bundle-switching bifurcation of the fourth-order at a point $\mathbf{p} = \mathbf{p}_1$.

(iii$_{5b}$) For $a_0 > 0$, the fixed-point of $x_k^* = a_{i_1}$ is monotonically unstable (a monotonic lower-saddle of the fourth-order, $d^4 f / dx_k^4|_{x_k^* = a_{i_1}} < 0$).

- Such a discrete flow is called a monotonic lower-saddle discrete flow of the fourth-order.
- The bifurcation of fixed-point at $x_k^* = a_{i_1}$ for two simple fixed-points switching to four simple fixed-points is called a monotonic lower-saddle-node flower-bundle-switching bifurcation of the fourth-order at a point $\mathbf{p} = \mathbf{p}_1$.

Based on the previous definition, the stability and bifurcations of fixed-points in the 1-dimensional, quartic nonlinear discrete system $(a_0 > 0)$ is presented in Fig. 3.14. In Fig. 3.14(i)–(iii), monotonic upper-saddle-node (mUSN) and monotonic lower-saddle-node (mLSN) switching bifurcations are at two locations for two simple fixed-points, and one monotonic upper-saddle-node (mUSN) appearing bifurcation is for two simple fixed-points. In Fig. 3.14(iv), a third-order monotonic sink (3rd mSI) pitchfork-switching bifurcation for a switching of one monotonic sink fixed-point to three monotonic simple fixed-points is presented, and one monotonic upper-saddle-node (mUSN) switching bifurcation for two monotonic simple fixed-points switching is also presented. In Fig. 3.14(v), a third-order source (3rd mSO) bundle-switching bifurcation for three fixed-point bundle-switching is presented, and a monotonic upper-saddle-node (mUSN) appearing bifurcation for two fixed-point onsets is also presented. In Fig. 3.14(vi), a fourth-order upper-saddle (4th mUS) flower-bundle switching bifurcation for four simple fixed-points are presented.

Similarly, the stability and bifurcations of fixed-points in the 1-dimensional, quartic nonlinear discrete system $(a_0 < 0)$ is presented in Fig. 3.15. In Fig. 3.15(i)–(iii), monotonic-lower-saddle-node (mLSN) and monotonic-upper-saddle-node (mUSN) switching bifurcations are at two locations for two simple fixed-points, and one monotonic-lower-saddle-node (mLSN) appearing bifurcation is for two simple fixed-points appearing. In Fig. 3.15(iv), a third-order monotonic source (3rd mSO) pitchfork-switching bifurcation for a switching of one monotonic source fixed-point to three monotonic simple fixed-points is presented, and one monotonic-lower-saddle-node (mLSN) switching bifurcation for two simple fixed-points switching is also presented. In Fig. 3.15(v), a third-order monotonic sink (3rd mSI) bundle-switching bifurcation for three fixed-point bundle-switching is presented, and a monotonic lower-saddle-node (mLSN) appearing bifurcation for two fixed-point onset is also presented. In Fig. 3.15(vi), a fourth-order order monotonic-lower-saddle (4th mLS) flower-bundle switching bifurcation for four simple fixed-points are presented. The period-2 fixed-points are also sketched for mSI-oSO or mSI-oSO-mSI.

For the further discussion on the switching bifurcation, the following definition is given for the 1-dimensional, quartic nonlinear discrete system.

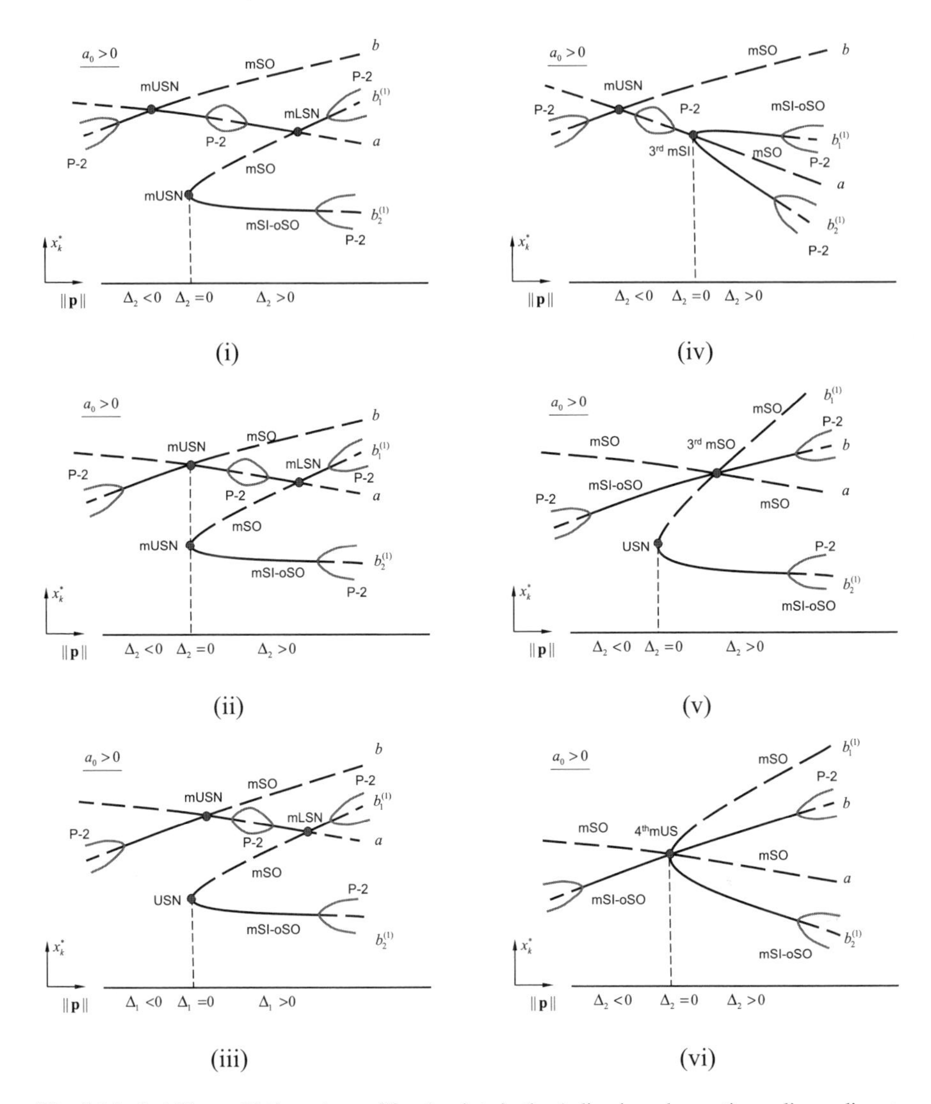

Fig. 3.14 Stability and bifurcations of fixed-points in the 1-dimeisonal, quartic nonlinear discrete system ($a_0 > 0$): (i)-(iii) Two (mUSN and mLSN) switching and one mUSN appearing bifurcations, (iv) 3rd mSI pitchfork-switching bifurcation, (v) 3rd mSO bundle-switching bifurcation, (vi) 4thm mUS flower-bundle switching bifurcation. mLSN: monotonic-lower-saddle-node, mUSN: monotonic-upper-saddle-node, mSI-oSO: monotonic-sink to oscillatory source, mSI-oSO-mSI: monotonic-sink to oscillatory source to monotonic sink, mSO: monotonic-source. Stable and unstable fixed-points are represented by solid and dashed curves, respectively. The bifurcation points are marked by circular symbols. P-2: Period-2 fixed-point. Open curves of P-2 are for mSI-oSO. Closed loop of P-2 are for mSI-oSO-mSI.

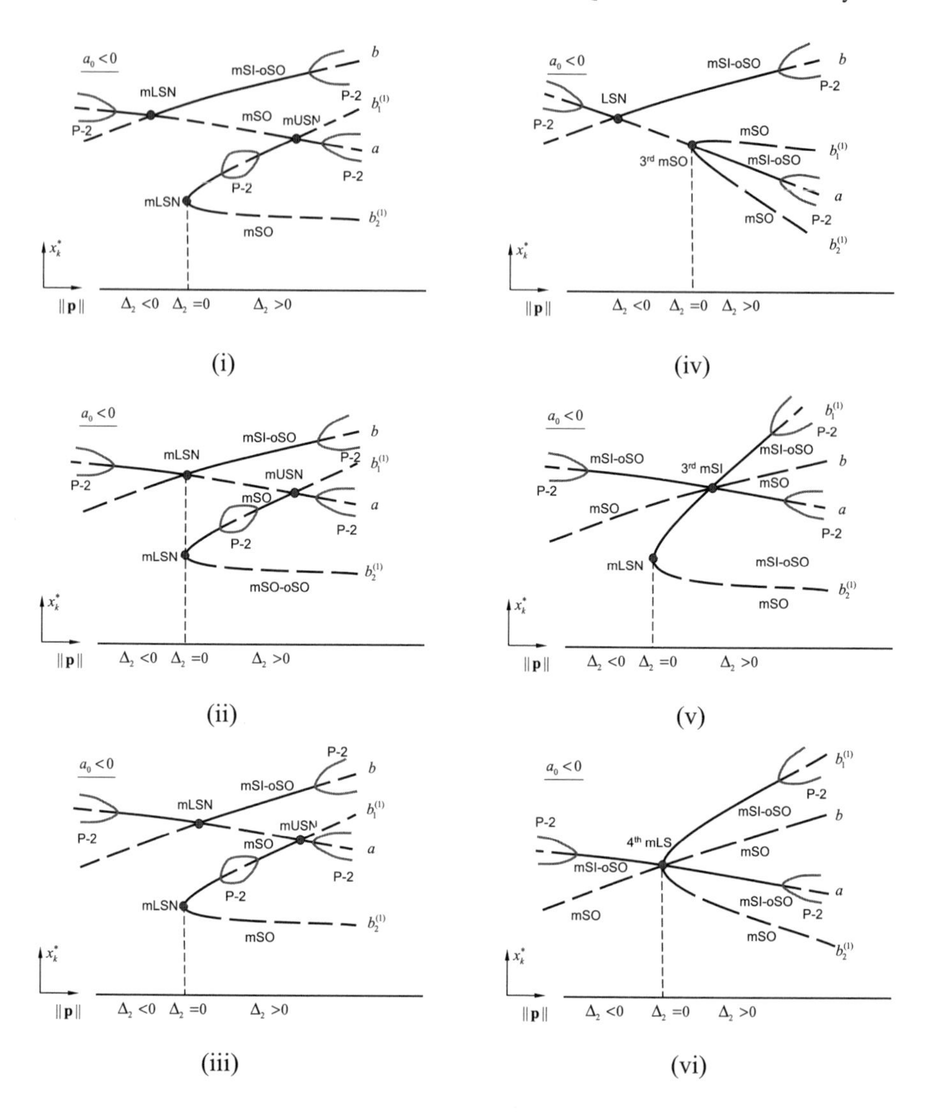

Fig. 3.15 Stability and bifurcations of fixed-points in the 1-dimeisonal, quartic nonlinear discrete system ($a_0 < 0$): (i)-(iii) Two (mLSN and mUSN) switching and one mLSN appearing bifurcations, (iv) 3rd mSO pitchfork switching bifurcation, (v) 3rd mSI bundle-switching bifurcation, (vi) 4th mLS flower-bundle switching bifurcation. mLSN: monotonic-lower-saddle-node, mUSN: monotonic-upper-saddle-node, mSI-oSO: monotonic-sink to oscillatory source, mSI-oSO: monotonic-sink to oscillatory source to monotonic sink, mSO: monotonic-source. Stable and unstable fixed-points are represented by solid and dashed curves, respectively. The bifurcation points are marked by circular symbols. Open curves of P-2 are for mSI-oSO. Closed loop of P-2 are for mSI-oSO-mSI.

Definition 3.5 Consider a 1-dimensional, quartic nonlinear discrete system

$$\begin{aligned} x_{k+1} &= x_k + f(x_k, \mathbf{p}) \\ &= x_k + A(\mathbf{p})x_k^4 + B(\mathbf{p})x_k^3 + C(\mathbf{p})x_k^2 + D(\mathbf{p})x_k + E(\mathbf{p}) \\ &= a_0(\mathbf{p})(x_k - a)(x_k - b)(x_k - c)(x_k - d) \end{aligned} \tag{3.119}$$

where $A(\mathbf{p}) \neq 0$, and

$$\mathbf{p} = (p_1, p_2, \ldots, p_m)^{\mathrm{T}}. \tag{3.120}$$

(i) If

$$\{a_1, a_2, a_3, a_4\} = \mathrm{sort}\{a, b, c, d\}, a_i \leq a_{i+1}, \tag{3.121}$$

the quartic nonlinear discrete system has any four simple fixed-points. The standard form is

$$\begin{aligned} x_{k+1} &= x_k + f(x_k, \mathbf{p}) \\ &= x_k + a_0(\mathbf{p})(x_k - a_1)(x_k - a_2)(x_k - a_2)(x_k - a_3). \end{aligned} \tag{3.122}$$

(i_1) For $a_0 > 0$, the fixed-point flow is a (mSI-oSO:mSO:mSI-oSO:mSO) flow.

(i_{1a}) The fixed-point of $x_k^* = a_{1,3}$ is

- monotonically stable (monotonic sink) if $df/dx_k|_{x_k^*=a_{1,3}} \in (-1, 0)$,
- invariantly stable (invariant sink) if $df/dx_k|_{x_k^*=a_{1,3}} = -1$,
- oscillatorilly stable (oscillatory sink) if $df/dx_k|_{x_k^*=a_{1,3}} \in (-2, -1)$,
- flipped if $df/dx_k|_{x_k^*=a_{1,3}} = -2$, which is
  - an oscillatory upper-saddle of the second-order for $d^2f/dx_k^2|_{x_k^*=a_{1,3}} > 0$;
  - an oscillatory lower-saddle of the second-order for $d^2f/dx_k^2|_{x_k^*=a_{1,3}} < 0$;
- oscillatorilly stable (oscillatory source) if $df/dx_k|_{x_k^*=a_{1,3}} \in (-\infty, -2)$.

(i_{1b}) The fixed-point of $x^* = a_{2,4}$ is monotonically unstable (a monotonic source) if $df/dx_k|_{x_k^*=a_{2,4}} \in (0, \infty)$.

(i_2) For $a_0 < 0$, the fixed-point flow is a (mSO:mSI-oSO:mSO:mSI-oSO) flow.

(i_{2a}) The fixed-point of $x_k^* = a_{1,3}$ is monotonically unstable (a monotonic source) if $df/dx_k|_{x_k^*=a_{1,3}} \in (0, \infty)$.

(i_{2b}) The fixed-point of $x_k^* = a_{2,4}$ is

- monotonically stable (monotonic sink) if $df/dx_k|_{x_k^*=a_{2,4}} \in (-1, 0)$,
- invariantly stable (invariant sink) if $df/dx_k|_{x_k^*=a_{2,4}} = -1$,
- oscillatorilly stable (oscillatory sink) if $df/dx_k|_{x_k^*=a_{2,4}} \in (-2, -1)$,
- flipped if $df/dx_k|_{x_k^*=a_{2,4}} = -2$, which is
  - an oscillatory upper-saddle of the second-order for $d^2f/dx_k^2|_{x_k^*=a_{2,4}} > 0$;
  - an oscillatory lower-saddle of the second-order for $d^2f/dx_k^2|_{x_k^*=a_{2,4}} < 0$;
- oscillatorilly stable (oscillatory source) if $df/dx_k|_{x_k^*=a_{2,4}} \in (-\infty, -2)$.

(ii) If

$$\Delta_{i_1 i_2} = (a_{i_1} - a_{i_2})^2 = 0 \text{ with } a_{i_1} = a_{i_2}; i_1, i_2 \in \{1, 2, 3, 4\}, \tag{3.123}$$

the quartic nonlinear discrete system has a standard form as

$$x_{k+1} = x_k + f(x_k, \mathbf{p}) = x_k + a_0(x_k - a_{i_1})^2(x_k - a_{i_3})(x_k - a_{i_4}). \tag{3.124}$$

(ii_1) The fixed-point of $x_k^* = a_{i_1}$ is monotonically unstable (a monotonic-upper-saddle of second-order, $d^2f/dx_k^2|_{x_k^*=a_{i_1}} > 0$).

- Such a discrete flow is called a monotonic-upper-saddle discrete flow of the second-order at $x_k^* = a_{i_1}$.
- The bifurcation of fixed-point at $x_k^* = a_{i_1}$ for two fixed-points switching of $x_k^* = a_{i_1}, a_{i_2}$ is called a monotonic-upper-saddle-node switching bifurcation of the second-order at a point $\mathbf{p} = \mathbf{p}_1$.

(ii_2) The fixed-point of $x_k^* = a_{i_1}$ is monotonically unstable (a monotonic-lower-saddle of second-order, $d^2f/dx_k^2|_{x_k^*=a_{i_1}} < 0$).

- Such a discrete flow is called a lower-saddle discrete flow of the second-order at $x_k^* = a_{i_1}$.
- The bifurcation of fixed-point at $x_k^* = a_{i_1}$ for two fixed-points switching of $x_k^* = a_{i_1}, a_{i_2}$ is called a lower-saddle-node switching bifurcation of the second-order at a point $\mathbf{p} = \mathbf{p}_1$.

(ii_3) The fixed-point of $x_k^* = a_j$ $(j = i_3, i_4)$ is

- monotonically unstable (monotonic source) if $df/dx_k|_{x_k^*=a_j} \in (0, \infty)$;
- monotonically stable (monotonic sink) if $df/dx_k|_{x_k^*=a_j} \in (-1, 0)$;
- invariantly stable (invariant sink) if $df/dx_k|_{x_k^*=a_j} = -1$;

- oscillatorilly stable (oscillatory sink) if $df/dx_k|_{x_k^*=a_j} \in (-2,-1)$;
- flipped if $df/dx_k|_{x_k^*=a_j} = -2$, which is
  - an oscillatory upper-saddle of the second-order for $d^2f/dx_k^2|_{x_k^*=a_j} > 0$;
  - an oscillatory lower-saddle of the second-order for $d^2f/dx_k^2|_{x_k^*=a_j} < 0$;
- oscillatorilly stable (oscillatory source) if $df/dx_k|_{x_k^*=a_j} \in (-\infty,-2)$.

(iii) If

$$\begin{aligned}&\Delta_{i_1 i_2} = (a_{i_1} - a_{i_2})^2 = 0, \Delta_{i_2 i_3} = (a_{i_2} - a_{i_3})^2 = 0,\\&a_{i_1} = a_{i_2} = a_3, i_1, i_2, i_3 \in \{1,2,3,4\}, i_1 \neq i_2 \neq i_3,\end{aligned} \tag{3.125}$$

the corresponding standard form is

$$\begin{aligned}&x_{k+1} = x_k + f(x_k, \mathbf{p}) = x_k + a_0(x_k - a_{i_1})^3(x_k - a_{i_4})\\&\quad i_\alpha \in \{1,2,3,4\}, \alpha = 1,4.\end{aligned} \tag{3.126}$$

(iii$_1$) The fixed-point of $x_k^* = a_{i_1}$ is monotonically unstable (a third-order monotonic source,$d^3f/dx_k^3|_{x_k^*=a_{i_1}} > 0$).

- Such a discrete flow is called a monotonic source discrete flow of the third-order at $x_k^* = a_{i_1}$.
- The bifurcation of fixed-point at $x_k^* = a_{i_1}$ for a bundle switching of three simple fixed-points of $x_k^* = a_{i_1}, a_{i_2}, a_{i_3}$ is called a third-order monotonic source bundle-switching bifurcation at a point $\mathbf{p} = \mathbf{p}_1$.

(iii$_2$) The fixed-point of $x_k^* = a_{i_1}$ is monotonically stable (a third-order monotonic sink, $d^3f/dx_k^3|_{x_k^*=a_{i_1}} < 0$).

- Such a discrete flow is called a monotonic sink discrete flow of the third-order at $x_k^* = a_{i_1}$.
- The bifurcation of fixed-point at $x_k^* = a_{i_1}$ for a bundle switching of three simple fixed-points of $x_k^* = a_{i_1}, a_{i_2}, a_{i_3}$ is called a monotonic sink bundle-switching bifurcation of the third-order at a point $\mathbf{p} = \mathbf{p}_1$.

(iii$_3$) The fixed-point of $x_k^* = a_{i_4}$ is

- monotonically unstable (monotonic source) if $df/dx_k|_{x_k^*=a_{i_4}} \in (0,\infty)$;
- monotonically stable (monotonic sink) if $df/dx_k|_{x_k^*=a_{i_4}} \in (-1,0)$;
- invariantly stable (invariant sink) if $df/dx_k|_{x_k^*=a_{i_4}} = -1$;

- oscillatorilly stable (oscillatory sink) if $df/dx_k|_{x_k^*=a_{i_4}} \in (-2,-1)$;
- flipped if $df/dx_k|_{x_k^*=a_{i_4}} = -2$, which is
  - an oscillatory upper-saddle of the second-order for $d^2f/dx_k^2|_{x_k^*=a_{i_4}} > 0$;
  - an oscillatory lower-saddle of the second-order for $d^2f/dx_k^2|_{x_k^*=a_{i_4}} < 0$;
- oscillatorilly stable (oscillatory source) if $df/dx_k|_{x_k^*=a_{i_4}} \in (-\infty,-2)$.

(iv) If

$$\begin{aligned} &\Delta_{i_1 i_2} = (a_{i_1} - a_{i_2})^2 = 0, \Delta_{i_2 i_3} = (a_{i_2} - a_{i_3})^2 = 0, \\ &\Delta_{i_3 i_4} = (a_{i_4} - a_{i_4})^2 = 0, a_{i_1} = a_{i_2} = a_{i_3} = a_{i_4}, \\ &i_1, i_2, i_3, i_4 \in \{1,2,3,4\}, i_1 \neq i_2 \neq i_3 \neq i_4, \end{aligned} \tag{3.127}$$

the corresponding standard form is

$$x_{k+1} = x_k + f(x_k, \mathbf{p}) = x_k + a_0(x_k - a_{i_1})^4. \tag{3.128}$$

(iv$_1$) The fixed-point of $x_k^* = a_{i_1}$ is monotonically unstable (a monotonic-upper-saddle of the fourth-order, $d^4f/dx_k^4|_{x_k^*=a_{i_1}} > 0$).

- Such a discrete flow is called a monotonic-upper-saddle discrete flow of the fourth-order at $x_k^* = a_{i_1}$.
- The bifurcation of fixed-point at $x_k^* = a_{i_1}$ for a bundle switching of four simple fixed-points of $x_k^* = a_{1,2,3,4}$ is called a monotonic-upper-saddle-node bundle-switching bifurcation of the fourth-order at a point $\mathbf{p} = \mathbf{p}_1$.

(iv$_2$) The fixed-point of $x_k^* = a_{i_1}$ is monotonically unstable (a 4th order monotonic-lower-saddle, $d^4f/dx_k^4|_{x_k^*=a_{i_1}} < 0$).

- Such a flow is called a monotonic-lower-saddle discrete flow of the fourth-order at $x_k^* = a_{i_1}$.
- The bifurcation of fixed-point at $x_k^* = a_{i_1}$ for a bundle switching of four simple fixed-points of $x_k^* = a_{1,2,3,4}$ is called a monotonic-lower-saddle bundle-switching bifurcation of the fourth-order at a point $\mathbf{p} = \mathbf{p}_1$.

From the previous definition, stability and bifurcations of fixed-points in the 1-dimensional, quartic nonlinear discrete system is presented in Fig. 3.16. For $a_0 > 0$, the bifurcations and stability of fixed-points are presented in Fig. 3.16(i)–(iii). In Fig. 3.16(i), four monotonic-upper-saddle-node (mUSN) and two monotonic-lower-saddle-node (mLSN) switching bifurcation network are presented for all possible switching bifurcation between two simple fixed-points. In Fig. 3.16(ii), a third-order monotonic-sink (3rd mSI) bundle-switching bifurcation for three simple fixed-point is presented, and there are three possible monotonic-upper-saddle-node (mUSN) and monotonic-lower-saddle-node (mLSN) switching bifurcations for two simple fixed-points. Figure 3.16(iii) a fourth-order monotonic-upper-saddle (4th mUS) bundle-switching bifurcation for four simple fixed-points are presented. Similarly, For $a_0 < 0$, the bifurcations and stability of fixed-points are presented in Fig. 3.16 (iv)–(vi). In Fig. 3.16(iv), four monotonic-lower-saddle-node (mLSN) and two monotonic-upper-saddle-node (mUSN) switching bifurcation network are presented for all possible switching bifurcation between two simple fixed-points. In Fig. 3.16 (v), a third-order monotonic source (mSO) bundle-switching bifurcation for three simple fixed-point is presented, and there are three possible monotonic-lower-saddle-node (mLSN) and monotonic-upper-saddle-node (mUSN) switching bifurcations for two simple fixed-points. Figure 3.16(vi) a fourth-order monotonic-lower-saddle (4th mLS) bundle-switching bifurcation for four simple fixed-points are presented. The corresponding period-2 fixed points are sketched as well.

For the switching bifurcation between the second-order and simple fixed-points, the following definition is given for the 1-dimensional, quartic nonlinear discrete system.

3.4.2 *Higher-Order Period-1 Switching Bifurcations*

Definition 3.6 Consider a 1-dimensional, quartic nonlinear discrete system

$$\begin{aligned} x_{k+1} &= x_k + f(x_k, \mathbf{p}) \\ &= x_k + A(\mathbf{p})x_k^4 + B(\mathbf{p})x_k^3 + C(\mathbf{p})x_k^2 + D(\mathbf{p})x_k + E(\mathbf{p}) \\ &= x_k + a_0(\mathbf{p})(x_k - a)^2(x_k - b)(x_k - c) \end{aligned} \tag{3.129}$$

where $A(\mathbf{p}) \neq 0$, and

$$\mathbf{p} = (p_1, p_2, \ldots, p_m)^{\mathrm{T}}. \tag{3.130}$$

(i) If

$$\begin{aligned} &\{a_1, a_2, a_3\} = \mathrm{sort}\{a, b, c\}, a_i < a_{i+1} \\ &i_1, i_2, i_3 \in \{1, 2, 3\} \end{aligned} \tag{3.131}$$

the quartic nonlinear discrete system has a standard form as

$$\begin{aligned} x_{k+1} &= x_k + f(x_k, \mathbf{p}) \\ &= x_k + a_0(x_k - a_{i_1})^2(x_k - a_{i_2})(x_k - a_{i_3}). \end{aligned} \tag{3.132}$$

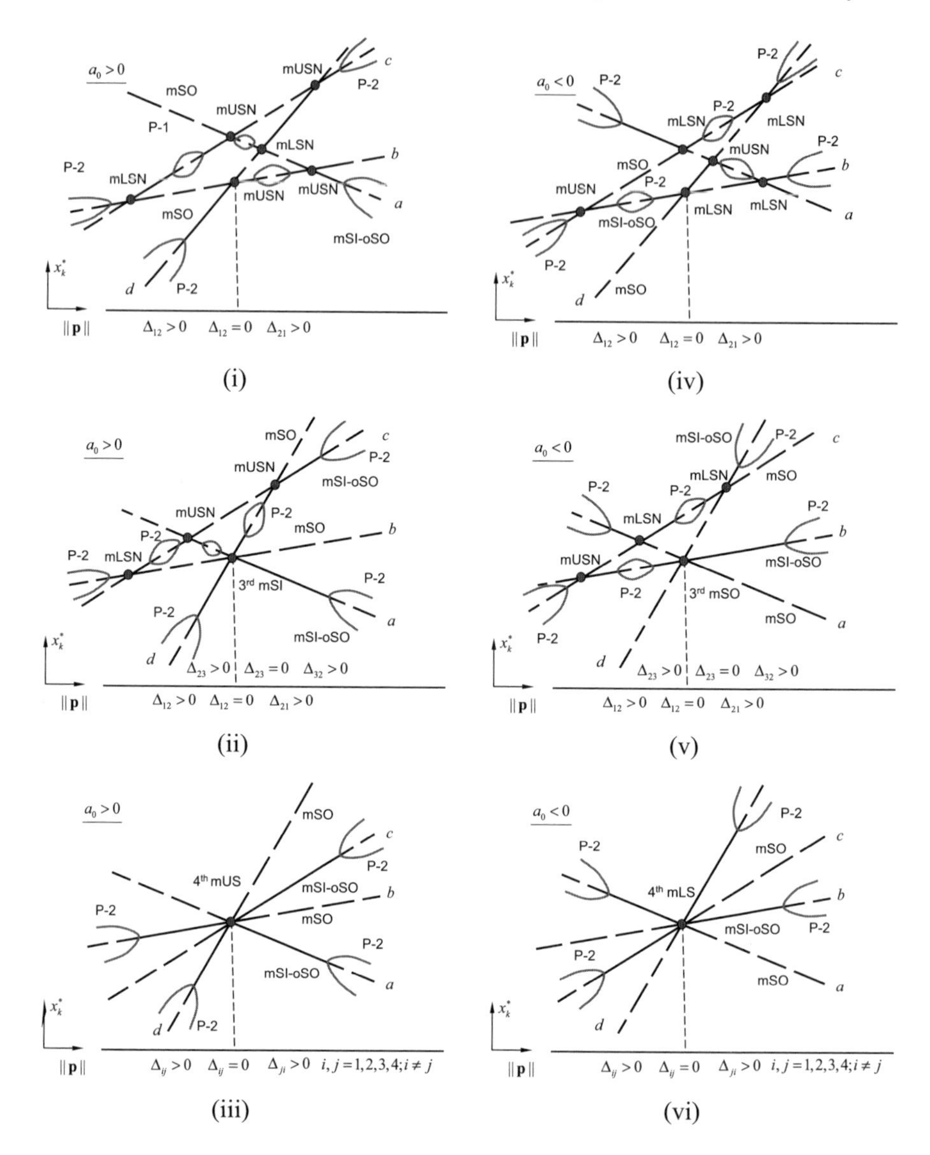

Fig. 3.16 Stability and bifurcations of fixed-points in the 1-dimensional, quartic nonlinear discrete system ($a_0 > 0$): (i) Four USN and two LSN switching bifurcation network, (ii) 3rd order mSI bundle-switching bifurcation, (iii) 4th order mUS bundle-switching bifurcation, ($a_0 < 0$): (iv) Four mLSN and two mUSN switching bifurcation network, (v) 3rd order mSO bundle-switching bifurcation, (vi) 4th order mLS bundle-switching bifurcation. mLSN: monotonic lower-saddle-node, mUSN: upper-saddle-node, mSI-oSO: sink, mSO: source. Stable and unstable fixed-points are represented by solid and dashed curves, respectively. The bifurcation points are marked by circular symbols. Open curves of P-2 are for mSI-oSO. Closed loop of P-2 are for mSI-oSO-mSI.

(i_{1a}) The fixed-point of $x_k^* = a_{i_1}$ is monotonically unstable (a monotonic-upper-saddle of second-order, $d^2f/dx_k^2|_{x_k^*=a_{i_1}} > 0$). Such a discrete flow is called a monotonic upper-saddle discrete flow of the second-order at $x_k^* = a_{i_1}$.

(i_{1b}) The fixed-point of $x_k^* = a_{i_1}$ is monotonically unstable (a monotonic-lower-saddle of second-order, $d^2f/dx_k^2|_{x_k^*=a_{i_1}} < 0$). Such a discrete flow is called a monotonic lower-saddle discrete flow of the second-order at $x_k^* = a_{i_1}$.

(i_{1c}) The fixed-point of $x_k^* = a_j$ ($j = i_2, i_3$) is

- monotonically unstable (monotonic source) if $df/dx_k|_{x_k^*=a_j} \in (0, \infty)$;
- monotonically stable (monotonic sink) if $df/dx_k|_{x_k^*=a_j} \in (-1, 0)$;
- invariantly stable (invariant sink) if $df/dx_k|_{x_k^*=a_j} = -1$;
- oscillatorilly stable (oscillatory sink) if $df/dx_k|_{x_k^*=a_j} \in (-2, -1)$;
- flipped if $df/dx_k|_{x_k^*=a_j} = -2$, which is
  - an oscillatory upper-saddle of the second-order for $d^2f/dx_k^2|_{x_k^*=a_j} > 0$;
  - an oscillatory lower-saddle of the second-order for $d^2f/dx_k^2|_{x_k^*=a_j} < 0$;
- oscillatorilly stable (oscillatory source) if $df/dx_k|_{x_k^*=a_j} \in (-\infty, -2)$.

(ii) If

$$\begin{aligned} &a = a_{i_1}, b = a_{i_2}, c = a_{i_3}; \\ &\Delta_{i_2 i_3} = (a_{i_2} - a_{i_3})^2 = 0, a_{i_2} = a_{i_3}; \\ &i_1, i_2, i_3 \in \{1, 2\}, i_1 \neq i_2 \neq i_3; \end{aligned} \tag{3.133}$$

the corresponding standard form is

$$\begin{aligned} x_{k+1} &= x_k + f(x_k, \mathbf{p}) = x_k + a_0(x - a_{i_1})^2 (x - a_{i_2})^2 \\ &i_\alpha \in \{1, 2\}, \alpha = 1, 2. \end{aligned} \tag{3.134}$$

The fixed-points of $x_k^* = a_{i_1}, a_{i_2}$ are monotonically unstable (a monotonic-upper-saddle of the second-order, $d^2f/dx_k^2|_{x_k^*=a_{i_1}\text{or}a_{i_2}} > 0$ or a monotonic-lower-saddle of the second-order, $d^2f/dx_k^2|_{x_k^*=a_{i_2}\text{or}a_{i_1}} < 0$).

- Such a discrete flow is called a (mUS:mUS) or (mLS:mLS) discrete flow.
- The bifurcation of fixed-point at $x_k^* = a_{i_2}$ for two simple fixed-points switching of $x_k^* = b, c$ is called a monotonic-upper-saddle or monotonic-lower-saddle switching bifurcation of the second-order at a point $\mathbf{p} = \mathbf{p}_1$.

(iii) If

$$\begin{aligned} & a = a_{i_1}; a_{i_3}, a_{i_2} \in \{b, c\}; \\ & \Delta_{i_1 i_3} = (a_{i_1} - a_{i_3})^2 = 0, a_{i_1} = a_{i_3}, a_{i_3} \neq a_{i_2}; \\ & i_1, i_2, i_3 \in \{1, 2, 3\}, i_1 \neq i_2 \neq i_3, \end{aligned} \tag{3.135}$$

the corresponding standard form is

$$\begin{aligned} x_{k+1} = x_k + f(x_k, \mathbf{p}) = x_k + a_0 (x - a_{i_1})^3 (x - a_{i_2}) \\ i_\alpha \in \{1, 2\}, \alpha = 1, 2. \end{aligned} \tag{3.136}$$

(iii_1) The fixed-point of $x_k^* = a_{i_1}$ is monotonically unstable (a 3rd order monotonic source, $d^3f/dx_k^3|_{x_k^*=a_{i_1}} > 0$).

- Such a discrete flow is called a monotonic source discrete flow of the third-order at $x_k^* = a_{i_1}$.
- The bifurcation of fixed-point at $x_k^* = a_{i_1}$ for a switching of second-order and simple fixed-points of $x_k^* = a_{i_1}, a_{i_2}$ is called a monotonic source switching bifurcation of the third-order at a point $\mathbf{p} = \mathbf{p}_1$.

(iii_2) The fixed-point of $x_k^* = a_{i_1}$ is monotonically stable (a third-order monotonic sink, $d^3f/dx_k^3|_{x_k^*=a_{i_1}} < 0$).

- Such a discrete flow is called a monotonic sink discrete flow of the third-order at $x_k^* = a_{i_1}$.
- The bifurcation of fixed-point at $x_k^* = a_{i_1}$ for a switching of one second-order and one simple fixed-points of $x_k^* = a_{i_1}, a_{i_2}$ is called a monotonic sink switching bifurcation of the third-order at a point $\mathbf{p} = \mathbf{p}_1$.

(iii_3) The fixed-point of $x_k^* = a_{i_2}$ is

- monotonically unstable (monotonic source) if $df/dx_k|_{x_k^*=a_{i_2}} \in (0, \infty)$;
- monotonically stable (monotonic sink) if $df/dx_k|_{x_k^*=a_{i_2}} \in (-1, 0)$;
- invariantly stable (invariant sink) if $df/dx_k|_{x_k^*=a_{i_2}} = -1$;
- oscillatorilly stable (oscillatory sink) if $df/dx_k|_{x_k^*=a_{i_2}} \in (-2, -1)$;
- flipped if $df/dx_k|_{x_k^*=a_{i_2}} = -2$, which is
  - an oscillatory upper-saddle of the second-order for $d^2f/dx_k^2|_{x_k^*=a_{i_2}} > 0$;
  - an oscillatory lower-saddle of the second-order for $d^2f/dx_k^2|_{x_k^*=a_{i_2}} < 0$;
- oscillatorilly stable (oscillatory source) if $df/dx_k|_{x_k^*=a_{i_2}} \in (-\infty, -2)$.

(iv) If

$$\begin{aligned} &\Delta_{i_1 i_2} = (a_{i_1} - a_{i_2})^2 = 0, \Delta_{i_2 i_3} = (a_{i_2} - a_{i_3})^2 = 0, \\ &i_1, i_2, i_3 \in \{1,2,3\}, i_1 \neq i_2 \neq i_3, \end{aligned} \tag{3.137}$$

the corresponding standard form is

$$x_{k+1} = x_k + f(x_k, \mathbf{p}) = x_k + a_0(x_k - a_{i_1})^4. \tag{3.138}$$

(iv_1) The fixed-point of $x_k^* = a_{i_1}$ is monotonically unstable (a fourth-order monotonic-upper-saddle, $d^4 f/dx_k^4|_{x_k^*=a_{i_1}} > 0$).

- Such a discrete flow is called a monotonic-upper-saddle discrete flow of the fourth-order at $x_k^* = a_{i_1}$.
- The bifurcation of fixed-point at $x_k^* = a_{i_1}$ for a bundle switching of one second-order and two simple fixed-points of $x_k^* = a_{1,2,3}$ is called a upper-saddle bundle-switching bifurcation of the fourth-order at a point $\mathbf{p} = \mathbf{p}_1$.

(iv_2) The fixed-point of $x_k^* = a_{i_1}$ is monotonically unstable (a fourth-order monotonic lower-saddle, $d^4 f/dx_k^4|_{x_k^*=a_{i_1}} < 0$).

- Such a discrete flow is called a monotonic-lower-saddle discrete flow of the fourth-order at $x_k^* = a_{i_1}$.
- The bifurcation of fixed-point at $x_k^* = a_{i_1}$ for a bundle switching of one second-order and two simple fixed-points of $x_k^* = a_{1,2,3}$ is called a monotonic-lower-saddle bundle-switching bifurcation of the fourth-order at a point $\mathbf{p} = \mathbf{p}_1$.

From Definition 3.6, stability and bifurcations of fixed-points in the 1-dimensional, quartic nonlinear discrete system is presented in Fig. 3.17. For $a_0 > 0$, the bifurcations and stability of fixed-points are presented in Fig. 3.17(i)–(iii). In Fig. 3.17(i), there is a switching bifurcation network with two third-order monotonic-source switching bifurcations and one monotonic upper-saddle-node bifurcation. The two third-order monotonic source (3rd mSO) switching bifurcations are for (mLS:mSO) switching to (mUS:mSO) fixed-points and for (mSO:mUS) switching to (mSO:mLS)-fixed-points. The upper-saddle-node (mUSN) switching bifurcation is for two simple fixed-points. In Fig. 3.17(ii), a fourth-order monotonic-upper-saddle (4th US) bundle-switching bifurcation for (mSI-oSO:mLS:mSO) switching to (mSO: mLS:mSI-oSO) fixed-points. In Fig. 3.17(iii), a fourth-order monotonic-upper-saddle (4th mUS) bundle-switching bifurcation for (mSI-oSO:mSO:mUS) switching to (mSO:mSI-oSO:mUS) fixed-points. Similarly, for $a_0 < 0$, the bifurcations and stability of fixed-points are presented in in Fig. 3.17(iv)–(vi). In Fig. 3.17(iv), the switching bifurcation network consists of two third-order monotonic sink switching

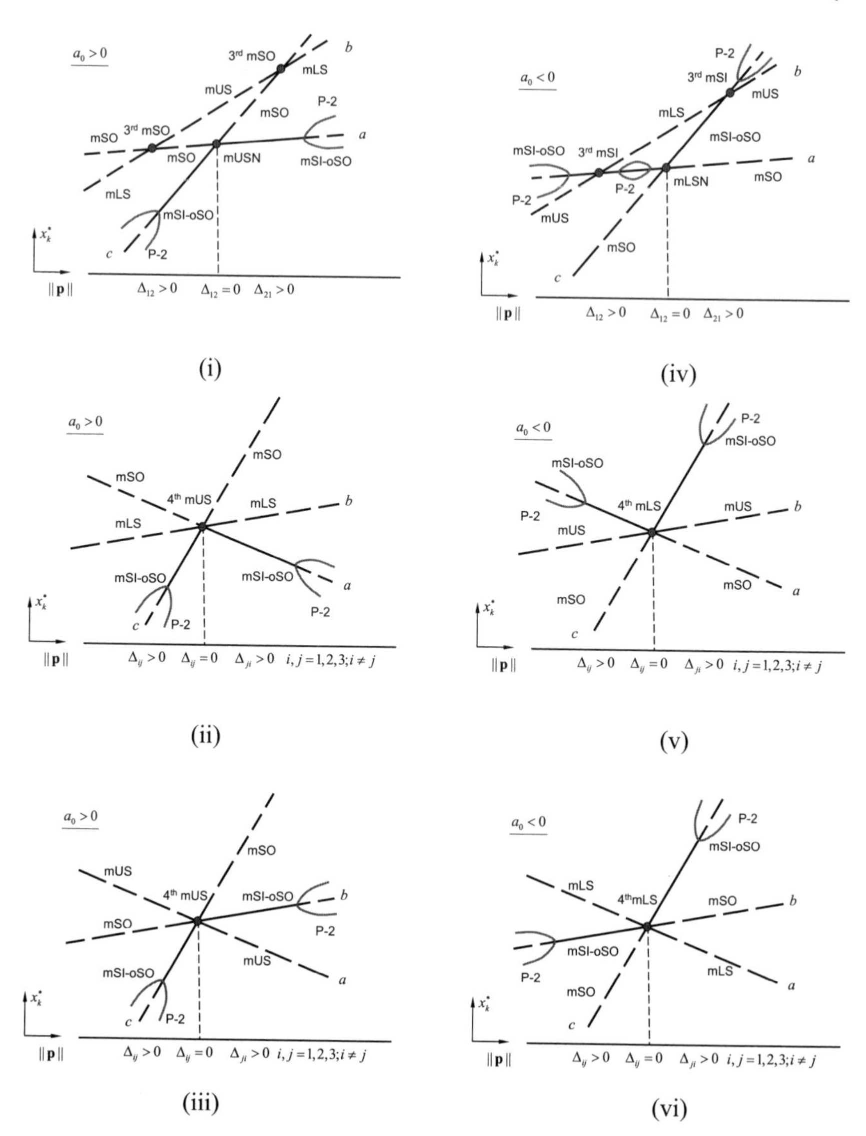

Fig. 3.17 Stability and bifurcations of fixed-points in the 1-dimeisonal, quartic nonlinear discrete system ($a_0 > 0$): (i) two 3rd order mSO and mUSN switching bifurcation network, (ii) 4th mUS bundle switching bifurcation, (iii) 4th order mUS bundle-switching bifurcation, ($a_0 < 0$): (iv) two 3rd order mSI and mLSN switching bifurcation network, (v) 4th order mLS bundle-switching bifurcation, (vi) 4th order mLS bundle switching. mLSN: monotonic-lower-saddle-node, mUSN: monotonic-upper-saddle-node, mSI-oSO: monotonic-sink to oscillatory source, mSI-oSO-mSI: monotonic-sink to oscillatory source to monotonic sink, mSO: monotonic-source. Stable and unstable fixed-points are represented by solid and dashed curves, respectively. The bifurcation points are marked by circular symbols. P-2: period-2 fixed-point. Open curves of P-2 are for mSI-oSO. Closed loop of P-2 are for mSI-oSO-mSI.

bifurcations and one monotonic lower-saddle-node bifurcation. The two 3rd order monotonic sink (3rd mSI) switching bifurcations are for the (mLS:mSO) switching to (mUS:mSO)-fixed-points and for the (mSO:mUS) switching to (mSO:mLS)-fixed-points. The monotonic upper-saddle-node (mUSN) switching bifurcation is for two simple fixed-points. In Fig. 3.17(v), a fourth-order monotonic lower-saddle (4th mLS) bundle-switching bifurcation for (mSO:mUS:mSI-oSO) switching to (mSI-oSO:mUS:mSO) fixed-points. In Fig. 3.17(vi), a fourth-order monotonic-lower-saddle (4th mLS) bundle-switching bifurcation for (mSO:mSI-oSO:mLS) switching to (mSI-oSO:mSO:mLS) fixed-points.

For the switching bifurcation between the third-order and simple fixed-points, the following definition is given for the 1-dimensional, quartic nonlinear discrete system.

Definition 3.7 Consider a 1-dimensional, quartic nonlinear discrete system

$$\begin{aligned} x_{k+1} &= x_k + f(x_k, \mathbf{p}) \\ &= x_k + A(\mathbf{p})x_k^4 + B(\mathbf{p})x_k^3 + C(\mathbf{p})x_k^2 + D(\mathbf{p})x_k + E(\mathbf{p}) \\ &= x_k + a_0(\mathbf{p})(x_k - a)^3(x_k - b) \end{aligned} \tag{3.139}$$

where $A(\mathbf{p}) \neq 0$, and

$$\mathbf{p} = (p_1, p_2, \ldots, p_m)^{\mathrm{T}}. \tag{3.140}$$

(i) If

$$\begin{aligned} &\{a_1, a_2\} = \mathrm{sort}\{a, b\}, a_i < a_{i+1} \\ &i_1, i_2 \in \{1, 2\}, \end{aligned} \tag{3.141}$$

the quartic nonlinear discrete system has a standard form as

$$x_{k+1} = x_k + f(x_k, \mathbf{p}) = x_k + a_0(x_k - a_{i_1})^3(x_k - a_{i_2}). \tag{3.142}$$

(i_{1a}) The fixed-point of $x_k^* = a_{i_1}$ is monotonically unstable (a third-order monotonic-source, $d^3f/dx_k^3|_{x_k^*=a_{i_1}} > 0$).

(i_{1b}) The fixed-point of $x_k^* = a_{i_2}$ is

- monotonically stable (monotonic sink) if $df/dx_k|_{x_k^*=a_{i_2}} \in (-1, 0)$,
- invariantly stable (invariant sink) if $df/dx_k|_{x_k^*=a_{i_2}} = -1$,
- oscillatorilly stable (oscillatory sink) if $df/dx_k|_{x_k^*=a_{i_2}} \in (-2, -1)$,
- flipped if $df/dx_k|_{x_k^*=a_{i_2}} = -2$,
- oscillatorilly stable (oscillatory source) if $df/dx_k|_{x_k^*=a_{i_2}} \in (-\infty, -2)$.

($\mathrm{i}_{1\mathrm{c}}$) Such a discrete flow is called a (3rd mSO:mSI-oSO) or (mSI-oSO:3rd mSO)-discrete flow.

($\mathrm{i}_{2\mathrm{a}}$) The fixed-point of $x^* = a_{i_1}$ is monotonically stable (a third-order monotonic-sink, $d^3f/dx_k^3|_{x_k^*=a_{i_1}} < 0$).

($\mathrm{i}_{2\mathrm{b}}$) The fixed-point of $x^* = a_{i_2}$ is monotonically unstable (a monotonic source, $df/dx_k|_{x_k^*=a_{i_2}} \in (0,\infty)$).

($\mathrm{i}_{2\mathrm{c}}$) Such a discrete flow is called a (3rd mSI:mSO) or (mSO:3rd mSI)-discrete flow.

(ii) If

$$
\begin{aligned}
&a = a_{i_1}, b = a_{i_2};\\
&\Delta_{i_1 i_2} = (a_{i_1} - a_{i_2})^2 = 0, a_{i_1} = a_{i_2};\\
&i_1, i_2 \in \{1,2\}, i_1 \neq i_2;
\end{aligned}
\tag{3.143}
$$

the corresponding standard form is

$$
x_{k+1} = x_k + f(x_k, \mathbf{p}) = x_k + a_0(x_k - a_{i_1})^4. \tag{3.144}
$$

(ii_1) The fixed-point of $x_k^* = a_{i_1}$ is monotonically unstable (a fourth-order monotonically upper-saddle, $d^4f/dx_k^4|_{x_k^*=a_{i_1}} > 0$).

- Such a discrete flow is called a fourth-order monotonic upper-saddle discrete flow.
- The bifurcation of fixed-point at $x_k^* = a_{i_1}$ for a switching of one third-order and one simple fixed-points of $x_k^* = a_{1,2}$ is called a fourth-order upper-saddle switching bifurcation at a point $\mathbf{p} = \mathbf{p}_1$.

(ii_2) The fixed-point of $x_k^* = a_{i_1}$ is monotonically unstable (a fourth-order monotonic-lower-saddle, $d^4f/dx_k^4|_{x_k^*=a_{i_1}} < 0$).

- Such a discrete flow is called a fourth-order monotonic-lower-saddle discrete flow.
- The bifurcation of fixed-point at $x_k^* = a_{i_1}$ for a switching of one third-order and one simple fixed-points of $x_k^* = a_{1,2}$ is called a fourth-order monotonic lower-saddle switching bifurcation at a point $\mathbf{p} = \mathbf{p}_1$.

From the Definition 3.7, the stability and bifurcations of fixed-points in the 1-dimensional, quartic nonlinear discrete system is presented in Fig. 3.18. In Fig. 3.18(i), the fourth-order monotonic upper-saddle (4th mUS) switching bifurcation for $a_0 > 0$ is presented for the third-order monotonic sink (3rd mSI) with a

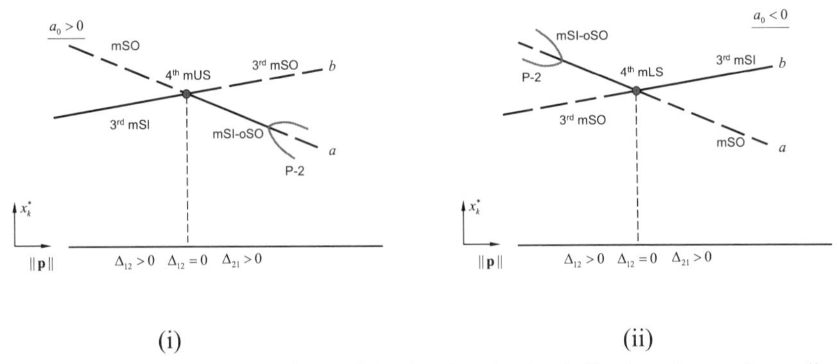

(i) (ii)

Fig. 3.18 Stability and bifurcations of fixed-points in the 1-dimeisonal, quartic nonlinear discrete system ($a_0 > 0$): (i) 4th order US switching bifurcation of (3rdmSI: mSO) to (mSI-oSO: 3rd mSO), ($a_0 < 0$):(ii) 4th order mLS switching bifurcation of (3rd mSO: mSI-oSO) to (mSO: 3rd mSI). mLS: monotonic-lower-saddle, mUS: monotonic-upper-saddle, mSI-oSO: monotonic sink to oscillatory source, mSO: monotonic-source. Stable and unstable fixed-points are represented by solid and dashed curves, respectively. The bifurcation points are marked by circular symbols. P-2: period-2 fixed-point. Open curves of P-2 are for mSI-oSO. Closed loop of P-2 are for mSI-oSO-mSI.

simple monotonic source (mSO) fixed-points (i.e., (3rd mSI:mSO)) to the third-order monotonic source (3rd mSO) with a simple sink to oscillatory source (mSI-oSO) fixed-points (i.e., (mSI-oSO:3rd mSO)). Similarly, in Fig. 3.18(ii), the fourth-order lower-saddle (4th mLS) switching bifurcation for $a_0 < 0$ is presented for the third-order source (3rd mSO) with a simple monotonic sink to oscillatory source (mSI-oSO) fixed-points (i.e., (3rd mSO:mSI-oSO)) to the 3rd order monotonic sink (3rd mSI) with a simple monotonic source (mSO) fixed-points (i.e., (mSO: 3rd mSI)).

For the switching bifurcation between the two second-order monotonic fixed-points, the following definition was presented in Definition 3.3, and the corresponding illustrations are presented in Fig. 3.12.

3.5 Forward Quartic Discrete Systems

In this section, the analytical bifurcation scenario of a quartic nonlinear discrete system will be discussed. The period-doubling bifurcation scenario will be discussed first through nonlinear renormalization techniques, and the bifurcation scenario based on the saddle-node bifurcation will be discussed, which is independent of period-1 fixed-points.

3.5.1 Period-2 Quartic Discrete Systems

After the period-doubling bifurcation of a period-1 fixed-point, the period-doubled fixed-points of a quartic nonlinear discrete system can be obtained. Consider the period-doubling solutions of a forward quartic nonlinear discrete system first.

Theorem 3.2 *Consider a 1-dimensional, forward, quartic nonlinear discrete system*

$$\begin{aligned} x_{k+1} &= x_k + A(\mathbf{p})x_k^4 + B(\mathbf{p})x_k^3 + C(\mathbf{p})x_k^2 + D(\mathbf{p})x_k + E(\mathbf{p}) \\ &= x_k + a_0(\mathbf{p})[x_k^2 + B_1(\mathbf{p})x_k + C_1(\mathbf{p})][x_k^2 + B_2(\mathbf{p})x_k + C_2(\mathbf{p})] \end{aligned} \tag{3.145}$$

where $a_0(\mathbf{p}) = A(\mathbf{p}) \neq 0$, *and*

$$\mathbf{p} = (p_1, p_2, \ldots, p_m)^{\mathrm{T}}. \tag{3.146}$$

Under

$$\begin{aligned} &\Delta_i = B_i^2 - 4C_i > 0, \text{ for } i = 1,2 \text{ with} \\ &a_{1,2} = \frac{1}{2}B_1 \pm \frac{1}{2}\sqrt{\Delta_1}; a_{3,4} = \frac{1}{2}B_2 \pm \frac{1}{2}\sqrt{\Delta_2}, \end{aligned} \tag{3.147}$$

the standard form of such a 1-dimensional system is

$$x_{k+1} = x_k + a_0(x_k - a_1)(x_k - a_2)(x_k - a_3)(x_k - a_4). \tag{3.148}$$

Thus, a general standard form of such a 1-dimensional quartic discrete system is

$$\begin{aligned} x_{k+1} &= x_k + Ax_k^4 + Bx_k^3 + Cx_k^2 + Dx_k + E \\ &= x_k + a_0 \prod\nolimits_{i=1}^4 (x_k - a_i^{(1)}) \end{aligned} \tag{3.149}$$

where

$$\begin{aligned} &b_{i,1}^{(1)} = -\tfrac{1}{2}(B_i^{(1)} + \sqrt{\Delta_i^{(1)}}),\ b_{i,2}^{(1)} = -\tfrac{1}{2}(B_i^{(1)} - \sqrt{\Delta_i^{(1)}}) \\ &\text{for } \Delta_i^{(1)} \geq 0,\ i \in \{1,2\}; \\ &\cup_{i=1}^4 a_i^{(1)} = \text{sort}\{\cup_{i=1}^2 \{b_{i,2}^{(1)}, b_{i,2}^{(1)}\}\}, a_i^{(1)} \leq a_{i+1}^{(i)}; \\ &\text{for } \Delta_i^{(1)} < 0,\ i \in \{1,2\},\ \mathbf{i} = \sqrt{-1}; \\ &a_1^{(1)} = b_{1,1}^{(1)}, a_2^{(1)} = b_{1,2}^{(1)}; a_3^{(1)} = b_{2,1}^{(1)}, a_4^{(1)} = b_{2,2}^{(1)}. \end{aligned} \tag{3.150}$$

(i) *Consider a forward period-2 discrete system of Eq.* (3.145) *as*

$$
\begin{aligned}
x_{k+2} &= x_k + [a_0 \prod_{i_1=1}^{4} (x_k - a_{i_1}^{(1)})]\{1 + \prod_{i_1=1}^{4} [1 + a_0 \prod_{i_2=1, i_2 \neq i_1}^{4} (x_k - a_{i_2}^{(1)})]\} \\
&= x_k + [a_0 \prod_{i_1=1}^{4} (x_k - a_{i_1}^{(1)})][a_0^4 \prod_{i_2=1}^{(4^2-4)/2} (x_k^2 + B_{i_2}^{(2)} x_k + C_{i_2}^{(2)})] \\
&= x_k + [a_0 \prod_{j_1=1}^{3} (x_k - a_{i_1}^{(1)})][a_0^4 \prod_{j_2=1}^{4^2-4} (x_k - b_{j_2}^{(2)})] \\
&= x_k + a_0^{1+4} \prod_{i=1}^{4^2} (x_k - a_i^{(2)})
\end{aligned}
\tag{3.151}
$$

where

$$
\begin{aligned}
& b_{i,1}^{(2)} = -\tfrac{1}{2}(B_i^{(2)} + \sqrt{\Delta^{(2)}}), b_{i,2}^{(2)} = -\tfrac{1}{2}(B_i^{(2)} - \sqrt{\Delta_i^{(2)}}), \\
& \Delta_i^{(2)} = (B_i^{(2)})^2 - 4C_i^{(2)} \geq 0,\ i \in \cup_{q_1=1}^{N_1} I_{q_1}^{(2^0)} \cup \cup_{q_2=1}^{N_2} I_{q_2}^{(2^2)} \\
& I_{q_1}^{(2^0)} = \{l_{(q_1-1)\times 2^0 \times m_1+1}, l_{(q_1-1)\times 2^0\times m_1+2}, \cdots, l_{q_1\times 2^0\times m_1}\} \\
& \qquad \subseteq \{1, 2, \cdots, M_1\} \cup \{\varnothing\}, \\
& q_1 \in \{1, 2, \cdots, N_1\}, M_1 = N_1 \times 2^0 \times m_1, m_1 \in \{1, 2\}; \\
& I_{q_2}^{(2^1)} = \{l_{(q_2-1)\times 2^1\times m_1+1}, l_{(q_2-1)\times 2^1\times m_1+2}, \cdots, l_{q_2\times 2^1\times m_1}\} \\
& \qquad \subseteq \{M_1+1, M_1+2, \cdots, M_2\} \cup \{\varnothing\}, \\
& q_2 \in \{1, 2, \cdots, N_2\}, M_2 = (4^2 - 4)/2; \\
& b_{i,1}^{(2)} = -\tfrac{1}{2}(B_i^{(2)} + \mathbf{i}\sqrt{\Delta^{(2)}}), b_{i,2}^{(2)} = -\tfrac{1}{2}(B_i^{(2)} - \mathbf{i}\sqrt{\Delta_i^{(2)}}), \\
& \mathbf{i} = \sqrt{-1}, \Delta_i^{(2)} = (B_i^{(2)})^2 - 4C_i^{(2)} < 0, \\
& i \in J^{(2^1)} = \{l_{N_2\times 2^1\times m_1+1}, l_{N_2\times 2^1\times m_1+2}, \cdots, l_{M_2}\} \\
& \qquad\qquad \subseteq \{M_1+1, M_1+2, \cdots, M_2\}
\end{aligned}
\tag{3.152}
$$

with fixed-points

$$
\begin{aligned}
& x_{k+2}^* = x_k^* = a_i^{(2)} (i = 1, 2, \ldots, 4^2), \\
& \cup_{i=1}^{4^2} \{a_i^{(2)}\} = \mathrm{sort}\{\cup_{j_1=1}^{4} \{a_{j_1}^{(1)}\}, \cup_{j_2=1}^{M_2} \{b_{j_2,1}^{(2)}, b_{j_2,2}^{(2)}\}\} \\
& \text{with } a_i^{(2)} < a_{i+1}^{(2)},\ M_2 = (4^2 - 4)/2.
\end{aligned}
\tag{3.153}
$$

(ii) *For a fixed-point of* $x_{k+1}^* = x_k^* = a_{i_1}^{(1)}$ $(i_1 \in \{1, 2, \ldots, 4\})$, *if*

$$
\frac{dx_{k+1}}{dx_k}\Big|_{x_k^* = a_{i_1}^{(1)}} = 1 + a_0 \prod_{i_2=1, i_2 \neq i_1}^{4} (a_{i_1}^{(1)} - a_{i_2}^{(1)}) = -1,
\tag{3.154}
$$

with

- *a* r^{th}*-order oscillatory upper-saddle-node bifurcation* $(d^r x_{k+1}/dx_k^r|_{x_k^*=a_{i_1}^{(1)}} > 0, r = 2l_1)$,
- *a* r^{th}*-order oscillatory lower-saddle-node bifurcation* $(d^r x_{k+1}/dx_k^r|_{x_k^*=a_{i_1}^{(1)}} < 0,\ r = 2l_1)$,
- *a* r^{th}*-order oscillatory sink bifurcation* $(d^r x_{k+1}/dx_k^r|_{x_k^*=a_{i_1}^{(1)}} > 0, r = 2l_1 + 1)$,
- *a* r^{th}*-order oscillatory source bifurcation* $(d^r x_{k+1}/dx_k^r|_{x_k^*=a_{i_1}^{(1)}} > 0, r = 2l_1 + 1)$,

then the following relations satisfy

$$a_{i_1}^{(1)} = -\frac{1}{2}B_{i_1}^{(2)}, \Delta_{i_1}^{(2)} = (B_{i_1}^{(2)})^2 - 4C_{i_1}^{(2)} = 0, \tag{3.155}$$

and there is a period-2 discrete system of the quartic discrete system in Eq. (3.145) *as*

$$x_{k+2} = x_k + a_0^5 \prod_{i_2 \in I_{q_1}^{(20)}} (x_k - a_{i_2}^{(1)})^3 \prod_{i_3=1}^{4^2} (x_k - a_{i_3}^{(2)})^{(1-\delta(i_2,i_3))}. \tag{3.156}$$

For $i_1 \in \{1, 2, \ldots, 4\}$, $i_1 \neq i_2$ *with*

$$\frac{dx_{k+2}}{dx_k}|_{x_k^*=a_{i_1}^{(1)}} = 1, \frac{d^2 x_{k+2}}{dx_k^2}|_{x_k^*=a_{i_1}^{(1)}} = 0; \tag{3.157}$$

- x_{k+2} at $x_k^* = a_{i_1}^{(1)}$ *is a monotonic sink of the third-order if*

$$\begin{aligned}\frac{d^3 x_{k+2}}{dx_k^3}|_{x_k^*=a_{i_1}^{(1)}} &= 6a_0^5 \prod_{i_2 \in I_{q_1}^{(20)}, i_2 \neq i_1} (a_{i_1}^{(1)} - a_{i_2}^{(1)})^3 \\ &\times \prod_{i_3=1}^{4^2} (a_{i_1}^{(1)} - a_{i_3}^{(2)})^{(1-\delta(i_2,i_3))} < 0,\end{aligned} \tag{3.158}$$

  and the corresponding bifurcations is a third-order monotonic sink bifurcation for the period-2 discrete system;
- x_{k+2} at $x_k^* = a_{i_1}^{(1)}$ *is a monotonic source of the third-order if*

$$\begin{aligned}\frac{d^3 x_{k+2}}{dx_k^3}|_{x_k^*=a_{i_1}^{(1)}} &= 6a_0^5 \prod_{i_2 \in I_{q_1}^{(20)}, i_2 \neq i_1} (a_{i_1}^{(1)} - a_{i_2}^{(1)})^3 \\ &\times \prod_{i_3=1}^{4^2} (a_{i_1}^{(1)} - a_{i_3}^{(2)})^{(1-\delta(i_2,i_3))} > 0,\end{aligned} \tag{3.159}$$

and the corresponding bifurcations is a third-order monotonic source bifurcation for the period-2 discrete system.

(ii$_1$) *The period-2 fixed-points are trivial and unstable if*

$$x_{k+2}^{*}=x_{k}^{*}=a_{i}^{(2)}\in\cup_{i_1=1}^{4}\{a_{i_1}^{(1)}\}. \tag{3.160}$$

(ii$_2$) *The period-2 fixed-points are non-trivial and stable if*

$$x_{k+2}^{*}=x_{k}^{*}=a_{i}^{(2)}\in\cup_{i_1=1}^{M_2}\{b_{i_1,1}^{(2)},b_{i_1,2}^{(2)}\}. \tag{3.161}$$

Proof The proof is straightforward through the simple algebraic manipulation. Following the proof of quadratic discrete system, this theorem is proved. ■

3.5.2 Period-Doubling Renormalization

The generalized cases of period-doublization of quartic discrete systems are presented through the following theorem. The analytical period-doubling bifurcation trees can be developed for quartic discrete systems.

Theorem 3.3 *Consider a 1-dimensional quartic nonlinear discrete system as*

$$\begin{aligned}x_{k+1}&=x_k+Ax_k^4+Bx_k^3+Cx_k^2+Dx_k+E\\&=x_k+a_0\prod_{i=1}^{4}(x_k-a_i^{(1)}).\end{aligned} \tag{3.162}$$

(i) *After* l*-times period-doubling bifurcations, a period-*2^l $(l=1,2,\ldots)$ *discrete system for the quartic discrete system in Eq.* (3.162) *is given through the nonlinear renormalization as*

$$\begin{aligned}x_{k+2^l}&=x_k+[a_0^{(2^{l-1})}\prod_{i_1=1}^{4^{2^{l-1}}}(x_k-a_{i_1}^{(2^{l-1})})]\\&\quad\times\{1+\prod_{i_1=1}^{4^{2^{l-1}}}[1+a_0^{(2^{l-1})}\prod_{i_2=1,i_2\neq i_1}^{4^{2^{l-1}}}(x_k-a_{i_2}^{(2^{l-1})})]\}\\&=x_k+[a_0^{(2^{l-1})}\prod_{i_1=1}^{4^{2^{l-1}}}(x_k-a_{i_1}^{(2^{l-1})})]\\&\quad\times[(a_0^{(2^{l-1})})^{4^{2^{l-1}}}\prod_{j_1=1}^{(4^{2^l}-4^{2^{l-1}})/2}(x_k^2+B_{j_2}^{(2^l)}x_k+C_{j_2}^{(2^l)})]\\&=x_k+[a_0^{(2^{l-1})}\prod_{i_1=1,}^{2^{2^{l-1}}}(x_k-a_{i_1}^{(2^{l-1})})]\\&\quad\times[(a_0^{(2^{l-1})})^{4^{2^{l-1}}}\prod_{i_2=1}^{(4^{2^l}-4^{2^{l-1}})/2}(x_k-b_{i_2,1}^{(2^l)})(x_k-b_{i_2,2}^{(2^l)})]\\&=x_k+(a_0^{(2^{l-1})})^{1+4^{2^{l-1}}}\prod_{i=1}^{4^{2^l}}(x_k-a_i^{(2^l)})\\&=x_k+a_0^{(2^l)}\prod_{i=1}^{4^{2^l}}(x_k-a_i^{(2^l)})\end{aligned} \tag{3.163}$$

with

$$\frac{dx_{k+2^l}}{dx_k} = 1 + a_0^{(2^l)} \sum_{i_1=1}^{4^{2^l}} \prod_{i_2=1, i_2\neq i_1}^{4^{2^l}} (x_k - a_{i_2}^{(2^l)}),$$

$$\frac{d^2 x_{k+2^l}}{dx_k^2} = a_0^{(2^l)} \sum_{i_1=1}^{4^{2^l}} \sum_{i_2=1, i_2\neq i_1}^{4^{2^l}} \prod_{i_3=1, i_3\neq i_1, i_2}^{4^{2^l}} (x_k - a_{i_3}^{(2^l)}),$$

$$\vdots$$

$$\frac{d^r x_{k+2^l}}{dx_k^r} = a_0^{(2^l)} \sum_{i_1=1}^{4^{2^l}} \cdots \sum_{i_r=1, i_3\neq i_1, i_2 \cdots i_{r-1}}^{4^{2^l}} \prod_{i_{r+1}=1, i_3\neq i_1, i_2\cdots, i_r}^{4^{2^l}} (x_k - a_{i_{r+1}}^{(2^l)}) \tag{3.164}$$

for $r \leq 4^{2^l}$ *where*

$$\begin{aligned}
& a_0^{(2)} = (a_0)^{1+4}, a_0^{(2^l)} = (a_0^{(2^{l-1})})^{1+4^{2^{l-1}}}, l = 1, 2, 3, \cdots; \\
& \cup_{i=1}^{4^{2^l}} \{a_i^{(2^l)}\} = \text{sort}\{\cup_{i_1=1}^{4^{2^{l-1}}} \{a_{i_1}^{(2^l)}\}, \cup_{i_2=1}^{M_2} \{b_{i_2,1}^{(2^l)}, b_{i_2,2}^{(2^l)}\}\}, a_i^{(2^l)} \leq a_{i+1}^{(2^l)}; \\
& b_{i,1}^{(2^l)} = -\tfrac{1}{2}(B_i^{(2^l)} + \sqrt{\Delta_i^{(2^l)}}), b_{i,2}^{(2^{l-1})} = -\tfrac{1}{2}(B_i^{(2^l)} - \sqrt{\Delta_i^{(2^l)}}), \\
& \Delta_i^{(2^l)} = (B_i^{(2^l)})^2 - 4C_i^{(2^l)} \geq 0 \text{ for } i \in \cup_{q_1=1}^{N_1} I_{q_1}^{(2^{l-1})} \cup \cup_{q_2=1}^{N_2} I_{q_2}^{(2^l)}, \\
& I_{q_1}^{(2^{l-1})} = \{l_{(q_1-1)\times 2^{l-1}\times m_1+1}, l_{(q_1-1)\times 2^{l-1}\times m_1+2}, \cdots, l_{q_1\times 2^{l-1}\times m_1}\} \\
& \qquad \subseteq \{1, 2, \cdots, M_1\} \cup \{\varnothing\}, \\
& \text{for } q_1 \in \{1, 2, \cdots, N_1\}, M_1 = N_1 \times 2^{l-1} \times m_1, m_1 \in \{1, 2\}; \\
& I_{q_2}^{(2^l)} = \{l_{(q_2-1)\times 2^l\times m_1+1}, l_{(q_2-1)\times 2^l\times m_1+2}, \cdots, l_{q_2\times 2^l\times m_1}\} \\
& \qquad \subseteq \{M_1+1, M_1+2, \cdots, M_2\} \cup \{\varnothing\}, \\
& \text{for } q_2 \in \{1, 2, \cdots, N_2\}, M_2 = (4^{2^l} - 4^{2^{l-1}})/2; \\
& b_{i,1}^{(2^l)} = -\tfrac{1}{2}(B_i^{(2^l)} + \mathbf{i}\sqrt{|\Delta_i^{(2^l)}|}), \ b_{i,2}^{(2^l)} = -\tfrac{1}{2}(B_i^{(2^l)} - \mathbf{i}\sqrt{|\Delta_i^{(2^l)}|}), \\
& \Delta_i^{(2^l)} = (B_i^{(2^l)})^2 - 4C_i^{(2^l)} < 0, \mathbf{i} = \sqrt{-1}, \\
& i \in J^{(2^l)} = \{l_{N_2\times 2^l\times m_1+1}, l_{N_2\times 2^l\times m_1+2}, \cdots, l_{M_2}\} \\
& \qquad \subset \{M_1+1, M_1+2, \cdots, M_2\} \cup \{\varnothing\};
\end{aligned} \tag{3.165}$$

with fixed-points

$$\begin{aligned}
& x_{k+2^l}^* = x_k^* = a_i^{(2^l)}, (i = 1, 2, \ldots, 4^{2^l}) \\
& \cup_{i=1}^{4^{2^l}} \{a_i^{(2^l)}\} = \text{sort}\{\cup_{i_1=1}^{4^{2^{l-1}}} \{a_{i_1}^{(2^{l-1})}\}, \cup_{i_2=1}^{M_2} \{b_{i_2,1}^{(2^l)}, b_{i_2,2}^{(2^l)}\}\} \\
& \text{with } a_i^{(2^l)} < a_{i+1}^{(2^l)}.
\end{aligned} \tag{3.166}$$

(ii) *For a fixed-point of* $x^*_{k+2^{l-1}} = x^*_k = a^{(2^{l-1})}_{i_1}$ $(i_1 \in I^{(2^{l-1})}_q)$, *if*

$$\begin{aligned}
&\frac{dx_{k+2^{l-1}}}{dx_k}\Big|_{x^*_k=a^{(2^{l-1})}_{i_1}} = 1 + a^{(2^{l-1})}_0 \prod_{i_2=1,i_2\neq i_1}^{4^{2^{l-1}}} (a^{(2^{l-1})}_{i_1} - a^{(2^{l-1})}_{i_2}) = -1,\\
&\frac{d^s x_{k+2^{l-1}}}{dx^s_k}\Big|_{x^*_k=a^{(2^{l-1})}_{i_1}} = 0, \text{ for } s = 2, \cdots, r-1;\\
&\frac{d^r x_{k+2^{l-1}}}{dx^r_k}\Big|_{x^*_k=a^{(2^{l-1})}_{i_1}} \neq 0 \text{ for } 1 < r \leq 4^{2^{l-1}},
\end{aligned} \tag{3.167}$$

with

- a rth-*order oscillatory sink for* $d^r x_{k+2^{l-1}}/dx^r_k|_{x^*_k=a^{(2^{l-1})}_{i_1}} > 0$ *and* $r = 2l_1+1$;
- a rth-*order oscillatory source for* $d^r x_{k+2^{l-1}}/dx^r_k|_{x^*_k=a^{(2^{l-1})}_{i_1}} < 0$ *and* $r = 2l_1+1$;
- a rth-*order oscillatory upper-saddle for* $d^r x_{k+2^{l-1}}/dx^r_k|_{x^*_k=a^{(2^{l-1})}_{i_1}} > 0$ *and* $r = 2l_1$;
- a rth-*order oscillatory lower-saddle for* $d^r x_{k+2^{l-1}}/dx^r_k|_{x^*_k=a^{(2^{l-1})}_{i_1}} < 0$ *and* $r = 2l_1$;

then there is a period-2^l *fixed-point discrete system*

$$x_{k+2^l} = x_k + a^{(2^l)}_0 \prod_{i_1\in I^{(2^l)}_q} (x_k - a^{(2^{l-1})}_{i_1})^3 \prod_{j_2=1}^{4^{2^l}} (x_k - a^{(2^l)}_{j_2})^{(1-\delta(i_1,j_2))} \tag{3.168}$$

where

$$\delta(i_1,j_2) = 1 \text{ if } a^{(2^l)}_{j_2} = a^{(2^{l-1})}_{i_1},\ \delta(i_1,j_2) = 0 \text{ if } a^{(2^l)}_{j_2} \neq a^{(2^{l-1})}_{i_1} \tag{3.169}$$

and

$$\frac{dx_{k+2^l}}{dx_k}\Big|_{x^*_k=a^{(2^{l-1})}_{i_1}} = 1,\ \frac{d^2 x_{k+2^l}}{dx^2_k}\Big|_{x^*_k=a^{(2^{l-1})}_{i_1}} = 0. \tag{3.170a}$$

- x_{k+2^l} at $x^*_k = a^{(2^{l-1})}_{i_1}$ *is a monotonic sink of the third-order if*

$$\begin{aligned}
\frac{d^3 x_{k+2^l}}{dx^3_k}\Big|_{x^*_k=a^{(2^{l-1})}_{i_1}} &= 6a^{(2^l)}_0 \prod_{i_2\in I^{(2^{l-1})}_q, i_2\neq i_1} (a^{(2^{l-1})}_{i_1} - a^{(2^{l-1})}_{i_2})^3\\
&\quad \times \prod_{j_2\neq 1}^{4^{2^l}} (a^{(2^{l-1})}_{i_1} - a^{(2^l)}_{j_2})^{(1-\delta(i_2,j_2))} < 0\\
(i_1 \in I^{(2^{l-1})}_q,\ &q \in \{1,2,\ldots,N_1\}),
\end{aligned} \tag{3.170b}$$

and such a bifurcation at $x^*_k = a^{(2^{l-1})}_{i_1}$ *is a third-order monotonic sink bifurcation.*

- x_{k+2^l} at $x^*_k = a^{(2^{l-1})}_{i_1}$ *is a monotonic source of the third-order if*

$$\begin{aligned}&\frac{d^3x_{k+2^l}}{dx_k^3}\Big|_{x_k^*=a_{i_1}^{(2^{l-1})}}=6a_0^{(2^l)}\prod\nolimits_{i_2\in I_q^{(2^{l-1})},i_2\neq i_1}(a_{i_1}^{(2^{l-1})}-a_{i_2}^{(2^{l-1})})^3\\&\qquad\times\prod\nolimits_{j_2\neq 1}^{4^{2^l}}(a_{i_1}^{(2^{l-1})}-a_{j_2}^{(2^l)})^{(1-\delta(i_2,j_2))}>0\\&(i_1\in I_q^{(2^{l-1})},q\in\{1,2,\ldots,N_1\})\end{aligned}\tag{3.171}$$

and such a bifurcation at $x_k^*=a_{i_1}^{(2^{l-1})}$ *is a third-order monotonic source bifurcation.*

(ii$_1$) *The period-*2^l *fixed-points are trivial if*

$$x_{k+2^l}^*=x_k^*=a_{i_1}^{(2^l)}\in\cup_{i_1}^{4^{2^{l-1}}}\{a_{i_1}^{(2^{l-1})}\};\tag{3.172}$$

(ii$_2$) *The period-*2^l *fixed-points are non-trivial if*

$$x_{k+2^l}^*=x_k^*=a_{i_1}^{(2^l)}\in\cup_{j_1=1}^{M_2}\{b_{j_1,1}^{(2^l)},b_{j_1,2}^{(2^l)}\}.\tag{3.173}$$

Such a period-2^l *fixed-point is*

- *monotonically unstable if* $dx_{k+2^l}/dx_k|_{x_k^*=a_{i_1}^{(2^l)}}\in(1,\infty)$;
- *monotonically invariant if* $dx_{k+2^l}/dx_k|_{x_k^*=a_{i_1}^{(2^l)}}=1$, *which is*
  - *a monotonic upper-saddle of the* $(2l_1)^{\text{th}}$ *order for* $d^{2l_1}x_{k+2^l}/dx_k^{2l_1}|_{x_k^*}>0$;
  - *a monotonic lower-saddle of the* $(2l_1)^{\text{th}}$ *order for* $d^{2l_1}x_{k+2^l}/dx_k^{2l_1}|_{x_k^*}<0$;
  - *a monotonic source of the* $(2l_1+1)^{\text{th}}$ *order for* $d^{2l_1+1}x_{k+2^l}/dx_k^{2l_1+1}|_{x_k^*}>0$;;
  - *a monotonic sink the* $(2l_1+1)^{\text{th}}$ *order for* $d^{2l_1+1}x_{k+2^l}/dx_k^{2l_1+1}|_{x_k^*}<0$;
- *monotonically stable if* $dx_{k+2^l}/dx_k|_{x_k^*=a_{i_1}^{(2^l)}}\in(0,1)$;
- *invariantly zero-stable if* $dx_{k+2^l}/dx_k|_{x_k^*=a_{i_1}^{(2^{l-1})}}=0$;
- *oscillatorilly stable if* $dx_{k+2^l}/dx_k|_{x_k^*=a_{i_1}^{(2^{l-1})}}\in(-1,0)$;
- *flipped if* $dx_{k+2^l}/dx_k|_{x_k^*=a_{i_1}^{(2^{l-1})}}=-1$, *which is*
  - *an oscillatory upper-saddle of the* $(2l_1)^{\text{th}}$ *order for* $d^{2l_1}x_{k+2^l}/dx_k^{2l_1}|_{x_k^*}>0$;
  - *an oscillatory lower-saddle the* $(2l_1)^{\text{th}}$ *order for* $d^{2l_1}x_{k+2^l}/dx_k^{2l_1}|_{x_k^*}<0$;
  - *an oscillatory source of the* $(2l_1+1)^{\text{th}}$ *order if* $d^{2l_1+1}x_{k+2^l}/dx_k^{2l_1+1}|_{x_k^*}<0$;
  - *an oscillatory sink the* $(2l_1+1)^{\text{th}}$ *order with* $d^{2l_1+1}x_{k+2^l}/dx_k^{2l_1+1}|_{x_k^*}>0$;
- *oscillatorilly unstable if* $dx_{k+2^l}/dx_k|_{x_k^*=a_{i_1}^{(2^l)}}\in(-\infty,-1)$.

Proof Through the nonlinear renormalization, this theorem can be proved. ■

3.5.3 Period-n Appearing and Period-Doublization

The forward period-n discrete system for the quartic nonlinear discrete systems will be discussed, and the period-doublization of period-n discrete systems is discussed through the nonlinear renormalization.

Theorem 3.4 *Consider a 1-dimensional quartic nonlinear discrete system*

$$\begin{aligned}x_{k+1} &= x_k + Ax_k^4 + Bx_k^3 + Cx_k^2 + Dx_k + E\\ &= x_k + a_0\prod_{i=1}^{4}(x_k - a_i^{(1)}).\end{aligned} \tag{3.174}$$

(i) *After* n*-times iterations, a period-*n *discrete system for the quartic discrete system in Eq.* (3.174) *is*

$$\begin{aligned}x_{k+n} &= x_k + a_0\prod_{i_1=1}^{4}(x_k - a_{i_2})\{1+\textstyle\sum_{j=1}^{n}Q_j\}\\ &= x_k + a_0^{(4^n-1)/3}\prod_{i_1=1}^{4}(x_k - a_{i_1})[\prod_{j_2=1}^{(4^n-4)/2}(x_k^2 + B_{j_2}^{(2^l)}x_k + C_{j_2}^{(2^l)})]\\ &= x_k + a_0^{(n)}\prod_{i=1}^{4^n}(x_k - a_i^{(n)})\end{aligned} \tag{3.175}$$

with

$$\begin{aligned}\frac{dx_{k+n}}{dx_k} &= 1 + a_0^{(n)}\textstyle\sum_{i_1=1}^{4^n}\prod_{i_2=1,i_2\neq i_1}^{4^n}(x_k - a_{i_2}^{(n)}),\\ \frac{d^2x_{k+n}}{dx_k^2} &= a_0^{(n)}\textstyle\sum_{i_1=1}^{4^n}\sum_{i_2=1,i_2\neq i_1}^{4^n}\prod_{i_3=1,i_3\neq i_1,i_2}^{4^n}(x_k - a_{i_3}^{(n)}),\\ &\vdots\\ \frac{d^rx_{k+n}}{dx_k^r} &= a_0^{(n)}\textstyle\sum_{i_1=1}^{4^n}\cdots\sum_{i_r=1,i_r\neq i_1,i_2\cdots,i_{r-1}}^{4^n}\prod_{i_{r+1}=1,i_{r+1}\neq i_1,i_2\cdots,i_r}^{4^n}(x_k - a_{i_{r+1}}^{(n)})\\ &\text{for } r\leq 4^n;\end{aligned} \tag{3.176}$$

where

$$\begin{aligned}&a_0^{(n)} = (a_0)^{(4^n-1)/3}, Q_1 = 0, Q_2 = \textstyle\prod_{i_2=1}^{4}[1 + a_0\prod_{i_1=1,i_1\neq i_2}^{4}(x_k - a_{i_1}^{(1)})],\\ &Q_n = \textstyle\prod_{i_n=1}^{4}[1 + a_0(1+Q_{n-1})\prod_{i_{n-1}=1,i_{n-1}\neq i_n}^{4}(x_k - a_{i_{n-1}}^{(1)})],\ n = 3,4,\cdots;\\ &\cup_{i=1}^{4^n}\{a_i^{(n)}\} = \text{sort}\{\cup_{i_1=1}^{4}\{a_{i_1}^{(1)}\}, \cup_{i_2=1}^{M}\{b_{i_2,1}^{(n)}, b_{i_2,2}^{(n)}\}\}\ ;\\ &b_{i_2,1}^{(n)} = -\tfrac{1}{2}(B_{i_2}^{(n)} + \sqrt{\Delta_{i_2}^{(n)}}), b_{i_2,2}^{(n)} = -\tfrac{1}{2}(B_{i_2}^{(n)} - \sqrt{\Delta_{i_2}^{(n)}}),\end{aligned}$$

$$
\begin{aligned}
&\Delta_{i_2}^{(n)} = (B_{i_2}^{(n)})^2 - 4C_{i_2}^{(n)} \geq 0 \text{ for } i_2 \in \cup_{q=1}^{N} I_q^{(n)},\\
&I_q^{(n)} = \{l_{(q-1)\times n+1}, l_{(q-1)\times n+2}, \cdots, l_{q\times n}\} \subseteq \{1,2,\cdots,M\} \cup \{\varnothing\},\\
&\text{for } q \in \{1,2,\cdots,N\}, M = (4^n - 4)/2;\\
&b_{i,1}^{(n)} = -\tfrac{1}{2}(B_i^{(n)} + \mathbf{i}\sqrt{|\Delta_i^{(n)}|}),\ b_{i,2}^{(n)} = -\tfrac{1}{2}(B_i^{(n)} - \mathbf{i}\sqrt{|\Delta_i^{(n)}|}),\\
&\Delta_i^{(n)} = (B_i^{(n)})^2 - 4C_i^{(n)} < 0,\ \mathbf{i} = \sqrt{-1}\\
&i \in \{l_{N\times n+1}, l_{N\times n+2}, \cdots, l_M\} \subset \{1,2,\cdots,M\} \cup \{\varnothing\};
\end{aligned}
\tag{3.177}
$$

with fixed-points

$$
\begin{aligned}
&x_{k+n}^* = x_k^* = a_i^{(n)}, (i = 1,2,\ldots,4^n)\\
&\cup_{i=1}^{4^n}\{a_i^{(n)}\} = \text{sort}\{\cup_{i_1=1}^{4}\{a_{i_1}^{(1)}\}, \cup_{i_2=1}^{M}\{b_{i_2,1}^{(n)}, b_{i_2,2}^{(n)}\}\}\\
&\text{with } a_i^{(n)} < a_{i+1}^{(n)}.
\end{aligned}
\tag{3.178}
$$

(ii) *For a fixed-point of* $x_{k+n}^* = x_k^* = a_{i_1}^{(n)}$ $(i_1 \in I_q^{(n)}, q \in \{1,2,\ldots,N\})$, *if*

$$
\frac{dx_{k+n}}{dx_k}\Big|_{x_k^*=a_{i_1}^{(n)}} = 1 + a_0^{(n)} \prod_{i_2=1, i_2\neq i_1}^{4^n} (a_{i_1}^{(n)} - a_{i_2}^{(n)}) = 1, \tag{3.179}
$$

with

$$
\frac{d^2x_{k+n}}{dx_k^2}\Big|_{x_k^*=a_{i_1}^{(n)}} = a_0^{(n)} \sum_{i_2=1, i_2\neq i_1}^{4^n} \prod_{i_3=1, i_3\neq i_1, i_2}^{4^n} (a_{i_1}^{(n)} - a_{i_3}^{(n)}) \neq 0, \tag{3.180}
$$

then there is a new discrete system for onset of the qth*- set of period-*n *fixed-points based on the second-order monotonic saddle-node bifurcation as*

$$
x_{k+n} = x_k + a_0^{(n)} \prod_{i_1 \in I_q^{(n)}} (x_k - a_{i_1}^{(n)})^2 \prod_{j_2=1}^{4^n} (x_k - a_{j_2}^{(n)})^{(1-\delta(i_1,j_2))} \tag{3.181}
$$

where

$$
\delta(i_1, j_2) = 1 \text{ if } a_{j_2}^{(n)} = a_{i_1}^{(n)},\ \delta(i_1, j_2) = 0 \text{ if } a_{j_2}^{(n)} \neq a_{i_1}^{(n)}. \tag{3.182}
$$

(ii_1) If

$$
\begin{aligned}
&\frac{dx_{k+n}}{dx_k}\Big|_{x_k^*=a_{i_1}^{(n)}} = 1\,(i_1 \in I_q^{(n)}),\\
&\frac{d^2x_{k+n}}{dx_k^2}\Big|_{x_k^*=a_{i_1}^{(n)}} = 2a_0^{(n)} \prod\nolimits_{i_1\in I_q^{(n)}, i_2\neq i_1} (a_{i_1}^{(n)} - a_{i_1}^{(n)})^2\\
&\qquad\qquad \times \prod\nolimits_{j_2=1}^{4^n} (a_{i_1}^{(n)} - a_{j_2}^{(n)})^{(1-\delta(i_2,j_2))} \neq 0
\end{aligned}
\tag{3.183}
$$

x_{k+n} at $x_k^* = a_{i_1}^{(n)}$ is

- *a monotonic lower-saddle of the second-order for* $d^2x_{k+n}/dx_k^2|_{x_k^*=a_{i_1}^{(n)}} < 0$;
- *a monotonic upper-saddle of the second-order for* $d^2x_{k+n}/dx_k^2|_{x_k^*=a_{i_1}^{(n)}} > 0$.

(ii_2) *The period-*n *fixed-points* $(n = 2^{n_1} \times m)$ *are trivial if*

$$
\left.\begin{aligned}
&x_k^* = x_{k+n}^* = a_{j_1}^{(n)} \in \{\cup_{i_j=1}^4 \{a_{i_1}^{(1)}\}, \cup_{i_2=1}^{4^{2^{n_1-1}m}} \{a_{i_2}^{(2^{n_1-1}m)}\}\}\\
&\text{for } n_1 = 1,2,\ldots; m = 2l_1+1; j_1 \in \{1,2,\ldots,4^n\}\cup\{\varnothing\}
\end{aligned}\right\}
\\
\text{for } n \neq 2^{n_2},
\\
\left.\begin{aligned}
&x_k^* = x_{k+n}^* = a_{j_1}^{(n)} \in \{\cup_{i_2=1}^{4^{2^{n_1-1}m}} \{a_{i_2}^{(2^{n_1-1}m)}\}\}\\
&\text{for } n_1 = 1,2,\ldots; m = 1; j_1 \in \{1,2,\ldots,4^n\}\cup\{\varnothing\}
\end{aligned}\right\}
\\
\text{for } n = 2^{n_2}.
\tag{3.184}
$$

(ii_3) *The period-*n *fixed-points* $(n = 2^{n_1} \times m)$ *are non-trivial if*

$$
\left.\begin{aligned}
&x_k^* = x_{k+n}^* = a_{j_1}^{(n)} \notin \{\cup_{i_j=1}^4 \{a_{i_1}^{(1)}\}, \cup_{i_2=1}^{4^{2^{n_1-1}m}} \{a_{i_2}^{(2^{n_1-1}m)}\}\}\\
&\text{for } n_1 = 1,2,\ldots; m = 2l_1+1; j_1 \in \{1,2,\ldots,4^n\}\cup\{\varnothing\}
\end{aligned}\right\}
\\
\text{for } n \neq 2^{n_2},
\\
\left.\begin{aligned}
&x_k^* = x_{k+n}^* = a_{j_1}^{(n)} \notin \{\cup_{i_2=1}^{4^{2^{n_1-1}m}} \{a_{i_2}^{(2^{n_1-1}m)}\}\}\\
&\text{for } n_1 = 1,2,\ldots; m = 1; j_1 \in \{1,2,\ldots,4^n\}\cup\{\varnothing\}
\end{aligned}\right\}
\\
\text{for } n = 2^{n_2}.
\tag{3.185}
$$

*Such a forward period-*n *fixed-point is*

- *monotonically unstable if* $dx_{k+n}/dx_k|_{x_k^*=a_{i_1}^{(n)}} \in (1,\infty)$;
- *monotonically invariant if* $dx_{k+n}/dx_k|_{x_k^*=a_{i_1}^{(n)}} = 1$, *which is*

- *a monotonic upper-saddle of the* $(2l_1)^{\text{th}}$ *order for* $d^{2l_1}x_{k+n}/dx_k^{2l_1}|_{x_k^*} > 0$;
- *a monotonic lower-saddle the* $(2l_1)^{\text{th}}$ *order for* $d^{2l_1}x_{k+n}/dx_k^{2l_1}|_{x_k^*} < 0$;
- *a monotonic source of the* $(2l_1+1)^{\text{th}}$ *order for* $d^{2l_1+1}x_{k+n}/dx_k^{2l_1+1}|_{x_k^*} > 0$;
- *a monotonic sink the* $(2l_1+1)^{\text{th}}$ *order for* $d^{2l_1+1}x_{k+n}/dx_k^{2l_1+1}|_{x_k^*} < 0$;

- *monotonically unstable if* $dx_{k+n}/dx_k|_{x_k^*=a_{i_1}^{(n)}} \in (0,1)$;
- *invariantly zero-stable if* $dx_{k+n}/dx_k|_{x_k^*=a_{i_1}^{(n)}} = 0$;
- *oscillatorilly stable if* $dx_{k+n}/dx_k|_{x_k^*=a_{i_1}^{(n)}} \in (-1,0)$;
- *flipped if* $dx_{k+n}/dx_k|_{x_k^*=a_{i_1}^{(n)}} = -1$, *which is*
  - *an oscillatory upper-saddle of the* $(2l_1)^{\text{th}}$ *order for* $d^{2l_1}x_{k+n}/dx_k^{2l_1}|_{x_k^*} > 0$;
  - *an oscillatory lower-saddle the* $(2l_1)^{\text{th}}$ *order for* $d^{2l_1}x_{k+n}/dx_k^{2l_1}|_{x_k^*} < 0$;
  - *an oscillatory source of the* $(2l_1+1)^{\text{th}}$ *order for* $d^{2l_1+1}x_{k+n}/dx_k^{2l_1+1}|_{x_k^*} < 0$;
  - *an oscillatory sink the* $(2l_1+1)^{\text{th}}$ *order for* $d^{2l_1+1}x_{k+n}/dx_k^{2l_1+1}|_{x_k^*} > 0$;
- *oscillatorilly unstable if* $dx_{k+n}/dx_k|_{x_k^*=a_{i_1}^{(n)}} \in (-\infty,-1)$.

(iii) *For a fixed-point of* $x_{k+n}^* = x_k^* = a_{i_1}^{(n)}$ $(i_1 \in I_q^{(n)}, q \in \{1,2,\ldots,N\})$, *there is a period-doubling of the* q^{th}*-set of period-*n *fixed-points if*

$$\begin{aligned}
&\frac{dx_{k+n}}{dx_k}|_{x_k^*=a_{i_1}^{(n)}} = 1 + a_0^{(n)} \prod_{j_2=1, j_2\neq i_1}^{4^n} (a_{i_1}^{(n)} - a_{j_2}^{(n)}) = -1,\\
&\frac{d^s x_{k+n}}{dx_k^s}|_{x_k^*=a_{i_1}^{(n)}} = 0, \text{ for } s = 2,\ldots, r-1;\\
&\frac{d^r x_{k+n}}{dx_k^r}|_{x_k^*=a_{i_1}^{(n)}} \neq 0 \text{ for } 1 < r \leq 4^n
\end{aligned} \tag{3.186}$$

with

- *a* r^{th} *-order oscillatory sink for* $d^r x_{k+n}/dx_k^r|_{x_k^*=a_{i_1}^{(n)}} > 0$ *and* $r = 2l_1+1$;
- *a* r^{th} *-order oscillatory source for* $d^r x_{k+n}/dx_k^r|_{x_k^*=a_{i_1}^{(n)}} < 0$ *and* $r = 2l_1+1$;
- *a* r^{th} *-order oscillatory upper-saddle for* $d^r x_{k+n}/dx_k^r|_{x_k^*=a_{i_1}^{(n)}} > 0$ *and* $r = 2l_1$;
- *a* r^{th} *-order oscillatory lower-saddle for* $d^r x_{k+n}/dx_k^r|_{x_k^*=a_{i_1}^{(n)}} < 0$ *and* $r = 2l_1$.

The corresponding period-$2\times n$ *discrete system of the quartic discrete system in Eq.* (3.174) *is*

$$x_{k+2\times n} = x_k + a_0^{(2\times n)} \prod_{i_1\in I_q^{(n)}} (x_k - a_{i_1}^{(n)})^3 \prod_{j_2=1}^{4^{2\times n}} (x_k - a_{j_2}^{(2\times n)})^{(1-\delta(i_1,j_2))} \tag{3.187}$$

with

$$\begin{aligned} &\frac{dx_{k+2\times n}}{dx_k}\Big|_{x_k^*=a_{i_1}^{(n)}} = 1, \quad \frac{d^2x_{k+2\times n}}{dx_k^2}\Big|_{x_k^*=a_{i_1}^{(n)}} = 0; \\ &\frac{d^3x_{k+2\times n}}{dx_k^3}\Big|_{x_k^*=a_{i_1}^{(n)}} = 6a_0^{(2\times n)} \prod_{i_1\in I_q^{(n)}, i_2\neq i_1} (a_{i_1}^{(n)} - a_{i_2}^{(n)})^3 \\ &\qquad\qquad \times \prod_{j_2=1}^{4^{2\times n}} (a_{i_1}^{(n)} - a_{j_2}^{(2\times n)})^{(1-\delta(i_1,j_2))}. \end{aligned} \tag{3.188}$$

Thus, $x_{k+2\times n}$ at $x_k^* = a_{i_1}^{(n)}$ for $i_1 \in I_q^{(n)}$, $q\in\{1,2,\cdots,N\}$ is

- *a monotonic sink of the third-order if* $d^3x_{k+2\times n}/dx_k^3|_{x_k^*=a_{i_1}^{(n)}} < 0$,
- *a monotonic source of the third-order if* $d^3x_{k+2\times n}/dx_k^3|_{x_k^*=a_{i_1}^{(n)}} > 0$.

(iv) *After* l*-times period-doubling bifurcations of period-*n *fixed points, a period-*$2^l\times n$ *discrete system of the quartic discrete system in Eq.* (3.174) *is*

$$\begin{aligned} x_{k+2^l\times n} &= x_k + [a_0^{(2^{l-1}\times n)} \prod_{i_1=1}^{4^{2^{l-1}\times n}} (x_k - a_{i_1}^{(2^{l-1}\times n)})] \\ &\quad \times \{1 + \prod_{i_1=1}^{4^{2^{l-1}\times n}} [1 + a_0^{(2^{l-1})} \prod_{i_2=1, i_2\neq i_1}^{4^{2^{l-1}\times n}} (x_k - a_{i_2}^{(2^{l-1}\times n)})]\} \\ &= x_k + [a_0^{(2^{l-1}\times n)} \prod_{i_1=1}^{4^{2^{l-1}\times n}} (x_k - a_{i_1}^{(2^{l-1}\times n)})] \\ &\quad \times [(a_0^{(2^{l-1}\times n)})^{4^{(2^{l-1}\times n)}} \prod_{j_1=1}^{(4^{2^l\times n}-4^{2^{l-1}\times n})/2} (x_k^2 + B_{j_2}^{(2^l\times n)} x_k + C_{j_2}^{(2^l\times n)})] \\ &= x_k + [a_0^{(2^{l-1}\times n)} \prod_{i_1=1}^{4^{2^{l-1}\times n}} (x_k - a_{i_1}^{(2^{l-1}\times n)})] \\ &\quad \times [(a_0^{(2^{l-1}\times n)})^{4^{2^{l-1}\times n}} \prod_{j_2=1}^{(4^{2^l\times n}-4^{2^{l-1}\times n})/2} (x_k - b_{j_2,1}^{(2^l\times n)})(x_k - b_{j_2,2}^{(2^l\times n)})] \\ &= x_k + (a_0^{(2^{l-1}\times n)})^{4^{2^{l-1}\times n}} \prod_{i=1}^{4^{2^l\times n}} (x_k - a_i^{(2^l\times n)}) \\ &= x_k + a_0^{(2^l\times n)} \prod_{i=1}^{4^{2^l\times n}} (x_k - a_i^{(2^l\times n)}) \end{aligned} \tag{3.189}$$

with

$$\begin{aligned}
\frac{dx_{k+2^l\times n}}{dx_k} &= 1 + a_0^{(2^l\times n)} \sum_{i_1=1}^{4^{2^l\times n}} \prod_{i_2=1,i_2\neq i_1}^{4^{2^l\times n}} (x_k - a_{i_2}^{(2^l\times n)}),\\
\frac{d^2x_{k+2^l\times n}}{dx_k^2} &= a_0^{(2^l\times n)} \sum_{i_1=1}^{4^{2^l\times n}} \sum_{i_2=1,i_2\neq i_1}^{4^{2^l\times n}} \prod_{i_3=1,i_3\neq i_1,i_2}^{4^{2^l\times n}} (x_k - a_{i_3}^{(2^l\times n)}),\\
&\vdots\\
\frac{d^r x_{k+2^l\times n}}{dx_k^r} &= a_0^{(2^l\times n)} \sum_{i_1=1}^{4^{2^l\times n}} \cdots \sum_{i_r=1,i_r\neq i_1,i_2\cdots,i_{r-1}}^{4^{2^l\times n}} \prod_{i_{r+1}=1,i_{r+1}\neq i_1,i_2\cdots,i_r}^{4^{2^l\times n}} (x_k - a_{i_{r+1}}^{(2^l\times n)})\\
&\text{for } r \le 4^{2^l\times n};
\end{aligned} \tag{3.190}$$

where

$$\begin{aligned}
&a_0^{(2\times n)} = (a_0^{(n)})^{1+4^{2\times n}}, a_0^{(2^l\times n)} = (a_0^{(2^{l-1}\times n)})^{1+4^{2^{l-1}\times n}}, l = 1,2,3,\ldots;\\
&\cup_{i=1}^{4^{2^l\times n}} \{a_i^{(2^l\times n)}\} = \text{sort}\{\cup_{i_1=1}^{4^{2^{l-1}\times n}} \{a_{i_1}^{(2^{l-1}\times n)}\}, \cup_{i_2=1}^{M_2}\{b_{i_2,1}^{(2^l\times n)}, b_{i_2,2}^{(2^l\times n)}\}\};\\
&b_{i,1}^{(2^l\times n)} = -\frac{1}{2}(B_i^{(2^l\times n)} + \sqrt{\Delta_i^{(2^l\times n)}}),\\
&b_{i,2}^{(2^l\times n)} = -\frac{1}{2}(B_i^{(2^l\times n)} - \sqrt{\Delta_i^{(2^l\times n)}}),\\
&\Delta_i^{(2^l\times n)} = (B_i^{(2^l\times n)})^2 - 4C_i^{(2^l\times n)} \ge 0\\
&\text{for } i \in \cup_{q_1=1}^{N_1} I_{q_1}^{(2^{l-1}\times n)} \cup \cup_{q_2=1}^{N_2} I_{q_2}^{(2^l\times n)}\\
&I_{q_1}^{(2^{l-1}\times n)} = \{l_{(q_1-1)\times(2^{l-1}\times n)+1}, l_{(q_1-1)\times(2^{l-1}\times n)+2}, \cdots, l_{q_1\times(2^{l-1}\times n)}\}\\
&\qquad \subseteq \{1,2,\ldots,M_1\} \cup \{\varnothing\},\\
&\text{for } q_1 \in \{1,2,\ldots,N_1\}, M_1 = N_1 \times (2^{l-1}\times n);\\
&I_{q_2}^{(2^l\times n)} = \{l_{(q_2-1)\times(2^l\times n)+1}, l_{(q_2-1)\times(2^l\times n)+2}, \cdots, l_{q_2\times(2^{l-1}\times n)}\}\\
&\qquad \subseteq \{M_1+1, M_1+2,\ldots,M_2\} \cup \{\varnothing\},\\
&\text{for } q_2 \in \{1,2,\ldots,N_2\}, M_2 = (4^{2^l\times n} - 4^{2^{l-1}\times n})/2;\\
&b_{i,1}^{(2^l\times n)} = -\frac{1}{2}(B_i^{(2^l\times n)} + \mathbf{i}\sqrt{|\Delta_i^{(2^l\times n)}|}),\\
&b_{i,2}^{(2^l\times n)} = -\frac{1}{2}(B_i^{(2^l\times n)} - \mathbf{i}\sqrt{|\Delta_i^{(2^l\times n)}|}),\\
&\Delta_i^{(2^l\times n)} = (B_i^{(2^l\times n)})^2 - 4C_i^{(2^l\times n)} < 0,\ \mathbf{i} = \sqrt{-1},\\
&i \in \{l_{N\times(2^l\times n)+1}, l_{N\times(2^l\times n)+2}, \cdots, l_{M_2}\} \subset \{1,2,\ldots,M_2\} \cup \{\varnothing\}
\end{aligned} \tag{3.191}$$

with fixed-points

$$
\begin{aligned}
& x^*_{k+2^l\times n} = x^*_k = a_i^{(2^l\times n)}, (i=1,2,\ldots,4^{2^l\times n}) \\
& \cup_{i=1}^{4^{2^l\times n}}\{a_i^{(2^l\times n)}\} = \text{sort}\{\cup_{i_1=1}^{4^{2^{l-1}\times n}}\{a_{i_1}^{(2^{l-1}\times n)}\}, \cup_{i_2=1}^{M_2}\{b_{i_2,1}^{(2^{l-1}\times n)}, b_{i_2,2}^{(2^{l-1}\times n)}\}\} \\
& \text{with } a_i^{(2^l\times n)} < a_{i+1}^{(2^l\times n)}.
\end{aligned}
\tag{3.192}
$$

(ii) *For a fixed-point of* $x^*_{k+(2^l\times n)} = x^*_k = a_{i_1}^{(2^{l-1}\times n)}$ $(i_1 \in I_q^{(2^{l-1}\times n)},\ q \in \{1,2,\ldots, N_1\})$, *there is a period-*$2^{l-1}\times n$ *discrete system if*

$$
\begin{aligned}
& \frac{dx_{k+2^{l-1}\times n}}{dx_k}\Big|_{x^*_k=a_{i_1}^{(2^{l-1}\times n)}} = 1 + a_0^{(2^{l-1}\times n)}\prod_{i_2=1, i_2\neq i_1}^{4^{2^{l-1}\times n}}(a_{i_1}^{(2^{l-1}\times n)} - a_{i_2}^{(2^{l-1}\times n)}) = -1, \\
& \frac{d^s x_{k+2^{l-1}\times n}}{dx_k^s}\Big|_{x^*_k=a_{i_1}^{(2^{l-1}\times n)}} = 0, \text{ for } s = 2,\ldots, r-1; \\
& \frac{d^r x_{k+2^{l-1}\times n}}{dx_k^r}\Big|_{x^*_k=a_{i_1}^{(2^{l-1}\times n)}} \neq 0 \text{ for } 1 < r \leq 4^{2^{l-1}\times n}
\end{aligned}
\tag{3.193}
$$

with

- *a* r^{th}*-order oscillatory sink for* $d^r x_{k+2^l\times n}/dx_k^r|_{x^*_k=a_{i_1}^{(2^l\times n)}} > 0$ *and* $r = 2l_1+1$;
- *a* r^{th}*-order oscillatory source for* $d^r x_{k+2^l\times n}/dx_k^r|_{x^*_k=a_{i_1}^{(2^l\times n)}} < 0$ *and* $r = 2l_1+1$;
- *a* r^{th}*-order oscillatory upper-saddle for* $d^r x_{k+2^l\times n}/dx_k^r|_{x^*_k=a_{i_1}^{(2^l\times n)}} > 0$ *and* $r = 2l_1$;
- *a* r^{th}*-order oscillatory lower-saddle for* $d^r x_{k+2^l\times n}/dx_k^r|_{x^*_k=a_{i_1}^{(2^l\times n)}} < 0$ *and* $r = 2l_1$.

The corresponding period-$2^l\times n$ *discrete system is*

$$
\begin{aligned}
x_{k+2^l\times n} = x_k + a_0^{(2^l\times n)} & \prod_{i_1\in I_q^{(2^{l-1}\times n)}}(x_k - a_{i_1}^{(2^{l-1}\times n)})^3 \\
& \times \prod_{j_2=1}^{4^{2^l\times n}}(x_k - a_{j_2}^{(2^l\times n)})^{(1-\delta(i_1,j_2))}
\end{aligned}
\tag{3.194}
$$

where

$$
\delta(i_1,j_2) = 1 \text{ if } a_{j_2}^{(2^l\times n)} = a_{i_1}^{(2^{l-1}\times n)}, \delta(i_1,j_2) = 0 \text{ if } a_{j_2}^{(2^l\times n)} \neq a_{i_1}^{(2^{l-1}\times n)} \tag{3.195}
$$

with

$$\begin{aligned}
&\frac{dx_{k+2^l\times n}}{dx_k}\Big|_{x_k^*=a_{i_1}^{(2^{l-1}\times n)}}=1,\ \frac{d^2x_{k+2^l\times n}}{dx_k^2}\Big|_{x_k^*=a_{i_1}^{(2^{l-1}\times n)}}=0;\\
&\frac{d^3x_{k+2^l\times n}}{dx_k^3}\Big|_{x_k^*=a_{i_1}^{(2^{l-1}\times n)}}=6a_0^{(2^l\times n)}\prod\nolimits_{i_2\in I_q^{(2^{l-1}\times n)},i_2\neq i_1}(a_{i_1}^{(2^{l-1}\times n)}-a_{i_2}^{(2^{l-1}\times n)})^3\\
&\qquad\qquad\times\prod\nolimits_{j_2=1}^{4^{2^l\times n}}(a_{i_1}^{(2^{l-1}\times n)}-a_{j_2}^{(2^l\times n)})^{(1-\delta(i_2,j_2))}\neq 0\\
&(i_1\in I_q^{(2^{l-1}\times n)},q\in\{1,2,\ldots,N_1\}).
\end{aligned}\tag{3.196}$$

Thus, $x_{k+2^l\times n}$ *at* $x_k^*=a_{i_1}^{(2^{l-1}\times n)}$ *is*

- *a monotonic sink of the third-order if* $d^3x_{k+2^l\times n}/dx_k^3|_{x_k^*=a_{i_1}^{(2^{l-1})}}<0$;
- *a monotonic source of the third-order if* $d^3x_{k+2^l\times n}/dx_k^3|_{x_k^*=a_{i_1}^{(2^{l-1})}}>0$.

(ii$_1$) *The period-*$2^l\times n$ *fixed-points are trivial if*

$$\left.\begin{aligned}
&x_{k+2^l\times n}^*=x_k^*=a_j^{(2^l\times n)}\in\{\cup_{i_i=1}^4\{a_{i_1}^{(1)}\},\cup_{i_2=1}^{4^{2^{l-1}\times n}}\{a_{i_2}^{(2^{l-1}\times n)}\}\}\\
&\text{for } j=1,2,\cdots,4^{2^l\times n}
\end{aligned}\right\}$$
$$\text{for } n\neq 2^{n_1},$$
$$\left.\begin{aligned}
&x_{k+2^l\times n}^*=x_k^*=a_j^{(2^l\times n)}\in\cup_{i_2=1}^{4^{2^{l-1}\times n}}\{a_{i_2}^{(2^{l-1}\times n)}\}\\
&\text{for } j=1,2,\cdots,4^{2^l\times n}
\end{aligned}\right\}\tag{3.197}$$
$$\text{for } n=2^{n_1}.$$

(ii$_2$) *The period-*$2^l\times n$ *fixed-points are non-trivial if*

$$\left.\begin{aligned}
&x_{k+2^l\times n}^*=x_k^*=a_j^{(2^l\times n)}\notin\{\cup_{i_i=1}^4\{a_{i_1}^{(1)}\},\cup_{i_2=1}^{4^{2^{l-1}\times n}}\{a_{i_2}^{(2^{l-1}\times n)}\}\}\\
&\text{for } j=1,2,\cdots,4^{(2^l\times n)}
\end{aligned}\right\}$$
$$\text{for } n\neq 2^{n_1},$$
$$\left.\begin{aligned}
&x_{k+2^l\times n}^*=x_k^*=a_j^{(2^l\times n)}\notin\cup_{i_2=1}^{4^{2^{l-1}\times n}}\{a_{i_2}^{(2^{l-1}\times n)}\}\\
&\text{for } j=1,2,\cdots,4^{2^l\times n}
\end{aligned}\right\}\tag{3.198}$$
$$\text{for } n=2^{n_1}.$$

Such a period-$2^l\times n$ *fixed-point is*

- *monotonically unstable if* $dx_{k+2^l\times n}/dx_k|_{x_k^*=a_{i_1}^{(2^l\times n)}}\in(1,\infty)$;
- *monotonically invariant if* $dx_{k+2^l\times n}/dx_k|_{x_k^*=a_{i_1}^{(2^l\times n)}}=1$, *which is*

  - *a monotonic upper-saddle of the* $(2l_1)^{\text{th}}$ *order for* $d^{2l_1}x_{k+2^l\times n}/dx_k^{2l_1}|_{x_k^*} > 0$ *(independent* $(2l_1)$*-branch appearance);*
  - *a monotonic lower-saddle the* $(2l_1)^{\text{th}}$ *order for* $d^{2l_1}x_{k+2^l\times n}/dx_k^{2l_1}|_{x_k^*} < 0$ *(independent* $(2l_1)$*-branch appearance)*
  - *a monotonic source of the* $(2l_1+1)^{\text{th}}$ *order for* $d^{2l_1+1}x_{k+2^l\times n}/\ dx_k^{2l_1+1}|_{x_k^*} > 0$ *(dependent* $(2l_1+1)$*-branch appearance from one branch);*
  - *a monotonic sink the* $(2l_1+1)^{\text{th}}$ *order for* $d^{2l_1+1}x_{k+2^l\times n}/dx_k^{2l_1+1}|_{x_k^*} < 0$ *(dependent* $(2l_1+1)$*-branch appearance from one branch);*

- *monotonically stable if* $dx_{k+2^l\times n}/dx_k|_{x_k^*=a_{i_1}^{(2^l\times n)}} \in (0,1)$*;*
- *invariantly zero-stable if* $dx_{k+2^l\times n}/dx_k|_{x_k^*=a_{i_1}^{(2^l\times n)}} = 0$*;*
- *oscillatorilly stable if* $dx_{k+2^l\times n}/dx_k|_{x_k^*=a_{i_1}^{(2^l\times n)}} \in (-1,0)$*;*
- *flipped if* $dx_{k+2^l\times n}/dx_k|_{x_k^*=a_{i_1}^{(2^l\times n)}} = -1$*, which is*
  - *an oscillatory upper-saddle of the* $(2l_1)^{\text{th}}$ *order for* $d^{2l_1}x_{k+2^l\times n}/dx_k^{2l_1}|_{x_k^*} > 0$*;*
  - *an oscillatory lower-saddle the* $(2l_1)^{\text{th}}$ *order for* $d^{2l_1}x_{k+2^l\times n}/dx_k^{2l_1}|_{x_k^*} <0$
  - *an oscillatory source of the* $(2l_1+1)^{\text{th}}$ *order for* $d^{2l_1+1}x_{k+2^l\times n}/dx_k^{2l_1+1}|_{x_k^*} < 0$*;*
  - *an oscillatory sink the* $(2l_1+1)^{\text{th}}$ *order for* $d^{2l_1+1}x_{k+2^l\times n}/dx_k^{2l_1+1}|_{x_k^*} > 0$*;*
- *oscillatorilly unstable if* $dx_{k+2^l\times n}/dx_k|_{x_k^*=a_{i_1}^{(2^l\times n)}} \in (-\infty,-1)$.

Proof Through the nonlinear renormalization, the proof of this theorem is similar to the proof of Theorem 1.11. This theorem can be easily proved. ■

3.6 Backward Quartic Discrete Systems

In this section, the analytical bifurcation scenario for backward quartic discrete systems will be discussed in a similar fashion through nonlinear renormalization techniques, and the backward bifurcation scenario based on the monotonic saddle-node bifurcations will be discussed, which is independent of period-1 fixed-points.

3.6.1 Backward Period-2 Quartic Discrete Systems

After the period-doubling bifurcation of a period-1 fixed-points in the backward quartic nonlinear discrete systems, the backward period-doubled fixed-point systems can be obtained. Consider the period-doubling fixed-points for a backward quartic nonlinear discrete system as follows.

Theorem 3.5 *Consider a 1-dimensional, backward, quartic nonlinear discrete system*

$$\begin{aligned} x_k &= x_{k+1} + f(x_{k+1}, \mathbf{p}) \\ &= x_{k+1} + A(\mathbf{p})x_{k+1}^4 + B(\mathbf{p})x_{k+1}^3 + C(\mathbf{p})x_{k+1}^2 + D(\mathbf{p})x_{k+1} + E(\mathbf{p}) \\ &= x_{k+1} + a_0(\mathbf{p})[x_{k+1}^2 + B_1(\mathbf{p})x_{k+1} + C_1(\mathbf{p})][x_{k+1}^2 + B_2(\mathbf{p})x_{k+1} + C_2(\mathbf{p})] \end{aligned} \tag{3.199}$$

where $a_0(\mathbf{p}) = A(\mathbf{p}) \neq 0$, *and*

$$\mathbf{p} = (p_1, p_2, \cdots, p_m)^{\mathrm{T}}. \tag{3.200}$$

Under

$$\begin{aligned} &\Delta_i = B_i^2 - 4C_i > 0, \text{ for } i = 1, 2 \text{ with} \\ &a_{1,2} = \frac{1}{2}(B_1 \pm \sqrt{\Delta_1}); a_{3,4} = \frac{1}{2}(B_2 \pm \sqrt{\Delta_2}), \end{aligned} \tag{3.201}$$

the standard form of such a backward discrete system is

$$x_k = x_{k+1} + a_0(x_{k+1} - a_1)(x_{k+1} - a_2)(x_{k+1} - a_3)(x_{k+1} - a_4). \tag{3.202}$$

Thus, a general standard form of such a backward quartic discrete system is

$$\begin{aligned} x_k &= x_{k+1} + Ax_{k+1}^4 + Bx_{k+1}^3 + Cx_{k+1}^2 + Dx_{k+1} + E \\ &= x_{k+1} + a_0 \textstyle\prod_{i=1}^4 (x_{k+1} - a_i^{(1)}) \end{aligned} \tag{3.203}$$

where

$$\begin{aligned} &b_{i,1}^{(1)} = -\tfrac{1}{2}(B_i^{(1)} + \sqrt{\Delta_i^{(1)}}),\ b_{i,2}^{(1)} = -\tfrac{1}{2}(B_i^{(1)} - \sqrt{\Delta_i^{(1)}}) \\ &\text{for } \Delta_i^{(1)} \geq 0,\ i \in \{1, 2\}; \\ &\cup_{i=1}^4 a_i^{(1)} = \text{sort}\{\cup_{i=1}^2 \{b_{i,2}^{(1)}, b_{i,2}^{(1)}\}\}, a_i^{(1)} \leq a_{i+1}^{(i)}; \\ &b_{i,1}^{(1)} = -\tfrac{1}{2}(B_i^{(1)} + \mathbf{i}\sqrt{|\Delta_i^{(1)}|}),\ b_{i,2}^{(1)} = -\tfrac{1}{2}(B_i^{(1)} - \mathbf{i}\sqrt{|\Delta_i^{(1)}|}) \\ &\text{for } \Delta_i^{(1)} < 0,\ i \in \{1, 2\},\ \mathbf{i} = \sqrt{-1}; \\ &a_1^{(1)} = b_{1,1}^{(1)}, a_2^{(1)} = b_{1,2}^{(1)}; a_3^{(1)} = b_{2,1}^{(1)}, a_4^{(1)} = b_{2,2}^{(1)}. \end{aligned} \tag{3.204}$$

(i) *Consider a backward period-2 discrete system of Eq.* (3.199) *as*

$$\begin{aligned} x_k &= x_{k+2} + [a_0 \textstyle\prod_{i_1=1}^{4}(x_{k+2} - a_{i_1}^{(1)})]\{1 + \prod_{i_1=1}^{4}[1 + a_0 \prod_{i_2=1, i_2 \neq i_1}^{4}(x_{k+2} - a_{i_2}^{(1)})]\} \\ &= x_{k+2} + [a_0 \textstyle\prod_{i_1=1}^{4}(x_{k+2} - a_{i_1}^{(1)})][a_0^4 \prod_{i_2=1}^{(4^2-4)/2}(x_{k+1}^2 + B_{i_2}^{(2)} x_{k+1} + C_{i_2}^{(2)})] \\ &= x_{k+2} + [a_0 \textstyle\prod_{j_1=1}^{4}(x_{k+2} - a_{i_1}^{(1)})][a_0^4 \prod_{j_2=1}^{4^2-4}(x_{k+2} - b_{j_2}^{(2)})] \\ &= x_{k+2} + a_0^{1+4} \textstyle\prod_{i=1}^{4^2}(x_{k+2} - a_i^{(2)}) \end{aligned} \tag{3.205}$$

where

$$\begin{aligned} & b_{i,1}^{(2)} = -\tfrac{1}{2}(B_i^{(2)} + \sqrt{\Delta^{(2)}}), b_{i,2}^{(2)} = -\tfrac{1}{2}(B_i^{(2)} - \sqrt{\Delta_i^{(2)}}), \\ & \Delta_i^{(2)} = (B_i^{(2)})^2 - 4C_i^{(2)} \geq 0,\ i \in \cup_{q_1=1}^{N_1} I_{q_1}^{(2^0)} \cup \cup_{q_2=1}^{N_2} I_{q_2}^{(2^2)} \\ & I_{q_1}^{(2^0)} = \{l_{(q_1-1)\times 2^0 \times m_1+1}, l_{(q_1-1)\times 2^0 \times m_1+2}, \cdots, l_{q_1 \times 2^0 \times m_1}\} \\ & \qquad \subseteq \{1, 2, \cdots, M_1\} \cup \{\varnothing\}, \\ & q_1 \in \{1, 2, \cdots, N_1\}, M_1 = N_1 \times 2^0 \times m_1, m_1 \in \{1, 2\}; \\ & I_{q_2}^{(2^1)} = \{l_{(q_2-1)\times 2^1 \times m_1+1}, l_{(q_2-1)\times 2^1 \times m_1+2}, \cdots, l_{q_2 \times 2^1 \times m_1}\} \\ & \qquad \subseteq \{M_1+1, M_1+2, \cdots, M_2\} \cup \{\varnothing\}, \\ & q_2 \in \{1, 2, \cdots, N_2\}, M_2 = (4^2 - 4)/2; \\ & b_{i,1}^{(2)} = -\tfrac{1}{2}(B_i^{(2)} + \mathbf{i}\sqrt{\Delta^{(2)}}), b_{i,2}^{(2)} = -\tfrac{1}{2}(B_i^{(2)} - \mathbf{i}\sqrt{\Delta_i^{(2)}}), \\ & \mathbf{i} = \sqrt{-1}, \Delta_i^{(2)} = (B_i^{(2)})^2 - 4C_i^{(2)} < 0, \\ & i \in J^{(2^1)} = \{l_{N_2 \times 2^1 \times m_1+1}, l_{N_2 \times 2^1 \times m_1+2}, \cdots, l_{M_2}\} \\ & \qquad \subseteq \{M_1+1, M_1+2, \cdots, M_2\} \end{aligned} \tag{3.206}$$

with fixed-points

$$\begin{aligned} & x_k^* = x_{k+2}^* = a_i^{(2)}, (i = 1, 2, \ldots, 4^2) \\ & \cup_{i=1}^{4^2}\{a_i^{(2)}\} = \mathrm{sort}\{\cup_{j_1=1}^{4}\{a_{j_1}^{(1)}\}, \cup_{j_2=1}^{M_2}\{b_{j_2,1}^{(2)}, b_{j_2,2}^{(2)}\}\} \\ & \text{with } a_i^{(2)} < a_{i+1}^{(2)}, M_2 = (4^2 - 4)/2. \end{aligned} \tag{3.207}$$

(ii) *For a fixed-point of* $x_k^* = x_{k+1}^* = a_{i_1}^{(1)}$ $(i_1 \in \{1, 2, \ldots, 4\})$, *if*

$$\frac{dx_k}{dx_{k+1}}\Big|_{x_{k+1}^* = a_{i_1}^{(1)}} = 1 + a_0 \textstyle\prod_{i_2=1, i_2 \neq i_1}^{4}(a_{i_1}^{(1)} - a_{i_2}^{(1)}) = -1, \tag{3.208}$$

with

- *a* r^{th}*-order oscillatory lower-saddle-node bifurcation* $(d^r x_k/dx_{k+1}^r|_{x_{k+1}^*} > 0, r = 2l_1)$,
- *a* r^{th}*-order oscillatory upper-saddle-node bifurcation* $(d^r x_k/dx_{k+1}^r|_{x_{k+1}^*} < 0,\ r = 2l_1)$,
- *a* r^{th}*-order oscillatory source bifurcation* $(d^r x_k/dx_{k+1}^r|_{x_{k+1}^*} > 0,\ r = 2l_1 + 1)$,
- *a* r^{th}*-order oscillatory sink bifurcation* $(d^r x_k/dx_{k+1}^r|_{x_{k+1}^*} < 0, r = 2l_1 + 1)$,

then the following relations satisfy

$$a_{i_1}^{(1)} = -\frac{1}{2}B_{i_1}^{(2)}, \Delta_{i_1}^{(2)} = (B_{i_1}^{(2)})^2 - 4C_{i_1}^{(2)} = 0, \tag{3.209}$$

and there is a period-2 discrete system of the backward quartic discrete system in Eq. (3.199) *as*

$$\begin{aligned} x_k = x_{k+2} + a_0^5 \textstyle\prod_{i_1 \in I_{q_1}^{(20)}} (x_{k+2} - a_{i_1}^{(1)})^3 \\ \times \textstyle\prod_{i_2=1}^{4^2} (x_{k+2} - a_{i_2}^{(2)})^{(1-\delta(i_1,i_2))} \end{aligned} \tag{3.210}$$

for $i_1 \in \{1, 2, \ldots, 4\},\ i_1 \neq i_2$ *with*

$$\frac{dx_k}{dx_{k+2}}|_{x_{k+2}^* = a_{i_1}^{(1)}} = 1,\ \frac{d^2 x_k}{dx_{k+2}^2}|_{x_{k+2}^* = a_{i_1}^{(1)}} = 0; \tag{3.211}$$

- x_k *at* $x_{k+2}^* = a_{i_1}^{(1)}$ *is a monotonic sink of the third-order if*

$$\begin{aligned} \frac{d^3 x_k}{dx_{k+2}^3}|_{x_{k+2}^* = a_{i_1}^{(1)}} = 6a_0^5 \textstyle\prod_{i_2 \in I_{q_1}^{(20)}, i_2 \neq i_1} (a_{i_1}^{(1)} - a_{i_2}^{(1)})^3 \\ \times \textstyle\prod_{i_3=1}^{4^2} (a_{i_1}^{(1)} - a_{i_3}^{(2)})^{(1-\delta(i_2,i_3))} < 0, \end{aligned} \tag{3.212}$$

and the corresponding bifurcations is a third-order monotonic source bifurcation for the period-2 discrete system;

- x_{k+2} at $x_k^* = a_{i_1}^{(1)}$ *is a monotonic source of the third-order if*

$$\begin{aligned} \frac{d^3 x_k}{dx_{k+2}^3}|_{x_{k+2}^* = a_{i_1}^{(1)}} = 6a_0^5 \textstyle\prod_{i_2 \in I_{q_1}^{(20)}, i_2 \neq i_1} (a_{i_1}^{(1)} - a_{i_2}^{(1)})^3 \\ \times \textstyle\prod_{i_3=1}^{4^2} (a_{i_1}^{(1)} - a_{i_3}^{(2)})^{(1-\delta(i_2,i_3))} < 0, \end{aligned} \tag{3.213}$$

and the corresponding bifurcations is a third-order monotonic sink bifurcation for the period-2 discrete system.

(ii_1) *The backward period-2 fixed-points are trivial and unstable if*

$$x_k^* = x_{k+2}^* \in \mathrm{U}_{i_1=1}^{4^1}\{\mathrm{a}_i^{(t)}\}. \tag{3.214}$$

(ii_2) *The backward period-2 fixed-points are non-trivial and stable if*

$$x_k^* = x_{k+2}^* = \in \mathrm{U}_{i_1=1}^{4^2}\{b_{i_1,1}^{(2)}, \{b_{i_1,2}^{(2)}\}. \tag{3.215}$$

Proof The proof is straightforward through the simple algebraic manipulation. Following the proof of quadratic discrete system, this theorem is proved. ■

3.6.2 Backward Period-Doubling Renormalization

The generalized case of period-doublization of a backward quartic discrete system is presented through the following theorem. The backward period-doubling bifurcation trees can be developed for backward quartic discrete systems.

Theorem 3.6 *Consider a 1-dimensional, backward quartic discrete system as*

$$\begin{aligned} x_k &= x_{k+1} + Ax_{k+1}^4 + Bx_{k+1}^3 + Cx_{k+1}^2 + Dx_{k+1} + E \\ &= x_k + a_0 \textstyle\prod_{i=1}^4 (x_{k+1} - a_i^{(1)}). \end{aligned} \tag{3.216}$$

(i) *After* l*-times period-doubling bifurcations, a period-*2^l ($l = 1, 2, \ldots$) *discrete system for the quartic discrete system in Eq.* (3.216) *is given through the nonlinear renormalization as*

$$\begin{aligned} x_k &= x_{k+2^l} + [a_0^{(2^{l-1})} \textstyle\prod_{i_1=1}^{4^{2^{l-1}}} (x_{k+2^l} - a_{i_1}^{(2^{l-1})})] \\ &\quad \times \{1 + \textstyle\prod_{i_1=1}^{4^{2^{l-1}}} [1 + a_0^{(2^{l-1})} \textstyle\prod_{i_2=1, i_2 \neq i_1}^{4^{2^{l-1}}} (x_{k+2^l} - a_{i_2}^{(2^{l-1})})]\} \\ &= x_{k+2^l} + [a_0^{(2^{l-1})} \textstyle\prod_{i_1=1}^{4^{2^{l-1}}} (x_{k+2^l} - a_{i_1}^{(2^{l-1})})] \\ &\quad \times [(a_0^{(2^{l-1})})^{4^{2^{l-1}}} \textstyle\prod_{j_1=1}^{(4^{2^l} - 4^{2^{l-1}})/2} (x_{k+2^l}^2 + B_{j_2}^{(2^l)} x_{k+2^l} + C_{j_2}^{(2^l)})] \\ &= x_{k+2^l} + [a_0^{(2^{l-1})} \textstyle\prod_{i_1=1,}^{2^{2^{l-1}}} (x_{k+2^l} - a_{i_1}^{(2^{l-1})})] \\ &\quad \times [(a_0^{(2^{l-1})})^{4^{2^{l-1}}} \textstyle\prod_{i_2=1}^{(4^{2^l} - 4^{2^{l-1}})/2} (x_{k+2^l} - b_{i_2,1}^{(2^l)})(x_{k+2^l} - b_{i_2,2}^{(2^l)})] \\ &= x_{k+2^l} + (a_0^{(2^{l-1})})^{1+4^{2^{l-1}}} \textstyle\prod_{i=1}^{4^{2^l}} (x_{k+2^l} - a_i^{(2^l)}) \\ &= x_{k+2^l} + a_0^{(2^l)} \textstyle\prod_{i=1}^{4^{2^l}} (x_{k+2^l} - a_i^{(2^l)}) \end{aligned} \tag{3.217}$$

with

$$
\begin{aligned}
&\frac{dx_k}{dx_{k+2^l}} = 1 + a_0^{(2^l)} \Sigma_{i_1=1}^{4^{2^l}} \prod_{i_2=1, i_2 \neq i_1}^{4^{2^l}} (x_{k+2^l} - a_{i_2}^{(2^l)}),\\
&\frac{d^2 x_k}{dx_{k+2^l}^2} = a_0^{(2^l)} \sum_{i_1=1}^{4^{2^l}} \sum_{i_2=1, i_2 \neq i_1}^{4^{2^l}} \prod_{i_3=1, i_3 \neq i_1, i_2}^{4^{2^l}} (x_{k+2^l} - a_{i_3}^{(2^l)}),\\
&\vdots\\
&\frac{d^r x_k}{dx_{k+2^l}^r} = a_0^{(2^l)} \sum_{i_1=1}^{4^{2^l}} \cdots \sum_{i_r=1, i_3 \neq i_1, i_2 \ldots i_{r-1}}^{4^{2^l}} \prod_{i_{r+1}=1, i_3 \neq i_1, i_2 \ldots, i_r}^{4^{2^l}} (x_{k+2^l} - a_{i_{r+1}}^{(2^l)})\\
&\text{for } r \leq 4^{2^l}.
\end{aligned} \tag{3.218}
$$

where

$$
\begin{aligned}
&a_0^{(2)} = (a_0)^{1+4}, a_0^{(2^l)} = (a_0^{(2^{l-1})})^{1+4^{2^{l-1}}}, l = 2, 3, \cdots;\\
&\cup_{i=1}^{4^{2^l}} \{a_i^{(2^l)}\} = \text{sort}\{\cup_{i_1=1}^{4^{2^{l-1}}} \{a_{i_1}^{(2^l)}\}, \cup_{i_2=1}^{M_2} \cup_{I_q^{(2^l)}} \{b_{i_2,1}^{(2^l)}, b_{i_2,2}^{(2^l)}\}\}\ , a_i^{(2^l)} \leq a_{i+1}^{(2^l)};\\
&b_{i,1}^{(2^l)} = -\tfrac{1}{2}(B_i^{(2^l)} + \sqrt{\Delta_i^{(2^l)}}), b_{i,2}^{(2^{l-1})} = -\tfrac{1}{2}(B_i^{(2^l)} - \sqrt{\Delta_i^{(2^l)}}),\\
&\Delta_i^{(2^l)} = (B_i^{(2^l)})^2 - 4C_i^{(2^l)} \geq 0 \ \text{ for } \ i \in \cup_{q_1=1}^{N_1} I_{q_1}^{(2^{l-1})} \cup \cup_{q_2=1}^{N_2} I_{q_2}^{(2^l)},\\
&I_{q_1}^{(2^{l-1})} = \{l_{(q_1-1)\times 2^{l-1}\times m_1+1}, l_{(q_1-1)\times 2^{l-1}\times m_1+2}, \cdots, l_{q_1\times 2^{l-1}\times m_1}\}\\
&\qquad \subseteq \{1, 2, \cdots, M_1\} \cup \{\varnothing\},\\
&\text{for } q_1 \in \{1, 2, \cdots, N_1\}, M_1 = N_1 \times 2^{l-1} \times m_1, m_1 \in \{1, 2\};\\
&I_{q_2}^{(2^l)} = \{l_{(q_2-1)\times 2^l\times m_1+1}, l_{(q_2-1)\times 2^l\times m_1+2}, \cdots, l_{q_2\times 2^l\times m_1}\}\\
&\qquad \subseteq \{M_1+1, M_1+2, \cdots, M_2\} \cup \{\varnothing\},\\
&\text{for } q_2 \in \{1, 2, \cdots, N_2\}, M_2 = (4^{2^l} - 4^{2^{l-1}})/2;\\
&b_{i,1}^{(2^l)} = -\tfrac{1}{2}(B_i^{(2^l)} + \mathbf{i}\sqrt{|\Delta_i^{(2^l)}|}),\ b_{i,2}^{(2^l)} = -\tfrac{1}{2}(B_i^{(2^l)} - \mathbf{i}\sqrt{|\Delta_i^{(2^l)}|}),\\
&\Delta_i^{(2^l)} = (B_i^{(2^l)})^2 - 4C_i^{(2^l)} < 0,\ \mathbf{i} = \sqrt{-1},\\
&i \in J^{(2^l)} = \{l_{N_2\times 2^l\times m_1+1}, l_{N_2\times 2^l\times m_1+2}, \cdots, l_{M_2}\}\\
&\qquad \subset \{M_1+1, M_1+2, \cdots, M_2\} \cup \{\varnothing\}
\end{aligned} \tag{3.219}
$$

with fixed-points

$$
\begin{aligned}
&x_k^* = x_{k+2^l}^* = a_i^{(2^l)}, (i = 1, 2, \ldots, 4^{2^l})\\
&\cup_{i=1}^{4^{2^l}} \{a_i^{(2^l)}\} = \text{sort}\{\cup_{i=1}^{4^{2^{l-1}}} \{a_i^{(2^{l-1})}\}, \cup_{i=1}^{M_2} \{b_{i,1}^{(2^l)}, b_{i,2}^{(2^l)}\}\}\\
&\text{with } a_i^{(2^l)} < a_{i+1}^{(2^l)}.
\end{aligned} \tag{3.220}
$$

(ii) *For a fixed-point of* $x_k^* = x_{k+2^{l-1}}^* = a_{i_1}^{(2^{l-1})}$ $(i_1 \in I_q^{(2^{l-1})}, q \in \{1, 2, \ldots, N_1\})$, *if*

$$\begin{aligned}
&\frac{dx_k}{dx_{k+2^{l-1}}}\Big|_{x_{k+2^{l-1}}^* = a_{i_1}^{(2^{l-1})}} = 1 + a_0 \prod_{i_2=1, i_2 \neq i_1}^{4^{2^{l-1}}} (a_{i_1}^{(2^{l-1})} - a_{i_2}^{(2^{l-1})}) = -1, \\
&\frac{d^s x_k}{dx_{k+2^{l-1}}^s}\Big|_{x_{k+2^{l-1}}^* = a_{i_1}^{(2^{l-1})}} = 0, \text{ for } s = 2, \ldots, r-1; \\
&\frac{d^r x_k}{dx_{k+2^{l-1}}^r}\Big|_{x_{k+2^{l-1}}^* = a_{i_1}^{(2^{l-1})}} \neq 0 \text{ for } 1 < r \leq 4^{2^{l-1}},
\end{aligned} \tag{3.221}$$

with

- *a* r*th-order oscillatory source for* $d^r x_k / dx_{k+2^{l-1}}^r|_{x_{k+2^{l-1}}^* = a_{i_1}^{(2^{l-1})}} > 0$ *and* $r = 2l_1 + 1$;
- *a* r*th-order oscillatory sink for* $d^r x_k / dx_{k+2^{l-1}}^r|_{x_{k+2^{l-1}}^* = a_{i_1}^{(2^{l-1})}} < 0$ *and* $r = 2l_1 + 1$;
- *a* r*th-order oscillatory lower-saddle for* $d^r x_k / dx_{k+2^{l-1}}^r|_{x_{k+2^{l-1}}^* = a_{i_1}^{(2^{l-1})}} > 0$ *and* $r = 2l_1$;
- *a* r*th-order oscillatory upper-saddle for* $d^r x_k / dx_{k+2^{l-1}}^r|_{x_{k+2^{l-1}}^* = a_{i_1}^{(2^{l-1})}} < 0$ *and* $r = 2l_1$;

then, there is a backward period-2^l *fixed-point discrete system*

$$\begin{aligned}
x_k = x_{k+2^l} &+ a_0^{(2^l)} \prod_{i_1 \in I_q^{(2^{l-1})}} (x_{k+2^l} - a_{i_1}^{(2^{l-1})})^3 \\
&\times \prod_{j_2=1}^{4^{2^l}} (x_{k+2^l} - a_{j_2}^{(2^l)})^{(1-\delta(i_1, j_2))}
\end{aligned} \tag{3.222}$$

where

$$\delta(i_1, j_2) = 1 \text{ if } a_{j_2}^{(2^l)} = a_{i_1}^{(2^{l-1})}, \ \delta(i_1, j_2) = 0 \text{ if } a_{j_2}^{(2^l)} \neq a_{i_1}^{(2^{l-1})} \tag{3.223}$$

and

$$\frac{dx_k}{dx_{k+2^l}}\Big|_{x_{k+2^l}^* = a_{i_1}^{(2^{l-1})}} = 1, \ \frac{d^2 x_k}{dx_{k+2^l}^2}\Big|_{x_{k+2^l}^* = a_{i_1}^{(2^{l-1})}} = 0. \tag{3.224}$$

- x_k at $x_{k+2^l}^* = a_{i_1}^{(2^{l-1})}$ *is a monotonic source of the third-order if*

$$\frac{d^3x_k}{dx^3_{k+2^l}}\Big|_{x^*_{k+2^l}=a^{(2^{l-1})}_{i_1}} = 6a_0^{(2^l)}\prod_{i_2\in I_q^{(2^{l-1})},i_2\neq i_1}(a_{i_1}^{(2^{l-1})}-a_{i_2}^{(2^{l-1})})^3 \times \prod_{j_2=1}^{4^{2^l}}(a_{i_1}^{(2^{l-1})}-a_{j_2}^{(2^l)})^{(1-\delta(i_2,j_2))}<0$$
$$(i_1\in I_q^{(2^{l-1})}, q\in\{1,2,\ldots,N_1\}), \tag{3.225}$$

and such a bifurcation at $x^*_{k+2^l}=a^{(2^{l-1})}_{i_1}$ *is a third-order monotonic source bifurcation.*

- x_k at $x^*_{k+2^l}=a^{(2^{l-1})}_{i_1}$ *is a monotonic sink of the third-order if*

$$\frac{d^3x_k}{dx^3_{k+2^l}}\Big|_{x^*_{k+2^l}=a^{(2^{l-1})}_{i_1}} = 6a_0^{(2^l)}\prod_{i_2\in I_q^{(2^{l-1})},i_2\neq i_1}(a_{i_1}^{(2^{l-1})}-a_{i_2}^{(2^{l-1})})^3 \times \prod_{j_2=1}^{4^{2^l}}(a_{i_1}^{(2^{l-1})}-a_{j_2}^{(2^l)})^{(1-\delta(i_2,j_2))}>0$$
$$(i_1\in I_q^{(2^{l-1})}, q\in\{1,2,\ldots,N_1\}) \tag{3.226}$$

and such a bifurcation at $x^*_{k+2^l}=a^{(2^{l-1})}_{i_1}$ *is a third-order monotonic sink bifurcation.*

(ii_1) *The period-*2^l *fixed-points are trivial if*

$$x_k^*=x^*_{k+2^l}=a_i^{(2^l)}\in\cup_{i_1=1}^{4^{2^{l-1}}}a_{i_1}^{(2^{l-1})}, \tag{3.227}$$

(ii_2) *The period-*2^l *fixed-points are non-trivial if*

$$x_k^*=x^*_{k+2^l}=a_i^{(2^l)}\in\cup_{j_1=1}^{M_2}\{b_{j_1,1}^{(2^l)},b_{j_1,2}^{(2^l)}\}. \tag{3.228}$$

Such a backward period-2^l *fixed-point is*

- *monotonically stable if* $dx_k/dx_{k+2^l}|_{x^*_{k+2^l}=a^{(2^l)}_{i_1}}\in(1,\infty)$;
- *monotonically invariant if* $dx_k/dx_{k+2^l}|_{x^*_{k+2^l}=a^{(2^l)}_{i_1}}=1$, *which is*
  - *a monotonic lower-saddle of the* $(2l_1)^{\text{th}}$ *order for* $d^{2l_1}x_k/dx^{2l_1}_{k+2^l}|_{x^*_{k+2^l}}>0$;
  - *a monotonic upper-saddle of the* $(2l_1)^{\text{th}}$ *order for* $d^{2l_1}x_k/dx^{2l_1}_{k+2^l}|_{x^*_{k+2^l}}<0$;
  - *a monotonic sink of the* $(2l_1+1)^{\text{th}}$ *order for* $d^{2l_1+1}x_k/dx^{2l_1+1}_{k+2^l}|_{x^*_{k+2^l}}>0$;
  - *a monotonic source the* $(2l_1+1)^{\text{th}}$ *order for* $d^{2l_1+1}x_k/dx^{2l_1+1}_{k+2^l}|_{x^*_{k+2^l}}<0$;

- *monotonically unstable if* $dx_k/dx_{k+2^l}|_{x^*_{k+2^l}=a^{(2^l)}_{i_1}} \in (0,1)$;
- *invariantly zero-stable if* $dx_k/dx_{k+2^l}|_{x^*_{k+2^l}=a^{(2^l)}_{i_1}} = 0$;
- *oscillatorilly unstable if* $dx_k/dx_{k+2^l}|_{x^*_{k+2^l}=a^{(2^l)}_{i_1}} \in (-1,0)$;
- *flipped if* $dx_k/dx_{k+2^l}|_{x^*_{k+2^l}=a^{(2^l)}_{i_1}} = -1$, *which is*
  - *an oscillatory lower-saddle of the* $(2l_1)^{\text{th}}$ *order for* $d^{2l_1}x_k/dx^{2l_1}_{k+2^l}|_{x^*_{k+2^l}} > 0$;
  - *an oscillatory upper-saddle of the* $(2l_1)^{\text{th}}$ *order for* $d^{2l_1}x_k/dx^{2l_1}_{k+2^l}|_{x^*_{k+2^l}} < 0$;
  - *an oscillatory source of the* $(2l_1+1)^{\text{th}}$ *order for* $d^{2l_1+1}x_k/dx^{2l_1+1}_{k+2^l}|_{x^*_{k+2^l}} > 0$;
  - *an oscillatory sink of the* $(2l_1+1)^{\text{th}}$ *order for* $d^{2l_1+1}x_k/dx^{2l_1+1}_{k+2^l}|_{x^*_{k+2^l}} < 0$;
- *oscillatorilly stable if* $dx_k/dx_{k+2^l}|_{x^*_{k+2^l}=a^{(2^l)}_{i_1}} \in (-\infty,-1)$.

Proof Through the nonlinear renormalization, this theorem can be proved. ■

3.6.3 Backward Period-n Appearing and Period-Doublization

The period-n discrete system for a backward quartic nonlinear discrete system will be discussed, and the period-doublization of a backward period-n discrete system is discussed through the nonlinear renormalization.

Theorem 3.7 *Consider a 1-dimensional, backward quartic discrete system as*

$$\begin{aligned} x_k &= x_{k+1} + Ax^4_{k+1} + Bx^3_{k+1} + Cx^2_{k+1} + Dx_{k+1} + E \\ &= x_{k+1} + a_0 \textstyle\prod_{i=1}^4 (x_{k+1} - a^{(1)}_i). \end{aligned} \tag{3.229}$$

(i) *After* n*-times iterations, a backward period-*n *discrete system for the backward quartic discrete system in Eq.* (3.229) *is*

$$\begin{aligned} x_k &= x_{k+n} + a_0 \textstyle\prod_{i_1=1}^4 (x_{k+n} - a_{i_2})[1 + \sum_{j=1}^n Q_j] \\ &= x_{k+n} + a_0^{(4^n-1)/3} \textstyle\prod_{i_1=1}^4 (x_{k+n} - a_{i_1}) \\ &\quad \times [\textstyle\prod_{j_2=1}^{(4^n-4)/2} (x^2_{k+n} + B^{(2^n)}_{j_2} x_{k+n} + C^{(2^n)}_{j_2})] \\ &= x_{k+n} + a_0^{(n)} \textstyle\prod_{i=1}^{4^n} (x_{k+n} - a^{(n)}_i) \end{aligned} \tag{3.230}$$

with

$$
\begin{aligned}
&\frac{dx_k}{dx_{k+n}} = 1 + a_0^{(n)} \sum_{i_1=1}^{4^n} \prod_{i_2=1, i_2 \neq i_1}^{4^n} (x_{k+n} - a_{i_2}^{(n)}),\\
&\frac{d^2 x_k}{dx_{k+n}^2} = a_0^{(n)} \sum_{i_1=1}^{4^n} \sum_{i_2=1, i_2 \neq i_1}^{4^n} \prod_{i_3=1, i_3 \neq i_1, i_2}^{4^n} (x_{k+n} - a_{i_3}^{(n)}),\\
&\vdots\\
&\frac{d^r x_k}{dx_{k+n}^r} = a_0^{(n)} \sum_{i_1=1}^{4^n} \cdots \sum_{i_r=1, i_r \neq i_1, i_2 \cdots, i_{r-1}}^{4^n} \prod_{i_{r+1}=1, i_{r+1} \neq i_1, i_2 \cdots, i_r}^{4^n} (x_{k+n} - a_{i_{r+1}}^{(n)})\\
&\text{for } r \leq 4^n;
\end{aligned}
\tag{3.231}
$$

where

$$
\begin{aligned}
&a_0^{(n)} = (a_0)^{(4^n-1)/2}, Q_1 = 0, Q_2 = \prod_{i_2=1}^{4} [1 + a_0 \prod_{i_1=1, i_1 \neq i_2}^{4} (x_{k+n} - a_{i_1}^{(1)})],\\
&Q_n = \prod_{i_n=1}^{4} [1 + a_0 (1 + Q_{n-1}) \prod_{i_{n-1}=1, i_{n-1} \neq i_n}^{4} (x_{k+n} - a_{i_{n-1}}^{(1)})],\ n = 3, 4, \ldots;\\
&\cup_{i=1}^{4^n} \{a_i^{(n)}\} = \text{sort}[\cup_{i_1=1}^{4} \{a_{i_1}^{(1)}\}, \cup_{q=1}^{N} \cup_{i_2 \in I_q^{(n)}} \{b_{i_2,1}^{(n)}, b_{i_2,2}^{(n)}\}];\\
&b_{i_2,1}^{(n)} = -\frac{1}{2}(B_{i_2}^{(n)} + \sqrt{\Delta_{i_2}^{(n)}}), b_{i_2,2}^{(n)} = -\frac{1}{2}(B_{i_2}^{(n)} - \sqrt{\Delta_{i_2}^{(n)}}),\\
&\Delta_{i_2}^{(n)} = (B_{i_2}^{(n)})^2 - 4C_{i_2}^{(n)} \geq 0 \ \text{ for } \ i_2 \in \cup_{q=1}^{N} I_q^{(n)},\\
&I_q^{(n)} = \{l_{(q-1)\times n+1}, l_{(q-1)\times n+2}, \cdots, l_{q\times n}\} \subseteq \{1, 2, \ldots, \} \cup \{\varnothing\},\\
&\text{for } q \in \{1, 2, \ldots, N\}, M = (4^n - 4)/2;\\
&b_{i,1}^{(n)} = -\frac{1}{2}(B_i^{(n)} + \mathbf{i}\sqrt{|\Delta_i^{(n)}|}),\ b_{i,2}^{(n)} = -\frac{1}{2}(B_i^{(n)} - \mathbf{i}\sqrt{|\Delta_i^{(n)}|}),\\
&\Delta_i^{(n)} = (B_i^{(n)})^2 - 4C_i^{(n)} < 0,\ \mathbf{i} = \sqrt{-1}\\
&i \in \{l_{N\times n+1}, l_{N\times n+2}, \ldots, l_M\} \subset \{1, 2, \ldots, M\} \cup \{\varnothing\};
\end{aligned}
\tag{3.232}
$$

with backward fixed-points

$$
\begin{aligned}
&x_k^* = x_{k+n}^* = a_i^{(n)}, (i = 1, 2, \ldots, 4^n)\\
&\cup_{i=1}^{4^n} \{a_i^{(n)}\} = \text{sort}\{\cup_{i_1=1}^{4} \{a_{i_1}^{(1)}\}, \cup_{i=1}^{M} \{b_{i_2,1}^{(n)}, b_{i_2,2}^{(n)}\}\}\\
&\text{with } a_i^{(n)} < a_{i+1}^{(n)}.
\end{aligned}
\tag{3.233}
$$

(ii) *For a backward fixed-point of* $x_k^* = x_{k+n}^* = a_{i_1}^{(n)}$ $(i_1 \in I_q^{(n)}, q \in \{1, 2, \ldots, N\})$, *if*

$$
\frac{dx_k}{dx_{k+n}}\Big|_{x_{k+n}^* = a_{i_1}^{(n)}} = 1 + a_0^{(n)} \prod_{i_2=1, i_2 \neq i_1}^{4^n} (a_{i_1}^{(n)} - a_{i_2}^{(2^l)}) = 1, \tag{3.234}
$$

with

$$\frac{d^2x_k}{dx_{k+n}^2}\Big|_{x_{k+n}^*=a_{i_1}^{(n)}} = a_0^{(n)}\sum_{i_2=1,i_2\neq i_1}^{4^n}\prod_{i_3=1,i_3\neq i_1,i_2}^{4^n}(a_{i_1}^{(n)}-a_{i_3}^{(2^l)})\neq 0, \tag{3.235}$$

then there is a new discrete system for onset of the qth*-set of period-*n *fixed-points based on the second-order monotonic saddle-node bifurcation as*

$$x_k = x_{k+n} + a_0^{(n)}\prod_{i_1\in I_q^{(n)}}(x_{k+n}-a_{i_1}^{(n)})^2\prod_{j_2=1}^{4^n}(x_{k+n}-a_{j_2}^{(n)})^{(1-\delta(i_1,j_2))} \tag{3.236}$$

where

$$\delta(i_1,j_2)=1 \text{ if } a_{j_2}^{(n)}=a_{i_1}^{(n)},\ \delta(i_1,j_2)=0 \text{ if } a_{j_2}^{(n)}\neq a_{i_1}^{(n)}. \tag{3.237}$$

(ii_1) If

$$\begin{aligned}
&\frac{dx_k}{dx_{k+n}}\Big|_{x_{k+n}^*=a_{i_1}^{(n)}} = 1\ \ (i_1\in I_q^{(n)}),\\
&\frac{d^2x_k}{dx_{k+n}^2}\Big|_{x_{k+n}^*=a_{i_1}^{(n)}} = 2a_0^{(n)}\prod_{i_2\in I_q^{(n)},i_2\neq i_1}(a_{i_1}^{(n)}-a_{i_2}^{(n)})^2\\
&\qquad\qquad\times\prod_{j_2=1}^{4^n}(a_{i_1}^{(n)}-a_{j_2}^{(n)})^{(1-\delta(i_2,j_2))}\neq 0
\end{aligned} \tag{3.238}$$

x_k at $x_{k+n}^*=a_{i_1}^{(n)}$ is

- *a monotonic upper-saddle of the second-order for* $d^2x_k/dx_{k+n}^2|_{x_{k+n}^*=a_{i_1}^{(n)}}<0;$
- *a monotonic lower-saddle of the second-order for* $d^2x_k/dx_{k+n}^2|_{x_{k+n}^*=a_{i_1}^{(n)}}>0.$

(ii_2) *The backward period-*n *fixed-points* $(n=2^{n_1}\times m)$ *are trivial if*

$$\left.\begin{aligned}
&x_{k+n}^*=x_k^*=a_{j_1}^{(n)}\in\{\cup_{i_i=1}^{4}\{a_{i_1}^{(1)}\},\cup_{i_2=1}^{4^{2^{n_1-1}m}}\{a_{i_2}^{(2^{n_1-1}m)}\}\}\\
&\text{for } n_1=1,2,\ldots;m=2l_1+1;j_1\in\{1,2,\ldots,4^n\}\cup\{\varnothing\}\\
&\text{for } n\neq 2^{n_2},\\
&x_{k+n}^*=x_k^*=a_{j_1}^{(n)}\in\{\cup_{i_2=1}^{4^{2^{n_1-1}m}}\{a_{i_2}^{(2^{n_1-1}m)}\}\}\\
&\text{for } n_1=1,2,\ldots;m=1;j_1\in\{1,2,\ldots,4^n\}\cup\{\varnothing\}\\
&\text{for } n=2^{n_2}.
\end{aligned}\right\} \tag{3.239}$$

(ii_3) *The period-*n *fixed-points* $(n=2^{n_1}\times m)$ *are non-trivial if*

$$\left.\begin{array}{l} x_{k+n}^{*}=x_{k}^{*}=a_{j_1}^{(n)}\notin\{\cup_{i_j=1}^{4}\{a_{i_1}^{(1)}\},\cup_{i_2=1}^{4^{2^{n_1-1}m}}\{a_{i_2}^{(2^{n_1-1}m)}\}\} \\ \text{for } n_1=1,2,\ldots;m=2l_1+1;j_1\in\{1,2,\ldots,4^n\}\cup\{\varnothing\} \end{array}\right\}$$
$$\text{for } n\neq 2^{n_2},$$
$$\left.\begin{array}{l} x_{k+n}^{*}=x_{k}^{*}=a_{j_1}^{(n)}\notin\{\cup_{i_2=1}^{4^{2^{n_1-1}m}}\{a_{i_2}^{(2^{n_1-1}m)}\}\} \\ \text{for } n_1=1,2,\ldots;m=1;j_1\in\{1,2,\ldots,4^n\}\cup\{\varnothing\} \end{array}\right\} \tag{3.240}$$
$$\text{for } n=2^{n_2}.$$

*Such a backward period-*n *fixed-point is*

- *monotonically stable if* $dx_k/dx_{k+n}|_{x_{k+n}^*=a_{i_1}^{(n)}}\in(1,\infty)$;
- *monotonically invariant if* $dx_k/dx_{k+n}|_{x_{k+n}^*=a_{i_1}^{(n)}}=1$, *which is*
  - *a monotonic lower-saddle of the* $(2l_1)^{\text{th}}$ *order for* $d^{2l_1}x_k/dx_{k+n}^{2l_1}|_{x_{k+n}^*}>0$;
  - *a monotonic upper-saddle the* $(2l_1)^{\text{th}}$ *order for* $d^{2l_1}x_k/dx_{k+n}^{2l_1}|_{x_{k+n}^*}<0$;
  - *a monotonic sink of the* $(2l_1+1)^{\text{th}}$ *order for* $d^{2l_1+1}x_k/dx_{k+n}^{2l_1+1}|_{x_{k+n}^*}>0$;
  - *a monotonic source of the* $(2l_1+1)^{\text{th}}$ *order for* $d^{2l_1+1}x_k/dx_{k+n}^{2l_1+1}|_{x_{k+n}^*}<0$;
- *monotonically unstable if* $dx_k/dx_{k+n}|_{x_{k+n}^*=a_{i_1}^{(n)}}\in(-1,0)$;
- *monotonically unstable with infinity eigenvalue if* $dx_k/dx_{k+n}|_{x_{k+n}^*=a_{i_1}^{(n)}}=0^+$;
- *oscillatorilly source with infinity eigenvalue if* $dx_k/dx_{k+n}|_{x_{k+n}^*=a_{i_1}^{(n)}}=0^-$;
- *oscillatorilly unstable if* $dx_k/dx_{k+n}|_{x_{k+n}^*=a_{i_1}^{(n)}}\in(-1,0)$;
- *flipped if* $dx_k/dx_{k+n}|_{x_{k+n}^*=a_{i_1}^{(n)}}=-1$, *which is*
  - *an oscillatory lower-saddle of the* $(2l_1)^{\text{th}}$ *order for* $d^{2l_1}x_k/dx_{k+n}^{2l_1}|_{x_{k+n}^*}>0$;
  - *an oscillatory upper-saddle the* $(2l_1)^{\text{th}}$ *order for* $d^{2l_1}x_k/dx_{k+n}^{2l_1}|_{x_{k+n}^*}<0$;
  - *an oscillatory source of the* $(2l_1+1)^{\text{th}}$ *order for* $d^{2l_1+1}x_k/dx_{k+n}^{2l_1+1}|_{x_{k+n}^*}>0$;
  - *an oscillatory sink the* $(2l_1+1)^{\text{th}}$ *order for* $d^{2l_1+1}x_k/dx_{k+n}^{2l_1+1}|_{x_{k+n}^*}<0$;
- *oscillatorilly stable if* $dx_k/dx_{k+n}|_{x_{k+n}^*=a_{i_1}^{(n)}}\in(-\infty,-1)$.

(iii) *For a backward fixed-point of* $x_k^*=x_{k+n}^*=a_{i_1}^{(n)}$ $(i_1\in I_q^{(n)}, q\in\{1,2,\ldots,N\})$, *there is a period-doubling of the* q^{th}*-set of period-*n *fixed-points if*

$$\begin{aligned}
&\frac{dx_k}{dx_{k+n}}|_{x^*_{k+n}=a^{(n)}_{i_1}} = 1 + a^{(n)}_0 \prod_{j_2=1,j_2\neq i_1}^{4^n}(a^{(n)}_{i_1} - a^{(n)}_{j_2}) = -1,\\
&\frac{d^s x_k}{dx^s_{k+n}}|_{x^*_{k+n}=a^{(n)}_{i_1}} = 0,\ \text{for } s = 2,\ldots,\ r-1;\\
&\frac{d^r x_k}{dx^r_{k+n}}|_{x^*_{k+n}=a^{(n)}_{i_1}} \neq 0 \text{ for } 1 < r \leq 4^n
\end{aligned} \tag{3.241}$$

with

- *a* r*th-order oscillatory sink for* $d^r x_k/dx^r_{k+n}|_{x^*_{k+n}=a^{(n)}_{i_1}} < 0$ *and* $r = 2l_1 + 1$;
- *a* r*th-order oscillatory source for* $d^r x_k/dx^r_{k+n}|_{x^*_{k+n}=a^{(n)}_{i_1}} > 0$ *and* $r = 2l_1 + 1$;
- *a* r*th-order oscillatory lower-saddle for* $d^r x_k/dx^r_{k+n}|_{x^*_{k+n}=a^{(n)}_{i_1}} > 0$ *and* $r = 2l_1$;
- *a* r*th-order oscillatory upper-saddle for* $d^r x_k/dx^r_{k+n}|_{x^*_{k+n}=a^{(n)}_{i_1}} < 0$ *and* $r = 2l_1$.

The corresponding period-$2\times n$ *discrete system of the backword quartic discrete system in Eq.* (3.229) *is*

$$\begin{aligned}
x_k = x_{k+2\times n} &+ a^{(2\times n)}_0 \prod_{i_1\in I^{(n)}_q}(x_{k+2\times n} - a^{(n)}_{i_1})^3\\
&\times \prod_{j_2=1}^{4^{2\times n}}(x_{k+2\times n} - a^{(2\times n)}_{j_2})^{(1-\delta(i_1,j_2))}
\end{aligned} \tag{3.242}$$

with

$$\begin{aligned}
&\frac{dx_k}{dx_{k+2\times n}}|_{x^*_{k+2\times n}=a^{(n)}_{i_1}} = 1,\ \frac{d^2 x_k}{dx^2_{k+2\times n}}|_{x^*_{k+2\times n}=a^{(n)}_{i_1}} = 0;\\
&\frac{d^3 x_k}{dx^3_{k+2\times n}}|_{x^*_{k+2\times n}=a^{(n)}_{i_1}} = 6a^{(2\times n)}_0 \prod_{i_1\in I^{(n)}_q, i_2\neq i_1}(a^{(n)}_{i_1} - a^{(n)}_{i_2})^3\\
&\qquad\qquad\times \prod_{j_2=1}^{4^{2\times n}}(a^{(n)}_{i_1} - a^{(2\times n)}_{j_2})^{(1-\delta(i_1,j_2))}.
\end{aligned} \tag{3.243}$$

Thus, x_k *at* $x^*_{k+2\times n} = a^{(n)}_{i_1}$ *for* $i_1 \in I^{(n)}_q$, $q \in \{1,2,\ldots,N\}$ *is*

- *a monotonic source of the third-order if* $d^3 x_k/dx^3_{k+2\times n}|_{x^*_{k+2\times n}=a^{(n)}_{i_1}} < 0$,
- *a monotonic sink of the third-order if* $d^3 x_k/dx^3_{k+2\times n}|_{x^*_{k+2\times n}=a^{(n)}_{i_1}} > 0$.

(iv) *After* l*-times period-doubling bifurcations of period-*n *fixed points, a period-*$2^l \times n$ *discrete system of the quartic discrete system in Eq.* (3.216) *is*

$$
\begin{aligned}
x_k &= x_{k+2^l\times n} + [a_0^{(2^{l-1}\times n)} \prod_{i_1=1}^{4^{2^{l-1}\times n}} (x_{k+2^l\times n} - a_{i_1}^{(2^{l-1}\times n)})] \\
&\quad \times \{1 + \prod_{i_1=1}^{4^{2^{l-1}\times n}} [1 + a_0^{(2^{l-1})} \prod_{i_2=1,i_2\neq i_1}^{4^{2^{l-1}\times n}} (x_{k+2^l\times n} - a_{i_2}^{(2^{l-1}\times n)})]\} \\
&= x_{k+2^l\times n} + [a_0^{(2^{l-1}\times n)} \prod_{i_1=1}^{4^{2^{l-1}\times n}} (x_{k+2^l\times n} - a_{i_1}^{(2^{l-1}\times n)})] \\
&\quad \times [(a_0^{(2^{l-1}\times n)})^{4^{(2^{l-1}\times n)}} \prod_{j_1=1}^{(4^{2^l\times n}-4^{2^{l-1}\times n})/2} (x_{k+2^l\times n}^2 + B_{j_2}^{(2^l\times n)} x_{k+2^l\times n} + C_{j_2}^{(2^l\times n)})] \\
&= x_{k+2^l\times n} + [a_0^{(2^{l-1}\times n)} \prod_{i_1=1}^{4^{(2^{l-1}\times n)}} (x_{k+2^l\times n} - a_{i_1}^{(2^{l-1}\times n)})] \\
&\quad \times [(a_0^{(2^{l-1}\times n)})^{4^{2^{l-1}\times n}} \prod_{j_2=1}^{(4^{2^l\times n}-4^{2^{l-1}\times n})/2} (x_{k+2^l\times n} - b_{j_2,1}^{(2^l\times n)})(x_{k+2^l\times n} - b_{j_2,2}^{(2^l\times n)})] \\
&= x_{k+2^l\times n} + (a_0^{(2^{l-1}\times n)})^{4^{2^{l-1}\times n}} \prod_{i=1}^{4^{2^l\times n}} (x_{k+2^l\times n} - a_i^{(2^l\times n)}) \\
&= x_{k+2^l\times n} + a_0^{(2^l\times n)} \prod_{i=1}^{4^{2^l\times n}} (x_{k+2^l\times n} - a_i^{(2^l\times n)})
\end{aligned}
\tag{3.244}
$$

with

$$
\begin{aligned}
\frac{dx_k}{dx_{k+2^l\times n}} &= 1 + a_0^{(2^l\times n)} \sum_{i_1=1}^{4^{2^l\times n}} \prod_{i_2=1,i_2\neq i_1}^{4^{2^l\times n}} (x_{k+2^l\times n} - a_{i_2}^{(2^l\times n)}), \\
\frac{d^2x_k}{dx_{k+2^l\times n}^2} &= a_0^{(2^l\times n)} \sum_{i_1=1}^{4^{2^l\times n}} \sum_{i_2=1,i_2\neq i_1}^{4^{2^l\times n}} \prod_{i_3=1,i_3\neq i_1,i_2}^{4^{2^l\times n}} (x_{k+2^l\times n} - a_{i_3}^{(2^l\times n)}), \\
&\vdots \\
\frac{d^rx_k}{dx_{k+2^l\times n}^r} &= a_0^{(2^l\times n)} \sum_{i_1=1}^{4^{2^l\times n}} \cdots \sum_{i_r=1,i_r\neq i_1,i_2\cdots,i_{r-1}}^{4^{2^l\times n}} \prod_{i_{r+1}=1,i_{r+1}\neq i_1,i_2\cdots,i_r}^{4^{2^l\times n}} (x_{k+2^l\times n} - a_{i_{r+1}}^{(2^l\times n)}) \\
&\text{for } r \leq 4^{2^l\times n};
\end{aligned}
\tag{3.245}
$$

where

$$
\begin{aligned}
a_0^{(2^l\times n)} &= (a_0^{(2^{l-1}\times n)})^{1+4^{2^{l-1}\times n}}, l = 1,2,3,\ldots; \\
\cup_{i=1}^{4^{2^l\times n}} \{a_i^{(2^l\times n)}\} &= \text{sort}\{\cup_{i_1=1}^{4^{2^{l-1}\times n}} \{a_{i_1}^{(2^{l-1}\times n)}\}, \cup_{i_2=1}^{M_2} \{b_{i_2,1}^{(2^l\times n)}, b_{i_2,2}^{(2^l\times n)}\}\}; \\
b_{i,1}^{(2^l\times n)} &= -\frac{1}{2}(B_i^{(2^l\times n)} + \sqrt{\Delta_i^{(2^l\times n)}}), \\
b_{i,2}^{(2^l\times n)} &= -\frac{1}{2}(B_i^{(2^l\times n)} - \sqrt{\Delta_i^{(2^l\times n)}}), \\
\Delta_i^{(2^l\times n)} &= (B_i^{(2^l\times n)})^2 - 4C_i^{(2^l\times n)} \geq 0 \\
\text{for } i &\in \cup_{q_1=1}^{N_1} I_{q_1}^{(2^{l-1}\times n)} \cup \cup_{q_2=1}^{N_2} I_{q_2}^{(2^l\times n)}, \\
I_{q_1}^{(2^{l-1}\times n)} &= \{l_{(q_1-1)\times(2^{l-1}\times n)+1}, l_{(q_1-1)\times(2^{l-1}\times n)+2}, \ldots, l_{q_1\times(2^{l-1}\times n)}\} \\
&\subseteq \{1,2,\ldots,M_1\} \cup \{\varnothing\},
\end{aligned}
$$

$$
\begin{aligned}
&\text{for } q_1 \in \{1,2,\ldots,N_1\}, M_1 = N_1 \times (2^{l-1} \times n);\\
&I_{q_2}^{(2^l \times n)} = \{l_{(q_2-1)\times(2^l\times n)+1}, l_{(q_2-1)\times(2^l\times n)+2}, \ldots, l_{q_2\times(2^l\times n)}\}\\
&\qquad \subseteq \{1,2,\ldots,M_1\} \cup \{\varnothing\},\\
&\text{for } q_2 \in \{1,2,\ldots,N_2\}, M_2 = (4^{2^l} - 4^{2^{l-1}})/2;\\
&b_{i,1}^{(2^l\times n)} = -\frac{1}{2}(B_i^{(2^l\times n)} + \mathbf{i}\sqrt{|\Delta_i^{(2^l\times n)}|}),\\
&b_{i,2}^{(2^l\times n)} = -\frac{1}{2}(B_i^{(2^l\times n)} - \mathbf{i}\sqrt{|\Delta_i^{(2^l\times n)}|}),\\
&\Delta_i^{(2^l\times n)} = (B_i^{(2^l\times n)})^2 - 4C_i^{(2^l\times n)} < 0,\ \mathbf{i} = \sqrt{-1},\\
&i \in J^{(2^l\times n)} = \{l_{N_2\times(2^l\times n)+1}, l_{N_2\times(2^{l-1}\times n)+2}, \ldots, l_{M_2}\}\\
&\qquad \subset \{1,2,\ldots,M_2\} \cup \{\varnothing\}
\end{aligned}
\tag{3.246}
$$

with backward fixed-points

$$
\begin{aligned}
&x_k^* = x_{k+2^l\times n}^* = a_i^{(2^l\times n)}, (i = 1,2,\ldots,4^{2^l\times n})\\
&\cup_{i=1}^{4^{2^l\times n}}\{a_i^{(2^l\times n)}\} = \text{sort}\{\cup_{i=1}^{4^{2^{l-1}\times n}}\{a_i^{(2^{l-1}\times n)}\}, \cup_{i=1}^{M_2}\{b_{i,1}^{(2^l\times n)}, b_{i,2}^{(2^l\times n)}\}\}\\
&\text{with } a_i^{(2^l\times n)} < a_{i+1}^{(2^l\times n)}.
\end{aligned}
\tag{3.247}
$$

(ii) *For a fixed-point of* $x_k^* = x_{k+2^l\times n}^* = a_{i_1}^{(2^{l-1}\times n)}$ $(i_1 \in I_q^{(2^{l-1}\times n)}, q \in \{1,2,\ldots,N_1\})$, *there is a backward period-*$2^l \times n$ *discrete system if*

$$
\begin{aligned}
&\frac{dx_k}{dx_{k+2^{l-1}\times n}}\Big|_{x_{k+2^{l-1}\times n}^* = a_{i_1}^{(2^{l-1}\times n)}} = 1 + a_0 \prod_{i_2=1, i_2\neq i_1}^{4^{2^{l-1}\times n}} (a_{i_1}^{(2^{l-1}\times n)} - a_{i_2}^{(2^{l-1}\times n)}) = -1,\\
&\frac{d^s x_k}{dx_{k+2^{l-1}\times n}^s}\Big|_{x_{k+2^{l-1}\times n}^* = a_{i_1}^{(2^{l-1}\times n)}} = 0,\ \text{for } s = 2,\ldots,\ r-1;\\
&\frac{d^r x_k}{dx_{k+2^{l-1}\times n}^r}\Big|_{x_{k+2^{l-1}\times n}^* = a_{i_1}^{(2^{l-1}\times n)}} \neq 0 \text{ for } 1 < r \leq 4^{2^{l-1}\times n}
\end{aligned}
\tag{3.248}
$$

with

- *a* r*th-order oscillatory source for* $d^r x_k/dx_{k+2^{l-1}\times n}^r|_{x_{k+2^{l-1}\times n}^* = a_{i_1}^{(2^{l-1}\times n)}} > 0$ *and* $r = 2l_1 + 1$;
- *a* r*th-order oscillatory sink for* $d^r x_k/dx_{k+2^{l-1}\times n}^r|_{x_{k+2^{l-1}\times n}^* = a_{i_1}^{(2^{l-1}\times n)}} < 0$ *and* $r = 2l_1 + 1$;
- *a* r*th-order oscillatory lower-saddle for* $d^r x_k/dx_{k+2^{l-1}\times n}^r|_{x_{k+2^{l-1}\times n}^* = a_{i_1}^{(2^{l-1}\times n)}} > 0$ *and* $r = 2l_1$;

- *a* r*th-order oscillatory upper-saddle for* $d^r x_k/dx^r_{k+2^{l-1}\times n}|_{x^*_{k+2^{l-1}\times n}=a^{(2^{l-1}\times n)}_{i_1}} < 0$ *and* $r = 2l_1$.

The corresponding period-$2^l \times n$ *discrete system is*

$$\begin{aligned} x_k = x_{k+2^l\times n} &+ a_0^{(2^l\times n)} \prod\nolimits_{i_1\in I_q^{(2^l\times n)}} (x_{k+2^l\times n} - a_{i_1}^{(2^{l-1}\times n)})^3 \\ &\times \prod\nolimits_{j_2=1}^{4^{2^l\times n}} (x_{k+2^l\times n} - a_{j_2}^{(2^l\times n)})^{(1-\delta(i_1,j_2))} \end{aligned} \tag{3.249}$$

where

$$\delta(i_1,j_2) = 1 \text{ if } a_{j_2}^{(2^l\times n)} = a_{i_1}^{(2^{l-1}\times n)},\ \delta(i_1,j_2) = 0 \text{ if } a_{j_2}^{(2^l\times n)} \neq a_{i_1}^{(2^{l-1}\times n)} \tag{3.250}$$

with

$$\begin{aligned} &\frac{dx_k}{dx_{k+2^l\times n}}|_{x^*_{k+2^l\times n}=a_{i_1}^{(2^{l-1}\times n)}} = 1,\ \frac{d^2x_k}{dx^2_{k+2^l\times n}}|_{x^*_{k+2^l\times n}=a_{i_1}^{(2^{l-1}\times n)}} = 0; \\ &\frac{d^3x_k}{dx^3_{k+2^l\times n}}|_{x^*_{k+2^l\times n}=a_{i_1}^{(2^{l-1})}} = 6a_0^{(2^l\times n)} \prod\nolimits_{i_2=1,i_2\neq i_1}^{4^{2^{l-1}\times n}} (a_{i_1}^{(2^{l-1}\times n)} - a_{i_2}^{(2^{l-1}\times n)})^3 \\ &\qquad\qquad \times \prod\nolimits_{j_2=1}^{4^{2^l\times n}} (a_{i_1}^{(2^{l-1}\times n)} - a_{j_2}^{(2^l\times n)})^{(1-\delta(i_2,j_2))} \neq 0 \\ &(i_1 \in I_q^{(2^{l-1}\times n)}, q\in\{1,2,\ldots,N_1\}). \end{aligned} \tag{3.251}$$

Thus, x_k *at* $x^*_{k+2^l\times n} = a_{i_1}^{(2^{l-1}\times n)}$ $x_{k+2^l\times n}$ *is*

- *a monotonic source of the third-order if* $d^3x_k/dx^3_{k+2^l\times n}|_{x^*_{k+2^l\times n}=a_{i_1}^{(2^{l-1})}} < 0$;
- *a monotonic sink of the third-order if* $d^3x_k/dx^3_{k+2^l\times n}|_{x^*_{k+2^l\times n}=a_{i_1}^{(2^{l-1})}} > 0$.

(ii$_1$) *The period-*$2^l \times n$ *fixed-points are trivial if*

$$\left.\begin{aligned} &x_k^* = x^*_{k+2^l\times n} = a_j^{(2^l\times n)} \in \{\cup_{i_l=1}^{4}\{a_{i_1}^{(1)}\}, \cup_{i_2=1}^{4^{2^{l-1}\times n}}\{a_{i_2}^{(2^{l-1}\times n)}\}\} \\ &\text{for } j = 1,2,\ldots,4^{(2^l\times n)} \end{aligned}\right\} \\ \text{for } n \neq 2^{n_1}, \\ \left.\begin{aligned} &x_k^* = x^*_{k+2^l\times n} = a_j^{(2^l\times n)} \in \cup_{i_2=1}^{4^{2^{l-1}\times n}}\{a_{i_2}^{(2^{l-1}\times n)}\} \\ &\text{for } j = 1,2,\ldots,4^{2^l\times n} \end{aligned}\right\} \\ \text{for } n = 2^{n_1}. \tag{3.252}$$

(ii$_2$) *The backward period-*$2^l \times n$ *fixed-points are non-trivial if*

$$\left.\begin{array}{l} x_k^* = x_{k+2^l\times n}^* = a_j^{(2^l\times n)} \notin \{\cup_{i_i=1}^{2}\{a_{i_1}^{(1)}\}, \cup_{i_2=1}^{2^{2^{l-1}\times n}}\{a_{i_2}^{(2^{l-1}\times n)}\}\} \\ \text{for } j=1,2,\ldots,2^{(2^l\times n)} \end{array}\right\}$$
$$\text{for } n \neq 2^{n_1},$$
$$\left.\begin{array}{l} x_k^* = x_{k+2^l\times n}^* = a_j^{(2^l\times n)} \notin \{\cup_{i_2=1}^{2^{2^{l-1}\times n}}\{a_{i_2}^{(2^{l-1}\times n)}\}\} \\ \text{for } j=1,2,\ldots,2^{2^l\times n} \end{array}\right\}$$
$$\text{for } n = 2^{n_1}. \tag{3.253}$$

Such a backward period-$2^l \times n$ fixed-point for the quartic discrete system is

- *monotonically unstable if* $dx_k/dx_{k+2^l\times n}|_{x_{k+2^l\times n}^*=a_{i_1}^{(2^l\times n)}} \in (1,\infty)$;
- *monotonically invariant if* $dx_k/dx_{k+2^l\times n}|_{x_{k+2^l\times n}^*=a_{i_1}^{(2^l\times n)}} = 1$, *which is*
  - *a monotonic lower-saddle of the* $(2l_1)^{\text{th}}$ *order for* $d^{2l_1}x_k/dx_{k+2^l\times n}^{2l_1}|_{x_{k+2^l\times n}^*} > 0$ *(independent* $(2l_1)$*-branch appearance);*
  - *a monotonic upper-saddle the* $(2l_1)^{\text{th}}$ *order for* $d^{2l_1}x_k/dx_{k+2^l\times n}^{2l_1}|_{x_{k+2^l\times n}^*} < 0$ *(independent* $(2l_1)$*-branch appearance);*
  - *a monotonic sink of the* $(2l_1+1)^{\text{th}}$ *order for* $d^{2l_1+1}x_k/dx_{k+2^l\times n}^{2l_1+1}|_{x_{k+2^l\times n}^*} > 0$ *(dependent* $(2l_1+1)$*-branch appearance from one branch);*
  - *a monotonic source the* $(2l_1+1)^{\text{th}}$ *order for* $d^{2l_1+1}x_k/dx_{k+2^l\times n}^{2l_1+1}|_{x_{k+2^l\times n}^*} < 0$ *(dependent* $(2l_1+1)$*-branch appearance from one branch);*
- *monotonically unstable if* $dx_k/dx_{k+2^l\times n}|_{x_{k+2^l\times n}^*=a_{i_1}^{(2^l\times n)}} \in (0,1)$;
- *monotonically infinity-unstable if* $dx_k/dx_{k+2^l\times n}|_{x_{k+2^l\times n}^*=a_{i_1}^{(2^l\times n)}} = 0^+$;
- *oscillatorilly infinity-unstable if* $dx_k/dx_{k+2^l\times n}|_{x_{k+2^l\times n}^*=a_{i_1}^{(2^l\times n)}} = 0^-$;
- *oscillatorilly unstable if* $dx_k/dx_{k+2^l\times n}|_{x_{k+2^l\times n}^*=a_{i_1}^{(2^l\times n)}} \in (-1,0)$;
- *flipped if* $dx_k/dx_{k+2^l\times n}|_{x_{k+2^l\times n}^*=a_{i_1}^{(2^l\times n)}} = -1$, *which is*
  - *an oscillatory lower-saddle of the* $(2l_1)^{\text{th}}$ *order for* $d^{2l_1}x_k/dx_{k+2^l\times n}^{2l_1}|_{x_{k+2^l\times n}^*} > 0$;
  - *an oscillatory upper-saddle of the* $(2l_1)^{\text{th}}$ *order for* $d^{2l_1}x_k/dx_{k+2^l\times n}^{2l_1}|_{x_{k+2^l\times n}^*} < 0$;
  - *an oscillatory sink of the* $(2l_1+1)^{\text{th}}$ *order for* $d^{2l_1+1}x_k/dx_{k+2^l\times n}^{2l_1+1}|_{x_{k+2^l\times n}^*} < 0$;
  - *an oscillatory source the* $(2l_1+1)^{\text{th}}$ *order for* $d^{2l_1+1}x_k/dx_{k+2^l\times n}^{2l_1+1}|_{x_{k+2^l\times n}^*} > 0$;
- *oscillatorilly stable if* $dx_k/dx_{k+2^l\times n}|_{x_{k+2^l\times n}^*=a_{i_1}^{(2^l\times n)}} \in (-\infty,-1)$.

Proof Through the nonlinear renormalization, this theorem can be easily proved. ■

Reference

Luo ACJ (2019) The stability and bifurcation of fixed-points in low-degree polynomial systems. J Vib Test Syst Dyn 3(4):403–451

Chapter 4 $(2m)^{\text{th}}$-Degree Polynomial Discrete Systems

In this Chapter, the global stability and bifurcations of period-1 fixed-points in the $(2m)^{\text{th}}$-degree polynomial discrete system are presented. The *parallel-appearing*, *spraying-appearing*, *sprinkler-spraying-appearing* bifurcations for simple and higher-order period-1 fixed-points are presented, and the *antenna-switching*, *straw-bundle-switching bifurcations* and *flower-bundle-switching bifurcations* for simple and higher-order period-1 fixed-points are presented. From the period-doubling bifurcation, the period-2 fixed-point solutions and the corresponding period-doubling renormalization of such a forwarded $(2m)^{\text{th}}$-degree polynomial discrete system are discussed. For multiple iterations, the appearing bifurcations of period-n fixed-points and the corresponding period-doublization of the forward $(2m)^{\text{th}}$-degree polynomial discrete system are presented as well.

4.1 Global Stability and Bifurcations

In a similar fashion in Chaps. 1–3, the global stability and bifurcation of fixed-points in the $(2m)^{\text{th}}$-degree polynomial nonlinear discrete systems are discussed as in Luo (2020a, b). The stability and bifurcation of each individual fixed-point are analyzed from the local analysis.

Definition 4.1 Consider a $(2m)^{\text{th}}$-degree polynomial nonlinear discrete system

$$\begin{aligned} x_{k+1} &= x_k + f(x_k, \mathbf{p}) \\ &= x_k + A_0(\mathbf{p})x_k^{2m} + A_1(\mathbf{p})x_k^{2m-1} + \cdots + A_{2m-2}(\mathbf{p})x_k^2 + A_{2m-1}(\mathbf{p})x_k + A_{2m}(\mathbf{p}) \\ &= x_k + a_0(\mathbf{p})[x_k^2 + B_1(\mathbf{p})x_k + C_1(\mathbf{p})] \cdots [x_k^2 + B_m(\mathbf{p})x_k + C_m(\mathbf{p})] \end{aligned} \tag{4.1}$$

A. C. J Luo, *Bifurcation Dynamics in Polynomial Discrete Systems*, Nonlinear Physical Science, https://doi.org/10.1007/978-981-15-5208-3_4

where $A_0(\mathbf{p}) \neq 0$, and

$$\mathbf{p} = (p_1, p_2, \ldots, p_{m_1})^{\mathrm{T}}. \tag{4.2}$$

(i) If

$$\Delta_i = B_i^2 - 4C_i < 0 \text{ for } i = 1, 2, \ldots, m, \tag{4.3}$$

the 1-dimensional nonlinear discrete system with a $(2m)^{\text{th}}$-degree polynomial does not have any period-1 fixed-point, and the corresponding standard form is

$$x_{k+1} = x_k + a_0[(x_k + \tfrac{1}{2}B_1)^2 + \tfrac{1}{4}(-\Delta_1)] \cdots [(x_k + \tfrac{1}{2}B_m)^2 + \tfrac{1}{4}(-\Delta_m)]. \tag{4.4}$$

The flow of such a discrete system without fixed-points is called a non-fixed-point discrete flow.

(a) If $a_0 > 0$, the non-fixed-point discrete flow is called a positive discrete flow.
(b) If $a_0 < 0$,the non-fixed-point discrete flow is called a negative discrete flow.

(ii) If

$$\begin{array}{l} \Delta_i = B_i^2 - 4C_i > 0, i = i_1, i_2, \ldots, i_l \in \{1, 2, \ldots, m\}, \\ \Delta_j = B_j^2 - 4C_j < 0, j = i_{l+1}, i_{l+2}, \ldots, i_m \in \{1, 2, \ldots, m\} \\ \text{with } l \in \{0, 1, \ldots, m\}, \end{array} \tag{4.5}$$

the 1-dimensional, $(2m)^{\text{th}}$-degree polynomial discrete system has $2l$-fixed-points as

$$\begin{array}{l} x_k^* = b_1^{(i)} = -\frac{1}{2}(B_i + \sqrt{\Delta_i}), x_k^* = b_2^{(i)} = -\frac{1}{2}(B_i - \sqrt{\Delta_i}) \\ i \in \{i_1, i_2, \ldots, i_l\} \subseteq \{1, 2, \ldots, m\}. \end{array} \tag{4.6}$$

(ii_1) If

$$\begin{array}{l} b_r^{(i)} \neq b_s^{(j)} \text{ for } r, s \in \{1, 2\}; i, j = 1, 2, \ldots, l \\ \{a_1, a_2 \ldots, a_{2l}\} = \text{sort}\{b_1^{(1)}, b_2^{(1)}, \ldots, b_1^{(l)}, b_2^{(l)}\}, a_s < a_{s+1}, \end{array} \tag{4.7}$$

then, the corresponding standard form is

$$x_{k+1} = x_k + a_0 \prod_{i=1}^{l} (x_k - a_{2i-1})(x_k - a_{2i}) \prod_{k=l+1}^{m} [(x_k + \tfrac{1}{2}B_{i_k})^2 + \tfrac{1}{4}(-\Delta_{i_k})]. \tag{4.8}$$

(a) If $a_0 > 0$, the simple fixed-point discrete flow is called a (mSI-oSO: mSO:. . . : mSI-oSO:mSO) discrete flow.

(b) If $a_0 < 0$, the simple fixed-point discrete flow is called a (mSO: mSI-oSO:... : mSO:mSI-oSO) discrete flow.

(ii$_2$) If

$$
\begin{aligned}
&\{a_1, a_2 ..., a_{2l}\} = \text{sort}\{b_1^{(1)}, b_2^{(1)}, ..., b_1^{(l)}, b_2^{(l)}\}, \\
&a_{i_1} \equiv a_1 = \cdots = a_{l_1}, \\
&a_{i_2} \equiv a_{l_1+1} = \cdots = a_{l_1+l_2}, \\
&\vdots \\
&a_{i_r} \equiv a_{\sum_{i=1}^{r-1} l_i + 1} = \cdots = a_{\sum_{i=1}^{r-1} l_i + l_r} = a_{2l} \\
&\text{with } \textstyle\sum_{s=1}^{r} l_s = 2l,
\end{aligned}
\tag{4.9}
$$

then, the corresponding standard form is

$$
x_{k+1} = x_k + a_0 \textstyle\prod_{s=1}^{r} (x_k - a_{i_s})^{l_s} \prod_{k=l+1}^{m} [(x_k + \frac{1}{2}B_{i_k})^2 + \frac{1}{4}(-\Delta_{i_k})]. \tag{4.10}
$$

The fixed-point discrete flow is called an (l_1th mXX : l_2th mXX : $\cdots$: l_rth mXX) discrete flow.

(a) For $a_0 > 0$ and $p = 1, 2, ..., r$,

$$
l_p\text{th mXX} = \begin{cases}
(2r_p - 1)^{\text{th}}\text{mSO} \equiv (2r_p - 1)^{\text{th}}\text{order monotonic source,} \\
\text{for } \alpha_p = 2M_p - 1, l_p = 2r_p - 1; \\
(2r_p - 1)^{\text{th}}\text{mSI} \equiv (2r_p - 1)^{\text{th}}\text{order monotonic sink,} \\
\text{for } \alpha_p = 2M_p, l_p = 2r_p - 1; \\
(2r_p)^{\text{th}}\text{mLS} \equiv (2r_p)^{\text{th}}\text{order monotonic lower-saddle,} \\
\text{for } \alpha_p = 2M_p - 1, l_p = 2r_p; \\
(2r_p - 1)^{\text{th}}\text{mUS} \equiv (2r_p)^{\text{th}}\text{order monotonic upper-saddle,} \\
\text{for } \alpha_p = 2M_p, l_p = 2r_p;
\end{cases}
\tag{4.11}
$$

where $(2r_p - 1)^{\text{th}}\text{mSI} \equiv \text{mSI-oSO}$ for $r_p = 1$ and

$$
\alpha_p = \textstyle\sum_{s=p}^{r} l_s. \tag{4.12}
$$

(b) For $a_0 < 0$ and $p = 1, 2, ..., r$,

$$l_p\text{th mXX} = \begin{cases} (2r_p-1)^{\text{th}}\text{mSI} \equiv (2r_p-1)^{\text{th}}\text{order monotonic sink,} \\ \text{for } \alpha_p = 2M_p - 1, l_p = 2r_p - 1; \\ (2r_p-1)^{\text{th}}\text{mSO} \equiv (2r_p-1)^{\text{th}}\text{order monotonic source,} \\ \text{for } \alpha_p = 2M_p, l_p = 2r_p - 1; \\ ((2r_p-1)^{\text{th}}\text{mUS} \equiv 2r_p)^{\text{th}}\text{order monotonic upper-saddle,} \\ \text{for } \alpha_p = 2M_p - 1, l_p = 2r_p; \\ (2r_p-1)^{\text{th}}\text{mLS} \equiv (2r_p)^{\text{th}}\text{order monotonic lower-saddle,} \\ \text{for } \alpha_p = 2M_p, l_p = 2r_p. \end{cases} \tag{4.13}$$

(c) The fixed-point of $x_k^* = a_{i_p}$ for $(l_p > 1)$-repeated fixed-points switching is called an l_pth mXX switching bifurcation of $(l_{p_1}\text{th mXX} : l_{p_2}\text{th mXX} : \ldots : l_{p_\beta}\text{th mXX})$ fixed-points at a point $\mathbf{p} = \mathbf{p}_1 \in \partial\Omega_{12}$, and the corresponding bifurcation condition is

$$\begin{aligned} &a_{i_p} \equiv a_{\sum_{i=1}^{p-1} l_i + 1} = \cdots = a_{\sum_{i=1}^{p-1} l_i + l_p}, \\ &a^{\pm}_{\sum_{i=1}^{p-1} l_i + 1} \neq \cdots \neq a^{\pm}_{\sum_{i=1}^{p-1} l_i + l_p}; l_p = \textstyle\sum_{i=1}^{\beta} l_{p_i}. \end{aligned} \tag{4.14}$$

(iii) If

$$\begin{aligned} &\Delta_{j_1} = B_{j_1}^2 - 4C_{j_1} = 0, \\ &j_1 \in \{i_{11}, i_{12}, \ldots, i_{1s_1}\} \subseteq \{i_1, i_2, \ldots, i_l\} \subseteq \{1, 2, \ldots, m\}; \\ &\Delta_{j_2} = B_{j_2}^2 - 4C_{j_2} > 0, \\ &j_2 \in \{i_{21}, i_{22}, \ldots, i_{2s_2}\} \subseteq \{i_1, i_2, \ldots, i_l\} \subseteq \{1, 2, \ldots, m\}; \\ &\Delta_{j_3} = B_{j_3}^2 - 4C_{j_3} < 0, \\ &j_3 \in \{i_{l+1}, i_{l+2}, \ldots, i_m\} \subseteq \{1, 2, \ldots, m\}; \end{aligned} \tag{4.15}$$

the 1-dimensional, $(2m)^{\text{th}}$-degree polynomial discrete system has $2l$-fixed-points as

$$\begin{aligned} &\left.\begin{aligned} x_k^* &= b_1^{(j_1)} = -\tfrac{1}{2}B_{j_1}, \\ x_k^* &= b_2^{(j_1)} = -\tfrac{1}{2}B_{j_1} \end{aligned}\right\} \text{for } j_1 \in \{i_{11}, i_{12}, \ldots, i_{1s_1}\}, \\ &\left.\begin{aligned} x_k^* &= b_1^{(j_2)} = -\tfrac{1}{2}(B_{j_2} + \sqrt{\Delta_{j_2}}), \\ x_k^* &= b_2^{(j_2)} = -\tfrac{1}{2}(B_{j_2} - \sqrt{\Delta_{j_2}}) \end{aligned}\right\} \text{for } j_2 \in \{i_{21}, i_{22}, \ldots, i_{2s_2}\}. \end{aligned} \tag{4.16}$$

If

$$
\begin{aligned}
&\{a_1, a_2 \ldots, a_{2l}\} = \text{sort}\{b_1^{(1)}, b_2^{(1)}, \ldots, b_1^{(l)}, b_2^{(l)}\},\\
&a_{i_1} \equiv a_1 = \cdots = a_{l_1},\\
&a_{i_2} \equiv a_{l_1+1} = \cdots = a_{l_1+l_2},\\
&\vdots\\
&a_{i_r} \equiv a_{\sum_{i=1}^{r-1} l_i+1} = \cdots = a_{\sum_{i=1}^{r-1} l_i + l_r} = a_{2l}\\
&\text{with } \textstyle\sum_{s=1}^{r} l_s = 2l,
\end{aligned}
\tag{4.17}
$$

then, the corresponding standard form is

$$
x_{k+1} = x_k + a_0 \textstyle\prod_{s=1}^{r} (x_k - a_{i_s})^{l_s} \prod_{j=l+1}^{m} [(x_k + \frac{1}{2}B_{i_j})^2 + \frac{1}{4}(-\Delta_{i_j})]. \tag{4.18}
$$

The fixed-point discrete flow is called an (l_1th mXX : l_2th mXX : $\cdots$: l_rth mXX) discrete flow.

(a) The fixed-point of $x_k^* = a_{i_p}$ for $(l_p > 1)$-repeated fixed-point appearance or vanishing is called an l_pth mXX appearing bifurcation of fixed-points at a point $\mathbf{p} = \mathbf{p}_1 \in \partial\Omega_{12}$, and the appearing bifurcation condition is

$$
\begin{aligned}
&a_{i_p} \equiv a_{\sum_{i=1}^{p-1} l_i+1} = \cdots = a_{\sum_{i=1}^{p-1} l_i + l_p} = -\tfrac{1}{2}B_{i_p},\\
&\text{with } \Delta_{i_p} = B_{i_p}^2 - 4C_{i_p} = 0 \; (i_p \in \{i_1, i_2, \ldots, i_l\}),\\
&a^{+}_{\sum_{i=1}^{p-1} l_i+1} \neq \cdots \neq a^{+}_{\sum_{i=1}^{p-1} l_i + l_p} \text{ or } a^{-}_{\sum_{i=1}^{p-1} l_i+1} \neq \cdots \neq a^{-}_{\sum_{i=1}^{p-1} l_i + l_p}.
\end{aligned}
\tag{4.19}
$$

(b) The fixed-point of $x_k^* = a_{i_q}$ for $(l_q > 1)$-repeated fixed-points switching is called an l_qth XX bifurcation of (l_{q_1}th mXX : l_{q_2}th mXX : $\cdots$: l_{q_β}th mXX) fixed-point switching at a point $\mathbf{p} = \mathbf{p}_1 \in \partial\Omega_{12}$, and the switching bifurcation condition is

$$
\begin{aligned}
&a_{i_q} \equiv a_{\sum_{i=1}^{q-1} l_i+1} = \cdots = a_{\sum_{i=1}^{q-1} l_i + l_q},\\
&a^{\pm}_{\sum_{i=1}^{q-1} l_i+1} \neq \cdots \neq a^{\pm}_{\sum_{i=1}^{q-1} l_i + l_q};\; l_q = \textstyle\sum_{i=1}^{\beta} l_{q_i}.
\end{aligned}
\tag{4.20}
$$

(c) The fixed-point of $x_k^* = a_{i_p}$ for $(l_{p_1} \geq 1)$-repeated fixed-points appearance/vanishing and $(l_{p_2} \geq 2)$ repeated fixed-points switching of ($l_{p_{21}}$th mXX : $l_{p_{22}}$th mXX : $\cdots$: $l_{p_{2\beta}}$th mXX)-fixed-point switching is called an l_pth mXX bifurcation of fixed-point at a point $\mathbf{p} = \mathbf{p}_1 \in \partial\Omega_{12}$, and the flower-switching bifurcation condition is

$$\begin{aligned}
&a_{i_p} \equiv a_{\sum_{i=1}^{p-1} l_i+1} = \cdots = a_{\sum_{i=1}^{p-1} l_i+l_p} \\
&\text{with } \Delta_{i_q} = B_{i_q}^2 - 4C_{i_q} = 0 \ (i_q \in \{i_1, i_2, \ldots, i_l\}) \\
&a^{+}_{\sum_{i=1}^{p-1} l_i+j_1} \neq \cdots \neq a^{+}_{\sum_{i=1}^{p-1} l_i+j_{p_1}} \text{ or } a^{-}_{\sum_{i=1}^{p_1-1} l_i+j_1} \neq \cdots \neq a^{-}_{\sum_{i=1}^{p_1-1} l_i+j_{p_1}}, \\
&\text{for } \{j_1, j_2, \ldots, j_{p_1}\} \subseteq \{1, 2, \ldots, l_p\}, \\
&a^{\pm}_{\sum_{i=1}^{p-1} l_i+k_1} \neq \cdots \neq a^{\pm}_{\sum_{i=1}^{p-1} l_i+k_{p_2}} \\
&\text{for } \{k_1, k_2, \ldots, k_{p_2}\} \subseteq \{1, 2, \ldots, l_p\}, \\
&\text{with } l_{p_1} + l_{p_2} = l_p; l_{p_2} = \textstyle\sum_{i=1}^{\beta} l_{p_{2i}}
\end{aligned} \tag{4.21}$$

(iv) If

$$\Delta_i = B_i^2 - 4C_i > 0 \text{ for } i = 1, 2, \ldots, m \tag{4.22}$$

the 1-dimensional, $(2m)^{\text{th}}$-degree polynomial discrete system has $(2m)$ fixed-points as

$$\left.\begin{aligned}
x_k^* = b_1^{(i)} = -\tfrac{1}{2}(B_i + \sqrt{\Delta_i}), \\
x_k^* = b_2^{(i)} = -\tfrac{1}{2}(B_i - \sqrt{\Delta_i})
\end{aligned}\right\} \text{ for } i = 1, 2, \ldots, m. \tag{4.23}$$

(iv_1) If

$$\begin{aligned}
&b_r^{(i)} \neq b_s^{(j)} \text{ for } r, s \in \{1, 2\}; i, j = 1, 2, \ldots, m \\
&\{a_1, a_2 \ldots, a_{2m}\} = \text{sort}\{b_1^{(1)}, b_2^{(1)}, \ldots, b_1^{(m)}, b_2^{(m)}\},\ a_s < a_{s+1}.
\end{aligned} \tag{4.24}$$

The corresponding standard form is

$$x_{k+1} = x_k + a_0(x_k - a_1)(x_k - a_2)(x_k - a_3)\ldots(x_k - a_{2m-1})(x_k - a_{2m}). \tag{4.25}$$

Such a discrate flow is formed with all the simple fixed-points.

(a) If $a_0 < 0$, the simple fixed-point discrete flow is called a (mSO : mSI-oSO : ... : mSO : mSI-oSO) discrete flow.
(b) If $a_0 > 0$, the simple fixed-point discrete flow is called a (mSI-oSO : mSO : ... : mSI-oSO : mSO) discrete flow.

(iv$_2$) If

$$\begin{aligned}&\{a_1,a_2\ldots,a_{2m}\}=\text{sort}\{b_1^{(1)},b_2^{(1)},\ldots,b_1^{(m)},b_2^{(m)}\},\\&a_{i_1}\equiv a_1=\ldots=a_{l_1},\\&a_{i_2}\equiv a_{l_1+1}=\cdots=a_{l_1+l_2},\\&\vdots\\&a_{i_r}\equiv a_{\sum_{i=1}^{r-1}l_i+1}=\cdots=a_{\sum_{i=1}^{r-1}l_i+l_r}=a_{2m}\\&\text{with }\textstyle\sum_{s=1}^{r}l_s=2m,\end{aligned}\tag{4.26}$$

then, the corresponding standard form is

$$x_{k+1}=x_k+a_0\textstyle\prod_{s=1}^{r}(x_k-a_{i_s})^{l_s}.\tag{4.27}$$

The fixed-point discrete flow is called an (l_1th mXX : l_2th mXX : $\cdots$: l_rth mXX)-discrete flow. The fixed-point of $x_k^*=a_{i_p}$ for l_p-repeated fixed-points switching is called an l_pth XX bifurcation of (l_{p_1}th mXX : l_{p_2}th mXX : $\cdots$: l_{p_β}th mXX) fixed-point switching at a point $\mathbf{p}=\mathbf{p}_1\in\partial\Omega_{12}$, and the switching bifurcation condition is

$$\begin{aligned}&a_{i_p}\equiv a_{\sum_{i=1}^{p-1}l_i+1}=\cdots=a_{\sum_{i=1}^{p-1}l_i+l_p},\\&a^{\pm}_{\sum_{i=1}^{p-1}l_i+1}\neq\cdots\neq a^{\pm}_{\sum_{i=1}^{p-1}l_i+l_p};l_p=\textstyle\sum_{i=1}^{\beta}l_{p_i}.\end{aligned}\tag{4.28}$$

Definition 4.2 Consider a 1-dimensional, $(2m)^{\text{th}}$-degree polynomial nonlinear forward discrete system as

$$\begin{aligned}x_{k+1}&=x_k+f(x_k,\mathbf{p})\\&=x_k+A_0(\mathbf{p})x_k^{2m}+A_1(\mathbf{p})x_k^{2m-1}+\cdots+A_{2m-2}(\mathbf{p})x_k^2+A_{2m-1}x_k+A_{2m}(\mathbf{p})\\&=a_0(\mathbf{p})\textstyle\prod_{i=1}^{n}[x_k^2+B_i(\mathbf{p})x_k+C_i(\mathbf{p})]^{q_i}\end{aligned}\tag{4.29}$$

where $A_0(\mathbf{p})\neq0$, and

$$\mathbf{p}=(p_1,p_2,\ldots,p_{m_1})^{\mathrm{T}},m=\textstyle\sum_{i=1}^{n}q_i.\tag{4.30}$$

(i) If

$$\Delta_i=B_i^2-4C_i<0\text{ for }i=1,2,\ldots,n,\tag{4.31}$$

the 1-dimensional nonlinear discrete system with a $(2m)^{\text{th}}$-degree polynomial does not have any fixed-point, and the corresponding standard form is

$$x_{k+1} = x_k + a_0 \prod_{i=1}^{n} [(x_k + \tfrac{1}{2}B_i)^2 + \tfrac{1}{4}(-\Delta_i)]^{q_i}. \tag{4.32}$$

The discrete flow of such a system without fixed-points is called a non-fixed-point discrete flow.

(a) If $a_0 > 0$, the non-fixed-point discrete flow is called a positive discrete flow.

(b) If $a_0 < 0$, the non-fixed-point discrete flow is called a negative discrete flow.

(ii) If

$$\begin{aligned} &\Delta_i = B_i^2 - 4C_i > 0,\ i \in \{i_1, i_2, \ldots, i_l\} \subseteq \{1, 2, \ldots, n\};\\ &\Delta_j = B_j^2 - 4C_j < 0,\ j \in \{i_{l+1}, i_{l+2}, \ldots, i_n\} \subseteq \{1, 2, \ldots, n\}, \end{aligned} \tag{4.33}$$

the 1-dimensional, $(2m)^{\text{th}}$-degree polynomial discrete system has $(2l)$ fixed-points as

$$\left.\begin{aligned} x_k^* = b_1^{(i)} &= -\tfrac{1}{2}(B_i + \sqrt{\Delta_i}),\\ x_k^* = b_2^{(i)} &= -\tfrac{1}{2}(B_i - \sqrt{\Delta_i}) \end{aligned}\right\} \tag{4.34}$$
$$i \in \{i_1, i_2, \ldots, i_l\} \subseteq \{1, 2, \ldots, n\}.$$

(ii$_1$) If

$$\begin{aligned} &b_r^{(i)} \neq b_s^{(j)} \text{ for } r, s \in \{1, 2\};\ i, j = 1, 2, \ldots, l;\\ &\{a_1, a_2 \ldots, a_{2l}\} = \text{sort}\{b_1^{(1)}, b_2^{(1)}, \ldots, b_1^{(l)}, b_2^{(l)}\},\ a_s < a_{s+1}, \end{aligned} \tag{4.35}$$

then, the corresponding standard form is

$$\begin{aligned} &x_{k+1} = x_k + a_0 \prod_{s=1}^{2l} (x_k - a_s)^{l_s} \prod_{k=l+1}^{n} [(x_k + \tfrac{1}{2}B_{i_k})^2 + \tfrac{1}{4}(-\Delta_{i_k})]^{q_{i_k}}\\ &\text{with } l_s \in \{q_{i_1}, q_{i_2}, \ldots, q_{i_l}\}. \end{aligned} \tag{4.36}$$

The fixed-point discrete flow is called an $(l_1\text{th mXX} : l_2\text{th mXX} : \cdots : l_{2l}\text{th mXX})$ discrete flow.

(a) For $a_0 > 0$ and $p = 1, 2, \ldots, 2l$,

$$l_p\text{th mXX} = \begin{cases} (2r_p-1)^{\text{th}}\text{mSO} \equiv (2r_p-1)^{\text{th}}\text{order monotonic source}, \\ \text{for } \alpha_p = 2M_p - 1, l_p = 2r_p - 1; \\ (2r_p-1)^{\text{th}}\text{mSI} \equiv (2r_p-1)^{\text{th}}\text{order monotonic sink}, \\ \text{for } \alpha_p = 2M_p, l_p = 2r_p - 1; \\ (2r_p-1)^{\text{th}}\text{mLS} \equiv (2r_p)^{\text{th}}\text{order monotonic lower-saddle}, \\ \text{for } \alpha_p = 2M_p - 1, l_p = 2r_p; \\ (2r_p-1)^{\text{th}}\text{mUS} \equiv (2r_p)^{\text{th}}\text{order monotonic upper-saddle}, \\ \text{for } \alpha_p = 2M_p, l_p = 2r_p; \end{cases} \tag{4.37}$$

where $(2r_p-1)^{\text{th}}\text{mSI} \equiv \text{mSI-oSO}$ for $r_p = 1$ and

$$\alpha_p = \sum_{s=p}^{2l} l_s. \tag{4.38}$$

(b) For $a_0 < 0$ and $p = 1, 2, \ldots, 2l$,

$$l_p\text{th mXX} = \begin{cases} (2r_p-1)^{\text{th}}\text{mSI} \equiv (2r_p-1)^{\text{th}}\text{order monotonic sink}, \\ \text{for } \alpha_p = 2M_p - 1, l_p = 2r_p - 1; \\ (2r_p-1)^{\text{th}}\text{mSO} \equiv (2r_p-1)^{\text{th}}\text{order monotonic source}, \\ \text{for } \alpha_p = 2M_p, l_p = 2r_p - 1; \\ (2r_p)^{\text{th}}\text{mUS} \equiv (2r_p)^{\text{th}}\text{order monotonic upper-saddle}, \\ \text{for } \alpha_p = 2M_p - 1, l_p = 2r_p; \\ (2r_p-1)^{\text{th}}\text{mLS} \equiv (2r_p)^{\text{th}}\text{order monotonic lower-saddle}, \\ \text{for } \alpha_p = 2M_p, l_p = 2r_p. \end{cases} \tag{4.39}$$

(ii$_2$) If

$$\left.\begin{array}{l} \{a_1, a_2 \ldots, a_{2l}\} = \text{sort}\{b_1^{(1)}, b_2^{(1)}, \ldots, b_1^{(l)}, b_2^{(l)}\}, \\ a_{i_1} \equiv a_1 = \cdots = a_{l_1}, \\ a_{i_2} \equiv a_{l_1+1} = \cdots = a_{l_1+l_2}, \\ \vdots \\ a_{i_r} \equiv a_{\sum_{i=1}^{r-1} l_i + 1} = \cdots = a_{\sum_{i=1}^{r-1} l_i + l_r} = a_{2l} \\ \text{with } \sum_{s=1}^{r} l_s = 2l, \end{array}\right. \tag{4.40}$$

then, the corresponding standard form is

$$x_{k+1} = x_k + a_0 \prod_{s=1}^{r} (x_k - a_{i_s})^{l_s} \prod_{j=l+1}^{n} [(x_k + \tfrac{1}{2}B_{i_j})^2 + \tfrac{1}{4}(-\Delta_{i_j})]^{q_{i_k}}. \quad (4.41)$$

The fixed-point discrete flow is called an $(l_1\text{th mXX} : l_2\text{th mXX} : \cdots : l_r\text{th mXX})$-discrete flow.

(a) For $a_0 > 0$ and $p = 1, 2, \ldots, r$,

$$l_p\text{th mXX} = \begin{cases} (2r_p-1)^{\text{th}}\text{mSO} \equiv (2r_p-1)^{\text{th}}\text{order monotonic source}, \\ \text{for } \alpha_p = 2M_p - 1, l_p = 2r_p - 1; \\ (2r_p-1)^{\text{th}}\text{mSI} \equiv (2r_p-1)^{\text{th}}\text{order monotonic sink}, \\ \text{for } \alpha_p = 2M_p, l_p = 2r_p - 1; \\ (2r_p-1)^{\text{th}}\text{mLS} \equiv (2r_p)^{\text{th}}\text{order monotonic lower-saddle}, \\ \text{for } \alpha_p = 2M_p - 1, l_p = 2r_p; \\ (2r_p-1)^{\text{th}}\text{mUS} \equiv (2r_p)^{\text{th}}\text{order monotonic upper-saddle}, \\ \text{for } \alpha_p = 2M_p, l_p = 2r_p; \end{cases} \quad (4.42)$$

where

$$\alpha_p = \sum_{s=p}^{r} l_s. \quad (4.43)$$

(b) For $a_0 < 0$ and $p = 1, 2, \ldots, r$,

$$l_p\text{th mXX} = \begin{cases} (2r_p-1)^{\text{th}}\text{mSI} \equiv (2r_p-1)^{\text{th}}\text{order monotonic sink}, \\ \text{for } \alpha_p = 2M_p - 1, l_p = 2r_p - 1; \\ (2r_p-1)^{\text{th}}\text{mSO} \equiv (2r_p-1)^{\text{th}}\text{order monotonic source}, \\ \text{for } \alpha_p = 2M_p, l_p = 2r_p - 1; \\ (2r_p-1)^{\text{th}}\text{mUS} \equiv (2r_p)^{\text{th}}\text{order monotonic upper-saddle}, \\ \text{for } \alpha_p = 2M_p - 1, l_p = 2r_p; \\ (2r_p-1)^{\text{th}}\text{mLS} \equiv (2r_p)^{\text{th}}\text{order monotonic lower-saddle}, \\ \text{for } \alpha_p = 2M_p, l_p = 2r_p. \end{cases} \quad (4.44)$$

(c) The fixed-point of $x_k^* = a_{i_p}$ for $(l_p > 1)$-repeated fixed-points switching is called an $l_p\text{th mXX}$ switching bifurcation of $(l_{p_1}\text{th mXX} : l_{p_2}\text{th mXX} : \cdots : l_{p_\beta}\text{th mXX})$ fixed point switching at a point $\mathbf{p} = \mathbf{p}_1 \in \partial\Omega_{12}$, and the corresponding bifurcation condition is

$$\begin{aligned} & a_{i_p} \equiv a_{\sum_{i=1}^{p-1} l_i + 1} = \cdots = a_{\sum_{i=1}^{p-1} l_i + l_p}, \\ & a^{\pm}_{\sum_{i=1}^{p-1} l_i + 1} \neq \cdots \neq a^{\pm}_{\sum_{i=1}^{p-1} l_i + l_p}; l_p = \sum_{i=1}^{\beta} l_{p_i}. \end{aligned} \tag{4.45}$$

(iii) If

$$\begin{aligned} & \Delta_i = B_i^2 - 4C_i = 0, \\ & i \in \{i_{11}, i_{12}, \ldots, i_{1s}\} \subseteq \{i_1, i_2, \ldots, i_l\} \subseteq \{1, 2, \ldots, n\}; \\ & \Delta_k = B_k^2 - 4C_k > 0, \\ & k \in \{i_{21}, i_{22}, \ldots, i_{2r}\} \subseteq \{i_1, i_2, \ldots, i_l\} \subseteq \{1, 2, \ldots, n\}; \\ & \Delta_j = B_j^2 - 4C_j < 0, \\ & j \in \{i_{l+1}, i_{l+2}, \ldots, i_n\} \subseteq \{1, 2, \ldots, n\} \text{with } i \neq j \neq k; \end{aligned} \tag{4.46}$$

the 1-dimensional, $(2m)^{\text{th}}$-degree polynomial discrete system has $(2l)$ fixed-points as

$$\left.\begin{aligned} & x_k^* = b_1^{(j_1)} = -\frac{1}{2}B_{j_1}, \\ & x_k^* = b_2^{(j_1)} = -\frac{1}{2}B_{j_1} \end{aligned}\right\} \text{for } j_1 \in \{i_{11}, i_{12}, \ldots, i_{1s}\},$$
$$\left.\begin{aligned} & x_k^* = b_1^{(j_2)} = -\frac{1}{2}(B_{j_2} + \sqrt{\Delta_{j_2}}), \\ & x_k^* = b_2^{(j_2)} = -\frac{1}{2}(B_{j_2} - \sqrt{\Delta_{j_2}}) \end{aligned}\right\} \text{for } j_2 \in \{i_{21}, i_{22}, \ldots, i_{2r}\}. \tag{4.47}$$

If

$$\begin{aligned} & \{a_1, a_2 \ldots, a_{2l}\} = \text{sort}\{b_1^{(1)}, b_2^{(1)}, \ldots, b_1^{(l)}, b_2^{(l)}\}, \\ & a_{i_1} \equiv a_1 = \cdots = a_{l_1}, \\ & a_{i_2} \equiv a_{l_1+1} = \cdots = a_{l_1+l_2}, \\ & \vdots \\ & a_{i_r} \equiv a_{\sum_{i=1}^{r-1} l_i + 1} = \cdots = a_{\sum_{i=1}^{r-1} l_i + l_r} = a_{2l} \\ & \text{with } \sum_{s=1}^{r} l_s = 2l, \end{aligned} \tag{4.48}$$

then, the corresponding standard form is

$$x_{k+1} = x_k + a_0 \prod_{s=1}^{r} (x_k - a_{i_s})^{l_s} \prod_{k=l+1}^{n} [(x_k + \frac{1}{2}B_{i_k})^2 + \frac{1}{4}(-\Delta_{i_k})]^{q_{i_k}}. \tag{4.49}$$

The fixed-point discrete flow is called an $(l_1\text{th mXX} : l_2\text{th mXX} : \cdots : l_r\text{th mXX})$-discrete flow.

(a) The fixed-point of $x_k^* = a_{i_p}$ for $(l_p > 1)$-repeated fixed-points appearance or vanishing is called an l_pth mXX appearing bifurcation of fixed-points at a point $\mathbf{p} = \mathbf{p}_1 \in \partial\Omega_{12}$, and the appearing bifurcation condition is

$$\begin{aligned} & a_{i_p} \equiv a_{\sum_{i=1}^{p-1} l_i + 1} = \cdots = a_{\sum_{i=1}^{p-1} l_i + l_p} = -\frac{1}{2} B_{i_p} \\ & \text{with } \Delta_{i_p} = B_{i_p}^2 - 4C_{i_p} = 0 \, (i_p \in \{i_1, i_2, \ldots, i_l\}), \\ & a^+_{\sum_{i=1}^{p-1} l_i + 1} \neq \cdots \neq a^+_{\sum_{i=1}^{p-1} l_i + l_p} \text{ or } a^-_{\sum_{i=1}^{p-1} l_i + 1} \neq \cdots \neq a^-_{\sum_{i=1}^{p-1} l_i + l_p}. \end{aligned} \tag{4.50}$$

(b) The fixed-point of $x_k^* = a_{i_p}$ for $(l_p > 1)$- repeated fixed-points switching is called an l_pth mXX switching bifurcation of $(l_{p_1}\text{th mXX} : l_{p_2}\text{th mXX} : \cdots : l_{p_\beta}\text{th mXX})$-fixed-points switching at a point $\mathbf{p} = \mathbf{p}_1 \in \partial\Omega_{12}$, and the switching bifurcation condition is

$$\begin{aligned} & a_{i_p} \equiv a_{\sum_{i=1}^{p-1} l_i + 1} = \cdots = a_{\sum_{i=1}^{p-1} l_i + l_p}, \\ & a^\pm_{\sum_{i=1}^{p-1} l_i + 1} \neq \cdots \neq a^\pm_{\sum_{i=1}^{p-1} l_i + l_p}; l_p = \textstyle\sum_{i=1}^{\beta} l_{p_i}. \end{aligned} \tag{4.51}$$

(iv) If

$$\Delta_i = B_i^2 - 4C_i > 0 \text{ for } i = 1, 2, \ldots, n \tag{4.52}$$

the 1-dimensional, $(2m)^{\text{th}}$-degree polynomial discrete system has $2n$-fixed-points as

$$\left. \begin{aligned} x_k^* = b_1^{(i)} = -\frac{1}{2}(B_i + \sqrt{\Delta_i}), \\ x_k^* = b_2^{(i)} = -\frac{1}{2}(B_i - \sqrt{\Delta_i}) \end{aligned} \right\} \text{ for } i = 1, 2, \ldots, n. \tag{4.53}$$

(iv_1) If

$$\begin{aligned} & b_r^{(i)} \neq b_s^{(j)} \text{ for } r, s \in \{1, 2\}; i, j = 1, 2, \ldots, n; \\ & \{a_1, a_2 \ldots, a_{2n}\} = \text{sort}\{b_1^{(1)}, b_2^{(1)}, \ldots, b_1^{(n)}, b_2^{(n)}\}, \; a_s < a_{s+1}, \end{aligned} \tag{4.54}$$

then the corresponding standard form is

$$x_{k+1} = x_k + a_0 \textstyle\prod_{s=1}^{2n} (x_k - a_s)^{l_s} \text{ with } l_s \in \{q_{i_1}, q_{i_2}, \ldots, q_{i_n}\}. \tag{4.55}$$

The fixed-point discrete flow is called an $(l_1\text{th mXX} : l_2\text{th mXX} : \cdots : l_{2n}\text{th mXX})$-discrete flow.

(a) For $a_0 > 0$ and $p = 1, 2, \ldots, 2n$,

$$
l_p\text{th mXX} = \begin{cases}
(2r_p - 1)^{\text{th}}\text{mSO} \equiv (2r_p - 1)^{\text{th}}\text{order monotonic source,} \\
\text{for } \alpha_p = 2M_p - 1, l_p = 2r_p - 1; \\
(2r_p - 1)^{\text{th}}\text{mSI} \equiv (2r_p - 1)^{\text{th}}\text{order monotonic sink,} \\
\text{for } \alpha_p = 2M_p, l_p = 2r_p - 1; \\
(2r_p)^{\text{th}}\text{mLS} \equiv (2r_p)^{\text{th}}\text{order lower-saddle,} \\
\text{for } \alpha_p = 2M_p - 1, l_p = 2r_p; \\
(2r_p)^{\text{th}}\text{mUS} \equiv (2r_p)^{\text{th}}\text{order upper-saddle,} \\
\text{for } \alpha_p = 2M_p, l_p = 2r_p;
\end{cases} \tag{4.56}
$$

where $(2r_p - 1)^{\text{th}}\text{mSI} \equiv \text{mSI-oSO}$ for $r_p = 1$, and

$$
\alpha_p = \sum\nolimits_{s=p}^{2n} l_s. \tag{4.57}
$$

(b) For $a_0 < 0$ and $p = 1, 2, \ldots, 2n$,

$$
l_p\text{th mXX} = \begin{cases}
(2r_p - 1)^{\text{th}}\text{mSI} \equiv (2r_p - 1)^{\text{th}}\text{order monotonic sink,} \\
\text{for } \alpha_p = 2M_p - 1, l_p = 2r_p - 1; \\
(2r_p - 1)^{\text{th}}\text{mSO} \equiv (2r_p - 1)^{\text{th}}\text{order monotonic source,} \\
\text{for } \alpha_p = 2M_p, l_p = 2r_p - 1; \\
(2r_p - 1)^{\text{th}}\text{mUS} \equiv (2r_p)^{\text{th}}\text{order montonic upper-saddle,} \\
\text{for } \alpha_p = 2M_p - 1, l_p = 2r_p; \\
(2r_p - 1)^{\text{th}}\text{mLS} \equiv (2r_p)^{\text{th}}\text{order monotonic lower-saddle,} \\
\text{for } \alpha_p = 2M_p, l_p = 2r_p.
\end{cases} \tag{4.58}
$$

(iv_2) If

$$
\begin{aligned}
&\{a_1, a_2 \ldots, a_{2n}\} = \text{sort}\{b_1^{(1)}, b_2^{(1)}, \ldots, b_1^{(n)}, b_2^{(n)}\}, \\
&a_{i_1} \equiv a_1 = \cdots = a_{l_1}, \\
&a_{i_2} \equiv a_{l_1+1} = \cdots = a_{l_1+l_2}, \\
&\vdots \\
&a_{i_r} \equiv a_{\sum_{i=1}^{r-1} l_i + 1} = \cdots = a_{\sum_{i=1}^{r-1} l_i + l_r} = a_{2n}, \\
&\text{with } \sum\nolimits_{s=1}^{r} l_s = 2n,
\end{aligned} \tag{4.59}
$$

then, the corresponding standard form is

$$x_{k+1} = x_k + a_0 \prod_{s=1}^{r} (x_k - a_{i_s})^{l_s}. \tag{4.60}$$

The fixed-point discrete flow is called an (l_1th mXX : l_2th mXX : ⋯ : l_rth mXX)-discrete flow. The fixed-point of $x_k^* = a_{i_p}$ for l_p- repeated fixed-points switching is called a l_pth mXX switching bifurcation of (l_{p_1}th mXX : l_{p_2}th mXX : ⋯ : l_{p_β}th mXX) fixed-point at a point $\mathbf{p} = \mathbf{p}_1 \in \partial\Omega_{12}$, and the corresponding bifurcation condition is

$$\begin{aligned} &a_{i_p} \equiv a_{\sum_{i=1}^{p-1} l_i + 1} = \cdots = a_{\sum_{i=1}^{p-1} l_i + l_p}, \\ &a^{\pm}_{\sum_{i=1}^{p-1} l_i + 1} \neq \cdots \neq a^{\pm}_{\sum_{i=1}^{p-1} l_i + l_p}; l_p = \sum_{i=1}^{\beta} l_{p_i}. \end{aligned} \tag{4.61}$$

Definition 4.3 Consider a 1-dimensional, $(2m)^{\text{th}}$-degree polynomial nonlinear discrete system

$$\begin{aligned} x_{k+1} &= x_k + f(x_k, \mathbf{p}) \\ &= x_k + A_0(\mathbf{p}) x_k^{2m} + A_1(\mathbf{p}) x_k^{2m-1} + \cdots + A_{2m-2}(\mathbf{p}) x_k^2 + A_{2m-1} x_k + A_{2m}(\mathbf{p}) \\ &= a_0(\mathbf{p}) \prod_{s=1}^{r} (x_k - c_{i_s}(\mathbf{p}))^{l_s} \prod_{i=r+1}^{n} [x_k^2 + B_i(\mathbf{p}) x_k + C_i(\mathbf{p})]^{q_i} \end{aligned} \tag{4.62}$$

where $A_0(\mathbf{p}) \neq 0$, and

$$\sum_{s=1}^{r} l_s = 2l,\ \sum_{i=r+1}^{n} q_i = (m - l),\ \mathbf{p} = (p_1, p_2, \ldots, p_{m_1})^{\mathrm{T}}. \tag{4.63}$$

(i) If

$$\begin{aligned} &\Delta_i = B_i^2 - 4C_i < 0 \text{ for } i = r+1, r+2, \ldots, n, \\ &\{a_1, a_2, \ldots, a_r\} = \text{sort}\{c_1, c_2, \ldots, c_r\} \text{ with } a_i < a_{i+1}, \end{aligned} \tag{4.64}$$

the 1-dimensional nonlinear discrete system with a $(2m)^{\text{th}}$-degree polynomial have a fixed-point of $x_k^* = a_{i_s}(\mathbf{p})$ $(s = 1, 2, \ldots, r)$, and the corresponding standard form is

$$x_{k+1} = x_k + a_0(\mathbf{p}) \prod_{s=1}^{r} (x_k - a_{i_s})^{l_s} \prod_{i=r+1}^{n} [(x_k + \tfrac{1}{2} B_i)^2 + \tfrac{1}{4}(-\Delta_i)]^{l_i}. \tag{4.65}$$

The fixed-point discrete flow is called an (l_1th mXX : l_2th mXX : ⋯ : l_rth mXX)-discrete flow.

(a) For $a_0 > 0$ and $s = 1, 2, \ldots, r$,

$$
l_p\text{th mXX} = \begin{cases}
(2r_p - 1)^{\text{th}}\text{mSO} \equiv (2r_p - 1)^{\text{th}}\text{order monotonic source,} \\
\text{for } \alpha_p = 2M_p - 1, l_p = 2r_p - 1; \\
(2r_p - 1)^{\text{th}}\text{mSI} \equiv (2r_p - 1)^{\text{th}}\text{order monotonic sink,} \\
\text{for } \alpha_p = 2M_p, l_p = 2r_p - 1; \\
(2r_p)^{\text{th}}\text{mLS} \equiv (2r_p)^{\text{th}}\text{order monotonic lower-saddle,} \\
\text{for } \alpha_p = 2M_p - 1, l_p = 2r_p; \\
(2r_p)^{\text{th}}\text{mUS} \equiv (2r_p)^{\text{th}}\text{order monotonic upper-saddle,} \\
\text{for } \alpha_p = 2M_p, l_p = 2r_p;
\end{cases}
\tag{4.66}
$$

where

$$
\alpha_p = \textstyle\sum_{s=p}^{r} l_s. \tag{4.67}
$$

(b) For $a_0 < 0$ and $p = 1, 2, \ldots, r$,

$$
l_p\text{th mXX} = \begin{cases}
(2r_p - 1)^{\text{th}}\text{mSI} \equiv (2r_p - 1)^{\text{th}}\text{order monotonic sink,} \\
\text{for } \alpha_p = 2M_p - 1, l_p = 2r_p - 1; \\
(2r_p - 1)^{\text{th}}\text{mSO} \equiv (2r_p - 1)^{\text{th}}\text{order monotonic source,} \\
\text{for } \alpha_p = 2M_p, l_p = 2r_p - 1; \\
(2r_p)^{\text{th}}\text{mUS} \equiv (2r_p)^{\text{th}}\text{order monotonic upper-saddle,} \\
\text{for } \alpha_p = 2M_p - 1, l_p = 2r_p; \\
(2r_p)^{\text{th}}\text{mLS} \equiv (2r_p)^{\text{th}}\text{order monotonic lower-saddle,} \\
\text{for } \alpha_p = 2M_p, l_p = 2r_p.
\end{cases}
\tag{4.68}
$$

(ii) If

$$
\begin{array}{l}
\Delta_i = B_i^2 - 4C_i > 0, i = j_1, j_2, \ldots, j_s \in \{l+1, l+2, \ldots, n\}; \\
\Delta_j = B_j^2 - 4C_j < 0, j = j_{s+1}, j_{s+2}, \ldots, j_n \in \{l+1, l+2, \ldots, n\} \\
\text{with } s \in \{1, \ldots, n-l\},
\end{array}
\tag{4.69}
$$

the 1-dimensional, $(2m)^{\text{th}}$-degree polynomial discrete system has $2n_2$-fixed-points as

$$
\begin{aligned}
&x_k^* = b_1^{(i)} = -\tfrac{1}{2}(B_i + \sqrt{\Delta_i}), x_k^* = b_2^{(i)} = -\tfrac{1}{2}(B_i - \sqrt{\Delta_i}) \\
&i \in \{j_1, j_2, \ldots, j_{n_1}\} \subseteq \{l+1, l+2, \ldots, n\}.
\end{aligned} \tag{4.70}
$$

If

$$
\begin{aligned}
&\{a_1, a_2 \ldots, a_{2n_2}\} = \text{sort}\{c_1, c_2 \ldots, c_{2l}, \underbrace{b_1^{(r+1)}, b_2^{(r+1)}}_{q_{r+1}\,\text{sets}}, \ldots, \underbrace{b_1^{(n_1)}, b_2^{(n_1)}}_{q_{n_1}\,\text{sets}}\}, \\
&a_{i_1} \equiv a_1 = \cdots = a_{l_1}, \\
&a_{i_2} \equiv a_{l_1+1} = \cdots = a_{l_1+l_2}, \\
&\vdots \\
&a_{i_{n_1}} \equiv a_{\sum_{i=1}^{n_1-1} l_i + 1} = \cdots = a_{\sum_{i=1}^{n_1-1} l_i + l_{n_1}} = a_{2n_2} \\
&\text{with } \textstyle\sum_{s=1}^{n_1} l_s = 2n_2,
\end{aligned} \tag{4.71}
$$

then, the corresponding standard form is

$$
x_{k+1} = x_k + a_0 \prod_{s=1}^{n_1} (x_k - a_{i_s})^{l_s} \prod_{i=n_2+1}^{n} [(x_k + \tfrac{1}{2}B_i)^2 + \tfrac{1}{4}(-\Delta_i)]^{q_i}. \tag{4.72}
$$

The fixed-point discrete flow is called an (l_1th mXX : l_2th mXX : $\cdots$: l_{n_1}th mXX)-discrete flow.

(a) For $a_0 > 0$ and $p = 1, 2, \ldots, r, r+1, \ldots, n_1$,

$$
l_p\text{th mXX} = \begin{cases}
(2r_p - 1)^{\text{th}}\text{mSO} \equiv (2r_p - 1)^{\text{th}}\text{order monotonc source}, \\
\text{for } \alpha_p = 2M_p - 1, l_p = 2r_p - 1; \\
(2r_p - 1)^{\text{th}}\text{mSI} \equiv (2r_p - 1)^{\text{th}}\text{order monotonic sink}, \\
\text{for } \alpha_p = 2M_p, l_p = 2r_p - 1; \\
(2r_p)^{\text{th}}\text{mLS} \equiv (2r_p)^{\text{th}}\text{order monotonic lower-saddle}, \\
\text{for } \alpha_p = 2M_p - 1, l_p = 2r_p; \\
(2r_p - 1)^{\text{th}}\text{mUS} \equiv (2r_p)^{\text{th}}\text{order monotonic upper-saddle}, \\
\text{for } \alpha_p = 2M_p, l_p = 2r_p;
\end{cases} \tag{4.73}
$$

where $(2r_p - 1)^{\text{th}}\text{mSI} \equiv \text{mSI-oSO}$ for $r_p = 1$, and

$$
\alpha_p = \textstyle\sum_{s=p}^{n_1} l_s. \tag{4.74}
$$

(b) For $a_0 < 0$ and $p = 1, 2, \ldots, r, r+1, \ldots, n_1$,

$$
l_p\text{th mXX} = \begin{cases} (2r_p-1)^{\text{th}}\text{mSI} \equiv (2r_p-1)^{\text{th}}\text{order monotonic sink}, \\ \text{for } \alpha_p = 2M_p - 1, l_p = 2r_p - 1; \\ (2r_p-1)^{\text{th}}\text{mSO} \equiv (2r_p-1)^{\text{th}}\text{order monotonic source}, \\ \text{for } \alpha_p = 2M_p, l_p = 2r_p - 1; \\ (2r_p)^{\text{th}}\text{mUS} \equiv (2r_p)^{\text{th}}\text{order monotonic upper-saddle}, \\ \text{for } \alpha_p = 2M_p - 1, l_p = 2r_p; \\ (2r_p)^{\text{th}}\text{mLS} \equiv (2r_p)^{\text{th}}\text{order monotonic lower-saddle}, \\ \text{for } \alpha_p = 2M_p, l_p = 2r_p. \end{cases} \tag{4.75}
$$

(c) The fixed-point of $x_k^* = a_{i_p}$ for $(l_p > 1)$-repeated fixed-points switching is called an l_pth mXX switching bifurcation of (l_{p_1}th mXX : l_{p_2}th mXX : $\cdots$: l_{p_β}th mXX) fixed-point at a point $\mathbf{p} = \mathbf{p}_1 \in \partial\Omega_{12}$, and the switching bifurcation condition is

$$
a_{i_p} \equiv a_{\sum_{i=1}^{p-1} l_i + 1} = \cdots = a_{\sum_{i=1}^{p-1} l_i + l_p}, \; a^{\pm}_{\sum_{i=1}^{p-1} l_i + 1} \neq \cdots \neq a^{\pm}_{\sum_{i=1}^{p-1} l_i + l_p}. \tag{4.76}
$$

(iii) If

$$
\begin{array}{l} \Delta_i = B_i^2 - 4C_i = 0, \\ \text{for } i \in \{i_{11}, i_{12}, \ldots, i_{1s}\} \subseteq \{i_{l+1}, i_{l+2}, \ldots, i_{n_2}\} \subseteq \{l+1, l+2, \ldots, n\}, \\ \Delta_k = B_k^2 - 4C_k > 0, \\ \text{for } k \in \{i_{21}, i_{22}, \ldots, i_{2r}\} \subseteq \{i_{l+1}, i_{l+2}, \ldots, i_{n_2}\} \subseteq \{l+1, l+2, \ldots, n\}, \\ \Delta_j = B_j^2 - 4C_j < 0, \\ \text{for } j \in \{i_{n_2+1}, i_{n_2+2}, \ldots, i_n\} \subseteq \{l+1, l+2, \ldots, n\}, \end{array} \tag{4.77}
$$

the 1-dimensional, $(2m)^{\text{th}}$-degree polynomial discrete system has $(2n_2)$-fixed-points as

$$
\left. \begin{array}{l} x_k^* = b_1^{(j_1)} = -\frac{1}{2} B_i, \\ x_k^* = b_2^{(j_1)} = -\frac{1}{2} B_i \end{array} \right\} \text{for } j_1 \in \{i_{11}, i_{12}, \ldots, i_{1s}\}; \\
\left. \begin{array}{l} x_k^* = b_1^{(j_2)} = -\frac{1}{2}(B_{j_2} + \sqrt{\Delta_{j_2}}), \\ x_k^* = b_2^{(j_2)} = -\frac{1}{2}(B_{j_2} - \sqrt{\Delta_{j_2}}) \end{array} \right\} \text{for } j_2 \in \{i_{21}, i_{22}, \ldots, i_{2r}\}. \tag{4.78}
$$

If

$$
\begin{aligned}
&\{a_1, a_2 \ldots, a_{2n_2}\} = \operatorname{sort}\{c_1, c_2 \ldots, c_{2l}, \underbrace{b_1^{(r+1)}, b_2^{(r+1)}}_{q_{r+1}\text{sets}}, \ldots, \underbrace{b_1^{(n_1)}, b_2^{(n_1)}}_{q_{n_1}\text{ sets}}\},\\
&a_{i_1} \equiv a_1 = \cdots = a_{l_1},\\
&a_{i_2} \equiv a_{l_1+1} = \cdots = a_{l_1+l_2},\\
&\vdots\\
&a_{i_{n_1}} \equiv a_{\sum_{i=1}^{n_1-1} l_i+1} = \cdots = a_{\sum_{i=1}^{n_1-1} l_i+l_{n_1}} = a_{2n_2}\\
&\text{with } \textstyle\sum_{s=1}^{n_1} l_s = 2n_2,
\end{aligned}
\tag{4.79}
$$

then, the corresponding standard form is

$$
x_{k+1} = x_k + a_0 \prod_{s=1}^{n_1} (x_k - a_{i_s})^{l_s} \prod_{i=n_2+1}^{n} [(x_k + \tfrac{1}{2}B_i)^2 + \tfrac{1}{4}(-\Delta_i)]^{q_i}. \tag{4.80}
$$

The fixed-point discrete flow is called an (l_1th mXX : l_2th mXX : $\cdots$: l_{n_1}th mXX)-discrete flow.

(a) The fixed-point of $x_k^* = a_{i_p}$ for ($l_p > 1$)-repeated fixed-points appearance or vanishing is called an l_pth mXX appearing bifurcation of fixed-point at a point $\mathbf{p} = \mathbf{p}_1 \in \partial\Omega_{12}$, and the appearing bifurcation condition is

$$
\begin{aligned}
&a_{i_p} \equiv a_{\sum_{i=1}^{p-1} l_i+1} = \cdots = a_{\sum_{i=1}^{p-1} l_i+l_p} = -\tfrac{1}{2}B_{i_p}\\
&\text{with } \Delta_{i_p} = B_{i_p}^2 - 4C_{i_p} = 0 \; (i_p \in \{i_1, i_2, \ldots, i_l\})\\
&a^{+}_{\sum_{i=1}^{p-1} l_i+1} \neq \cdots \neq a^{+}_{\sum_{i=1}^{p-1} l_i+l_p} \text{ or } a^{-}_{\sum_{i=1}^{p-1} l_i+1} \neq \cdots \neq a^{-}_{\sum_{i=1}^{p-1} l_i+l_p}.
\end{aligned}
\tag{4.81}
$$

(b) The fixed-point of $x_k^* = a_{i_p}$ for ($l_p > 1$)- repeated fixed-points switching is called an l_pth mXX switching bifurcation of (l_{p_1}th mXX : l_{p_2}th mXX : $\cdots$: l_{p_β}th mXX) fixed-point at a point $\mathbf{p} = \mathbf{p}_1 \in \partial\Omega_{12}$, and the switching bifurcation condition is

$$
\begin{aligned}
&a_{i_p} \equiv a_{\sum_{i=1}^{p-1} l_i+1} = \cdots = a_{\sum_{i=1}^{p-1} l_i+l_p},\\
&a^{\pm}_{\sum_{i=1}^{p-1} l_i+1} \neq \cdots \neq a^{\pm}_{\sum_{i=1}^{p-1} l_i+l_p}, l_p = \textstyle\sum_{i=1}^{\beta} l_{p_i}.
\end{aligned}
\tag{4.82}
$$

(c) The fixed-point of $x_k^* = a_{i_p}$ for ($l_{p_1} \geq 1$)-repeated fixed-points appearance/vanishing and ($l_{p_2} \geq 2$) repeated fixed-points switching of ($l_{p_{21}}$th mmXX : $l_{p_{22}}$th mXX : $\cdots$: $l_{p_{2\beta}}$th mXX) is called an l_pth mXX switching bifurcation of fixed-point at a point $\mathbf{p} = \mathbf{p}_1 \in \partial\Omega_{12}$, and the flower-bundle witching bifurcation condition is

$$
\begin{aligned}
& a_{i_p} \equiv a_{\sum_{i=1}^{p-1} q_i + 1} = \cdots = a_{\sum_{i=1}^{p-1} q_i + q_p} \\
& \text{with } \Delta_{i_q} = B_{i_q}^2 - 4C_{i_q} = 0\,(i_q \in \{i_1, i_2, \ldots, i_l\}) \\
& a^{+}_{\sum_{i=1}^{p-1} l_i + j_1} \neq \cdots \neq a^{+}_{\sum_{i=1}^{p-1} l_i + j_{p_1}} \text{ or } a^{-}_{\sum_{i=1}^{p_1-1} l_i + j_1} \neq \cdots \neq a^{-}_{\sum_{i=1}^{p_1-1} l_i + j_{p_1}}, \\
& \text{for } \{j_1, j_2, \ldots, j_{p_1}\} \subseteq \{1, 2, \ldots, l_p\}, \\
& a^{\pm}_{\sum_{i=1}^{p-1} l_i + k_1} \neq \cdots \neq a^{\pm}_{\sum_{i=1}^{p-1} l_i + k_{p_2}} \\
& \text{for } \{k_1, k_2, \ldots, k_{p_2}\} \subseteq \{1, 2, \ldots, l_p\}, \\
& \text{with } l_{p_1} + l_{p_2} = l_p.
\end{aligned}
\tag{4.83}
$$

(iv) If

$$
\Delta_i = B_i^2 - 4C_i > 0 \ \text{ for } \ i = l+1, l+2, \ldots, n \tag{4.84}
$$

the 1-dimensional, $(2m)^{\text{th}}$-degree polynomial discrete system has $(2m)$ fixed-points as

$$
\left.
\begin{aligned}
x_k^* = b_1^{(i)} = -\frac{1}{2}(B_i + \sqrt{\Delta_i}), \\
x_k^* = b_2^{(i)} = -\frac{1}{2}(B_i - \sqrt{\Delta_i})
\end{aligned}
\right\} \text{for } i = l+1, l+2, \ldots, n. \tag{4.85}
$$

If

$$
\begin{aligned}
& \{a_1, a_2 \ldots, a_{2m}\} = \text{sort}\{c_1, c_2 \ldots, c_{2l}, \underbrace{b_1^{(r+1)}, b_2^{(r+1)}}_{q_{r+1}\text{sets}}, \ldots, \underbrace{b_1^{(n)}, b_2^{(n)}}_{q_n\text{ sets}}\}, \\
& a_{i_1} \equiv a_1 = \cdots = a_{l_1}, \\
& a_{i_2} \equiv a_{l_1+1} = \cdots = a_{l_1+l_2}, \\
& \vdots \\
& a_{i_n} \equiv a_{\sum_{i=1}^{n-1} l_i + 1} = \cdots = a_{\sum_{i=1}^{n-1} l_i + l_r} = a_{2m} \\
& \text{with } \textstyle\sum_{s=1}^{n} l_s = 2m,
\end{aligned}
\tag{4.86}
$$

then, the corresponding standard form is

$$
x_{k+1} = x_k + a_0 \textstyle\prod_{s=1}^{n} (x_k - a_{i_s})^{l_s}. \tag{4.87}
$$

The fixed-point discrete flow is called an $(l_1\text{th mXX} : l_2\text{th mXX} : \cdots : l_n\text{th mXX})$-discrete flow. The fixed-point of $x_k^* = a_{i_p}$ for l_p-repeated fixed-points switching is called an l_pth mXX switching bifurcation of $(l_{p_1}\text{th mXX} : l_{p_2}\text{th mXX} : \cdots : l_{p_\beta}\text{th mXX})$ fixed-point switching at a point $\mathbf{p} = \mathbf{p}_1 \in \partial\Omega_{12}$, and the switching bifurcation condition is

$$\begin{array}{l} a_{i_p} \equiv a_{\sum_{i=1}^{p-1} l_i + 1} = \cdots = a_{\sum_{i=1}^{p-1} l_i + l_p}, \\ a^{\pm}_{\sum_{i=1}^{p-1} l_i + 1} \neq \cdots \neq a^{\pm}_{\sum_{i=1}^{p-1} l_i + l_p}; l_p = \sum_{i=1}^{\beta} l_{p_i}. \end{array} \tag{4.88}$$

4.2 Simple Fixed-Point Bifurcations

From the global analysis, the bifurcations of simple fixed-points in the $(2m)^{\text{th}}$-degree polynomial discrete systems are discussed, which include appearing/vanishing bifurcations, switching bifurcations, and switching and appearing bifurcations.

4.2.1 *Appearing Bifurcations*

Consider a $(2m)^{\text{th}}$-degree polynomial discrete system in a form of

$$x_{k+1} = x_k + a_0 Q(x_k) \prod_{i=1}^{n} (x_k^2 + B_i x_k + C_i). \tag{4.89}$$

Without loss of generality, a function of $Q(x_k) > 0$ is either a polynomial function or a non-polynomial function. The roots of $x_k^2 + B_i x_k + C_i = 0$ are

$$\begin{array}{l} b_{1,2}^{(i)} = -\frac{1}{2} B_i \pm \frac{1}{2} \sqrt{\Delta_i}, \Delta_i = B_i^2 - 4C_i \geq 0 (i = 1, 2, \ldots, n) \\ \{a_1, a_2, \ldots, a_{2l}\} \subset \text{sort}\{b_1^{(1)}, b_2^{(1)}, b_1^{(2)}, b_2^{(2)}, \ldots, b_1^{(n)}, b_2^{(n)}\}, a_s \leq a_{s+1} \\ \left. \begin{array}{l} B_i \neq B_j (i, j = 1, 2, \ldots, n; i \neq j) \\ \Delta_i = 0 (i = 1, 2, \ldots, n) \end{array} \right\} \text{at bifurcation.} \end{array} \tag{4.90}$$

The second-order singularity bifurcation is for the birth of a pair of simple fixed points with monotonic sink and monotonic source. There are two *appearing* bifurcations for $i \in \{1, 2, \ldots, n\}$

$$2^{\text{nd}} \text{ order mUS} \xrightarrow[\text{appearing bifurcation}]{i^{\text{th}} \text{ quadratic factor}} \begin{cases} \text{mSO}, \text{ for } x_k^* = a_{2i}; \\ \text{mSI-oSO}, \text{ for } x_k^* = a_{2i-1}. \end{cases} \tag{4.91}$$

$$2^{\text{nd}} \text{ order mLS} \xrightarrow[\text{appearing bifurcation}]{i^{\text{th}} \text{ quadratic factor}} \begin{cases} \text{mSI-oSO}, \text{ for } x_k^* = a_{2i}; \\ \text{mSO}, \text{ for } x_k^* = a_{2i-1}. \end{cases} \tag{4.92}$$

If $Q(x_k) = 1$ and $n = m$, a set of paralleled different simple monotonic upper-saddle *appearing* bifurcations in the $(2m)^{\text{th}}$ degree polynomial nonlinear discrete system is called an m-monotonic-upper-saddle-node (m-mUSN) *parallel appearing* bifurcation. Such a bifurcation is also called an m-monotonic-upper-saddle-node (m-mUSN) *teethcomb appearing* bifurcation. At the *appearing* bifurcation point, $\Delta_i = 0 (i = 1, 2, \ldots, m)$, and the m-mUSN *teethcomb appearing* bifurcation structure is

$$
m\text{-mUSN}\begin{cases}
\text{mUS}\xrightarrow[\text{appearing}]{m^{\text{th}}\text{bifurcation}}\begin{cases}\text{mSO, for } x_k^* = a_{2m},\\ \text{mSI-oSO, for } x_k^* = a_{2m-1};\end{cases}\\
\quad\vdots\\
\text{mUS}\xrightarrow[\text{appearing}]{i^{\text{th}}\text{bifurcation}}\begin{cases}\text{mSO, for } x_k^* = a_{2i},\\ \text{mSI-oSO, for } x_k^* = a_{2i-1};\end{cases}\\
\quad\vdots\\
\text{mUS}\xrightarrow[\text{appearing}]{1^{\text{st}}\text{bifurcation}}\begin{cases}\text{mSO, for } x_k^* = a_2,\\ \text{mSI-oSO, for } x_k^* = a_1.\end{cases}
\end{cases}
\tag{4.93}
$$

Similarly, a set of paralleled different simple monotonic lower-saddle *appearing* bifurcations is called an m-monotonic-lower-saddle-node (m-mLSN) parallel *appearing* bifurcation for the $(2m)^{\text{th}}$ degree polynomial nonlinear discrete system. The monotonic lower-saddle-node bifurcation is called the m-monotonic-lower-saddle-node (m-mLSN) *teethcomb appearing* bifurcation. At the bifurcation point, $\Delta_i = 0 (i = 1, 2, \ldots, m)$, and the m-mLSN *appearing* bifurcation structure is

$$
m\text{-mLSN}\begin{cases}
\text{mLS}\xrightarrow[\text{appearing}]{m^{\text{th}}\text{bifurcation}}\begin{cases}\text{mSI-oSO, for } x_k^* = a_{2m},\\ \text{mSO, for } x_k^* = a_{2m-1};\end{cases}\\
\quad\vdots\\
\text{mLS}\xrightarrow[\text{appearing}]{i^{\text{th}}\text{bifurcation}}\begin{cases}\text{mSI-oSO, for } x_k^* = a_{2i},\\ \text{mSO, for } x_k^* = a_{2i-1};\end{cases}\\
\quad\vdots\\
\text{mLS}\xrightarrow[\text{appearing}]{1^{\text{st}}\text{bifurcation}}\begin{cases}\text{mSI-oSO, for } x_k^* = a_2,\\ \text{mSO, for } x_k^* = a_1.\end{cases}
\end{cases}
\tag{4.94}
$$

Consider an appearing bifurcation for a cluster of fixed-points with monotonic sink to oscillatory source and monotonic source with the following conditions.

$$\left.\begin{array}{l} B_i = B_j \ (i,j \in \{1,2,\ldots,n\}; i \neq j) \\ \Delta_i = 0 \ (i = 1,2,\ldots,n) \end{array}\right\} \text{at bifurcation.} \tag{4.95}$$

Thus, the $(2l)^{\text{th}}$-order appearing bifurcation is for a cluster of fixed-points with simple monotonic sinks to oscillatory sources and monotonic sources. Two $(2l)^{\text{th}}$ order *appearing* bifurcations for $l \in \{1,2,\ldots,s\}$ are

$$(2l)^{\text{th}}\text{order mUS} \xrightarrow[\text{appearing bifurcation}]{\text{cluster of } l\text{-quadratics}} \begin{cases} \text{mSO, for } x_k^* = a_{2s_l}, \\ \text{mSI-oSO, for } x_k^* = a_{2s_l-1}, \\ \vdots \\ \text{mSO, for } x_k^* = a_{2s_1}, \\ \text{mSI-oSO, for } x_k^* = a_{2s_1-1}. \end{cases} \tag{4.96}$$

$$(2l)^{\text{th}}\text{order mLS} \xrightarrow[\text{appearing bifurcation}]{\text{cluster of } l\text{-quadratics}} \begin{cases} \text{mSI-oSO, for } x_k^* = a_{2s_l}, \\ \text{mSO, for } x_k^* = a_{2s_l-1}, \\ \vdots \\ \text{mSI-oSO, for } x_k^* = a_{2s_1}, \\ \text{mSO, for } x_k^* = a_{2s_1-1}. \end{cases} \tag{4.97}$$

A set of paralleled, different, higher-order upper-saddle-node bifurcations with multiplicity is a $((2l_1)^{\text{th}}\text{mUS} : (2l_2)^{\text{th}}\text{mUS} : \cdots : (2l_s)^{\text{th}}\text{mUS})$ *parallel appearing* bifurcation in the $(2m)^{\text{th}}$ degree polynomial discrete system. $(2l_i)^{\text{th}}$mUS for $(i = 1,2,\ldots,s)$ with monotonic sources and monotonic sinks to oscillatory source is the $(2l_i)^{\text{th}}$-order monotonic upper-saddle with l_i-pairs of simple monotonic source and monotonic sink to oscillatory source fixed-points. With different orders of l_i-pairs of fixed points with simple monotonic sources and monotonic sinks to oscillatory sources, the $(2l_i)^{\text{th}}$mUSN bifurcation possesses different *spraying appearing clusters* of fixed points with monotonic sinks to oscillatory sources and monotonic sources. $\sum_{i=1}^{s} l_i = n \leq m$ where $s, l_i \in \{0,1,2,\ldots,m\}$. If $l_i = 1$ for $i = 1,2,\ldots,m$ with $n = m$, the monotonic upper-saddle-node parallel bifurcation or the monotonic upper-saddle-node *teethcomb appearing* bifurcation is recovered. Introduce

$$((2l_1)^{\text{th}}\text{mUS}:(2l_2)^{\text{th}}\text{mUS}:\cdots:(2l_s)^{\text{th}}\text{mUS}) \equiv (2l_1 : 2l_2 : \cdots : 2l_s)^{\text{th}}\text{mUS}. \tag{4.98}$$

At the *sprinkler-spraying appearing* bifurcation, $\Delta_i = 0 (i = 1,2,\ldots,s)$ and $B_i = B_j \ \ (i,j \in \{1,2,\ldots,s\}; i \neq j)$. The *sprinkler-spraying* mUSN *appearing* bifurcation is

$$(2l_1 : 2l_2 : \cdots : 2l_s)^{\text{th}}\text{mUS} = \begin{cases} (2l_s)^{\text{th}} \text{ order mUS}, \\ \vdots \\ (2l_2)^{\text{th}} \text{ order mUS}, \\ (2l_1)^{\text{th}} \text{ order mUS}. \end{cases} \tag{4.99}$$

Thus, the $(2l_1 : 2l_2 : \cdots : 2l_s)^{\text{th}}$ mUS appearing (or vanishing) bifurcation is called a $(2l_1 : 2l_2 : \cdots : 2l_s)^{\text{th}}$ mUSN *sprinkler-spraying appearing (or vanishing)* bifurcation.

Similarly, a set of paralleled different lower-saddle *appearing* bifurcations with multiplicity is the $((2l_1)^{\text{th}}\text{mLS}:(2l_2)^{\text{th}}\text{mLS}:\cdots:(2l_s)^{\text{th}}\text{mLS})$ *appearing* bifurcation in the $(2m)^{\text{th}}$ degree polynomial system. Thus, the $(2l_1:2l_2:\cdots:2l_s)^{\text{th}}\text{mLS}$ *appearing (or vanishing)* bifurcation is also called a $(2l_1:2l_2:\cdots:2l_s)^{\text{th}}\text{mLS}$ *sprinkler-spraying appearing (or vanishing)* bifurcation. Again, at the mLS *sprinkler-spraying* bifurcation, $\Delta_i = 0\ (i = 1, 2, \ldots, n)$ and $B_i = B_j\ (i, j \in \{1, 2, \ldots, n\}; i \neq j)$. Thus, the *sprinkler-spraying* mLSN *appearing* bifurcation is

$$(2l_1 : 2l_2 : \cdots : 2l_s)^{\text{th}}\text{mLS} = \begin{cases} (2l_s)^{\text{th}}\text{order mLS}, \\ \vdots \\ (2l_2)^{\text{th}}\text{order mLS}, \\ (2l_1)^{\text{th}}\text{order mLS}. \end{cases} \tag{4.100}$$

Two m-mUSN and m-mLSN *teethcomb appearing* bifurcations are presented in Fig. 4.1(i) and (ii) for $a_0 > 0$ and $a_0 < 0$, respectively. The set of paralleled $(4^{\text{th}}\text{mUS}:\cdots:(2r)^{th}\text{mUS}:\cdots:4^{\text{th}}\text{mUS}:6^{\text{th}}\text{mUS})$ appearing bifurcations for simple monotonic sinks to oscillatory sources and monotonic sources is presented in Fig. 4.1(iii) for $a_0 > 0$, where $l_1 = 2, \ldots, l_i = r, \ldots, l_{s-1} = 2$, $l_s = 3$ with $\sum_{i=1}^{s} l_i = m$. The $(4 : \cdots : 2r : \cdots : 4 : 6)^{\text{th}}$-mUSN *appearing* bifurcation is a mUSN *sprinkler-spraying appearing* bifurcation. However, for $a_0 < 0$, the $(4^{\text{th}}\text{mLS}:\cdots:(2r)^{\text{th}}\text{mLS}:\cdots:4^{\text{th}}\text{mLS}:6^{\text{th}}\text{mLS})$ appearing bifurcations for simple sources and sinks is presented in Fig. 4.1(iv). The $(4 : \cdots : 2r : \cdots : 4 : 6)^{\text{th}}$-mLSN *appearing* bifurcation is a mLSN *sprinkler-spraying appearing* bifurcation.

For a cluster of m-quadratics, $B_i = B_j (i, j \in \{1, 2, \ldots, m\}; i \neq j)$ and $\Delta_i = 0$ $(i = 1, 2, \ldots, m)$. The $(2m)^{\text{th}}$ order monotonic upper-saddle-node appearing bifurcation for m-pairs of fixed points with monotonic sink to oscillatory source and monotonic sources is

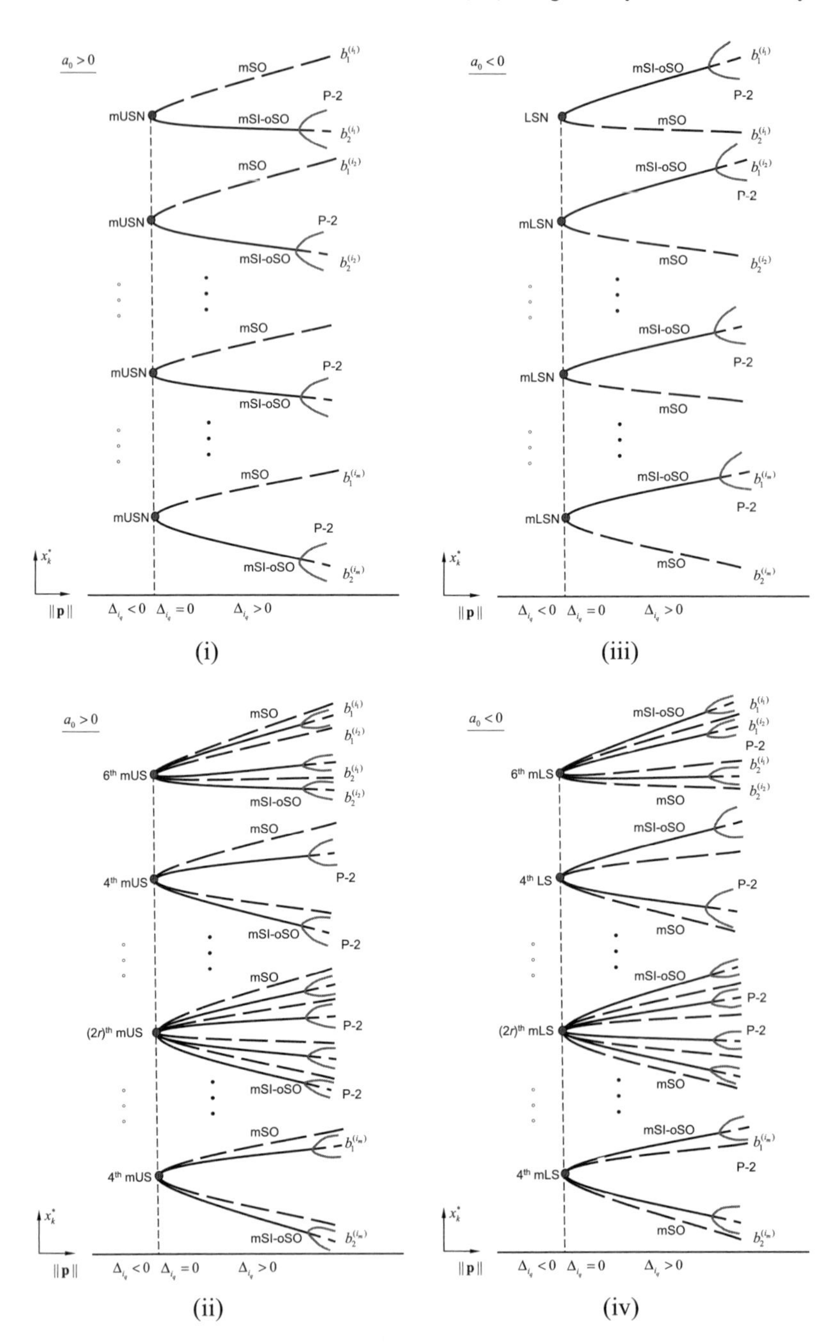

Fig. 4.1 (i) m-mUSN parallel bifurcation $(a_0 > 0)$, (ii) m-mLSN parallel bifurcation $(a_0 < 0)$, (iii) $((2l_1)^{\text{th}}$mUS:$(2l_2)^{\text{th}}$mUS:$\cdots$: $(2l_s)^{\text{th}}$mUS) parallel bifurcation $(a_0 > 0)$. (iv) $((2l_1)^{\text{th}}$mLS: $(2l_2)^{\text{th}}$mLS:$\cdots$: $(2l_s)^{\text{th}}$mLS) parallel bifurcation $(a_0 < 0)$ in a $(2m)^{\text{th}}$-degree polynomial discrete system. mLS: monotonic-lower-saddle, mUS: monotonic-upper-saddle, mSI-oSO: monotonic sink to oscillatory source, mSO: monotonic source. Stable and unstable fixed-points are represented by solid and dashed curves, respectively. The bifurcation points are marked by circular symbols.

$$(2m)^{\text{th}}\text{order mUS} \xrightarrow[\text{appearing bifurcation}]{\text{cluster of } m\text{-quadratics}} \begin{cases} \text{mSO}, \text{ for } x_k^* = a_{2m}, \\ \text{mSI-oSO}, \text{ for } x_k^* = a_{2m-1}, \\ \vdots \\ \text{mSO}, \text{ for } x_k^* = a_2, \\ \text{mSI-oSO}, \text{ for } x_k^* = a_1. \end{cases} \quad (4.101)$$

The $(2m)^{\text{th}}$ order lower-saddle-node appearing bifurcation for m-pairs of fixed-points with monotonic sink to oscillatory source, and monotonic source is

$$(2m)^{\text{th}}\text{order mLS} \xrightarrow[\text{appearing bifurcation}]{\text{cluster of } m\text{-quadratics}} \begin{cases} \text{mSI-oSO}, \text{ for } x_k^* = a_{2m}, \\ \text{mSO}, \text{ for } x_k^* = a_{2m-1}, \\ \vdots \\ \text{mSI-oSO}, \text{ for } x_k^* = a_2, \\ \text{mSO}, \text{ for } x_k^* = a_1. \end{cases} \quad (4.102)$$

The $(2m)^{\text{th}}$ order monotonic upper-saddle-node appearing bifurcation with m-pairs of fixed-points with monotonic sources and monotonic sinks to oscillatory sources is a *sprinkler-spraying cluster* of the m-pairs of fixed-points with monotonic sources and monotonic sinks to oscillatory sources. The $(2m)^{\text{th}}$ order monotonic lower-saddle-node *appearing* bifurcation with m-pairs of fixed-points is also a *sprinkler-spraying cluster* of the m-pairs of monotonic sources and monotonic sinks to oscillatory sources. Thus, the $(2m)^{\text{th}}$ order mUSN appearing bifurcation $(a_0 > 0)$ and $(2m)^{\text{th}}$ order mLSN bifurcation $(a_0 < 0)$ are presented in Fig. 4.2(i) and (ii), respectively. The $(2m)^{\text{th}}$ order monotonic-upper-saddle-node appearing bifurcation is named a $(2m)^{\text{th}}$ order mUSN *sprinkler-spaying* appearing bifurcation, and the $(2m)^{\text{th}}$ order monotonic-lower-saddle-node appearing bifurcation is named a $(2m)^{\text{th}}$ order mLSN *sprinkler-spraying* appearing bifurcation.

A series of the monotonic saddle-node bifurcations is aligned up with varying with parameters, which is formed a special pattern. For m-quadratics in the $(2m)^{\text{th}}$ order polynomial discrete system, the following conditions should be satisfied.

$$\begin{aligned} &B_i \approx B_j \, i,j \in \{1,2,\ldots,n\}; i \neq j, \\ &\Delta_i > \Delta_{i+1} \ (i = 1,2,\ldots,n; n \leq m), \\ &\Delta_i = 0 \text{ with } \|\mathbf{p}_i\| < \|\mathbf{p}_{i+1}\|. \end{aligned} \quad (4.103)$$

Thus, a series of m-(mUSN-mLSN-mUSN-...) *appearing* bifurcations $(a_0 > 0)$ and a series of m-(mLSN-mUSN-mLSN-...) *appearing* bifurcations $(a_0 < 0)$ are presented in Fig. 4.3(i) and (ii), respectively. The bifurcation scenario is formed by the swapping pattern of mUSN and mLSN appearing bifurcations. Such a bifurcation scenario is like the fish-scale. Thus, such a bifurcation swapping pattern of the mUSN and mLSN is called a *fish-scale appearing* bifurcation in the $(2m)^{\text{th}}$

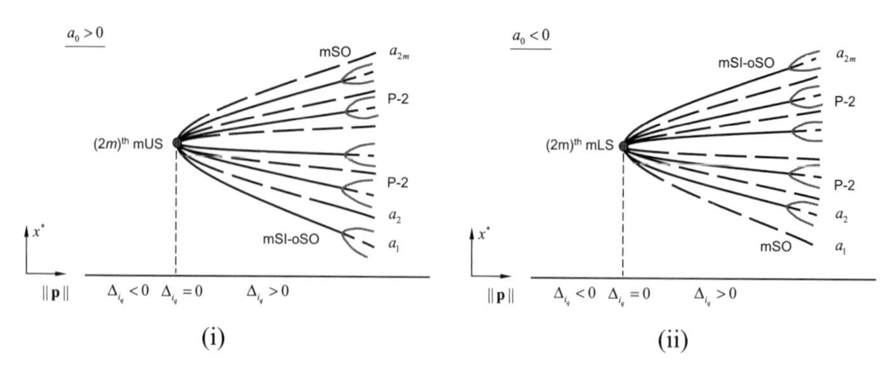

Fig. 4.2 (i) $(2m)^{\text{th}}$ order mUSN bifurcation $(a_0 > 0)$, (ii) $(2m)^{\text{th}}$ order mLSN bifurcation $(a_0 < 0)$ in the $(2m)^{\text{th}}$ polynomial system. mLS: monotonic-lower-saddle, mUS: monotonic-upper-saddle, mSI-oSO: monotonic sink to oscillatory source, mSO: monotonic source. Stable and unstable fixed-points are represented by solid and dashed curves, respectively. The bifurcation points are marked by circular symbols.

degree polynomial nonlinear discrete system. There are two swapping bifurcations: (i) the USN-LSN *fish-scale appearing* bifurcation and (ii) the mLSN-mUSN *fish-scale*, appearing bifurcation.

4.2.2 Switching Bifurcations

Consider the roots of $x_k^2 + B_i x_k + C_i = 0$ as

$$\begin{aligned}
&B_i = -(b_1^{(i)} + b_2^{(i)}),\ \Delta_i = (b_1^{(i)} - b_2^{(i)})^2 \geq 0,\\
&x_{k;1,2}^{(i)} = b_{1,2}^{(i)},\ \Delta_i > 0 \text{ if } b_1^{(i)} \neq b_2^{(i)} (i = 1, 2, \ldots, n);\\
&\left.\begin{aligned}&B_i \neq B_j (i, j = 1, 2, \ldots, n; i \neq j)\\ &\Delta_i = 0 \text{ at } b_1^{(i)} = b_2^{(i)} (i = 1, 2, \ldots, n)\end{aligned}\right\}\text{at bifurcation.}
\end{aligned} \tag{4.104}$$

The 2nd order singularity bifurcation is for the *switching* of a pair of fixed point with simple monotonic sink to oscillatory source and monotonic source. There are two *switching* bifurcations for $i \in \{1, 2, \ldots, n\}$

$$2^{\text{nd}}\text{order mUS} \xrightarrow[\text{switching bifurcation}]{i^{\text{th}}\text{quadratic factor}} \begin{cases} \text{mSO, for } a_{2i} = b_2^{(i)} \to b_1^{(i)}, \\ \text{mSI-oSO, for } a_{2i-1} = b_1^{(i)} \to b_2^{(i)}. \end{cases} \tag{4.105}$$

$$2^{\text{nd}}\text{order mLS} \xrightarrow[\text{switching bifurcation}]{i^{\text{th}}\text{quadratic factor}} \begin{cases} \text{mSI-oSO, for } a_{2i} = b_2^{(i)} \to b_1^{(i)}, \\ \text{mSO, for } a_{2i-1} = b_1^{(i)} \to b_2^{(i)}. \end{cases} \tag{4.106}$$

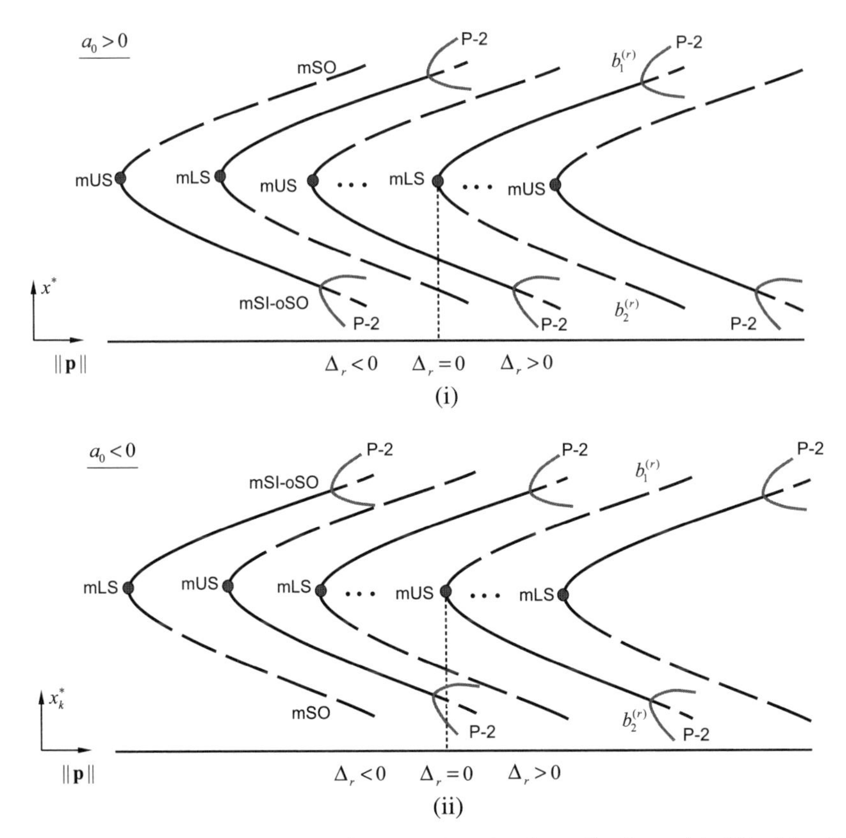

Fig. 4.3 (i) m-(mUS-mLS-mUS-...) series bifurcation $(a_0 > 0)$, (ii) m–(-(mUS-mLS-mUS-...) series bifurcation $(a_0 > 0)$ in the $(2m)^{\text{th}}$-degree polynomial discrete system. mLS: monotonic-lower-saddle, mUS: monotonic-upper-saddle, mSI-oSO: monotonic sink to oscillatory source, mSO: monotonic source. Stable and unstable fixed-points are represented by solid and dashed curves, respectively. The bifurcation points are marked by circular symbols.

A set of m-paralleled-pairs of different simple-monotonic-upper-saddle-node *switching* bifurcations in the $(2m)^{\text{th}}$degree polynomial nonlinear discrete system is called an m-monotonic-upper-saddle-node (m-mUSN) *parallel switching* bifurcation. Such a bifurcation is also called an m-monotonic-upper-saddle-node (m-mUSN) *antenna switching* bifurcation. For non-switching point, $\Delta_i > 0$ at $b_1^{(i)} \neq b_2^{(i)}$ $(i = 1, 2, \ldots, n)$. At the bifurcation point, $\Delta_i = 0$ at $b_1^{(i)} = b_2^{(i)}$ $(i = 1, 2, \ldots, n)$. The m-mUSN *parallel switching* bifurcation is

$$
m\text{-mUSN}\left\{\begin{array}{l}
\text{mUS}\xrightarrow[\text{switching bifurcation}]{m^{\text{th}}\text{bifurcation}}\left\{\begin{array}{l}\text{mSO}\downarrow\text{mSI-oSO},\ \text{for}\ b_2^{(m)}=a_{2m}\downarrow a_{2m-1},\\ \text{mSI-oSO}\uparrow m\text{SO},\ \text{for}\ b_1^{(m)}=a_{2m-1}\uparrow a_{2m};\end{array}\right.\\
\vdots\\
\text{mUS}\xrightarrow[\text{switching bifurcation}]{i^{\text{th}}\text{bifurcation}}\left\{\begin{array}{l}\text{mSO}\downarrow\text{mSI-oSO},\ \text{for}\ b_2^{(i)}=a_{2i}\downarrow a_{2i-1},\\ \text{mSI-oSO}\uparrow\text{mSO},\ \text{for}\ b_1^{(i)}=a_{2i-1}\uparrow a_{2i};\end{array}\right.\\
\vdots\\
\text{mUS}\xrightarrow[\text{switching bifurcation}]{1^{\text{st}}\text{bifurcation}}\left\{\begin{array}{l}\text{mSO}\downarrow\text{mSI-oSO},\ \text{for}\ b_2^{(1)}=a_2\downarrow a_1,\\ \text{mSI-oSO}\uparrow\text{mSO},\ \text{for}\ b_1^{(1)}=a_1\uparrow a_2.\end{array}\right.
\end{array}\right.
\tag{4.107}
$$

Similarly, a set of paralleled different simple monotonic-lower-saddle bifurcations is called an m-monotonic-lower-saddle-node (m-mLSN) *parallel switching* bifurcation for the $(2m)^{\text{th}}$ degree polynomial nonlinear system. The monotonic-lower-saddle-node *switching* bifurcation is also called an m-monotonic-lower-saddle-node (m-mLSN) *antenna switching* bifurcation. For non-switching point, $\Delta_i > 0$ at $b_1^{(i)} \neq b_2^{(i)}$ $(i = 1, 2, \ldots, n)$. At the bifurcation point, $\Delta_i = 0$ at $b_1^{(i)} = b_2^{(i)}$ $(i = 1, 2, \ldots, n)$. The m-mLSN *antenna switching* bifurcation is

$$
m\text{-mLSN}\left\{\begin{array}{l}
\text{mLS}\xrightarrow[\text{switching bifurcation}]{m^{\text{th}}\text{bifurcation}}\left\{\begin{array}{l}\text{mSI-oSO}\downarrow\text{mSO, for}\ b_2^{(m)}=a_{2m}\downarrow a_{2m-1},\\ \text{mSO}\uparrow\text{mSI-oSO, for}\ b_1^{(m)}=a_{2m-1}\uparrow a_{2m};\end{array}\right.\\
\vdots\\
\text{mLS}\xrightarrow[\text{switching bifurcation}]{i^{\text{th}}\text{bifurcation}}\left\{\begin{array}{l}\text{mSI-oSO}\downarrow\text{mSO, for}\ b_2^{(i)}=a_{2i}\downarrow a_{2i-1},\\ \text{mSO}\uparrow\text{mSI-oSO, for}\ b_1^{(i)}=a_{2i-1}\uparrow a_{2i};\end{array}\right.\\
\vdots\\
\text{mLS}\xrightarrow[\text{switching bifurcation}]{1^{\text{st}}\text{bifurcation}}\left\{\begin{array}{l}\text{mSI-oSO}\downarrow\text{mSO, for}\ b_2^{(1)}=a_2\downarrow a_1,\\ \text{mSO}\uparrow\text{mSI-oSO, for}\ b_1^{(1)}=a_1\uparrow a_2.\end{array}\right.
\end{array}\right.
\tag{4.108}
$$

Consider a switching bifurcation for a bundle of fixed-points with monotonic sink to oscillatory source and monotonic source with the following conditions,

$$
\begin{aligned}
&B_i = -(b_1^{(i)} + b_2^{(i)}),\ \Delta_i = (b_1^{(i)} - b_2^{(i)})^2 \geq 0, \\
&x_{k;1,2}^{(i)} = b_{1,2}^{(i)}, \Delta_i > 0 \text{ if } b_1^{(i)} \neq b_2^{(i)} (i = 1, 2, \ldots, n); \\
&\left.\begin{array}{l} B_i = B_j\ (i, j \in \{1, 2, \ldots, n\}; i \neq j) \\ \Delta_i = 0 \text{ at } b_1^{(i)} = b_2^{(i)}\ (i = 1, 2, \ldots, n) \end{array}\right\} \text{at bifurcation.}
\end{aligned}
\tag{4.109}
$$

Thus, the $(2l)^{\text{th}}$ order *switching* bifurcation can be for a bundle of simple monotonic sinks to oscillatory sources, and monotonic sources. Two $(2l)^{\text{th}}$ order monotonic upper- and lower-saddle switching bifurcations for $l \in \{1, 2, \ldots, s\}$ are

$$
(2l)^{\text{th}}\text{order mUS} \xrightarrow[\text{switching bifurcation}]{\text{a bundle of } (2l)\text{-fixed points}} \begin{cases} \text{mSO, for } a_{2s_l} \to b_{2s_l}, \\ \text{mSI-oSO, for } a_{2s_l-1} \to b_{2s_l-1}, \\ \vdots \\ \text{mSO, for } a_{2s_1} \to b_{2s_1}, \\ \text{mSI-oSO, for } a_{2s_1-1} \to b_{2s_1-1}. \end{cases} \tag{4.110}
$$

$$
(2l)^{\text{th}}\text{order mLS} \xrightarrow[\text{switching bifurcation}]{\text{a bundle of } (2l)\text{-fixed points}} \begin{cases} \text{mSI-oSO, for } a_{2s_l} \to b_{2s_l}, \\ \text{mSO, for } a_{2s_l-1} \to b_{2s_l-1}, \\ \vdots \\ \text{mSI-oSO, for } a_{2s_1} \to b_{2s_1}, \\ \text{mSO, for } a_{2s_1-1} \to b_{2s_1-1}. \end{cases} \tag{4.111}
$$

where $\Delta_{ij} = (a_i - a_j)^2 = (b_i - b_j)^2 = 0$ with $B_i = B_j(i, j = 2s_1 - 1, 2s_1, \ldots, 2s_l - 1, 2s_l)$ and

$$
\begin{aligned}
&\{a_{2s_1-1}, a_{2s_1}, \ldots, a_{2s_l-1}, a_{2s_l}\} \underset{\text{before bifurcation}}{\subset} \text{sort}\{b_1^{(1)}, b_2^{(1)}, \ldots b_1^{(n)}, b_2^{(n)}\}, \\
&\{b_{2s_1-1}, b_{2s_1}, \ldots, b_{2s_l-1}, b_{2s_l}\} \underset{\text{after bifurcation}}{\subset} \text{sort}\{b_1^{(1)}, b_2^{(1)}, \ldots b_1^{(n)}, b_2^{(n)}\}.
\end{aligned}
\tag{4.112}
$$

The $(2l-1)^{\text{th}}$ order *switching* bifurcation can be for a bundle of simple fixed-points with monotonic-sinks to oscillatory-sources and monotonic-sources. Two $(2l-1)^{\text{th}}$ order monotonic sink and monotonic source switching bifurcations for $l \in \{1, 2, \ldots, s\}$ are

$$
(2l-1)^{\text{th}}\text{order mSO} \xrightarrow[\text{switching bifurcation}]{\text{abundle of } (2l-1)\text{-fixed points}} \begin{cases} \text{mSO, for } a_{2s_l-1} \to b_{2s_l-1}, \\ \vdots \\ \text{mSI-oSO, for } a_{2s_1} \to b_{2s_1}, \\ \text{mSO, for } a_{2s_1-1} \to b_{2s_1-1}. \end{cases} \tag{4.113}
$$

$$
(2l-1)^{\text{th}}\text{order mSI}\xrightarrow[\text{switching bifurcation}]{\text{a bundle of }(2l-1)\text{-fixed points}}\begin{cases}\text{mSI-oSO, for } a_{2s_l-1}\to b_{2s_l-1},\\ \vdots\\ \text{mSO, for } a_{2s_1}\to b_{2s_1},\\ \text{mSI-oSO, for } a_{2s_1-1}\to b_{2s_1-1}.\end{cases}
\tag{4.114}
$$

where $\Delta_{ij}=(a_i-a_j)^2=(b_i-b_j)^2=0$ with $B_i=B_j(i,j=2s_1-1,2s_1,\ldots,2s_l-1)$ and

$$
\begin{aligned}
&\{a_{2s_1-1},a_{2s_1},\ldots,a_{2s_l-1}\}\underset{\text{before bifurcation}}{\subset}\text{sort}\{b_1^{(1)},b_2^{(1)},\ldots b_1^{(n)},b_2^{(n)}\},\\
&\{b_{2s_1-1},b_{2s_1},\ldots,b_{2s_l-1}\}\underset{\text{after bifurcation}}{\subset}\text{sort}\{b_1^{(1)},b_2^{(1)},\ldots b_1^{(n)},b_2^{(n)}\}.
\end{aligned}
\tag{4.115}
$$

A set of paralleled, different, higher-order upper-saddle-node switching bifurcations with multiplicity is the $((\alpha_1)^{\text{th}}\text{mXX}:(\alpha_2)^{\text{th}}\text{mXX}:\cdots:(\alpha_s)^{\text{th}}\text{mXX})$ *parallel switching* bifurcation in the $(2m)^{\text{th}}$ degree polynomial discrete system. At the *straw-bundle switching* bifurcation, $\Delta_i=0(i=1,2,\ldots,n)$ and $B_i=B_j(i,j\in\{1,2,\ldots,n\};\ i\neq j)$. Thus, the *parallel straw-bundle switching* bifurcation is

$$
\begin{aligned}
&((\alpha_1)^{\text{th}}\text{mXX}:(\alpha_2)^{\text{th}}\text{mXX}:\cdots:(\alpha_s)^{\text{th}}\text{mXX})\text{-switching}\\
&=\begin{cases}(\alpha_s)^{\text{th}}\text{order mXX switching},\\ \vdots\\ (\alpha_2)^{\text{th}}\text{order mXX switching},\\ (\alpha_1)^{\text{th}}\text{order mXX switching};\end{cases}
\end{aligned}
\tag{4.116}
$$

where

$$
\begin{aligned}
&\alpha_i\in\{2l_i,2l_i-1\}\text{ with}\textstyle\sum_{i=1}^{s}\alpha_i=2m,\\
&\text{and XX}\in\{\text{US},\text{LS},\text{SO},\text{SI}\}.
\end{aligned}
\tag{4.117}
$$

The $(2l_i)^{\text{th}}$mUS for $(i=1,2,\ldots,s)$ with monotonic-sinks to oscillatory sources, and monotonic-sources is the $(2l_i)^{\text{th}}$ order monotonic-upper-saddle for a switching of l_i-pairs of simple monotonic-sinks to oscillatory-sources, and monotonic-sources. With different orders of l_i-pairs of simple monotonic-sinks to oscillatory-sources, and monotonic-sources, the $(2l_i)^{\text{th}}$mUSN switching bifurcation possesses different *straw-bundle switching* for a bundle of stable and unstable fixed-points. The $(2l_1:$

$2l_2 : \cdots : 2l_s)^{\text{th}}$ mUSN bifurcation is called the $(2l_1 : 2l_2 : \cdots : 2l_s)^{\text{th}}$ mUSN *straw-bundle switching* bifurcation.

$$
(2l_1 : 2l_2 : \cdots : 2l_s)^{\text{th}}\text{mUSN switching} = \begin{cases} (2l_s)^{\text{th}}\text{order mUSN switching}, \\ \vdots \\ (2l_2)^{\text{th}}\text{order mUSN switching}, \\ (2l_1)^{\text{th}}\text{order mUSN switching}. \end{cases} \tag{4.118}
$$

If $l_i = 1$ for $i = 1, 2, \ldots, m$ with $n = m$, the simple upper-saddle-node parallel switching bifurcation or the upper-saddle-node *antenna switching* bifurcation is recovered.

Similarly, a set of paralleled different monotonic lower-saddle switching bifurcations with multiplicity is a $((2l_1)^{\text{th}}\text{mLS}{:}(2l_2)^{\text{th}}\text{mLS}{:} \cdots : (2l_s)^{\text{th}}\text{mLS})$ parallel switching bifurcation in the $(2m)^{\text{th}}$ degree polynomial discrete system. Thus, the $(2l_1 : 2l_2 : \cdots : 2l_s)^{\text{th}}$mLSN switching bifurcation is also called a $(2l_1 : 2l_2 : \cdots : 2l_s)^{\text{th}}$mLSN *straw-bundle* switching bifurcation. Again, at the mLSN *straw-bundle switching* bifurcation, $\Delta_i = 0 (i = 1, 2, \ldots, n)$ and $B_i = B_j (i, j \in \{1, 2, \ldots, n\}; i \neq j)$. Thus, the mLSN *straw-bundle* switching bifurcation is

$$
(2l_1 : 2l_2 : \cdots : 2l_s)^{\text{th}}\text{mLSN switching} = \begin{cases} (2l_s)^{\text{th}}\text{order mLSN switching}, \\ \vdots \\ (2l_2)^{\text{th}}\text{order mLSN switching}, \\ (2l_1)^{\text{th}}\text{order mLSN switching}. \end{cases} \tag{4.119}
$$

The set of m-monotonic-upper-saddle-node (m-mUSN) *parallel switching* bifurcation is equivalent to the set of $(2 : 2 : \cdots : 2)^{\text{nd}}$-mUSN bifurcations. The set of m-lower-saddle-node (m-mLSN) *parallel switching* bifurcation is equivalent to the set of $(2 : 2 : \cdots : 2)^{\text{nd}}$-mLSN bifurcations. Such two sets of parallel switching bifurcations are presented in Fig. 4.4(i) and (ii) for $a_0 > 0$ and $a_0 < 0$, respectively. A set of paralleled $(3^{\text{rd}}\text{mSO}{:}2^{\text{nd}}\text{mLS}{:} \cdots : 4^{\text{th}}\text{mLS}{:} \cdots : 3^{\text{rd}}\text{mSI})$ *switching* bifurcations for mSI-oSO and mSO fixed-points is presented in Fig. 4.4(iii) for $a_0 > 0$. However, for $a_0 < 0$, the set of $(3^{\text{rd}}\text{mSI}{:}2^{\text{nd}}\text{mUS}{:} \cdots : 4^{\text{th}}\text{mUS}{:} \cdots : 3^{\text{rd}}\text{mSI})$ *switching* bifurcations for monotonic-sources and monotonic-sink-to-oscillatory-sources is presented in Fig. 4.4(iv).

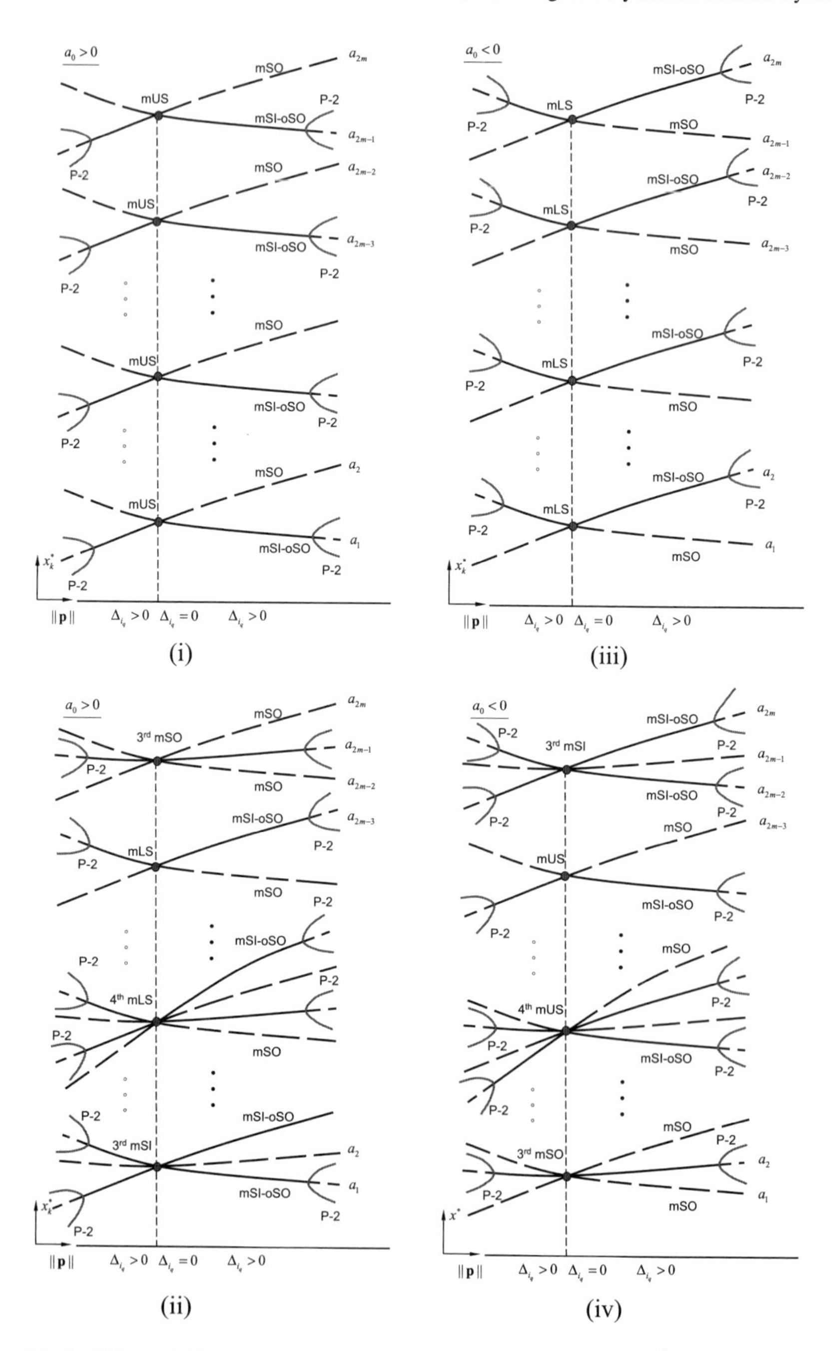

Fig. 4.4 Stability and bifurcations of fixed-points in a 1-dimensional, $(2m)^{\text{th}}$-degree polynomial discrete system: (i) m-mUSN parallel switching bifurcation ($a_0 > 0$), (ii) m-mLSN parallel switching bifurcation ($a_0 < 0$), (iii) (3^{rd}mSO:2^{nd}mLS: $\cdots$: 3^{rd}mSI) parallel switching bifurcation ($a_0 > 0$). (iv) (3^{rd}mSI: 2^{nd}mUS: $\cdots$: 3^{rd}mSO) parallel switching bifurcation. mLS: monotonic-lower-saddle, mUS: monotonic-upper-saddle, mSI-oSO: monotonic sink to oscillatory source, mSO: monotonic source. Stable and unstable fixed-points are represented by solid and dashed curves, respectively. The bifurcation points are marked by circular symbols.

4.2.3 *Switching-Appearing Bifurcations*

Consider a $(2m)^{\text{th}}$ degree 1-dimensional polynomial discrete system in a form of

$$x_{k+1} = x_k + a_0 Q(x_k) \prod_{i=1}^{2n_1} (x_k - c_i) \prod_{j=1}^{n_2} (x_k^2 + B_j x_k + C_j). \tag{4.120}$$

Without loss of generality, a function of $Q(x_k) > 0$ is either a polynomial function or a non-polynomial function. The roots of $x_k^2 + B_j x_k + C_j = 0$ are

$$b_{1,2}^{(j)} = -\frac{1}{2}B_j \pm \frac{1}{2}\sqrt{\Delta_j}, \Delta_j = B_j^2 - 4C_j \geq 0 (j = 1, 2, \ldots, n_2); \tag{4.121}$$

either

$$\begin{aligned}
&\{a_1^-, a_2^-, \ldots, a_{2n_1}^-\} = \text{sort}\{c_1, c_2 \ldots, c_{2n_1}\}, a_s^- \leq a_{s+1}^- \text{ before bifurcation} \\
&\{a_1^+, a_2^+, \ldots, a_{2n_3}^+\} = \text{sort}\{c_1, \ldots, c_{2n_1}; b_1^{(1)}, b_2^{(1)}, \ldots, b_1^{(n_2)}, b_2^{(n_2)}\}, \\
&a_s^+ \leq a_{s+1}^+, \ n_3 = n_1 + n_2 \text{ after bifurcation;}
\end{aligned} \tag{4.122}$$

or

$$\begin{aligned}
&\{a_1^-, a_2^-, \ldots, a_{2n_3}^-\} = \text{sort}\{c_1, c_2 \ldots, c_{2n_1}; b_1^{(1)}, b_2^{(1)}, \ldots, b_1^{(n_2)}, b_2^{(n_2)}\}, \\
&a_s^- \leq a_{s+1}^-, n_3 = n_1 + n_2 \text{ before bifurcation;} \\
&\{a_1^+, a_2^+, \ldots, a_{2n_1}^+\} = \text{sort}\{c_1, \ldots, c_{2n_1}\}, \ a_s^+ \leq a_{s+1}^+ \text{ after bifurcation;}
\end{aligned} \tag{4.123}$$

and

$$\left.\begin{aligned}
&B_{j_1} = B_{j_2} = \cdots = B_{j_s} \ (j_{k_1} \in \{1, 2, \ldots, n\}; j_{k_1} \neq j_{k_2}) \\
&(k_1, k_2 \in \{1, 2, \ldots, s\}; k_1 \neq k_2) \\
&\Delta_j = 0 \ (j \in U \subset \{1, 2, \ldots, n_2\} \\
&c_i \neq -\tfrac{1}{2}B_j \ (i = 1, 2, \ldots, 2n_1, j = 1, 2, \ldots, n_2)
\end{aligned}\right\} \text{at bifurcation.} \tag{4.124}$$

Consider a just before bifurcation of $((\alpha_1^-)^{\text{th}}\text{mXX}_1^- : (\alpha_2^-)^{\text{th}}\text{mXX}_2^- : \ldots : (\alpha_{s_1}^-)^{\text{th}}\text{mXX}_{s_1}^-)$ for simple sources and sinks. For $\alpha_i^- = 2l_i^- - 1$, $\text{mXX}_i^- \in \{\text{mSO}, \text{mSI}\}$ and for $\alpha_i^- = 2l_i^-$, $\text{mXX}_i^- \in \{\text{mUS, mLS}\} (i = 1, 2, \ldots, s_1)$. The detailed structures are as follows.

$$
\left.\begin{array}{l}
\text{mSI-oSO} \\
\text{mSO} \\
\vdots \\
\text{mSO} \\
\text{mSI-oSO}
\end{array}\right\} \to (2l_i^- - 1)^{\text{th}}\text{mSI, and }
\left.\begin{array}{l}
\text{mSO} \\
\text{mSI-oSO} \\
\vdots \\
\text{mSI-oSO} \\
\text{mSO}
\end{array}\right\} \to (2l_i^- - 1)^{\text{th}}\text{mSO;}
$$

$$
\left.\begin{array}{l}
\text{mSO} \\
\text{mSI-oSO} \\
\vdots \\
\text{mSO} \\
\text{mSI-oSO}
\end{array}\right\} \to (2l_i^-)^{\text{th}}\text{mUS,} \quad \text{and }
\left.\begin{array}{l}
\text{mSI-oSO} \\
\text{mSO} \\
\vdots \\
\text{mSI-oSO} \\
\text{mSO}
\end{array}\right\} \to (2l_i^-)^{\text{th}}\text{mLS.} \tag{4.125}
$$

The bifurcation set of $((\alpha_1^-)^{\text{th}}\text{mXX}_1^-,:(\alpha_2^-)^{\text{th}}\text{mXX}_2^- : \ldots : (\alpha_{s_1}^-)^{\text{th}}\text{mXX}_{s_1}^-)$ at the same parameter point is called a *left-parallel-bundle* switching bifurcation

Consider a just after bifurcation of $((\alpha_1^+)^{\text{th}}\text{mXX}_1^+:(\alpha_2^+)^{\text{th}}\text{mXX}_2^+:\cdots:(\alpha_{s_2}^+)^{\text{th}}\text{mXX}_{s_2}^+)$ for simple fixed-points with monotonic sources and monotonic sinks to oscillatory sources. For $\alpha_i^+ = 2l_i^+ - 1, \text{mXX}_i^+ \in \{\text{mSO}, \text{mSI}\}$ and for $\alpha_i^+ = 2l_i^+$, $\text{mXX}_i^- \in \{\text{mUS, mLS}\}$. The four detailed structures are as follows.

$$
(2l_i^+ - 1)^{\text{th}}\text{mSI} \to \left\{\begin{array}{l}
\text{mSI-oSO} \\
\text{mSO} \\
\vdots \\
\text{mSO} \\
\text{mSI-oSO}
\end{array}\right., \text{ and } (2l_i^+ - 1)^{\text{th}}\text{mSO} \to \left\{\begin{array}{l}
\text{mSO} \\
\text{mSI-oSO} \\
\vdots \\
\text{mSI-oSO} \\
\text{mSO}
\end{array}\right.;
$$

$$
(2l_i^+)^{\text{th}}\text{mUS} \to \left\{\begin{array}{l}
\text{mSO} \\
\text{mSI-oSO} \\
\vdots \\
\text{mSO} \\
\text{mSI-oSO}
\end{array}\right., \quad \text{and} \quad (2l_i^+)^{\text{th}}\text{mLS} \to \left\{\begin{array}{l}
\text{mSI-oSO} \\
\text{mSO} \\
\vdots \\
\text{mSI-oSO} \\
\text{mSO}
\end{array}\right.. \tag{4.126}
$$

The bifurcation set of $((\alpha_1^+)^{\text{th}}\text{mXX}_1^+ : (\alpha_2^+)^{\text{th}}\text{mXX}_2^+ : \ldots : (\alpha_{s_2}^+)^{\text{th}}\text{mXX}_{s_2}^+)$ at the same parameter point is called a *right-parallel-bundle* switching bifurcation

(i) For the just before and after bifurcation structure, if there exists a relation of

$$\begin{aligned}&(\alpha_i^-)^{\text{th}}\text{mXX}_i^- = (\alpha_j^+)^{\text{th}}\text{mXX}_j^+ = \alpha^{\text{th}}\text{mXX, for } x_k^* = a_i^- = a_j^+ \\ &(i \in \{1,2,\ldots,s_1\}, j \in \{1,2,\ldots,s_2\}),\ \text{XX} \in \{\text{US, LS, SO, SI}\}\end{aligned} \tag{4.127}$$

then the bifurcation is a α^{th}mXX *switching* bifurcation for simple fixed-points.

(ii) Just for the just before bifurcation structure, if there exists a relation of

$$\begin{aligned}&(2l_i^-)^{\text{th}}\text{mXX}_i^- = (2l)^{\text{th}}\text{mXX, for } x_k^* = a_i^- = a_i \\ &(i \in \{1,2,\ldots,s_1\}, \text{XX} \in \{\text{US, LS}\}\end{aligned} \tag{4.128}$$

then, the bifurcation is a $(2l)^{\text{th}}$mXX *left appearing* (or *right vanishing*) bifurcation for simple fixed-points.

(iii) Just for the just after bifurcation structure, if there exists a relation of

$$\begin{aligned}&(2l_i^+)^{\text{th}}\text{mXX}_i^+ = (2l)^{\text{th}}\text{mXX, for } x_k^* = a_i^+ = a_i \\ &(i \in \{1,2,\ldots,s_1\}), \text{XX} \in \{\text{US, LS}\}\end{aligned} \tag{4.129}$$

then, the bifurcation is a $(2l)^{\text{th}}$mXX *right appearing* (or *left vanishing*) bifurcation for simple fixed-points.

(iv) For the just before and after bifurcation structure, if there exists a relation of

$$\begin{aligned}&(\alpha_i^-)^{\text{th}}\text{mXX}_i^- \neq (\alpha_j^+)^{\text{th}}\text{mXX}_j^+ \text{ for } x_k^* = a_i^- = a_j^+ \\ &\text{XX}_i^-,\ \text{XX}_j^+ \in \{\text{US, LS, SO, SI}\} \\ &(i \in \{1,2,\ldots,s_1\}, j \in \{1,2,\ldots,s_2\}),\end{aligned} \tag{4.130}$$

then, there are two *flower-bundle* switching bifurcations of simple fixed-points:

(iv$_1$) for $\alpha_j = \alpha_i + 2l$, the bifurcation is called a α_j^{th}mXX right *flower-bundle* switching bifurcation for α_i to α_j-simple fixed-points with the appearance (birth) of $2l$-simple fixed-points.

(iv$_2$) for $\alpha_j = \alpha_i - 2l$, the bifurcation is called a α_i^{th}mXX left *flower-bundle* switching bifurcation for α_i to α_j-simple fixed-points with the vanishing (death) of $2l$-simple fixed-points.

A general parallel switching bifurcation is

$$\begin{aligned}&((\alpha_1^-)^{\text{th}}\text{mXX}_1^- : (\alpha_2^-)^{\text{th}}\text{mXX}_2^- : \cdots : (\alpha_{s_1}^-)^{\text{th}}\text{mXX}_{s_1}^-) \xrightarrow[\text{bifurcation}]{\text{switching}} \\ &((\alpha_1^+)^{\text{th}}\text{mXX}_1^+ : (\alpha_2^+)^{\text{th}}\text{mXX}_2^+ : \cdots : (\alpha_{s_2}^+)^{\text{th}}\text{mXX}_{s_2}^+).\end{aligned} \tag{4.131}$$

Such a general, parallel switching bifurcation consists of the *left* and *right* parallel-bundle switching bifurcations.

If the *left* and *right* parallel-bundle switching bifurcations are same in a parallel *flower-bundle* switching bifurcation, i.e.,

$$\begin{aligned}&(\alpha_i^-)^{\text{th}}\text{mXX}_i^- = (\alpha_i^+)^{\text{th}}\text{mXX}_i^+ = \alpha^{\text{th}}\text{mXX},\\&\text{for } x_k^* = a_i^- = a_i^+ \ (i = 1, 2, \ldots, s\},\end{aligned} \tag{4.132}$$

then the parallel *flower-bundle* switching bifurcation becomes a parallel *straw-bundle switching* bifurcation of $((\alpha_1)^{\text{th}}\text{mXX}{:}(\alpha_2)^{\text{th}}\text{mXX}{:}\cdots{:}\ (\alpha_s)^{\text{th}}\ \text{mXX})$.

If the *left* and *right* parallel-bundle switching bifurcations are different in a parallel *flower-bundle* switching bifurcation, i.e.,

$$\begin{aligned}&(\alpha_i^-)^{\text{th}}\text{mXX}_i^- = (2l_i^-)^{\text{th}}\text{mXX},\ (\alpha_j^+)^{\text{th}}\text{mXX}_j^+ = (2l_j^+)^{\text{th}}\text{mYY},\\&\text{for } x_k^* = a_i^- \neq a_i^+ \ (i = 1, 2, \ldots, s\}\\&\text{mXX} \in \{\text{mUS}, \text{mLS}\}, \text{mYY} \in \{\text{mUS}, \text{mLS}\},\end{aligned} \tag{4.133}$$

then the parallel *flower-bundle* switching bifurcation becomes a combination of two independent left and right parallel appearing bifurcations:

(i) a $((2l_1^-)^{\text{th}}\text{mXX}_1^- : (2l_2^-)^{\text{th}}\text{mXX}_2^- : \cdots : (2l_{s_1}^-)^{\text{th}}\text{mXX}_{s_1}^-)$*-left parallel sprinkler-spraying appearing (or right vanishing)* bifurcation and
(ii) a $((2l_1^+)^{\text{th}}\text{mXX}_1^+ : (2l_2^+)^{\text{th}}\text{mXX}_2^+ : \cdots : (2l_{s_2}^+)^{\text{th}}\text{mXX}_{s_2}^+)$*-right parallel sprinkler-spraying appearing (or left vanishing)* bifurcation.

The $(6^{\text{th}}\text{mUS}{:}4^{\text{th}}\text{mLS}{:}\cdots{:}\ 4^{\text{th}}\text{mUS}{:}\text{mSI-oSO})$ *appearing* bifurcation for $a_0 > 0$ is presented in Fig. 4.5(i). Compared to the case of $a_0 > 0$, the bifurcation and stability conditions of fixed-points for $a_0 > 0$ will be swapped. The $(6^{\text{th}}\text{mLS}{:}\ 4^{\text{th}}\text{mUS}{:}\cdots{:}\ 4^{\text{th}}\text{mLS}{:}\text{mSO})$ *parallel appearing* bifurcation is shown in Fig. 4.5(ii).

Such a kind of bifurcation is like a *waterfall appearing* bifurcation. The *switching* and *appearing* bifurcations of fixed-points exist at the same parameter. A set of paralleled, different *switching and appearing* bifurcations of higher-order fixed-points is also named an $(l_1^{\text{th}}\text{mXX}{:}l_2^{\text{th}}\text{mXX}{:}\cdots{:}\ l_s^{\text{th}}\text{mXX})$ parallel *switching and appearing* bifurcation in the $(2m)^{\text{th}}$ degree polynomial discrete system. The $l_i^{\text{th}}\text{mXX}$ *switching* and *appearing* bifurcation possesses different *clusters* of stable and unstable fixed-points *before* and *after* the bifurcation.

The set of $(5^{\text{th}}\text{mSI}{:}\cdots{:}\ \text{mSO}{:}6^{\text{th}}\text{mUS})$ *flower-bundle switching* bifurcation for mSI-oSO and mSO fixed-points is presented in Fig. 4.5(iii) for $a_0 > 0$. Such a *flower-bundle* switching bifurcation is from $(m\text{SI-oSO}{:}\ mSO : mSI - oSO : mSO)$ to $(5^{\text{th}}\text{mSI}{:}\cdots{:}\ \text{mSO}{:}6^{\text{th}}\text{mUS})$ with a *waterfall appearing*. The set of $(5^{\text{th}}\text{mSO}{:}\cdots{:}\ \text{mSI-oSO}{:}6^{\text{th}}\text{mLS})$ *flower-bundle switching* bifurcation for mSI-oSO and mSO

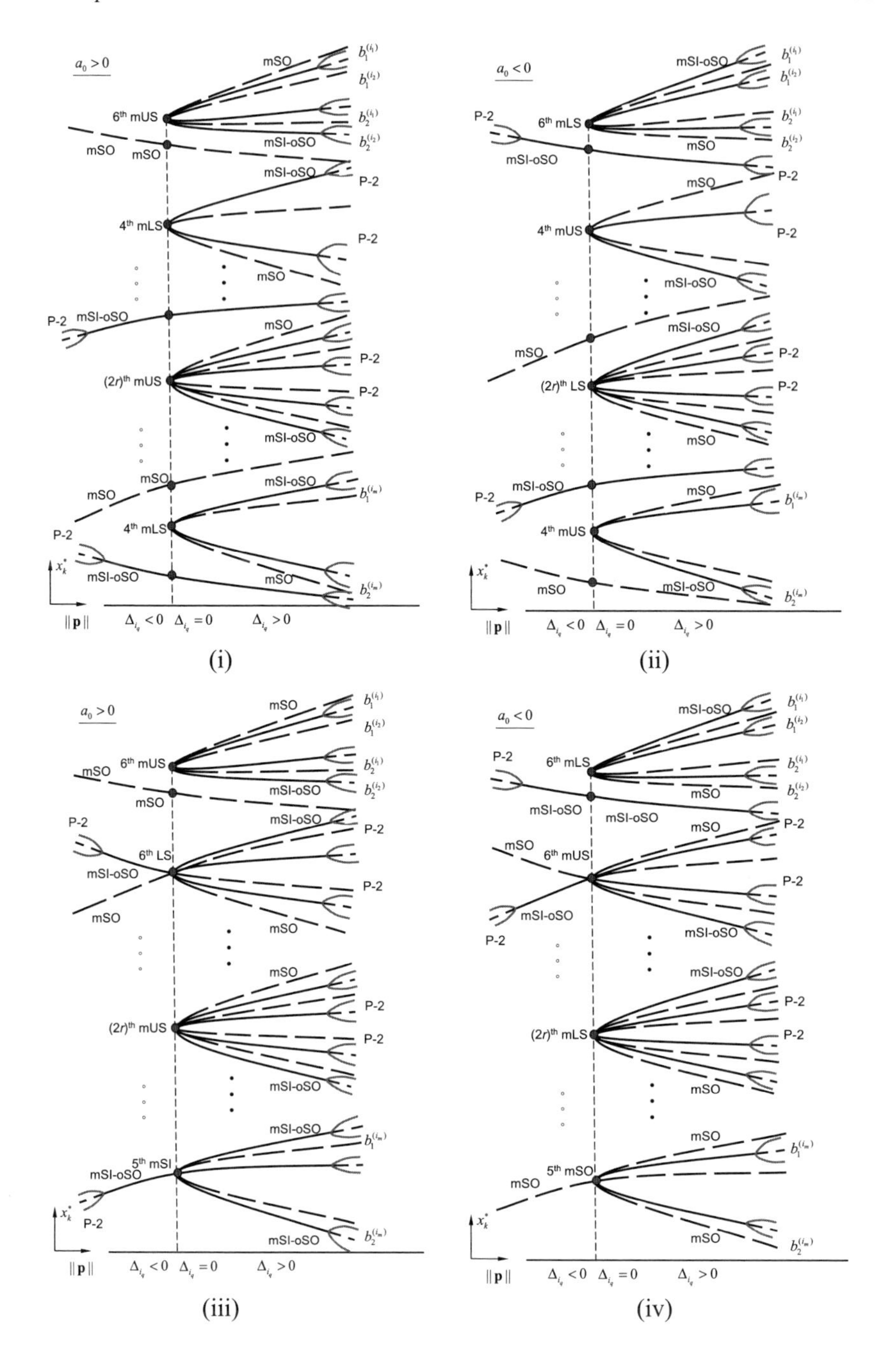

Fig. 4.5 Stability and bifurcation. (i)(6^{th}mUS:mSO:4^{th}mLS: $\cdots$: mSI-oSO) appearing bifurcation ($a_0 > 0$). (ii) (6^{th}mLS:mSI-oSO:4^{th}mUS: $\cdots$: mSO) appearing bifurcation ($a_0 < 0$). (iii) (6^{th}mUS:6^{th}mLS: $\cdots$: 5^{th}mSI) switching-appearing bifurcation ($a_0 > 0$). (iv) (6^{th}mLS:mSI-*oSO* : 6^{th}US: $\cdots$: 5^{th}mSO) switching-appearing bifurcation in a $(2m)^{th}$-degree polynomial discrete system. mLS: monotonic-lower-saddle, mUS: monotonic-upper-saddle, mSI-oSO: monotonic sink to oscillatory source, mSO: monotonic source. Stable and unstable fixed-points are represented by solid and dashed curves, respectively. The bifurcation points are marked by circular symbols.

fixed-points is presented in Fig. 4.5(iv) for $a_0 < 0$. Such a *flower-bundle* switching bifurcation is from (mSO:mSI-oSO:mSO:mSI-oSO) to (5^{th}mSI : $\cdots$: mSO:6th mUS) with a *waterfall appearing*. After the bifurcation, the *waterfall* fixed-points birth can be observed. The fixed-points before such a bifurcation are much less than after the bifurcation.

4.3 Higher-Order Fixed-Point Bifurcations

The afore-discussed appearing and switching bifurcations in the $(2m)^{\text{th}}$ degree polynomial system are relative to simple monotonic sources and monotonic sinks. As in Luo (2020a), the higher-order singularity bifurcations in the $(2m)^{\text{th}}$ degree polynomial discrete system can be for higher-order fixed-points (i.e., monotonic sinks, monotonic sources, monotonic upper-saddles, monotonic lower-saddles).

4.3.1 Appearing Bifurcations

Consider a $(2m)^{\text{th}}$ degree polynomial discrete system as

$$x_{k+1} = x_k + a_0 Q(x_k) \prod_{i=1}^{s} (x_k^2 + B_i x_k + C_i)^{\alpha_i}, \tag{4.134}$$

where $\alpha_i \in \{2l-1, 2l\}$. Without loss of generality, a function of $Q(x_k) > 0$ is either a polynomial function or a non-polynomial function. The roots of $x_k^2 + B_i x_k + C_i = 0$ are

$$\begin{aligned} &b_{1,2}^{(i)} = -\tfrac{1}{2}B_i \pm \tfrac{1}{2}\sqrt{\Delta_i}, \Delta_i = B_i^2 - 4C_i \geq 0; \\ &\{a_1, a_2, \ldots, a_{2s-1}, a_{2s}\} = \text{sort}\{b_1^{(1)}, b_2^{(1)}, \ldots, b_1^{(s)}, b_2^{(s)}\}, a_j \leq a_{j+1}. \end{aligned} \tag{4.135}$$

There are four higher-order bifurcations as follows:

$$\begin{aligned} &(2(2l_i-1))^{\text{th}}\text{order mUS} \xrightarrow[\text{appearing bifurcation}]{(2l_i-1)^{\text{th}}\text{order quadratics}} \\ &\qquad \begin{cases} (2l_i-1)^{\text{th}}\text{order mSO}, \ x_k^* = a_{2i}, \\ (2l_i-1)^{\text{th}}\text{order mSI}, \ x_k^* = a_{2i-1}; \end{cases} \end{aligned} \tag{4.136}$$

$$
\begin{aligned}
&(2(2l_i-1))^{\text{th}}\text{ order mLS}\xrightarrow[\text{appearing bifurcation}]{(2l_i-1)^{\text{th}}\text{order quadratics}}\\
&\qquad\begin{cases}(2l_i-1)^{\text{th}}\text{ order mSI}, x_k^*=a_{2i},\\(2l_i-1)^{\text{th}}\text{ order mSO}, x_k^*=a_{2i-1};\end{cases}
\end{aligned}
\tag{4.137}
$$

$$
\begin{aligned}
&(2(2l_i))^{\text{th}}\text{order mUS}\xrightarrow[\text{switching bifurcation}]{(2l_i)^{\text{th}}\text{ order quadratics}}\\
&\qquad\begin{cases}(2l_i)^{\text{th}}\text{order mUS}, x_k^*=a_{2i},\\(2l_i)^{\text{th}}\text{order mUS}, x_k^*=a_{2i-1};\end{cases}
\end{aligned}
\tag{4.138}
$$

$$
\begin{aligned}
&(2(2l_i))^{\text{th}}\text{order mLS}\xrightarrow[\text{switching bifurcation}]{(2l_i)^{\text{th}}\text{order quadratics}}\\
&\qquad\begin{cases}(2l_i)^{\text{th}}\text{order mLS}, x_k^*=a_{2i},\\(2l_i)^{\text{th}}\text{order mLS}, x_k^*=a_{2i-1}.\end{cases}
\end{aligned}
\tag{4.139}
$$

(i) For $\alpha_i = 2l_i - 1$, the $(2(2l_i-1))^{\text{th}}$ order monotonic upper-saddle (mUS) *appearing* bifurcation is for the onset of the $(2l_i-1)^{\text{th}}$ order monotonic source (mSO) $(x^* = a_{2i})$ and the $(2l-1)^{\text{th}}$ order monotonic sink (mSI) $(x^* = a_{2i-1})$ with $a_{2i} > a_{2i-1}$ for $a_0 > 0$.

(ii) For $\alpha_i = 2l_i - 1$, the $(2(2l_i-1))^{\text{th}}$ order monotonic lower-saddle (mLS) *appearing* bifurcation is for the onset of the $(2l_i-1)^{\text{th}}$ order monotonic sink (mSI) $(x^* = a_{2i})$ and the $(2l_i-1)^{\text{th}}$ order monotonic source (mSO) $(x^* = a_{2i-1})$ with $a_{2i} > a_{2i-1}$ for $a_0 < 0$.

(iii) For $\alpha_i = 2l_i$, the $(2(2l_i))^{\text{th}}$ order monotonic upper-saddle (mUS) *appearing* bifurcation is for the onset of two $(2l_i)^{\text{th}}$ order monotonic upper-saddles (mUS) $(x^* = a_{2i-1}, a_{2i})$ with $a_{2i} \neq a_{2i-1}$ for $a_0 > 0$.

(iv) For $\alpha_i = 2l_i$, the $(2(2l_i))^{\text{th}}$ order monotonic lower-saddle (mLS) *appearing* bifurcation is for the onset of two $(2l_i)^{\text{th}}$ order monotonic lower-saddles (mLS) $(x^* = a_{2i-1}, a_{2i})$ with $a_{2i} \neq a_{2i-1}$ for $a_0 < 0$.

From the higher-order singular bifurcation conditions, in a $(2m)^{\text{th}}$ degree polynomial discrete system, the higher-order saddle-node bifurcations for appearing and switching of the higher-order fixed-points are discussed herein. A set of paralleled different higher order monotonic upper-saddle *appearing* bifurcations in the $(2m)^{\text{th}}$ degree polynomial nonlinear discrete system is called a $((2\alpha_1)^{\text{th}}\text{mUS}:(2\alpha_2)^{\text{th}}\text{mUS}:\cdots:(2\alpha_s)^{\text{th}}\text{mUS})$ *parallel appearing* bifurcation for $a_0 > 0$.

Define

$$((2\alpha_1)^{\text{th}}\text{mUS}:(2\alpha_2)^{\text{th}}\text{mUS}:\cdots:(2\alpha_s)^{\text{th}}\text{mUS}) = (2\alpha_1:2\alpha_2:\cdots:2\alpha_s)^{\text{th}}\text{mUS} \tag{4.140}$$

where $\alpha_i \in \{2l_i - 1, 2l_i\}$ for $i = 1, 2, \ldots, s$. Such an *appearing* bifurcation is called a $(2\alpha_1 : 2\alpha_2 : \cdots : 2\alpha_s)^{\text{th}}$mUS *teethcomb appearing* bifurcation.

Similarly, a set of paralleled different higher order lower-saddle *appearing* bifurcations in the $(2m)^{\text{th}}$ degree polynomial nonlinear system is called a $((2\alpha_1)^{\text{th}}\text{mLS}:(2\alpha_2)^{\text{th}}\text{mLS}:\cdots:(2\alpha_s)^{\text{th}}\text{mLS})$ *parallel appearing* bifurcation for $a_0 < 0$. Define

$$((2\alpha_1)^{\text{th}}\text{mLS}:(2\alpha_2)^{\text{th}}\text{mLS}:\cdots:(2\alpha_s)^{\text{th}}\text{mLS}) = (2\alpha_1:2\alpha_2:\cdots:2\alpha_s)^{\text{th}}\text{mLS} \tag{4.141}$$

where $\alpha_i \in \{2l_i - 1, 2l_i\}$ for $i = 1, 2, \ldots, s$. Such an appearing bifurcation is called a $(2\alpha_1 : 2\alpha_2 : \cdots : 2\alpha_s)^{\text{th}}$mLS *teethcomb appearing* bifurcation.

Consider a 1-dimensional polynomial system as

$$x_{k+1} = x_k + a_0 Q(x_k) \prod\nolimits_{i=1}^{n} (x_k^2 + B_i x_k + C_i)^{\alpha_i} \tag{4.142}$$

where $\alpha_i \in \{2r_i - 1, 2r_i\}$ $(i = 1, 2, \ldots, n)$. Without loss of generality, a function of $Q(x_k) > 0$ is either a polynomial function or a non-polynomial function. The roots of $x_k^2 + B_i x_k + C_i = 0$ are

$$\begin{aligned}
&x_{k;1,2}^{(i)} = -\frac{1}{2}B_i \pm \frac{1}{2}\sqrt{\Delta_i},\ \Delta_i = B_i^2 - 4C_i \geq 0;\\
&B_i = B_j\ (i,j = 1, 2, \ldots, n; i \neq j)\\
&\{a_1, a_2, \ldots, a_{2l}\} \in \text{sort}\{x_{k,1}^{(1)}, x_{k,2}^{(1)}, x_{k,1}^{(2)}, x_{k,2}^{(2)}, \ldots, x_{k,1}^{(r)}, x_{k,2}^{(r)}\}, a_i \leq a_{i+1}.
\end{aligned} \tag{4.143}$$

The higher-order singularity bifurcation can be for a cluster of higher-order monotonic sinks, monotonic sources, monotonic upper-saddles, and monotonic lower-saddles. There are four higher-order bifurcations as follows:

For the higher-order upper-saddle *appearing* bifurcation, the cluster of higher-order monotonic sinks, monotonic sources, monotonic upper-saddles and monotonic lower saddles is given by the following two cases:

(i) The $(2(2l-1))^{\text{th}}$order US *spraying appearing* bifurcation for a cluster of fixed points with higher-order monotonic sinks, monotonic sources, monotonic upper-saddles and monotonic-lower-saddles is

$$(2(2l-1))^{\text{th}}\text{order mUS} \xrightarrow[\text{appearing bifurcation}]{\text{a cluster of } 2n\text{-mXX}} \begin{cases} (\alpha_{2n})^{\text{th}}\text{order mXX for } x_k^* = a_{2n}, \\ (\alpha_{2n-1})^{\text{th}}\text{order mXX for } x_k^* = a_{2n-1}, \\ \vdots \\ (\alpha_1)^{\text{th}}\text{order mXX for } x_k^* = a_1; \end{cases} \tag{4.144}$$

where $2(2l-1) = \Sigma_{i=1}^{n}\alpha_i$ and the minimum and maximum fixed-points satisfy

$$\begin{aligned} (\alpha_{2n})^{\text{th}}\text{order mXX} &= \begin{cases} (2r_{2n})^{\text{th}}\text{order mUS, for } \alpha_{2n} = 2r_n, \\ (2r_{2n}-1)^{\text{th}}\text{order mSO, for } \alpha_{2n} = 2r_n - 1; \end{cases} \\ (\alpha_1)^{\text{th}}\text{order mXX} &= \begin{cases} (2r_1)^{\text{th}}\text{order mUS, for } \alpha_1 = 2r_1, \\ (2r_1-1)^{\text{th}}\text{order mSI, for } \alpha_1 = 2r_1 - 1. \end{cases} \end{aligned} \tag{4.145}$$

(ii) The $(2(2l))^{\text{th}}$order mUS *spraying-appearing* bifurcation for a cluster of fixed points with higher-order monotonic sinks, monotonic sources, monotonic upper-saddles and monotonic lower-saddles is

$$(2(2l))^{\text{th}}\text{order mUS} \xrightarrow[\text{appearing bifurcation}]{\text{a cluster of } 2n\text{-mXX}} \begin{cases} (\alpha_{2n})^{\text{th}}\text{order mXX for } x_k^* = a_{2n}, \\ (\alpha_{2n-1})^{\text{th}}\text{order mXX for } x_k^* = a_{2n-1}, \\ \vdots \\ (\alpha_1)^{\text{th}}\text{order mXX for } x_k^* = a_1; \end{cases} \tag{4.146}$$

where $2(2l) = \Sigma_{i=1}^{n}\alpha_i$ and the minimum and maximum fixed-points satisfy

$$\begin{aligned} (\alpha_{2n})^{\text{th}}\text{order mXX} &= \begin{cases} (2r_{2n})^{\text{th}}\text{order mUS, for } \alpha_{2n} = 2r_n, \\ (2r_{2n}-1)^{\text{th}}\text{order mSO, for } \alpha_{2n} = 2r_n - 1; \end{cases} \\ (\alpha_1)^{\text{th}}\text{order mXX} &= \begin{cases} (2r_1)^{\text{th}}\text{order mUS, for } \alpha_1 = 2r_1, \\ (2r_1-1)^{\text{th}}\text{order mSI, for } \alpha_1 = 2r_1 - 1. \end{cases} \end{aligned} \tag{4.147}$$

For the higher-order monotonic lower-saddle bifurcation, the cluster of the higher-order fixed-points is given by the following two cases.

(iii) The $(2(2l-1))^{\text{th}}$order mLS *spraying-appearing* bifurcation for a cluster of fixed-points with higher-order monotonic sinks, monotonic sources, monotonic upper-saddles and monotonic lower-saddles is

$$
\begin{aligned}
&(2(2l-1))^{\text{th}}\text{order mLS} \xrightarrow[\text{appearing bifurcation}]{\text{a cluster of } 2n\text{-mXX}} \\
&\qquad\qquad \begin{cases}
(\alpha_{2n})^{\text{th}}\text{order mXX, for } x_k^* = a_{2n}, \\
(\alpha_{2n-1})^{\text{th}}\text{order mXX, for } x_k^* = a_{2n-1}, \\
\vdots \\
(\alpha_1)^{\text{th}}\text{order mXX, for } x_k^* = a_1;
\end{cases}
\end{aligned} \tag{4.148}
$$

where $2(2l-1) = \Sigma_{i=1}^{n}\alpha_i$ and the minimum and maximum fixed-points satisfy

$$
\begin{aligned}
&(\alpha_{2n})^{\text{th}}\text{order mXX} = \begin{cases}
(2r_{2n})^{\text{th}}\text{order mLS, for } \alpha_{2n} = 2r_n, \\
(2r_{2n}-1)^{\text{th}}\text{order mSI, for } \alpha_{2n} = 2r_n - 1;
\end{cases} \\
&(\alpha_1)^{\text{th}}\text{order mXX} = \begin{cases}
(2r_1)^{\text{th}}\text{order mLS, for } \alpha_1 = 2r_1, \\
(2r_1-1)^{\text{th}}\text{order mSO, for } \alpha_1 = 2r_1 - 1.
\end{cases}
\end{aligned} \tag{4.149}
$$

(iv) The $(2(2l))^{\text{th}}$order LS *spraying-appearing* bifurcation for a cluster of fixed-points with higher-order monotonic sinks, monotonic sources, monotonic upper-saddles and monotonic lower-saddles is

$$
\begin{aligned}
&(2(2l))^{\text{th}}\text{order mLS} \xrightarrow[\text{appearing bifurcation}]{\text{a cluster of } 2n\text{-mXX}} \\
&\qquad\qquad \begin{cases}
(\alpha_{2n})^{\text{th}}\text{order mXX, for } x_k^* = a_{2n}, \\
(\alpha_{2n-1})^{\text{th}}\text{order mXX, for } x_k^* = a_{2n-1}, \\
\vdots \\
(\alpha_1)^{\text{th}}\text{order mXX, for } x_k^* = a_1;
\end{cases}
\end{aligned} \tag{4.150}
$$

where $2(2l) = \Sigma_{i=1}^{n}\alpha_i$ and the minimum and maximum fixed-points satisfy

$$(\alpha_{2n})^{\text{th}}\text{ order mXX} = \begin{cases} (2r_{2n})^{\text{th}}\text{order mLS, for } \alpha_{2n} = 2r_n, \\ (2r_{2n}-1)^{\text{th}}\text{order mSI, for } \alpha_{2n} = 2r_n - 1; \end{cases}$$
$$(\alpha_1)^{\text{th}}\text{ order mXX} = \begin{cases} (2r_1)^{\text{th}}\text{order mLS, for } \alpha_1 = 2r_1, \\ (2r_1-1)^{\text{th}}\text{order mSO, for } \alpha_1 = 2r_1 - 1. \end{cases} \tag{4.151}$$

A set of paralleled, different, higher-order monotonic upper-saddle-node *appearing* bifurcations in the $(2m)^{\text{th}}$-degree polynomial system is the $((2\beta_1)^{\text{th}}\text{mUS}:(2\beta_2)^{\text{th}}\text{mUS}:\cdots:(2\beta_s)^{\text{th}}\text{mUS})$ parallel *appearing* bifurcation for clusters of fixed-points with higher-order monotonic sinks, monotonic sources, monotonic upper-saddles and monotonic lower saddles. For the $(2\beta_i)^{\text{th}}\text{mUS}$

$$(2\beta_i)^{\text{th}}\text{mUS} = \begin{cases} (2(2l_i-1))^{\text{th}}\text{order mUS, for } \beta_i = 2l_i - 1, \\ (2(2l_i))^{\text{th}}\text{order mUS, for } \beta_i = 2l_i. \end{cases} \tag{4.152}$$

Similarly, the following notation is introduced as

$$((2\beta_1)^{\text{th}}\text{mUS}:(2\beta_2)^{\text{th}}\text{mUS}:\cdots:(2\beta_s)^{\text{th}}\text{mUS}) = (2\beta_1:2\beta_2:\cdots:2\beta_s)^{\text{th}}\text{mUS}. \tag{4.153}$$

Thus, the paralleled $(2\beta_1:2\beta_2:\cdots:2\beta_s)^{\text{th}}\text{mUS}$ spraying *appearing* bifurcation is called the $(2\beta_1:2\beta_2:\cdots:2\beta_s)^{\text{th}}\text{mUS}$ *sprinkler-spraying appearing* bifurcation for the higher-order fixed-points. Similarly, a set of paralleled different lower-saddle appearing bifurcations for higher-order singularity of fixed-points is called the $((2\beta_1)^{\text{th}}\text{mLS}:(2\beta_2)^{\text{th}}\text{mLS}:\cdots:(2\beta_s)^{\text{th}}\text{mLS})$ parallel appearing bifurcation in the $(2m)^{\text{th}}$-degree polynomial discrete system. Thus, the paralleled $(2\beta_1:2\beta_2:\cdots:2\beta_s)^{\text{th}}$ mLS bifurcation is also called the $(2\beta_1:2\beta_2:\cdots:2\beta_s)^{\text{th}}\text{mLS}$ *sprinkler-spraying* appearing bifurcation for higher-order fixed-points.

The $(2\alpha_1:2\alpha_2:\cdots:2\alpha_n)^{\text{th}}\text{mUS}$ and $(2\alpha_1:2\alpha_2:\cdots:2\alpha_n)^{\text{th}}\text{mLS}$ *teethcomb appearing* bifurcations for the higher-order singularity of fixed-points are presented in Figs. 4.6(i) and (ii) for $a_0 > 0$ and $a_0 < 0$, respectively. The components of the *teethcomb appearing* bifurcation are

$$(2\alpha_{j_1})^{\text{th}}\text{mUS} \xrightarrow[\text{appearing}]{\alpha_{j_1}=2r_{j_1}} \begin{cases} (2r_{j_1})^{\text{th}}\text{mUS} \\ (2r_{j_1})^{\text{th}}\text{mUS} \end{cases} (j_1 = i, n-1, \cdots),$$
$$(2\alpha_{j_2})^{\text{th}}\text{mUS} \xrightarrow[\text{appearing}]{\alpha_{j_2}=2r_{j_2}-1} \begin{cases} (2r_{j_2}-1)^{\text{th}}\text{mSO} \\ (2r_{j_2}-1)^{\text{th}}\text{mSI} \end{cases} (j_2 = 1, n, \cdots); \tag{4.154}$$

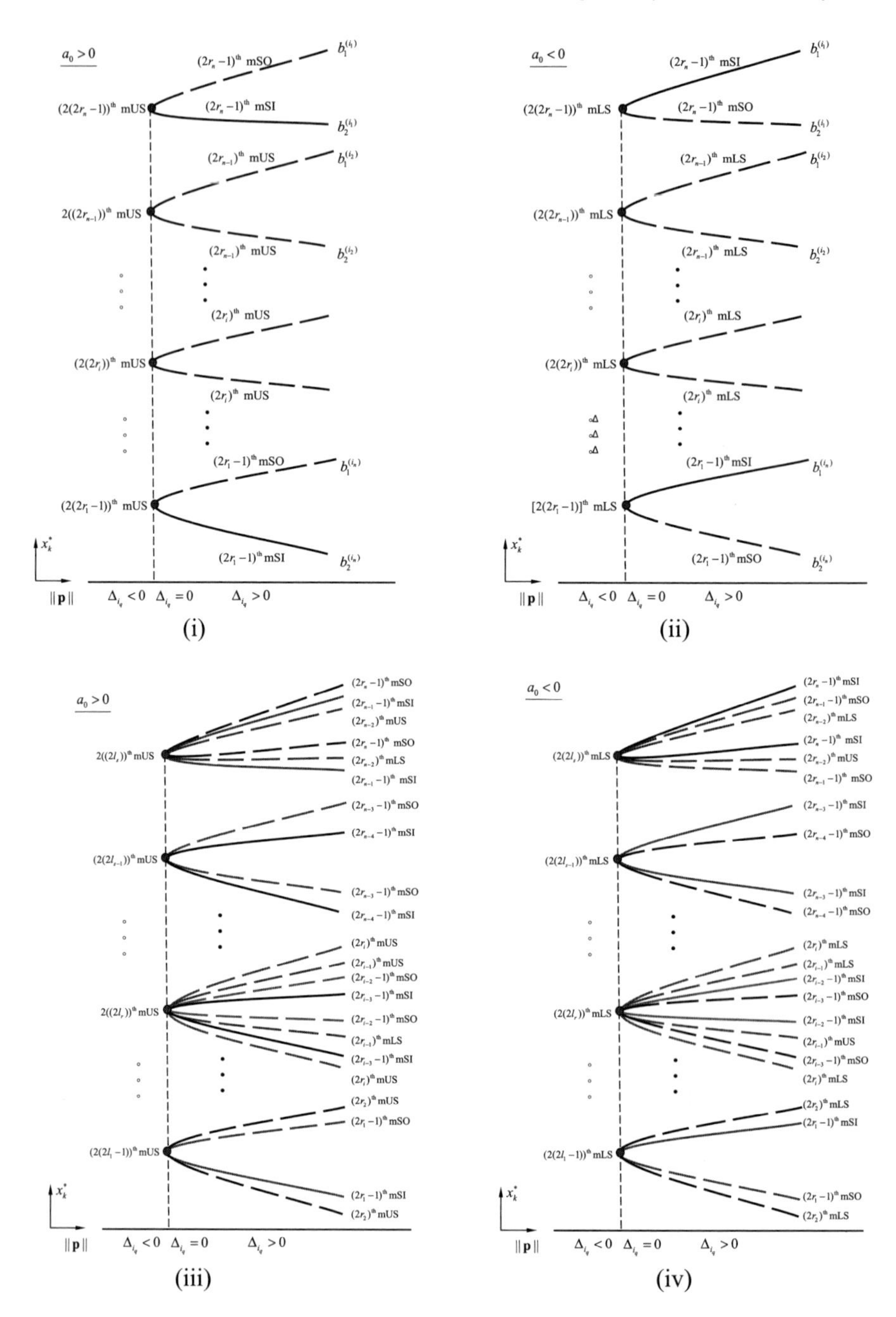

Fig. 4.6 The *teethcomb appearing* bifurcations of $(2(2r_1 - 1) : \cdots : 2(2r_{n-1}) : 2(2r_n - 1))^{\text{th}}$mXX: (i) XX = US($a_0 > 0$) and (ii) XX = LS($a_0 < 0$). The *sprinkler-spraying appearing* bifurcations of $(2(2l_1 - 1) : \cdots : (2(2l_{n-1}) : 2(2l_n))^{\text{th}}$mXX: (iii) XX = US($a_0 > 0$) and (iv) XX = LS($a_0 < 0$). mLS: monotonic-lower-saddle, mUS: monotonic-upper-saddle, mSI: monotonic sink, mSO: monotonic source. Stable and unstable fixed-points are represented by solid and dashed curves, respectively. The bifurcation points are marked by circular symbols.

and

$$
\begin{aligned}
&(2\alpha_{j_1})^{\text{th}}\text{mLS} \xrightarrow[\text{appearing}]{\alpha_{j_1}=2r_{j_1}} \begin{cases} (2r_{j_1})^{\text{th}}\text{mLS} \\ (2r_{j_1})^{\text{th}}\text{mLS} \end{cases} (j_1 = i, n-1, \cdots), \\
&(2\alpha_{j_2})^{\text{th}}\text{mLS} \xrightarrow[\text{appearing}]{\alpha_{j_2}=2r_{j_2}-1} \begin{cases} (2r_{j_2}-1)^{\text{th}}\text{mSI} \\ (2r_{j_2}-1)^{\text{th}}\text{mSO} \end{cases} (j_2 = 1, n, \cdots).
\end{aligned}
\tag{4.155}
$$

The $(2\beta_1 : 2\beta_2 : \cdots : 2\beta_n)^{\text{th}}$mUS and $(2\beta_1 : 2\beta_2 : \cdots : 2\beta_s)^{\text{th}}$mLS *sprinkler-spraying appearing* bifurcations for the higher-order singularity of fixed-points are presented in Figs. 4.6(iii) and (iv) for $a_0 > 0$ and $a_0 < 0$, respectively. The components of the *sprinkler-spraying appearing* bifurcation are

$$
\begin{aligned}
&(2\beta_1 : 2\beta_2 : \cdots : 2\beta_n)^{\text{th}}\text{mUS} \\
&= ((2(2l_1-1) : \cdots : 2(2l_i) : \cdots : 2(2l_{n-1}) : 2(2l_n))^{\text{th}}\text{mUS}
\end{aligned}
\tag{4.156}
$$

and

$$
\begin{aligned}
&(2\beta_1 : 2\beta_2 : \cdots : 2\beta_n)^{\text{th}}\text{mLS} \\
&= ((2(2l_1-1) : \cdots : 2(2l_i) : \cdots : 2(2l_{n-1}) : 2(2l_n))^{\text{th}}\text{mLS}.
\end{aligned}
\tag{4.157}
$$

For a cluster of m-quadratics, $B_i = B_j$ $(i, j \in \{1, 2, \ldots, n\}; i \neq j)$ and $\Delta_i = 0$ $(i = 1, 2, \ldots, n)$. The $(2m)^{\text{th}}$ order monotonic upper-saddle *appearing* bifurcation for n-pairs of the higher-order singularity of fixed-points is

$$
(2m)^{\text{th}}\text{order mUS} \xrightarrow[\text{appearing bifurcation}]{\text{a cluster of } 2n\text{-mXX}} \begin{cases} (\alpha_{2n})^{\text{th}}\text{order mXX for } x_k^* = a_{2n}, \\ (\alpha_{2n-1})^{\text{th}}\text{order mXX for } x_k^* = a_{2n-1}, \\ \vdots \\ (\alpha_1)^{\text{th}}\text{order mXX for } x_k^* = a_1; \end{cases}
\tag{4.158}
$$

where

$$
2m = 2(2l) = \textstyle\sum_{i=1}^{2n} \alpha_i, 2m = 2(2l-1) = \sum_{i=1}^{2n} \alpha_i. \tag{4.159}
$$

The $(2m)^{\text{th}}$ order monotonic lower-saddle-node appearing bifurcation for higher-order fixed-points is

$$(2m)^{\text{th}}\text{order mLS} \xrightarrow[\text{appearing bifurcation}]{\text{a cluster of } 2n\text{-mXX}} \begin{cases} (\alpha_{2n})^{\text{th}}\text{order mXX for } x_k^* = a_{2n}, \\ (\alpha_{2n-1})^{\text{th}}\text{order mXX for } x_k^* = a_{2n-1}, \\ \vdots \\ (\alpha_1)^{\text{th}}\text{order mXX for } x_k^* = a_1. \end{cases} \tag{4.160}$$

The $(2m)^{\text{th}}$ order upper-saddle appearing bifurcation with n-pairs of higher-order singularity of fixed-points is a *sprinkler-spraying cluster* of the n-pairs of higher-order singularity of fixed-points. The $(2m)^{\text{th}}$order monotonic lower-saddle *appearing* bifurcation with n-pairs of fixed-points is also a *sprinkler-spraying cluster* of the n-pairs of higher-order singularity of fixed-points.

Thus, the $(2m)^{\text{th}}$ order mUS bifurcation $(a_0 > 0)$ and $(2m)^{\text{th}}$ order mLS bifurcation $(a_0 < 0)$ are presented in Fig. 4.7(i)–(iv), respectively. The $(2m)^{\text{th}}$ order monotonic upper-saddle appearing bifurcation for higher-order singularity of fixed-points is called the $(2m)^{\text{th}}$ order mUS *sprinkler-spaying appearing* bifurcation, and the $(2m)^{\text{th}}$ order monotonic lower saddle-node *appearing* bifurcation for higher-order singularity of fixed-points is also called the $(2m)^{\text{th}}$ order mLS *sprinkler-spraying* appearing bifurcation.

A series of the monotonic saddle-node bifurcations for higher-order singularity of fixed-points is aligned up with varying with parameters, which is formed a special pattern. For n-quadratics in the $(2m)^{\text{th}}$ order polynomial discrete systems, the following conditions should be satisfied.

$$\begin{aligned} & B_i \approx B_j \; i,j \in \{1,2,\ldots,s\}; i \neq j, \\ & \Delta_i > \Delta_{i+1} \; (i = 1,2,\ldots,s; s \leq n < m), \\ & \Delta_i = 0 \text{ with } ||\mathbf{p}_i|| < ||\mathbf{p}_{i+1}||. \end{aligned} \tag{4.161}$$

The two series of the fish-scale switching bifurcations in Fig. 4.8(i) and (iii) for $a_0 < 0$ have the following detailed structures.

$$\begin{cases} (2(2r_1-1))^{\text{th}}\text{mUS} \to \begin{cases} (2r_1-1)^{\text{th}}\text{mSO}, \\ (2r_1-1)^{\text{th}}\text{mSI}; \end{cases} \\ (2(2r_2))^{\text{th}}\text{mLS} \to \begin{cases} (2r_2)^{\text{th}}\text{mLS}, \\ (2r_2)^{\text{th}}\text{mLS}; \end{cases} \\ \vdots \\ (2(2r_n-1))^{\text{th}}\text{mUS} \to \begin{cases} (2r_n-1)^{\text{th}}\text{mSO}, \\ (2r_n-1)^{\text{th}}\text{mSI}; \end{cases} \end{cases} \tag{4.162}$$

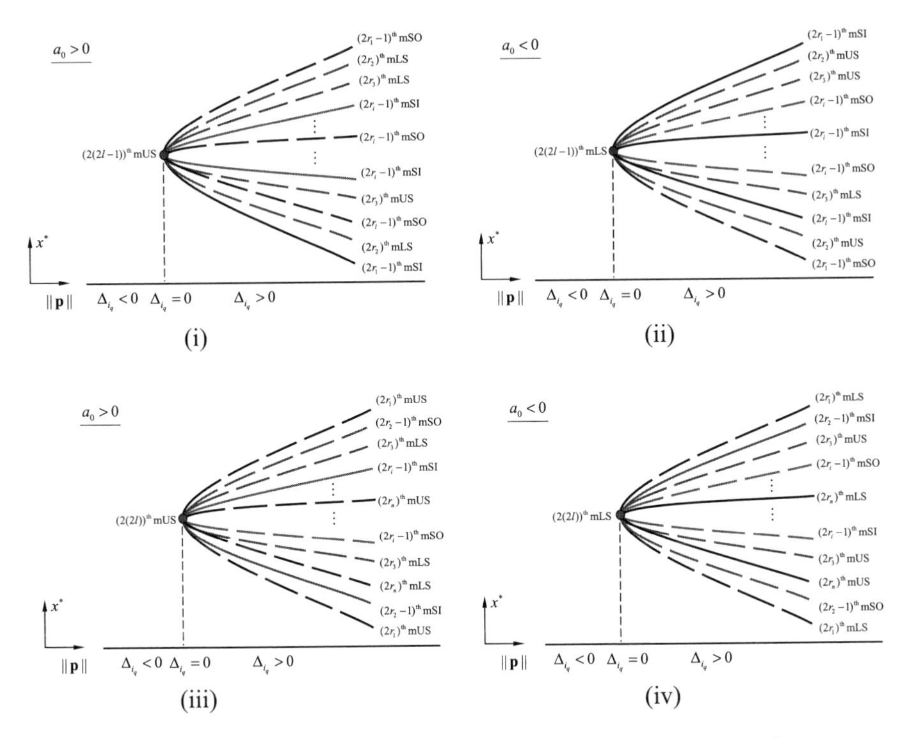

Fig. 4.7 Spraying appearing bifurcations for higher-order fixed-points in the $(2m)^{\text{th}}$ polynomial system: (i) $(2(2l-1))^{\text{th}}$mUS spraying-appearing bifurcation $(a_0 > 0)$, (ii) $(2(2l-1))^{\text{th}}$mLS spraying appearing bifurcation $(a_0 < 0)$, (iii) $(2(2l))^{\text{th}}$mUS spraying appearing bifurcation $(a_0 > 0)$, (iv)$(2(2l))^{\text{th}}$mUS spraying appearing bifurcation $(a_0 < 0)$. mLS: monotonic-lower-saddle, mUS: monotonic-upper-saddle, mSI: monotonic sink, mSO: monotonic source. Stable and unstable fixed-points are represented by solid and dashed curves, respectively. The bifurcation points are marked by circular symbols.

and

$$\begin{cases} (2(2r_1))^{\text{th}}\text{mUS} \rightarrow \begin{cases} (2r_1)^{\text{th}}\text{mUS}, \\ (2r_1)^{\text{th}}\text{mUS}; \end{cases} \\ (2(2r_2-1))^{\text{th}}\text{mUS} \rightarrow \begin{cases} (2r_2-1)^{\text{th}}\text{mSO}, \\ (2r_2-1)^{\text{th}}\text{mSI}; \end{cases} \\ \vdots \\ (2(2r_n-1))^{\text{th}}\text{mUS} \rightarrow \begin{cases} (2r_n-1)^{\text{th}}\text{mSO}, \\ (2r_n-1)^{\text{th}}\text{mSI}. \end{cases} \end{cases} \tag{4.163}$$

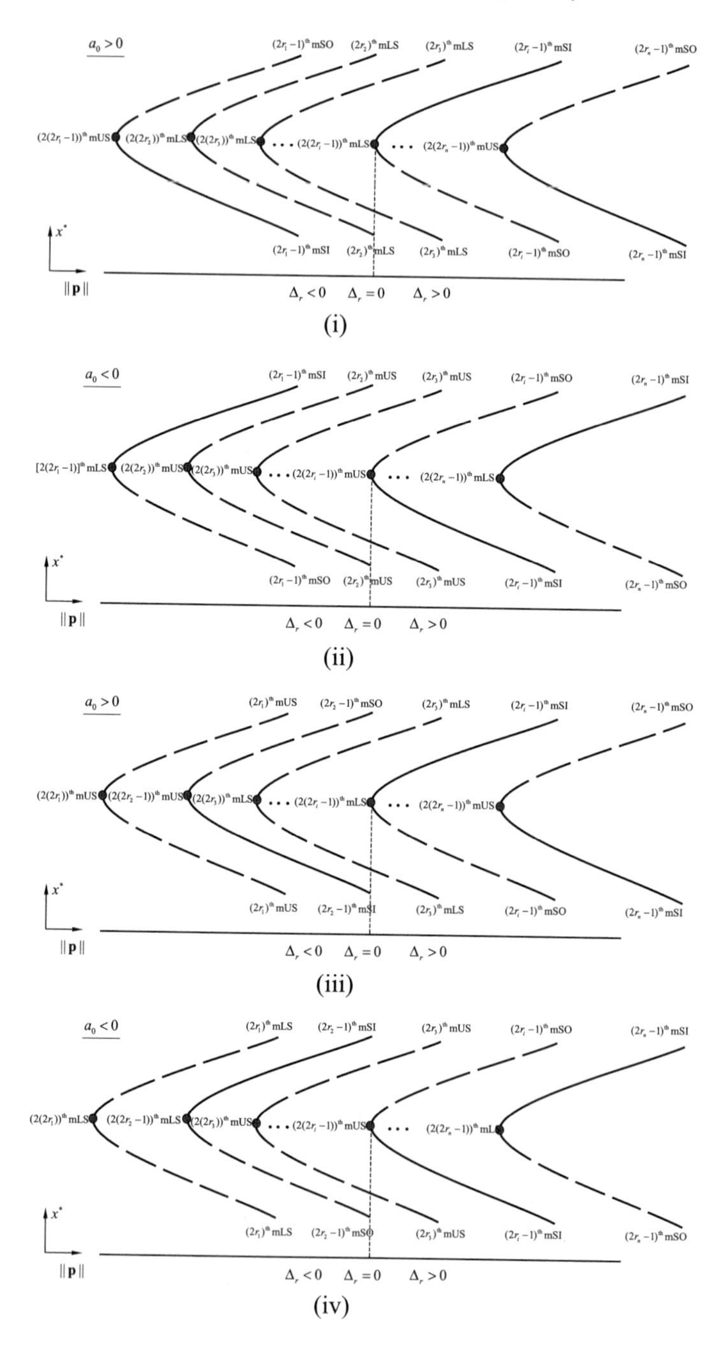

Fig. 4.8 The *fish-scale appearing* bifurcation patterns in a $(2m)^{\text{th}}$-degree polynomial discrete system: (i) $(2(2r_1-1))^{\text{th}}$mUS - $(2(2r_2))^{\text{th}}$mLS - $\ldots(a_0>0)$, (ii) $(2(2r_1-1))^{\text{th}}$mLS - $(2(2r_2))^{\text{th}}$ mUS - $\ldots(a_0<0)$, (iii) $(2(2r_1))^{\text{th}}$mUS - $(2(2r_2))^{\text{th}}$mUS - $\ldots(a_0>0)$, (iv)$(2(2r_1))^{\text{th}}$mLS - $(2(2r_2-1))^{\text{th}}$mLS - $\ldots(a_0>0)$. mLS: monotonic-lower-saddle, mUS: monotonic-upper-saddle, mSI: monotonic sink, mSO: monotonic source. Stable and unstable fixed-points are represented by solid and dashed curves, respectively. The bifurcation points are marked by circular symbols.

Two series of fish-scale *appearing* bifurcations in Fig. 4.8(ii) and (iv) for $a_0 < 0$ have the following structures as

$$\begin{cases} (2(2r_1-1))^{\text{th}}\text{mLS} \to \begin{cases} (2r_1-1)^{\text{th}}\text{mSI}, \\ (2r_1-1)^{\text{th}}\text{mSO}; \end{cases} \\ (2(2r_2))^{\text{th}}\text{mUS} \to \begin{cases} (2r_2)^{\text{th}}\text{mUS}, \\ (2r_2)^{\text{th}}\text{mUS}; \end{cases} \\ \vdots \\ (2(2r_n-1))^{\text{th}}\text{mLS} \to \begin{cases} (2r_n-1)^{\text{th}}\text{mSI}, \\ (2r_n-1)^{\text{th}}\text{mSO}; \end{cases} \end{cases} \quad (4.164)$$

and

$$\begin{cases} (2(2r_1))^{\text{th}}\text{mLS} \to \begin{cases} (2r_1)^{\text{th}}\text{mLS}, \\ (2r_1)^{\text{th}}\text{mLS}; \end{cases} \\ (2(2r_2-1))^{\text{th}}\text{mLS} \to \begin{cases} (2r_2-1)^{\text{th}}\text{mSI}, \\ (2r_2-1)^{\text{th}}\text{mSO}; \end{cases} \\ \vdots \\ (2(2r_n-1))^{\text{th}}\text{mLS} \to \begin{cases} (2r_n-1)^{\text{th}}\text{mSI}, \\ (2r_n-1)^{\text{th}}\text{mSO}. \end{cases} \end{cases} \quad (4.165)$$

The four *fish-scale appearing* bifurcation patterns for higher-order fixed-points are different from the fish-scale appearing bifurcation patterns for simple fixed-points.

4.3.2 Switching Bifurcations

Consider the roots of $(x_k^2 + B_i x_k + C_i)^{\alpha_i} = 0$ as

$$
\left.\begin{array}{l}
B_i = -(b_1^{(i)} + b_2^{(i)}),\ \Delta_i = (b_1^{(i)} - b_2^{(i)})^2 \geq 0, \\
x_{k;1,2}^{(i)} = b_{1,2}^{(i)},\ \Delta_i > 0 \text{ if } b_1^{(i)} \neq b_2^{(i)} (i = 1,2,\ldots,n); \\
\left.\begin{array}{l}
B_i \neq B_j\ (i,j = 1,2,\ldots,n; i \neq j) \\
\Delta_i = 0 \text{ at } b_1^{(i)} = b_2^{(i)} (i = 1,2,\ldots,n)
\end{array}\right\} \text{at bifurcation.}
\end{array}\right. \tag{4.166}
$$

The α_i^{th}-order singularity bifurcation is for the *switching* of a pair of higher order fixed-points (i.e., monotonic sinks, monotonic sources, monotonic-upper-saddles and monotonic-lower-saddles). There are six *switching* bifurcations for $i \in \{1,2,\ldots,n\}$

$$
(2l_i)^{\text{th}}\text{order mUS} \xrightarrow[\text{switching bifurcation}]{l_i = r_1^{(i)} + r_2^{(i)} - 1}
\begin{cases}
(2r_2^{(i)} - 1)^{\text{th}}\text{order mSO} \downarrow \text{mSI, for } b_2^{(i)} = a_{2i} \downarrow a_{2i-1}, \\
(2r_1^{(i)} - 1)^{\text{th}}\text{order mSI} \uparrow \text{mSO, for } b_1^{(i)} = a_{2i-1} \uparrow a_{2i};
\end{cases} \tag{4.167}
$$

$$
(2l_i)^{\text{th}}\text{order mLS} \xrightarrow[\text{switching bifurcation}]{l_i = r_1^{(i)} + r_2^{(i)} - 1}
\begin{cases}
(2r_2^{(i)} - 1)^{\text{th}}\text{order mSI} \downarrow \text{mSO, for } b_2^{(i)} = a_{2i} \downarrow a_{2i-1}, \\
(2r_1^{(i)} - 1)^{\text{th}}\text{order mSO} \uparrow \text{mSI, for } b_1^{(i)} = a_{2i-1} \uparrow a_{2i};
\end{cases} \tag{4.168}
$$

$$
(2l_i)^{\text{th}}\text{order mUS} \xrightarrow[\text{switching bifurcation}]{l_i = r_1^{(i)} + r_2^{(i)}}
\begin{cases}
(2r_2^{(i)})^{\text{th}}\text{order mUS} \downarrow \text{mUS, for } b_2^{(i)} = a_{2i} \downarrow a_{2i-1}, \\
(2r_1^{(i)})^{\text{th}}\text{order mUS} \uparrow \text{mUS for } b_1^{(i)} = a_{2i-1} \uparrow a_{2i};
\end{cases} \tag{4.169}
$$

$$
(2l_i)^{\text{th}}\text{order mLS} \xrightarrow[\text{switching bifurcation}]{l_i = r_1^{(i)} + r_2^{(i)}}
\begin{cases}
(2r_2^{(i)})^{\text{th}}\text{order mLS} \downarrow \text{mLS, for } b_2^{(i)} = a_{2i} \downarrow a_{2i-1}, \\
(2r_1^{(i)})^{\text{th}}\text{order mLS} \uparrow \text{mLS for } b_1^{(i)} = a_{2i-1} \uparrow a_{2i};
\end{cases} \tag{4.170}
$$

$$
(2l_i - 1)^{\text{th}}\text{order mSO} \xrightarrow[\text{switching bifurcation}]{l_i = r_1^{(i)} + r_2^{(i)}}
\begin{cases}
(2r_2^{(i)} - 1)^{\text{th}}\text{order mSO} \downarrow \text{mSO, for } b_2^{(i)} = a_{2i} \downarrow a_{2i-1}, \\
(2r_1^{(i)})^{\text{th}}\text{order mLS} \uparrow \text{mUS for } b_1^{(i)} = a_{2i-1} \uparrow a_{2i};
\end{cases} \tag{4.171}
$$

$$
\begin{aligned}
&(2l_i-1)^{\text{th}}\text{order mSI}\xrightarrow[\text{switching bifurcation}]{l_i=r_1^{(i)}+r_2^{(i)}}\\
&\begin{cases}
(2r_2^{(i)}-1)^{\text{th}}\text{order mSI}\downarrow\text{mSI, for } b_2^{(i)}=a_{2i}\downarrow a_{2i-1},\\
2r_1^{(i)})^{\text{th}}\text{order mUS}\uparrow\text{mLS for } b_1^{(i)}=a_{2i-1}\uparrow a_{2i}.
\end{cases}
\end{aligned}
\tag{4.172}
$$

A set of n-paralleled higher-order mXX *switching* bifurcations is called a $((\alpha_1)^{\text{th}}\text{mXX}:(\alpha_2)^{\text{th}}\text{mXX}:\cdots:(\alpha_n)^{\text{th}}\text{mXX})$ *parallel switching* bifurcation in the $(2m)^{\text{th}}$ degree polynomial nonlinear discrete system. Such a bifurcation is also called a $((\alpha_1)^{\text{th}}\text{mXX}:(\alpha_2)^{\text{th}}\text{mXX}:\cdots:(\alpha_n)^{\text{th}}\text{mXX})$ *antenna switching* bifurcation. $\alpha_i\in\{2l_i,2l_i-1\}$ and $\text{XX}\in\{\text{SO, SI, US, LS}\}$. For non-switching points, $\Delta_i>0$ at $b_1^{(i)}\neq b_2^{(i)}$ $(i=1,2,\ldots,n)$. At the bifurcation point, $\Delta_i=0$ at $b_1^{(i)}=b_2^{(i)}$ $(i=1,2,\ldots,n)$. The $((\alpha_1)^{\text{th}}\text{mXX}:(\alpha_2)^{\text{th}}\text{mXX}:\cdots:(\alpha_n)^{\text{th}}\text{mXX})$ *parallel antenna switching* bifurcation is

$$
\begin{cases}
\alpha_n^{\text{th}}\text{mXX}_n\xrightarrow[\text{switiching}]{n^{\text{th}}\text{bifurcation}}\begin{cases}(r_2^{(n)})^{\text{th}}\text{mXX}_2^{(n)}\downarrow\text{mYY}_1^{(n)},\text{ for } b_2^{(n)}=a_{2n}\downarrow a_{2n-1},\\(r_1^{(n)})^{\text{th}}\text{mXX}_1^{(n)}\uparrow\text{mYY}_2^{(n)},\text{ for } b_1^{(n)}=a_{2n-1}\uparrow a_{2n};\end{cases}\\
\vdots\\
\alpha_2^{\text{th}}\text{mXX}_2\xrightarrow[\text{switiching}]{2^{\text{nd}}\text{ bifurcation}}\begin{cases}(r_2^{(2)})^{\text{th}}\text{mXX}_2^{(2)}\downarrow\text{mYY}_1^{(2)},\text{ for } b_2^{(2)}=a_4\downarrow a_3,\\(r_1^{(2)})^{\text{th}}\text{mXX}_1^{(2)}\uparrow\text{mYY}_2^{(2)},\text{ for } b_1^{(2)}=a_3\uparrow a_4;\end{cases}\\
\alpha_1^{\text{th}}\text{mXX}_1\xrightarrow[\text{switiching}]{1^{\text{st}}\text{ bifurcation}}\begin{cases}(r_2^{(1)})^{\text{th}}\text{mXX}_2^{(1)}\downarrow\text{mYY}_1^{(1)},\text{ for } b_2^{(1)}=a_2\downarrow a_1,\\(r_1^{(1)})^{\text{th}}\text{mXX}_1^{(1)}\uparrow\text{mYY}_2^{(1)},\text{ for } b_1^{(1)}=a_1\uparrow a_2.\end{cases}
\end{cases}
\tag{4.173}
$$

Such eight sets of parallel *switching* bifurcations of $((\alpha_1)^{\text{th}}\text{mXX}:(\alpha_1)^{\text{th}}\text{mXX}:\cdots:(\alpha_n)^{\text{th}}\text{mXX})$ are presented in Fig. 4.9(i, iii, v, vii) and (ii, iv, vi, viii) for $a_0>0$ and $a_0<0$, respectively. The eight switching bifurcation structures are as follows:

(i) $((2l_1)^{\text{th}}\text{mUS}:\cdots:(2l_{n-1}-1)^{\text{th}}\text{mSO}:(2l_n)^{\text{th}}\text{mUS})$ for $a_0>0$,
(ii) $((2l_1)^{\text{th}}\text{mLS}:\cdots:(2l_{n-1}-1)^{\text{th}}\text{mSI}:(2l_n)^{\text{th}}\text{mLS})$ for $a_0<0$,
(iii) $((2l_1)^{\text{th}}\text{mLS}:\cdots:(2l_{n-1}-1)^{\text{th}}\text{mSI}:(2l_n-1)^{\text{th}}\text{mSO})$ for $a_0>0$,
(iv) $((2l_1)^{\text{th}}\text{mUS}:\cdots:(2l_{n-1}-1)^{\text{th}}\text{mSO}:(2l_n-1)^{\text{th}}\text{mSI})$ for $a_0<0$,
(v) $((2l_1)^{\text{th}}\text{mLS}:\cdots:(2l_{n-1}-1)^{\text{th}}\text{mSI}:(2l_n-1)^{\text{th}}\text{mSO})$ for $a_0>0$,
(vi) $((2l_1)^{\text{th}}\text{mUS}:\cdots:(2l_{n-1}-1)^{\text{th}}\text{mSI}:(2l_n-1)^{\text{th}}\text{mSI})$ for $a_0<0$,
(vii) $((2l_1)^{\text{th}}\text{mUS}:\cdots:(2l_{n-1}-1)^{\text{th}}\text{mSO}:(2l_n)^{\text{th}}\text{mUS})$ for $a_0>0$,
(viii) $((2l_1)^{\text{th}}\text{mLS}:\cdots:(2l_{n-1}-1)^{\text{th}}\text{mSI}:(2l_n)^{\text{th}}\text{mLS})$ for $a_0<0$.

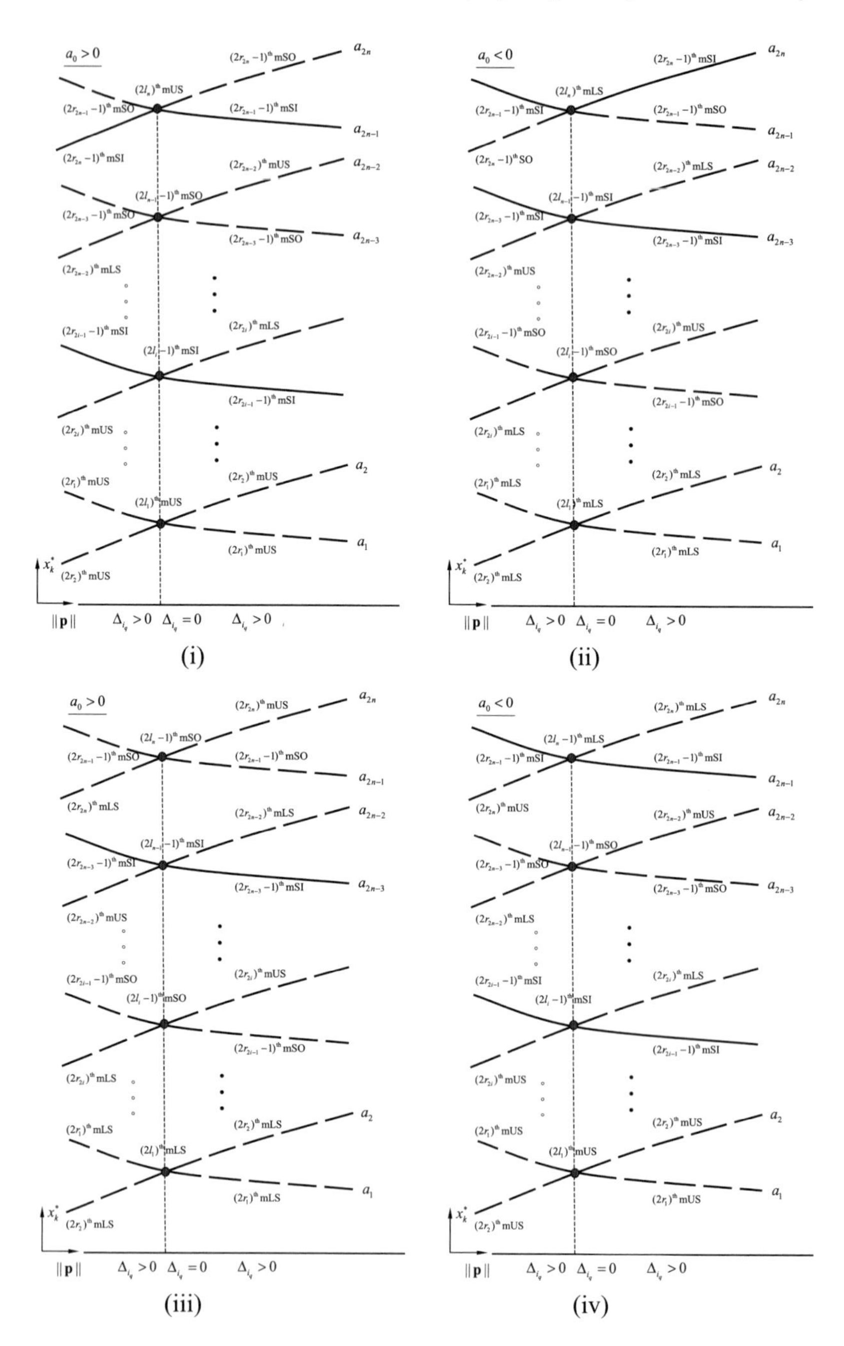

Fig. 4.9 Parallel *antenna* switching bifurcations for high-order fixed-points in a $(2m)^{\text{th}}$-degree polynomial discrete system. (α_1^{th}mXX:α_2^{th}mXX:... : α_n^{th}mXX): (i, iii, v, vii) for $a_0 > 0$. (ii, iv, vi, viii) for $a_0 < 0$. mLS: monotonic-lower-saddle, mUS: monotonic-upper-saddle, mSI: monotonic sink, mSO: monotonic source. Stable and unstable fixed-points are represented by solid and dashed curves, respectively. The bifurcation points are marked by circular symbols. Continued

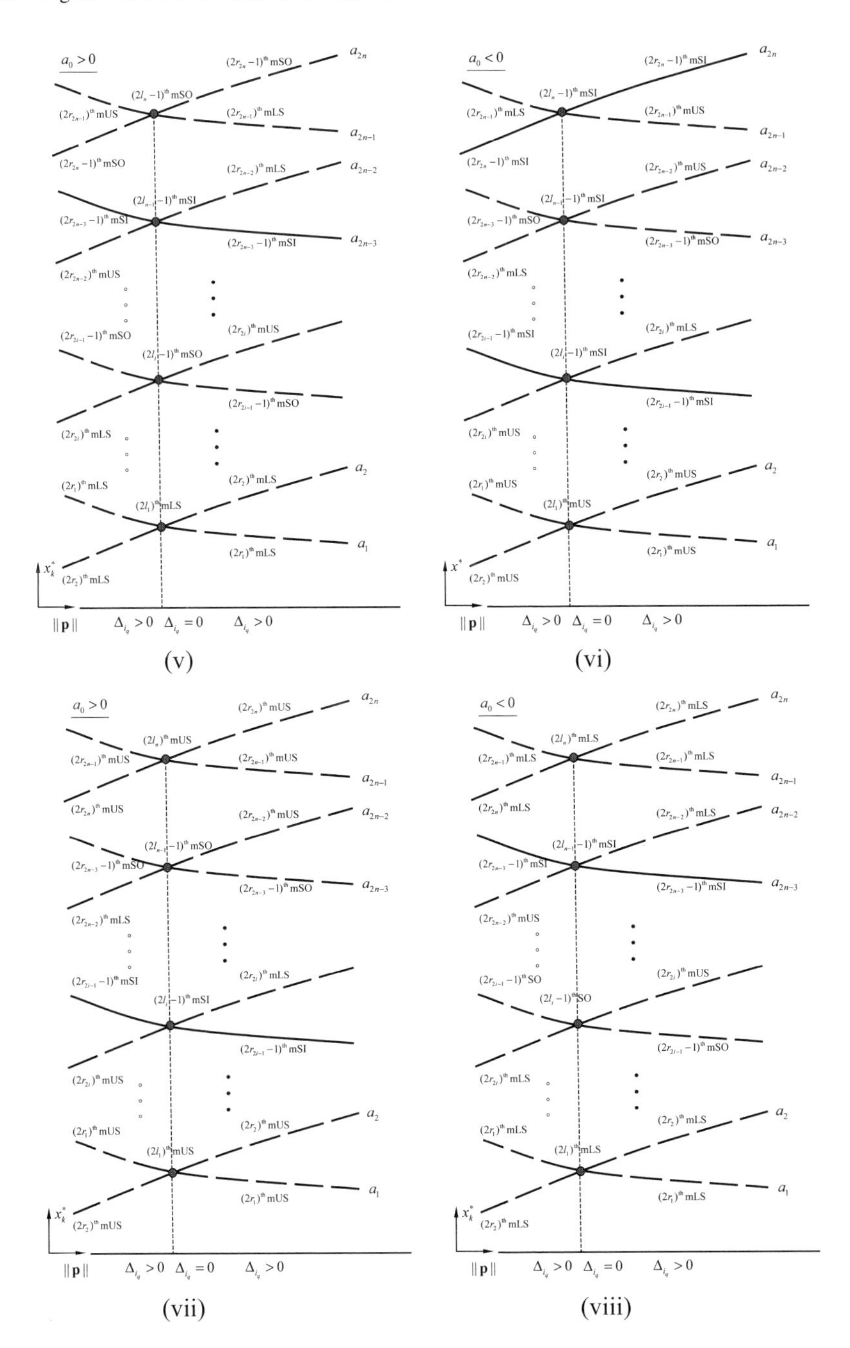

Fig. 4.9 (continued)

The same switching bifurcations with different higher-order fixed-points are illustrated, which is different from the m-mUSN and m-mLSN for simple monotonic sinks and monotonic sources.

Consider a switching bifurcation for a cluster of higher-order fixed-points with the following conditions,

$$\left.\begin{array}{l}B_i = -(b_1^{(i)} + b_2^{(i)}),\ \Delta_i = (b_1^{(i)} - b_2^{(i)})^2 \geq 0, \\ x_{k;1,2}^{(i)} = b_{1,2}^{(i)}, \Delta_i > 0 \text{ if } b_1^{(i)} \neq b_2^{(i)} (i = 1, 2, \ldots, n); \\ \left.\begin{array}{l} B_i = B_j (i, j \in \{1, 2, \ldots, n\}; i \neq j) \\ \Delta_i = 0 \text{ at } b_1^{(i)} = b_2^{(i)} (i = 1, 2, \ldots, n) \end{array}\right\} \text{at bifurcation.}\end{array}\right. \tag{4.174}$$

Thus, the $(\alpha_i)^{\text{th}}$ order *switching* bifurcation can be for a cluster of higher-order fixed-points. The $(\alpha_i)^{\text{th}}$ order switching bifurcations for $i \in \{1, 2, \ldots, s\}$ are

$$(\alpha_i)^{\text{th}}\text{order mXX} \xrightarrow[\text{switching bifurcation}]{\alpha_i = \sum_{j=1}^{l_i} r_j^{(i)}} \begin{cases} (r_s^{(i)})^{\text{th}}\text{order mXX}_{l_i}^{(i)} \downarrow \text{mYY}_{l_i}^{(i)}, \text{ for } b_{l_i}^{(i)} \downarrow a_{l_i}^{(i)}, \\ \vdots \\ (r_j^{(i)})^{\text{th}}\text{order mXX}_j^{(i)} \downarrow \text{mYY}_j^{(i)}, \text{ for } b_j^{(i)} \downarrow a_s^{(i)}, \\ \vdots \\ (r_1^{(i)})^{\text{th}}\text{order mXX}_1^{(i)} \uparrow \text{mYY}_1^{(i)}, \text{ for } b_1^{(i)} \downarrow a_s^{(i)}; \end{cases} \tag{4.175}$$

where

$$\begin{array}{l}\{a_1^{(i)}, a_2^{(i)}, \ldots, a_{l_i-1}^{(i)}, a_{l_i}^{(i)}\} \underset{\text{before bifurcation}}{\subset} \text{sort}\{b_1^{(1)}, b_2^{(1)}, \ldots, b_1^{(n)}, b_2^{(n)}\}, \\ \{b_1^{(i)}, b_2^{(i)}, \ldots, b_{l_i-1}^{(i)}, b_{l_i}^{(i)}\} \underset{\text{After bifurcation}}{\subset} \text{sort}\{b_1^{(1)}, b_2^{(1)}, \ldots, b_1^{(n)}, b_2^{(n)}\}.\end{array} \tag{4.176}$$

A set of paralleled, different, higher-order upper-saddle-node switching bifurcations with multiplicity is the $((\alpha_1)^{\text{th}}\text{mXX}:(\alpha_2)^{\text{th}}\text{mXX}:\cdots:(\alpha_s)^{\text{th}}\text{mXX})$ *parallel switching* bifurcation in the $(2m)^{\text{th}}$ degree polynomial discrete system. At the *straw-bundle switching* bifurcation, $\Delta_i = 0$ $(i = 1, 2, \ldots, n)$ and $B_i = B_j$ $(i, j \in \{1, 2, \ldots, n\};\ i \neq j)$. The *parallel straw-bundle switching* bifurcation for higher order fixed-points is

$$\begin{aligned}
&((\alpha_1)^{\text{th}}\text{mXX}:(\alpha_2)^{\text{th}}\text{mXX}:\cdots:(\alpha_s)^{\text{th}}\text{mXX})\text{-switching} \\
&= \begin{cases} (\alpha_s)^{\text{th}}\text{order mXX switching}, \\ \vdots \\ (\alpha_2)^{\text{th}}\text{order mXX switching}, \\ (\alpha_1)^{\text{th}}\text{order mXX switching}; \end{cases}
\end{aligned} \tag{4.177}$$

where

$$\alpha_i \in \{2l_i, 2l_i - 1\} \text{ and } \text{mXX} \in \{\text{mUS}, \text{mLS}, \text{mSO}, \text{mSI}\}. \tag{4.178}$$

Eight *parallel straw-bundle switching* bifurcations of (α_1^{th}mXX:α_2^{th}mXX:$\cdots$: α_n^{th}mXX) are presented in Fig. 4.10 and Fig. 4.11 for $a_0 > 0$ and $a_0 < 0$, respectively.

4.3.3 Switching-Appearing Bifurcations

Consider a $(2m)^{\text{th}}$ degree polynomial discrete system in a form of

$$x_{k+1} = x_k + a_0 Q(x_k) \prod_{i=1}^{2n_1} (x_k - c_i)^{\alpha_i} \prod_{j=1}^{n_2} (x_k^2 + B_j x_k + C_j)^{\alpha_j}. \tag{4.179}$$

Without loss of generality, a function of $Q(x_k) > 0$ is either a polynomial function or a non-polynomial function. The roots of $x_k^2 + B_j x_k + C_j = 0$ are

$$b_{1,2}^{(j)} = -\tfrac{1}{2}B_j \pm \tfrac{1}{2}\sqrt{\Delta_j}, \Delta_j = B_j^2 - 4C_j \geq 0 \quad (j = 1, 2, \ldots, n_2); \tag{4.180}$$

either

$$\begin{aligned}
&\{a_1^-, a_2^-, \ldots, a_{2n_1}^-\} = \text{sort}\{c_1, c_2 \ldots, c_{2n_1}\}, a_s^- \leq a_{s+1}^- \text{ before bifurcation} \\
&\{a_1^+, a_2^+, \ldots, a_{2n_3}^+\} = \text{sort}\{c_1, \ldots, c_{2n_1}; b_1^{(1)}, b_2^{(1)}, \ldots, b_1^{(n_2)}, b_2^{(n_2)}\}, \\
&a_s^+ \leq a_{s+1}^+,\ n_3 = n_1 + n_2 \text{ after bifurcation};
\end{aligned} \tag{4.181}$$

or

$$\begin{aligned}
&\{a_1^-, a_2^-, \ldots, a_{2n_3}^-\} = \text{sort}\{c_1, c_2 \ldots, c_{2n_1}; b_1^{(1)}, b_2^{(1)}, \ldots, b_1^{(n_2)}, b_2^{(n_2)}\}, \\
&a_s^- \leq a_{s+1}^-, n_3 = n_1 + n_2 \text{ before bifurcation}; \\
&\{a_1^+, a_2^+, \ldots, a_{2n_1}^+\} = \text{sort}\{c_1, \ldots, c_{2n_1}\},\ a_s^+ \leq a_{s+1}^+ \text{ after bifurcation};
\end{aligned} \tag{4.182}$$

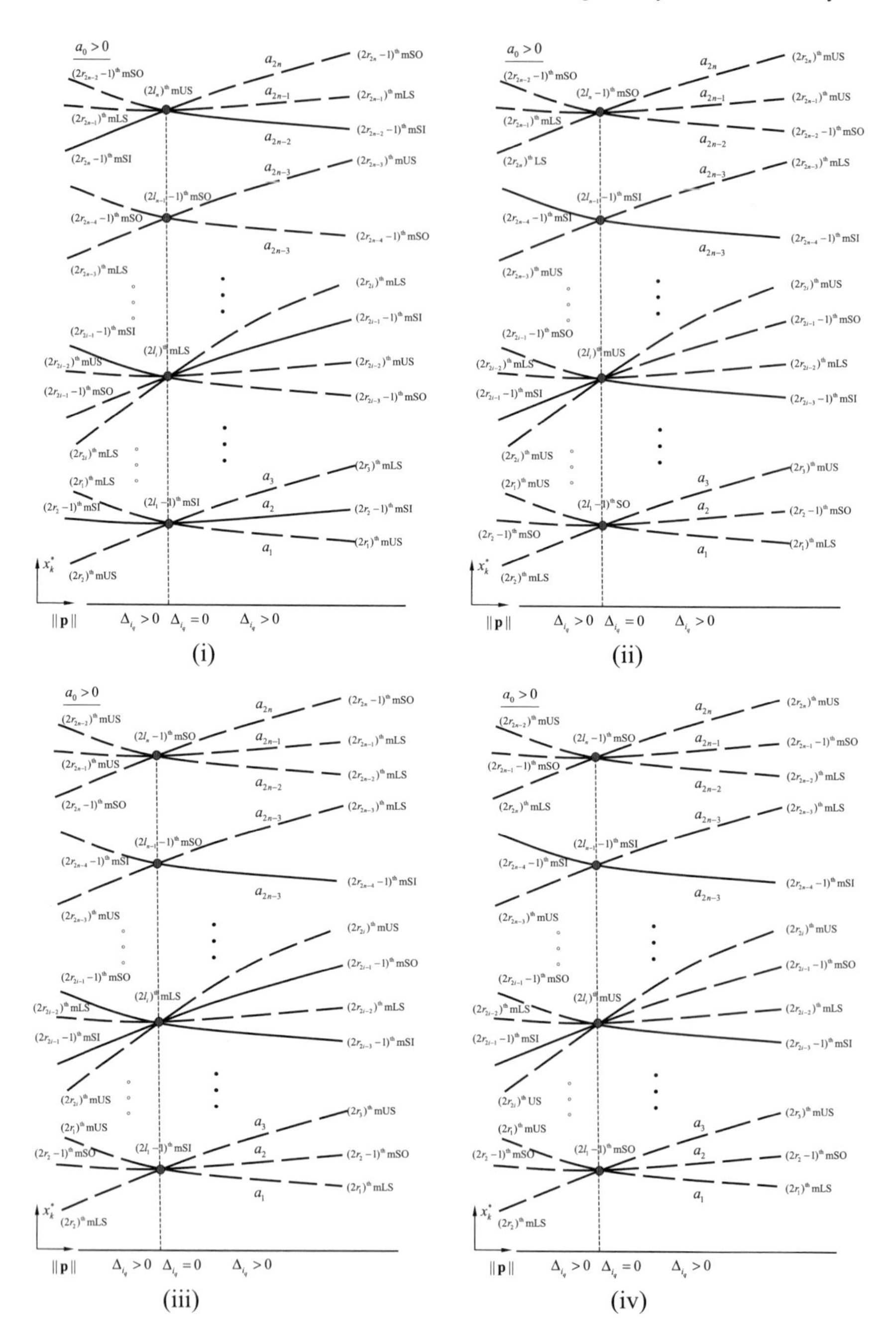

Fig. 4.10 (i)–(iv) Four types of $(r_1\text{th mXX}{:}r_2\text{th mXX}{:}\cdots{:}\,r_m\text{th mXX})$ parallel switching bifurcation for $a_0 > 0$ in the $(2m)^{\text{th}}$-degree polynomial system. mLS: monotonic-lower-saddle, mUS: monotonic-upper-saddle, mSI: monotonic sink, mSO: monotonic source. Stable and unstable fixed-points are represented by solid and dashed curves, respectively. The bifurcation points are marked by circular symbols.

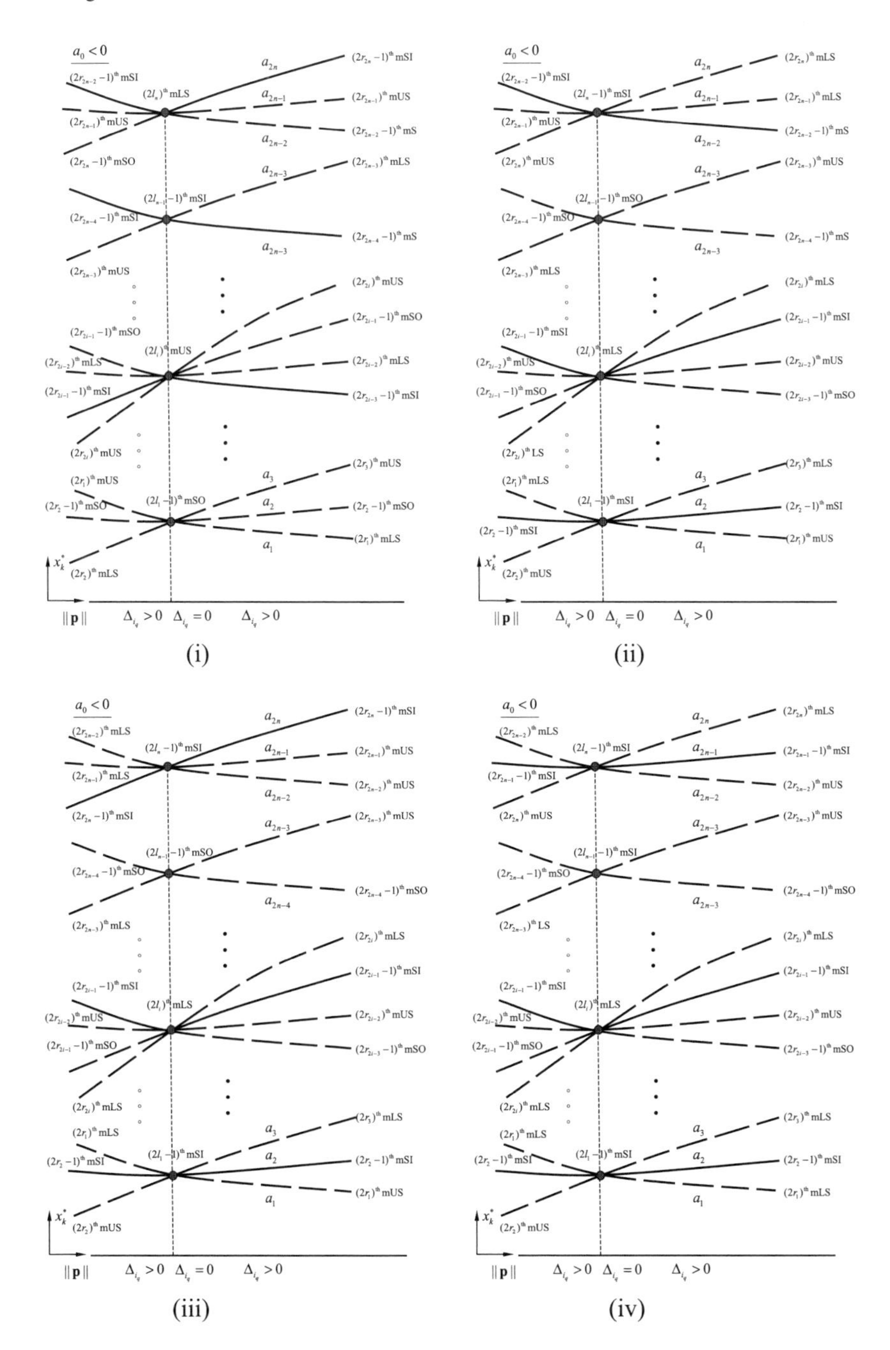

Fig. 4.11 (i)–(iv) Four types of $(r_1\text{th mXX}{:}r_2\text{th mXX}{:}\cdots{:}r_m\text{th mXX})$ parallel switching bifurcation for $a_0 > 0$ in the $(2m)^{\text{th}}$-degree polynomial discrete system. mLS: monotonic-lower-saddle, mUS: monotonic-upper-saddle, mSI: monotonic sink, mSO: monotonic source. Stable and unstable fixed-points are represented by solid and dashed curves, respectively. The bifurcation points are marked by circular symbols.

and

$$\left.\begin{array}{l} B_{j_1} = B_{j_2} = \cdots = B_{j_s}\ (j_{k_1} \in \{1,2,\ldots,n\}; j_{k_1} \neq j_{k_2}) \\ (k_1,k_2 \in \{1,2,\ldots,s\}; k_1 \neq k_2) \\ \Delta_j = 0\ (j \in U \subset \{1,2,\ldots,n_2\} \\ c_i \neq -\frac{1}{2}B_j\ (i=1,2,\ldots,2n_1, j=1,2,\ldots,n_2) \end{array}\right\} \text{at bifurcation.} \tag{4.183}$$

Consider a just before bifurcation of $((\beta_1^-)^{\text{th}}\text{mXX}_1^- : (\beta_2^-)^{\text{th}}\text{mXX}_2^- : \cdots : (\beta_{s_1}^-)^{\text{th}}\text{mXX}_{s_1}^-)$ for higher-order fixed-points. For $\beta_i^- = 2l_i^- - 1, \text{XX}_i^- \in \{\text{SO,SI}\}$ and for $\alpha_i^- = 2l_i^-, \text{XX}_i^- \in \{\text{US,LS}\}$ $(i = 1,2,\ldots,s_1)$. The detailed structures are as follows.

$$\left.\begin{array}{l} (r_{s_i}^{(i)-})^{\text{th}} \text{ order mXX}_{s}^{(i)-}, x_k^* = a_{s_i}^{(i)-}, \\ \vdots \\ (r_j^{(i)-})^{\text{th}} \text{ order mXX}_j^{(i)-}, x_k^* = a_j^{(i)-} \\ \vdots \\ (r_1^{(i)-})^{\text{th}} \text{ order mXX}_1^{(i)-}, x_k^* = a_1^{(i)-} \end{array}\right\} \xrightarrow[\text{switching bifurcation}]{\beta_i^- = \sum_{j=1}^{s_i} r_j^{(i)-}} (\beta_i^-)^{\text{th}} \text{order mXX}^{(i)-} \tag{4.184}$$

The bifurcation set of $((\beta_1^-)^{\text{th}}\text{mXX}_1^- : (\beta_2^-)^{\text{th}}\text{mXX}_2^- : \cdots : (\beta_{s_1}^-)^{\text{th}}\text{mXX}_{s_1}^-)$ at the same parameter point is called a *left-parallel-straw-bundle* switching bifurcation.

Consider a just after bifurcation of $((\beta_1^+)^{\text{th}}\text{mXX}_1^+ : (\beta_2^+)^{\text{th}}\text{mXX}_2^+ : \cdots : (\beta_{s_2}^+)^{\text{th}}\text{mXX}_{s_2}^+)$ for monotonic sources and monotonic sinks. For $\beta_i^+ = 2l_i^+ - 1$, $\text{XX}_i^+ \in \{\text{SO,SI}\}$ and for $\beta_i^+ = 2l_i^+$, $\text{XX}_i^+ \in \{\text{US,LS}\}$. The detailed structures are as follows.

$$(\beta_i^+)^{\text{th}} \text{order mXX}^{(i)+} \xrightarrow[\text{switching bifurcation}]{\beta_i^- = \sum_{j=1}^{s_i} r_j^{(i)+}} \left\{\begin{array}{l} (r_{s_i}^{(i)+})^{\text{th}} \text{order mXX}_{s_i}^{(i)+}, x_k^* = a_{s_i}^{(i)+}, \\ \vdots \\ (r_j^{(i)+})^{\text{th}} \text{order mXX}_j^{(i)+}, x_k^* = a_j^{(i)+} \\ \vdots \\ (r_1^{(i)+})^{\text{th}} \text{order mXX}_1^{(i)+}, x_k^* = a_1^{(i)+}. \end{array}\right. \tag{4.185}$$

The bifurcation set of $((\beta_1^+)^{\text{th}}\text{mXX}_1^+ : (\beta_2^+)^{\text{th}}\text{mXX}_2^+ : \cdots : (\beta_{s_2}^+)^{\text{th}}\text{mXX}_{s_2}^+)$ at the same parameter point is called a *right-parallel-straw-bundle* switching bifurcation

(i) For the just before and after bifurcation structure, if there exists a relation of

$$\begin{aligned}&(\beta_i^-)^{\text{th}}\text{mXX}_i^- = (\beta_j^+)^{\text{th}}\text{mXX}_j^+ = \beta_j^{\text{th}}\text{mXX, for } x_k^* = a_i^- = a_j^+\\&(i,j \in \{1,2,\ldots,k\}),\ \text{XX} \in \{\text{US,LS,SO,SI}\}\end{aligned} \tag{4.186}$$

then the bifurcation is a $(\beta_j)^{\text{th}}\text{mXX}_j$ *switching* bifurcation for higher-order fixed-points.

(ii) Just for the just before bifurcation structure, if there exists a relation of

$$\begin{aligned}&(2l_i^-)^{\text{th}}\text{mXX}_i^- = (2l_i)^{\text{th}}\text{mXX, for } x_k^* = a_i^- = a_i\\&(i \in \{1,2,\ldots,s_1\}, \text{mXX} \in \{\text{mUS,mLS}\}\end{aligned} \tag{4.187}$$

then, the bifurcation is a $(2l)^{\text{th}}\text{mXX}$ *left appearing* (or *right vanishing*) bifurcation for higher-order fixed-points.

(iii) Just for the just after bifurcation structure, if there exists a relation of

$$\begin{aligned}&(2l_i^+)^{\text{th}}\text{mXX}_i^+ = (2l_i)^{\text{th}}\text{mXX, for } x_k^* = a_i^+ = a_i\\&(i \in \{1,2,\ldots,s_1\}), \text{XX} \in \{\text{US,LS}\}\end{aligned} \tag{4.188}$$

then, the bifurcation is a $(2l)^{\text{th}}\text{mXX}$ *right appearing* (or *left vanishing*) bifurcation for higher-order fixed-points.

(iv) For the just before and after bifurcation structure, if there exists a relation of

$$\begin{aligned}&(\beta_i^-)^{\text{th}}\text{mXX}_i^- \neq (\beta_j^+)^{\text{th}}\text{mXX}_j^+ \text{ for } x_k^* = a_i^- = a_j^+\\&\text{XX}_i^-,\ \text{XX}_j^+ \in \{\text{US,LS, SO,SI}\}\\&(i \in \{1,2,\ldots,s_1\}, j \in \{1,2,\ldots,s_2\}),\end{aligned} \tag{4.189}$$

then, two *flower-bundle* switching bifurcations of higher-order fixed-points are as follows.

(iv$_1$) For $\beta_j = \beta_i + 2l$, the bifurcation is called a $\beta_j^{\text{th}}\text{mXX}$ right *flower-bundle* switching bifurcation for the $\beta_i^{\text{th}}\text{mXX}$ to $\beta_j^{\text{th}}\text{mXX}$ switching of *higher-order* fixed-points with the appearance (or birth) of $(2l)^{\text{th}}\text{mXX}$ right appearing (or left vanishing) bifurcation.

(iv$_2$) For $\beta_j = \beta_i - 2l$, the bifurcation is called a $\beta_i^{\text{th}}\text{mXX}$ left *flower-bundle* switching bifurcation for the $\beta_i^{\text{th}}\text{mXX}$ to $\beta_j^{\text{th}}\text{mXX}$ switching of higher-order fixed-points with the vanishing (or death) of $(2l)^{\text{th}}$mXX *left appearing* (or right vanishing) bifurcation.

A general parallel switching bifurcation is

$$
\begin{aligned}
&((\beta_1^-)^{\text{th}}\text{mXX}_1^- : (\beta_2^-)^{\text{th}}\text{mXX}_2^- : \cdots : (\beta_{s_1}^-)^{\text{th}}\text{mXX}_{s_1}^-) \xrightarrow[\text{bifurcation}]{\text{switching}} \\
&((\beta_1^+)^{\text{th}}\text{mXX}_1^+ : (\beta_2^+)^{\text{th}}\text{mXX}_2^+ : \cdots : (\beta_{s_2}^+)^{\text{th}}\text{mXX}_{s_2}^+).
\end{aligned}
\tag{4.190}
$$

Such a general, parallel switching bifurcation consists of the *left* and *right* parallel-bundle switching bifurcations for higher-order fixed-points.

If the *left* and *right* parallel-bundle switching bifurcations are same in a parallel *flower-bundle* switching bifurcation, i.e.,

$$
\begin{aligned}
&(\beta_i^-)^{\text{th}}\text{mXX}_i^- = (\beta_i^+)^{\text{th}}\text{mXX}_i^+ = \beta^{\text{th}}\text{mXX}, \\
&\text{for } x_k^* = a_i^- = a_i^+ \, (i = 1, 2, \ldots, s\}
\end{aligned}
\tag{4.191}
$$

then the parallel *flower-bundle* switching bifurcation becomes a parallel *straw-bundle switching* bifurcation of $((\alpha_1)^{\text{th}}\text{mXX}{:}(\beta_2)^{\text{th}}\text{mXX}{:} \cdots : (\beta_s)^{\text{th}}\text{mXX})$.

If the *left* and *right* parallel-bundle switching bifurcations are different in a parallel *flower-bundle* switching bifurcation, i.e.,

$$
\begin{aligned}
&(\alpha_i^-)^{\text{th}}\text{mXX}_i^- = (2l_i^-)^{\text{th}}\text{mXX}, \; (\alpha_j^+)^{\text{th}}\text{mXX}_j^+ = (2l_j^+)^{\text{th}}\text{mYY}, \\
&\text{for } x_k^* = a_i^- \neq a_j^+ \; (i = 1, 2, \ldots, s_1; j = 1, 2, \ldots, s_2), \\
&\text{XX} \in \{\text{US}, \text{LS}\}, \text{YY} \in \{\text{US}, \text{LS}\}.
\end{aligned}
\tag{4.192}
$$

then the parallel *flower-bundle* switching bifurcation for higher-order fixed-points becomes a combination of two independent left and right parallel appearing bifurcations:

(i) a $((2l_1^-)^{\text{th}}\text{mXX}_1^- : (2l_2^-)^{\text{th}}\text{mXX}_2^- : \cdots : (2l_{s_1}^-)^{\text{th}}\text{mXX}_{s_1}^-)$-*left parallel sprinkler-spraying appearing (or right vanishing)* bifurcation and

(ii) a $((2l_1^+)^{\text{th}}\text{mXX}_1^+ : (2l_2^+)^{\text{th}}\text{mXX}_2^+ : \cdots : (2l_{s_2}^+)^{\text{th}}\text{mXX}_{s_2}^+)$-*right parallel sprinkler-spraying appearing (or left vanishing)* bifurcation.

The parallel switching and appearing bifurcations for higher-order fixed-points are presented in Fig. 4.12(i)–(iv). The *waterfall* appearing bifurcations and the *flower-bundle* switching bifurcations for higher-order fixed-points are presented.

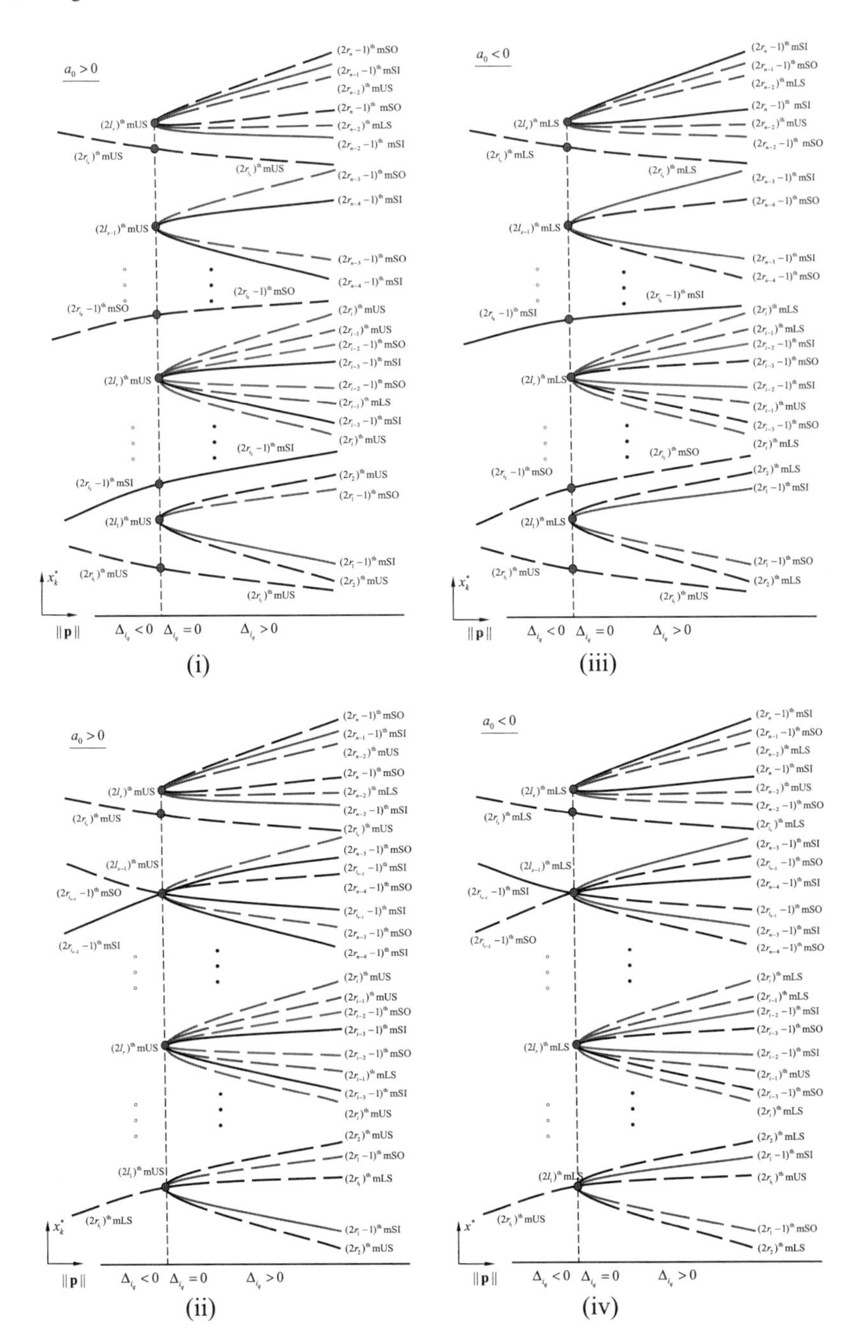

Fig. 4.12 (r_1th mXX:r_2th mXX:$\cdots$: r_nth mXX) parallel switching-appearing bifurcations. ($a_0 > 0$): (i) without switching, and (ii) with switching. ($a_0 > 0$): (iii) without switching, and (vi) with switching. mLS: monotonic-lower-saddle, mUS: monotonic-upper-saddle, mSI: monotonic sink, mSO: monotonic source. Stable and unstable fixed-points are represented by solid and dashed curves, respectively. The bifurcation points are marked by circular symbols.

4.4 Forward Bifurcation Trees

In this section, the analytical bifurcation scenario of a $(2m)^{\text{th}}$-degree polynomial nonlinear discrete system will be discussed. The period-doubling bifurcation scenario will be discussed first through nonlinear renormalization techniques, and the bifurcation scenario based on the monotonic saddle-node bifurcation will be discussed, which is independent of period-1 fixed-points.

4.4.1 *Period-Doubled* $(2\mathrm{m})^{\text{th}}$*-Degree Polynomial Discrete Systems*

After the period-doubling bifurcation of a period-1 fixed-point, the period-doubled fixed-points of a $(2m)^{\text{th}}$-degree polynomial nonlinear discrete system can be obtained. Consider the period-doubling solutions of a forward $(2\mathrm{m})^{\text{th}}$-degree polynomial nonlinear discrete system.

Theorem 4.1 *Consider a* $(2m)^{th}$*-degree polynomial nonlinear discrete system*

$$\begin{aligned} x_{k+1} &= x_k + A_0(\mathbf{p})x_k^{2m} + A_1(\mathbf{p})x_k^{2m-1} + \cdots + A_{2m-2}(\mathbf{p})x_k^2 + A_{2m-1}(\mathbf{p})x_k + A_{2m}(\mathbf{p}) \\ &= x_k + a_0(\mathbf{p})[x_k^2 + B_1(\mathbf{p})x_k + C_1(\mathbf{p})]\cdots[x_k^2 + B_m(\mathbf{p})x_k + C_m(\mathbf{p})] \end{aligned} \tag{4.193}$$

where $A_0(\mathbf{p}) \neq 0$, *and*

$$\mathbf{p} = (p_1, p_2, \ldots, p_{m_1})^{\mathrm{T}}. \tag{4.194}$$

If

$$\begin{aligned} &\Delta_i = B_i^2 - 4C_i > 0,\ i = i_1, i_2, \ldots, i_l \in \{1, 2, \ldots, m\} \cup \{\varnothing\}, \\ &\Delta_j = B_j^2 - 4C_j < 0,\ j = i_{l+1}, i_{l+2}, \ldots, i_m \in \{1, 2, \ldots, m\} \cup \{\varnothing\} \\ &\text{with } l \in \{0, 1, \ldots, m\}, \end{aligned} \tag{4.195}$$

then, the corresponding standard form is

$$x_{k+1} = x_k + a_0 \textstyle\prod_{i=1}^{2m}(x_k - a_i). \tag{4.196}$$

where

$$\begin{aligned}
&b_{i,1}^{(1)}=-\tfrac{1}{2}(B_i^{(1)}+\sqrt{\Delta_i^{(1)}}),\ b_{i,2}^{(1)}=-\tfrac{1}{2}(B_i^{(1)}-\sqrt{\Delta_i^{(1)}})\\
&\text{for }\Delta_i^{(1)}\geq 0,\ i\in\{1,2,\ldots,l\}\cup\{\varnothing\};\\
&\cup_{i=1}^{2l}\{a_i^{(1)}\}=\text{sort}\{\cup_{i_1=1}^{l}\{b_{i_1,2}^{(1)},b_{i_1,2}^{(1)}\}\},a_i^{(1)}\leq a_{i+1}^{(i)};\\
&b_{i,1}^{(1)}=-\tfrac{1}{2}(B_i^{(1)}+\mathbf{i}\sqrt{|\Delta_i^{(1)}|}),\ b_{i,2}^{(1)}=-\tfrac{1}{2}(B_i^{(1)}-\mathbf{i}\sqrt{|\Delta_i^{(1)}|})\\
&\text{for }\Delta_i^{(1)}<0,\ i\in\{l+1,l+2,\ldots,m\}\cup\{\varnothing\},\ \mathbf{i}=\sqrt{-1};\\
&\cup_{i=2l+1}^{2m}\{a_i^{(1)}\}=\{\cup_{i_1=l+1}^{m}\{b_{i_1,2}^{(1)},b_{i_1,2}^{(1)}\}\}.
\end{aligned}\tag{4.197}$$

(i) *Consider a forward period-2 discrete system of Eq. (4.193) as*

$$\begin{aligned}
x_{k+2}&=x_k+[a_0\textstyle\prod_{i_1=1}^{2m}(x_k-a_{i_1}^{(1)})]\{1+\prod_{i_1=1}^{2m}[1+a_0\prod_{i_2=1,i_2\neq i_1}^{2m}(x_k-a_{i_2}^{(1)})]\}\\
&=x_k+[a_0\textstyle\prod_{i_1=1}^{2m}(x_k-a_{i_1}^{(1)})][a_0^{2m}\prod_{i_2=1}^{((2m)^2-2m)/2}(x_k^2+B_{i_2}^{(2)}x_k+C_{i_2}^{(2)})]\\
&=x_k+[a_0\textstyle\prod_{j_1=1}^{3}(x_k-a_{i_1}^{(1)})][a_0^{2m}\prod_{j_2=1}^{(2m)^2-2m}(x_k-b_{j_2}^{(2)})]\\
&=x_k+a_0^{1+2m}\textstyle\prod_{i=1}^{(2m)^2}(x_k-a_i^{(2)})
\end{aligned}\tag{4.198}$$

where

$$\begin{aligned}
&b_{i,1}^{(2)}=-\frac{1}{2}(B_i^{(2)}+\sqrt{\Delta^{(2)}}),b_{i,2}^{(2)}=-\frac{1}{2}(B_i^{(2)}-\sqrt{\Delta_i^{(2)}}),\\
&\Delta_i^{(2)}=(B_i^{(2)})^2-4C_i^{(2)}\geq 0,\ i\in\cup_{q_1=1}^{N_1}I_{q_1}^{(2^0)}\cup\cup_{q_2=1}^{N_2}I_{q_2}^{(2^2)}\\
&I_{q_1}^{(2^0)}=\{l_{(q_1-1)\times 2^0\times m_1+1},l_{(q_1-1)\times 2^0\times m_1+2},\cdots,l_{q_1\times 2^0\times m_1}\}\\
&\qquad\subseteq\{1,2,\cdots,M_1\}\cup\{\varnothing\},\\
&\quad q_1\in\{1,2,\cdots,N_1\},M_1=N_1\times 2^0\times m_1,m_1\in\{1,2,\cdots,m\};\\
&I_{q_2}^{(2^1)}=\{l_{(q_2-1)\times 2^1\times m_1+1},l_{(q_2-1)\times 2^1\times m_1+2},\cdots,l_{q_2\times 2^1\times m_1}\}\\
&\qquad\subseteq\{M_1+1,M_1+2,\cdots,M_2\}\cup\{\varnothing\},\\
&\quad q_2\in\{1,2,\cdots,N_2\},M_2=((2m)^2-2m)/2;\\
&\quad b_{i,1}^{(2)}=-\frac{1}{2}(B_i^{(2)}+\mathbf{i}\sqrt{|\Delta^{(2)}|}),b_{i,2}^{(2)}=-\frac{1}{2}(B_i^{(2)}-\mathbf{i}\sqrt{|\Delta_i^{(2)}|}),\\
&\qquad\mathbf{i}=\sqrt{-1},\Delta_i^{(2)}=(B_i^{(2)})^2-4C_i^{(2)}<0,\\
&\qquad i\in J^{(2^1)}=\{l_{N_2\times 2^1\times m_1+1},l_{N_2\times 2^1\times m_1+2},\cdots,l_{M_2}\}\\
&\qquad\qquad\subseteq\{M_1+1,M_1+2,\cdots,M_2\}
\end{aligned}\tag{4.199}$$

with fixed-points

$$\begin{array}{l} x_{k+2}^{*} = x_{k}^{*} = a_{i}^{(2)}, (i = 1, 2, \ldots, (2m)^{2}) \\ \cup_{i=1}^{(2m)^{2}} \{a_{i}^{(2)}\} = \text{sort}\{\cup_{j_1=1}^{2m}\{a_{j_1}^{(1)}\}, \cup_{j_2=1}^{M}\{b_{j_2,1}^{(2)}, b_{j_2,2}^{(2)}\}\} \\ \text{with } a_{i}^{(2)} < a_{i+1}^{(2)}, \ M = ((2m)^{2} - 2m)/2. \end{array} \tag{4.200}$$

(ii) *For a fixed-point of* $x_{k+1}^{*} = x_{k}^{*} = a_{i_1}^{(1)}$ $(i_1 \in \{1, 2, \ldots, 2m\})$, *if*

$$\frac{dx_{k+1}}{dx_k}\Big|_{x_k^*=a_{i_1}^{(1)}} = 1 + a_0 \prod_{i_2=1, i_2 \neq i_1}^{2m} (a_{i_1}^{(1)} - a_{i_2}^{(1)}) = -1, \tag{4.201}$$

with

- *a* r^{th}*-order oscillatory upper-saddle-node bifurcation* $(d^r x_{k+1}/dx_k^r|_{x_k^*=a_{i_1}^{(1)}} > 0,\ r = 2l_1)$,
- a rth*-order oscillatory lower-saddle-node bifurcation* $(d^r x_{k+1}/dx_k^r|_{x_k^*=a_{i_1}^{(1)}} < 0, r = 2l_1)$,
- a rth*-order oscillatory sink bifurcation* $(d^r x_{k+1}/dx_k^r|_{x_k^*=a_{i_1}^{(1)}} < 0, r = 2l_1 + 1)$,
- a rth*-order oscillatory source bifurcation* $(d^r x_{k+1}/dx_k^r|_{x_k^*=a_{i_1}^{(1)}} > 0, r = 2l_1 + 1)$,

then the following relations satisfy

$$a_{i_1}^{(1)} = -\frac{1}{2} B_{i_1}^{(2)}, \Delta_{i_1}^{(2)} = (B_{i_1}^{(2)})^2 - 4C_{i_1}^{(2)} = 0, \tag{4.202}$$

and there is a period-2 discrete system of the quartic discrete system in Eq. (4.193) as

$$x_{k+2} = x_k + a_0^{1+(2m)} \prod_{i_2 \in I_{q_1}^{(20)}} (x_k - a_{i_2}^{(1)})^3 \prod_{i_3=1}^{(2m)^2} (x_k - a_{i_3}^{(2)})^{(1-\delta(i_2, i_3))} \tag{4.203}$$

for $i_1 \in \{1, 2, \ldots, 2m\}$, $i_1 \neq i_2$ *with*

$$\frac{dx_{k+2}}{dx_k}\Big|_{x_k^*=a_{i_1}^{(1)}} = 1, \ \frac{d^2 x_{k+2}}{dx_k^2}\Big|_{x_k^*=a_{i_1}^{(1)}} = 0; \tag{4.204}$$

- x_{k+2} *at* $x_k^* = a_{i_1}^{(1)}$ *is a monotonic sink of the third-order if*

$$\frac{d^3 x_{k+2}}{dx_k^3}\Big|_{x_k^*=a_{i_1}^{(1)}} = 6a_0^{1+2m} \prod\nolimits_{i_2 \in I_{q_1}^{(2^0)}, i_2 \neq i_1} (a_{i_1}^{(1)} - a_{i_2}^{(1)})^3 \times \prod\nolimits_{i_3=1}^{(2m)^2} (a_{i_1}^{(1)} - a_{i_3}^{(2)})^{(1-\delta(i_2,i_3))} < 0, \tag{4.205}$$

and the corresponding bifurcations is a third-order monotonic sink bifurcation for the period-2 discrete system;

- x_{k+2} *at* $x_k^* = a_{i_1}^{(1)}$ *is a monotonic source of the third-order if*

$$\frac{d^3 x_{k+2}}{dx_k^3}\Big|_{x_k^*=a_{i_1}^{(1)}} = 6a_0^{1+2m} \prod\nolimits_{i_2 \in I_{q_1}^{(2^0)}, i_2 \neq i_1} (a_{i_1}^{(1)} - a_{i_2}^{(1)})^3 \times \prod\nolimits_{i_3=1}^{(2m)^2} (a_{i_1}^{(1)} - a_{i_3}^{(2)})^{(1-\delta(i_2,i_3))} > 0, \tag{4.206}$$

and the corresponding bifurcations is a third-order monotonic source bifurcation for the period-2 discrete system.

(ii_1) *The period-2 fixed-points are trivial and unstable if*

$$x_{k+2}^* = x_k^* = a_i^{(2)} \in \cup_{i_1}^{2m} \{a_{i_1}^{(1)}\}. \tag{4.207}$$

(ii_2) *The period-2 fixed-points are non-trivial and stable if*

$$x_{k+2}^* = x_k^* = a_i^{(2)} \in \cup_{i_1=1}^{M_2} \{b_{i_1,1}^{(2)}, b_{i_1,2}^{(2)}\}. \tag{4.208}$$

Proof The proof is straightforward through the simple algebraic manipulation. Following the proof of quadratic discrete system, this theorem is proved. ■

4.4.2 Renormalization and Period-Doubling

The generalized case of period-doublization of a $(2m)^{\text{th}}$-degree polynomial discrete system is presented through the following theorem. The analytical period-doubling bifurcation trees can be developed for such a $(2m)^{\text{th}}$-degree polynomial discrete systems.

Theorem 4.2 *Consider a 1-dimensional $(2m)^{th}$-degree polynomial discrete system as*

$$\begin{aligned} x_{k+1} &= x_k + A_0 x_k^{2m} + A_1 x_k^{2m-1} + \cdots + A_{2m-2} x_k^2 + A_{2m-1} x_k + A_{2m} \\ &= x_k + a_0 \prod_{i=1}^{2m} (x_k - a_i). \end{aligned} \tag{4.209}$$

(i) *After l-times period-doubling bifurcations, a period-2l discrete system ($l = 1, 2, \ldots$) for the $(2m)^{th}$-degree polynomial discrete system in Eq. (4.209) is given through the nonlinear renormalization as*

$$\begin{aligned} x_{k+2^l} &= x_k + [a_0^{(2^{l-1})} \prod_{i_1=1}^{(2m)^{2^{l-1}}} (x_k - a_{i_1}^{(2^{l-1})})] \\ &\quad \times \{1 + \prod_{i_1=1}^{(2m)^{2^{l-1}}} [1 + a_0^{(2^{l-1})} \prod_{i_2=1, i_2 \neq i_1}^{(2m)^{2^{l-1}}} (x_k - a_{i_2}^{(2^{l-1})})]\} \\ &= x_k + [a_0^{(2^{l-1})} \prod_{i_1=1}^{(2m)^{2^{l-1}}} (x_k - a_{i_1}^{(2^{l-1})})] \\ &\quad \times [(a_0^{(2^{l-1})})^{4^{2^{l-1}}} \prod_{j_1=1}^{((2m)^{2^l} - (2m)^{2^{l-1}})/2} (x_k^2 + B_{j_2}^{(2^l)} x_k + C_{j_2}^{(2^l)})] \\ &= x_k + [a_0^{(2^{l-1})} \prod_{i_1=1,}^{(2m)^{2^{l-1}}} (x_k - a_{i_1}^{(2^{l-1})})] \\ &\quad \times [(a_0^{(2^{l-1})})^{4^{2^{l-1}}} \prod_{i_2=1}^{((2m)^{2^l} - (2m)^{2^{l-1}})/2} (x_k - b_{i_2,1}^{(2^l)})(x_k - b_{i_2,2}^{(2^l)})] \\ &= x_k + (a_0^{(2^{l-1})})^{1+(2m)^{2^{l-1}}} \prod_{i=1}^{(2m)^{2^l}} (x_k - a_i^{(2^l)}) \\ &= x_k + a_0^{(2^l)} \prod_{i=1}^{(2m)^{2^l}} (x_k - a_i^{(2^l)}) \end{aligned} \tag{4.210}$$

with

$$\begin{aligned} \frac{dx_{k+2^l}}{dx_k} &= 1 + a_0^{(2^l)} \sum_{i_1=1}^{(2m)^{2^l}} \prod_{i_2=1, i_2 \neq i_1}^{(2m)^{2^l}} (x_k - a_{i_2}^{(2^l)}), \\ \frac{d^2 x_{k+2^l}}{dx_k^2} &= a_0^{(2^l)} \sum_{i_1=1}^{(2m)^{2^l}} \sum_{i_2=1, i_2 \neq i_1}^{(2m)^{2^l}} \prod_{i_3=1, i_3 \neq i_1, i_2}^{(2m)^{2^l}} (x_k - a_{i_3}^{(2^l)}), \\ &\vdots \\ \frac{d^r x_{k+2^l}}{dx_k^r} &= a_0^{(2^l)} \sum_{i_1=1}^{(2m)^{2^l}} \cdots \sum_{i_r=1, i_3 \neq i_1, i_2 \ldots i_{r-1}}^{(2m)^{2^l}} \prod_{i_{r+1}=1, i_{r+1} \neq i_1, i_2 \ldots, i_r}^{(2m)^{2^l}} (x_k - a_{i_{r+1}}^{(2^l)}) \end{aligned} \tag{4.211}$$

for $r \leq (2m)^{2^l}$ *where*

$$
\begin{aligned}
& a_0^{(2)} = (a_0)^{1+2m}, a_0^{(2^l)} = (a_0^{(2^{l-1})})^{1+(2m)^{2^{l-1}}}, l = 1,2,3,\cdots; \\
& \cup_{i=1}^{(2m)^{2^l}} \{a_i^{(2^l)}\} = \mathrm{sort}\{\cup_{i_1=1}^{(2m)^{2^{l-1}}} \{a_{i_1}^{(2^l)}\}, \cup_{i_2=1}^{M_2}\{b_{i_2,1}^{(2^l)}, b_{i_2,2}^{(2^l)}\}\}, a_i^{(2^l)} \le a_{i+1}^{(2^l)}; \\
& b_{i,1}^{(2^l)} = -\tfrac{1}{2}(B_i^{(2^l)} + \sqrt{\Delta_i^{(2^l)}}), b_{i,2}^{(2^{l-1})} = -\tfrac{1}{2}(B_i^{(2^l)} - \sqrt{\Delta_i^{(2^l)}}), \\
& \Delta_i^{(2^l)} = (B_i^{(2^l)})^2 - 4C_i^{(2^l)} \ge 0 \text{ for } i \in \cup_{q_1=1}^{N_1} I_{q_1}^{(2^{l-1})} \bigcup \cup_{q_2=1}^{N_2} I_{q_2}^{(2^l)}, \\
& I_{q_1}^{(2^{l-1})} = \{l_{(q_1-1)\times 2^{l-1}\times m_1+1}, l_{(q_1-1)\times 2^{l-1}\times m_1+2}, \cdots, l_{q_1\times 2^{l-1}\times m_1}\} \\
& \qquad \subseteq \{1,2,\cdots,M_1\} \cup \{\varnothing\}, \\
& \text{for } q_1 \in \{1,2,\cdots,N_1\}, M_1 = N_1 \times 2^{l-1} \times m_1, m_1 \in \{1,2,\cdots,m\}; \\
& I_{q_2}^{(2^l)} = \{l_{(q_2-1)\times 2^l\times m_1+1}, l_{(q_2-1)\times 2^l\times m_1+2}, \cdots, l_{q_2\times 2^l\times m_1}\} \\
& \qquad \subseteq \{M_1+1, M_1+2, \cdots, M_2\} \cup \{\varnothing\}, \\
& \text{for } q_2 \in \{1,2,\cdots,N_2\}, M_2 = ((2m)^{2^l} - (2m)^{2^{l-1}})/2; \\
& b_{i,1}^{(2^l)} = -\tfrac{1}{2}(B_i^{(2^l)} + \mathbf{i}\sqrt{|\Delta_i^{(2^l)}|}), \ b_{i,2}^{(2^l)} = -\tfrac{1}{2}(B_i^{(2^l)} - \mathbf{i}\sqrt{|\Delta_i^{(2^l)}|}), \\
& \Delta_i^{(2^l)} = (B_i^{(2^l)})^2 - 4C_i^{(2^l)} < 0, \mathbf{i} = \sqrt{-1}, \\
& i \in J^{(2^l)} = \{l_{N_2\times 2^l\times m_1+1}, l_{N_2\times 2^l\times m_1+2}, \cdots, l_{M_2}\} \\
& \qquad \subset \{M_1+1, M_1+2, \cdots, M_2\} \cup \{\varnothing\};
\end{aligned} \tag{4.212}
$$

with fixed-points

$$
\begin{aligned}
& x_{k+2^l}^* = x_k^* = a_i^{(2^l)}, (i = 1,2,\ldots,(2m)^{2^l}) \\
& \cup_{i=1}^{(2m)^{2^l}} \{a_i^{(2^l)}\} = \mathrm{sort}\{\cup_{i_1=1}^{(2m)^{2^{l-1}}} \{a_{i_1}^{(2^{l-1})}\}, \cup_{i_2=1}^{M_2}\{b_{i_2,1}^{(2^l)}, b_{i_2,2}^{(2^l)}\}\} \\
& \text{with } a_i^{(2^l)} < a_{i+1}^{(2^l)}.
\end{aligned} \tag{4.213}
$$

(ii) *For a fixed-point of* $x_{k+2^{l-1}}^* = x_k^* = a_{i_1}^{(2^{l-1})}$ $(i_1 \in I_q^{(2^{l-1})})$, *if*

$$
\begin{aligned}
& \frac{dx_{k+2^{l-1}}}{dx_k}\Big|_{x_k^* = a_{i_1}^{(2^{l-1})}} = 1 + a_0^{(2^{l-1})} \prod_{i_2=1, i_2\ne i_1}^{(2m)^{2^{l-1}}} (a_{i_1}^{(2^{l-1})} - a_{i_2}^{(2^{l-1})}) = -1, \\
& \frac{d^s x_{k+2^{l-1}}}{dx_k^s}\Big|_{x_k^* = a_{i_1}^{(2^{l-1})}} = 0, \ \text{ for } s = 2,\cdots,r-1; \\
& \frac{d^r x_{k+2^{l-1}}}{dx_k^r}\Big|_{x_k^* = a_{i_1}^{(2^{l-1})}} \ne 0 \text{ for } 1 < r \le (2m)^{2^{l-1}},
\end{aligned} \tag{4.214}
$$

with

- *a rth-order oscillatory sink for* $d^r x_{k+2^{l-1}}/dx_k^r|_{x_k^*=a_{i_1}^{(2^{l-1})}} > 0$ and $r = 2l_1 + 1$;
- *a rth-order oscillatory source for* $d^r x_{k+2^{l-1}}/dx_k^r|_{x_k^*=a_{i_1}^{(2^{l-1})}} < 0$ *and* $r = 2l_1 + 1$;
- *a rth -order oscillatory upper-saddle for* $d^r x_{k+2^{l-1}}/dx_k^r|_{x_k^*=a_{i_1}^{(2^{l-1})}} > 0$ *and* $r = 2l_1$;
- *a rth-order oscillatory lower-saddle for* $d^r x_{k+2^{l-1}}/dx_k^r|_{x_k^*=a_{i_1}^{(2^{l-1})}} < 0$ and $r = 2l_1$;

then there is a period- 2^l *fixed-point discrete system*

$$\begin{aligned} x_{k+2^l} = x_k + a_0^{(2^l)} \prod\nolimits_{i_1 \in I_{q_1}^{(2^{l-1})}} (x_k - a_{i_1}^{(2^{l-1})})^3 \\ \times \prod\nolimits_{j_2=1}^{(2m)^{2^l}} (x_k - a_{j_2}^{(2^l)})^{(1-\delta(i_1,j_2))} \end{aligned} \tag{4.215}$$

where

$$\delta(i_1,j_2) = 1 \text{ if } a_{j_2}^{(2^l)} = a_{i_1}^{(2^{l-1})},\ \delta(i_1,j_2) = 0 \text{ if } a_{j_2}^{(2^l)} \neq a_{i_1}^{(2^{l-1})} \tag{4.216}$$

and

$$\frac{dx_{k+2^l}}{dx_k}\Big|_{x_k^*=a_{i_1}^{(2^{l-1})}} = 1,\ \frac{d^2 x_{k+2^l}}{dx_k^2}\Big|_{x_k^*=a_{i_1}^{(2^{l-1})}} = 0. \tag{4.217}$$

- x_{k+2^l} at $x_k^* = a_{i_1}^{(2^{l-1})}$ *is a monotonic sink of the third-order if*

$$\begin{aligned} \frac{d^3 x_{k+2^l}}{dx_k^3}\Big|_{x_k^*=a_{i_1}^{(2^{l-1})}} = 6a_0^{(2^l)} \prod\nolimits_{i_2 \in I_q^{(2^{l-1})}, i_2 \neq i_1} (a_{i_1}^{(2^{l-1})} - a_{i_2}^{(2^{l-1})})^3 \\ \times \prod\nolimits_{j_2=1}^{(2m)^{2^l}} (a_{i_1}^{(2^{l-1})} - a_{j_2}^{(2^l)})^{(1-\delta(i_2,j_2))} < 0 \\ (i_1 \in I_q^{(2^{l-1})}, q \in \{1,2,\ldots,N_1\}), \end{aligned} \tag{4.218}$$

and such a bifurcation at $x_k^* = a_{i_1}^{(2^{l-1})}$ *is a third-order monotonic sink bifurcation.*

- x_{k+2^l} *at* $x_k^* = a_{i_1}^{(2^{l-1})}$ *is a monotonic source of the third-order if*

$$\begin{aligned} \frac{d^3 x_{k+2^l}}{dx_k^3}\Big|_{x_k^*=a_{i_1}^{(2^{l-1})}} &= 6a_0^{(2^l)} \prod\nolimits_{i_2 \in I_q^{(2^{l-1})}, i_2 \neq i_1} (a_{i_1}^{(2^{l-1})} - a_{i_2}^{(2^{l-1})})^3 \\ &\times \prod\nolimits_{j_2=1}^{(2m)^{2^l}} (a_{i_1}^{(2^{l-1})} - a_{j_2}^{(2^l)})^{(1-\delta(i_2,j_2))} > 0 \end{aligned} \tag{4.219}$$
$$(i_1 \in I_q^{(2^{l-1})}, q \in \{1,2,\ldots,N_1\})$$

and such a bifurcation at $x_k^* = a_{i_1}^{(2^{l-1})}$ *is a third-order monotonic source bifurcation.*

(ii$_1$) *The period-* 2^l *fixed-points are trivial if*

$$x_{k+2^l}^* = x_k^* = a_i^{(2^l)} \in \cup_{i_1=1}^{(2m)^{2^{l-1}}} \{a_{i_1}^{(2^{l-1})}\} \text{ for } i_1 = 1,2,\ldots,(2m)^{2^{l-1}}. \tag{4.220}$$

(ii$_2$) *The period-* 2^l *fixed-points are non-trivial if*

$$\begin{aligned} & x_{k+2^l}^* = x_k^* = a_i^{(2^l)} \in \cup_{i_1=1}^{(2m)^{2^{l-1}}} \{b_{i_1,1}^{(2^l)}, b_{i_1,2}^{(2^l)}\} \\ & j_1 \in \{1,2,\ldots,M_2\} \cup \{\varnothing\} \end{aligned}. \tag{4.221}$$

Such a period-2^l *fixed-point is*

- *monotonically unstable if* $dx_{k+2^l}/dx_k|_{x_k^*=a_{i_1}^{(2^l)}} \in (1,\infty)$;
- *monotonically invariant if* $dx_{k+2^l}/dx_k|_{x_k^*=a_{i_1}^{(2^l)}} = 1$, which is
  - *a monotonic upper-saddle of the* $(2l_1)^{\text{th}}$ *order for* $d^{2l_1}x_{k+2^l}/\,dx_k^{2l_1}|_{x_k^*} > 0$;
  - *a monotonic lower-saddle of the* $(2l_1)^{\text{th}}$ *order for* $d^{2l_1}x_{k+2^l}/dx_k^{2l_1}|_{x_k^*} < 0$;
  - *a monotonic source of the* $(2l_1+1)^{\text{th}}$ *order for* $d^{2l_1+1}x_{k+2^l}/\,dx_k^{2l_1+1}|_{x_k^*} > 0$;
  - *a monotonic sink the* $(2l_1+1)^{\text{th}}$ *order for* $d^{2l_1+1}x_{k+2^l}/dx_k^{2l_1+1}|_{x_k^*} < 0$;
- *monotonically stable if* $dx_{k+2^l}/dx_k|_{x_k^*=a_{i_1}^{(2^l)}} \in (0,1)$;
- *invariantly zero-stable if* $dx_{k+2^l}/dx_k|_{x_k^*=a_{i_1}^{(2^{l-1})}} = 0$;
- *oscillatorilly stable if* $dx_{k+2^l}/dx_k|_{x_k^*=a_{i_1}^{(2^{l-1})}} \in (-1,0)$;
- *flipped if* $dx_{k+2^l}/dx_k|_{x_k^*=a_{i_1}^{(2^{l-1})}} = -1$, *which is*
  - *an oscillatory upper-saddle of the* $(2l_1)^{\text{th}}$ *order for* $d^{2l_1}x_{k+2^l}/\,dx_k^{2l_1}|_{x_k^*} > 0$;
  - *an oscillatory lower-saddle the* $(2l_1)^{\text{th}}$ *order for* $d^{2l_1}x_{k+2^l}/dx_k^{2l_1}|_{x_k^*} < 0$;

- *an oscillatory source of the* $(2l_1+1)^{\text{th}}$ *order if* $d^{2l_1+1}x_{k+2^l}/\,dx_k^{2l_1+1}|_{x_k^*}<0$;
- *an oscillatory sink the* $(2l_1+1)^{\text{th}}$ *order with* $d^{2l_1+1}x_{k+2^l}/\,dx_k^{2l_1+1}|_{x_k^*}>0$;

• *oscillatorilly unstable if* $dx_{k+2^l}/dx_k|_{x_k^*=a_{i_1}^{(2^l)}}\in(-\infty,-1)$.

Proof Through the nonlinear renormalization, this theorem can be proved. ■

4.4.3 *Period-*n *Appearing and Period-Doublization*

The forward period-*n* discrete system for the $(2m)^{th}$-degree polynomial quartic nonlinear discrete systems will be discussed, and the period-doublization of period-*n* discrete systems is discussed through the nonlinear renormalization.

Theorem 4.3 *Consider a 1-dimensional* $(2m)^{th}$-degree polynomial discrete system as

$$\begin{aligned} x_{k+1} &= x_k + A_0x_k^{2m} + A_1x_k^{2m-1} + \cdots + A_{2m-2}x_k^2 + A_{2m-1}x_k + A_{2m} \\ &= x_k + a_0\prod_{i=1}^{2m}(x_k-a_i). \end{aligned} \tag{4.222}$$

(i) *After n-times iterations, a period-n discrete system for the quartic discrete system in Eq. (4.222) is*

$$\begin{aligned} x_{k+n} &= x_k + a_0\prod_{i_1=1}^{2m}(x_k-a_{i_2})[1+\sum_{j=1}^{n}Q_j] \\ &= x_k + a_0^{((2m)^n-1)/(2m-1)}\prod_{i_1=1}^{2m}(x_k-a_{i_1})[\prod_{j_2=1}^{((2m)^n-2m)/2}(x_k^2+B_{j_2}^{(n)}x_k+C_{j_2}^{(n)})] \\ &= x_k + a_0^{(n)}\prod_{i=1}^{(2m)^n}(x_k-a_i^{(n)}) \end{aligned} \tag{4.223}$$

with

$$\begin{aligned} \frac{dx_{k+n}}{dx_k} &= 1 + a_0^{(n)}\sum_{i_1=1}^{(2m)^n}\prod_{i_2=1,i_2\neq i_1}^{(2m)^n}(x_k-a_{i_2}^{(n)}), \\ \frac{d^2x_{k+n}}{dx_k^2} &= a_0^{(n)}\sum_{i_1=1}^{(2m)^n}\sum_{i_2=1,i_2\neq i_1}^{(2m)^n}\prod_{i_3=1,i_3\neq i_1,i_2}^{(2m)^n}(x_k-a_{i_3}^{(n)}), \\ &\vdots \\ \frac{d^rx_{k+n}}{dx_k^r} &= a_0^{(n)}\sum_{i_1=1}^{(2m)^n}\cdots\sum_{i_r=1,i_r\neq i_1,i_2\cdots,i_{r-1}}^{(2m)^n}\prod_{i_{r+1}=1,i_{r+1}\neq i_1,i_2\cdots,i_r}^{(2m)^n}(x_k-a_{i_{r+1}}^{(n)}) \\ &\text{for } r\leq(2m)^n; \end{aligned} \tag{4.224}$$

where

$$
\begin{aligned}
&a_0^{(n)} = (a_0)^{((2m)^n-1)/(2m-1)}, Q_1 = 0, Q_2 = \prod_{i_2=1}^{2m}[1 + a_0 \prod_{i_1=1, i_1 \neq i_2}^{2m}(x_k - a_{i_1}^{(1)})],\\
&Q_n = \prod_{i_n=1}^{2m}[1 + a_0(1 + Q_{n-1}) \prod_{i_{n-1}=1, i_{n-1} \neq i_n}^{2m}(x_k - a_{i_{n-1}}^{(1)})],\ n = 3, 4, \cdots;\\
&\cup_{i=1}^{(2m)^n}\{a_i^{(n)}\} = \mathrm{sort}\{\cup_{i_1=1}^{2m}\{a_{i_1}^{(1)}\}, \cup_{i_2=1}^{M}\{b_{i_2,1}^{(n)}, b_{i_2,2}^{(n)}\}\}\ ;\\
&b_{i_2,1}^{(n)} = -\frac{1}{2}(B_{i_2}^{(n)} + \sqrt{\Delta_{i_2}^{(n)}}), b_{i_2,2}^{(n)} = -\frac{1}{2}(B_{i_2}^{(n)} - \sqrt{\Delta_{i_2}^{(n)}}),\\
&\Delta_{i_2}^{(n)} = (B_{i_2}^{(n)})^2 - 4C_{i_2}^{(n)} \geq 0 \ \text{ for }\ i_2 \in \cup_{q=1}^{N} I_q^{(n)},\\
&I_q^{(n)} = \{l_{(q-1)\times n+1}, l_{(q-1)\times n+2}, \cdots, l_{q\times n}\} \subseteq \{1, 2, \cdots, M\} \cup \{\varnothing\},\\
&\text{for } q \in \{1, 2, \cdots, N\}, M = ((2m)^n - 2m)/2;\\
&b_{i,1}^{(n)} = -\frac{1}{2}(B_i^{(n)} + \mathbf{i}\sqrt{|\Delta_i^{(n)}|}),\ b_{i,2}^{(n)} = -\frac{1}{2}(B_i^{(n)} - \mathbf{i}\sqrt{|\Delta_i^{(n)}|}),\\
&\Delta_i^{(n)} = (B_i^{(n)})^2 - 4C_i^{(n)} < 0,\ \mathbf{i} = \sqrt{-1}\\
&i \in \{l_{N\times n+1}, l_{N\times n+2}, \cdots, l_M\} \subset \{1, 2, \cdots, M\} \cup \{\varnothing\};
\end{aligned}
\tag{4.225}
$$

with fixed-points

$$
\begin{aligned}
&x_{k+n}^* = x_k^* = a_i^{(n)}, (i = 1, 2, \ldots, (2m)^n)\\
&\cup_{i=1}^{(2m)^n}\{a_i^{(n)}\} = \mathrm{sort}\{\cup_{i_1=1}^{2m}\{a_{i_1}^{(1)}\}, \cup_{i_2=1}^{M}\{b_{i_2,1}^{(n)}, b_{i_2,2}^{(n)}\}\}\\
&\text{with } a_i^{(n)} < a_{i+1}^{(n)}.
\end{aligned}
\tag{4.226}
$$

(ii) *For a fixed-point of* $x_{k+n}^* = x_k^* = a_{i_1}^{(n)}$ $(i_1 \in I_q^{(n)}, q \in \{1, 2, \ldots, N\})$, *if*

$$
\frac{dx_{k+n}}{dx_k}\Big|_{x_k^* = a_{i_1}^{(n)}} = 1 + a_0^{(n)} \prod_{i_2=1, i_2 \neq i_1}^{(2m)^n}(a_{i_1}^{(n)} - a_{i_2}^{(n)}) = 1, \tag{4.227}
$$

with

$$
\frac{d^2 x_{k+n}}{dx_k^2}\Big|_{x_k^* = a_{i_1}^{(n)}} = a_0^{(n)} \sum_{i_2=1, i_2 \neq i_1}^{(2m)^n} \prod_{i_3=1, i_3 \neq i_1, i_2}^{(2m)^n}(a_{i_1}^{(n)} - a_{i_3}^{(n)}) \neq 0, \tag{4.228}
$$

then there is a new discrete system for onset of the qth*-set of period-n fixed-points based on the second-order monotonic saddle-node bifurcation as*

$$
x_{k+n} = x_k + a_0^{(n)} \prod_{i_1 \in I_q^{(n)}}(x_k - a_{i_1}^{(n)})^2 \prod_{j_2=1}^{(2m)^n}(x_k - a_{j_2}^{(n)})^{(1-\delta(i_1, j_2))} \tag{4.229}
$$

where

$$\delta(i_1,j_2)=1 \text{ if } a_{j_2}^{(n)}=a_{i_1}^{(n)},\ \delta(i_1,j_2)=0 \text{ if } a_{j_2}^{(n)}\neq a_{i_1}^{(n)}. \tag{4.230}$$

(ii$_1$) *If*

$$\begin{aligned}&\frac{dx_{k+n}}{dx_k}\Big|_{x_k^*=a_{i_1}^{(n)}}=1\ (i_1\in I_q^{(n)}),\\&\frac{d^2x_{k+n}}{dx_k^2}\Big|_{x_k^*=a_{i_1}^{(n)}}=2a_0^{(n)}\prod\nolimits_{i_1\in I_q^{(n)},i_2\neq i_1}(a_{i_1}^{(n)}-a_{i_1}^{(n)})^2\\&\qquad\times\prod\nolimits_{j_2=1}^{(2m)^n}(a_{i_1}^{(n)}-a_{j_2}^{(n)})^{(1-\delta(i_2,j_2))}\neq 0\end{aligned} \tag{4.231}$$

x_{k+n} *at* $x_k^*=a_{i_1}^{(n)}$ *is*

- *a monotonic lower-saddle of the second-order for* $d^2x_{k+n}/dx_k^2|_{x_k^*=a_{i_1}^{(n)}}<0$;
- *a monotonic upper-saddle of the second-order for* $d^2x_{k+n}/dx_k^2|_{x_k^*=a_{i_1}^{(n)}}>0$.

(ii$_2$) *The period-n fixed-points* $(n=2^{n_1}\times s)$ *are trivial if*

$$\left.\begin{aligned}&x_k^*=x_{k+n}^*=a_{j_1}^{(n)}\in\{\cup_{i_i=1}^{2m}\{a_{i_1}^{(1)}\},\cup_{i_2=1}^{(2m)^{2^{n_1-1}s}}\{a_{i_2}^{(2^{n_1-1}s)}\}\}\\&\text{for } n_1=1,2,\ldots;s=2l_1+1;j_1\in\{1,2,\ldots,(2m)^n\}\cup\{\varnothing\}\end{aligned}\right\}$$
$$\text{for } n\neq 2^{n_2},$$
$$\left.\begin{aligned}&x_k^*=x_{k+n}^*=a_{j_1}^{(n)}\in\cup_{i_2=1}^{(2m)^{2^{n_1-1}s}}\{a_{i_2}^{(2^{n_1-1}s)}\}\\&\text{for } n_1=1,2,\ldots;s=1;j_1\in\{1,2,\ldots,(2m)^n\}\cup\{\varnothing\}\end{aligned}\right\}$$
$$\text{for } n=2^{n_2}. \tag{4.232}$$

(ii$_3$) *The period-n fixed-points* $(n=2^{n_1}\times s)$ *are non-trivial if*

$$\left.\begin{aligned}&x_k^*=x_{k+n}^*=a_{j_1}^{(n)}\notin\{\cup_{i_i=1}^{2m}\{a_{i_1}^{(1)}\},\cup_{i_2=1}^{(2m)^{2^{n_1-1}s}}\{a_{i_2}^{(2^{n_1-1}s)}\}\}\\&\text{for } n_1=1,2,\ldots;s=2l_1+1;j_1\in\{1,2,\ldots,(2m)^n\}\cup\{\varnothing\}\end{aligned}\right\}$$
$$\text{for } n\neq 2^{n_2},$$
$$\left.\begin{aligned}&x_k^*=x_{k+n}^*=a_{j_1}^{(n)}\notin\cup_{i_2=1}^{(2m)^{2^{n_1-1}s}}\{a_{i_2}^{(2^{n_1-1}s)}\}\\&\text{for } n_1=1,2,\ldots;s=1;j_1\in\{1,2,\ldots,(2m)^n\}\cup\{\varnothing\}\end{aligned}\right\}$$
$$\text{for } n=2^{n_2}. \tag{4.233}$$

Such a forward period-n fixed-point is

- *monotonically unstable if* $dx_{k+n}/dx_k|_{x_k^*=a_{i_1}^{(n)}} \in (1,\infty)$;
- *monotonically invariant if* $dx_{k+n}/dx_k|_{x_k^*=a_{i_1}^{(n)}} = 1$, *which is*
  - *a monotonic upper-saddle of the* $(2l_1)^{\text{th}}$ *order for* $d^{2l_1}x_{k+n}/dx_k^{2l_1}|_{x_k^*} > 0$;
  - *a monotonic lower-saddle the* $(2l_1)^{\text{th}}$ *order for* $d^{2l_1}x_{k+n}/dx_k^{2l_1}|_{x_k^*} < 0$;
  - *a monotonic source of the* $(2l_1+1)^{\text{th}}$ *order for* $d^{2l_1+1}x_{k+n}/dx_k^{2l_1+1}|_{x_k^*} > 0$;
  - *a monotonic sink the* $(2l_1+1)^{\text{th}}$ *order for* $d^{2l_1+1}x_{k+n}/dx_k^{2l_1+1}|_{x_k^*} < 0$;
- *monotonically unstable if* $dx_{k+n}/dx_k|_{x_k^*=a_{i_1}^{(n)}} \in (0,1)$;
- *invariantly zero-stable if* $dx_{k+n}/dx_k|_{x_k^*=a_{i_1}^{(n)}} = 0$;
- *oscillatorilly stable if* $dx_{k+n}/dx_k|_{x_k^*=a_{i_1}^{(n)}} \in (-1,0)$;
- *flipped if* $dx_{k+n}/dx_k|_{x_k^*=a_{i_1}^{(n)}} = -1$, *which is*
  - *an oscillatory upper-saddle of the* $(2l_1)^{\text{th}}$ *order for* $d^{2l_1}x_{k+n}/dx_k^{2l_1}|_{x_k^*} > 0$;
  - *an oscillatory lower-saddle the* $(2l_1)^{\text{th}}$ *order for* $d^{2l_1}x_{k+n}/dx_k^{2l_1}|_{x_k^*} < 0$;
  - *an oscillatory source of the* $(2l_1+1)^{\text{th}}$ *order for* $d^{2l_1+1}x_{k+n}/dx_k^{2l_1+1}|_{x_k^*} < 0$;
  - *an oscillatory sink the* $(2l_1+1)^{\text{th}}$ *order for* $d^{2l_1+1}x_{k+n}/dx_k^{2l_1+1}|_{x_k^*} > 0$;
- *oscillatorilly unstable if* $dx_{k+n}/dx_k|_{x_k^*=a_{i_1}^{(n)}} \in (-\infty,-1)$.

(iii) *For a fixed-point of* $x_{k+n}^* = x_k^* = a_{i_1}^{(n)}$ $(i_1 \in I_q^{(n)}, q \in \{1,2,\ldots,N\})$, *there is a period-doubling of the* q^{th}*-set of period-n fixed-points if*

$$\begin{aligned}
&\frac{dx_{k+n}}{dx_k}|_{x_k^*=a_{i_1}^{(n)}} = 1 + a_0^{(n)} \prod_{j_2=1,j_2\neq i_1}^{(2m)^n} (a_{i_1}^{(n)} - a_{j_2}^{(n)}) = -1,\\
&\frac{d^s x_{k+n}}{dx_k^s}|_{x_k^*=a_{i_1}^{(n)}} = 0, \text{ for } s = 2,\ldots, r-1;\\
&\frac{d^r x_{k+n}}{dx_k^r}|_{x_k^*=a_{i_1}^{(n)}} \neq 0 \text{ for } 1 < r \leq (2m)^n
\end{aligned} \tag{4.234}$$

with

- a rth*-order oscillatory sink for* $d^r x_{k+n}/dx_k^r|_{x_k^*=a_{i_1}^{(n)}} > 0$ *and* $r = 2l_1+1$;
- *a* r^{th}*-order oscillatory source for* $d^r x_{k+n}/dx_k^r|_{x_k^*=a_{i_1}^{(n)}} < 0$ *and* $r = 2l_1+1$;
- *a* r^{th}*-order oscillatory upper-saddle for* $d^r x_{k+n}/dx_k^r|_{x_k^*=a_{i_1}^{(n)}} > 0$ *and* $r = 2l_1$;
- *a* r^{th}*-order oscillatory lower-saddle for* $d^r x_{k+n}/dx_k^r|_{x_k^*=a_{i_1}^{(n)}} < 0$ *and* $r = 2l_1$.

The corresponding period-$2\times n$ discrete system of the $(2m)^{th}$-degree polynomial discrete system in Eq. (4.222) is

$$x_{k+2\times n}=x_k+a_0^{(2\times n)}\prod\nolimits_{i_1\in I_q^{(n)}}(x_k-a_{i_1}^{(n)})^3\prod\nolimits_{j_2=1}^{(2m)^{2\times n}}(x_k-a_{j_2}^{(2\times n)})^{(1-\delta(i_1,j_2))} \tag{4.235}$$

with

$$\begin{aligned}&\frac{dx_{k+2\times n}}{dx_k}\Big|_{x_k^*=a_{i_1}^{(n)}}=1,\ \frac{d^2x_{k+2\times n}}{dx_k^2}\Big|_{x_k^*=a_{i_1}^{(n)}}=0;\\&\frac{d^3x_{k+2\times n}}{dx_k^3}\Big|_{x_k^*=a_{i_1}^{(n)}}=6a_0^{(2\times n)}\prod\nolimits_{i_1\in I_q^{(n)},i_2\neq i_1}(a_{i_1}^{(n)}-a_{i_2}^{(n)})^3\\&\qquad\times\prod\nolimits_{j_2=1}^{(2m)^{2\times n}}(a_{i_1}^{(n)}-a_{j_2}^{(2\times n)})^{(1-\delta(i_1,j_2))}.\end{aligned} \tag{4.236}$$

Thus, $x_{k+2\times n}$ at $x_k^*=a_{i_1}^{(n)}$ for $i_1\in I_q^{(n)}$, $q\in\{1,2,\ldots,N\}$ is

- *a monotonic sink of the third-order if $d^3x_{k+2\times n}/dx_k^3|_{x_k^*=a_{i_1}^{(n)}}<0$,*
- *a monotonic source of the third-order if $d^3x_{k+2\times n}/dx_k^3|_{x_k^*=a_{i_1}^{(n)}}>0$.*

(iv) *After l-times period-doubling bifurcations of period-n fixed points, a period-$2^l\times n$ discrete system of the $(2m)^{th}$-degree polynomial discrete system in Eq. (4.222) is*

$$\begin{aligned}x_{k+2^l\times n}&=x_k+[a_0^{(2^{l-1}\times n)}\prod\nolimits_{i_1=1}^{(2m)^{2^{l-1}\times n}}(x_k-a_{i_1}^{(2^{l-1}\times n)})]\\&\quad\times\{1+\prod\nolimits_{i_1=1}^{(2m)^{2^{l-1}\times n}}[1+a_0^{(2^{l-1})}\prod\nolimits_{i_2=1,i_2\neq i_1}^{(2m)^{2^{l-1}\times n}}(x_k-a_{i_2}^{(2^{l-1}\times n)})]\}\\&=x_k+[a_0^{(2^{l-1}\times n)}\prod\nolimits_{i_1=1}^{(2m)^{2^{l-1}\times n}}(x_k-a_{i_1}^{(2^{l-1}\times n)})]\\&\quad\times[(a_0^{(2^{l-1}\times n)})^{(2m)^{(2^{l-1}\times n)}}\prod\nolimits_{j_1=1}^{((2m)^{2^l\times n}-(2m)^{2^{l-1}\times n})/2}(x_k^2+B_{j_2}^{(2^l\times n)}x_k+C_{j_2}^{(2^l\times n)})]\\&=x_k+[a_0^{(2^{l-1}\times n)}\prod\nolimits_{i_1=1}^{(2m)^{2^{l-1}\times n}}(x_k-a_{i_1}^{(2^{l-1}\times n)})]\\&\quad\times[(a_0^{(2^{l-1}\times n)})^{(2m)^{2^{l-1}\times n}}\prod\nolimits_{j_2=1}^{((2m)^{2^l\times n}-(2m)^{2^{l-1}\times n})/2}(x_k-b_{j_2,1}^{(2^l\times n)})(x_k-b_{j_2,2}^{(2^l\times n)})]\\&=x_k+(a_0^{(2^{l-1}\times n)})^{(2m)^{2^{l-1}\times n}}\prod\nolimits_{i=1}^{(2m)^{2^l\times n}}(x_k-a_i^{(2^l\times n)})\\&=x_k+a_0^{(2^l\times n)}\prod\nolimits_{i=1}^{(2m)^{2^l\times n}}(x_k-a_i^{(2^l\times n)})\end{aligned} \tag{4.237}$$

with

$$
\begin{aligned}
&\frac{dx_{k+2^l\times n}}{dx_k} = 1 + a_0^{(2^l\times n)} \sum_{i_1=1}^{(2m)^{2^l\times n}} \prod_{i_2=1, i_2\neq i_1}^{(2m)^{2^l\times n}} (x_k - a_{i_2}^{(2^l\times n)}),\\
&\frac{d^2x_{k+2^l\times n}}{dx_k^2} = a_0^{(2^l\times n)} \sum_{i_1=1}^{(2m)^{2^l\times n}} \sum_{i_2=1, i_2\neq i_1}^{(2m)^{2^l\times n}} \prod_{i_3=1, i_3\neq i_1, i_2}^{(2m)^{2^l\times n}} (x_k - a_{i_3}^{(2^l\times n)}),\\
&\vdots\\
&\frac{d^r x_{k+2^l\times n}}{dx_k^r} = a_0^{(2^l\times n)} \sum_{i_1=1}^{(2m)^{2^l\times n}} \cdots \sum_{i_r=1, i_r\neq i_1, i_2\cdots, i_{r-1}}^{(2m)^{2^l\times n}} \prod_{i_{r+1}=1, i_{r+1}\neq i_1, i_2\cdots, i_r}^{(2m)^{2^l\times n}} (x_k - a_{i_{r+1}}^{(2^l\times n)})\\
&\text{for } r \leq (2m)^{2^l\times n};
\end{aligned}
\tag{4.238}
$$

where

$$
\begin{aligned}
&a_0^{(2\times n)} = (a_0^{(n)})^{1+(2m)^{2\times n}}, a_0^{(2^l\times n)} = (a_0^{(2^{l-1}\times n)})^{1+(2m)^{2^{l-1}\times n}}, l = 1, 2, 3, \ldots;\\
&\cup_{i=1}^{(2m)^{2^l\times n}} \{a_i^{(2^l\times n)}\} = \text{sort}\{\cup_{i_1=1}^{(2m)^{2^{l-1}\times n}} \{a_{i_1}^{(2^{l-1}\times n)}\}, \cup_{i_2=1}^{M_2} \{b_{i_2,1}^{(2^l\times n)}, b_{i_2,2}^{(2^l\times n)}\}\};\\
&b_{i,1}^{(2^l\times n)} = -\tfrac{1}{2}(B_i^{(2^l\times n)} + \sqrt{\Delta_i^{(2^l\times n)}}),\\
&b_{i,2}^{(2^l\times n)} = -\tfrac{1}{2}(B_i^{(2^l\times n)} - \sqrt{\Delta_i^{(2^l\times n)}}),\\
&\Delta_i^{(2^l\times n)} = (B_i^{(2^l\times n)})^2 - 4C_i^{(2^l\times n)} \geq 0\\
&\text{for } i \in \cup_{q_1=1}^{N_1} I_{q_1}^{(2^{l-1}\times n)} \bigcup \cup_{q_2=1}^{N_2} I_{q_2}^{(2^l\times n)}\\
&I_{q_1}^{(2^{l-1}\times n)} = \{l_{(q_1-1)\times(2^{l-1}\times n)+1}, l_{(q_1-1)\times(2^{l-1}\times n)+2}, \cdots, l_{q_1\times(2^{l-1}\times n)}\}\\
&\qquad \subseteq \{1, 2, \ldots, M_1\} \cup \{\varnothing\},\\
&\text{for } q_1 \in \{1, 2, \ldots, N_1\}, M_1 = N_1 \times (2^{l-1} \times n);\\
&I_{q_2}^{(2^l\times n)} = \{l_{(q_2-1)\times(2^l\times n)+1}, l_{(q_2-1)\times(2^l\times n)+2}, \cdots, l_{q_2\times(2^{l-1}\times n)}\}\\
&\qquad \subseteq \{M_1+1, M_1+2, \ldots, M_2\} \cup \{\varnothing\},\\
&\text{for } q_2 \in \{1, 2, \ldots, N_2\}, M_2 = ((2m)^{2^l\times n} - (2m)^{2^{l-1}\times n})/2;\\
&b_{i,1}^{(2^l\times n)} = -\tfrac{1}{2}(B_i^{(2^l\times n)} + \mathbf{i}\sqrt{|\Delta_i^{(2^l\times n)}|}),\\
&b_{i,2}^{(2^l\times n)} = -\tfrac{1}{2}(B_i^{(2^l\times n)} - \mathbf{i}\sqrt{|\Delta_i^{(2^l\times n)}|}),\\
&\Delta_i^{(2^l\times n)} = (B_i^{(2^l\times n)})^2 - 4C_i^{(2^l\times n)} < 0,\ \mathbf{i} = \sqrt{-1},\\
&i \in \{l_{N\times(2^l\times n)+1}, l_{N\times(2^l\times n)+2}, \cdots, l_{M_2}\} \subset \{1, 2, \ldots, M_2\} \cup \{\varnothing\}
\end{aligned}
\tag{4.239}
$$

with fixed-points

$$
\begin{aligned}
&x_{k+2^l\times n}^* = x_k^* = a_i^{(2^l\times n)}, (i = 1, 2, \cdots, (2m)^{2^l\times n})\\
&\cup_{i=1}^{(2m)^{2^l\times n}} \{a_i^{(2^l\times n)}\} = \text{sort}\{\cup_{i_1=1}^{(2m)^{2^{l-1}\times n}} \{a_{i_1}^{(2^{l-1}\times n)}\}, \cup_{i_2=1}^{M_2} \{b_{i_2,1}^{(2\times n)}, b_{i_2,2}^{(2\times n)}\}\}\\
&\text{with } a_i^{(2^l\times n)} < a_{i+1}^{(2^l\times n)}.
\end{aligned}
\tag{4.240}
$$

(v) *For a fixed-point of* $x_{k+(2^l\times n)}^* = x_k^* = a_{i_1}^{(2^{l-1}\times n)}$ $(i_1 \in I_q^{(2^{l-1}\times n)}, q \in \{1, 2, \ldots, N_1\})$, *there is a period-* $2^{l-1} \times n$ *discrete system if*

$$\begin{aligned}
&\frac{dx_{k+2^{l-1}\times n}}{dx_k}\Big|_{x_k^*=a_{i_1}^{(2^{l-1}\times n)}} = 1 + a_0^{(2^{l-1}\times n)} \prod_{i_2=1,i_2\neq i_1}^{(2m)^{2^{l-1}\times n}} (a_{i_1}^{(2^{l-1}\times n)} - a_{i_2}^{(2^{l-1}\times n)}) = -1,\\
&\frac{d^s x_{k+2^{l-1}\times n}}{dx_k^s}\Big|_{x_k^*=a_{i_1}^{(2^{l-1}\times n)}} = 0,\ \text{for } s = 2, \ldots, r-1;\\
&\frac{d^r x_{k+2^{l-1}\times n}}{dx_k^r}\Big|_{x_k^*=a_{i_1}^{(2^{l-1}\times n)}} \neq 0 \text{ for } 1 < r \leq (2m)^{2^{l-1}\times n}
\end{aligned} \tag{4.241}$$

with

- *a* $r^{\rm th}$*-order oscillatory sink for* $d^r x_{k+2^{l-1}\times n}/dx_k^r|_{x_k^*=a_{i_1}^{(2^{l-1}\times n)}} > 0$ and $r = 2l_1 + 1$;
- *a* $r^{\rm th}$*-order oscillatory source for* $d^r x_{k+2^{l-1}\times n}/dx_k^r|_{x_k^*=a_{i_1}^{(2^{l-1}\times n)}} < 0$ *and* $r = 2l_1 + 1$;
- *a* $r^{\rm th}$*-order oscillatory upper-saddle for* $d^r x_{k+2^{l-1}\times n}/dx_k^r|_{x_k^*=a_{i_1}^{(2^{l-1}\times n)}} > 0$ *and* $r = 2l_1$;
- *a* $r^{\rm th}$*-order oscillatory lower-saddle for* $d^r x_{k+2^{l-1}\times n}/dx_k^r|_{x_k^*=a_{i_1}^{(2^{l-1}\times n)}} < 0$ and $r = 2l_1$.

The corresponding period- $2^{l-1}\times n$ *discrete system is*

$$\begin{aligned}
x_{k+2^l\times n} = {} & x_k + a_0^{(2^l\times n)} \prod_{i_1\in I_q^{(2^{l-1}\times n)}} (x_k - a_{i_1}^{(2^{l-1}\times n)})^3\\
& \times \prod_{j_2=1}^{(2m)^{2^l\times n}} (x_k - a_{j_2}^{(2^l\times n)})^{(1-\delta(i_1,j_2))}
\end{aligned} \tag{4.242}$$

where

$$\delta(i_1,j_2) = 1 \text{ if } a_{j_2}^{(2^l\times n)} = a_{i_1}^{(2^{l-1}\times n)},\ \delta(i_1,j_2) = 0 \text{ if } a_{j_2}^{(2^l\times n)} \neq a_{i_1}^{(2^{l-1}\times n)} \tag{4.243}$$

with

$$\begin{aligned}
&\frac{dx_{k+2^l\times n}}{dx_k}\Big|_{x_k^*=a_{i_1}^{(2^{l-1}\times n)}} = 1,\ \frac{d^2 x_{k+2^l\times n}}{dx_k^2}\Big|_{x_k^*=a_{i_1}^{(2^{l-1}\times n)}} = 0;\\
&\frac{d^3 x_{k+2^l\times n}}{dx_k^3}\Big|_{x_k^*=a_{i_1}^{(2^{l-1})}} = 6a_0^{(2^l\times n)} \prod_{i_2\in I_q^{(2^{l-1}\times n)},i_2\neq i_1} (a_{i_1}^{(2^{l-1}\times n)} - a_{i_2}^{(2^{l-1}\times n)})^3\\
&\qquad\qquad \times \prod_{j_2=1}^{(2m)^{(2^l\times n)}} (a_{i_1}^{(2^{l-1}\times n)} - a_{j_2}^{(2^l\times n)})^{(1-\delta(i_2,j_2))} \neq 0\\
&(i_1 \in I_q^{(2^{l-1}\times n)}, q \in \{1,2,\ldots,N_1\})
\end{aligned} \tag{4.244}$$

Thus, $x_{k+2^l\times n}$ at $x_k^*=a_{i_1}^{(2^{l-1}\times n)}$ *is*

- *a monotonic sink of the third-order if* $d^3x_{k+2^l\times n}/dx_k^3|_{x_k^*=a_{i_1}^{(2^{l-1})}}<0$;
- *a monotonic source of the third-order if* $d^3x_{k+2^l\times n}/dx_k^3|_{x_k^*=a_{i_1}^{(2^{l-1})}}>0$.

(v_1) *The period-* $2^l\times n$ *fixed-points are trivial if*

$$\left.\begin{array}{l} x_{k+2^l\times n}^*=x_k^*=a_j^{(2^l\times n)}\in\{\cup_{i_i=1}^{(2m)}\{a_{i_1}^{(1)}\},\cup_{i_2=1}^{(2m)^{2^{l-1}\times n}}\{a_{i_2}^{(2^{l-1}\times n)}\}\} \\ \text{for } j=1,2,\ldots,(2m)^{(2^l\times n)} \\ \text{for } n\neq 2^{n_1} \end{array}\right\}$$
$$\left.\begin{array}{l} x_{k+2^l\times n}^*=x_k^*=a_j^{(2^l\times n)}\in\{\cup_{i_2=1}^{(2m)^{2^{l-1}\times n}}\{a_{i_2}^{(2^{l-1}\times n)}\} \\ \text{for } j=1,2,\ldots,(2m)^{2^l\times n} \\ \text{for } n=2^{n_1}. \end{array}\right\} \tag{4.245}$$

(v_2) *The period-* $2^l\times n$ *fixed-points are non-trivial if*

$$\left.\begin{array}{l} x_{k+2^l\times n}^*=x_k^*=a_j^{(2^l\times n)}\notin\{\cup_{i_i=1}^{2m}\{a_{i_1}^{(1)}\},\cup_{i_2=1}^{(2m)^{2^{l-1}\times n}}\{a_{i_2}^{(2^{l-1}\times n)}\}\} \\ \text{for } j=1,2,\ldots,(2m)^{2^l\times n} \\ \text{for } n\neq 2^{n_1} \end{array}\right\}$$
$$\left.\begin{array}{l} x_{k+2^l\times n}^*=x_k^*=a_j^{(2^l\times n)}\notin\{\cup_{i_2=1}^{(2m)^{2^{l-1}\times n}}\{a_{i_2}^{(2^{l-1}\times n)}\} \\ \text{for } j=1,2,\ldots,(2m)^{2^l\times n} \\ \text{for } n=2^{n_1}. \end{array}\right\} \tag{4.246}$$

Such a period- $2^l\times n$ *fixed-point is*

- *monotonically unstable if* $dx_{k+2^l\times n}/dx_k|_{x_k^*=a_{i_1}^{(2^l\times n)}}\in(1,\infty)$;
- *monotonically invariant if* $dx_{k+2^l\times n}/dx_k|_{x_k^*=a_{i_1}^{(2^l\times n)}}=1$, *which is*
  - *a monotonic upper-saddle of the* $(2l_1)^{\text{th}}$ *order for* $d^{2l_1}x_{k+2^l\times n}/dx_k^{2l_1}|_{x_k^*}>0$ *(independent* $(2l_1)$*-branch appearance);*
  - *a monotonic lower-saddle the* $(2l_1)^{\text{th}}$ *order for* $d^{2l_1}x_{k+2^l\times n}/dx_k^{2l_1}|_{x_k^*}<0$ *(independent* $(2l_1)$*-branch appearance)*
  - *a monotonic source of the* $(2l_1+1)^{\text{th}}$ *order for* $d^{2l_1+1}x_{k+2^l\times n}/dx_k^{2l_1+1}|_{x_k^*}>0$ *(dependent* $(2l_1+1)$*-branch appearance from one branch);*
  - *a monotonic sink the* $(2l_1+1)^{\text{th}}$ *order for* $d^{2l_1+1}x_{k+2^l\times n}/dx_k^{2l_1+1}|_{x_k^*}<0$ *(dependent* $(2l_1+1)$*-branch appearance from one branch);*

- *monotonically stable if* $dx_{k+2^l\times n}/dx_k|_{x_k^*=a_{i_1}^{(2^l\times n)}} \in (0,1)$;
- *invariantly zero-stable if* $dx_{k+2^l\times n}/dx_k|_{x_k^*=a_{i_1}^{(2^l\times n)}} = 0$;
- *oscillatorilly stable if* $dx_{k+2^l\times n}/dx_k|_{x_k^*=a_{i_1}^{(2^l\times n)}} \in (-1,0)$;
- *flipped if* $dx_{k+2^l\times n}/dx_k|_{x_k^*=a_{i_1}^{(2^l\times n)}} = -1$, which is
  - *an oscillatory upper-saddle of the* $(2l_1)^{\rm th}$ *order for* $d^{2l_1}x_{k+2^l\times n}/dx_k^{2l_1}|_{x_k^*} > 0$;
  - *an oscillatory lower-saddle the* $(2l_1)^{\rm th}$ *order for* $d^{2l_1}x_{k+2^l\times n}/dx_k^{2l_1}|_{x_k^*} < 0$
  - *an oscillatory source of the* $(2l_1+1)^{\rm th}$ *order for* $d^{2l_1}x_{k+2^l\times n}/dx_k^{2l_1}|_{x_k^*} < 0$;
  - *an oscillatory sink the* $(2l_1+1)^{\rm th}$ *order for* $d^{2l_1+1}x_{k+2^l\times n}/dx_k^{2l_1+1}|_{x_k^*} > 0$
- *oscillatorilly unstable if* $dx_{k+2^l\times n}/dx_k|_{x_k^*=a_{i_1}^{(2^l\times n)}} \in (-\infty,-1)$.

Proof Through the nonlinear renormalization, the proof of this theorem is similar to the proof of Theorem 5.11. This theorem can be easily proved. ■

References

Luo ACJ (2020a) The stability and bifurcations of the $(2m)^{\rm th}$ degree polynomial systems. J Vib Test Syst Dyn 4(1):1–42

Luo ACJ (2020b) Bifurcation and stability in nonlinear dynamical systems. Springer, New York

Chapter 5
$(2m + 1)^{\text{th}}$-Degree Polynomial Discrete Systems

In this chapter, the global stability and bifurcations of period-1 fixed-points in a forward $(2m+1)^{\text{th}}$-degree polynomial discrete system are presented. The *broom-appearing*, *broom-spraying-appearing* and *broom-sprinkler-spraying-appearing* bifurcations for simple and higher-order period-1 fixed-points are discussed, and the *antenna switching*, *straw-bundle-switching* and *flower-bundle-switching* bifurcations for simple and higher-order period-1 fixed-points are also presented. As in cubic nonlinear discrete systems, the period-2 fixed-point solutions and the corresponding period-doubling renormalization of such a forwarded $(2m+1)^{\text{th}}$-degree polynomial discrete system are discussed. For multiple iterations, the period-n appearing and period-doublization of the forward $(2m+1)^{\text{th}}$-degree polynomial discrete system are discussed.

5.1 Global Stability and Bifurcations

In a similar fashion of low-degree polynomial discrete systems, the global stability and bifurcation of fixed-points in the $(2m+1)^{\text{th}}$-degree polynomial nonlinear discrete systems are discussed as in Luo (2020a, b). The stability and bifurcation of each individual fixed-point are analyzed from the local analysis.

Definition 5.1 Consider a $(2m+1)^{\text{th}}$-degree polynomial nonlinear forward discrete system

$$\begin{aligned} x_{k+1} &= x_k + f(x_k, \mathbf{p}) \\ &= x_k + A_0(\mathbf{p})x_k^{2m+1} + A_1(\mathbf{p})x_k^{2m} + \cdots + A_{2m-1}(\mathbf{p})x_k^2 + A_{2m}x_k + A_{2m+1}(\mathbf{p}) \\ &= x_k + a_0(\mathbf{p})(x_k - a(\mathbf{p}))[x_k^2 + B_1(\mathbf{p})x_k + C_1(\mathbf{p})] \\ &\quad \cdots [x_k^2 + B_m(\mathbf{p})x_k + C_m(\mathbf{p})] \end{aligned} \tag{5.1}$$

A. C. J Luo, *Bifurcation Dynamics in Polynomial Discrete Systems*, Nonlinear Physical Science, https://doi.org/10.1007/978-981-15-5208-3_5

where $A_0(\mathbf{p}) \neq 0$, and

$$\mathbf{p} = (p_1, p_2, \cdots, p_m)^{\mathrm{T}}. \tag{5.2}$$

(i) If

$$\Delta_i = B_i^2 - 4C_i < 0 \text{ for } i = 1, 2, \ldots, m, \tag{5.3}$$

the $(2m+1)^{\text{th}}$-degree polynomial discrete system has one fixed-point of $x_k^* = a$, and the corresponding standard form is

$$x_{k+1} = x_k + a_0(x_k - a)[(x_k + \tfrac{1}{2}B_1)^2 + \tfrac{1}{4}(-\Delta_1)] \cdots [(x_k + \tfrac{1}{2}B_m)^2 + \tfrac{1}{4}(-\Delta_m)]. \tag{5.4}$$

The discrete flow of such a system with one fixed-point is called a single-fixed-point flow.

(a) If $a_0 > 0$, the fixed-point discrete flow with $x_k^* = a$ is called a monotonic source discrete flow for $df/dx_k|_{x_k^*=a} \in (0, \infty)$.
(b) If $a_0 < 0$, the fixed-point discrete flow with $x^* = a$ is called

- a monotonic sink discrete flow for $df/dx_k|_{x_k^*=a} \in (-1, 0)$,
- an invariant sink discrete flow for $df/dx_k|_{x_k^*=a} = -1$,
- an oscillatory sink discrete flow for $df/dx_k|_{x_k^*=a} \in (-2, -1)$,
- a flipped discrete flow for $df/dx_k|_{x_k^*=a} = -2$
  - of the oscillatory upper-saddle $(d^2f/dx_k^2|_{x_k^*=a} > 0)$,
  - of the oscillatory lower-saddle $(d^2f/dx_k^2|_{x_k^*=a} < 0)$.
- an oscillatory source discrete flow for $df/dx_k|_{x_k^*=a} \in (-\infty, -2)$.

(ii) If

$$\begin{aligned} &\Delta_i = B_i^2 - 4C_i > 0, i = i_1, i_2, \cdots, i_l \in \{1, 2, \ldots, m\}, \\ &\Delta_j = B_j^2 - 4C_j < 0, j = i_{l+1}, i_{l+2}, \cdots, i_m \in \{1, 2, \ldots, m\} \\ &\text{with } l \in \{0, 1, \ldots, m\}, \end{aligned} \tag{5.5}$$

the $(2m+1)^{\text{th}}$-degree polynomial nonlinear discrete system has $(2l+1)$-fixed-points as

$$\left.\begin{aligned} x_k^* = b_1^{(i)} = -\tfrac{1}{2}(B_i + \sqrt{\Delta_i}), \\ x_k^* = b_2^{(i)} = -\tfrac{1}{2}(B_i - \sqrt{\Delta_i}) \end{aligned}\right\} \\ \text{for } i \in \{i_1, i_2, \cdots, i_l\} \subseteq \{1, 2, \cdots, m\}. \tag{5.6}$$

(ii$_1$) If

$$\begin{aligned}&b_r^{(i)}\neq b_s^{(j)}\ \text{ for }\ r,s\in\{1,2\};i,j=1,2,\ldots,l\\&\{a_1,a_2\cdots,a_{2l}\}=\text{sort}\{a,b_1^{(1)},b_2^{(1)},\cdots,b_1^{(l)},b_2^{(l)}\},a_s<a_{s+1},\end{aligned}\tag{5.7}$$

then, the corresponding standard form is

$$x_{k+1}=x_k+a_0\prod\nolimits_{i_1=1}^{2l+1}(x_k-a_{i_1})\prod\nolimits_{k=l+1}^{m}[(x_k+\tfrac{1}{2}B_{i_k})^2+\tfrac{1}{4}(-\Delta_{i_k})].\tag{5.8}$$

(a) If $a_0>0$, the simple fixed-point discrete flow is called a (mSI-oSO: mSO:$\cdots$:mSO : mSI-oSO)-discrete flow.
(b) If $a_0<0$, the simple fixed-point discrete flow is called a (mSO:mSI-oSO : $\cdots$:mSI-oSO:mSO)-discrete flow.

(ii$_2$) If

$$\begin{aligned}&\{a_1,a_2\cdots,a_{2l+1}\}=\text{sort}\{a,b_1^{(1)},b_2^{(1)},\cdots,b_1^{(l)},b_2^{(l)}\},\\&a_{i_1}\equiv a_1=\cdots=a_{l_1},\\&a_{i_2}\equiv a_{l_1+1}=\cdots=a_{l_1+l_2},\\&\vdots\\&a_{i_r}\equiv a_{\Sigma_{i=1}^{r-1}l_i+1}=\cdots=a_{\Sigma_{i=1}^{r-1}l_i+l_r}=a_{2l+1}\\&\text{with }\Sigma_{s=1}^{r}l_s=2l+1,\end{aligned}\tag{5.9}$$

then, the corresponding standard form is

$$x_{k+1}=x_k+a_0\prod\nolimits_{s=1}^{r}(x_k-a_{i_s})^{l_s}\prod\nolimits_{k=l+1}^{m}[(x_k+\tfrac{1}{2}B_{i_k})^2+\tfrac{1}{4}(-\Delta_{i_k})].\tag{5.10}$$

The fixed-point discrete flow is called an (l_1th mXX:l_2th mXX:$\cdots$:l_rth mXX)-discrete flow.

(a) for $a_0>0$ and $p=1,2,\ldots,r$,

$$l_p\text{th mXX}=\begin{cases}(2r_p-1)^{\text{th}}\text{ order monotonic source,}\\\text{for }\alpha_p=2M_p-1,l_p=2r_p-1;\\(2r_p-1)^{\text{th}}\text{ order monotonic sink,}\\\text{for }\alpha_p=2M_p,l_p=2r_p-1;\\(2r_p)^{\text{th}}\text{ order monotonic lower-saddle,}\\\text{for }\alpha_p=2M_p-1,l_p=2r_p;\\(2r_p)^{\text{th}}\text{ order monotonic upper-saddle,}\\\text{for }\alpha_p=2M_p,l_p=2r_p;\end{cases}\tag{5.11}$$

where

$$\alpha_p = \sum_{s=p}^{r} l_s. \tag{5.12}$$

(b) for $a_0 < 0$ and $p = 1, 2, \ldots, r$,

$$l_p\text{th mXX} = \begin{cases} (2r_p - 1)^{\text{th}} \text{ order monotonic sink,} \\ \text{for } \alpha_p = 2M_p - 1, l_p = 2r_p - 1; \\ (2r_p - 1)^{\text{th}} \text{order monotonic source,} \\ \text{for } \alpha_p = 2M_p, l_p = 2r_p - 1; \\ (2r_p)^{\text{th}} \text{ order monotonic upper-saddle,} \\ \text{for } \alpha_p = 2M_p - 1, l_p = 2r_p; \\ (2r_p)^{\text{th}} \text{ order monotonic lower-saddle,} \\ \text{for } \alpha_p = 2M_p, l_p = 2r_p. \end{cases} \tag{5.13}$$

(c) The fixed-point of $x_k^* = a_{i_p}$ for $(l_p > 1)$-repeated fixed-points switching is called an l_pth mXX switching bifurcation of (l_{p_1}th mXX:l_{p_2}th mXX:$\cdots$: l_{p_β}th mXX) fixed-point at a point $\mathbf{p} = \mathbf{p}_1 \in \partial\Omega_{12}$, and the corresponding switching bifurcation condition is

$$\begin{aligned} & a_{i_p} \equiv a_{\sum_{i=1}^{p-1} l_i + 1} = \cdots = a_{\sum_{i=1}^{p-1} l_i + l_p}, \\ & a^{\pm}_{\sum_{i=1}^{p-1} l_i + 1} \neq \cdots \neq a^{\pm}_{\sum_{i=1}^{p-1} l_i + l_p}; l_p = \sum_{i=1}^{\beta} l_{p_i}. \end{aligned} \tag{5.14}$$

(iii) If

$$\begin{aligned} & \Delta_i = B_i^2 - 4C_i = 0,\ i \in \{i_{11}, i_{12}, \cdots, i_{1s}\} \subseteq \{i_1, i_2, \cdots, i_l\} \subseteq \{1, 2, \cdots, m\}, \\ & \Delta_i = B_i^2 - 4C_i > 0,\ i \in \{i_{21}, i_{22}, \cdots, i_{2r}\} \subseteq \{i_1, i_2, \cdots, i_l\} \subseteq \{1, 2, \cdots, m\}, \\ & \Delta_i = B_i^2 - 4C_i < 0,\ i \in \{i_{l+1}, i_{l+2}, \cdots, i_m\} \subseteq \{1, 2, \cdots, m\}, \end{aligned} \tag{5.15}$$

the $(2m+1)^{\text{th}}$-degree polynomial nonlinear discrete system has $(2l+1)$-fixed-points as

$$\begin{aligned} & \left.\begin{aligned} x_k^* = b_1^{(i)} = -\tfrac{1}{2}B_i, \\ x_k^* = b_2^{(i)} = -\tfrac{1}{2}B_i \end{aligned}\right\} \text{for } i \in \{i_{11}, i_{12}, \cdots, i_{1s}\}, \\ & \left.\begin{aligned} x_k^* = b_1^{(i)} = -\tfrac{1}{2}(B_i + \sqrt{\Delta_i}), \\ x_k^* = b_2^{(i)} = -\tfrac{1}{2}(B_i - \sqrt{\Delta_i}) \end{aligned}\right\} \text{for } i \in \{i_{21}, i_{22}, \cdots, i_{2r}\}. \end{aligned} \tag{5.16}$$

If

$$
\begin{aligned}
&\{a_1, a_2 \cdots, a_{2l+1}\} = \text{sort}\{a, b_1^{(1)}, b_2^{(1)}, \cdots, b_1^{(l)}, b_2^{(l)}\},\\
&a_{i_1} \equiv a_1 = \cdots = a_{l_1},\\
&a_{i_2} \equiv a_{l_1+1} = \cdots = a_{l_1+l_2},\\
&\vdots\\
&a_{i_r} \equiv a_{\Sigma_{i=1}^{r-1} l_i+1} = \cdots = a_{\Sigma_{i=1}^{r-1} l_i+l_r} = a_{2l+1}\\
&\text{with } \Sigma_{s=1}^{r} l_s = 2l+1,
\end{aligned}
\tag{5.17}
$$

then the corresponding standard form is

$$
x_{k+1} = x_k + a_0 \prod_{s=1}^{r} (x_k - a_{i_s})^{l_s} \prod_{k=l+1}^{m} [(x_k + \tfrac{1}{2}B_{i_k})^2 + \tfrac{1}{4}(-\Delta_{i_k})]. \tag{5.18}
$$

The fixed-point discrete flow is called an (l_1th mXX: l_2th mXX:$\cdots$:l_rth mXX)-discrete flow.

(a) The fixed-point of $x_k^* = a_{i_p}$ for $(l_p > 1)$-repeated fixed-points appearance or vanishing is called an l_pth mXX appearing bifurcation of fixed-point at a point $\mathbf{p} = \mathbf{p}_1 \in \partial\Omega_{12}$, and the corresponding bifurcation condition is

$$
\begin{aligned}
&a_{i_p} \equiv a_{\Sigma_{i=1}^{p-1} l_i+1} = \cdots = a_{\Sigma_{i=1}^{p-1} l_i+l_p} = -\tfrac{1}{2}B_{i_p},\\
&\text{with } \Delta_{i_p} = B_{i_p}^2 - 4C_{i_p} = 0\,(i_p \in \{i_1, i_2, \cdots, i_l\}),\\
&a^{+}_{\Sigma_{i=1}^{p-1} l_i+1} \neq \cdots \neq a^{+}_{\Sigma_{i=1}^{p-1} l_i+l_p} \text{ or } a^{-}_{\Sigma_{i=1}^{p-1} l_i+1} \neq \cdots \neq a^{-}_{\Sigma_{i=1}^{p-1} l_i+l_p}.
\end{aligned}
\tag{5.19}
$$

(b) The fixed-point of $x_k^* = a_{i_q}$ for $(l_q > 1)$-repeated fixed-points switching is called an l_qth mXX switching bifurcation of (l_{q_1}th mXX:l_{q_2}th mXX:$\cdots$: l_{q_β}th mXX) fixed-point at a point $\mathbf{p} = \mathbf{p}_1 \in \partial\Omega_{12}$, and the switching bifurcation condition is

$$
\begin{aligned}
&a_{i_q} \equiv a_{\Sigma_{i=1}^{q-1} l_i+1} = \cdots = a_{\Sigma_{i=1}^{q-1} l_i+l_q},\\
&a^{\pm}_{\Sigma_{i=1}^{q-1} l_i+1} \neq \cdots \neq a^{\pm}_{\Sigma_{i=1}^{q-1} l_i+l_q};\, l_q = \textstyle\sum_{i=1}^{\beta} l_{q_i}.
\end{aligned}
\tag{5.20}
$$

(c) The fixed-point of $x_k^* = a_{i_p}$ for $(l_n > 1)$-repeated fixed-points appearance or vanishing and $(l_{p_2} \geq 2)$ repeated fixed-points switching of ($l_{p_{11}}$th mXX : $l_{p_{22}}$th mXX:$\cdots$:$l_{p_{2\beta}}$th mXX)-fixed-point switching is called an l_pth mXX bifurcation of fixed-point at a point $\mathbf{p} = \mathbf{p}_1 \in \partial\Omega_{12}$, and the flower-switching bifurcation condition is

$$\begin{array}{l}
a_{i_p} \equiv a_{\Sigma_{i=1}^{p-1} l_i + 1} = \cdots = a_{\Sigma_{i=1}^{p-1} + l_i + l_p} \\
\text{with } \Delta_{i_p} = B_{i_p}^2 - 4C_{i_p} = 0\left(i_p \in \{i_1, i_2, \cdots, i_l\}\right) \\
\text{for } \left\{j_1, j_2, \cdots, j_{p_1}\right\} \subseteq \left\{1, 2, \ldots, l_p\right\} \\
\text{for } \left\{k_1, k_2, \cdots, k_{p_2}\right\} \subseteq \left\{1, 2, \ldots, l_p\right\} \\
\text{with } l_{p_1} + l_{p_2} = l_p; l_{p_2} = \sum_{i=1}^{\beta} l_{p_{2i}}
\end{array} \tag{5.21}$$

(iv) If

$$\Delta_i = B_i^2 - 4C_i > 0 \text{ for } i = 1, 2, \ldots, m \tag{5.22}$$

the $(2m+1)^{\text{th}}$-degree polynomial nonlinear discrete system has $(2m+1)$-fixed-points as

$$\left.\begin{array}{l}
x_k^* = b_1^{(i)} = -\frac{1}{2}(B_i + \sqrt{\Delta_i}), \\
x_k^* = b_2^{(i)} = -\frac{1}{2}(B_i - \sqrt{\Delta_i})
\end{array}\right\} \text{for } i = 1, 2, \ldots, m. \tag{5.23}$$

(iv_1) If

$$\begin{array}{l}
b_r^{(i)} \neq b_s^{(j)} \text{ for } r, s \in \{1, 2\}; i, j = 1, 2, \ldots, m \\
\{a_1, a_2 \cdots, a_{2m}\} = \text{sort}\{a, b_1^{(1)}, b_2^{(1)}, \cdots, b_1^{(m)}, b_2^{(m)}\}(a_s < a_{s+1}),
\end{array} \tag{5.24}$$

then, the corresponding standard form is

$$x_{k+1} = x_k + a_0(x_k - a_1)(x_k - a_2) \cdots (x_k - a_{2m})(x_k - a_{2m+1}). \tag{5.25}$$

This discrete flow is formed with all the simple fixed-points.

(a) If $a_0 > 0$, the fixed-point discrete flow with $(2m+1)$ fixed-points is called a (mSO:mSI-oSO:$\cdots$:mSI-oSO:mSO)-discrete flow.
(b) If $a_0 < 0$, the fixed-point discrete flow with $(2m+1)$ fixed-points is called a (mSI-oSO:mSO:$\cdots$:mSO : mSI-oSO)-discrete flow.

(iv$_2$) If

$$\begin{aligned}
&\{a_1, a_2 \cdots, a_{2m+1}\} = \text{sort}\{a, b_1^{(1)}, b_2^{(1)}, \cdots, b_1^{(m)}, b_2^{(m)}\},\\
&a_{i_1} \equiv a_1 = \cdots = a_{l_1},\\
&a_{i_2} \equiv a_{l_1+1} = \cdots = a_{l_1+l_2},\\
&\vdots\\
&a_{i_r} \equiv a_{\Sigma_{i=1}^{r-1} l_i+1} = \cdots = a_{\Sigma_{i=1}^{r-1} l_i + l_r} = a_{2m+1}\\
&\text{with } \Sigma_{s=1}^{r} l_s = 2m+1,
\end{aligned} \tag{5.26}$$

then, the corresponding standard form is

$$x_{k+1} = x_k + a_0 \prod_{s=1}^{r} (x_k - a_{i_s})^{l_s}. \tag{5.27}$$

The fixed-point discrete flow is called an (l_1th mXX:l_2th $mXX : \cdots :$ l_rth mXX)-discrete flow. The fixed-point of $x_k^* = a_{i_p}$ for l_p-repeated fixed-points switching is called an l_pth mXX bifurcation of (l_{p_1}th mXX: l_{p_2}th mXX: $\cdots$:l_{p_β}th mXX) fixed-point switching at a point $\mathbf{p} = \mathbf{p}_1$ $\in \partial\Omega_{12}$, and the switching bifurcation condition is

$$\begin{aligned}
&a_{i_p} \equiv a_{\Sigma_{i=1}^{p-1} l_i+1} = \cdots = a_{\Sigma_{i=1}^{p-1} l_i + l_p},\\
&a^{\pm}_{\Sigma_{i=1}^{p-1} l_i+1} \neq \cdots \neq a^{\pm}_{\Sigma_{i=1}^{p-1} l_i + l_p}; l_p = \sum_{i=1}^{\beta} l_{p_i}.
\end{aligned} \tag{5.28}$$

Definition 5.2 Consider a $(2m+1)^{\text{th}}$-degree polynomial nonlinear discrete system as

$$\begin{aligned}
x_{k+1} &= x_k + f(x_k, \mathbf{p})\\
&= x_k + A_0(\mathbf{p})x_k^{2m+1} + A_1(\mathbf{p})x_k^{2m} + \cdots + A_{2m-1}(\mathbf{p})x_k^2 + A_{2m}x_k + A_{2m+1}(\mathbf{p})\\
&= a_0(\mathbf{p})(x_k - a(\mathbf{p})) \prod_{i=1}^{n} [x_k^2 + B_i(\mathbf{p})x_k + C_i(\mathbf{p})]^{q_i}
\end{aligned} \tag{5.29}$$

where $A_0(\mathbf{p}) \neq 0$, and

$$\mathbf{p} = (p_1, p_2, \cdots, p_m)^{\mathrm{T}}, m = \sum_{i=1}^{n} q_i. \tag{5.30}$$

(i) If

$$\Delta_i = B_i^2 - 4C_i < 0 \text{ for } i = 1, 2, \ldots, n \tag{5.31}$$

the $(2m+1)^{\text{th}}$-degree polynomial nonlinear system has one fixed-point of $x_k^* = a$, and the corresponding standard form is

$$x_{k+1} = x_k + a_0(x_k - a)\prod_{i=1}^{n}[(x_k + \tfrac{1}{2}B_i)^2 + \tfrac{1}{4}(-\Delta_i)]^{q_i}. \tag{5.32}$$

The discrete flow of such a system with one fixed-point is called a single fixed-point discrete flow.

(a) If $a_0 > 0$, the fixed-point discrete flow of $x_k^* = a$ is called a monotonic source discrete flow for $df/dx_k|_{x_k^*=a} \in (0, \infty)$.

(b) If $a_0 < 0$, the fixed-point discrete flow of $x_k^* = a$ is called

- a monotonic sink discrete flow for $df/dx_k|_{x_k^*=a} \in (-1, 0)$,
- an invariant sink discrete flow for $df/dx_k|_{x_k^*=a} = -1$,
- an oscillatory sink discrete flow for $df/dx_k|_{x_k^*=a} \in (-2, -1)$,
- a flipped discrete flow for $df/dx_k|_{x_k^*=a} = -2$ with
  - an oscillatory upper-saddle $(d^2f/dx_k^2|_{x_k^*=a} > 0)$,
  - an oscillatory lower-saddle $(d^2f/dx_k^2|_{x_k^*=a} < 0)$,
- an oscillatory source discrete flow for $df/dx_k|_{x_k^*=a} \in (-\infty, -2)$.

(ii) If

$$\left.\begin{aligned} &\Delta_i = B_i^2 - 4C_i > 0, i \in \{i_1, i_2, \cdots, i_l\} \subseteq \{1, 2, \ldots, n\}, \\ &\Delta_j = B_j^2 - 4C_j < 0, j \in \{i_{l+1}, i_{l+2}, \cdots, i_n\} \subseteq \{1, 2, \ldots, n\} \end{aligned}\right. \tag{5.33}$$

the $(2m+1)^{\text{th}}$-degree polynomial nonlinear discrete system has $(2l+1)-$fixed-points as

$$\left.\begin{aligned} x_k^* = b_1^{(i)} &= -\tfrac{1}{2}(B_i + \sqrt{\Delta_i}), \\ x_k^* = b_2^{(i)} &= -\tfrac{1}{2}(B_i - \sqrt{\Delta_i}) \end{aligned}\right\} \text{for } i \in \{i_1, i_2, \cdots, i_l\} \subseteq \{1, 2, \ldots, n\}. \tag{5.34}$$

(ii$_1$) If

$$\begin{aligned}&b_r^{(i)} \neq b_s^{(j)} \text{ for } r,s \in \{1,2\}; i,j = 1,2,\ldots,l\\&\{a_1, a_2 \cdots, a_{2l+1}\} = \text{sort}\{a, \underbrace{b_1^{(1)}, b_2^{(1)}}_{q_1 \text{ sets}}, \cdots, \underbrace{b_1^{(r)}, b_2^{(r)}}_{q_r \text{ sets}}\},\\&a_s \leq a_{s+1},\end{aligned} \tag{5.35}$$

then, the corresponding standard form is

$$\begin{aligned}&x_{k+1} = x_k + a_0 \prod_{s=1}^{2l+1} (x_k - a_s)^{l_s} \prod_{k=l+1}^{n} [(x_k + \tfrac{1}{2}B_{i_k})^2 + \tfrac{1}{4}(-\Delta_{i_k})]^{q_{i_k}}\\&\text{with } l_s \in \{q_{i_1}, q_{i_2}, \cdots, q_{i_l}, 1\}.\end{aligned} \tag{5.36}$$

The fixed-point discrete flow is called an (l_1th mXX:l_2th mXX:$\cdots$:l_{2l+1}th mXX)-discrete flow.

(a) For $a_0 > 0$ and $p = 1, 2, \ldots, 2l+1$,

$$l_p\text{th mXX} = \begin{cases}(2r_p-1)^{\text{th}}\text{order montonic source,}\\ \text{for } \alpha_p = 2M_p - 1, l_p = 2r_p - 1;\\(2r_p-1)^{\text{th}}\text{ order monotonic sink,}\\ \text{for } \alpha_p = 2M_p, l_p = 2r_p - 1;\\(2r_p)^{\text{th}}\text{ order monotonic lower-saddle,}\\ \text{for } \alpha_p = 2M_p - 1, l_p = 2r_p;\\(2r_p)^{\text{th}}\text{ order monotonic upper-saddle,}\\ \text{for } \alpha_p = 2M_p, l_p = 2r_p;\end{cases} \tag{5.37}$$

where

$$\alpha_p = \sum_{s=p}^{2l+1} l_s. \tag{5.38}$$

(b) For $a_0 < 0$ and $p = 1, 2, \ldots, 2l+1$,

$$l_p\text{th mXX} = \begin{cases}(2r_p-1)^{\text{th}}\text{ order monotonic sink,}\\ \text{for } \alpha_p = 2M_p - 1, l_p = 2r_p - 1;\\(2r_p-1)^{\text{th}}\text{ order monotonic source,}\\ \text{for } \alpha_p = 2M_p, l_p = 2r_p - 1;\\(2r_p)^{\text{th}}\text{ order monotonic upper-saddle,}\\ \text{for } \alpha_p = 2M_p - 1, l_p = 2r_p;\\(2r_p)^{\text{th}}\text{ order monotonic lower-saddle,}\\ \text{for } \alpha_p = 2M_p, l_p = 2r_p.\end{cases} \tag{5.39}$$

(ii$_2$) If

$$
\begin{aligned}
&\{a_1, a_2 \cdots, a_{2l+1}\} = \text{sort}\{a, \underbrace{b_1^{(1)}, b_2^{(1)}}_{q_1 \textit{ sets}}, \cdots, \underbrace{b_1^{(r)}, b_2^{(r)}}_{q_r \textit{ sets}}\},\\
&a_{i_1} \equiv a_1 = \cdots = a_{l_1},\\
&a_{i_2} \equiv a_{l_1+1} = \cdots = a_{l_1+l_2},\\
&\vdots\\
&a_{i_r} \equiv a_{\Sigma_{i=1}^{r-1} l_i+1} = \cdots = a_{\Sigma_{i=1}^{r-1} l_i+l_r} = a_{2l+1}\\
&\text{with } \Sigma_{s=1}^{r} l_s = 2l+1,
\end{aligned}
\tag{5.40}
$$

then, the corresponding standard form is

$$
x_{k+1} = x_k + a_0 \prod_{s=1}^{r} (x_k - a_{i_s})^{l_s} \prod_{k=l+1}^{n} [(x_k + \tfrac{1}{2}B_{i_k})^2 + \tfrac{1}{4}(-\Delta_{i_k})]^{q_{i_k}}. \tag{5.41}
$$

The fixed-point discrete flow is called an (l_1th mXX:l_2th mXX:$\cdots$: l_rth mXX)-discrete flow.

(a) For $a_0 > 0$ and $s = 1, 2, \ldots, r$,

$$
l_p\text{th mXX} = \begin{cases}
(2r_p-1)^{\text{th}} \text{ order monotonic source,}\\
\text{for } \alpha_p = 2M_p - 1, l_p = 2r_p - 1;\\
(2r_p-1)^{\text{th}} \text{ order monotonic sink,}\\
\text{for } \alpha_p = 2M_p, l_p = 2r_p - 1;\\
(2r_p)^{\text{th}} \text{ order monotonic lower-saddle,}\\
\text{for } \alpha_p = 2M_p - 1, l_p = 2r_p;\\
(2r_p)^{\text{th}} \text{ order monotonic upper-saddle,}\\
\text{for } \alpha_p = 2M_p, l_p = 2r_p;
\end{cases}
\tag{5.42}
$$

where

$$
\alpha_p = \textstyle\sum_{s=p}^{r} l_s. \tag{5.43}
$$

(b) For $a_0 < 0$ and $p = 1, 2, \ldots, r$,

$$
l_p\text{th mXX} = \begin{cases}
(2r_p-1)^{\text{th}} \text{order monotonic sink,}\\
\text{for } \alpha_p = 2M_p - 1, l_p = 2r_p - 1;\\
(2r_p-1)^{\text{th}} \text{ order monotonic source,}\\
\text{for } \alpha_p = 2M_p, l_p = 2r_p - 1;\\
(2r_p)^{\text{th}} \text{ order monotonic upper-saddle,}\\
\text{for } \alpha_p = 2M_p - 1, l_p = 2r_p;\\
(2r_p)^{\text{th}} \text{ order monotonic lower-saddle,}\\
\text{for } \alpha_p = 2M_p, l_p = 2r_p.
\end{cases}
\tag{5.44}
$$

(c) The fixed-point of $x_k^* = a_{i_p}$ for ($l_p > 1$)-repeated fixed-points switching is called an l_pth mXX switching bifurcation of (l_{p_1}th mXX:l_{p_2}th XmX: $\cdots$:l_{p_β}th mXX) fixed-point at a point $\mathbf{p} = \mathbf{p}_1 \in \partial\Omega_{12}$, and the switching bifurcation condition is

$$\begin{aligned} &a_{i_p} \equiv a_{\Sigma_{i=1}^{p-1} l_i + 1} = \cdots = a_{\Sigma_{i=1}^{p-1} l_i + l_p}, \\ &a^{\pm}_{\Sigma_{i=1}^{p-1} l_i + 1} \neq \cdots \neq a^{\pm}_{\Sigma_{i=1}^{p-1} l_i + l_p}; l_p = \Sigma_{i=1}^{\beta} l_{p_i}. \end{aligned} \tag{5.45}$$

(iii) If

$$\begin{aligned} &\Delta_i = B_i^2 - 4C_i = 0,\ i \in \{i_{11}, i_{12}, \cdots, i_{1s}\} \subseteq \{i_1, i_2, \cdots, i_l\} \subseteq \{1, 2, \ldots, n\}, \\ &\Delta_k = B_k^2 - 4C_k > 0,\ k \in \{i_{21}, i_{22}, \cdots, i_{2r}\} \subseteq \{i_1, i_2, \cdots, i_l\} \subseteq \{1, 2, \ldots, n\}, \\ &\Delta_j = B_j^2 - 4C_j < 0,\ j \in \{i_{l+1}, i_{l+2}, \cdots, i_n\} \subseteq \{1, 2, \ldots, n\} \text{ with } i \neq j \neq k, \end{aligned} \tag{5.46}$$

the $(2m+1)^{\text{th}}$-degree polynomial nonlinear system has $(2l+1)$ - fixed-points as

$$\begin{aligned} &\left.\begin{aligned} x_k^* = b_1^{(i)} = -\tfrac{1}{2}B_i, \\ x_k^* = b_2^{(i)} = -\tfrac{1}{2}B_i \end{aligned}\right\} \text{for } i \in \{i_{11}, i_{12}, \cdots, i_{1s}\}, \\ &\left.\begin{aligned} x_k^* = b_1^{(k)} = -\tfrac{1}{2}(B_k + \sqrt{\Delta_k}), \\ x_k^* = b_2^{(k)} = -\tfrac{1}{2}(B_k - \sqrt{\Delta_k}) \end{aligned}\right\} \text{for } i \in \{i_{21}, i_{22}, \cdots, i_{2r}\}. \end{aligned} \tag{5.47}$$

If

$$\begin{aligned} &\{a_1, a_2 \cdots, a_{2l+1}\} = \text{sort}\{a, b_1^{(1)}, b_2^{(1)}, \cdots, b_1^{(l)}, b_2^{(l)}\}, \\ &a_{i_1} \equiv a_1 = \cdots = a_{l_1}, \\ &a_{i_2} \equiv a_{l_1+1} = \cdots = a_{l_1+l_2}, \\ &\vdots \\ &a_{i_r} \equiv a_{\Sigma_{i=1}^{r-1} l_i + 1} = \cdots = a_{\Sigma_{i=1}^{r-1} l_i + l_r} = a_{2l+1} \\ &\text{with } \Sigma_{s=1}^{r} l_s = 2l+1, \end{aligned} \tag{5.48}$$

then, the corresponding standard form is

$$x_{k+1} = x_k + a_0 \prod_{s=1}^{r} (x_k - a_{i_s})^{l_s} \prod_{k=l+1}^{n} [(x_k + \frac{1}{2}B_{i_k})^2 + \frac{1}{4}(-\Delta_{i_k})]^{q_{i_k}}. \tag{5.49}$$

The fixed-point discrete flow is called an (l_1th mXX:l_2th mXX: $\cdots$:l_rth mXX)-discrete flow.

(a) The fixed-point of $x_k^* = a_{i_p}$ for $(l_p > 1)$-repeated fixed-points appearance or vanishing is called an l_pth mXX appearing bifurcation of fixed-point at a point $\mathbf{p} = \mathbf{p}_1 \in \partial\Omega_{12}$, and the corresponding bifurcation condition is

$$\begin{aligned} & a_{i_p} \equiv a_{\Sigma_{i=1}^{p-1} l_i + 1} = \cdots = a_{\Sigma_{i=1}^{p-1} l_i + l_p} = -\frac{1}{2}B_{i_p} \\ & \text{with } \Delta_{i_p} = B_{i_p}^2 - 4C_{i_p} = 0 \, (i_p \in \{i_1, i_2, \cdots, i_l\}), \\ & a^{+}_{\Sigma_{i=1}^{p-1} l_i + 1} \neq \cdots \neq a^{+}_{\Sigma_{i=1}^{p-1} l_i + l_p} \text{ or } a^{-}_{\Sigma_{i=1}^{p-1} l_i + 1} \neq \cdots \neq a^{-}_{\Sigma_{i=1}^{p-1} l_i + l_p}. \end{aligned} \tag{5.50}$$

(b) The fixed-point of $x_k^* = a_{i_p}$ for $(l_p > 1)$-repeated fixed-points switching is called an l_pth mXX switching bifurcation l_pth mXX of (l_{p_1}th mXX: l_{p_2}th mXX : $\cdots$:l_{p_β}th mXX) fixed-point at a point $\mathbf{p} = \mathbf{p}_1 \in \partial\Omega_{12}$, and the corresponding switching bifurcation condition is

$$\begin{aligned} & a_{i_p} \equiv a_{\Sigma_{i=1}^{p-1} l_i + 1} = \cdots = a_{\Sigma_{i=1}^{p-1} l_i + l_p}, \\ & a^{\pm}_{\Sigma_{i=1}^{p-1} l_i + 1} \neq \cdots \neq a^{\pm}_{\Sigma_{i=1}^{p-1} l_i + l_p} \\ & l_p = \Sigma_{i=1}^{\beta} l_{p_i}. \end{aligned} \tag{5.51}$$

(iv) If

$$\Delta_i = B_i^2 - 4C_i > 0 \text{ for } i = 1, 2, \ldots, n \tag{5.52}$$

the $(2m + 1)^{\text{th}}$-degree polynomial nonlinear system has $(2n + 1)$-fixed-points as

$$\left. \begin{aligned} x_k^* = b_1^{(i)} = -\tfrac{1}{2}(B_i + \sqrt{\Delta_i}), \\ x_k^* = b_2^{(i)} = -\tfrac{1}{2}(B_i - \sqrt{\Delta_i}) \end{aligned} \right\} \text{for } i = 1, 2, \ldots, n. \tag{5.53}$$

(iv$_1$) If

$$\begin{aligned} & b_r^{(i)} \neq b_s^{(j)} \text{ for } r,s \in \{1,2\}, (i,j=1,2,\ldots,n); \\ & \{a_1, a_2 \cdots, a_{2n+1}\} = \text{sort}\{a, \underbrace{b_1^{(1)}, b_2^{(1)}}_{q_1 \text{sets}}, \cdots, \underbrace{b_1^{(n)}, b_2^{(n)}}_{q_n \text{sets}}\} \\ & a_s \le a_{s+1}, \end{aligned} \tag{5.54}$$

then, the corresponding standard form is

$$x_{k+1} = x_k + a_0 \prod_{s=1}^{2n+1} (x_k - a_s)^{l_s} \text{ with } l_s \in \{q_{i_1}, q_{i_2}, \cdots, q_{i_n}, 1\}. \tag{5.55}$$

The fixed-point discrete flow is called an (l_1th mXX:l_2th mXX:$\cdots$: l_{2n+1}th mXX)-discrete flow.

(a) For $a_0 > 0$ and $p = 1, 2, \ldots, 2n+1$,

$$l_p\text{th mXX} = \begin{cases} (2r_p-1)^{\text{th}} \text{ order monotonic source,} \\ \text{for } \alpha_p = 2M_p - 1, l_p = 2r_p - 1; \\ (2r_p-1)^{\text{th}} \text{ order monotonic sink,} \\ \text{for } \alpha_p = 2M_p, l_p = 2r_p - 1; \\ (2r_p)^{\text{th}} \text{ order monotonic lower-saddle,} \\ \text{for } \alpha_p = 2M_p - 1, l_p = 2r_p; \\ (2r_p)^{\text{th}} \text{ order monotonic upper-saddle,} \\ \text{for } \alpha_p = 2M_p, l_p = 2r_p; \end{cases} \tag{5.56}$$

where

$$\alpha_p = \sum_{s=p}^{2n+1} l_s. \tag{5.57}$$

(b) For $a_0 < 0$ and $p = 1, 2, \ldots, 2n+1$,

$$l_p\text{th mXX} = \begin{cases} (2r_p-1)^{\text{th}} \text{ order montonic sink,} \\ \text{for } \alpha_p = 2M_p - 1, l_p = 2r_p - 1; \\ (2r_p-1)^{\text{th}} \text{ order monotonic source,} \\ \text{for } \alpha_p = 2M_p, l_p = 2r_p - 1; \\ (2r_p)^{\text{th}} \text{ order montonic upper-saddle,} \\ \text{for } \alpha_p = 2M_p - 1, l_p = 2r_p; \\ (2r_p)^{\text{th}} \text{ order monotonic lower-saddle,} \\ \text{for } \alpha_p = 2M_p, l_p = 2r_p. \end{cases} \tag{5.58}$$

(iv$_2$) If

$$
\begin{aligned}
&\{a_1, a_2 \cdots, a_{2n+1}\} = sort\{a, \underbrace{b_1^{(1)}, b_2^{(1)}}_{q_1\text{sets}}, \cdots, \underbrace{b_1^{(n)}, b_2^{(n)}}_{q_n\text{sets}}\},\\
&a_{i_1} \equiv a_1 = \cdots = a_{l_1},\\
&a_{i_2} \equiv a_{l_1+1} = \cdots = a_{l_1+l_2},\\
&\vdots\\
&a_{i_r} \equiv a_{\Sigma_{i=1}^{r-1} l_i+1} = \cdots = a_{\Sigma_{i=1}^{r-1} l_i+l_r} = a_{2n+1},\\
&\text{with } \Sigma_{s=1}^{r} l_s = 2n+1,
\end{aligned}
\tag{5.59}
$$

then, the corresponding standard form is

$$
x_{k+1} = x_k + a_0 \prod_{s=1}^{r} (x_k - a_{i_s})^{l_s}. \tag{5.60}
$$

The fixed-point discrete flow is called an (l_1th mXX: l_2th mXX: $\cdots$:l_rth mXX)-discrete flow. The fixed-point of $x_k^* = a_{i_p}$ for l_p-repeated fixed-points switching is called an l_pth XX switching bifurcation of (l_{p_1}th mXX:l_{p_2}th mXX:$\cdots$:l_{p_β}th mXX) fixed-point at a point $\mathbf{p} = \mathbf{p}_1 \in \partial\Omega_{12}$, and the switching bifurcation condition is

$$
\begin{aligned}
&a_{i_p} \equiv a_{\Sigma_{i=1}^{p-1} l_i+1} = \cdots = a_{\Sigma_{i=1}^{p-1} l_i+l_p},\\
&a^{\pm}_{\Sigma_{i=1}^{p-1} l_i+1} \neq \cdots \neq a^{\pm}_{\Sigma_{i=1}^{p-1} l_i+l_p}; l_p = \Sigma_{i=1}^{\beta} l_{p_i}.
\end{aligned}
\tag{5.61}
$$

Definition 5.3 Consider a 1-dimensional, $(2m+1)^{\text{th}}$-degree polynomial nonlinear discrete system

$$
\begin{aligned}
x_{k+1} &= x_k + f(x_k, \mathbf{p})\\
&= x_k + A_0(\mathbf{p})x_k^{2m+1} + A_1(\mathbf{p})x_k^{2m} + \cdots + A_{2m-1}(\mathbf{p})x_k^2 + A_{2m}x_k + A_{2m+1}(\mathbf{p})\\
&= x_k + a_0(\mathbf{p}) \prod_{s=1}^{r} (x_k - c_{i_s}(\mathbf{p}))^{l_s} \prod_{i=r+1}^{n} [x_k^2 + B_i(\mathbf{p})x_k + C_i(\mathbf{p})]^{q_i}
\end{aligned}
\tag{5.62}
$$

where $A_0(\mathbf{p}) \neq 0$, and

$$
\Sigma_{s=1}^{r} l_s = 2l+1,\ \Sigma_{i=r+1}^{n} q_i = (m-l),\ \mathbf{p} = (p_1, p_2, \cdots, p_m)^{\mathrm{T}}. \tag{5.63}
$$

(i) If

$$\begin{aligned}&\Delta_i = B_i^2 - 4C_i < 0 \text{ for } i = r+1, r+2, \ldots, n,\\ &\{a_1, a_2, \ldots, a_r\} = \text{sort}\{c_1, c_2, \ldots, c_r\} \text{ with } a_i < a_{i+1}\end{aligned} \tag{5.64}$$

the $(2m+1)^{\text{th}}$-degree polynomial discrete system has fixed-points of $x_k^* = a_{i_s}(\mathbf{p})$ $(s = 1, 2, \ldots, r)$, and the corresponding standard form is

$$x_{k+1} = x_k + a_0(\mathbf{p}) \prod_{j=1}^{r} (x_k - a_{i_j})^{l_j} \prod_{i=r+1}^{n} [(x_k + \tfrac{1}{2}B_i)^2 + \tfrac{1}{4}(-\Delta_i)]^{l_i}. \tag{5.65}$$

The fixed-point discrete flow is called an (l_1th mXX:l_2th mXX:$\cdots$: l_rth mXX)-discrete flow.

(a) For $a_0 > 0$ and $s = 1, 2, \ldots, r$,

$$l_p\text{th mXX} = \begin{cases} (2r_p - 1)^{\text{th}} \text{ order monotonic source,} \\ \text{for } \alpha_p = 2M_p - 1, l_p = 2r_p - 1; \\ (2r_p - 1)^{\text{th}} \text{ order monotonic sink,} \\ \text{for } \alpha_p = 2M_p, l_p = 2r_p - 1; \\ (2r_p)^{\text{th}} \text{ order monotonic lower-saddle,} \\ \text{for } \alpha_p = 2M_p - 1, l_p = 2r_p; \\ (2r_p)^{\text{th}} \text{ order monotonic upper-saddle,} \\ \text{for } \alpha_p = 2M_p, l_p = 2r_p; \end{cases} \tag{5.66}$$

where

$$\alpha_p = \sum_{s=p}^{r} l_s. \tag{5.67}$$

(b) For $a_0 < 0$ and $p = 1, 2, \ldots, r$,

$$l_p\text{th mXX} = \begin{cases} (2r_p - 1)^{\text{th}} \text{ order monotonic sink,} \\ \text{for } \alpha_p = 2M_p - 1, l_p = 2r_p - 1; \\ (2r_p - 1)^{\text{th}} \text{ order monotonic source,} \\ \text{for } \alpha_p = 2M_p, l_p = 2r_p - 1; \\ (2r_p)^{\text{th}} \text{ order monotonic upper-saddle,} \\ \text{for } \alpha_p = 2M_p - 1, l_p = 2r_p; \\ (2r_p)^{\text{th}} \text{ order monotonic lower-saddle,} \\ \text{for } \alpha_p = 2M_p, l_p = 2r_p. \end{cases} \tag{5.68}$$

(ii) If

$$\begin{aligned}&\Delta_i = B_i^2 - 4C_i > 0, i = j_1, j_2, \cdots, j_s \in \{l+1, l+2, \ldots, n\},\\ &\Delta_j = B_j^2 - 4C_j < 0, j = j_{s+1}, j_{s+2}, \cdots, j_n \in \{l+1, l+2, \ldots, n\}\\ &\text{with s} \in \{1, \ldots, n-l\},\end{aligned} \tag{5.69}$$

the $(2m+1)^{\text{th}}$-degree polynomial nonlinear discrete system has $2n_2$-fixed-points as

$$\left.\begin{aligned} x_k^* &= b_1^{(i)} = -\tfrac{1}{2}(B_i + \sqrt{\Delta_i}), \\ x_k^* &= b_2^{(i)} = -\tfrac{1}{2}(B_i - \sqrt{\Delta_i}) \end{aligned}\right\} \\ \text{for } i \in \{j_1, j_2, \cdots, j_{n_1}\} \subseteq \{l+1, l+2, \ldots, n\}. \tag{5.70}$$

If

$$\begin{aligned} &\{a_1, a_2 \cdots, a_{2n_2+1}\} = \text{sort}\{c_1, c_2 \cdots, c_{2l+1}, \underbrace{b_1^{(r+1)}, b_2^{(r+1)}}_{q_{r+1}\ sets}, \cdots, \underbrace{b_1^{(n_1)}, b_2^{(n_1)}}_{q_{n_1}\ sets}\}, \\ &a_{i_1} \equiv a_1 = \cdots = a_{l_1}, \\ &a_{i_2} \equiv a_{l_1+1} = \cdots = a_{l_1+l_2}, \\ &\vdots \\ &a_{i_{n_1}} \equiv a_{\Sigma_{i=1}^{n_1-1} l_i + 1} = \cdots = a_{\Sigma_{i=1}^{n_1-1} l_i + l_{n_1}} = a_{2n_2+1} \\ &\text{with } \Sigma_{s=1}^{n_1} l_s = 2n_2 + 1, \end{aligned} \tag{5.71}$$

then, the corresponding standard form is

$$x_{k+1} = x_k + a_0 \prod_{s=1}^{n_1} (x_k - a_{i_s})^{l_s} \prod_{i=n_2+1}^{n} [(x_k + \frac{1}{2} B_i)^2 + \frac{1}{4}(-\Delta_i)]^{q_i}. \tag{5.72}$$

The fixed-point discrete flow is called an (l_1th mXX:l_2th mXX:$\cdots$:l_{n_1}th mXX)-discrete flow.

(a) For $a_0 > 0$ and $p = 1, 2, \ldots, r, r+1, \ldots, n_1$,

$$l_p\text{th mXX} = \begin{cases} (2r_p - 1)^{\text{th}} \text{ order monotonic source,} \\ \text{for } \alpha_p = 2M_p - 1, l_p = 2r_p - 1; \\ (2r_p - 1)^{\text{th}} \text{ order monotonic sink,} \\ \text{for } \alpha_p = 2M_p, l_p = 2r_p - 1; \\ (2r_p)^{\text{th}} \text{ order monotonic lower-saddle,} \\ \text{for } \alpha_p = 2M_p - 1, l_p = 2r_p; \\ (2r_p)^{\text{th}} \text{ order monotonic upper-saddle,} \\ \text{for } \alpha_p = 2M_p, l_p = 2r_p; \end{cases} \tag{5.73}$$

where

$$\alpha_p = \Sigma_{s=p}^{n_1} l_s. \tag{5.74}$$

(b) For $a_0 < 0$ and $p = 1, 2, \ldots, r, r+1, \ldots, n_1$,

$$l_p\text{th mXX} = \begin{cases} (2r_p - 1)^{\text{th}} \text{ order monotonic sink,} \\ \text{for } \alpha_p = 2M_p - 1, l_p = 2r_p - 1; \\ (2r_p - 1)^{\text{th}} \text{ order monotonic source,} \\ \text{for } \alpha_p = 2M_p, l_p = 2r_p - 1; \\ (2r_p)^{\text{th}} \text{ order monotonic upper-saddle,} \\ \text{for } \alpha_p = 2M_p - 1, l_p = 2r_p; \\ (2r_p)^{\text{th}} \text{ order monotonic lower-saddle,} \\ \text{for } \alpha_p = 2M_p, l_p = 2r_p. \end{cases} \tag{5.75}$$

(c) The fixed-point of $x_k^* = a_{i_p}$ for $(l_p > 1)$-repeated fixed-points switching is called an l_pth mXX switching bifurcation of (l_{p_1}th mXX:l_{p_2}th mXX:$\cdots$:l_{p_β}th mXX) fixed-point at $\mathbf{p} = \mathbf{p}_1 \in \partial\Omega_{12}$, and the corresponding switching bifurcation condition is

$$\begin{aligned} &a_{i_p} \equiv a_{\Sigma_{i=1}^{p-1} l_i + 1} = \cdots = a_{\Sigma_{i=1}^{p-1} l_i + l_p}, \\ &a^{\pm}_{\Sigma_{i=1}^{p-1} l_i + 1} \neq \cdots \neq a^{\pm}_{\Sigma_{i=1}^{p-1} l_i + l_p}; l_p = \textstyle\sum_{i=1}^{\beta} l_{p_i}. \end{aligned} \tag{5.76}$$

(iii) If

$$\begin{aligned} &\Delta_i = B_i^2 - 4C_i = 0, \\ &\text{for } i \in \{i_{11}, i_{12}, \cdots, i_{1s}\} \subseteq \{i_{l+1}, i_{l+2}, \cdots, i_{n_2}\} \subseteq \{l+1, l+2, \ldots, n\}, \\ &\Delta_k = B_k^2 - 4C_k > 0, \\ &\text{for } k \in \{i_{21}, i_{22}, \ldots, i_{2r}\} \subseteq \{i_{l+1}, i_{l+2}, \cdots, i_{n_2}\} \subseteq \{l+1, l+2, \ldots, n\}, \\ &\Delta_j = B_j^2 - 4C_j < 0, \\ &\text{for } j \in \{i_{n_2+1}, i_{n_2+2}, \ldots, i_n\} \subseteq \{l+1, l+2, \ldots, n\}, \end{aligned} \tag{5.77}$$

the $(2m+1)^{\text{th}}$-degree polynomial nonlinear discrete system has $(2n_2+1)$-fixed-points as

$$\begin{aligned} &\left.\begin{aligned} x_k^* &= b_1^{(i)} = -\tfrac{1}{2} B_i, \\ x_k^* &= b_2^{(i)} = -\tfrac{1}{2} B_i \end{aligned}\right\} \text{for } i \in \{i_{11}, i_{12}, \ldots, i_{1s}\}, \\ &\left.\begin{aligned} x_k^* &= b_1^{(k)} = -\tfrac{1}{2}(B_k + \sqrt{\Delta_k}), \\ x_k^* &= b_2^{(k)} = -\tfrac{1}{2}(B_k - \sqrt{\Delta_k}) \end{aligned}\right\} \text{for } i \in \{i_{21}, i_{22}, \ldots, i_{2r}\}. \end{aligned} \tag{5.78}$$

If

$$\begin{aligned}
&\{a_1, a_2 \cdots, a_{2n_2+1}\} = \text{sort}\{a, c_1, c_2 \cdots, c_{2l}, \underbrace{b_1^{(r)}, b_2^{(r)}}_{q_r \text{sets}}, \cdots, \underbrace{b_1^{(n_1)}, b_2^{(n_1)}}_{q_{n_1} \text{ sets}}\}, \\
&a_{i_1} \equiv a_1 = \cdots = a_{l_1}, \\
&a_{i_2} \equiv a_{l_1+1} = \cdots = a_{l_1+l_2}, \\
&\vdots \\
&a_{i_{n_1}} \equiv a_{\Sigma_{i=1}^{n_1-1} l_i+1} = \cdots = a_{\Sigma_{i=1}^{n_1-1} l_i+l_{n_1}} = a_{2n_2+1} \\
&\text{with } \Sigma_{s=1}^{n_1} l_s = 2n_2+1,
\end{aligned} \tag{5.79}$$

then, the corresponding standard form is

$$x_{k+1} = a_0 \prod_{s=1}^{n_1} (x_k - a_{i_s})^{l_s} \prod_{i=n_2+1}^{n} [(x_k + \tfrac{1}{2}B_i)^2 + \tfrac{1}{4}(-\Delta_i)]^{q_i}. \tag{5.80}$$

The fixed-point discrete flow is called an l_1th mXX:l_2th mXX:$\cdots$:l_{n_1}th mXX)-discrete flow.

(a) The fixed-point of $x^* = a_{i_p}$ for $(l_p > 1)$-repeated fixed-points appearance (or vanishing) is called an l_pth mXX appearing bifurcation of fixed-point at a point $\mathbf{p} = \mathbf{p}_1 \in \partial\Omega_{12}$, and the corresponding bifurcation condition is

$$\begin{aligned}
&a_{i_p} \equiv a_{\Sigma_{i=1}^{p-1} l_i+1} = \cdots = a_{\Sigma_{i=1}^{p-1} l_i+l_p} = -\frac{1}{2}B_{i_q} \\
&\text{with } \Delta_{i_p} = B_{i_p}^2 - 4C_{i_p} = 0\,(i_p \in \{i_1, i_2, \cdots, i_l\}) \\
&a^+_{\Sigma_{i=1}^{p-1} l_i+1} \neq \cdots \neq a^+_{\Sigma_{i=1}^{p-1} l_i+l_p} \text{ or } a^-_{\Sigma_{i=1}^{p-1} l_i+1} \neq \cdots \neq a^-_{\Sigma_{i=1}^{p-1} l_i+l_p}.
\end{aligned} \tag{5.81}$$

(b) The fixed-point of $x_k^* = a_{i_p}$ for $(l_p > 1)$-repeated fixed-points switching is called an l_pth mXX switching bifurcation of (l_{p_1}th mXX:l_{p_2} th mXX: $\cdots$:l_{p_β}th mXX) fixed-point at a point $\mathbf{p} = \mathbf{p}_1 \in \partial\Omega_{12}$, and the corresponding switching bifurcation condition is

$$\begin{aligned}
&a_{i_p} \equiv a_{\Sigma_{i=1}^{p-1} l_i+1} = \cdots = a_{\Sigma_{i=1}^{p-1} l_i+l_p}, \\
&a^\pm_{\Sigma_{i=1}^{p-1} l_i+1} \neq \cdots \neq a^\pm_{\Sigma_{i=1}^{p-1} l_i+l_p}, \\
&l_p = \textstyle\sum_{i=1}^{\beta} l_{p_i}.
\end{aligned} \tag{5.82}$$

(c) The fixed-point of $x_k^* = a_{i_p}$ for $(l_{p_1} \geq 1)$-repeated fixed-points appearance (or vanishing) and $(l_{p_2} \geq 2)$ repeated fixed-points switching of ($l_{p_{21}}$th mXX: $l_{p_{22}}$th mXX : $\cdots$:$l_{p_{2\beta}}$th mXX) is called an l_pth mXX switching bifurcation of fixed-point at a point $\mathbf{p} = \mathbf{p}_1 \in \partial\Omega_{12}$, and the corresponding bifurcation condition is

$$
\begin{aligned}
& a_{i_p} \equiv a_{\Sigma_{i=1}^{p-1} l_i + 1} = \cdots = a_{\Sigma_{i=1}^{p-1} l_i + l_p} \\
& \text{with } \Delta_{i_p} = B_{i_p}^2 - 4C_{i_p} = 0 \, (i_p \in \{i_1, i_2, \cdots, i_l\}) \\
& a^{+}_{\Sigma_{i=1}^{p-1} l_i + j_1} \neq \cdots \neq a^{+}_{\Sigma_{i=1}^{p-1} l_i + j_{p_1}} \text{ or } a^{-}_{\Sigma_{i=1}^{p_1-1} l_i + j_1} \neq \cdots \neq a^{-}_{\Sigma_{i=1}^{p_1-1} l_i + j_{p_1}}, \\
& \text{for } \{j_1, j_2, \cdots, j_{p_1}\} \subseteq \{1, 2, \ldots, l_p\}, \\
& a^{\pm}_{\Sigma_{i=1}^{p-1} l_i + k_1} \neq \cdots \neq a^{\pm}_{\Sigma_{i=1}^{p-1} l_i + k_{p_2}} \\
& \text{for } \{k_1, k_2, \ldots, k_{p_2}\} \subseteq \{1, 2, \ldots, l_p\}, \\
& \text{with } l_{p_1} + l_{p_2} = l_p.
\end{aligned}
\tag{5.83}
$$

(iv) If

$$
\Delta_i = B_i^2 - 4C_i > 0 \text{ for } i = l+1, l+2, \ldots, n \tag{5.84}
$$

the $(2m+1)^{\text{th}}$-degree polynomial nonlinear discrete system has $(2m+1)$ fixed-points as

$$
\left.
\begin{aligned}
x_k^* = b_1^{(i)} = -\tfrac{1}{2}(B_i + \sqrt{\Delta_i}), \\
x_k^* = b_2^{(i)} = -\tfrac{1}{2}(B_i - \sqrt{\Delta_i})
\end{aligned}
\right\} \text{for } i = l+1, l+2, \ldots, n. \tag{5.85}
$$

If

$$
\begin{aligned}
& \{a_1, a_2 \cdots, a_{2m+1}\} = \text{sort}\{c_1, c_2 \cdots, c_{2l+1}, \underbrace{b_1^{(r+1)}, b_2^{(r+1)}}_{q_{r+1}\text{sets}}, \cdots, \underbrace{b_1^{(n)}, b_2^{(n)}}_{q_n \text{ sets}}\}, \\
& a_{i_1} \equiv a_1 = \cdots = a_{l_1}, \\
& a_{i_2} \equiv a_{l_1+1} = \cdots = a_{l_1+l_2}, \\
& \vdots \\
& a_{i_n} \equiv a_{\Sigma_{i=1}^{n-1} l_i + 1} = \cdots = a_{\Sigma_{i=1}^{n-1} l_i + l_r} = a_{2m+1} \\
& \text{with } \Sigma_{s=1}^{n} l_s = 2m+1,
\end{aligned}
\tag{5.86}
$$

then, the corresponding standard form is

$$
x_{k+1} = x_k + a_0 \prod_{s=1}^{r} (x_k - a_{i_s})^{l_s}. \tag{5.87}
$$

The fixed-point discrete flow is called an (l_1th mXX: l_2th mXX:$\cdots$:l_rth mXX)-discrete flow. The fixed-point of $x_k^* = a_{i_p}$ for l_p-repeated fixed-points switching is called an l_pth mXX switching bifurcation of (l_{p_1}th mXX:l_{p_2}th mXX : $\cdots$:

l_{p_β}th mXX) fixed-point at $\mathbf{p} = \mathbf{p}_1 \in \partial\Omega_{12}$, and the corresponding switching bifurcation condition is

$$\begin{aligned} & a_{i_p} \equiv a_{\Sigma_{i=1}^{p-1} l_i + 1} = \cdots = a_{\Sigma_{i=1}^{p-1} l_i + l_p}, \\ & a^{\pm}_{\Sigma_{i=1}^{p-1} l_i + 1} \neq \cdots \neq a^{\pm}_{\Sigma_{i=1}^{p-1} l_i + l_p}; l_p = \Sigma_{i=1}^{\beta} l_{p_i}. \end{aligned} \tag{5.88}$$

5.2 Simple Fixed-Point Bifurcations

To illustrate the bifurcations in the $(2m+1)^{\text{th}}$-degree polynomial discrete system, the detailed discussion with graphical illustrations will be presented as follows.

5.2.1 *Appearing Bifurcations*

Consider a $(2m+1)^{\text{th}}$-degree polynomial nonlinear discrete system as

$$x_{k+1} = x_k + a_0 Q(x_k)(x_k - a)\prod_{i=1}^{n}(x_k^2 + B_i x_k + C_i). \tag{5.89}$$

without loss of generality, a function of $Q(x_k) > 0$ is either a polynomial function or a non-polynomial function. The roots of $x_k^2 + B_i x_k + C_i = 0$ are

$$\begin{aligned} & b_{1,2}^{(i)} = -\tfrac{1}{2}B_i \pm \tfrac{1}{2}\sqrt{\Delta_i}, \Delta_i = B_i^2 - 4C_i \geq 0 (i = 1, 2, \ldots, n); \\ & \{a_1, a_2, \cdots, a_{2l}\} \subset \text{sort}\{b_1^{(1)}, b_2^{(1)}, b_1^{(2)}, b_2^{(2)}, \cdots, b_1^{(n)}, b_2^{(n)}\}, a_s \leq a_{s+1}; \\ & \left.\begin{array}{l} B_i \neq B_j (i, j = 1, 2, \ldots, n; i \neq j) \\ \Delta_i = 0 (i = 1, 2, \ldots, n) \end{array}\right\} \text{at bifurcation.} \end{aligned} \tag{5.90}$$

The 2nd order singularity bifurcation is for the birth of a pair of simple monotonic sink to oscillatory source, and simple monotonic source. There are two *appearing* bifurcations for $i \in \{1, 2, \ldots, n\}$

$$2^{\text{nd}}\text{order mUS} \xrightarrow[\text{appearing bifurcation}]{i^{\text{th}}\text{quadratic factor}} \begin{cases} \text{mSO}, & \text{for } x_k^* = a_{2i}, \\ \text{mSI-oSO}, & \text{for } x_k^* = a_{2i-1}; \end{cases} \tag{5.91}$$

$$2^{\text{nd}}\text{order mLS} \xrightarrow[\text{appearing bifurcation}]{i^{\text{th}}\text{quadratic factor}} \begin{cases} \text{mSI-oSO}, & \text{for } x_k^* = a_{2i}, \\ \text{mSO}, & \text{for } x_k^* = a_{2i-1}. \end{cases} \tag{5.92}$$

If $x_k^* = a \neq -\frac{1}{2}B_i (i \in \{1, 2, \ldots, m\})$, the fixed-point of $x_k^* = a$ breaks a cluster of *teethcomb* appearing bifurcations to two parts. The *teethcomb* appearing bifurcation generated by the m-pairs of quadratics becomes a *broom appearing* bifurcation. The two *broom appearing* bifurcations are

$$
\begin{aligned}
&\text{mSO } (x_k^* = a) \xrightarrow[\text{appearing bifurcation}]{l_1 + l_2 = m} (l_1\text{-mLSN:mSO:}l_2\text{-mUSN}) \\
&= \begin{cases} l_2\text{-mUSN}, \text{ for } x^* \in \{a_{2j}, a_{2j+1}, j = l_1 + 1, \cdots, l_1 + l_2\}, \\ \text{mSO}, \text{ for } x^* = a = a_{2(l_1+1)-1} \\ l_1\text{-mLSN}, \text{ for } x^* \in \{a_{2i-1}, a_{2i}, i = 1, 2, \ldots, l_1\} \end{cases}
\end{aligned}
\tag{5.93}
$$

and

$$
\begin{aligned}
&\text{mSI-oSO}(x_k^* = a) \xrightarrow[\text{appearing bifurcation}]{l_1 + l_2 = m} (l_1\text{-mUSN:mSI-oSO:}l_2\text{-mLSN}) \\
&= \begin{cases} l_2\text{-mLSN}, \text{ for } x_k^* \in \{a_{2j}, a_{2j+1} : j = l_1 + 1, \cdots, l_1 + l_2\}, \\ \text{mSI-oSO}, \text{ for } x_k^* = a = a_{2(l_1+1)-1} \\ l_1\text{-mUSN}, \text{ for } x_k^* \in \{a_{2i-1}, a_{2i} : i = 1, 2, \ldots, l_1\} \end{cases}
\end{aligned}
\tag{5.94}
$$

where the l_j-mLSN and l_j-mUSN $(j = 1, 2)$ are

$$
l_j\text{-mUSN} \begin{cases} \text{mUS} \xrightarrow[\text{appearing}]{(l_j + s_j)^{\text{th}} \text{ bifurcation}} \begin{cases} \text{mSO}, \text{ for } x_k^* = a_{2(s_j + l_j) + \delta_j^2}, \\ \text{mSI-oSO}, \text{ for } x_k^* = a_{2(s_j + l_j) - 1 + \delta_j^2}; \end{cases} \\ \vdots \\ \text{mUS} \xrightarrow[\text{appearing}]{(s_j)^{\text{th}} \text{ bifurcation}} \begin{cases} \text{mSO}, \text{ for } x_k^* = a_{2s_j + \delta_j^2}, \\ \text{mSI-oSO}, \text{ for } x_k^* = a_{2s_j - 1 + \delta_j^2}. \end{cases} \end{cases}
\tag{5.95}
$$

$$
l_j\text{-mLSN} \begin{cases} \text{mLS} \xrightarrow[\text{appearing}]{(l_j + s_j)^{\text{th}} \text{ bifurcation}} \begin{cases} \text{mSI-oSO}, \text{ for } x_k^* = a_{2(s_j + l_j) + \delta_j^2}, \\ \text{mSO}, \text{ for } x_k^* = a_{2(s_j + l_j) - 1 + \delta_j^2}; \end{cases} \\ \vdots \\ \text{mLS} \xrightarrow[\text{appearing}]{(s_j)^{\text{th}} \text{ bifurcation}} \begin{cases} \text{mSI-oSO}, \text{ for } x_k^* = a_{2s_j + \delta_j^2}, \\ \text{mSO}, \text{ for } x_k^* = a_{2s_j - 1 + \delta_j^2}. \end{cases} \end{cases}
\tag{5.96}
$$

for $s_j \in \{0, 1, 2, \ldots, m\}$ and $0 \leq l_j \leq m$ with $0 \leq l_j \leq m$.

Four special broom appearing bifurcations are

$$
\text{mSO } (x_k^* = a) \to \begin{cases} \text{mSO} \to m\text{SO}, \ \text{for } x_k^* = a = a_{2m+1}, \\ m\text{-mLSN} \begin{cases} \text{mLS} \xrightarrow[\text{appearing}]{m^{\text{th}} \text{ bifurcation}} \begin{cases} \text{mSI-oSO}, \\ \text{for } x_k^* = a_{2m}, \\ \text{mSO}, \\ \text{for } x_k^* = a_{2m-1}; \end{cases} \\ \vdots \\ \text{mUS} \xrightarrow[\text{appearing}]{1^{\text{st}} \text{ bifurcation}} \begin{cases} \text{mSI-oSO}, \\ \text{for } x_k^* = a_2, \\ \text{mSO}, \\ \text{for } x_k^* = a_1; \end{cases} \end{cases} \end{cases} \tag{5.97}
$$

$$
\text{mSI-oSO } (x_k^* = a) \to \begin{cases} \text{mSI} \to m\text{SI}, \ \text{for } x_k^* = a = a_{2m+1}, \\ m\text{-mUSN} \begin{cases} \text{mUS} \xrightarrow[\text{appearing}]{m^{\text{th}} \text{ bifurcation}} \begin{cases} \text{mSO}, \\ \text{for } x_k^* = a_{2m}, \\ \text{mSI-oSO}, \\ \text{for } x_k^* = a_{2m-1}; \end{cases} \\ \vdots \\ \text{mUS} \xrightarrow[\text{appearing}]{1^{\text{st}} \text{ bifurcation}} \begin{cases} \text{mSO}, \\ \text{for } x_k^* = a_2, \\ \text{mSI-oSO}, \\ \text{for } x_k^* = a_1; \end{cases} \end{cases} \end{cases} \tag{5.98}
$$

and

$$
\text{mSO } (x_k^* = a) \to \begin{cases} m\text{-mUSN} \begin{cases} \text{mUS} \xrightarrow[\text{appearing}]{m^{\text{th}} \text{ bifurcation}} \begin{cases} \text{mSO}, \\ \text{for } x_k^* = a_{2m+1}, \\ \text{mSI-oSO}, \\ \text{for } x_k^* = a_{2m}; \end{cases} \\ \vdots \\ \text{mUS} \xrightarrow[\text{appearing}]{1^{\text{st}} \text{ bifurcation}} \begin{cases} \text{mSO}, \\ \text{for } x_k^* = a_3, \\ \text{mSI-oSO}, \\ \text{for } x_k^* = a_2; \end{cases} \end{cases} \\ \text{mSO} \to \text{mSO}, \ \text{for } x_k^* = a = a_1, \end{cases} \tag{5.99}
$$

$$\text{mSI-oSO } (x_k^* = a) \to \begin{cases} m\text{-mLSN} \begin{cases} \text{mLS} \xrightarrow[\text{appearing}]{m^{\text{th}}\text{bifurcation}} \begin{cases} \text{mSI-oSO}, \\ \text{for } x_k^* = a_{2m+1}, \\ \text{mSO}, \\ \text{for } x_k^* = a_{2m}; \end{cases} \\ \vdots \\ \text{mLS} \xrightarrow[\text{appearing}]{1^{\text{st}}\text{ bifurcation}} \begin{cases} \text{mSI-oSO}, \\ \text{for } x_k^* = a_3, \\ \text{mSO}, \\ \text{for } x_k^* = a_2; \end{cases} \end{cases} \\ \text{mSI-oSO} \to m\text{SI-oSO}, \text{ for } x_k^* = a = a_1. \end{cases} \tag{5.100}$$

If $x_k^* = a = -\frac{1}{2}B_i$ $(i \in \{1, 2, \ldots, m\})$, the fixed-point of $x_k^* = a$ possess a third-order mSI or mSO switching bifurcation (or pitchfork bifurcation). The teethcomb appearing bifurcation generated by the m-pairs of quadratics becomes a *broom appearing* bifurcation. The two *broom appearing* bifurcations are

$$\begin{aligned} &\text{mSO } (x_k^* = a) \xrightarrow[\text{appearing bifurcation}]{m=l_1+l_2+1} (l_1\text{-mLSN:3}^{\text{rd}}\text{mSO:}l_2\text{-mUSN}) \\ &= \begin{cases} l_2\text{-mUSN, for } x_k^* \in \{a_{2j}, a_{2j+1}, i = l_1+2, \cdots, l_1+l_2\}, \\ 3^{\text{rd}}\text{mSO} \to \begin{cases} \text{mSO, for } x_k^* = a_{2(l_1+2)-1} \\ \text{mSI-oSO, for } x_k^* = a = a_{2(l_1+1)} \\ \text{mSO, for } x_k^* = a_{2(l_1+1)-1} \end{cases} \\ l_1\text{-mLSN, for } x_k^* \in \{a_{2i-1}, a_{2i}, i = 1, 2, \ldots, l_1\} \end{cases} \end{aligned} \tag{5.101}$$

and

$$\begin{aligned} &\text{mSI-oSO } (x_k^* = a) \xrightarrow[\text{appearing bifurcation}]{m=l_1+l_2+1} (l_1\text{-mUSN:3}^{\text{rd}}\text{mSI:}l_2\text{-mLSN}) \\ &= \begin{cases} l_2\text{-mLSN, for } x_k^* \in \{a_{2j}, a_{2j+1}, j = l_1+2, \cdots, l_1+l_2\}, \\ 3^{\text{rd}}\text{mSI} \to \begin{cases} \text{mSI-oSO, for } x_k^* = a_{2(l_1+2)-1} \\ \text{mSO, for } x_k^* = a = a_{2(l_1+1)} \\ \text{mSI-oSO, for } x_k^* = a_{2(l_1+1)-1} \end{cases} \\ l_1\text{-mUSN, for } x_k^* \in \{a_{2i-1}, a_{2i}, i = 1, 2, \ldots, l_1\}. \end{cases} \end{aligned} \tag{5.102}$$

Consider an appearing bifurcation for a cluster of fixed-points with monotonic sink to oscillatory source, and monotonic source with the following conditions.

$$\left. \begin{array}{l} B_i = B_j(i, j \in \{1, 2, \ldots, n\}, \; i \neq j) \\ \Delta_j = 0(i = 1, 2, \ldots, n) \end{array} \right\} \text{at bifurcation.} \tag{5.103}$$

Thus, the $(2l)^{\text{th}}$-order appearing bifurcation is for a cluster of simple monotonic sinks to monotonic-sources and monotonic sources. Two $(2l)^{\text{th}}$ order *appearing* bifurcations for $l \in \{1, 2, \ldots, s\}$ are

$$(2l)^{\text{th}}\ \text{order mUSN} \xrightarrow[\text{appearing bifurcation}]{\text{cluster of } l\text{-quadratics}} \begin{cases} \text{mSO, for } x_k^* = a_{2s_l}, \\ \text{mSI-oSO, for } x_k^* = a_{2s_l-1}, \\ \vdots \\ \text{mSO, for } x_k^* = a_{2s_1}, \\ \text{mSI-oSO, for } x_k^* = a_{2s_1-1}. \end{cases} \tag{5.104}$$

$$(2l)^{\text{th}}\ \text{order mLSN} \xrightarrow[\text{appearing bifurcation}]{\text{cluster of } l\text{-quadratics}} \begin{cases} \text{mSI-oSO, for } x_k^* = a_{2s_l}, \\ \text{mSO, for } x_k^* = a_{2s_l-1}, \\ \vdots \\ \text{mSI-oSO, for } x_k^* = a_{2s_1}, \\ \text{mSO, for } x_k^* = a_{2s_1-1}. \end{cases} \tag{5.105}$$

If $x_k^* = a \neq -\frac{1}{2}B_i$ $(i \in \{1, 2, \ldots, n\})$, the fixed-point of $x_k^* = a$ breaks a cluster of *sprinkler-spraying* appearing bifurcations to two parts. The sprinkler-spraying appearing bifurcation generated by the m-pairs of quadratics becomes a *broom-sprinkler-spraying appearing* bifurcation. The two broom-sprinkler-spraying *appearing* bifurcations are

$$\begin{aligned} &\text{mSO}\ (x_k^* = a) \xrightarrow[\text{appearing bifurcation}]{m = m_1 + m_2} (r_1\text{-mLSG:}m\text{SO:}r_2\text{-mUSG}) \\ &= \begin{cases} r_2\text{-mUSG} \to \begin{cases} (2l_{r_2}^{(2)})^{\text{th}}\ \text{mUSN}\ (x_k^* = a_{r_2+1}), \\ \vdots \\ (2l_1^{(2)})^{\text{th}}\ \text{mUSN}\ (x_k^* = a_{r_1+2}), \end{cases} \\ \text{mSO}\ (a = a_{r_1+1}) \to m\text{SO}\ (a = a_{2(m_1+1)-1}), \\ r_1\text{-mLSG} \to \begin{cases} (2l_{r_1}^{(1)})^{\text{th}}\ \text{mLSN}\ (x_k^* = a_{r_1}), \\ \vdots \\ (2l_1^{(1)})^{\text{th}}\ \text{mLSN}\ (x_k^* = a_1); \end{cases} \end{cases} \end{aligned} \tag{5.106}$$

and

$$\begin{aligned} &\text{mSI-oSO}\ (x_k^* = a) \xrightarrow[\text{appearing bifurcation}]{m = m_1 + m_2} (r_1\text{-mUSG:mSI-oSO:}r_2\text{-mLSG}) \\ &= \begin{cases} r_2\text{-mLSG} \to \begin{cases} (2l_{r_2}^{(2)})^{\text{th}}\ \text{mLSN}\ (x_k^* = a_{r_2+1}), \\ \vdots \\ (2l_1^{(2)})^{\text{th}}\ \text{mLSN}\ (x_k^* = a_{r_1+2}), \end{cases} \\ \text{mSI-oSO}\ (a = a_{r_1+1}) \to \text{mSI-oSO}\ (a = a_{2(m_1+1)-1}), \\ r_1\text{-mUSG} \to \begin{cases} (2l_{r_1}^{(1)})^{\text{th}}\ \text{mUSN}\ (x_k^* = a_{r_1}), \\ \vdots \\ (2l_1^{(1)})^{\text{th}}\ \text{mUSN}\ (x_k^* = a_1); \end{cases} \end{cases} \end{aligned} \tag{5.107}$$

for $m_1 = \sum_{i=1}^{r_1} l_i^{(1)}, m_2 = \sum_{j=1}^{r_2} l_j^{(2)}$; and the acronyms USG and LSG are the upper-saddle-node and lower-saddle-node bifurcation groups, respectively.

Four special broom-sprinkler-spraying appearing bifurcations are

$$\text{mSO}\ (x_k^* = a) \xrightarrow[\text{appearing bifurcation}]{m=\sum_{i=1}^{r} l_i} \begin{cases} \text{mSO}\ (a = a_{2r+1}) \to \text{SmO}\ (a = a_{2m+1}), \\ r\text{-mLSG} \to \begin{cases} (2l_r)^{\text{th}}\ \text{mLSN}\ (x_k^* = a_r), \\ \vdots \\ (2l_1)^{\text{th}}\ \text{mLSN}\ (x_k^* = a_1), \end{cases} \end{cases} \tag{5.108}$$

$$\text{mSI-oSO}\ (x_k^* = a) \xrightarrow[\text{appearing bifurcation}]{m=\sum_{i=1}^{r} l_i} \begin{cases} \text{mSI-oSO}\ (a = a_{2r+1}) \to \text{mSI-oSO}\ (a = a_{2m+1}), \\ r\text{-mUSG} \to \begin{cases} (2l_r)^{\text{th}}\ \text{mUSN}\ (x_k^* = a_r), \\ \vdots \\ (2l_1)^{\text{th}}\ \text{mUSN}\ (x_k^* = a_1), \end{cases} \end{cases} \tag{5.109}$$

and

$$\text{mSO}\ (x_k^* = a) \xrightarrow[\text{appearing bifurcation}]{m=\sum_{i=1}^{r} l_i} \begin{cases} r\text{-mUSG} \to \begin{cases} (2l_r)^{\text{th}}\ \text{mUSN}\ (x_k^* = a_{r+1}), \\ \vdots \\ (2l_1)^{\text{th}}\ \text{mUSN}\ (x_k^* = a_2), \end{cases} \\ \text{mSO}\ (a = a_1) \to m\text{SO}\ (a = a_1), \end{cases} \tag{5.110}$$

$$\text{mSI-oSO}\ (x_k^* = a) \xrightarrow[\text{appearing bifurcation}]{m=\sum_{i=1}^{r} l_i} \begin{cases} r\text{-mLSG} \to \begin{cases} (2l_r)^{\text{th}}\ \text{mLSN}\ (x_k^* = a_{r+1}), \\ \vdots \\ (2l_1)^{\text{th}}\ \text{mLSN}\ (x_k^* = a_2), \end{cases} \\ \text{mSI-oSO}\ (a = a_1) \to \text{mSI-oSO}\ (a = a_1), \end{cases} \tag{5.111}$$

If $x_k^* = a = -\frac{1}{2}B_i$ $(i \in \{1, 2, \ldots, l\})$, the fixed-point of $x_k^* = a$ possesses a $(2l+1)^{\text{th}}$-order mSI or mSO switching bifurcation (or broom-switching bifurcation). The sprinkler-spraying appearing bifurcation generated by the m-pairs of quadratics becomes a *broom-sprinkler-spraying switching* bifurcation. The two *broom switching* bifurcations are

$$
\begin{aligned}
&\text{mSO}\ (x_k^* = a) \xrightarrow[\text{switching bifurcation}]{m=m_1+m_2+l} (r_1\text{-mLSG}:(2l+1)^{\text{rd}}\text{mSO}:r_2\text{-mUSG}) \\
&= \begin{cases}
r_2\text{-mUSG} \to \begin{cases}
(2l_{r_2}^{(2)})^{\text{th}}\text{mUSN}\ (x_k^* = a_{r_2+r_1+1}), \\
\vdots \\
(2l_1^{(2)})^{\text{th}}\text{mUSN}\ (x_k^* = a_{r_1+2}),
\end{cases} \\
(2l+1)^{\text{th}}\text{mSO}\ (a = a_{r_1+1}); \\
r_1\text{-mLSG} \to \begin{cases}
(2l_{r_1}^{(1)})^{\text{th}}\text{mLSN}\ (x_k^* = a_{r_1}), \\
\vdots \\
(2l_1^{(1)})^{\text{th}}\text{mLSN}\ (x_k^* = a_1);
\end{cases}
\end{cases}
\end{aligned}
\tag{5.112}
$$

and

$$
\begin{aligned}
&\text{mSI-oSO}\ (x_k^* = a) \xrightarrow[\text{switching bifurcation}]{m=m_1+m_2+l} (r_1\text{-mUSG}:(2l+1)^{th}\text{mSI}:r_2\text{-mLSG}) \\
&= \begin{cases}
r_2\text{-mLSG} \to \begin{cases}
(2l_{r_2}^{(2)})^{\text{th}}\text{mLSN}\ (x_k^* = a_{r_2+r_1+1}), \\
\vdots \\
(2l_1^{(2)})^{\text{th}}\text{mLSN}\ (x_k^* = a_{r_1+2}),
\end{cases} \\
(2l+1)^{th}\text{mSI}\ (a = a_{r_1+1}); \\
r_1\text{-mUSG} \to \begin{cases}
(2l_{r_1}^{(1)})^{\text{th}}\text{mUSN}\ (x_k^* = a_{r_1}), \\
\vdots \\
(2l_1^{(1)})^{\text{th}}\text{mUSN}\ (x_k^* = a_1),
\end{cases}
\end{cases}
\end{aligned}
\tag{5.113}
$$

where

$$
(2l+1)^{\text{th}}\ \text{order mSO}\ (x_k^* = a) \xrightarrow[\text{appearing bifurcation}]{\text{cluster of } l\text{-quadratics}} \begin{cases}
\text{mSO, for } x_k^* = a_{2s_l+1}, \\
\text{mSI-oSO, for } x_k^* = a_{2s_l}, \\
\vdots \\
\text{mSI-oSO, for } x_k^* = a_{2s_1}, \\
\text{mSO, for } x_k^* = a_{2s_1-1}.
\end{cases}
\tag{5.114}
$$

and

$$
(2l+1)^{\text{th}} \text{ order mSI } (x_k^* = a) \xrightarrow[\text{appearing bifurcation}]{\text{cluster of } l\text{-quadratics}} \begin{cases} \text{mSI-oSO, for } x_k^* = a_{2s_l+1}, \\ \text{mSO, for } x_k^* = a_{2s_l}, \\ \vdots \\ \text{mSO, for } x_k^* = a_{2s_1}, \\ \text{mSI-oSO, for } x_k^* = a_{2s_1-1} \end{cases} \tag{5.115}
$$

where $x_k^* = a \in \{a_{2s_1-1}, \cdots, a_{2s_l}, a_{2s_l+1}\}$.

In Fig. 5.1i and ii, the simple switching with two *teethcomb appearing* bifurcations are presented for $a_0 > 0$ and $a_0 < 0$, respectively. The two bifurcation structures are:

(i) mSO $\to$ (l_1-mLSN:mSO:l_2-mUSN),

(ii) mSI-oSO $\to$ (l_1-mUSN:mSI-oSO:l_2-mLSN)

with $l_1 + l_2 = m$. In Fig. 5.1iii and iv, the 3rd-order pitchfork switching bifurcation with two *teethcomb appearing* bifurcations are presented for $a_0 > 0$ and $a_0 < 0$, respectively. The two bifurcation structures are:

(iii) mSO $\to$ (l_1-mLSN:3rdmSO:l_2-mUSN),

(iv) mSI-oSO $\to$ (l_1-mUSN:3rdmSI:l_2-mLSN)

with $l_1 + l_2 = m - 1$. The period-2 fixed points are sketched as well through red curves.

In Fig. 5.2i and ii, the simple switching with two *sprinkler-spraying appearing* bifurcations are presented for $a_0 > 0$ and $a_0 < 0$, respectively. The two bifurcation structures are:

(i) mSO $\to$ (r_1-mLSG:mSO:r_2-mUSG),

(ii) mSI-oSO $\to$ (r_1-mUSG:mSI-oSO:r_2-mLSG)

with $r_1 + r_2 + 1 = n$ and $m_1 + m_2 = m$ where $m_1 = \Sigma_{i=1}^{r_1} l_i^{(1)}, m_2 = \Sigma_{j=1}^{r_2} l_j^{(2)}$. In Fig. 5.2iii and iv, the $(2l+1)^{\text{th}}$ order broom-switching with two *sprinkler-spraying appearing* bifurcations are presented for $a_0 > 0$ and $a_0 < 0$, respectively. The two bifurcation structures are:

(iii) mSO $\to$ (r_1-mLSG:$(2l+1)^{\text{th}}$mSO:r_2-mUSG),

(iv) mSI-oSO $\to$ (r_1-mUSG:$(2l+1)^{\text{th}}$mSI:r_2-mLSG)

with $r_1 + r_2 + 1 = n$ and $m_1 + m_2 + l = m$ where $m_1 = \Sigma_{i=1}^{r_1} l_i^{(1)}, m_2 = \Sigma_{j=1}^{r_2} l_j^{(2)}$. The period-2 fixed points are sketched as well through red curves.

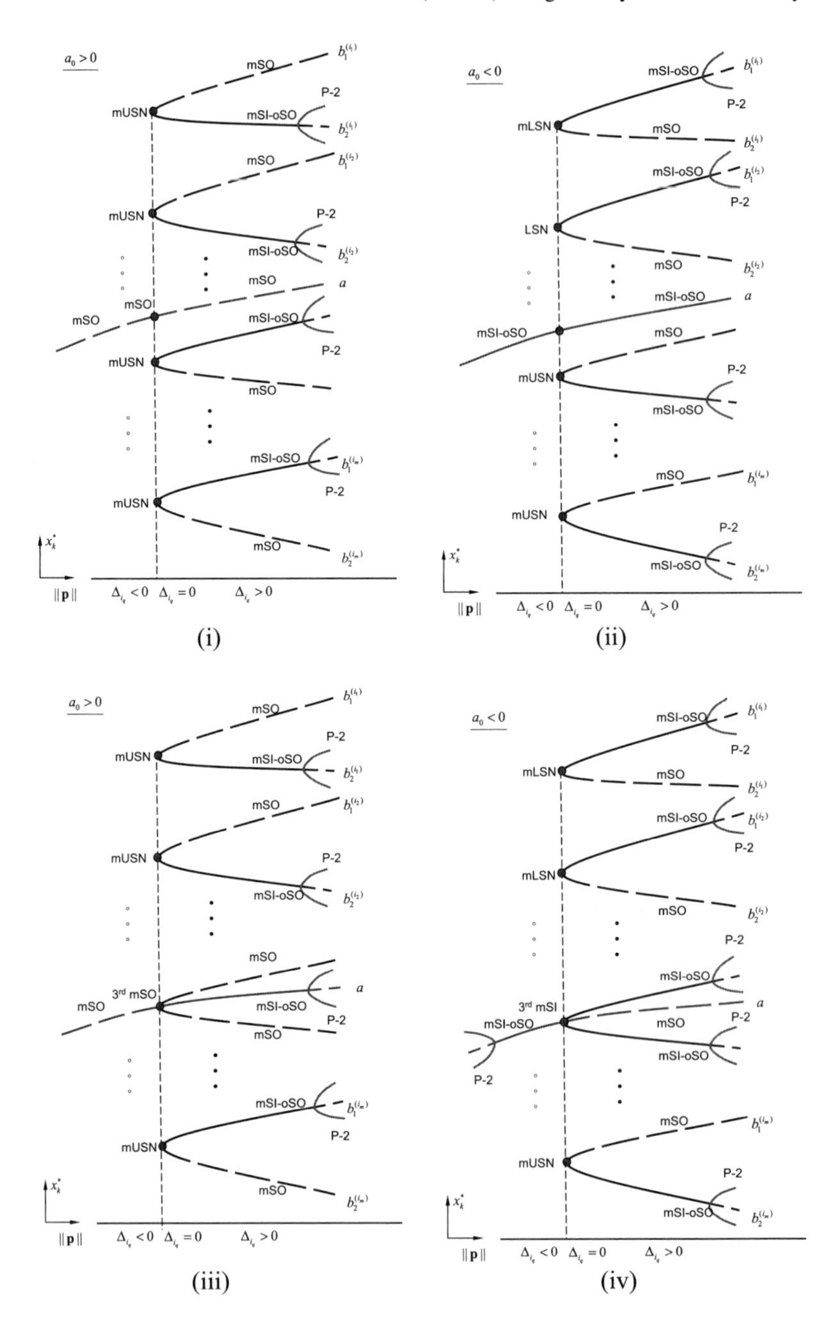

Fig. 5.1 Simple broom switching bifurcations: (i) (mUS:$\cdots$:mUS : *m*SO:mLS:$\cdots$:mLS)($a_0 > 0$), (ii) (mLS:$\cdots$:mLS:*m*SI-oSO:mUS:$\cdots$: *mUS*)($a_0 < 0$), (iii) (mUS:$\cdots$:mUS : 3rdmSO:mLS : $\cdots$:mLS) ($a_0 > 0$). (vi) (LS:$\cdots$:LS : 3rdSI:US:$\cdots$:US) ($a_0 < 0$) in a $(2m+1)^{\text{th}}$-degree polynomial system. mLS: monotonic lower saddle, mUS: monotonic upper saddle, mSI-oSO: monotonic sink to oscillatory source, mSO: monotonic source. Stable and unstable fixed-points are represented by solid and dashed curves, respectively. The bifurcation points are marked by circular symbols. P-2: period-2 fixed points.

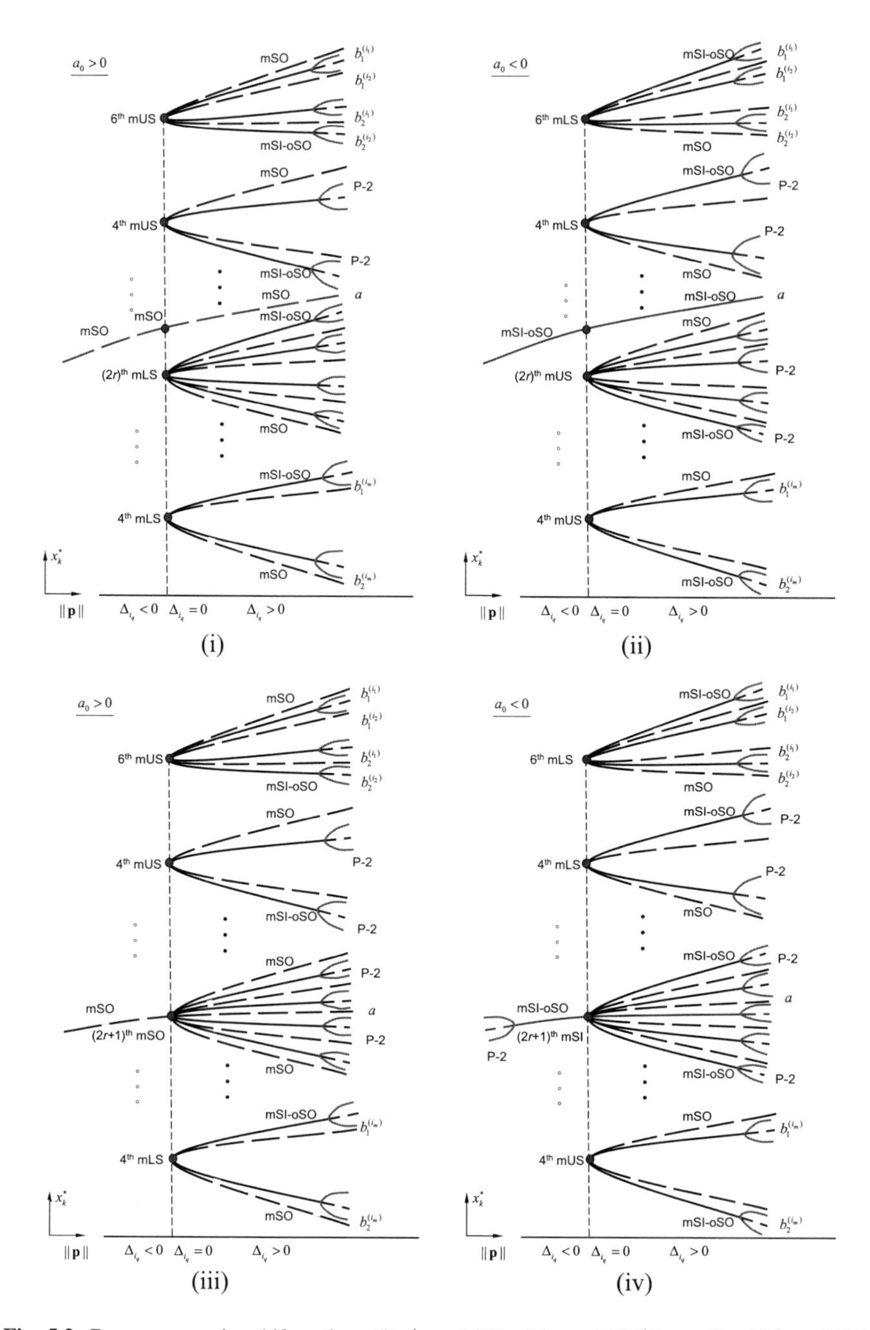

Fig. 5.2 Broom appearing bifurcation: (i) (r_1-mLSN:mSO:r_2-mUSN)($a_0 > 0$); (ii)(r_1-mUSN: mSI-oSO:r_2-mLSN) ($a_0 < 0$); broom-sprinkler-spraying switching bifurcation: (iii)(r_1- $mUSG$: $(2l_k + 1)^{\text{th}}$ mSO:r_2-mLSG) ($a_0 > 0$). (iv)(r_1-mLSG:$(2l_k + 1)^{\text{th}}$mSI:r_2-mUSG)($a_0 < 0$) in a $(2m + 1)^{\text{th}}$-degree polynomial system. mLS: monotonic lower-saddle, mUS: monotonic-upper-saddle, mSI-oSO: monotonic sink to oscillatory source, mSO: monotonic source. Stable and unstable fixed-points are represented by solid and dashed curves, respectively. The bifurcation points are marked by circular symbols. P-2: period-2 fixed points.

For a cluster of m-quadratics, $B_i = B_j$ $(i,j \in \{1,2,\ldots,m\}; i \neq j)$ and $\Delta_i = 0$ $(i \in \{1,2,\ldots,m\})$. The $(2m)^{\text{th}}$-order upper-saddle-node appearing bifurcation for m-pairs of fixed-points with monotonic sink to oscillatory source and monotonic source is

$$
(2m)^{\text{th}} \text{ order mUS} \xrightarrow[\text{appearing bifurcation}]{\text{cluster of } m\text{-quadratics}} \begin{cases} \text{mSO, for } x_k^* = a_{2m}, \\ \text{mSI-oSO, for } x_k^* = a_{2m-1}; \\ \vdots \\ \text{mSO, for } x_k^* = a_2, \\ \text{mSI-oSO, for } x_k^* = a_1. \end{cases} \tag{5.116}
$$

The $(2m)^{\text{th}}$-order lower-saddle-node appearing bifurcation for m-pairs of fixed-points with monotonic sink to oscillatory source and monotonic source is

$$
(2m)^{\text{th}} \text{ order mLS} \xrightarrow[\text{appearing bifurcation}]{\text{cluster of } m\text{-quadratics}} \begin{cases} \text{mSI-oSO, for } x_k^* = a_{2m}, \\ \text{mSO, for } x_k^* = a_{2m-1}; \\ \vdots \\ \text{mSI-oSO, for } x_k^* = a_2, \\ \text{mSO, for } x_k^* = a_1. \end{cases} \tag{5.117}
$$

There are four simple switching and $(2m)^{\text{th}}$-order saddle-node appearing bifurcations: The two switching bifurcations of mSO $\to$ $((2m)^{\text{th}}$mUS:mSO) and mSI-oSO $\to$ $((2m)^{\text{th}}$mLS:mSI) with two $(2m)^{\text{th}}$-order mUSN and mLSN spraying appearing bifurcations are

$$
\text{mSO } (x_k^* = a) \to \begin{cases} \text{mSO} \to mSO, \text{ for } x_k^* = a = a_{2m+1} \\ (2m)^{\text{th}} \text{ order mLSN} \to \begin{cases} \text{mSI-oSO, for } x_k^* = a_{2m}, \\ \text{mSO, for } x_k^* = a_{2m-1}; \\ \vdots \\ \text{mSI-oSO, for } x_k^* = a_2, \\ \text{mSO, for } x_k^* = a_1; \end{cases} \end{cases} \tag{5.118}
$$

$$
\text{mSI-oSO } (x_k^* = a) \to \begin{cases} \text{mSI-oSO} \to \text{mSI-oSO, for } x_k^* = a = a_{2m+1} \\ (2m)^{\text{th}} \text{ order mUSN} \to \begin{cases} \text{mSO, for } x_k^* = a_{2m}, \\ \text{mSI-oSO, for } x_k^* = a_{2m-1}; \\ \vdots \\ \text{mSO, for } x_k^* = a_2, \\ \text{mSI-oSO, for } x_k^* = a_1. \end{cases} \end{cases} \tag{5.119}
$$

and the two switching bifurcations of mSO $\to$ (mSO : $(2m)^{\text{th}}$ mUS) and mSI-oSO $\to$ (mSI-oSO : $(2m)^{\text{th}}$ mLS) with two $(2m)^{\text{th}}$-order mUSN and mLSN spraying appearing bifurcations are

$$\text{mSO } (x_k^* = a) \to \begin{cases} (2m)^{\text{th}} \text{ order mUSN} \to \begin{cases} \text{mSO, for } x_k^* = a_{2m+1}, \\ \text{mSI-oSO, for } x_k^* = a_{2m}; \\ \vdots \\ \text{mSO, for } x_k^* = a_3, \\ \text{mSI-oSO, for } x_k^* = a_2; \end{cases} \\ \text{mSO} \to \text{mSO, for } x_k^* = a = a_1 \end{cases} \tag{5.120}$$

$$\text{mSI-oSO } (x_k^* = a) \to \begin{cases} (2m)^{\text{th}} \text{ order mLSN} \to \begin{cases} \text{mSI-oSO, for } x_k^* = a_{2m+1}, \\ \text{mSO, for } x_k^* = a_{2m}; \\ \vdots \\ \text{mSI-oSO, for } x_k^* = a_3, \\ \text{mSO, for } x_k^* = a_2; \end{cases} \\ \text{mSI-oSO} \to \text{mSI-oSO, for } x_k^* = a = a_1. \end{cases} \tag{5.121}$$

The $(2m+1)^{\text{th}}$ order monotonic source broom switching bifurcation is

$$\text{mSO}(x_k^* = a) \xrightarrow{\text{switching}} (2m+1)^{\text{th}} \text{ order mSO} \begin{cases} \text{mSO, for } x_k^* = a_{2m+1}, \\ \text{mSI-oSO, for } x_k^* = a_{2m}, \\ \vdots \\ \text{mSI-oSO, for } x_k^* = a_2, \\ \text{mSO, for } x_k^* = a_1. \end{cases} \tag{5.122}$$

The $(2m+1)^{\text{th}}$ order monotonic sink *broom-switching* bifurcation is

$$\text{mSI-oSO } (x_k^* = a_1) \xrightarrow{\text{switching}} (2m+1)^{\text{th}} \text{ order mSI} \begin{cases} \text{mSI-oSO, for } x_k^* = a_{2m+1}, \\ \text{mSO, for } x_k^* = a_{2m}, \\ \vdots \\ \text{mSO, for } x_k^* = a_2, \\ \text{mSI-oSO, for } x_k^* = a_1. \end{cases} \tag{5.123}$$

The switching bifurcation consist of a simple switching and the $(2m)^{\text{th}}$ order monotonic saddle-node appearing bifurcation with m-pairs of monotonic sources and monotonic sink to oscillatory sources. The $(2m)^{\text{th}}$ order monotonic saddle-node appearing bifurcation is a *sprinkler-spraying cluster* of the m-pairs of monotonic sources and monotonic sinks to oscillatory sources. Thus, the four switching bifurcations of

$$
\begin{aligned}
&\text{mSO} \rightarrow ((2m)^{\text{th}}\,\text{mLS:mSO}) \text{ for } a_0 > 0,\\
&\text{mSI-oSO} \rightarrow ((2m)^{\text{th}}\,\text{mUS:mSI-oSO}) \text{ for } a_0 < 0,\\
&\text{mSO} \rightarrow (\text{mSO} : (2m)^{\text{th}}\,\text{mUS}) \text{ for } a_0 > 0,\\
&\text{mSI-oSO} \rightarrow (\text{mSI-oSO:}(2m)^{\text{th}}\,\text{mLS}) \text{ for } a_0 < 0,
\end{aligned}
$$

are presented in Fig. 5.3i–iv, respectively. The $(2m+1)^{\text{th}}$-order monotonic source switching bifurcation is named the $(2m+1)^{\text{th}}$ mSO broom switching bifurcation and the $(2m+1)^{\text{th}}$-order monotonic sink switching bifurcation is named the $(2m+1)^{\text{th}}$ mSI broom switching bifurcation. Such a $(2m+1)^{\text{th}}$ mXX *broom-switching* bifurcation is from simple fixed-point to a $(2m+1)^{\text{th}}$ mXX *broom-*switching bifurcation. The two broom-switching bifurcations of

$$
\begin{aligned}
&\text{mSO} \rightarrow (2m+1)^{\text{th}}\,\text{mSO for } a_0 > 0,\\
&\text{mSI-oSO} \rightarrow (2m+1)^{\text{th}}\,\text{mSI for } a_0 < 0,
\end{aligned}
$$

are presented in Fig. 5.3v–vi, respectively. The period-2 fixed points are sketched as well through red curves.

A series of the third-order monotonic source and monotonic sink bifurcations is aligned up with varying with parameters. Such a special pattern is from m-quadratics in the $(2m+1)^{\text{th}}$ order polynomial systems, the following conditions should be satisfied.

$$
\begin{aligned}
&a(\mathbf{p}_i) = -\frac{1}{2}B_i \text{ and } a(\mathbf{p}_j) = -\frac{1}{2}B_j\\
&B_i \approx B_j\, i,j \in \{1,2,\ldots,n\}; i \neq j,\\
&\Delta_i > \Delta_{i+1} (i = 1,2,\ldots,n; n \leq m),\\
&\Delta_i = 0 \text{ with } ||\mathbf{p}_i|| < ||\mathbf{p}_{i+1}||.
\end{aligned}
\tag{5.124}
$$

Thus, a series of m-(3^{rd}mSO-3^{rd}mSI-$\cdots$) *switching* bifurcations ($a_0 > 0$) and a series of m-(3^{rd}mSI-3^{rd}mSO-$\cdots$) *switching* bifurcations ($a_0 < 0$) are presented in Fig. 5.4i and ii. The bifurcation scenario is formed by the swapping pattern of 3rd mSI and 3rd mSO switching bifurcations. Such a bifurcation scenario is like the fish-bone. Thus, such a bifurcation swapping pattern of 3^{rd}mSI and 3^{rd}mSO switching bifurcations is called the *fish-bone switching* bifurcation in the $(2m+1)^{\text{th}}$

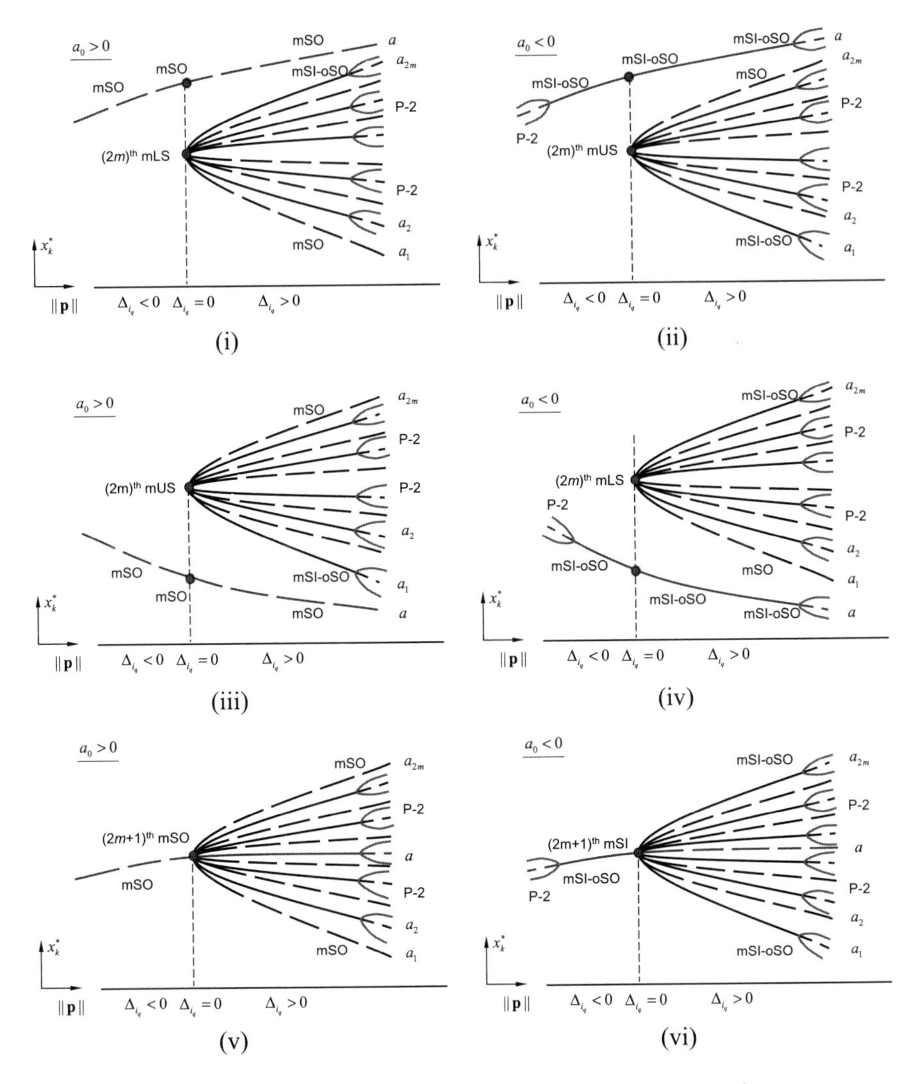

Fig. 5.3 (i) $((2m)^{\text{th}}$mLS:mSO)-switching bifurcation $(a_0 > 0)$, (ii) $((2m)^{\text{th}}$mUS:mSI-oSO)-switching bifurcation $(a_0 < 0)$, (iii) (mSO:$(2m)^{\text{th}}$mUS)-switching bifurcation $(a_0 > 0)$, (iv) (mSI-oSO:$(2m)^{\text{th}}$mLS)-switching bifurcation $(a_0 < 0)$, (v) $(2m+1)^{\text{th}}$mSO broom appearing bifurcation $((a_0 > 0)$, (vi) $(2m+1)^{\text{th}}$mSI-oSO broom appearing bifurcation $(a_0 < 0)$ in the $(2m+1)^{\text{th}}$ degree polynomial system. mLS: monotonic lower saddle, mUS: monotonic upper saddle, mSI-oSO: monotonic sink to oscillatory source, mSO: monotonic source. Stable and unstable fixed-points are represented by solid and dashed curves, respectively. The bifurcation points are marked by circular symbols. P-2 is for period-2 fixed points, which are sketched though red curves.

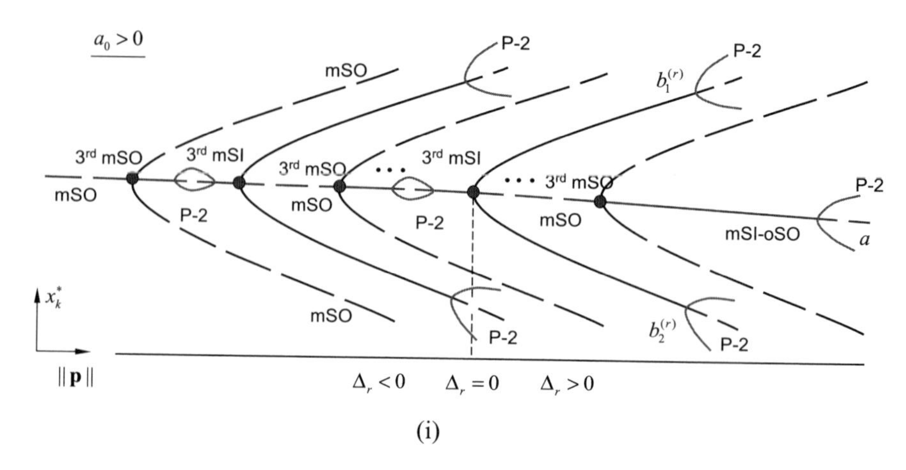

(i)

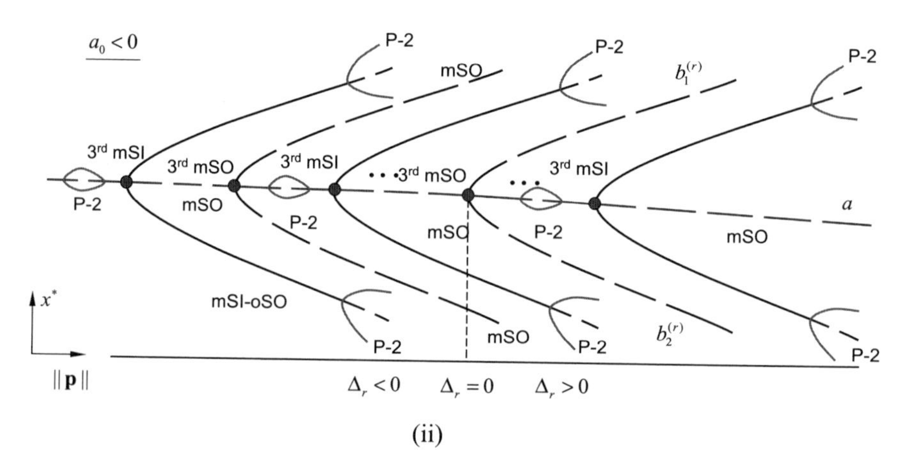

(ii)

Fig. 5.4 (i) m-(3^{rd}mSO-3^{rd}mSI-$\cdots$) series bifurcation ($a_0 > 0$), (ii) m-(3^{rd}mSI-3^{rd}mSO-$\cdots$) series switching bifurcation ($a_0 < 0$) in the $(2m+1)^{\text{th}}$-degree polynomial system. mSI: monotonic sink, mSO: monotonic source. Stable and unstable fixed-points are represented by solid and dashed curves, respectively. The bifurcation points are marked by circular symbols. P-2 is for period-2 fixed points, which are sketched though red curves.

degree polynomial nonlinear system. There are two swaps of the 3^{rd}mSI and 3^{rd}mSO bifurcations: (i) the 3^{rd}mSO-3^{rd}mSI *fish-bone switching* bifurcation and (ii) the 3^{rd}mSI-3^{rd}mSO *fish-bone, switching* bifurcation. The period-2 fixed-points are presented by P-2, which is sketched by red curves. The period-2 fixed-points are relative to the monotonic sink to the oscillatory source (mSI-oSO), and the monotonic sink to oscillatory source and back to monotonic sink (mSI-oSO-mSI).

5.2.2 Switching Bifurcations

In the $(2m+1)^{\text{th}}$ order polynomial discrete system, among the possible $(2m+1)$ roots, there are two roots to satisfy $x_k^2 + B_i x_k + C_i = 0$ with

$$\begin{aligned} & B_i = -(b_1^{(i)} + b_2^{(i)}),\ \Delta_i = (b_1^{(i)} - b_2^{(i)})^2 \geq 0, \\ & x_{k;1,2}^{(i)} = b_{1,2}^{(i)}, \Delta_i > 0 \text{ if } b_1^{(i)} \neq b_2^{(i)} (i = 1, 2, \cdots, n); \\ & \left.\begin{aligned} & B_i \neq B_j (i, j = 1, 2, \cdots, n; i \neq j) \\ & \Delta_i = 0 \text{ at } b_1^{(i)} = b_2^{(i)} (i = 1, 2, \cdots, n) \end{aligned}\right\} \text{at bifurcation.} \end{aligned} \tag{5.125}$$

The second-order singularity bifurcation is for the *switching* of a pair of fixed-points with a simple monotonic sink to oscillatory source, and monotonic source. There are two *switching* bifurcations for $i \in \{1, 2, \ldots, n\}$

$$2^{\text{nd}}\text{order mUS} \xrightarrow[\text{appearing bifurcation}]{i^{\text{th}}\text{quadratic factor}} \begin{cases} \text{mSO, for } a_{2i} = b_2^{(i)} \to b_1^{(i)}, \\ \text{mSI-oSO, for } a_{2i-1} = b_1^{(i)} \to b_2^{(i)}; \end{cases} \tag{5.126}$$

$$2^{\text{nd}}\text{order mLS} \xrightarrow[\text{switching bifurcation}]{i^{\text{th}}\text{quadratic factor}} \begin{cases} \text{mSI-oSO, for } a_{2i} = b_2^{(i)} \to b_1^{(i)}, \\ \text{mSO, for } a_{2i-1} = b_1^{(i)} \to b_2^{(i)}. \end{cases} \tag{5.127}$$

For non-switching point, $\Delta_i > 0$ at $b_1^{(i)} \neq b_2^{(i)}$ $(i = 1, 2, \ldots, n)$. At the bifurcation point, $\Delta_i = 0$ at $b_1^{(i)} = b_2^{(i)}$ $(i = 1, 2, \ldots, n)$. The l-mUSN *antenna switching* bifurcation for $s_i \in \{0, 1, \ldots, m\}$ $(i = 1, 2, \ldots, l)$ is

$$l\text{-mUSN} \begin{cases} \text{mUS} \xrightarrow[\text{switching}]{s_l^{\text{th}}\text{bifurcation}} \begin{cases} \text{mSO} \downarrow \text{mSI-oSO, for } b_2^{(s_l)} = a_{2s_l} \downarrow a_{2s_l-1}, \\ \text{mSI-oSO} \uparrow \text{mSO, for } b_1^{(s_l)} = a_{2s_l-1} \uparrow a_{2s_l}; \end{cases} \\ \vdots \\ \text{mUS} \xrightarrow[\text{switching}]{s_l^{\text{th}}\text{bifurcation}} \begin{cases} \text{mSO} \downarrow \text{mSI-oSO, for } b_2^{(s_1)} = a_{2s_1} \downarrow a_{2s_1-1}, \\ \text{mSI-oSO} \uparrow \text{mSO, for } b_1^{(s_1)} = a_{2s_1-1} \uparrow a_{2s_1}. \end{cases} \end{cases} \tag{5.128}$$

The l-mLSN *antenna switching* bifurcation for $s_i \in \{0, 1, \ldots, m\}$ $(i = 1, 2, \ldots, l)$ is

$$
l\text{-mLSN}\begin{cases}
\text{mLS}\xrightarrow[\text{switching}]{s_l^{\text{th}}\text{bifurcation}}\begin{cases}\text{mSI-oSO}\downarrow\text{mSO}, & \text{for } b_2^{(s_l)}=a_{2s_l}\downarrow a_{2s_l-1},\\ \text{mSO}\uparrow\text{mSI-oSO}, & \text{for } b_1^{(s_l)}=a_{2s_l-1}\uparrow a_{2s_l};\end{cases}\\
\vdots\\
\text{mLS}\xrightarrow[\text{switching}]{s_l^{\text{th}}\text{bifurcation}}\begin{cases}\text{mSI-oSO}\downarrow\text{mSO}, & \text{for } b_2^{(s_1)}=a_{2s_1}\downarrow a_{2s_1-1},\\ \text{mSO}\uparrow\text{mSI-oSO}, & \text{for } b_1^{(s_1)}=a_{2s_1-1}\uparrow a_{2s_1}.\end{cases}
\end{cases}
\tag{5.129}
$$

Two *antenna* switching bifurcation structures exist for the $(2m+1)^{\text{th}}$-order polynomial discrete system. The (l_1-mLSN:mSO:l_2-mUSN)-switching bifurcation for $a_0>0$ is

$$
(l_1\text{-mLSN}{:}m\text{SO}{:}l_2\text{-mUSN})\xrightarrow{l_1+l_2=m}\begin{cases}l_2\text{-mUSN}\\ \text{mSO}\to\text{mSO},\\ l_1\text{-mLSN};\end{cases}
\tag{5.130}
$$

and the (l_1-mUSN:mSI-oSO:l_2-mLSN)-switching bifurcation for $a_0<0$ is

$$
(l_1\text{-mUSN}{:}\text{mSI-oSO}{:}l_2\text{-mLSN})\xrightarrow{l_1+l_2=m}\begin{cases}l_2\text{-mLSN},\\ \text{mSI-oSO}\to\text{mSI-oSO},\\ l_1\text{-USN}.\end{cases}
\tag{5.131}
$$

As in the $(2m+1)^{\text{th}}$-order polynomial system, consider a switching bifurcation for a bundle of fixed-points with monotonic-sink-to-oscillatory-source and monotonic-source with the following conditions,

$$
\left.\begin{array}{l}B_i=B_j(i,j\in\{1,2,\ldots,n\};i\neq j)\\ \Delta_i=0 \text{ at } b_1^{(i)}=\mathrm{b}_2^{(i)}(i=1,2,\ldots,n)\end{array}\right\}\text{at bifurcation.}
\tag{5.132}
$$

Two $(2l)^{\text{th}}$ order switching bifurcations for $l\in\{1,2,\ldots,s\}$ are

$$
(2l)^{\text{th}}\text{order mUS}\xrightarrow[\text{switching bifurcation}]{\text{a bundle of }(2l)\text{-fixed-points}}\begin{cases}\text{mSO}, & \text{for } a_{2s_l}\to b_{2s_l},\\ \text{mSI-oSO}, & \text{for } a_{2s_l-1}\to b_{2s_l-1},\\ \vdots\\ \text{mSO}, & \text{for } a_{2s_1}\to b_{2s_1},\\ \text{mSI-oSO}, & \text{for } a_{2s_1-1}\to b_{2s_1-1}.\end{cases}
\tag{5.133}
$$

$$
(2l)^{\text{th}}\text{order mLS}\xrightarrow[\text{switching bifurcation}]{\text{a bundle of }(2l)\text{-fixed-points}}\begin{cases}\text{mSI-oSO}, & \text{for } a_{2s_l}\to b_{2s_l},\\ \text{mSO}, & \text{for } a_{2s_l-1}\to b_{2s_l-1},\\ \vdots\\ \text{mSI-oSO}, & \text{for } a_{2s_1}\to b_{2s_1},\\ \text{mSO}, & \text{for } a_{2s_1-1}\to b_{2s_1-1}.\end{cases}
\tag{5.134}
$$

where $\Delta_{ij} = (a_i - a_j)^2 = (b_i - b_j)^2 = 0$ with $B_i = B_j(i,j = 2s_1 - 1, 2s_1, \cdots, 2s_l - 1, 2s_l)$ and

$$\begin{aligned}
&\{a_{2s_1-1}, a_{2s_1}, \cdots, a_{2s_l-1}, a_{2s_l}\} \underset{\text{before bifurcation}}{\subset} \text{sort}\{b_1^{(1)}, b_2^{(1)}, \cdots b_1^{(n)}, b_2^{(n)}, a\},\\
&\{b_{2s_1-1}, b_{2s_1}, \cdots, b_{2s_l-1}, b_{2s_l}\} \underset{\text{after bifurcation}}{\subset} \text{sort}\{b_1^{(1)}, b_2^{(1)}, \cdots b_1^{(n)}, b_2^{(n)}, a\}.
\end{aligned} \tag{5.135}$$

Two $(2l+1)^{\text{th}}$ order switching bifurcations for $l \in \{1, 2, \ldots, s\}$ are

$$(2l+1)^{\text{th}}\text{order mSO} \xrightarrow[\text{switching bifurcation}]{\text{a bundle of }(2l+1)\text{-fixed-points}} \begin{cases} \text{mSO, for } a_{2s_l+1} \to b_{2s_l+1},\\ \vdots \\ \text{mSI-oSO, for } a_{2s_1} \to b_{2s_1},\\ \text{mSO, for } a_{2s_1-1} \to b_{2s_1-1}. \end{cases} \tag{5.136}$$

$$(2l+1)^{\text{th}}\text{order mSI} \xrightarrow[\text{switching bifurcation}]{\text{a bundle of }(2l+1)\text{-fixed-points}} \begin{cases} \text{mSI-oSO, for } a_{2s_l+1} \to b_{2s_l+1},\\ \vdots \\ \text{mSO, for } a_{2s_1} \to b_{2s_1},\\ \text{mSI-oSO, for } a_{2s_1-1} \to b_{2s_1-1}. \end{cases} \tag{5.137}$$

where $\Delta_{ij} = (a_i - a_j)^2 = (b_i - b_j)^2 = 0$ with $B_i = B_j(i,j = 2s_1 - 1, 2s_1, \cdots, 2s_l - 1)$ and

$$\begin{aligned}
&\{a_{2s_1-1}, a_{2s_1}, \cdots, a_{2s_l+1}\} \underset{\text{before bifurcation}}{\subset} \text{sort}\{b_1^{(1)}, b_2^{(1)}, \cdots b_1^{(n)}, b_2^{(n)}, a\},\\
&\{b_{2s_1-1}, b_{2s_1}, \cdots, b_{2s_l+1}\} \underset{\text{After bifurcation}}{\subset} \text{sort}\{b_1^{(1)}, b_2^{(1)}, \cdots b_1^{(n)}, b_2^{(n)}, a\}.
\end{aligned} \tag{5.138}$$

A set of paralleled, different, higher-order, monotonic upper-saddle-node switching bifurcations is the $((\alpha_1)^{\text{th}}\text{mXX}:(\alpha_2)^{\text{th}}\text{mXX}:\cdots:(\alpha_s)^{\text{th}}\text{ mXX})$ *parallel switching* bifurcation in the $(2m+1)^{\text{th}}$-degree polynomial discrete system. At the *straw-bundle switching* bifurcation, $\Delta_i = 0 (i = 1, 2, \ldots, n)$ and $B_i = B_j$ $(i,j \in \{1, 2, \ldots, n\};\ i \neq j)$. Thus, the *parallel straw-bundle switching* bifurcation is

$$\begin{aligned}
&((\alpha_1)^{\text{th}}\text{mXX}:(\alpha_2)^{\text{th}}\text{mXX}:\cdots:(\alpha_s)^{\text{th}}\text{ mXX})\text{-switching}\\
&= \begin{cases} (\alpha_s)^{\text{th}} \text{ order mXX switching},\\ \vdots \\ (\alpha_2)^{\text{th}} \text{ order mXX switching},\\ (\alpha_1)^{\text{th}} \text{ order mXX switching}; \end{cases}
\end{aligned} \tag{5.139}$$

where

$$\begin{aligned}&\alpha_i \in \{2l_i, 2l_i - 1\} \text{with} \textstyle\sum_{i=1}^{s} \alpha_i = 2m+1;\\ &\text{and } \text{mXX} \in \{\text{mUS}, \text{mLS}, \text{mSO}, \text{mSI}\}.\end{aligned} \tag{5.140}$$

The $(2l_1^{(j)} : 2l_2^{(j)} : \cdots : 2l_s^{(j)})^{\text{th}}$ mUSN parallel switching bifurcation is called a $(2l_1^{(j)} : 2l_2^{(j)} : \cdots : 2l_s^{(j)})^{\text{th}}$ mUSN parallel *straw-bundle switching* bifurcation.

$$\begin{aligned}s_j\text{-mUSG} &= (2l_1^{(j)} : 2l_2^{(j)} : \cdots : 2l_{s_j}^{(j)})^{\text{th}} \text{ mUSN switching}\\ &= \begin{cases}(2l_s^{(j)})^{\text{th}} \text{ order mUSN switching,}\\ \vdots\\ (2l_2^{(j)})^{\text{th}} \text{ order mUSN switching,}\\ (2l_1^{(j)})^{\text{th}} \text{ order mUSN switching.}\end{cases}\end{aligned} \tag{5.141}$$

The $(2l_1^{(j)} : 2l_2^{(j)} : \cdots : 2l_s^{(j)})^{\text{th}}$ mLSN parallel switching bifurcation is called a $(2l_1^{(j)} : 2l_2^{(j)} : \cdots : 2l_s^{(j)})^{\text{th}}$ mLSN parallel *straw-bundle switching* bifurcation.

$$\begin{aligned}s_j\text{-mLSG} &= (2l_1^{(j)} : 2l_2^{(j)} : \cdots : 2l_{s_j}^{(j)})^{\text{th}} \text{ mLSN switching}\\ &= \begin{cases}(2l_s^{(j)})^{\text{th}} \text{ order mLSN switching,}\\ \vdots\\ (2l_2^{(j)})^{\text{th}} \text{order mLSN switching,}\\ (2l_1^{(j)})^{\text{th}} \text{ order mLSN switching.}\end{cases}\end{aligned} \tag{5.142}$$

The $(s_1\text{-mLSG:}(2l_1^{(2)}+1)^{\text{th}}\text{mSO:}s_3\text{-mUSG})$-switching bifurcation for $a_0 > 0$ is

$$(s_1\text{-mLSG}, (2l_1^{(2)}+1)^{\text{th}} \text{ mSO}, s_3\text{-mUSG}) = \begin{cases}(2l_1^{(3)}, \cdots, 2l_{s_2}^{(3)})^{\text{th}} \text{-mUSN,}\\ (2l_1^{(2)}+1)^{\text{th}} \text{ mSO,}\\ (2l_1^{(1)}, \cdots, 2l_{s_1}^{(1)})^{\text{th}} \text{-mLSN;}\end{cases} \tag{5.143}$$

and the $(s_1\text{-mUSG:}(2l_1^{(2)}+1)^{\text{th}}\text{mSI:}s_3\text{-mLSG})$-switching bifurcation for $a_0 < 0$ is

$$(s_1\text{-mUSG:}(2l_1^{(2)}+1)^{\text{th}}\text{mSI:}s_3\text{-mLSG}) \to \begin{cases}(2l_1^{(3)}, \cdots, 2l_{s_3}^{(3)})^{\text{th}} \text{-mLSN,}\\ (2l_1^{(2)}+1)^{\text{th}} \text{ mSI,}\\ (2l_1^{(1)}, \cdots, 2l_{s_1}^{(1)})^{\text{th}} \text{-mUSN;}\end{cases} \tag{5.144}$$

The two (l_1-mUSN:*m*SO:l_2-mLSN) and (l_1-mLSN:mSI-oSO:l_2-mUSN) parallel-switching bifurcations ($l_1 + l_2 = m$) are presented in Fig. 5.5i and ii for $a_0 > 0$ and

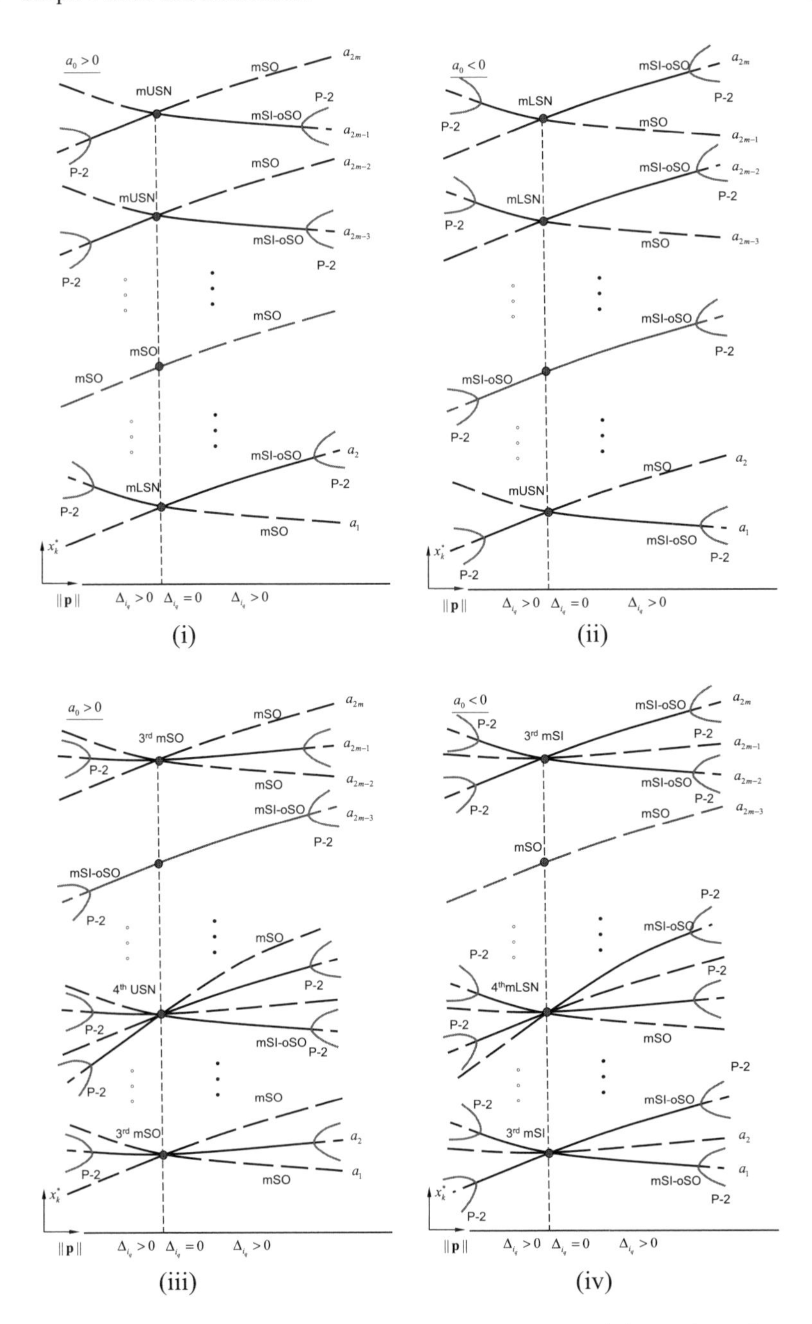

Fig. 5.5 Parallel switching bifurcations: (i) (l_1-mUSG:mSO:l_2-mUSG) ($a_0>0$), (ii) (l_1-mUSG: $mSI-oSO: l_2$-mLSG) ($a_0<0$); (iii)(3^{rd}mSI:$\cdots$:mUSN:3^{rd}mSO) ($a_0>0$), (vi) (3^{rd}mSO:$\cdots$: $mLSN: 3^{\text{rd}}$mSI) (($a_0<0$) in the $(2m+1)^{\text{th}}$-degree polynomial nonlinear system. mLSN: monotonic lower saddle-node, mUSN: monotonic upper saddle-node, mSI-oSO: monotonic sink, mSO: monotonic source. Stable and unstable fixed-points are represented by solid and dashed curves, respectively. The bifurcation points are marked by circular symbols. P-2: period-2 fixed points which is sketched through red curves.

$a_0 < 0$, respectively. A set of (3^{rd}mSO:$\cdots$:mSI-oSO:3^{rd}mSO) parallel, *switching* bifurcations for mSI-oSO and mSO fixed-points is presented in Fig. 5.5iii for $a_0 > 0$. However, for $a_0 < 0$, the set of (3^{rd}mSI:$\cdots$:*m*SO:3^{rd}mSI) *switching* bifurcations for monotonic sources and monotonic sinks is presented in Fig. 5.5iv. The period-2 fixed-points are sketched through red curves, which is relative to the monotonic sink to the oscillatory source.

5.2.3 Switching-Appearing Bifurcations

Consider a $(2m+1)^{\text{th}}$ degree polynomial discrete system in a form of

$$x_{k+1} = x_k + a_0 Q(x_k) \prod_{i=1}^{2n_1+1} (x_k - c_i) \prod_{j=1}^{n_2} (x_k^2 + B_j x_k + C_j). \tag{5.145}$$

Without loss of generality, a function of $Q(x_k) > 0$ is either a polynomial function or a non-polynomial function. The roots of $x_k^2 + B_j x_k + C_j = 0$ are

$$b_{1,2}^{(j)} = -\tfrac{1}{2}B_j \pm \tfrac{1}{2}\sqrt{\Delta_j}, \Delta_j = B_j^2 - 4C_j \geq 0 (j = 1, 2, \ldots, n_2); \tag{5.146}$$

either

$$\begin{array}{l} \{a_1^-, a_2^-, \cdots, a_{2n_1+1}^-\} = \text{sort}\{c_1, c_2 \cdots, c_{2n_1+1}\}, \\ a_s^- \leq a_{s+1}^- \text{ before bifurcation;} \\ \{a_1^+, a_2^+, \cdots, a_{2n_3+1}^+\} = \text{sort}\{c_1, \cdots, c_{2n_1+1}; b_1^{(1)}, b_2^{(1)}, \cdots, b_1^{(n_2)}, b_2^{(n_2)}\}, \\ a_s^+ \leq a_{s+1}^+, n_3 = n_1 + n_2 \text{ after bifurcation;} \end{array} \tag{5.147}$$

or

$$\begin{array}{l} \{a_1^-, a_2^-, \cdots, a_{2n_3+1}^-\} = \text{sort}\{c_1, c_2 \cdots, c_{2n_1}; b_1^{(1)}, b_2^{(1)}, \cdots, b_1^{(n_2)}, b_2^{(n_2)}, a\}, \\ a_s^- \leq a_{s+1}^-, n_3 = n_1 + n_2 \text{ before bifurcation;} \\ \{a_1^+, a_2^+, \cdots, a_{2n_1+1}^+\} = \text{sort}\{c_1, \cdots, c_{2n_1}, a\}, \\ a_s^+ \leq a_{s+1}^+ \text{ after bifurcation;} \end{array} \tag{5.148}$$

and

$$\left.\begin{array}{l} B_{j_1} = B_{j_2} = \cdots = B_{j_s} (j_{k_1} \in \{1, 2, \ldots, n\}; j_{k_1} \neq j_{k_2}) \\ (k_1, k_2 \in \{1, 2, \ldots, s\}; k_1 \neq k_2) \\ \Delta_j = 0 (j \in U \subset \{1, 2, \ldots, n_2\} \\ c_i \neq -\tfrac{1}{2}B_j (i = 1, 2, \ldots, 2n_1, j = 1, 2, \ldots, n_2) \end{array}\right\} \text{at bifurcation.} \tag{5.149}$$

Consider a just before bifurcation of $((\alpha_1^-)^{\text{th}}$ mXX$_1^-$:$(\alpha_2^-)^{\text{th}}$ mXX$_2^-$:$\cdots$: $(\alpha_{s_1}^-)^{\text{th}}$ mXX$_{s_1}^-$) with $\Sigma_{i=1}^{s_1} \alpha_i = 2m_1 + 1$ for simple monotonic-sources and monotonic-sink-to-oscillatory-sources in the $(2m+1)^{\text{th}}$ degree polynomial nonlinear discrete system. For $\alpha_i^- = 2l_i^- - 1$, $\text{mXX}_i^- \in \{m\text{SO,mSI}\}$ and for $\alpha_i^- = 2l_i^-$, $\text{mXX}_i^- \in \{\text{mUS,mLS}\}$ $(i = 1, 2, \ldots, s_1)$. The detailed structures are as follows.

$$\left.\begin{array}{l}\text{mSI-oSO}\\ \text{mSO}\\ \vdots\\ \text{mSO}\\ \text{mSI-oSO}\end{array}\right\} \to (2l_i^- - 1)^{\text{th}}\text{mSI}, \text{ and } \left.\begin{array}{l}\text{mSO}\\ \text{mSI-oSO}\\ \vdots\\ \text{mSI-oSO}\\ \text{mSO}\end{array}\right\} \to (2l_i^- - 1)^{\text{th}}\text{mSO};$$
$$\left.\begin{array}{l}\text{mSO}\\ \text{mSI-oSO}\\ \vdots\\ \text{mSO}\\ \text{mSI-oSO}\end{array}\right\} \to (2l_i^-)^{\text{th}}\,\text{mUS}, \quad \text{and } \left.\begin{array}{l}\text{mSI-oSO}\\ \text{mSO}\\ \vdots\\ \text{mSI-oSO}\\ \text{SO}\end{array}\right\} \to (2l_i^-)^{\text{th}}\,\text{mLS}. \tag{5.150}$$

The bifurcation set of $((\alpha_1^-)^{\text{th}}\,\text{mXX}_1^-{:}(\alpha_2^-)^{\text{th}}\,\text{mXX}_2^-{:}\cdots{:}(\alpha_{s_1}^-)^{\text{th}}\,\text{mXX}_{s_1}^-)$ at the same parameter point is called a *left-parallel-bundle* switching bifurcation

Consider a just after bifurcation of $((\alpha_1^+)^{\text{th}}\,\text{mXX}_1^+{:}(\alpha_2^+)^{\text{th}}\,\text{mXX}_2^+{:}\cdots{:}\,(\alpha_{s_2}^+)^{\text{th}}$ $\text{mXX}_{s_2}^+)$ with $\sum_{i=1}^{s_2}\alpha_i^+ = 2m_2+1$ for simple monotonic sources and monotonic sinks to oscillatory sources in the $(2m+1)^{\text{th}}$ degree polynomial nonlinear discrete system. $\text{mXX}_i^+ \in \{\text{mSO,mSI}\}$ for $\alpha_i^+ = 2l_i^+ - 1$, and $\text{mXX}_i^- \in \{\text{mUS,mLS}\}$ for $\alpha_i^+ = 2l_i^+$. The four detailed structures are as follows.

$$(2l_i^+ - 1)^{\text{th}}\text{mSI} \to \left\{\begin{array}{l}\text{mSI-oSO}\\ \text{mSO}\\ \vdots\\ \text{mSO}\\ \text{mSI-oSO}\end{array}\right., \text{ and } (2l_i^+ - 1)^{\text{th}}\,\text{mSO} \to \left\{\begin{array}{l}\text{mSO}\\ \text{mSI-oSO}\\ \vdots\\ \text{mSI-oSO}\\ \text{mSO}\end{array}\right.;$$
$$(2l_i^+)^{\text{th}}\,\text{mUS} \to \left\{\begin{array}{l}\text{mSO}\\ \text{mSI-oSO}\\ \vdots\\ \text{mSO}\\ \text{mSI-oSO}\end{array}\right., \text{ and } (2l_i^+)^{\text{th}}\,\text{mLS} \to \left\{\begin{array}{l}\text{mSI-oSO}\\ \text{mSO}\\ \vdots\\ \text{mSI-oSO}\\ \text{mSO}\end{array}\right.. \tag{5.151}$$

The bifurcation set of $((\alpha_1^+)^{\text{th}}\,\text{mXX}_1^+{:}(\alpha_2^+)^{\text{th}}\,\text{mXX}_2^+{:}\cdots{:}(\alpha_{s_2}^+)^{\text{th}}\,\text{mXX}_{s_2}^+)$ at the same parameter point is called a *right-parallel-bundle* switching bifurcation.

(i) For the just before and after bifurcation structure, if there exists a relation of

$$
\begin{gathered}
(\alpha_i^-)^{\text{th}}\text{mXX}_i^- = (\alpha_j^+)^{\text{th}}\text{mXX}_j^+ = \alpha^{\text{th}}\text{mXX}, \text{ for } x_k^* = a_i^- = a_j^+ \\
(i \in \{1,2,\cdots,s_1\}, j \in \{1,2,\cdots,s_2\}), \text{ mXX} \in \{m\text{US,mLS,mSO,mSI}\}
\end{gathered} \tag{5.152}
$$

then the bifurcation is a αth mXX *switching* bifurcation for simple fixed-points.

(ii) Just for the just before bifurcation structure, if there exists a relation of

$$
\begin{gathered}
(2l_i^-)^{\text{th}}\text{mXX}_i^- = (2l)^{\text{th}}\text{mXX}, \text{ for } x_k^* = a_i^- = a_i \\
(i \in \{1,2,\cdots,s_1\}, m\text{XX} \in \{m\text{US,mLS}\}
\end{gathered} \tag{5.153}
$$

then, the bifurcation is a $(2l)^{\text{th}}$ mXX *left appearing* (or *right vanishing*) bifurcation for simple fixed-points.

(iii) Just for the just after bifurcation structure, if there exists a relation of

$$
\begin{gathered}
(2l_i^+)^{\text{th}}\text{mXX}_i^+ = (2l)^{\text{th}}\text{mXX}, \text{ for } x_k^* = a_i^+ = a_i \\
(i \in \{1,2,\cdots,s_1\}), \text{mXX} \in \{\text{mUS,mLS}\}
\end{gathered} \tag{5.154}
$$

then, the bifurcation is a $(2l)^{\text{th}}$ mXX *right appearing* (or *left vanishing*) bifurcation for simple fixed-points.

(iv) For the just before and after bifurcation structure, if there exists a relation of

$$
\begin{gathered}
(\alpha_i^-)^{\text{th}}\text{mXX}_i^- \neq (\alpha_j^+)^{\text{th}}\text{mXX}_j^+ \text{ for } x_k^* = a_i^- = a_j^+ \\
\text{XX}_i^-, \text{XX}_j^+ \in \{\text{mUS,mLS, mSO,mSI}\} \\
(i \in \{1,2,\cdots,s_1\}, j \in \{1,2,\cdots,s_2\}),
\end{gathered} \tag{5.155}
$$

then, there are two *flower-bundle* switching bifurcations of simple fixed-points:

(iv$_1$) for $\alpha_j = \alpha_i + 2l$, the bifurcation is called a α_j^{th} mXX right *flower-bundle* switching bifurcation for α_i to α_j-simple fixed-points with the appearance (birth) of $2l$-simple fixed-points.

(iv$_2$) for $\alpha_j = \alpha_i - 2l$, the bifurcation is called a α_i^{th} mXX left *flower-bundle* switching bifurcation for α_i to α_j-simple fixed-points with the vanishing (death) of $2l$-simple fixed-points.

A general parallel switching bifurcation is

$$((\alpha_1^-)^{\text{th}}\,\text{mXX}_1^-,(\alpha_2^-)^{\text{th}}\,\text{mXX}_2^-,\cdots,(\alpha_{s_1}^-)^{\text{th}}\,\text{mXX}_{s_1}^-)\xrightarrow[\text{bifucation}]{\text{switching}}$$
$$((\alpha_1^+)^{\text{th}}\,\text{mXX}_1^+,(\alpha_2^+)^{\text{th}}\,\text{mXX}_2^+,\cdots,(\alpha_{s_2}^+)^{\text{th}}\,\text{mXX}_{s_2}^+). \tag{5.156}$$

Such a general, parallel switching bifurcation consists of the *left* and *right* parallel-bundle switching bifurcations.

If the *left* and *right* parallel-bundle switching bifurcations are same in a parallel *flower-bundle* switching bifurcation, i.e.,

$$\begin{array}{l}(\alpha_i^-)^{\text{th}}\text{mXX}_i^- = (\alpha_i^+)^{\text{th}}\text{mXX}_i^+ = (\alpha_i)^{\text{th}}\text{mXX}_i,\\ \text{for } x_k^* = a_i^- = a_i^+ \ (i = 1,2,\cdots,s\}\end{array} \tag{5.157}$$

then the parallel *flower-bundle* switching bifurcation becomes a parallel *straw-bundle switching* bifurcation of $((\alpha_1)^{\text{th}}\,\text{mXX}:(\alpha_2)^{\text{th}}\,\text{mXX}:\cdots:(\alpha_s)^{\text{th}}\,\text{mXX})$.

If the *left* and *right* parallel-bundle switching bifurcations are different in a parallel *flower-bundle* switching bifurcation, i.e.,

$$\begin{array}{l}(\alpha_i^-)^{\text{th}}\text{mXX}_i^- = (2l_i^-)^{\text{th}}\text{mXX},\ (\alpha_j^+)^{\text{th}}\text{mXX}_j^+ = (2l_j^+)^{\text{th}}\text{mYY},\\ \text{for } x_k^* = a_i^- \neq a_i^+ \ (i = 1,2,\cdots,s\}\\ \text{mXX} \in \{\ \text{mUS,mLS}\}\ \text{,mYY} \in \{\ \text{mUS,mLS}\}\end{array} \tag{5.158}$$

then the parallel *flower-bundle* switching bifurcation becomes a combination of two independent left and right parallel appearing bifurcations:

(i) a $((2l_1^-)^{\text{th}}\,\text{mXX}_1^- : (2l_2^-)^{\text{th}}\,\text{mXX}_2^- : \cdots : (2l_{s_1}^-)^{\text{th}}\,\text{mXX}_{s_1}^-)$-*left parallel sprinkler-spraying appearing (or right vanishing)* bifurcation and
(ii) a $((2l_1^+)^{\text{th}}\,\text{mXX}_1^+ : (2l_2^+)^{\text{th}}\,\text{mXX}_2^+ : \cdots : (2l_{s_2}^+)^{\text{th}}\,\text{mXX}_{s_2}^+)$-*right parallel sprinkler-spraying appearing (or left vanishing)* bifurcation.

The (4^{th}mLS:$\cdots$:mSO:m6^{th} US) *parallel appearing* bifurcation for $a_0 > 0$ is presented in Fig. 5.6i. The (4th mUS:$\cdots$:mSI-oSO:6th mLS) *parallel appearing* bifurcation for $a_0 < 0$ is shown in Fig. 5.6ii. Such a kind of bifurcation is also like a *waterfall appearing* bifurcation. The (5^{th}mSO:$\cdots$:6^{th}mUS:6^{th}mUS) parallel, *flower-bundle switching* bifurcation for mSI and mSO fixed-points is presented in Fig. 5.6iii for $a_0 > 0$. Such a parallel *flower-bundle* switching bifurcation is from (mSO:mSI-oSO:mSO) to (5^{th}mSO:$\cdots$:6^{th}mUS:6^{th}mUS) with a *waterfall appearance.* The set of (5^{th}mSI:$\cdots$:6^{th}mLS:6^{th}mLS) *flower-bundle switching* bifurcation for mSI and mSO fixed-points is presented in Fig. 5.6iv for $a_0 < 0$. Such a parallel *flower-bundle* switching bifurcation is from (mSI-oSO:mSO:mSI-oSO) to (5^{th}mSI:$\cdots$:6^{th}mLS:6^{th}mLS) with a *waterfall appearance.* After the bifurcation, the *waterfall* fixed-points birth can be observed. The fixed-points before such a bifurcation are much less than after the bifurcation.

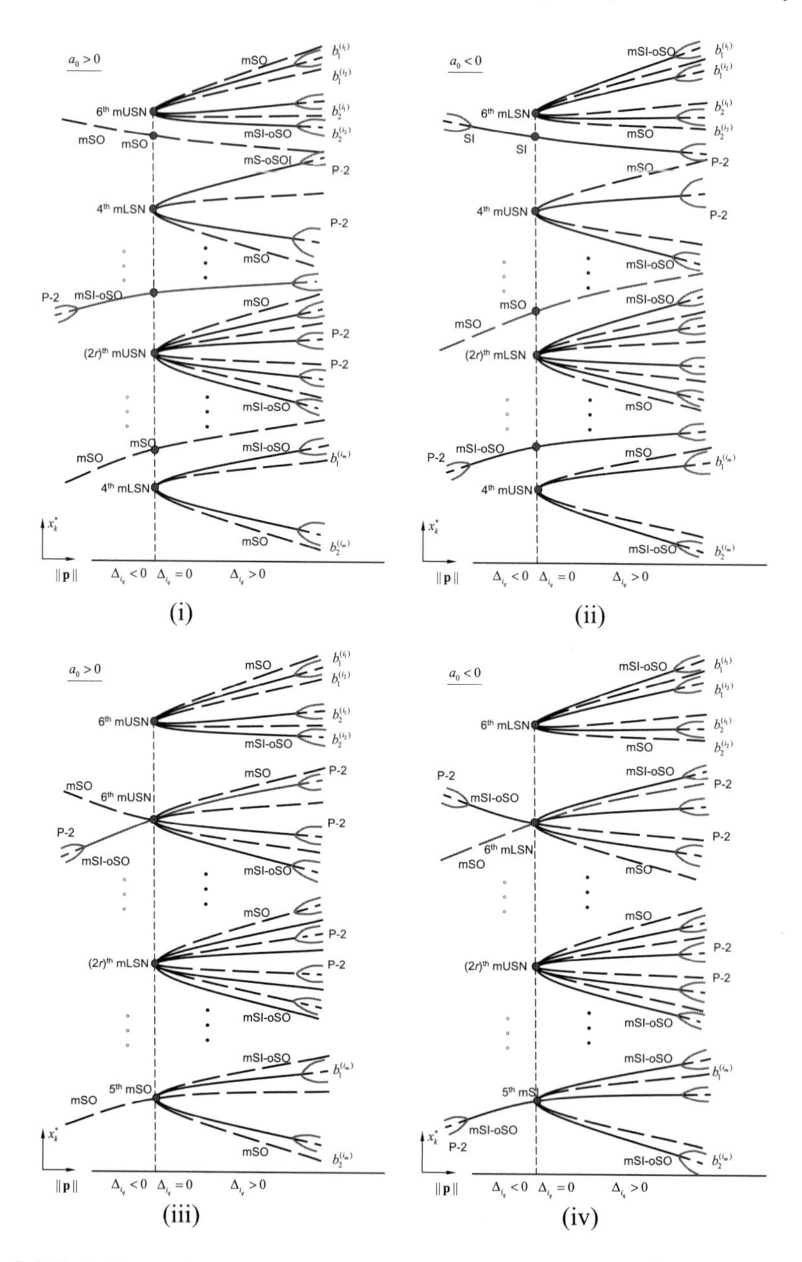

Fig. 5.6 Switching and appearing bifurcations. Simple switching: (i)(4^{th}mLSN: $\cdots$: mSO : 6^{th}mUSN) ($a_0 > 0$), (ii) (4^{th}mUSN: $\cdots$:mSI-oSO:6^{th}mLSN) ($a_0 < 0$). Higher-order switching: (iii) (5^{th}mSI: $\cdots$:6^{th}mUSN : 6^{th}mUSN) ($a_0 > 0$)), (vi) (5^{th}mSO: $\cdots$:6^{th}mLSN : 6^{th}mLSN) ($a_0 < 0$) in the $(2m+1)^{\text{th}}$-degree polynomial nonlinear system. mLSN: monotonic lower saddle-node, mUSN: monotonic upper saddle-node, mSI-oSO: monotonic sink to oscillatory source, mSO: monotonic source. Stable and unstable fixed-points are represented by solid and dashed curves, respectively. The bifurcation points are marked by circular symbols. The period-2 fixed points are represented by P-2, which are sketched through red curves.

5.3 Higher-Order Fixed-Point Bifurcations

The afore-discussed appearing and switching bifurcations in the $(2m+1)^{\text{th}}$ degree polynomial discrete system are relative to the simple monotonic sources and monotonic sinks to oscillatory sources. As similar to the $(2m)^{\text{th}}$ degree polynomial nonlinear discrete system, the higher-order singularity bifurcations in the $(2m+1)^{\text{th}}$ degree polynomial discrete system can be for higher-order monotonic sinks, monotonic sources, monotonic upper-saddles, and monotonic lower-saddles.

5.3.1 *Higher-Order Fixed-Point Bifurcations*

Consider a $(2m+1)^{\text{th}}$ degree polynomial nonlinear discrete system as

$$x_{k+1} = x_k + a_0 Q(x_k)(x_k - a)\prod_{i=1}^{s}(x_k^2 + B_i x_k + C_i)^{\alpha_i}, \tag{5.159}$$

where $\alpha_i \in \{2l_i - 1, 2l_i\}$. Without loss of generality, a function of $Q(x_k) > 0$ is either a polynomial function or a non-polynomial function. The roots of $x_k^2 + B_i x_k + C_i = 0$ are

$$\begin{aligned} &b_{1,2}^{(i)} = -\tfrac{1}{2}B_i \pm \tfrac{1}{2}\sqrt{\Delta_i},\ \Delta_i = B_i^2 - 4C_i \geq 0; \\ &\{a_1, a_2, \cdots, a_{2s-1}, a_{2s}, a_{2s+1}\} = \text{sort}\{b_1^{(1)}, b_2^{(1)}, \cdots, b_1^{(s)}, b_2^{(s)}, a\}, \\ &a_j \leq a_{j+1}. \end{aligned} \tag{5.160}$$

For $a \neq -\frac{1}{2}B_i (i = 1, 2, \ldots, s)$, there are four higher-order bifurcations as follows:

$$\begin{aligned} &(2(2l_i - 1))^{\text{th}} \text{ order mUS} \xrightarrow[\text{appearing bifurcation}]{(2l_i-1)^{\text{th}} \text{ order quadratics}} \\ &\quad \begin{cases} (2l_i - 1)^{\text{th}} \text{ order mSO}, & x_k^* = b_2^{(i)}, \\ (2l_i - 1)^{\text{th}} \text{ order mSI}, & x_k^* = b_1^{(i)}; \end{cases} \end{aligned} \tag{5.161}$$

$$\begin{aligned} &(2(2l_i - 1))^{\text{th}} \text{ order mLS} \xrightarrow[\text{appearing bifurcation}]{(2l_i-1)^{\text{th}} \text{ order quadratics}} \\ &\quad \begin{cases} (2l_i - 1)^{\text{th}} \text{ order mSI}, & x_k^* = b_2^{(i)}, \\ (2l_i - 1)^{\text{th}} \text{ order mSO}, & x_k^* = b_1^{(i)}; \end{cases} \end{aligned} \tag{5.162}$$

$$
\begin{aligned}
&(2(2l_i))^{\text{th}} \text{ order mUS} \xrightarrow[\text{appearing bifurcation}]{(2l_i)^{\text{th}}\text{-order power of quadratics}} \\
&\qquad\qquad \begin{cases} (2l_i)^{\text{th}} \text{ order mUS}, \; x_k^* = b_2^{(i)}, \\ (2l_i)^{\text{th}} \text{ order mUS}, \; x_k^* = b_1^{(i)}; \end{cases}
\end{aligned} \tag{5.163}
$$

$$
\begin{aligned}
&(2(2l_i))^{\text{th}} \text{ order mLS} \xrightarrow[\text{appearing bifurcation}]{(2l_i)^{\text{th}}\text{-order quadratics}} \\
&\qquad\qquad \begin{cases} (2l_i)^{\text{th}} \text{ order mLS}, \; x_k^* = b_2^{(i)}, \\ (2l_i)^{\text{th}} \text{ order mLS}, \; x_k^* = b_1^{(i)}. \end{cases}
\end{aligned} \tag{5.164}
$$

(i) For $\alpha_i = 2l_i - 1$, the $(2(2l_i - 1))^{\text{th}}$-order monotonic upper-saddle (mUS) *appearing* bifurcation is for the onset of the $(2l_i - 1)^{\text{th}}$-order monotonic source (mSO) $(x_k^* = b_2^{(i)})$ and the $(2l_i - 1)^{\text{th}}$-order monotonic sink (mSI) $(x_k^* = b_1^{(i)})$ with $b_2^{(i)} > b_1^{(i)}$.

(ii) For $\alpha_i = 2l_i - 1$, the $(2(2l_i - 1))^{\text{th}}$-order monotonic lower-saddle (mLS) *appearing* bifurcation is for the onset of the $(2l_i - 1)^{\text{th}}$-order monotonic sink (mSI) $(x_k^* = b_2^{(i)})$ and the $(2l_i - 1)^{\text{th}}$-order monotonic source (mSO) $(x_k^* = b_1^{(i)})$ with $b_2^{(i)} > b_1^{(i)}$.

(iii) For $\alpha_i = 2l_i$, the $(2(2l_i))^{\text{th}}$-order monotonic upper-saddle (mUS) *appearing* bifurcation is for the onset of two $(2l_i)^{\text{th}}$-order monotonic upper-saddles (mUS) $(x_k^* = b_1^{(i)}, b_2^{(i)})$ with $b_2^{(i)} > b_1^{(i)}$.

(iv) For $\alpha_i = 2l_i$, the $(2(2l_i))^{\text{th}}$ order monotonic lower-saddle (mLS) *appearing* bifurcation is for the onset of two $(2l_i)^{\text{th}}$-order monotonic lower-saddles (mLS) $(x_k^* = b_1^{(i)}, b_2^{(i)})$ with $b_2^{(i)} > b_1^{(i)}$.

The fixed-point of $x_k^* = a \neq -\frac{1}{2}B_i$ $(i = 1, 2, \ldots, s)$ breaks a cluster of *teethcomb* appearing bifurcations of higher order fixed-point to two parts. The *teethcomb* appearing bifurcation generated by the s-pairs of quadratics becomes a *broom appearing* bifurcation for higher-order fixed-points. The two *broom appearing* bifurcations for higher-order fixed-points are

$$
\text{mSO } (x_k^* = a) \xrightarrow[\text{appearing bifurcation}]{\sum_{j=1}^{2}\sum_{i=1}^{s_j}\alpha_i^{(j)}=m} \begin{cases} ((2\alpha_1^{(2)})^{\text{th}}\text{mUS}, \cdots, (2\alpha_{s_2}^{(2)})^{\text{th}}\text{mUS}), \\ \text{mSO}, \;\; \text{for } x_k^* = a = a_{2(s_1+1)-1}, \\ ((2\alpha_1^{(1)})^{\text{th}}\text{mLS}, \cdots, (2\alpha_{s_1}^{(1)})^{\text{th}}\text{mLS}); \end{cases} \tag{5.165}
$$

and

$$\text{mSI-oSO}\ (\ x_k^* = a) \xrightarrow[\text{appearing bifurcation}]{\sum_{j=1}^{2}\sum_{i=1}^{s_j}\alpha_i^{(j)}=m} \begin{cases} ((2\alpha_1^{(2)})^{\text{th}}\ \text{mLS}, \cdots, (2\alpha_{s_2}^{(2)})^{\text{th}}\ \text{mLS}), \\ \text{mSI-oSO}, \ \ \text{for}\ \ x_k^* = a = a_{2(s_1+1)-1}, \\ ((2\alpha_1^{(1)})^{\text{th}}\ \text{mUS}, \cdots, (2\alpha_{s_1}^{(1)})^{\text{th}}\ \text{mUS}); \end{cases} \tag{5.166}$$

where

$$((2\alpha_1^{(j)})^{\text{th}}\ \text{mUS}, \cdots, (2\alpha_{s_j}^{(j)})^{\text{th}}\ \text{mUS}) = \begin{cases} (2\alpha_{s_j}^{(j)})^{\text{th}}\ \text{mUS} \rightarrow \begin{cases} (\alpha_{s_j}^{(j)})^{\text{th}}\ \text{mXX}, \\ (\alpha_{s_j}^{(j)})^{\text{th}}\ \text{mXX}; \end{cases} \\ \vdots \\ (2\alpha_1^{(j)})^{\text{th}}\ \text{mUS} \rightarrow \begin{cases} (\alpha_1^{(j)})^{\text{th}}\ \text{mXX}, \\ (\alpha_1^{(j)})^{\text{th}}\ \text{mXX}; \end{cases} \end{cases} \tag{5.167}$$

$$((2\alpha_1^{(j)})^{\text{th}}\ \text{mLS}, \cdots, (2\alpha_{s_j}^{(j)})^{\text{th}}\ \text{mLS}) = \begin{cases} (2\alpha_{s_j}^{(j)})^{\text{th}}\ \text{mLS} \rightarrow \begin{cases} (\alpha_{s_j}^{(j)})^{\text{th}}\ \text{mXX}, \\ (\alpha_{s_j}^{(j)})^{\text{th}}\ \text{mXX}; \end{cases} \\ \vdots \\ (2\alpha_1^{(j)})^{\text{th}}\ \text{mLS} \rightarrow \begin{cases} (\alpha_1^{(j)})^{\text{th}}\ \text{mXX}, \\ (\alpha_1^{(j)})^{\text{th}}\ \text{mXX}; \end{cases} \end{cases} \tag{5.168}$$

for $j = 1, 2$.

Four special broom appearing bifurcations for higher-order fixed-points are

$$\text{mSO}\ (\ x_k^* = a) \xrightarrow[\text{appearing bifurcation}]{\sum_{i=1}^{s}\alpha_i=m} \begin{cases} \text{mSO}, \ \ \text{for}\ \ x_k^* = a = a_{2s+1}, \\ ((2\alpha_1)^{\text{th}}\ \text{mLS}, \cdots, (2\alpha_s)^{\text{th}}\ \text{mLS}); \end{cases} \tag{5.169}$$

$$\text{mSI-oSO}\ (\ x_k^* = a) \xrightarrow[\text{appearing bifurcation}]{\sum_{i=1}^{s}\alpha_i=m} \begin{cases} \text{mSo-oSO}, \ \ \text{for}\ \ x_k^* = a = a_{2s+1}, \\ ((2\alpha_1)^{\text{th}}\ \text{mUS}, \cdots, (2\alpha_s)^{\text{th}}\ \text{mUS}) \end{cases} \tag{5.170}$$

and

$$\text{mSO}\ (\ x_k^* = a) \xrightarrow[\text{appearing bifurcation}]{\sum_{i=1}^{s}\alpha_i=m} \begin{cases} ((2\alpha_1)^{\text{th}}\ \text{mUS}, \cdots, (2\alpha_s)^{\text{th}}\ \text{mUS}), \\ \text{mSO}, \ \ \text{for}\ \ x_k^* = a = a_1; \end{cases} \tag{5.171}$$

$$\text{mSI-oSO}\ (\ x_k^* = a) \xrightarrow[\text{appearing bifurcation}]{\sum_{i=1}^{s}\alpha_i=m} \begin{cases} ((2\alpha_1)^{\text{th}}\ \text{mLS}, \cdots, (2\alpha_s)^{\text{th}}\ \text{mLS}), \\ \text{mSI-oSO}, \ \ \text{for}\ \ x_k^* = a = a_1. \end{cases} \tag{5.172}$$

For $a = -\frac{1}{2}B_i$ $(i \in \{1, 2, \ldots, s\})$, there are four higher-order bifurcations as follows:

$$\text{mSO } (x_k^* = a) \to (2(2l_i - 1) + 1)^{\text{th}} \text{ mSO}$$
$$= \begin{cases} (2l_i - 1)^{\text{th}} \text{ order mSO}, \ x_k^* = b_2^{(i)}, \\ \text{mSI-oSO}, \ x_k^* = a, \\ (2l_i - 1)^{\text{th}} \text{ order SO}, \ x_k^* = b_1^{(i)}; \end{cases} \tag{5.173}$$

$$\text{mSI-oSO } (x_k^* = a) \to (2(2l_i - 1) + 1)^{\text{th}} \text{mSI}$$
$$= \begin{cases} (2l_i - 1)^{\text{th}} \text{ order mSI}, \ x_k^* = b_2^{(i)}, \\ \text{mSO}, \ x^* = a, \\ (2l_i - 1)^{\text{th}} \text{ order mSI}, \ x_k^* = b_1^{(i)}; \end{cases} \tag{5.174}$$

$$\text{mSI-oSO } (x_k^* = a) \to (2(2l_i) + 1)^{\text{th}} \text{ mSO}$$
$$= \begin{cases} (2l_i)^{\text{th}} \text{ order mUS}, \ x_k^* = b_2^{(i)}, \\ \text{mSO}, \ x_k^* = a, \\ (2l_i)^{\text{th}} \text{ order mLS}, \ x_k^* = b_1^{(i)}; \end{cases} \tag{5.175}$$

$$\text{mSI-oSO } (x_k^* = a) \to (2(2l_i) + 1)^{\text{th}} \text{ mSI}$$
$$= \begin{cases} (2l_i)^{\text{th}} \text{ order mLS}, \ x_k^* = b_2^{(i)}, \\ \text{mSI-oSO}, \ x^* = a, \\ (2l_i)^{\text{th}} \text{ order mUS}, \ x_k^* = b_1^{(i)}. \end{cases} \tag{5.176}$$

(i) For $\alpha_i = 2l_i - 1$, the $(2(2l_i - 1) + 1)^{\text{th}}$ order monotonic source (mSO) *switching* bifurcation is with the $(2l_i - 1)^{\text{th}}$ order monotonic source (mSO) $(x_k^* = b_2^{(i)})$ and the $(2l_i - 1)^{\text{th}}$ order monotonic sink (mSI) $(x_k^* = b_1^{(i)})$ with $b_2^{(i)} > a > b_1^{(i)}$.
(ii) For $\alpha_i = 2l_i - 1$ the $(2(2l_i - 1) + 1)^{\text{th}}$ order monotonic sink (mSI) *switching* bifurcation is with the $(2l_i - 1)^{\text{th}}$ order monotonic sink (mSI) $(x_k^* = b_2^{(i)})$ and the $(2l_i - 1)^{\text{th}}$ order monotonic source (mSO) $(x_k^* = b_1^{(i)})$ with $b_2^{(i)} > a > b_1^{(i)}$.
(iii) For $\alpha_i = 2l_i$ the $(2(2l_i) + 1)^{\text{th}}$ order monotonic source (mSO) *switching* bifurcation is with the $(2l_i)^{\text{th}}$ order monotonic upper-saddle (mUS) $(x_k^* = b_2^{(i)})$ and the $(2l_i)^{\text{th}}$ order monotonic upper-saddles (mLS) $(x_k^* = b_1^{(i)})$ with $b_2^{(i)} > a > b_1^{(i)}$.

(iv) For $\alpha_i = 2l_i$ the $(2(2l_i)+1)^{\text{th}}$ order monotonic sink (mSI) *switching* bifurcation is with the $(2l_i)^{\text{th}}$ order monotonic upper-saddle (mLS) $(x_k^* = b_2^{(i)})$ and the $(2l_i)^{\text{th}}$ order monotonic upper-saddles (mUS) $(x_k^* = b_1^{(i)})$ with $b_2^{(i)} > a > b_1^{(i)}$.

If $x_k^* = a = -\frac{1}{2}B_i$ $(i = 1, 2, \ldots, m)$, the fixed-point of $x_k^* = a$ possesses a $(2(2l_i - 1)+1)^{\text{th}}$ and $(2(2l_i)+1)^{\text{th}}$-order mSI or mSO switching bifurcations (or pitchfork bifurcations) for higher-order fixed-points. The teethcomb appearing bifurcation generated by the m-pairs of quadratics becomes a *broom switching* bifurcation. Such a *broom switching* bifurcation consists of a *pitchfork switching* bifurcation and two teethcomb appearing bifurcations in the $(2m+1)^{\text{th}}$-degree polynomial system. Four *broom switching* bifurcations for higher-order fixed-points are

$$\text{mSO } (x_k^* = a) \xrightarrow[\text{appearing bifurcation}]{\sum_{j=1}^{2}\sum_{i=1}^{s_j}\alpha_i^{(j)} + 2l_{s_1+1} = m} \left\{ \begin{array}{l} ((2\alpha_1^{(2)})^{\text{th}}\,\text{mUS}, \cdots, (2\alpha_{s_2}^{(2)})^{\text{th}}\,\text{mUS}), \\ (2(2l_{s_1+1})+1)^{\text{th}}\,\text{mSO} \left\{ \begin{array}{l} (2l_{s_1+1})^{\text{th}}\,\text{mUS}, \\ \text{mSO}, x^* = a, \\ (2l_{s_1+1})^{\text{th}}\,\text{mLS}, \end{array} \right. \\ ((2\alpha_1^{(1)})^{\text{th}}\,\text{mLS}, \cdots, (2\alpha_{s_1}^{(1)})^{\text{th}}\,\text{mLS}); \end{array} \right. \tag{5.177}$$

$$\text{mSO } (x_k^* = a) \xrightarrow[\text{appearing bifurcation}]{\sum_{j=1}^{2}\sum_{i=1}^{s_j}\alpha_i^{(j)} + 2l_{s_1+1} - 1 = m} \left\{ \begin{array}{l} ((2\alpha_1^{(2)})^{\text{th}}\,\text{mUS}, \cdots, (2\alpha_{s_2}^{(2)})^{\text{th}}\,\text{mUS}), \\ (2(2l_{s_1+1}-1)+1)^{\text{th}}\,\text{mSO} \left\{ \begin{array}{l} (2l_{s_1+1}-1)^{\text{th}}\,\text{mSO}, \\ \text{mSI-oSO}, x_k^* = a, \\ (2l_{s_1+1}-1)^{\text{th}}\,\text{mSO}, \end{array} \right. \\ ((2\alpha_1^{(1)})^{\text{th}}\,\text{mLS}, \cdots, (2\alpha_{s_1}^{(1)})^{\text{th}}\,\text{mLS}); \end{array} \right. \tag{5.178}$$

and

$$\text{mSI-oSO } (x_k^* = a) \xrightarrow[\text{appearing bifurcation}]{\sum_{j=1}^{2}\sum_{i=1}^{s_j}\alpha_i^{(j)} + 2l_{s_1+1} = m} \left\{ \begin{array}{l} ((2\alpha_1^{(2)})^{\text{th}}\,\text{mLS}, \cdots, (2\alpha_{s_2}^{(2)})^{\text{th}}\,\text{mLS}), \\ (2(2l_{s_1+1})+1)^{\text{th}}\,\text{mSI} \left\{ \begin{array}{l} (2l_{s_1+1})^{\text{th}}\,\text{mLS}, \\ \text{mSI-oSO}, x_k^* = a, \\ (2l_{s_1+1})^{\text{th}}\,\text{mUS}, \end{array} \right. \\ ((2\alpha_1^{(1)})^{\text{th}}\,\text{mUS}, \cdots, (2\alpha_{s_1}^{(1)})^{\text{th}}\,\text{mUS}); \end{array} \right. \tag{5.179}$$

$$\text{mSI-oSO}\ (\ x_k^* = a) \xrightarrow[\text{appearing bifurcation}]{\sum_{j=1}^{2}\sum_{i=1}^{s_j}\alpha_i^{(j)}+2l_{s_1+1}-1=m} \begin{cases} ((2\alpha_1^{(2)})^{\text{th}}\ \text{mLS}, \cdots, (2\alpha_{s_2}^{(2)})^{\text{th}}\ \text{mLS}), \\ (2(2l_{s_1+1}-1)+1)^{\text{th}}\ \text{mSI} \begin{cases} (2l_{s_1+1}-1)^{\text{th}}\ \text{mSI}, \\ \text{mSO},\ x_k^* = a, \\ (2l_{s_1+1}-1)^{\text{th}}\ \text{mSI}; \end{cases} \\ ((2\alpha_1^{(1)})^{\text{th}}\ \text{mUS}, \cdots, (2\alpha_{s_1}^{(1)})^{\text{th}}\ \text{mUS}). \end{cases} \tag{5.180}$$

Consider a $(2m+1)^{\text{th}}$ degree polynomial nonlinear discrete system as

$$x_{k+1} = x_k + a_0 Q(x_k)(x_k - a)\prod_{i=1}^{n}(x_k^2 + B_i x_k + C_i)^{\alpha_i} \tag{5.181}$$

where $\alpha_i \in \{2r_i - 1, 2r_i\}$ $(i = 1, 2, \ldots, n)$. Without loss of generality, a function of $Q(x_k) > 0$ is either a polynomial function or a non-polynomial function. The roots of $x_k^2 + B_i x_k + C_i = 0$ are

$$\begin{aligned} & b_{1,2}^{(i)} = -\tfrac{1}{2}B_i \pm \tfrac{1}{2}\sqrt{\Delta_i}, \Delta_i = B_i^2 - 4C_i \geq 0; \\ & B_i = B_j (i, j \in \{1, 2, \ldots, n\}; i \neq j) \\ & \{a_1, a_2, \cdots, a_{2n+1}\} \subset \text{sort}\{b_1^{(1)}, b_2^{(1)}, b_1^{(2)}, b_2^{(2)}, \cdots, b_1^{(n)}, b_2^{(n)}, a\}, a_i \leq a_{i+1}. \end{aligned} \tag{5.182}$$

The higher-order singularity bifurcation can be for a cluster of higher-order fixed-points. There are four higher-order bifurcations as follows:

(i) The $(2(2l-1))^{\text{th}}$ order monotonic upper-saddle (mUS) *spraying appearing* bifurcation for a cluster of higher-order monotonic sinks, monotonic sources, monotonic upper-saddles and monotonic lower-saddles is

$$\begin{aligned} & (2\beta)^{\text{th}}\ \text{mUS} = (2(2l-1))^{\text{th}}\ \text{order mUS} \\ & \xrightarrow[\text{appearing bifurcation}]{\text{a cluster of } 2n\text{-mXX}} \begin{cases} (\alpha_{2n})^{\text{th}}\ \text{order mXX for } x_k^* = a_{2n}, \\ (\alpha_{2n-1})^{\text{th}}\ \text{order mXX for } x_k^* = a_{2n-1}, \\ \vdots \\ (\alpha_1)^{\text{th}}\ \text{order mXX for } x_k^* = a_1; \end{cases} \end{aligned} \tag{5.183}$$

where $2(2l-1) = \sum_{i=1}^{n}\alpha_i$ and

$$
\begin{aligned}
(\alpha_{2n})^{\text{th}} \text{ order mXX} &= \begin{cases} (2r_{2n})^{\text{th}} \text{ order mUS, for } \alpha_{2n} = 2r_n, \\ (2r_{2n}-1)^{\text{th}} \text{ order mSO, for } \alpha_{2n} = 2r_n - 1; \end{cases} \\
(\alpha_1)^{\text{th}} \text{ order mXX} &= \begin{cases} (2r_1)^{\text{th}} \text{ order mUS, for } \alpha_1 = 2r_1, \\ (2r_1-1)^{\text{th}} \text{ order mSI, for } \alpha_1 = 2r_1 - 1. \end{cases}
\end{aligned}
\tag{5.184}
$$

(ii) The $(2(2l))^{\text{th}}$ order monotonic upper-saddle (mUS) *spraying-appearing* bifurcation for a cluster of higher-order monotonic sinks, monotonic sources, monotonic upper-saddles and monotonic lower-saddles is

$$
\begin{aligned}
&(2\beta)^{\text{th}} \text{ mUS} = (2(2l))^{\text{th}} \text{ order mUS} \\
&\xrightarrow[\text{appearing bifurcation}]{\text{a cluster of } 2n\text{-mXX}} \begin{cases} (\alpha_{2n})^{\text{th}} \text{ order mXX for } x_k^* = a_{2n}, \\ (\alpha_{2n-1})^{\text{th}} \text{ order mXX for } x_k^* = a_{2n-1}, \\ \vdots \\ (\alpha_1)^{\text{th}} \text{ order mXX for } x_k^* = a_1; \end{cases}
\end{aligned}
\tag{5.185}
$$

where $2(2l) = \Sigma_{i=1}^{n} \alpha_i$ and

$$
\begin{aligned}
(\alpha_{2n})^{\text{th}} \text{ order mXX} &= \begin{cases} (2r_{2n})^{\text{th}} \text{ order mUS, for } \alpha_{2n} = 2r_n, \\ (2r_{2n}-1)^{\text{th}} \text{ order mSO, for } \alpha_{2n} = 2r_n - 1; \end{cases} \\
(\alpha_1)^{\text{th}} \text{ order mXX} &= \begin{cases} (2r_1)^{\text{th}} \text{ order mUS, for } \alpha_1 = 2r_1, \\ (2r_1-1)^{\text{th}} \text{ order mSI, for } \alpha_1 = 2r_1 - 1. \end{cases}
\end{aligned}
\tag{5.186}
$$

For the higher-order monotonic lower-saddle bifurcation, the cluster of the higher-order fixed-points is given by the following two cases.

(iii) The $(2(2l-1))^{\text{th}}$ order monotonic lower-saddle (mLS) *spraying-appearing* bifurcation for a cluster of higher-order monotonic sinks, monotonic sources, monotonic upper-saddles and monotonic lower-saddles is

$$
\begin{aligned}
&(2\beta)^{\text{th}} \text{ mLS} = (2(2l-1))^{\text{th}} \text{ order mLS} \\
&\xrightarrow[\text{appearing bifurcation}]{\text{a cluster of } 2n\text{-mXX}} \begin{cases} (\alpha_{2n})^{\text{th}} \text{ order mXX, for } x_k^* = a_{2n}, \\ (\alpha_{2n-1})^{\text{th}} \text{ order mXX, for } x_k^* = a_{2n-1}, \\ \vdots \\ (\alpha_1)^{\text{th}} \text{ order mXX, for } x_k^* = a_1; \end{cases}
\end{aligned}
\tag{5.187}
$$

where $2(2l-1)=\Sigma_{i=1}^{n}\alpha_i$ and

$$
\begin{aligned}
(\alpha_{2n})^{\text{th}}\text{ order mXX} &= \begin{cases} (2r_{2n})^{\text{th}}\text{ order mLS, for } \alpha_{2n}=2r_n, \\ (2r_{2n}-1)^{\text{th}}\text{ order mSI, for } \alpha_{2n}=2r_n-1; \end{cases} \\
(\alpha_1)^{\text{th}}\text{ order mXX} &= \begin{cases} (2r_1)^{\text{th}}\text{ order mLS, for } \alpha_1=2r_1, \\ (2r_1-1)^{\text{th}}\text{ order mSO, for } \alpha_1=2r_1-1. \end{cases}
\end{aligned}
\tag{5.188}
$$

(iv) The $(2(2l))^{\text{th}}$-order lower-order *spraying-appearing* bifurcation for a cluster of higher-order sinks, sources, upper-saddles and lower-saddles is

$$
(2\beta)^{\text{th}}\text{ mLS} = (2(2l))^{\text{th}}\text{ order mLS}
\xrightarrow[\text{appearing bifurcation}]{\text{a cluster of } 2n\text{-mXX}}
\begin{cases} (\alpha_{2n})^{\text{th}}\text{ order mXX, for } x_k^*=a_{2n}, \\ (\alpha_{2n-1})^{\text{th}}\text{ order mXX, for } x_k^*=a_{2n-1}, \\ \vdots \\ (\alpha_1)^{\text{th}}\text{ order mXX, for } x_k^*=a_1; \end{cases}
\tag{5.189}
$$

where $2(2l)=\Sigma_{i=1}^{n}\alpha_i$ and

$$
\begin{aligned}
(\alpha_{2n})^{\text{th}}\text{ order mXX} &= \begin{cases} (2r_{2n})^{\text{th}}\text{ order mLS, for } \alpha_{2n}=2r_n, \\ (2r_{2n}-1)^{\text{th}}\text{ order mSI, for } \alpha_{2n}=2r_n-1; \end{cases} \\
(\alpha_1)^{\text{th}}\text{ order mXX} &= \begin{cases} (2r_1)^{\text{th}}\text{ order mLS, for } \alpha_1=2r_1, \\ (2r_1-1)^{\text{th}}\text{ order mSO, for } \alpha_1=2r_1-1. \end{cases}
\end{aligned}
\tag{5.190}
$$

If $x_k^*=a\neq-\frac{1}{2}B_i$ $(i=1,2,\ldots,n)$, the fixed-point of $x_k^*=a$ breaks a cluster of *sprinkler-spraying* appearing bifurcations for higher-order fixed-points to two parts. The sprinkler-spraying appearing bifurcation generated by the m-pairs of quadratics becomes a *broom-sprinkler-spraying appearing* bifurcation. The two *broom-sprinkler*-spraying *appearing* bifurcations in the $(2m+1)^{\text{th}}$-degree polynomial system are

$$
\text{mSO }(x^*=a)\xrightarrow[\text{appearing bifurcation}]{m=m_1+m_2}
\begin{cases} ((2\beta_1^{(2)})^{\text{th}}\text{mUS}:\cdots:(2\beta_{r_2}^{(2)})^{\text{th}}\text{mUS}), \\ \text{mSO }(a=a_{r_1+1})\to\text{mSO }(a=a_{2(m_1+1)-1}), \\ ((2\beta_1^{(1)})^{\text{th}}\text{mLS}:\cdots:(2\beta_{r_1}^{(1)})^{\text{th}}\text{mLS}); \end{cases}
\tag{5.191}
$$

and

$$
\begin{array}{l}
\text{mSI-oSO}\ (\ x_k^* = a) \xrightarrow[\text{appearing bifurcation}]{m=m_1+m_2} \\
\qquad \begin{cases} ((2\beta_1^{(2)})^{\text{th}}\text{mLS}:\cdots:(2\beta_{r_2}^{(2)})^{\text{th}}\text{mLS}), \\ \text{mSI-oSO}\ (a = a_{r_1+1}) \to \text{mSO}\ (a = a_{2(m_1+1)-1}), \\ ((2\beta_1^{(1)})^{\text{th}}\text{mUS}:\cdots:(2\beta_{r_1}^{(1)})^{\text{th}}\text{mUS}); \end{cases}
\end{array}
\tag{5.192}
$$

for $m_1 = \sum_{i=1}^{r_1} \beta_i^{(1)}, m_2 = \sum_{j=1}^{r_2} \beta_j^{(2)}$.

Four special broom-sprinkler-spraying appearing bifurcations the $(2m+1)^{\text{th}}$-degree polynomial nonlinear discrete system are

$$
\begin{array}{l}
\text{mSO}\ (\ x_k^* = a) \xrightarrow[\text{appearing bifurcation}]{m=\sum_{i=1}^{r}\beta_i} \\
\qquad \begin{cases} \text{mSO}\ (a = a_{2m+1}) \to m\text{SO}\ (a = a_{2m+1}), \\ ((2\beta_1)^{\text{th}}\text{mLS},\cdots,(2\beta_r)^{\text{th}}\text{mLS}); \end{cases}
\end{array}
\tag{5.193}
$$

$$
\begin{array}{l}
\text{mSI-oSO}\ (\ x_k^* = a) \xrightarrow[\text{appearing bifurcation}]{m=\sum_{i=1}^{r}\beta_i} \\
\qquad \begin{cases} \text{mSI-oSO}\ (a = a_{2m+1}) \to \text{mSI-oSO}\ (a = a_{2m+1}), \\ ((2\beta_1)^{\text{th}}\ \text{mUS},\cdots,(2\beta_r)^{\text{th}}\ \text{mUS}); \end{cases}
\end{array}
\tag{5.194}
$$

and

$$
\begin{array}{l}
\text{mSO}\ (\ x_k^* = a) \xrightarrow[\text{appearing bifurcation}]{m=\sum_{i=1}^{r}\beta_i} \\
\qquad \begin{cases} ((2\beta_1)^{\text{th}}\ \text{mUS},\cdots,(2\beta_r)^{\text{th}}\ \text{mUS}), \\ \text{mSO}\ (a = a_1) \to \text{mSO}\ (a = a_1); \end{cases}
\end{array}
\tag{5.195}
$$

$$
\begin{array}{l}
\text{mSI-oSO}(\ x_k^* = a) \xrightarrow[\text{appearing bifurcation}]{m=\sum_{i=1}^{r}\beta_i} \\
\qquad \begin{cases} ((2\beta_1)^{\text{th}}\ \text{mLS},\cdots,(2\beta_r)^{\text{th}}\ \text{mLS}), \\ \text{mSI-oSO}\ (a = a_1) \to \text{mSI-oSO}\ (a = a_1). \end{cases}
\end{array}
\tag{5.196}
$$

If $x_k^* = a = -\frac{1}{2}B_i$ $((i = 1, 2, \ldots, l)$, the fixed-point of $x_k^* = a$ possesses a $(2l+1)^{\text{th}}$-order mSI or mSO switching bifurcation (or broom-switching bifurcation) for higher-order fixed-points. The sprinkler-spraying appearing bifurcation generated by the m-pairs of quadratics becomes a *broom-sprinkler-spraying switching* bifurcation. The two *broom switching* bifurcations in the $(2m+1)^{\text{th}}$-degree polynomial system are

$$
\text{mSO } (x_k^* = a) \xrightarrow[\text{switching bifurcation}]{m=m_1+m_2+\beta}
\begin{cases}
(2\beta_1^{(2)}, \cdots, 2\beta_{r_2}^{(2)})^{\text{th}} \text{ mUS} \\
\xrightarrow[\text{appearing}]{m_2=\sum_{j=1}^{r_2}\beta_j^{(2)}}
\begin{cases}
(2\beta_{r_2}^{(2)})^{\text{th}} \text{ mUS } (x_k^* = a_{r_1+r_2+1}), \\
\vdots \\
(2\beta_1^{(2)})^{\text{th}} \text{ mUS } (x_k^* = a_{r_1+2});
\end{cases} \\
(2\beta+1)^{\text{th}} \text{ mSO } (a = a_{r_1+1}); \\
(2\beta_1^{(1)}, \cdots, 2\beta_{r_1}^{(1)})^{\text{th}} \text{mLS} \\
\xrightarrow[\text{appearing}]{m_1=\sum_{i=1}^{r_1}\beta_i^{(1)}}
\begin{cases}
(2\beta_{r_1}^{(1)})^{\text{th}} \text{ mLS } (x_k^* = a_{r_1}), \\
\vdots \\
(2\beta_1^{(1)})^{\text{th}} \text{ mLS } (x_k^* = a_1);
\end{cases}
\end{cases}
\tag{5.197}
$$

and

$$
\text{mSI-oSO } (x_k^* = a) \xrightarrow[\text{switching bifurcation}]{m=m_1+m_2+\beta}
\begin{cases}
(2\beta_1^{(2)}, \cdots, 2\beta_{r_2}^{(2)})^{\text{th}} \text{ mLS} \\
\xrightarrow[\text{appearing}]{m_2=\sum_{j=1}^{r_2}\beta_j^{(2)}}
\begin{cases}
(2\beta_{r_2}^{(2)})^{\text{th}} \text{ mLS } (x_k^* = a_{r_1+r_2+1}), \\
\vdots \\
(2\beta_1^{(2)})^{\text{th}} \text{ mLS } (x_k^* = a_{r_1+2});
\end{cases} \\
(2\beta+1)^{\text{th}} \text{ mSI } (a = a_{r_1+1}); \\
(2\beta_1^{(1)}, \cdots, 2\beta_{r_2}^{(1)})^{\text{th}} \text{ mUS} \\
\xrightarrow[\text{appearing}]{m_1=\sum_{i=1}^{r_1}\beta_i^{(1)}}
\begin{cases}
(2\beta_{r_1}^{(1)})^{\text{th}} \text{ mUS } (x_k^* = a_{r_1}), \\
\vdots \\
(2\beta_1^{(1)})^{\text{th}} \text{ mUS } (x_k^* = a_1);
\end{cases}
\end{cases}
\tag{5.198}
$$

where

$$
(2\beta+1)^{\text{th}} \text{ order mSO}(x_k^* = a) \xrightarrow[\text{appearing bifurcation}]{\text{cluster of } l\text{-quadratics}}
\begin{cases}
(\alpha_{2s_l+1})^{\text{th}} \text{ mXX}, \text{ for } x_k^* = a_{2s_l+1}, \\
(\alpha_{2s_l})^{\text{th}} \text{ mXX}, \text{ for } x_k^* = a_{2s_l}, \\
\vdots \\
(\alpha_{2s_1})^{\text{th}} \text{ mXX}, \text{ for } x_k^* = a_{2s_1}, \\
(\alpha_{2s_1-1})^{\text{th}} \text{ mXX}, \text{ for } x_k^* = a_{2s_1-1};
\end{cases} \tag{5.199}
$$

$$
(2\beta+1)^{\text{th}} \text{ order mSI } (x_k^* = a) \xrightarrow[\text{appearing bifurcation}]{\text{cluster of } l\text{-quadratics}}
\begin{cases}
(\alpha_{2s_l+1})^{\text{th}} \text{ mXX}, \text{ for } x_k^* = a_{2s_l+1}, \\
(\alpha_{2s_l})^{\text{th}} \text{ mXX}, \text{ for } x_k^* = a_{2s_l}, \\
\vdots \\
(\alpha_{2s_1})^{\text{th}} \text{ mXX}, \text{ for } x_k^* = a_{2s_1}, \\
(\alpha_{2s_1-1})^{\text{th}} \text{ mXX}, \text{ for } x_k^* = a_{2s_1-1}.
\end{cases} \tag{5.200}
$$

where $x_k^* = a \in \{a_{2s_1-1}, \cdots, a_{2s_l}, a_{2s_l+1}\}$ and $2\beta+1 = \sum_{i=1}^{l} \alpha_{2s_i-1} + \alpha_{2s_i} + \alpha_{2s_l+1}$. The two appearing bifurcations for the higher-order singularity of fixed-points are

(i) mSO $\to ((2\alpha_1)^{\text{th}}$mLS:$\cdots$:$(2\alpha_i)^{\text{th}}$mLS:mSO:$\cdots$:$(2\alpha_{n-1})^{\text{th}}$mUS:$(2\alpha_n)^{\text{th}}$mUS), and

(ii) mSO $\to ((2\alpha_1)^{\text{th}}$mUS:$\cdots$:$(2\alpha_i)^{\text{th}}$UmS:mSO:$\cdots$:$(2\alpha_{n-1})^{\text{th}}$mLS:$(2\alpha_n)^{\text{th}}$mLS), as presented in Fig. 5.7i and ii for $a_0 > 0$ and $a_0 < 0$, respectively. The *broom appearing* bifurcation for the higher-order fixed-points are illustrated. The components of the *broom appearing* bifurcation are

$$
\begin{aligned}
&(2\alpha_{j_1})^{\text{th}} \text{ mUS} \xrightarrow[\text{appearing}]{\alpha_{j_1}=2r_{j_1}} \begin{cases} (2r_{j_1})^{\text{th}} \text{ mUS} \\ (2r_{j_1})^{\text{th}} \text{ mUS} \end{cases} (j_1 = i, n-1, \ldots), \\
&(2\alpha_{j_2})^{\text{th}} \text{ mUS} \xrightarrow[\text{appearing}]{\alpha_{j_2}=2r_{j_2}-1} \begin{cases} (2r_{j_2}-1)^{\text{th}} \text{ mSO} \\ (2r_{j_2}-1)^{\text{th}} \text{ mSI} \end{cases} (j_2 = 1, n, \ldots);
\end{aligned} \tag{5.201}
$$

and

$$
\begin{aligned}
&(2\alpha_{j_1})^{\text{th}} \text{ mLS} \xrightarrow[\text{appearing}]{\alpha_{j_1}=2r_{j_1}} \begin{cases} (2r_{j_1})^{\text{th}} \text{ mLS} \\ (2r_{j_1})^{\text{th}} \text{ mLS} \end{cases} (j_1 = i, n-1, \cdots), \\
&(2\alpha_{j_2})^{\text{th}} \text{ mLS} \xrightarrow[\text{appearing}]{\alpha_{j_2}=2r_{j_2}-1} \begin{cases} (2r_{j_2}-1)^{\text{th}} \text{ mSI} \\ (2r_{j_2}-1)^{\text{th}} \text{ mSO} \end{cases} (j_2 = 1, n, \ldots);
\end{aligned} \tag{5.202}
$$

The simple fixed-point does not interact with the bifurcation points. The four switching and appearing switching bifurcation

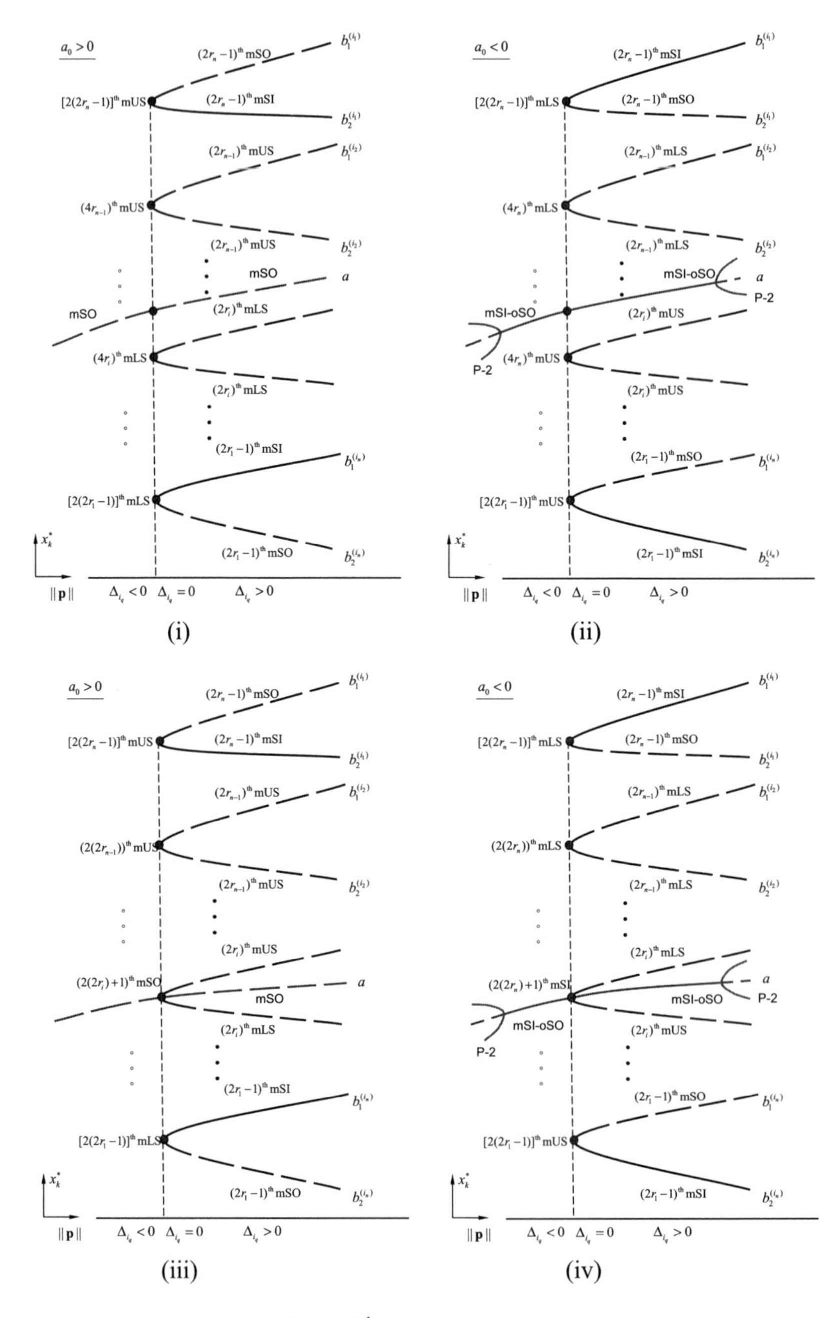

Fig. 5.7 Six bifurcations in a $(2m+1)^{\text{th}}$-degree polynomial system. (i) and (ii) two broom appearing bifurcations. (iii)–(vi) broom-switching bifurcations. mLS: monotonic lower-saddle, mUS: monotonic upper-saddle, mSI-oSO: monotonic sink to oscillatory source, mSO: monotonic source. Stable and unstable fixed-points are represented by solid and dashed curves, respectively. The bifurcation points are marked by circular symbols. P-2 is for period-2 fixed points, which sketched by red curves. Continued

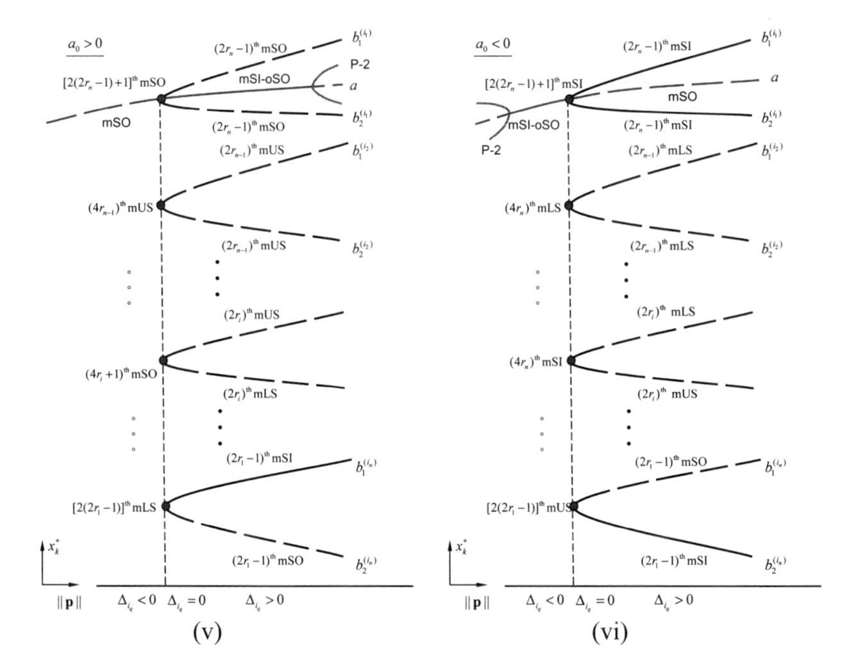

Fig. 5.7 (continued)

The two *broom-sprinkler-spraying switching* bifurcations for the higher-order singularity of fixed-points are

(iii) $((2(2r_1 - 1)^{\text{th}}\text{mLS}:\cdots:(2(2r_{n-1}))^{\text{th}}\text{mUS} : (2(2r_n - 1))^{\text{th}}\text{mUS})$

(iv) $((2(2r_1 - 1)^{\text{th}}\text{mUS}:\cdots:(2(2r_{n-1}))^{\text{th}}\text{mLS} : (2(2r_n - 1))^{\text{th}}\text{mLS})$

(v) $((2(2r_1 - 1)^{\text{th}}\text{mLS}:\cdots:(2(2r_{n-1}))^{\text{th}}\text{mLS} : (2(2r_n - 1) + 1)^{\text{th}}\text{mSO})$

(vi) $(2(2r_1 - 1)^{\text{th}}\text{mUS}:\cdots:(2(2r_{n-1}))^{\text{th}}\text{mUS} : (2(2r_n - 1) + 1)^{\text{th}}\text{mSI})$

as presented in Fig. 5.7iii, v and iv, vi for $a_0 > 0$ and $a_0 < 0$, respectively. The $(2(2r_i) + 1)^{\text{th}}$ mSO and $(2(2r_i) + 1)^{\text{th}}$ mSI switching bifurcations are

$$
\begin{aligned}
&(2(2r_i) + 1)^{\text{th}}\,\text{mSO} \xrightarrow[\text{appearing}]{\alpha_j = 2r_j} \begin{cases} (2r_i)^{\text{th}}\,\text{mUS}, \\ \text{mSO}, \\ (2r_i)^{\text{th}}\,\text{mLS}, \end{cases} \\
&(2(2r_i) + 1)^{\text{th}}\,\text{mSI} \xrightarrow[\text{appearing}]{\alpha_j = 2r_j} \begin{cases} (2r_i)^{\text{th}}\,\text{mLS}, \\ \text{mSI-oSO} \\ (2r_i)^{\text{th}}\,\text{mUS}, \end{cases}
\end{aligned}
\tag{5.203}
$$

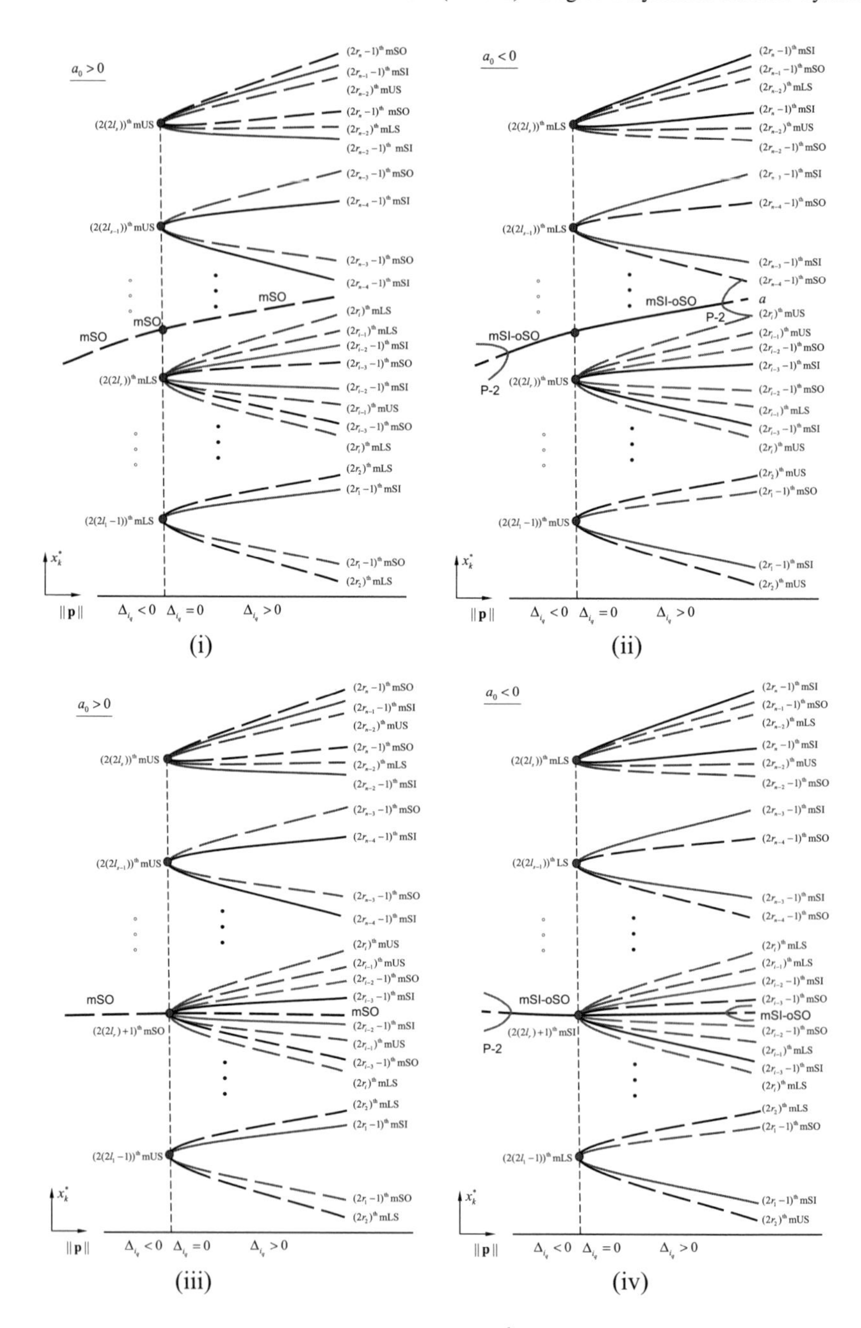

Fig. 5.8 Six types of bifurcations in a $(2m+1)^{\text{th}}$-degree polynomial system. (i) and (ii) broom-sprinkler–spraying appearing bifurcations, (iii)–(vi) broom-spraying switching bifurcations with fixed-points clusters. mLS: monotonic lower-saddle, mUS: monotonic upper-saddle, mSI-oSO: monotonic sink to oscillatory source, mSO: monotonic source. Stable and unstable fixed-points are represented by solid and dashed curves, respectively. The bifurcation points are marked by circular symbols. P-2 is for period-2 fixed-points, sketched by red curves and such period-2 fixed points are relative to the monotonic sinks to oscillatory sources. Continued

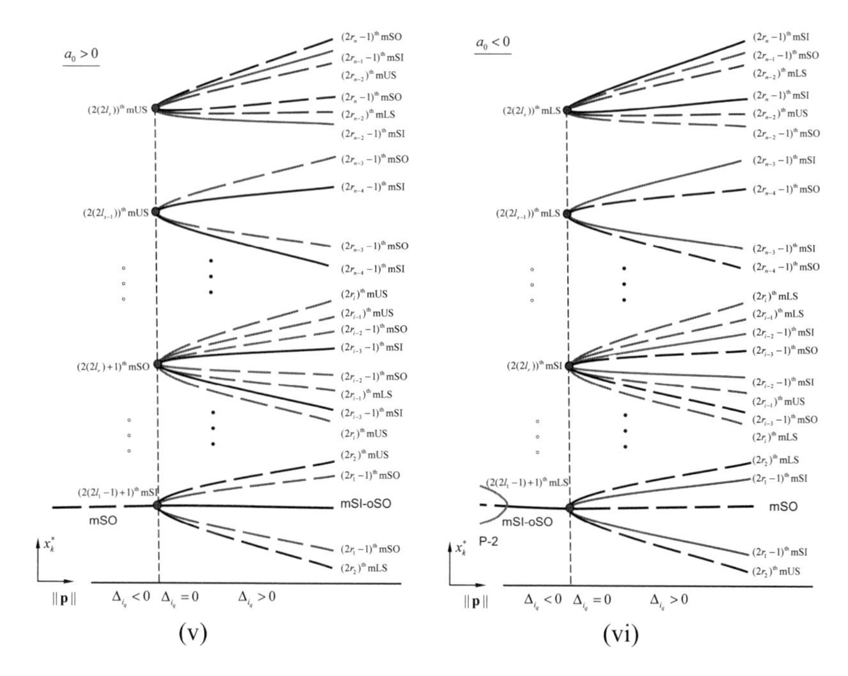

Fig. 5.8 (continued)

and the $(2(2r_n-1)+1)^{\text{th}}$ mSO and $(2(2r_n-1)+1)^{\text{th}}$ mSI switching bifurcations are

$$
\begin{aligned}
&(2(2r_n-1)+1)^{\text{th}}\,\text{mSO}\xrightarrow[\text{appearing}]{\alpha_j=2r_j}\begin{cases}(2r_n-1)^{\text{th}}\,\text{mSO},\\ \text{mSI-oSO},\\ (2r_n-1)^{\text{th}}\,\text{mSO},\end{cases}\\
&(2(2r_n-1)+1)^{\text{th}}\,\text{mSI}\xrightarrow[\text{appearing}]{\alpha_j=2r_j}\begin{cases}(2r_n-1)^{\text{th}}\,\text{mSI},\\ \text{mSO}\\ (2r_n-1)^{\text{th}}\,\text{mSI}.\end{cases}
\end{aligned}
\tag{5.204}
$$

In Fig. 5.8i and ii, the simple switching with two *sprinkler-spraying appearing* bifurcations are presented for $a_0>0$ and $a_0<0$, respectively. The two bifurcation structures are:

(i) $\text{mSO}\rightarrow((2\beta_1)^{\text{th}}\text{mLS}:\cdots:(2\beta_{n-1})^{\text{th}}\text{mUS}:(2\beta_n)^{\text{th}}\text{mUS})$,

(ii) $\text{mSI}\rightarrow((2\beta_1)^{\text{th}}\text{mUS}:\cdots:(2\beta_{n-1})^{\text{th}}\text{mLS}:(2\beta_n)^{\text{th}}\text{mLS})$,

where $m=\sum_{i=1}^{n}\beta_i$,$\beta_1=(2l_1-1),\cdots,\beta_i=2l_i,\cdots,\beta_{n-1}=2l_{n-1}$,$\beta_n=2l_n$. In Fig. 5.8iii, v and vi, iv, the $(2l+1)^{\text{th}}$-order *broom-switching* with two *sprinkler-*

spraying appearing bifurcations are presented for $a_0 > 0$ and $a_0 < 0$, respectively. The two bifurcation structures are:

(iii) mSO $\to$ $((2\beta_1)^{\text{th}}\text{mUS}:\cdots:(2\beta_{n-1})^{\text{th}}\text{mUS}:(2\beta_n)^{\text{th}}\text{mUS})$,
(iv) mSI-oSO $\to$ $((2\beta_1)^{\text{th}}\text{mLS}:\cdots:(2\beta_{n-1})^{\text{th}}\text{mLS}:(2\beta_n)^{\text{th}}\text{mLS})$,
(v) mSO $\to$ $((2\beta_1+1)^{\text{th}}\text{mSO}:\cdots:(2\beta_{n-1})^{\text{th}}\text{mUS}:(2\beta_n)^{\text{th}}\text{mUS})$,
(vi) mSI-oSO $\to$ $((2\beta_1+1)^{\text{th}}\text{mSI}:\cdots:(2\beta_{n-1})^{\text{th}}\text{mLS}:(2\beta_n)^{\text{th}}\text{mLS})$.

For a cluster of m-quadratics, $B_i = B_j$ $(i,j \in \{1,2,\ldots,m\}, i \neq j)$ and $\Delta_i = 0$ $(i \in \{1,2,\ldots,m\})$. The $(2m)^{\text{th}}$-order monotonic upper-saddle-node appearing bifurcation for s-pairs of higher-order fixed-points is

$$(2m)^{\text{th}}\text{ order mUS}\xrightarrow[\text{appearing bifurcation}]{\text{cluster of }s\text{-quadratics}}\begin{cases}(\alpha_{2s})^{\text{th}}\text{mXX}, & \text{for } x_k^* = a_{2s},\\ (\alpha_{2s-1})^{\text{th}}\text{mXX}, & \text{for } x_k^* = a_{2s-1},\\ \vdots & \\ (\alpha_1)^{\text{th}}\text{mXX}, & \text{for } x_k^* = a_1,\end{cases} \tag{5.205}$$

where $2m = \sum_{j=1}^{2s}\alpha_j$ and $2m = 2(2l-1), 2(2l)$.

$$\begin{aligned}(\alpha_1)^{\text{th}}\,\text{mXX} &= \begin{cases}(2l_1)^{\text{th}}\text{mUS}, & \text{for } \alpha_1 = 2l_1,\\ (2l_1-1)^{\text{th}}\text{mSI}, & \text{for } \alpha_1 = 2l_1-1;\end{cases}\\ (\alpha_{2s})^{\text{th}}\,\text{mXX} &= \begin{cases}(2l_{2s})^{\text{th}}\text{mUS}, & \text{for } \alpha_{2s} = 2l_{2s},\\ (2l_{2s}-1)^{\text{th}}\text{mSO}, & \text{for } \alpha_{2s} = 2l_{2s}-1.\end{cases}\end{aligned} \tag{5.206}$$

The $(2m)^{\text{th}}$-order monotonic lower-saddle-node appearing bifurcation for m-pairs of higher-order fixed points is

$$(2m)^{\text{th}}\text{ order mLS}\xrightarrow[\text{appearing bifurcation}]{\text{cluster of }s\text{-quadratics}}\begin{cases}(\alpha_{2s})^{\text{th}}\,\text{mXX}, & \text{for } x_k^* = a_{2s},\\ (\alpha_{2s-1})^{\text{th}}\,\text{mXX}, & \text{for } x_k^* = a_{2s-1};\\ \vdots & \\ (\alpha_1)^{\text{th}}\,\text{mXX}, & \text{for } x_k^* = a_1.\end{cases} \tag{5.207}$$

where

$$\begin{aligned}(\alpha_1)^{\text{th}}\,\text{mXX} &= \begin{cases}(2l_1)^{\text{th}}\,\text{mLS}, & \text{for } \alpha_1 = 2l_1,\\ (2l_1-1)^{\text{th}}\,\text{mSO}, & \text{for } \alpha_1 = 2l_1-1;\end{cases}\\ (\alpha_{2s})^{\text{th}}\,\text{mXX} &= \begin{cases}(2l_{2s})^{\text{th}}\,\text{mLS}, & \text{for } \alpha_{2s} = 2l_{2s},\\ (2l_{2s}-1)^{\text{th}}\,\text{mSI}, & \text{for } \alpha_{2s} = 2l_{2s}-1.\end{cases}\end{aligned} \tag{5.208}$$

There are four simple switching and $(2m)^{\text{th}}$-order saddle-node appearing bifurcations for higher-order fixed-points: The two switching bifurcations of mSO $\to$ $((2m)^{\text{th}}$mUS:mSO) and mSI-oSO $\to$ $((2m)^{\text{th}}$mLS:mSI-oSO) with two $(2m)^{\text{th}}$-order mUSN and mLSN spraying appearing bifurcations in the $(2m+1)^{\text{th}}$-degree polynomial system are

$$\text{mSO } (x_k^* = a) \to \begin{cases} \text{mSO} \to \text{mSO}, \text{ for } x_k^* = a = a_{2m+1}, \\ (2m)^{\text{th}} \text{ order mLS.} \end{cases} \tag{5.209}$$

$$\text{mSI-oSO } (x_k^* = a) \to \begin{cases} \text{mSI-oSO} \to \text{mSI-oSO}, \text{ for } x_k^* = a = a_{2m+1}, \\ (2m)^{\text{th}} \text{ order mUS.} \end{cases} \tag{5.210}$$

and the two switching bifurcations of mSO $\to$ (mSO:$(2m)^{\text{th}}$mUS) and mSI-oSO $\to$ (mSI-oSO:$(2m)^{\text{th}}$mLS) with two $(2m)^{\text{th}}$-order mUSN and mLSN spraying appearing bifurcations in the $(2m+1)^{\text{th}}$-degree polynomial nonlinear discrete system are

$$\text{mSO } (x_k^* = a) \to \begin{cases} (2m)^{\text{th}} \text{ order mUS}, \\ \text{mSO} \to \text{mSO}, \text{ for } x_k^* = a = a_1. \end{cases} \tag{5.211}$$

$$\text{mSI-oSO } (x_k^* = a) \to \begin{cases} (2m)^{\text{th}} \text{ order mLS}, \\ \text{mSI-oSO} \to \text{mSI-oSO}, \text{ for } x_k^* = a = a_1. \end{cases} \tag{5.212}$$

The $(2m+1)^{\text{th}}$ order source broom-switching bifurcation for higher-order fixed-points is

$$\text{mSO}(x_k^* = a) \xrightarrow{\text{switching}} (2m+1)^{\text{th}} \text{ order mSO} \begin{cases} (\alpha_{2s+1})^{\text{th}}\text{mXX}, \text{ for } x_k^* = a_{2s+1}, \\ (\alpha_{2s})^{\text{th}}\text{mXX}, \text{ for } x_k^* = a_{2s}, \\ \vdots \\ (\alpha_1)^{\text{th}}\text{mXX}, \text{ for } x_k^* = a_1; \end{cases} \tag{5.213}$$

where $m+1 = \sum_{j=1}^{2s+1} \alpha_j$ and $m = 2(2l-1), 2(2l)$.

$$\begin{aligned} (\alpha_1)^{\text{th}}\,\text{mXX} &= \begin{cases} (2l_1)^{\text{th}}\,\text{mLS}, \text{ for } \alpha_1 = 2l_1, \\ (2l_1-1)^{\text{th}}\,\text{mSI}, \text{ for } \alpha_1 = 2l_1 - 1; \end{cases} \\ (\alpha_{2s})^{\text{th}}\,\text{mXX} &= \begin{cases} (2l_{2s})^{\text{th}}\,\text{mUS}, \text{ for } \alpha_{2s} = 2l_{2s}, \\ (2l_{2s}-1)^{\text{th}}\,\text{mSO}, \text{ for } \alpha_{2s} = 2l_{2s} - 1. \end{cases} \end{aligned} \tag{5.214}$$

The $(2m+1)^{\text{th}}$ order monotonic sink *broom-switching* bifurcation is

$$\text{mSI-oSO}(x_k^* = a) \xrightarrow{\text{switching}} (2m+1)^{\text{th}} \text{ order mSI} \begin{cases} (\alpha_{2s+1})^{\text{th}}\text{mXX}, \text{ for } x_k^* = a_{2s+1}, \\ (\alpha_{2s})^{\text{th}}\text{mXX}, \text{ for } x_k^* = a_{2s}, \\ \vdots \\ (\alpha_1)^{\text{th}}\text{mXX}, \text{ for } x_k^* = a_1; \end{cases} \tag{5.215}$$

where

$$\begin{aligned} &(\alpha_1)^{\text{th}}\,\text{mXX} = \begin{cases} (2l_1)^{\text{th}}\,\text{mUS}, \text{ for } \alpha_1 = 2l_1, \\ (2l_1 - 1)^{\text{th}}\,\text{mSO}, \text{ for } \alpha_1 = 2l_1 - 1; \end{cases} \\ &(\alpha_{2s})^{\text{th}}\,\text{mXX} = \begin{cases} (2l_{2s})^{\text{th}}\,\text{mLS}, \text{ for } \alpha_{2s} = 2l_{2s}, \\ (2l_{2s} - 1)^{\text{th}}\,\text{mSI}, \text{ for } \alpha_{2s} = 2l_{2s} - 1. \end{cases} \end{aligned} \tag{5.216}$$

The switching bifurcation consist of a simple switching and the $(2m)^{\text{th}}$ order saddle-node appearing bifurcation with s-pairs of fixed-points. The $(2m)^{\text{th}}$ order saddle-node appearing bifurcation is a *sprinkler-spraying cluster* of the s-pairs of higher-order fixed-points. Thus, the four switching bifurcations of

mSO $\to$ $((2m)^{\text{th}}$mLS:mSO) for higher order fixed-points for $a_0 > 0$,
mSI-oSO $\to$ $((2m)^{\text{th}}$mUS:mSI-oSO) for higher order fixed-points for $a_0 < 0$,
mSO $\to$ (mSO:$(2m)^{\text{th}}$mUS) for higher order fixed-points for $a_0 > 0$,
mSI-oSO $\to$ (mSI-oSO:$(2m)^{\text{th}}$mLS) for higher order fixed-point for $a_0 < 0$

are presented in Fig. 5.9i–iv, respectively. The $(2m+1)^{\text{th}}$-order monotonic source switching bifurcation is named a $(2m+1)^{\text{th}}$mSO broom-sprinkle-spraying switching bifurcation, and the $(2m+1)^{\text{th}}$-order monotonic sink switching bifurcation is named a $(2m+1)^{\text{th}}$mSI broom switching bifurcation. Such a $(2m+1)^{\text{th}}$ mXX *broom-switching* bifurcation is from simple fixed-point to a $(2m+1)^{\text{th}}$ mXX *broom-*switching. The two broom-switching bifurcations for higher-order fixed-points of

$$\text{mSO} \to (2m+1)^{\text{th}}\text{mSO}$$
$$\text{mSI-oSO} \to (2m+1))^{\text{th}}\text{mSI}$$

are presented in Fig. 5.9v–vi, respectively.

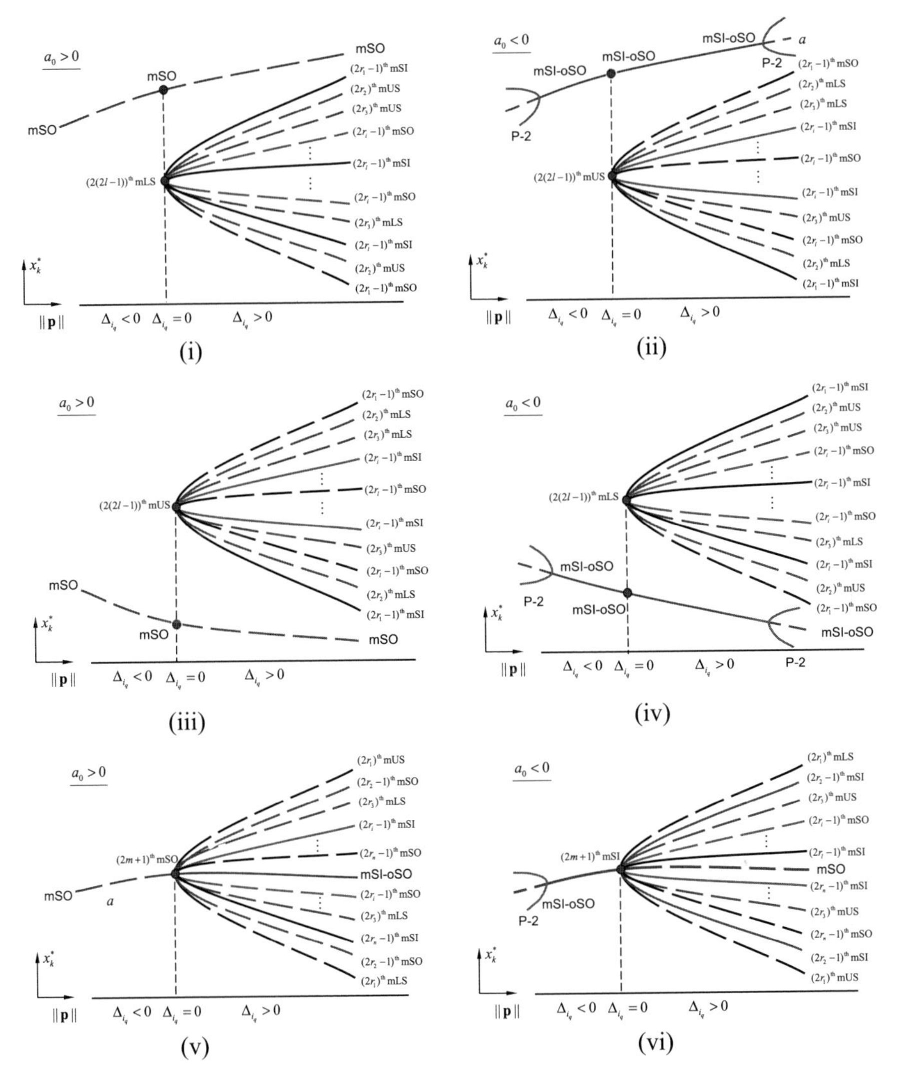

Fig. 5.9 Broom-switching bifurcations of fixed-points in $(2m+1)^{\text{th}}$ polynomial discrete system: (i) $((2m)^{\text{th}}$mLS:mSO)-appearing bifurcation $(a_0>0)$, (ii) $((2m)^{\text{th}}$mUS:mSI-oSO)-appearing bifurcation $(a_0<0)$, (iii) (mSO : $(2m)^{\text{th}}$mUS)-appearing bifurcation $(a_0>0)$, (iv) (mSI-oSO : $(2m)^{\text{th}}$mLS)-appearing bifurcation $(a_0>0)$. (v) $(2m+1)^{\text{th}}$mSO switching-appearing bifurcation $(a_0>0)$, (vi) $(2m+1)^{\text{th}}$mSI switching-appearing bifurcation $(a_0<0)$. mLS: monotonic lower-saddle, mUS: monotonic upper-saddle, mSI-oSO: monotonic sink to oscillatory source, mSO: monotonic source. Stable and unstable fixed-points are represented by solid and dashed curves, respectively. The bifurcation points are marked by circular symbols. P-2 is for period-2 fixed-points, sketched by red curves and such period-2 fixed points are relative to the monotonic sinks to oscillatory sources.

A series of the $(2\alpha_i+1)^{\text{th}}$-order monotonic source and monotonic sink bifurcations is aligned up with varying with parameters, which is formed a special pattern. Such a special pattern is from n-quadratics in the $(2m+1)^{\text{th}}$ degree polynomial nonlinear discrete system, and the following conditions should be satisfied.

$$
\begin{aligned}
& a(\mathbf{p}_i) = -\frac{1}{2}B_i \text{ and } a(\mathbf{p}_j) = -\frac{1}{2}B_j \\
& B_i \approx B_j \, i,j \in \{1,2,\ldots,s\}; i \neq j, \\
& \Delta_i > \Delta_{i+1} (i = 1,2,\ldots,s; s \leq n < m), \\
& \Delta_i = 0 \text{ with } ||\mathbf{p}_i|| < ||\mathbf{p}_{i+1}||.
\end{aligned}
\tag{5.217}
$$

Four series of *switching* bifurcations in the $(2m+1)^{\text{th}}$ degree polynomial nonlinear discrete system are

(i) $(2(2r_1-1)+1)^{\text{th}}\text{mSO-}(2(2r_2)+1)^{\text{th}}\text{mSI-}\cdots-(2(2r_n-1)+1)^{\text{th}}\text{mSO}),$
(ii) $(2(2r_1-1)+1)^{\text{th}}\text{mSI-}(2(2r_2)+1)^{\text{th}}\text{mSO-}\cdots-(2(2r_n-1)+1)^{\text{th}}\text{mSI}),$
(iii) $(2(2r_1)+1)^{\text{th}}\text{mSO-}(2(2r_2-1)+1)^{\text{th}}\text{mSO-}\cdots-(2(2r_n-1))^{\text{th}}\text{mSO}),$
(iv) $(2(2r_1)+1)^{\text{th}}\text{mSI-}(2(2r_2-1)+1)^{\text{th}}\text{mSI-}\cdots-(2(2r_n-1)+1)^{\text{th}}\text{mSI}),$

as presented in Fig. 5.10i, iii–ii, vi for $(a_0>0)$ and $(a_0<0)$, respectively. The swapping pattern of higher-order sinks and sources switching bifurcations cannot be observed. Such a bifurcation scenario is like the fish-bone for the higher-order switching bifurcations of higher-order fixed-points.

5.3.2 *Switching Bifurcations*

Consider the roots of $(x_k^2+B_ix_k+C_i)^{\alpha_i}=0$ as

$$
\begin{aligned}
& B_i = -(b_1^{(i)}+b_2^{(i)}),\ \Delta_i = (b_1^{(i)}-b_2^{(i)})^2 \geq 0, \\
& x_{k;1,2}^{(i)} = b_{1,2}^{(i)}, \Delta_i > 0 \text{ if } b_1^{(i)} \neq b_2^{(i)} (i=1,2,\ldots,n); \\
& \left.\begin{aligned} & B_i \neq B_j (i,j=1,2,\ldots,n; i\neq j) \\ & \Delta_i = 0 \text{ at } b_1^{(i)} = b_2^{(i)} (i=1,2,\ldots,n) \end{aligned}\right\} \text{at bifurcation.}
\end{aligned}
\tag{5.218}
$$

The $(\alpha_i)^{\text{th}}$-order singularity bifurcation is for the *switching* of a pair of higher order fixed-points (i.e., monotonic sinks, monotonic sources, monotonic upper-saddles and monotonic lower-saddles). There are six *switching* bifurcations for $i\in\{1,2,\ldots,n\}$

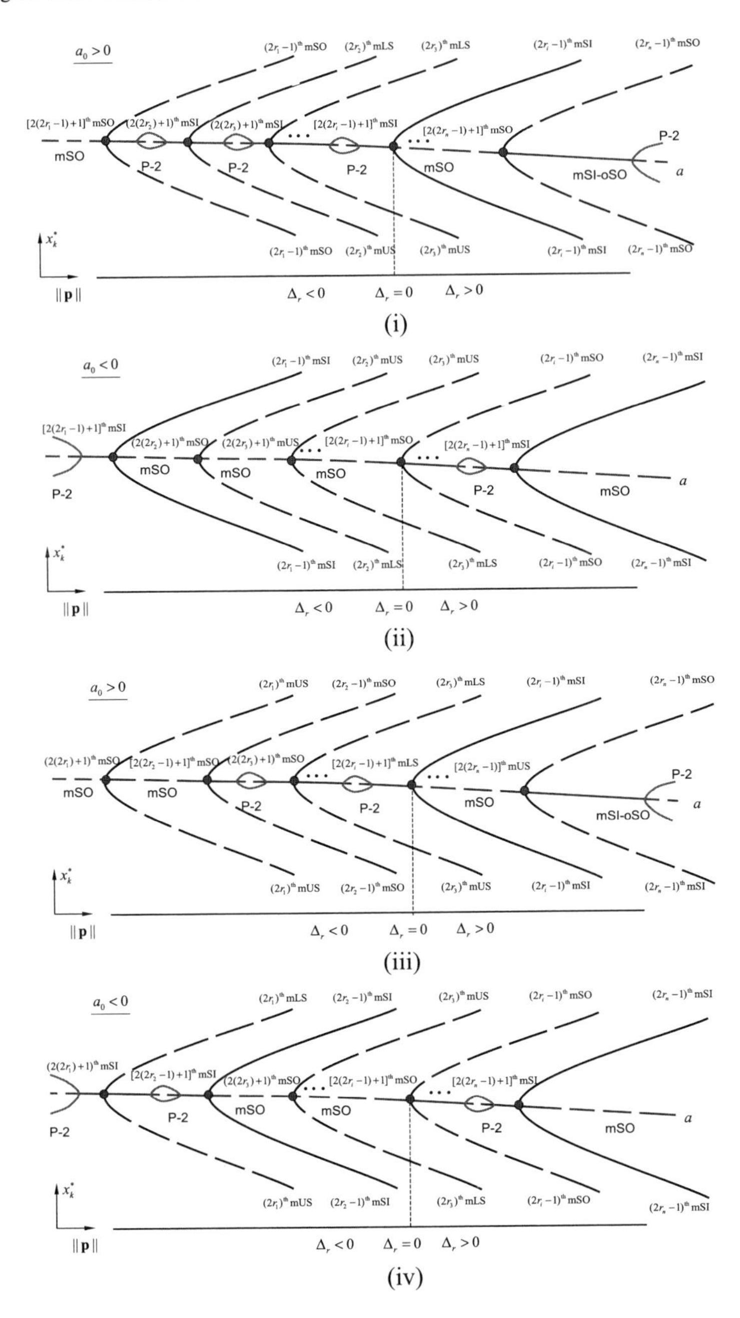

Fig. 5.10 Four series switching bifurcations of fixed-points in a $(2m+1)^{\text{th}}$ polynomial system: (i, iii) for $a_0 > 0$, (ii, iv) for $a_0 < 0$. mLS: monotonic lower-saddle, mUS: monotonic upper-saddle, mSI-oSO: monotonic sink to oscillatory source, mSO: monotonic source. Stable and unstable fixed-points are represented by solid and dashed curves, respectively. The bifurcation points are marked by circular symbols. P-2 is for period-2 fixed-points, sketched by red curves and such period-2 fixed points are relative to the monotonic sinks to oscillatory sources or the monotonic sinks to oscillatory source back to monotonic sinks.

$$(2l_i)^{\text{th}}\text{ order mUS}\xrightarrow[\text{switching bifurcation}]{l_i=r_1^{(i)}+r_2^{(i)}-1}\begin{cases}(2r_2^{(i)}-1)^{\text{th}}\text{ order mSO}\downarrow m\text{SI},\ \text{for}\ b_2^{(i)}=a_{2i}\downarrow a_{2i-1},\\(2r_1^{(i)}-1)^{\text{th}}\text{ order mSI}\uparrow m\text{SO},\ \text{for}\ b_1^{(i)}=a_{2i-1}\uparrow a_{2i};\end{cases}\tag{5.219}$$

$$(2l_i)^{\text{th}}\text{ order mLS}\xrightarrow[\text{switching bifurcation}]{l_i=r_1^{(i)}+r_2^{(i)}-1}\begin{cases}(2r_2^{(i)}-1)^{\text{th}}\text{ order mSI}\downarrow m\text{SO},\ \text{for}\ b_2^{(i)}=a_{2i}\downarrow a_{2i-1},\\(2r_1^{(i)}-1)^{\text{th}}\text{ order mSO}\uparrow m\text{SI},\ \text{for}\ b_1^{(i)}=a_{2i-1}\uparrow a_{2i};\end{cases}\tag{5.220}$$

$$(2l_i)^{\text{th}}\text{ order mUS}\xrightarrow[\text{switching bifurcation}]{l_i=r_1^{(i)}+r_2^{(i)}}\begin{cases}(2r_2^{(i)})^{\text{th}}\text{ order mUS}\downarrow \text{mUS},\ \text{for}\ b_2^{(i)}=a_{2i}\downarrow a_{2i-1},\\(2r_1^{(i)})^{\text{th}}\text{ order mUS}\uparrow \text{mUS}\ \text{for}\ b_1^{(i)}=a_{2i-1}\uparrow a_{2i};\end{cases}\tag{5.221}$$

$$(2l_i)^{\text{th}}\text{ order mLS}\xrightarrow[\text{switching bifurcation}]{l_i=r_1^{(i)}+r_2^{(i)}}\begin{cases}(2r_2^{(i)})^{\text{th}}\text{ order mLS}\downarrow \text{mLS},\ \text{for}\ b_2^{(i)}=a_{2i}\downarrow a_{2i-1},\\(2r_1^{(i)})^{\text{th}}\text{ order mLS}\uparrow \text{mLS}\ \text{for}\ b_1^{(i)}=a_{2i-1}\uparrow a_{2i};\end{cases}\tag{5.222}$$

$$(2l_i-1)^{\text{th}}\text{ order mSO}\xrightarrow[\text{switching bifurcation}]{l_i=r_1^{(i)}+r_2^{(i)}}\begin{cases}(2r_2^{(i)}-1)^{\text{th}}\text{ order mSO}\downarrow \text{mSO},\ \text{for}\ b_2^{(i)}=a_{2i}\downarrow a_{2i-1},\\(2r_1^{(i)})^{\text{th}}\text{ order mLS}\uparrow \text{mUS}\ \text{for}\ b_1^{(i)}=a_{2i-1}\uparrow a_{2i};\end{cases}\tag{5.223}$$

$$(2l_i-1)^{\text{th}}\text{ order mSI}\xrightarrow[\text{switching bifurcation}]{l_i=r_1^{(i)}+r_2^{(i)}}\begin{cases}(2r_2^{(i)}-1)^{\text{th}}\text{ order mSI}\downarrow \text{mSI},\ \text{for}\ b_2^{(i)}=a_{2i}\downarrow a_{2i-1},\\(2r_1^{(i)})^{\text{th}}\text{ order mUS}\uparrow \text{mLS}\ \text{for}\ b_1^{(i)}=a_{2i-1}\uparrow a_{2i}.\end{cases}\tag{5.224}$$

A set of n-paralleled higher-order mXX *switching* bifurcations is called the $(\alpha_1^{\text{th}}\,\text{mXX},\alpha_2^{\text{th}}\,\text{mXX},\cdots,\alpha_n^{\text{th}}\,\text{mXX})$ *parallel switching* bifurcation in the $(2m+1)^{\text{th}}$ degree polynomial nonlinear discrete system. Such a bifurcation is also called a $(\alpha_1^{\text{th}}\,\text{mXX},\alpha_2^{\text{th}}\,\text{mXX},\cdots,\alpha_n^{\text{th}}\,\text{mXX})$ *antenna switching* bifurcation. $\alpha_i\in\{2l_i,2l_i-1\}$ and mXX $\in\{m\text{SO, mSI, mUS, mLS}\}$. For non-switching points, $\Delta_i>0$ at $b_1^{(i)}\neq b_2^{(i)}$ $(i=1,2,\ldots,n)$. At the bifurcation point, $\Delta_i=0$ at $b_1^{(i)}=b_2^{(i)}$ $(i=1,2,\ldots,n)$.

The *parallel antenna switching* bifurcation for higher-order fixed-points in the $(2m+1)^{\text{th}}$ degree polynomial nonlinear discrete system is

$$\begin{cases} ((\alpha_1^{(2)})^{\text{th}}\text{mXX}:\cdots:(\alpha_{l_2}^{(2)})^{\text{th}}\,\text{mXX}) \\ \text{mSI-oSO (or mSO), for } x_k^* = a \\ ((\alpha_1^{(1)})^{\text{th}}\text{mXX}:\cdots:(\alpha_{l_1}^{(1)})^{\text{th}}\,\text{mXX}) \end{cases} \tag{5.225}$$

where

$$(\alpha_{s_i}^{(i)})^{\text{th}}\,\text{mXX}_{s_i} \xrightarrow[\text{switching}]{s_i^{\text{th}}\text{ bifurcation}} \begin{cases} (r_2^{(s_i)})^{\text{th}}\,\text{mXX}_2^{(s_i)} \downarrow \text{mYY}_1^{(s_i)}, & \text{for } b_2^{(s_i)} = a_{2s_i}^{(i)} \downarrow a_{2s_i-1}^{(i)}, \\ (r_1^{(s_i)})^{\text{th}}\,\text{mXX}_1^{(s_i)} \uparrow \text{mYY}_2^{(s_i)}, & \text{for } b_1^{(s_i)} = a_{2s_i-1}^{(i)} \uparrow a_{2s_i}^{(i)}; \end{cases} \tag{5.226}$$

$(s_i = 1, 2, \ldots, l_i, i = 1, 2)$

Such eight sets of parallel *switching* bifurcations for higher-order fixed-points are presented in Fig. 5.11i, iii, v, vii and ii, iv, vi, viii for $a_0 > 0$ and $a_0 < 0$, respectively. The eight switching bifurcation structures are as follows:

(i) $((2l_1)^{\text{th}}\text{mUS}:\cdots:\text{mSI-oSO}:\cdots:(2l_{n-1}-1)^{\text{th}}\text{mSO}:(2l_n)^{\text{th}}\text{mUS})$ for $a_0 > 0$,
(ii) $((2l_1)^{\text{th}}\text{mLS}\cdots:\text{mSO}:\cdots(2l_{n-1}-1)^{\text{th}}\text{mSI}:(2l_n)^{\text{th}}\text{mLS})$ for $a_0 < 0$,
(iii) $((2l_1)^{\text{th}}\text{mLS}\cdots:\text{mSO}:\cdots(2l_{n-1}-1)^{\text{th}}\text{mSI}:(2l_n-1)^{\text{th}}\text{mSO})$ for $a_0 > 0$,
(iv) $((2l_1)^{\text{th}}\text{mUS}\cdots:\text{mSI}:\cdots(2l_{n-1}-1)^{\text{th}}\text{mSO}:(2l_n-1)^{\text{th}}\text{mSI})$ for $a_0 < 0$,
(v) $((2l_1)^{\text{th}}\text{mLS}\cdots:\text{mSO}:\cdots(2l_{n-1}-1)^{\text{th}}\text{mSI}:(2l_n-1)^{\text{th}}\text{mSO})$ for $a_0 > 0$,
(vi) $((2l_1)^{\text{th}}\text{mUS}\cdots:\text{mSI-oSO}:\cdots(2l_{n-1}-1)^{\text{th}}\text{mSI}:(2l_n-1)^{\text{th}}\text{mSI})$ for $a_0 < 0$,
(vii) $((2l_1)^{\text{th}}\text{mUS}\cdots:\text{mSI}:\cdots(2l_{n-1}-1)^{\text{th}}\text{mSO}:(2l_n)^{\text{th}}\text{mUS})$ for $a_0 > 0$,
(viii) $((2l_1)^{\text{th}}\text{mLS}\cdots:\text{mSO}:\cdots(2l_{n-1}-1)^{\text{th}}\text{mSI}:(2l_n)^{\text{th}}\text{mLS})$ for $a_0 < 0$.

The switching bifurcations with different higher-order fixed-points are similar to the (l_1-mLSN:mSO:l_2-mUSN) and (l_1-mUSN:mSI-oSO:l_2-mLSN) switching bifurcations for simple sinks and sources.

Consider a switching bifurcation for a cluster of higher-order fixed-points with the following conditions,

$$\left.\begin{array}{l} B_i = B_j(i,j \in \{1,2,\ldots,n\}; i \neq j) \\ \Delta_i = 0 \text{ at } b_1^{(i)} = \text{b}_2^{(i)}(i = 1,2,\ldots,n) \end{array}\right\} \text{at bifurcation.} \tag{5.227}$$

Thus, the $(\alpha_i)^{\text{th}}$ order *switching* bifurcation can be for a cluster of higher-order fixed-points. The $(\alpha_i)^{\text{th}}$ order switching bifurcations for $i \in \{1, 2, \ldots, s\}$ are

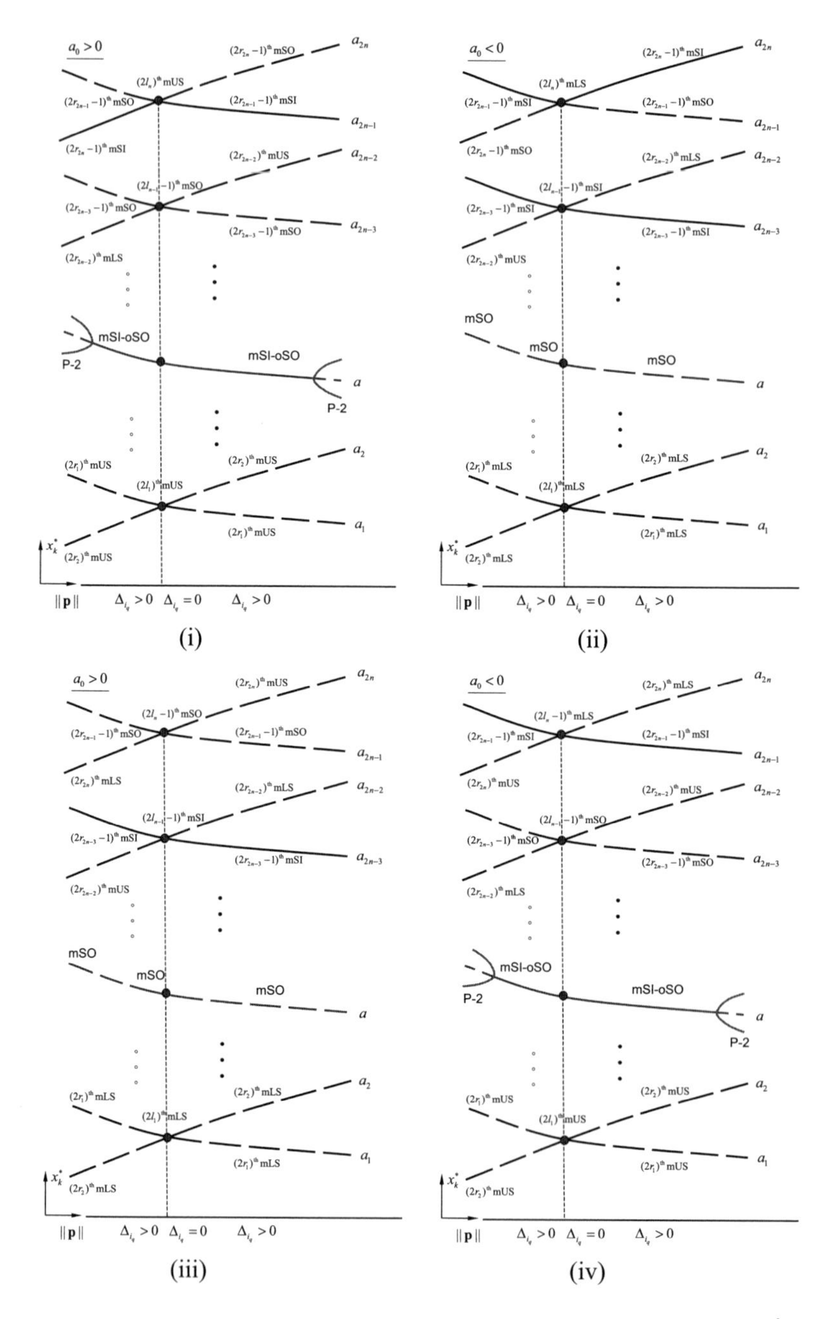

Fig. 5.11 Antenna parallel switching bifurcation of fixed-points for a $(2m+1)^{\text{th}}$-degree polynomial nonlinear discrete system. (i, iii, vi, vii) four parallel bifurcations for $a_0 > 0$. (ii, iv, vi, viii) four parallel bifurcations for $a_0 < 0$. mLS: monotonic lower-saddle, mUS: monotonic upper-saddle, mSI: monotonic sink, mSO: monotonic source. mSI-oSO: monotonic sink to oscillatory source. Stable and unstable fixed-points are represented by solid and dashed curves, respectively. The bifurcation points are marked by circular symbols. P-2 is for period-2 fixed points, sketched by red curves. The period-2 fixed-points are relative to the monotonic sinks to oscillatory sources. Continued

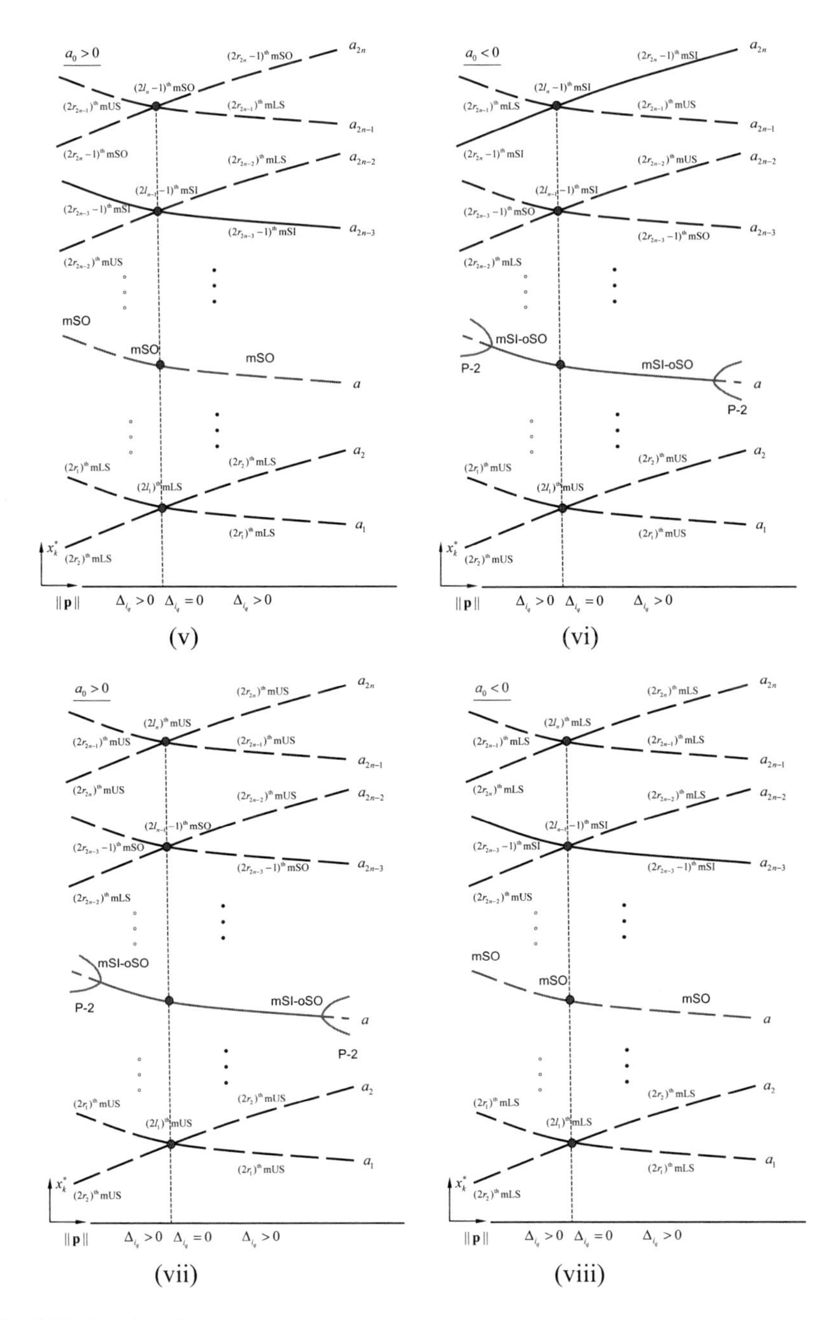

Fig. 5.11 (continued)

$$
(\alpha_i)^{\text{th}}\text{order mXX} \xrightarrow[\text{switching appearing}]{\alpha_i=\sum_{j=1}^{l_i} r_j^{(i)}} \begin{cases} (r_s^{(i)})^{\text{th}} \text{ order } \text{mXX}_{l_i}^{(i)} \downarrow \text{mYY}_{l_i}^{(i)}, \text{ for } b_{l_i}^{(i)} \downarrow a_{l_i}^{(i)}, \\ \vdots \\ (r_j^{(i)})^{\text{th}} \text{ order } \text{mXX}_j^{(i)} \downarrow \text{mYY}_j^{(i)}, \text{ for } b_j^{(i)} \downarrow a_s^{(i)}, \\ \vdots \\ (r_1^{(i)})^{\text{th}} \text{ order } \text{mXX}_1^{(i)} \uparrow \text{mYY}_1^{(i)}, \text{ for } b_1^{(i)} \downarrow a_s^{(i)}; \end{cases} \tag{5.228}
$$

where

$$
\begin{aligned}
&\{a_1^{(i)}, a_2^{(i)}, \cdots, a_{l_{i-1}}^{(i)}, a_{l_i}^{(i)}\} \underset{\text{before bifurcation}}{\subset} \text{sort}\{b_1^{(1)}, b_2^{(1)}, \cdots, b_1^{(n)}, b_2^{(n)}\}, \\
&\{b_1^{(i)}, b_2^{(i)}, \cdots, b_{l_{i-1}}^{(i)}, b_{l_i}^{(i)}\} \underset{\text{After bifurcation}}{\subset} \text{sort}\{b_1^{(1)}, b_2^{(1)}, \cdots, b_1^{(n)}, b_2^{(n)}\}.
\end{aligned} \tag{5.229}
$$

A set of paralleled, different, higher-order monotonic upper-saddle-node switching bifurcations with multiplicity is a $((\alpha_1)^{\text{th}}\text{mXX}:(\alpha_2)^{\text{th}}\text{mXX}:\cdots:(\alpha_s)^{\text{th}}\text{ mXX})$ *parallel switching* bifurcation in the $(2m+1)^{\text{th}}$ degree polynomial discrete system. At the *straw-bundle switching* bifurcation, $\Delta_i = 0$ $(i = 1, 2, \ldots, n)$ and $B_i = B_j$ $(i, j \in \{1, 2, \ldots, n\}$ $i \neq j)$. The *parallel straw-bundle switching* bifurcation for higher order fixed-points is

$$
\begin{aligned}
&((\alpha_1)^{\text{th}}\text{mXX}:(\alpha_2)^{\text{th}}\text{mXX}:\cdots:(\alpha_s)^{\text{th}}\text{ mXX})\text{-switching} \\
&= \begin{cases} (\alpha_s)^{\text{th}} \text{ order mXX switching}, \\ \vdots \\ (\alpha_2)^{\text{th}} \text{ order mXX switching}, \\ (\alpha_1)^{\text{th}} \text{ order mXX switching}; \end{cases}
\end{aligned} \tag{5.230}
$$

where

$$
\alpha_i \in \{2l_i, 2l_i - 1\} \textit{ and } mXX \in \{m\text{US, mLS, mSO, mSI}\}. \tag{5.231}
$$

Thus,

$$
\begin{cases} ((\alpha_1^{(2)})^{\text{th}}\text{mXX}:(\alpha_2^{(2)})^{\text{th}}\text{mXX}:\cdots:(\alpha_{s_2}^{(2)})^{\text{th}}\text{ mXX}) \\ \text{mSI-oSO (or mSO)} \\ ((\alpha_1^{(1)})^{\text{th}}\text{mXX}:(\alpha_2^{(1)})^{\text{th}}\text{mXX}:\cdots:(\alpha_{s_1}^{(1)})^{\text{th}}\text{ mXX}) \end{cases} \tag{5.232}
$$

Eight *parallel straw-bundle switching* bifurcations of $(\alpha_1^{\text{th}}\text{mXX} : \alpha_2^{\text{th}}\text{mXX}:\cdots : \alpha_n^{\text{th}}\text{ mXX})$ are presented in Figs. 5.12 and 5.13 for $a_0 > 0$ and $a_0 < 0$, respectively.

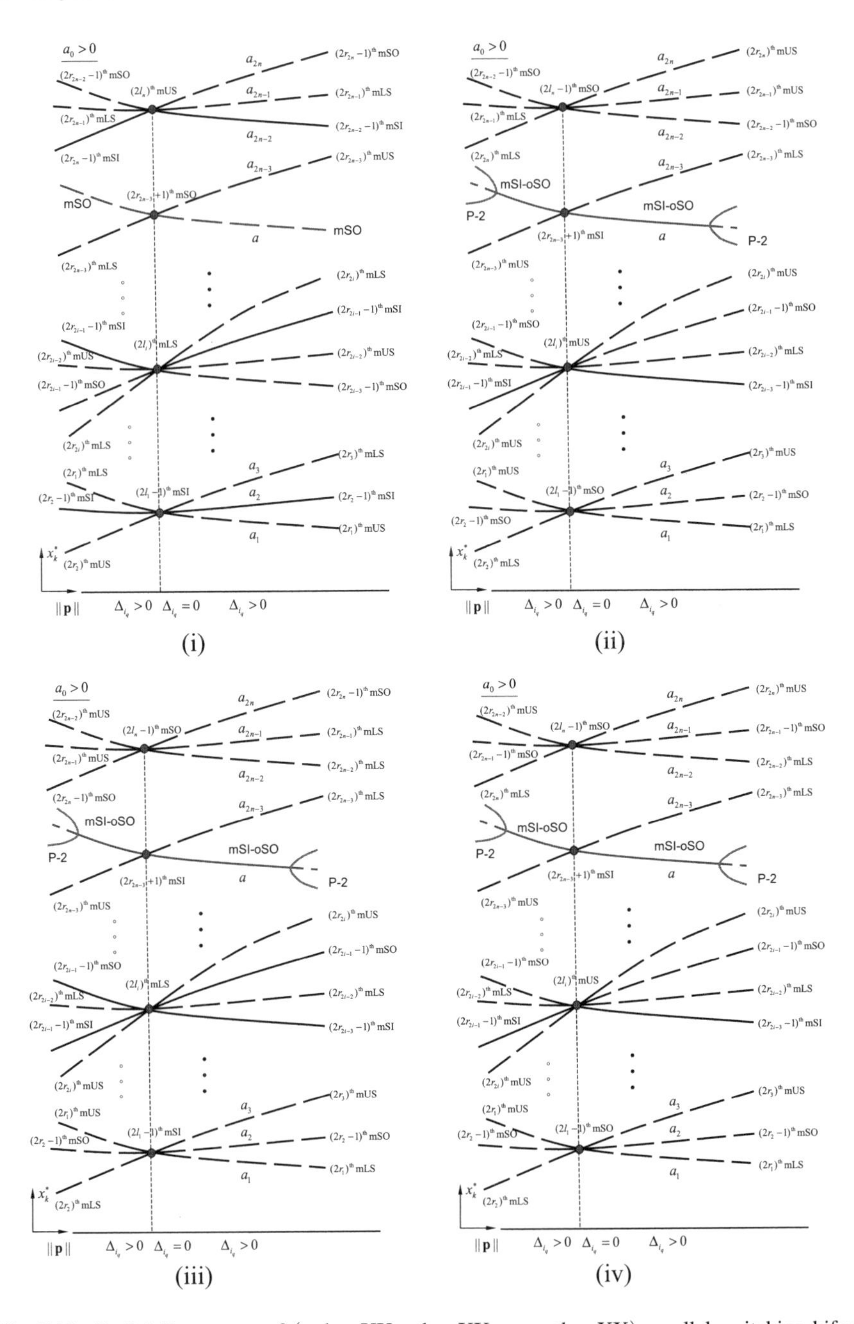

Fig. 5.12 (i)–(iv) Four types of (r_1th mXX:r_2th mXX:$\cdots$:r_mth mXX) parallel switching bifurcation for $a_0 > 0$ in the $(2m+1)^{\text{th}}$-degree polynomial discrete system. mLS: monotonic lower-saddle, mUS: monotonic upper-saddle, mSI: monotonic sink, mSO: monotonic source, mSI-oSO: monotonic sink to oscillatory source. Stable and unstable fixed-points are represented by solid and dashed curves, respectively. The bifurcation points are marked by circular symbols. P-2 is for period-2 fixed points, sketched by red curves, which are relative to the monotonic sinks to oscillatory sources.

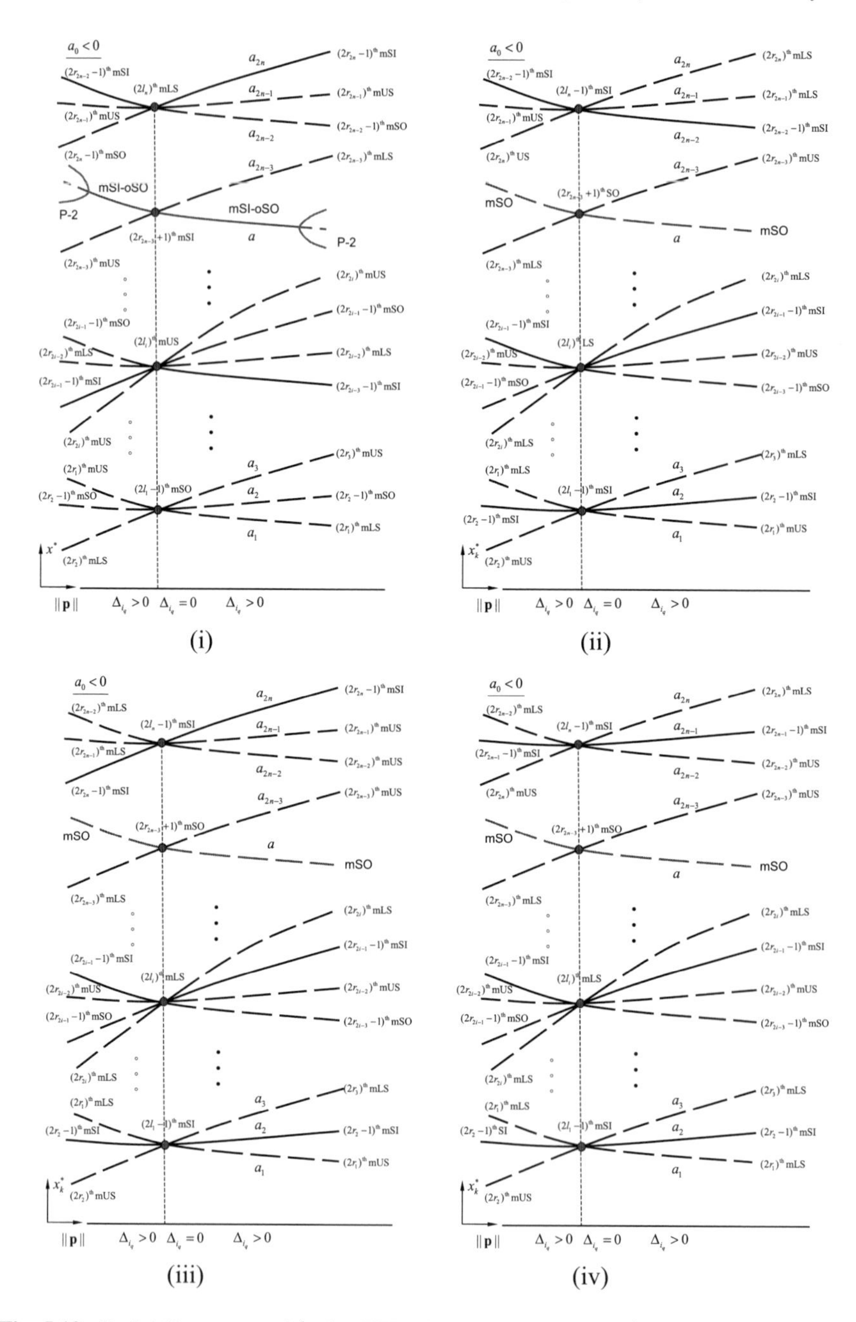

Fig. 5.13 (i)–(iv) Four types of (r_1th mXX:r_2th mXX:$\cdots$:r_mth mXX) parallel switching bifurcation for $a_0 < 0$ in the $(2m+1)^{\text{th}}$-degree polynomial discrete system. mLS: monotonic lower-saddle, mUS: monotonic upper-saddle, mSI: monotonic sink, mSO: monotonic source, mSI-oSO: monotoic sink to oscillatory source. Stable and unstable fixed-points are represented by solid and dashed curves, respectively. The bifurcation points are marked by circular symbols. P-2 is for period-2 fixed points, sketched by red curves. The period-2 fixed-points are relative to the monotonic sinks to oscillatory sources.

5.3.3 Switching-Appearing Bifurcations

Consider a $(2m+1)^{\text{th}}$ degree 1-dimensional polynomial nonlinear discrete system in a form of

$$x_{k+1} = x_k + a_0 Q(x_k) \prod_{i=1}^{n_1} (x_k - c_i)^{\alpha_i} \prod_{j=1}^{n_2} (x_k^2 + B_j x_k + C_j)^{\alpha_j}. \tag{5.233}$$

where $\sum_{i=1}^{n_1} \alpha_i = 2s_1 + 1$. Without loss of generality, a function of $Q(x_k) > 0$ is either a polynomial function or a non-polynomial function. The roots of $x_k^2 + B_j x_k + C_j = 0$ are

$$b_{1,2}^{(j)} = -\tfrac{1}{2}B_j \pm \tfrac{1}{2}\sqrt{\Delta_j}, \Delta_j = B_j^2 - 4C_j \geq 0 (j = 1, 2, \ldots, n_2); \tag{5.234}$$

either

$$\begin{aligned}
&\{a_1^-, a_2^-, \cdots, a_{2n_1+1}^-\} = \text{sort}\{c_1, c_2 \cdots, c_{2n_1}, a\}, a_s^- \leq a_{s+1}^- \text{ before bifurcation} \\
&\{a_1^+, a_2^+, \cdots, a_{2n_3+1}^+\} = \text{sort}\{c_1, \cdots, c_{2n_1}, a; b_1^{(1)}, b_2^{(1)}, \cdots, b_1^{(n_2)}, b_2^{(n_2)}\}, \\
&a_s^+ \leq a_{s+1}^+, n_3 = n_1 + n_2 \text{ after bifurcation;}
\end{aligned} \tag{5.235}$$

or

$$\begin{aligned}
&\{a_1^-, a_2^-, \cdots, a_{2n_3+1}^-\} = \text{sort}\{c_1, c_2 \cdots, c_{2n_1}, a; b_1^{(1)}, b_2^{(1)}, \cdots, b_1^{(n_2)}, b_2^{(n_2)}\}, \\
&a_s^- \leq a_{s+1}^-, n_3 = n_1 + n_2 \text{ before bifurcation;} \\
&\{a_1^+, a_2^+, \cdots, a_{2n_1+1}^+\} = \text{sort}\{c_1, \cdots, c_{2n_1}, a\}, a_s^+ \leq a_{s+1}^+ \text{ after bifurcation;}
\end{aligned} \tag{5.236}$$

and

$$\left.\begin{array}{l}
B_{j_1} = B_{j_2} = \cdots = B_{j_s} (j_{k_1} \in \{1, 2, \ldots, n\}; j_{k_1} \neq j_{k_2}) \\
(k_1, k_2 \in \{1, 2, \ldots, s\}; k_1 \neq k_2) \\
\Delta_j = 0 (j \in \{1, 2, \ldots, n_2\} \\
c_i \neq -\tfrac{1}{2}B_j (i = 1, 2, \ldots, 2n_1, j = 1, 2, \ldots, n_2)
\end{array}\right\} \text{at bifurcation.} \tag{5.237}$$

A just before bifurcation of $((\beta_1^-)^{\text{th}}\text{mXX}_1^- : (\beta_2^-)^{\text{th}}\text{mXX}_2^- : \cdots : (\beta_{3_1}^-)^{\text{th}}\text{mXX}_{\text{si}}^-)$ for higher-order fixed-points is considered. For $\beta_i^- = 2l_i^- - 1$ $\text{mXX}_i^- \in \{\text{mSO,mSI}\}$ and for $\alpha_i^- = 2l_i^-$, $\text{mXX}_i^- \in \{m\text{US,mLS}\}$ $(i = 1, 2, \ldots, s_1)$, the detailed structures are as follows.

$$\left.\begin{array}{l}(r_{s_i}^{(i)-})^{\text{th}}\ \text{order mXX}_s^{(i)-}, x_k^* = a_{s_i}^{(i)-},\\ \vdots\\ (r_j^{(i)-})^{\text{th}}\ \text{order mXX}_j^{(i)-}, x_k^* = a_j^{(i)-}\\ \vdots\\ (r_1^{(i)-})^{\text{th}}\ \text{order mXX}_1^{(i)-}, x_k^* = a_1^{(i)-}\end{array}\right\} \xrightarrow[\text{switching bifurcation}]{\beta_i^- = \sum_{j=1}^{s_i} r_j^{(i)-}} (\beta_i^-)^{\text{th}}\text{order mXX}^{(i)-} \tag{5.238}$$

The bifurcation set of $((\beta_1^-)^{\text{th}}\,\text{mXX}_1^- : (\beta_2^-)^{\text{th}}\,\text{mXX}_2^- : \cdots : (\beta_{s_1}^-)^{\text{th}}\,\text{mXX}_{s_1}^-)$ at the same parameter point is called a *left-parallel-straw-bundle* switching bifurcation

A just after bifurcation of $((\beta_1^+)^{\text{th}}\text{mXX}_1^+ : (\beta_2^+)^{\text{th}}\text{mXX}_2^+ : \cdots : (\beta_{s_2}^+)^{\text{th}}\text{mXX}_{s_2}^+)$ for higher-order singularity fixed-points is considered. For $\beta_j^+ = 2l_j^+ - 1$, $\text{mXX}_i^+ \in \{\text{mSO,mSI}\}$ and for $\beta^+ = 2l^+$, $\text{mXX}_i^+ \in \{m\text{US,mLS}\}$. The detailed structures are as follows.

$$(\beta_i^+)^{\text{th}}\text{order mXX}^{(i)+} \xrightarrow[\text{switching bifurcation}]{\beta_i^- = \sum_{j=1}^{s_i} r_j^{(i)+}} \left\{\begin{array}{l}(r_{s_i}^{(i)+})^{\text{th}}\ \text{order mXX}_{s_i}^{(i)+}, x_k^* = a_{s_i}^{(i)+},\\ \vdots\\ (r_j^{(i)+})^{\text{th}}\ \text{order mXX}_j^{(i)+}, x_k^* = a_j^{(i)+}\\ \vdots\\ (r_1^{(i)+})^{\text{th}}\ \text{order mXX}_1^{(i)+}, x_k^* = a_1^{(i)+}.\end{array}\right. \tag{5.239}$$

The bifurcation set of $((\beta_1^+)^{\text{th}}\,\text{mXX}_1^+ : (\beta_2^+)^{\text{th}}\,\text{mXX}_2^+ : \cdots : (\beta_{s_2}^+)^{\text{th}}\,\text{mXX}_{s_2}^+)$ at the same parameter point is called a *right-parallel-straw-bundle* switching bifurcation

(i) For the just before and after bifurcation structure, if there exists a relation of

$$\begin{array}{l}(\beta_i^-)^{\text{th}}\,\text{mXX}_i^- = (\beta_j^+)^{\text{th}}\,\text{mXX}_j^+ = \beta_j^{\text{th}}\text{mXX},\ \text{for}\ x^* = a_i^- = a_j^+\\ (i,j \in \{1,2,\ldots,k\}),\ \text{mXX} \in \{\text{mUS,mLS,mSO,mSI}\}\end{array} \tag{5.240}$$

then the bifurcation is a $(\beta_j)^{\text{th}}\,\text{mXX}_j$ *switching* bifurcation for higher-order fixed-points.

(ii) Just for the just before bifurcation structure, if there exists a relation of

$$\begin{aligned}&(2l_i^-)^{\text{th}}\,\text{mXX}_i^- = (2l_i)^{\text{th}}\text{mXX}, \text{ for } x^* = a_i^- = a_i\\&(i \in \{1,2,\ldots,s_1\}, \text{mXX} \in \{\text{mUS,mLS}\}\end{aligned} \tag{5.241}$$

then, the bifurcation is a $(2l)^{\text{th}}$ mXX *left appearing* (or *right vanishing*) bifurcation for higher-order fixed-points.

(iii) Just for the just after bifurcation structure, if there exists a relation of

$$\begin{aligned}&(2l_i^+)^{\text{th}}\,\text{mXX}_i^+ = (2l_i)^{\text{th}}\text{mXX}, \text{ for } x^* = a_i^+ = a_i\\&(i \in \{1,2,\ldots,s_1\}), \text{mXX} \in \{\text{mUS,mLS}\}\end{aligned} \tag{5.242}$$

then, the bifurcation is a $(2l)^{\text{th}}$ mXX *right appearing* (or *left vanishing*) bifurcation for higher-order fixed-points.

(iv) For the just before and after bifurcation structure, if there exists a relation of

$$\begin{aligned}&(\beta_i^-)^{\text{th}}\,\text{mXX}_i^- \neq (\beta_j^+)^{\text{th}}\,\text{mXX}_j^+ \text{ for } x^* = a_i^- = a_j^+\\&\text{mXX}_i^-,\ \text{mXX}_j^+ \in \{m\text{US,mLS, mSO,mSI}\}\\&(i \in \{1,2,\ldots,s_1\}, j \in \{1,2,\ldots,s_2\}),\end{aligned} \tag{5.243}$$

then, two *flower-bundle* switching bifurcations of higher-order fixed-points are as follows.

(iv$_1$) For $\beta_j = \beta_i + 2l$, the bifurcation is called a $(\beta_j)^{\text{th}}$ mXX right *flower-bundle* switching bifurcation for the $(\beta_i)^{\text{th}}$ mXX to $(\beta_j)^{\text{th}}$ mXX switching of *higher-order* fixed-points with the appearance (or birth) of $(2l)^{\text{th}}$ mXX right appearing (or left vanishing) bifurcation.

(iv$_2$) For $\beta_j = \beta_i - 2l$, the bifurcation is called a $(\beta_i)^{\text{th}}$ mXX left *flower-bundle* switching bifurcation for the $(\beta_i)^{\text{th}}$ mXX to $(\beta_j)^{\text{th}}$ mXX switching of higher-order fixed-points with the vanishing (or death) of $(2l)^{\text{th}}$ mXX *left appearing* (or right vanishing) bifurcation.

A general parallel switching bifurcation is

$$\begin{aligned}&((\beta_1^-)^{\text{th}}\,\text{mXX}_1^- : (\beta_2^-)^{\text{th}}\,\text{mXX}_2^- : \cdots : (\beta_{s_1}^-)^{\text{th}}\,\text{mXX}_{s_1}^-) \xrightarrow[\text{bifurcation}]{\text{switching}}\\&((\beta_1^+)^{\text{th}}\,\text{mXX}_1^+ : (\beta_2^+)^{\text{th}}\,\text{mXX}_2^+ : \cdots : (\beta_{s_2}^+)^{\text{th}}\,\text{mXX}_{s_2}^+).\end{aligned} \tag{5.244}$$

Such a general, parallel switching bifurcation consists of the *left* and *right* parallel-bundle switching bifurcations for higher-order fixed-points.

If the *left* and *right* parallel-bundle switching bifurcations are same in a parallel *flower-bundle* switching bifurcation, i.e.,

$$\begin{aligned}&(\beta_i^-)^{\text{th}}\,\text{mXX}_i^- = (\beta_i^+)^{\text{th}}\,\text{mXX}_i^+ = (\beta_i)^{\text{th}}\,\text{mXX}_i\\&\text{for } x_k^* = a_i^- = a_i^+\,(i = 1, 2, \ldots, s).\end{aligned} \tag{5.245}$$

then the parallel *flower-bundle* switching bifurcation becomes a parallel *straw-bundle switching* bifurcation of $((\beta_1)^{\text{th}}\,\text{mXX}_1 : (\beta_2)^{\text{th}}\,\text{mXX}_2 : \cdots : (\beta_{s_1})^{\text{th}}\,\text{mXX}_s)$.

If the *left* and *right* parallel-bundle switching bifurcations are different in a parallel *flower-bundle* switching bifurcation, i.e.,

$$\begin{aligned}&(\alpha_i^-)^{\text{th}}\,\text{mXX}_i^- = (2l_i^-)^{\text{th}}\,\text{mXX}_i^-, (\alpha_j^+)^{\text{th}}\,\text{mXX}_j^+ = (2l_j^+)^{\text{th}}\,\text{mXX}_j^+\\&\text{for } x_k^* = a_i^- \neq a_j^+\,(i = 1, 2, \ldots, s_1; j = 1, 2, \ldots, s_2),\\&\text{mXX}_i^- \in \{\text{mUS,mLS}\}, \text{mXX}_j^+ \in \{\text{mUS,mLS}\},\end{aligned} \tag{5.246}$$

then the parallel *flower-bundle* switching bifurcation for higher-order fixed-points becomes a combination of two independent left and right parallel appearing bifurcations:

(i) a $(2l_1^-)^{\text{th}}\,\text{mXX}_1^- : (2l_2^-)^{\text{th}}\,\text{mXX}_2^- : \cdots : (2l_{s_1}^-)^{\text{th}}\,\text{mXX}_{s_1}^-$-*left parallel sprinkler-spraying appearing (or right vanishing)* bifurcation and
(ii) a $(2l_1^+)^{\text{th}}\,\text{mXX}_1^+ : (2l_2^+)^{\text{th}}\,\text{mXX}_2^+ : \cdots : (2l_{s_2}^+)^{\text{th}}\,\text{mXX}_{s_1}^+$-*right parallel sprinkler-spraying appearing (or left vanishing)* bifurcation.

The parallel switching and appearing bifurcations for higher-order fixed-points are presented in Fig. 5.14i–iv. The *waterfall* appearing bifurcations and the *flower-bundle* switching bifurcations for higher-order fixed-points are presented. The period-2 fixed point is presented through red curves.

5.4 Forward Bifurcation Trees

In this section, the analytical bifurcation scenario of a $(2m+1)^{\text{th}}$-degree polynomial nonlinear discrete system will be discussed. The period-doubling bifurcation scenario will be discussed first through nonlinear renormalization techniques, and the bifurcation scenario based on the saddle-node bifurcation will be discussed, which is independent of period-1 fixed-points.

5.4.1 *Period-Doubled* $(2\text{m}+1)^{\text{th}}$*-Degree Polynomial Systems*

After the period-doubling bifurcation of a period-1 fixed-point, the period-doubled fixed-points of a $(2m+1)^{\text{th}}$-degree polynomial nonlinear discrete system can be obtained. Consider the period-doubling solutions of a forward quartic nonlinear discrete system first.

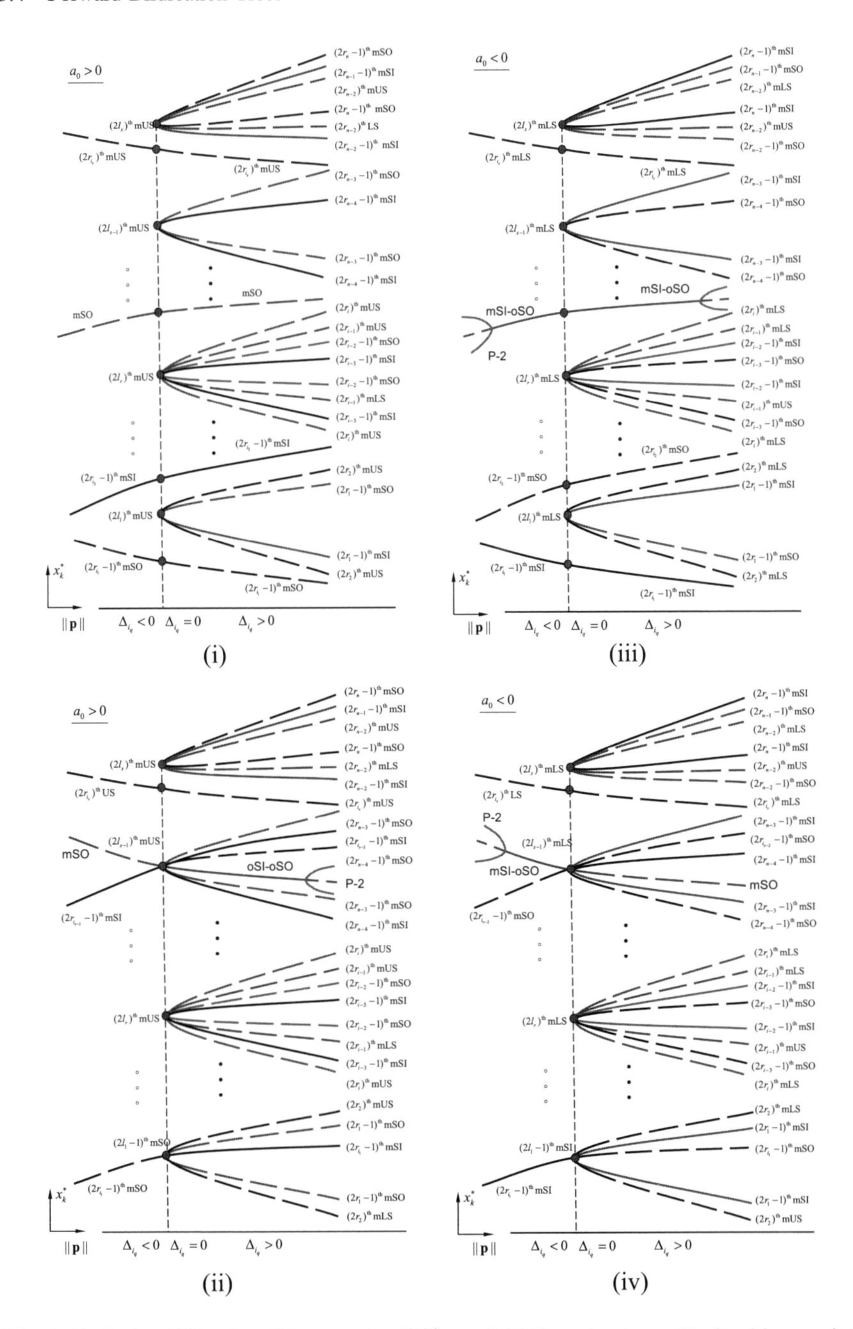

Fig. 5.14 (r_1th mXX:r_2th mXX:$\cdots$:r_nth mXX) parallel bifurcation ($a_0 > 0$): (i) without switching, and (ii) with switching. The (r_1th mXX:r_2th mXX:$\cdots$:r_nth mXX) parallel bifurcation ($a_0 < 0$): (iii) without switching, and (vi) with switching. mLS: monotonic lower-saddle, mUS: monotonic upper-saddle, mSI: monotonic sink, mSO: monotonic source, mSI-oSO: monotonic sink to oscillatory source. Stable and unstable fixed-points are represented by solid and dashed curves, respectively. The bifurcation points are marked by circular symbols. P-2 is for period-2 fixed points, sketched by red curves. The period-2 fixed-points are relative to the monotonic sinks to oscillatory sources.

Theorem 5.1 *Consider a* $(2m+1)^{\text{th}}$*-degree polynomial nonlinear discrete system*

$$\begin{aligned} x_{k+1} &= x_k + A_0(\mathbf{p})x_k^{2m+1} + A_1(\mathbf{p})x_k^{2m} + \cdots + A_{2m-1}(\mathbf{p})x_k^2 + A_{2m}x_k + A_{2m+1}(\mathbf{p}) \\ &= x_k + a_0(\mathbf{p})(x_k - a(\mathbf{p}))[x_k^2 + B_1(\mathbf{p})x_k + C_1(\mathbf{p})] \\ &\quad \cdots [x_k^2 + B_m(\mathbf{p})x_k + C_m(\mathbf{p})] \end{aligned} \tag{5.247}$$

where $A_0(\mathbf{p}) \neq 0$ *and*

$$\mathbf{p} = (p_1, p_2, \cdots, p_{m_1})^{\mathrm{T}}. \tag{5.248}$$

If

$$\begin{aligned} &\Delta_i = B_i^2 - 4C_i > 0, i = i_1, i_2, \ldots, i_l \in \{1, 2, \ldots, m\} \cup \{\varnothing\} \\ &\Delta_j = B_j^2 - 4C_j < 0, j = i_{l+1}, i_{l+2}, \cdots, i_m \in \{1, 2, \ldots, m\} \cup \{\varnothing\} \\ &\text{with } l \in \{0, 1, \ldots, m\} \end{aligned} \tag{5.249}$$

then, the corresponding standard form is

$$x_{k+1} = x_k + a_0 \prod\nolimits_{i=1}^{2m+1} (x_k - a_i^{(1)}). \tag{5.250}$$

where

$$\begin{aligned} &b_{i,1}^{(1)} = -\tfrac{1}{2}(B_i^{(1)} + \sqrt{\Delta_i^{(1)}}),\ b_{i,2}^{(1)} = -\tfrac{1}{2}(B_i^{(1)} - \sqrt{\Delta_i^{(1)}}) \\ &\text{for } \Delta_i^{(1)} \geq 0,\ i \in \{1, 2, \ldots, l\} \cup \{\varnothing\}; \\ &\cup_{i=1}^{2l}\{a_i^{(1)}\} = \mathrm{sort}\{\cup_{i_1=1}^{l}\{b_{i_1}^{(1)}, b_{i_1,2}^{(1)}\}\}, a_i^{(1)} \leq a_{i+1}^{(1)}; \\ &b_{i,1}^{(1)} = -\tfrac{1}{2}(B_i^{(1)} + \mathbf{i}\sqrt{|\Delta_i^{(1)}|}),\ b_{i,2}^{(1)} = -\tfrac{1}{2}(B_i^{(1)} - \mathbf{i}\sqrt{|\Delta_i^{(1)}|}) \\ &\text{for } \Delta_i^{(1)} < 0,\ i \in \{l+1, l+2, \ldots, m\} \cup \{\varnothing\},\ \mathbf{i} = \sqrt{-1}; \\ &\cup_{i=2l+1}^{2m+1}\{a_i^{(1)}\} = \{\cup_{i_1=l+1}^{m}\{b_{i_1}^{(1)}, b_{i_1,2}^{(1)}\}, a\}. \end{aligned} \tag{5.251}$$

(i) *Consider a forward period-2 discrete system of Eq. (5.247) as*

$$\begin{aligned} x_{k+2} &= x_k + [a_0 \prod\nolimits_{i_1=1}^{2m+1} (x_k - a_{i_1}^{(1)})]\{1 + \prod\nolimits_{i_1=1}^{2m+1} [1 + a_0 \prod\nolimits_{i_2=1, i_2 \neq i_1}^{2m+1} (x_k - a_{i_2}^{(1)})]\} \\ &= x_k + [a_0 \prod\nolimits_{i_1=1}^{2m+1} (x_k - a_{i_1}^{(1)})] \\ &\quad \times [a_0^{2m+1} \prod\nolimits_{i_2=1}^{((2m+1)^2 - (2m+1))/2} (x_k^2 + B_{i_2}^{(2)} x_k + C_{i_2}^{(2)})] \\ &= x_k + [a_0 \prod\nolimits_{j_1=1}^{2m+1} (x_k - a_{i_1}^{(1)})][a_0^{2m+1} \prod\nolimits_{j_2=1}^{(2m+1)^2 - (2m+1)} (x_k - b_{j_2}^{(2)})] \\ &= x_k + a_0^{1+(2m+1)} \prod\nolimits_{i=1}^{(2m+1)^2} (x_k - a_i^{(2)}) \end{aligned} \tag{5.252}$$

where

$$
\begin{aligned}
& b_{i,1}^{(2)} = -\tfrac{1}{2}(B_i^{(2)} + \sqrt{\Delta^{(2)}}), b_{i,2}^{(2)} = -\tfrac{1}{2}(B_i^{(2)} - \sqrt{\Delta_i^{(2)}}),\\
& \Delta_i^{(2)} = (B_i^{(2)})^2 - 4C_i^{(2)} \geq 0,\ i \in \cup_{q_1=1}^{N_1} I_{q_1}^{(2^0)} \cup \cup_{q_2=1}^{N_2} I_{q_2}^{(2^2)}\\
& I_{q_1}^{(2^0)} = \{l_{(q_1-1)\times 2^0 \times m_1+1}, l_{(q_1-1)\times 2^0\times m_1+2}, \cdots, l_{q_1\times 2^0\times m_1}\}\\
& \qquad \subseteq \{1,2,\cdots,M_1\} \cup \{\varnothing\},\\
& q_1 \in \{1,2,\cdots,N_1\}, M_1 = N_1 \times 2^0 \times m_1, m_1 \in \{1,2,\cdots,m\};\\
& I_{q_2}^{(2^1)} = \{l_{(q_2-1)\times 2^1 \times m_1+1}, l_{(q_2-1)\times 2^1\times m_1+2}, \cdots, l_{q_2\times 2^1\times m_1}\}\\
& \qquad \subseteq \{M_1+1, M_1+2,\cdots,M_2\} \cup \{\varnothing\},\\
& q_2 \in \{1,2,\cdots,N_2\}, M_2 = ((2m)^2 - 2m)/2;\\
& b_{i,1}^{(2)} = -\tfrac{1}{2}(B_i^{(2)} + \mathbf{i}\sqrt{\Delta^{(2)}}), b_{i,2}^{(2)} = -\tfrac{1}{2}(B_i^{(2)} - \mathbf{i}\sqrt{\Delta_i^{(2)}}),\\
& \mathbf{i} = \sqrt{-1}, \Delta_i^{(2)} = (B_i^{(2)})^2 - 4C_i^{(2)} < 0,\\
& i \in J^{(2^1)} = \{l_{N_2\times 2^1\times m_1+1}, l_{N_2\times 2^1\times m_1+2}, \cdots, l_{M_2}\}\\
& \qquad\quad \subseteq \{M_1+1, M_1+2, \cdots, M_2\}
\end{aligned}
\tag{5.253}
$$

with fixed-points

$$
\begin{aligned}
& x_{k+2}^* = x_k^* = a_i^{(2)}, (i = 1,2,\ldots,(2m+1)^2)\\
& \cup_{i=1}^{(2m+1)^2}\{a_i^{(2)}\} = \mathrm{sort}\{\cup_{j_1=1}^{2m+1}\{a_{j_1}^{(1)}\}, \cup_{j_2=1}^{M}\{b_{j_2,1}^{(2)}, b_{j_2,2}^{(2)}\}\}\\
& \text{with } a_i^{(2)} < a_{i+1}^{(2)},\ M = ((2m+1)^2 - (2m+1))/2.
\end{aligned}
\tag{5.254}
$$

(ii) *For a fixed-point of* $x_{k+1}^* = x_k^* = a_{i_1}^{(1)}$ $(i_1 \in \{1,2,\ldots,2m+1\})$, *if*

$$
\frac{dx_{k+1}}{dx_k}\Big|_{x_k^*=a_{i_1}^{(1)}} = 1 + a_0 \textstyle\prod_{i_2=1, i_2\neq i_1}^{2m+1}(a_{i_1}^{(1)} - a_{i_2}^{(1)}) = -1,
\tag{5.255}
$$

with

- *a* r^{th}*-order oscillatory upper-saddle-node bifurcation* $(d^r x_{k+1}/dx_k^r|_{x_k^*=a_{i_1}^{(1)}} > 0,\ r = 2l_1)$,
- *a* r^{th}*-order oscillatory lower-saddle-node bifurcation* $(d^r x_{k+1}/dx_k^r|_{x_k^*=a_{i_1}^{(1)}} < 0,\ r = 2l_1)$,
- *a* r^{th}*-order oscillatory source bifurcation* $(d^r x_{k+1}/dx_k^r|_{x_k^*=a_{i_1}^{(1)}} < 0,\ r = 2l_1+1)$,

- *a rth-order oscillatory sink bifurcation* $(d^r x_{k+1}/dx_k^r|_{x_k^*=a_{i_1}^{(1)}} > 0,\ r = 2l_1 + 1)$,

then the following relations satisfy

$$a_{i_1}^{(1)} = -\frac{1}{2}B_i^{(2)}, \Delta_{i_1}^{(2)} = (B_i^{(2)})^2 - 4C_{i_1}^{(2)} = 0, \tag{5.256}$$

and there is a period-2 discrete system of the quartic discrete system in Eq. (5.247) as

$$\begin{aligned} x_{k+2} = x_k + a_0^{1+(2m+1)} & \prod\nolimits_{i_2 \in I_{q_1}^{(2^0)}} (x_k - a_{i_2}^{(1)})^3 \\ & \times \prod\nolimits_{i_3=1}^{(2m+1)^2} (x_k - a_{i_3}^{(2)})^{(1-\delta(i_2,i_3))} \end{aligned} \tag{5.257}$$

for $i_1 \in I_{q1}^{(2^0)} \subset \{1, 2, \ldots, 2m+1\}$, $i_1 \neq i_2$ *with*

$$\frac{dx_{k+2}}{dx_k}|_{x_k^*=a_{i_1}^{(1)}} = 1, \frac{d^2 x_{k+2}}{dx_k^2}|_{x_k^*=a_{i_1}^{(1)}} = 0; \tag{5.258}$$

- x_{k+2} at $x_k^* = a_{i_1}^{(1)}$ *is a monotonic sink of the third-order if*

$$\begin{aligned} \frac{d^3 x_{k+2}}{dx_k^3}|_{x_k^*=a_{i_1}^{(1)}} = 6a_0^{1+2m} & \prod\nolimits_{i_2 \in I_{q_1}^{(2^0)}, i_2 \neq i_1} (a_{i_1}^{(1)} - a_{i_2}^{(1)})^3 \\ & \times \prod\nolimits_{i_3=1}^{(2m+1)^2} (a_{i_1}^{(1)} - a_{i_3}^{(2)})^{(1-\delta(i_2,i_3))} < 0 \end{aligned} \tag{5.259}$$

  and the corresponding bifurcations is a third-order monotonic sink bifurcation for the period-2 discrete system;

- x_{k+2} at $x_k^* = a_{i_1}^{(1)}$ is *a monotonic source of the third-order if*

$$\begin{aligned} \frac{d^3 x_{k+2}}{dx_k^3}|_{x_k^*=a_{i_1}^{(1)}} = 6a_0^{1+2m} & \prod\nolimits_{i_2 \in I_{q_1}^{(2^0)}, i_2 \neq i_1} (a_{i_1}^{(1)} - a_{i_2}^{(1)})^3 \\ & \times \prod\nolimits_{i_3=1}^{(2m+1)^2} (a_{i_1}^{(1)} - a_{i_3}^{(2)})^{(1-\delta(i_2,i_3))} > 0 \end{aligned} \tag{5.260}$$

  and the corresponding bifurcations is a third-order monotonic source bifurcation for the period-2 discrete system.

(ii$_1$) *The period-2 fixed-points are trivial and unstable if*

$$x^*_{k+2} = x^*_k = a_i^{(2)} \in \cup_{i_1=1}^{2m+1}\{a_{i_1}^{(1)}\} \ . \tag{5.261}$$

(ii$_2$) *The period-2 fixed-points are non-trivial and stable if*

$$x^*_{k+2} = x^*_k = a_i^{(2)} \in \cup_{i_1=1}^{M_2}\{b_{i_1,1}^{(2)}, b_{i_1,2}^{(2)}\} \ . \tag{5.262}$$

Proof The proof is straightforward through the simple algebraic manipulation. Following the proof of quadratic discrete system, this theorem is proved. ■

5.4.2 Renormalization and Period-Doubling

The generalized cases of period-doublization of a $(2m+1)^{\text{th}}$-degree polynomial discrete system are presented through the following theorem. The analytical period-doubling bifurcation trees can be developed for such a $(2m+1)^{\text{th}}$-degree polynomial discrete systems.

Theorem 5.2 *Consider a 1-dimensional $(2m+1)^{\text{th}}$-degree polynomial discrete system as*

$$\begin{aligned} x_{k+1} &= x_k + A_0 x_k^{2m+1} + A_1 x_k^{2m} + \cdots + A_{2m-1} x_k^2 + A_{2m} x_k + A_{2m+1} \\ &= x_k + a_0 \prod_{i=1}^{2m+1} (x_k - a_i). \end{aligned} \tag{5.263}$$

(i) *After l-times period-doubling bifurcations, a period- 2^l discrete system ($l = 1, 2, \ldots$) for the $(2m+1)^{\text{th}}$-degree polynomial discrete system in Eq. (5.263) is given through the nonlinear renormalization as*

$$\begin{aligned} x_{k+2^l} &= x_k + [a_0^{(2^{l-1})} \prod_{i_1=1}^{(2m+1)^{2^{l-1}}} (x_k - a_{i_1}^{(2^{l-1})})] \\ &\quad \times \{1 + \prod_{i_1=1}^{(2m+1)^{2^{l-1}}} [1 + a_0^{(2^{l-1})} \prod_{i_2=1, i_2\neq i_1}^{(2m+1)^{2^{l-1}}} (x_k - a_{i_2}^{(2^{l-1})})]\} \\ &= x_k + [a_0^{(2^{l-1})} \prod_{i_1=1}^{(2m+1)^{2^{l-1}}} (x_k - a_{i_1}^{(2^{l-1})})] \\ &\quad \times [(a_0^{(2^{l-1})})^{4^{2^{l-1}}} \prod_{j_1=1}^{((2m+1)^{2^l} - (2m+1)^{2^{l-1}})/2} (x_k^2 + B_{j_2}^{(2^l)} x_k + C_{j_2}^{(2^l)})] \\ &= x_k + [a_0^{(2^{l-1})} \prod_{i_1=1,}^{(2m+1)^{2^{l-1}}} (x_k - a_{i_1}^{(2^{l-1})})] \\ &\quad \times [(a_0^{(2^{l-1})})^{4^{2^{l-1}}} \prod_{i_2=1}^{((2m+1)^{2^l} - (2m+1)^{2^{l-1}})/2} (x_k - b_{i_2,1}^{(2^l)})(x_k - b_{i_2,2}^{(2^l)})] \\ &= x_k + (a_0^{(2^{l-1})})^{1+(2m+1)^{2^{l-1}}} \prod_{i=1}^{(2m+1)^{2^l}} (x_k - a_i^{(2^l)}) \\ &= x_k + a_0^{(2^l)} \prod_{i=1}^{(2m+1)^{2^l}} (x_k - a_i^{(2^l)}) \end{aligned} \tag{5.264}$$

with

$$
\begin{aligned}
&\frac{dx_{k+2^l}}{dx_k} = 1 + a_0^{(2^l)} \sum_{i_1=1}^{(2m+1)^{2^l}} \prod_{i_2=1, i_2 \neq i_1}^{(2m+1)^{2^l}} (x_k - a_{i_2}^{(2^l)}),\\
&\frac{d^2 x_{k+2^l}}{dx_k^2} = a_0^{(2^l)} \sum_{i_1=1}^{(2m+1)^{2^l}} \sum_{i_2=1, i_2 \neq i_1}^{(2m+1)^{2^l}} \prod_{i_3=1, i_3 \neq i_1, i_2}^{(2m+1)^{2^l}} (x_k - a_{i_3}^{(2^l)}),\\
&\vdots\\
&\frac{d^r x_{k+2^l}}{dx_k^r} = a_0^{(2^l)} \sum_{i_1=1}^{(2m+1)^{2^l}} \cdots \sum_{i_r=1, i_3 \neq i_1, i_2 \cdots i_{r-1}}^{(2m+1)^{2^l}} \prod_{i_{r+1}=1, i_{r+1} \neq i_1, i_2 \cdots, i_r}^{(2m+1)^{2^l}} (x_k - a_{i_{r+1}}^{(2^l)})\\
&\text{for } r \leq (2m+1)^{2^l}
\end{aligned}
\tag{5.265}
$$

where

$$
\begin{aligned}
&a_0^{(2)} = (a_0)^{1+(2m+1)}, a_0^{(2^l)} = (a_0^{(2^{l-1})})^{1+(2m+1)^{2^{l-1}}}, l = 1, 2, 3, \cdots;\\
&\cup_{i=1}^{(2m+1)^{2^l}} \{a_i^{(2^l)}\} = \text{sort}\{\cup_{i_1=1}^{(2m+1)^{2^{l-1}}} \{a_{i_1}^{(2^l)}\}, \cup_{i_2=1}^{M_2} \{b_{i_2,1}^{(2^l)}, b_{i_2,2}^{(2^l)}\}\}, a_i^{(2^l)} \leq a_{i+1}^{(2^l)};\\
&b_{i,1}^{(2^l)} = -\frac{1}{2}(B_i^{(2^l)} + \sqrt{\Delta_i^{(2^l)}}), b_{i,2}^{(2^{l-1})} = -\frac{1}{2}(B_i^{(2^l)} - \sqrt{\Delta_i^{(2^l)}}),\\
&\Delta_i^{(2^l)} = (B_i^{(2^l)})^2 - 4C_i^{(2^l)} \geq 0 \text{ for } i \in \cup_{q_1=1}^{N_1} I_{q_1}^{(2^{l-1})} \cup \cup_{q_2=1}^{N_2} I_{q_2}^{(2^l)},\\
&I_{q_1}^{(2^{l-1})} = \{l_{(q_1-1)\times 2^{l-1}\times m_1+1}, l_{(q_1-1)\times 2^{l-1}\times m_1+2}, \cdots, l_{q_1\times 2^{l-1}\times m_1}\}\\
&\qquad \subseteq \{1, 2, \cdots, M_1\} \cup \{\varnothing\},\\
&\text{for } q_1 \in \{1, 2, \cdots, N_1\}, M_1 = N_1 \times 2^{l-1} \times m_1, m_1 \in \{1, 2, \cdots, m\};\\
&I_{q_2}^{(2^l)} = \{l_{(q_2-1)\times 2^l\times m_1+1}, l_{(q_2-1)\times 2^l\times m_1+2}, \cdots, l_{q_2\times 2^l\times m_1}\}\\
&\qquad \subseteq \{M_1+1, M_1+2, \cdots, M_2\} \cup \{\varnothing\},\\
&\text{for } q_2 \in \{1, 2, \cdots, N_2\}, M_2 = ((2m+1)^{2^l} - (2m+1)^{2^{l-1}})/2;\\
&b_{i,1}^{(2^l)} = -\frac{1}{2}(B_i^{(2^l)} + \mathbf{i}\sqrt{|\Delta_i^{(2^l)}|}), b_{i,2}^{(2^l)} = -\frac{1}{2}(B_i^{(2^l)} - \mathbf{i}\sqrt{|\Delta_i^{(2^l)}|}),\\
&\Delta_i^{(2^l)} = (B_i^{(2^l)})^2 - 4C_i^{(2^l)} < 0, \mathbf{i} = \sqrt{-1},\\
&i \in J^{(2^l)} = \{l_{N\times 2^l\times m_1+1}, l_{N\times 2^l\times m_1+2}, \cdots, l_{M_2}\}\\
&\qquad \subset \{M_1+1, M_1+2, \cdots, M_2\} \cup \{\varnothing\};
\end{aligned}
\tag{5.266}
$$

with fixed-points

$$
\begin{aligned}
& x_{k+2^l}^* = x_k^* = a_i^{(2^l)}, (i = 1, 2, \cdots, (2m+1)^{2^l}) \\
& \cup_{i=1}^{(2m+1)^{2^l}} \{a_i^{(2^l)}\} = \mathrm{sort}\{\cup_{i_1=1}^{(2m+1)^{2^{l-1}}} \{a_{i_1}^{(2^{l-1})}\}, \cup_{i_2=1}^{M_2} \{b_{i_2,1}^{(2^l)}, b_{i_2,2}^{(2^l)}\}\} \\
& \text{with } a_i^{(2^l)} < a_{i+1}^{(2^l)}.
\end{aligned}
\tag{5.267}
$$

(ii) *For a fixed-point of* $x_{k+2^{l-1}}^* = x_k^* = a_{i_1}^{(2^{l-1})}$ $(i_1 \in I_q^{(2^{l-1})})$, *if*

$$
\begin{aligned}
& \frac{dx_{k+2^{l-1}}}{dx_k}\Big|_{x_k^*=a_{i_1}^{(2^{l-1})}} = 1 + a_0^{(2^{l-1})} \prod_{i_2=1, i_2 \neq i_1}^{(2m+1)^{2^{l-1}}} (a_{i_1}^{(2^{l-1})} - a_{i_2}^{(2^{l-1})}) = -1, \\
& \frac{d^s x_{k+2^{l-1}}}{dx_k^s}\Big|_{x_k^*=a_{i_1}^{(2^{l-1})}} = 0, \text{ for } s = 2, \cdots, r-1; \\
& \frac{d^r x_{k+2^{l-1}}}{dx_k^r}\Big|_{x_k^*=a_{i_1}^{(2^{l-1})}} \neq 0 \text{ for } 1 < r \leq (2m+1)^{2^{l-1}},
\end{aligned}
\tag{5.268}
$$

with

- *a* r^{th}*-order oscillatory sink for* $d^r x_{k+2^{l-1}}/dx_k^r\big|_{x_k^*=a_{i_1}^{(2^{l-1})}} > 0$ *and* $r = 2l_1 + 1$;
- *a* r^{th}*-order oscillatory source for* $d^r x_{k+2^{l-1}}/dx_k^r\big|_{x_k^*=a_{i_1}^{(2^{l-1})}} < 0$ *and* $r = 2l_1 + 1$;
- *a* r^{th}*-order oscillatory upper-saddle for* $d^r x_{k+2^{l-1}}/dx_k^r\big|_{x_k^*=a_{i_1}^{(2^{l-1})}} > 0$ *and* $r = 2l_1$;
- *a* r^{th}*-order oscillatory lower-saddle for* $d^r x_{k+2^{l-1}}/dx_k^r\big|_{x_k^*=a_{i_1}^{(2^{l-1})}} < 0$ *and* $r = 2l_1$;

then there is a period- 2^l *fixed-point discrete system*

$$
\begin{aligned}
x_{k+2^l} = x_k + a_0^{(2^l)} & \prod_{i_1 \in I_{q_1}^{(2^{l-1})}} (x_k - a_{i_1}^{(2^{l-1})})^3 \\
& \times \prod_{j_2=1}^{(2m+1)^{2^l}} (x_k - a_{j_2}^{(2^l)})^{(1-\delta(i_1, j_2))}
\end{aligned}
\tag{5.269}
$$

where

$$
\delta(i_1, j_2) = 1 \text{ if } a_{j_2}^{(2^l)} = a_{i_1}^{(2^{l-1})}, \ \delta(i_1, j_2) = 0 \text{ if } a_{j_2}^{(2^l)} \neq a_{i_1}^{(2^{l-1})}
\tag{5.270}
$$

and

$$
\frac{dx_{k+2^l}}{dx_k}\Big|_{x_k^*=a_{i_1}^{(2^{l-1})}} = 1, \ \frac{d^2 x_{k+2^l}}{dx_k^2}\Big|_{x_k^*=a_{i_1}^{(2^{l-1})}} = 0.
\tag{5.271}
$$

- x_{k+2^l} *at* $x_k^* = a_i^{(2^{l-1})}$ *is a monotonic sink of the third-order if*

$$\frac{d^3 x_{k+2^l}}{dx_k^3}\Big|_{x_k^*=a_{i_1}^{(2^{l-1})}} = 6a_0^{(2^l)} \prod\nolimits_{i_2 \in I_{q_1}^{(2^{l-1})}, i_2 \neq i_1} (a_{i_1}^{(2^{l-1})} - a_{i_2}^{(2^{l-1})})^3 \times \prod\nolimits_{j_2=1}^{(2m+1)^{2^l}} (a_{i_1}^{(2^{l-1})} - a_{j_2}^{(2^l)})^{(1-\delta(i_2,j_2))} < 0 \tag{5.272}$$

$(i_1 \in I_{q_1}^{(2^{l-1})}, q_1 \in \{1, 2, \cdots, N_1\})$,

and such a bifurcation at $x_k^* = a_i^{(2^{l-1})}$ *is a third-order monotonic sink bifurcation.*

- x_{k+2^l} *at* $x_k^* = a_i^{(2^{l-1})}$ *is a monotonic source of the third-order if*

$$\frac{d^3 x_{k+2^l}}{dx_k^3}\Big|_{x_k^*=a_{i_1}^{(2^{l-1})}} = 6a_0^{(2^l)} \prod\nolimits_{i_2 \in I_{q_1}^{(2^{l-1})}, i_2 \neq i_1} (a_{i_1}^{(2^{l-1})} - a_{i_2}^{(2^{l-1})})^3 \times \prod\nolimits_{j_2=1}^{(2m+1)^{2^l}} (a_{i_1}^{(2^{l-1})} - a_{j_2}^{(2^l)})^{(1-\delta(i_2,j_2))} > 0 \tag{5.273}$$

$(i_1 \in I_{q_1}^{(2^{l-1})}, q_1 \in \{1, 2, \cdots, N_1\})$

and such a bifurcation at $x_k^* = a_{i_1}^{(2^{l-1})}$ *is a third-order monotonic source bifurcation.*

(ii$_1$) *The period-* 2^l *fixed-points are trivial if*

$$x_{k+2^l}^* = x_k^* = a_i^{(2^{l-1})} \in \cup_{i_1=1}^{(2m+1)^{2^{l-1}}} \{a_{i_1}^{(2^{l-1})}\}\ . \tag{5.274}$$

(ii$_2$) *The period-* 2^l *fixed-points are non-trivial if*

$$x_{k+2^l}^* = x_k^* = a_i^{(2^{l-1})} \in \cup_{i_1=1}^{M_2} \{b_{j_1,1}^{(2^l)}, b_{j_1,2}^{(2^l)}\}. \tag{5.275}$$

Such a period- 2^l *fixed-point is*

- *monotonically unstable if* $dx_{k+2^l}/dx_k|_{x_k^*=a_{i_1}^{(2^l)}} \in (1, \infty)$;
- *monotonically invariant if* $dx_{k+2^l}/dx_k|_{x_k^*=a_{i_1}^{(2^l)}} = 1$, *which is*
  - *a monotonic upper-saddle of the* $(2l_1)^{\text{th}}$ *order for* $d^{2l_1} x_{k+2^l}/dx_k^{2l_1}|_{x_k^*} > 0$;
  - *a monotonic lower-saddle of the* $(2l_1)^{\text{th}}$ *order for* $d^{2l_1} x_{k+2^l}/dx_k^{2l_1}|_{x_k^*} < 0$;
  - *a monotonic source of the* $(2l_1 + 1)^{\text{th}}$ *order for* $d^{2l_1+1} x_{k+2^l}/dx_k^{2l_1+1}|_{x_k^*} > 0$;
  - *a monotonic sink the* $(2l_1 + 1)^{\text{th}}$ *order for* $d^{2l_1+1} x_{k+2^l}/dx_k^{2l_1+1}|_{x_k^*} < 0$;

- *monotonically stable if* $dx_{k+2^l}/dx_k|_{x_k^*=a_{i_1}^{(2^l)}} \in (0,1)$;
- *invariantly zero-stable if* $dx_{k+2^l}/dx_k|_{x_k^*=a_{i_1}^{(2^{l-1})}} = 0$;
- *oscillatorilly stable if* $dx_{k+2^l}/dx_k|_{x_k^*=a_{i_1}^{(2^{l-1})}} \in (-1,0)$;
- *flipped if* $dx_{k+2^l}/dx_k|_{x_k^*=a_{i_1}^{(2^{l-1})}} = -1$, *which is*
  - *an oscillatory upper-saddle of the* $(2l_1)^{\text{th}}$ *order for* $d^{2l_1}x_{k+2^l}/dx_k^{2l_1}|_{x_k^*} > 0$;
  - *an oscillatory lower-saddle the* $(2l_1)^{\text{th}}$ *order for* $d^{2l_1}x_{k+2^l}/dx_k^{2l_1}|_{x_i^*} < 0$
  - *an oscillatory source of the* $(2l_1+1)^{\text{th}}$ *order if* $d^{2l_1+1}x_{k+2^l}/dx_k^{2l_1+1}|_{x_k^*} < 0$;
  - *an oscillatory sink the* $(2l_1+1)^{\text{th}}$ *order with* $d^{2l_1+1}x_{k+2^l}/dx_k^{2l_1+1}|_{x_k^*} > 0$;
- *oscillatorilly unstable if* $dx_{k+2^l}/dx_k|_{x_k^*=a_{i_1}^{(2^l)}} \in (-\infty,-1)$.

Proof Through the nonlinear renormalization, this theorem can be proved. ■

5.4.3 *Period-*n *Appearing and Period-Doublization*

The forward period-n discrete system for the quartic nonlinear discrete systems will be discussed, and the period-doublization of period-n discrete systems is discussed through the nonlinear renormalization.

Theorem 5.3 *Consider a 1-dimensional* $(2m+1)^{\text{th}}$*-degree polynomial discrete system as*

$$\begin{aligned} x_{k+1} &= x_k + A_0 x_k^{2m+1} + A_1 x_k^{2m} + \cdots + A_{2m-1}x_k^2 + A_{2m}x_k + A_{2m+1} \\ &= x_k + a_0 \textstyle\prod_{i=1}^{2m+1}(x_k - a_i). \end{aligned} \tag{5.276}$$

(i) *After n-times iterations, a period-n discrete system for the quartic discrete system in Eq. (5.276) is*

$$\begin{aligned} x_{k+n} &= x_k + a_0 \textstyle\prod_{i_1=1}^{2m+1}(x_k - a_{i_1}^{(1)})\{1 + \Sigma_{j=1}^{n} Q_j\} \\ &= x_k + a_0^{((2m+1)^n-1)/(2m)} \textstyle\prod_{i_1=1}^{2m+1}(x_k - a_{i_1}^{(1)}) \\ &\quad \times [\textstyle\prod_{j_2=1}^{((2m+1)^n-(2m+1))/2}(x_k^2 + B_{j_2}^{(n)}x_k + C_{j_2}^{(n)})] \\ &= x_k + a_0^{(n)} \textstyle\prod_{i=1}^{(2m+1)^n}(x_k - a_i^{(n)}) \end{aligned} \tag{5.277}$$

with

$$\begin{aligned}
&\frac{dx_{k+n}}{dx_k} = 1 + a_0^{(n)} \sum_{i_1=1}^{(2m+1)^n} \prod_{i_2=1, i_2 \neq i_1}^{(2m+1)^n} (x_k - a_{i_2}^{(n)}),\\
&\frac{d^2 x_{k+n}}{dx_k^2} = a_0^{(n)} \sum_{i_1=1}^{(2m+1)^n} \sum_{i_2=1, i_2 \neq i_1}^{(2m+1)^n} \prod_{i_3=1, i_3 \neq i_1, i_2}^{(2m+1)^n} (x_k - a_{i_3}^{(n)}),\\
&\vdots\\
&\frac{d^r x_{k+n}}{dx_k^r} = a_0^{(n)} \sum_{i_1=1}^{(2m+1)^n} \cdots \sum_{i_r=1, i_r \neq i_1, i_2 \cdots, i_{r-1}}^{(2m+1)^n} \prod_{i_{r+1}=1, i_{r+1} \neq i_1, i_2 \cdots, i_r}^{(2m+1)^n} (x_k - a_{i_{r+1}}^{(n)})\\
&\text{for } r \leq (2m+1)^n;
\end{aligned} \tag{5.278}$$

where

$$\begin{aligned}
&a_0^{(n)} = (a_0)^{((2m+1)^n - 1)/(2m)}, Q_1 = 0, Q_2 = \prod_{i_2=1}^{2m+1} [1 + a_0 \prod_{i_1=1, i_1 \neq i_2}^{2m+1} (x_k - a_{i_1}^{(1)})],\\
&Q_n = \prod_{i_n=1}^{2m+1} [1 + a_0 (1 + Q_{n-1}) \prod_{i_{n-1}=1, i_{n-1} \neq i_n}^{2m+1} (x_k - a_{i_{n-1}}^{(1)})],\ n = 3, 4, \cdots;\\
&\cup_{i=1}^{(2m+1)^n} \{a_i^{(n)}\} = \text{sort}\{\cup_{i_1=1}^{2m+1} \{a_{i_1}^{(1)}\}, \cup_{i_2=1}^{M} \{b_{i_2,1}^{(n)}, b_{i_2,2}^{(n)}\}\}\ ;\\
&b_{i_2,1}^{(n)} = -\tfrac{1}{2}(B_{i_2}^{(n)} + \sqrt{\Delta_{i_2}^{(n)}}), b_{i_2,2}^{(n)} = -\tfrac{1}{2}(B_{i_2}^{(n)} - \sqrt{\Delta_{i_2}^{(n)}}),\\
&\Delta_{i_2}^{(n)} = (B_{i_2}^{(n)})^2 - 4C_{i_2}^{(n)} \geq 0 \text{ for } i_2 \in \cup_{q=1}^{N} I_q^{(n)},\\
&I_q^{(n)} = \{l_{(q-1)\times n+1}, l_{(q-1)\times n+2}, \cdots, l_{q\times n}\} \subseteq \{1, 2, \cdots, M\} \cup \{\varnothing\},\\
&\text{for } q \in \{1, 2, \cdots, N\}, M = ((2m+1)^n - (2m+1))/2;\\
&b_{i,1}^{(n)} = -\tfrac{1}{2}(B_i^{(n)} + \mathbf{i}\sqrt{|\Delta_i^{(n)}|}),\ b_{i,2}^{(n)} = -\tfrac{1}{2}(B_i^{(n)} - \mathbf{i}\sqrt{|\Delta_i^{(n)}|}),\\
&\Delta_i^{(n)} = (B_i^{(n)})^2 - 4C_i^{(n)} < 0, \mathbf{i} = \sqrt{-1}\\
&i \in \{l_{N\times n+1}, l_{N\times n+2}, \cdots, l_M\} \subset \{1, 2, \cdots, M\} \cup \{\varnothing\};
\end{aligned} \tag{5.279}$$

with fixed-points

$$\begin{aligned}
&x_{k+n}^* = x_k^* = a_i^{(n)}, (i = 1, 2, \ldots, (2m+1)^n)\\
&\cup_{i=1}^{(2m+1)^n} \{a_i^{(n)}\} = \text{sort}\{\cup_{i_1=1}^{2m+1} \{a_{i_1}^{(1)}\}, \cup_{i_2=1}^{M} \{b_{i_2,1}^{(n)}, b_{i_2,2}^{(n)}\}\}\\
&\text{with } a_i^{(n)} < a_{i+1}^{(n)}.
\end{aligned} \tag{5.280}$$

(ii) *For a fixed-point of* $x_{k+n}^* = x_k^* = a_{i_1}^{(n)}$ $(i_1 \in I_q^{(n)}, q \in \{1, 2, \ldots, N\})$, if

$$\frac{dx_{k+n}}{dx_k}\Big|_{x_k^* = a_{i_1}^{(n)}} = 1 + a_0^{(n)} \prod_{i_2=1, i_2 \neq i_1}^{(2m+1)^n} (a_{i_1}^{(n)} - a_{i_2}^{(n)}) = 1, \tag{5.281}$$

with

$$\frac{d^2 x_{k+n}}{dx_k^2}\Big|_{x_k^*=a_{i_1}^{(n)}} = a_0^{(n)} \sum_{i_2=1,i_2\neq i_1}^{(2m+1)^n} \prod_{i_3=1,i_3\neq i_1,i_2}^{(2m+1)^n} (a_{i_1}^{(n)} - a_{i_3}^{(n)}) \neq 0, \tag{5.282}$$

then there is a new discrete system for onset of the qth-set of period-n fixed-points based on the second-order monotonic saddle-node bifurcation as

$$x_{k+n} = x_k + a_0^{(n)} \prod_{i_1\in I_q^{(n)}} (x_k - a_{i_1}^{(n)})^2 \prod_{j_2=1}^{(2m+1)^n} (x_k - a_{j_2}^{(n)})^{(1-\delta(i_1,j_2))} \tag{5.283}$$

where

$$\delta(i_1,j_2) = 1 \text{ if } a_{j_2}^{(n)} = a_{i_1}^{(n)},\ \delta(i_1,j_2) = 0 \text{ if } a_{j_2}^{(n)} \neq a_{i_1}^{(n)}. \tag{5.284}$$

(ii_1) *If*

$$\begin{aligned} &\frac{dx_{k+n}}{dx_k}\Big|_{x_k^*=a_{i_1}^{(n)}} = 1\ (i_1 \in I_q^{(n)}),\\ &\frac{d^2 x_{k+n}}{dx_k^2}\Big|_{x_k^*=a_{i_1}^{(n)}} = 2a_0^{(n)} \prod_{i_2\in I_q^{(n)},i_2\neq i_1} (a_{i_1}^{(n)} - a_{i_2}^{(n)})^2\\ &\qquad\times \prod_{j_2=1}^{(2m+1)^n} (a_{i_1}^{(n)} - a_{j_2}^{(n)})^{(1-\delta(i_2,j_2))} \neq 0 \end{aligned} \tag{5.285}$$

x_{k+n} at $x_k^* = a_{i_1}^{(n)}$ *is*

- *a monotonic lower-saddle of the second-order for* $d^2x_{k+n}/dx_k^2|_{x_k^*=a_{i_1}^{(n)}} < 0$;
- a monotonic upper-saddle of the second-order for $d^2x_{k+n}/dx_k^2|_{x_k^*=a_{i_1}^{(n)}} > 0$.

(ii_2) *The period-n fixed-points* $(n = 2^{n_1} \times s)$ *are trivial*

$$\left.\begin{aligned} &x_k^* = x_{k+n}^* = a_{j_1}^{(n)} \in \left\{\cup_{i_i=1}^{2m+1}\{a_{i_1}^{(1)}\},\ \cup_{i_2=1}^{(2m+1)^{2^{n_1-1}s}}\{a_{i_2}^{(2^{n_1-1}s)}\}\right\}\\ &\text{for } n_1 = 1,2,\ldots; s = 2l_1+1; j_1 \in \{1,2,\ldots,(2m+1)^n\}\cup\{\varnothing\} \end{aligned}\right\}$$
$$\text{for } n \neq 2^{n_2},$$
$$\left.\begin{aligned} &x_k^* = x_{k+n}^* = a_{j_1}^{(n)} \in \cup_{i_2=1}^{(2m+1)^{2^{n_1-1}s}}\{a_{i_2}^{(2^{n_1-1}s)}\}\\ &\text{for } n_1 = 1,2,\ldots; s = 1; j_1 \in \{1,2,\ldots,(2m+1)^n\}\cup\{\varnothing\} \end{aligned}\right\}$$
$$\text{for } n = 2^{n_2}. \tag{5.286}$$

(ii_3) *The period-n fixed-points* ($n = 2^{n_1} \times s$) *are non-trivial if*

$$\left.\begin{array}{l} x_k^* = x_{k+n}^* = a_{j_1}^{(n)} \notin \{\cup_{i_i=1}^{2m+1}\{a_{i_1}^{(1)}\}, \cup_{i_2=1}^{(2m+1)^{2^{n_1-1}s}}\{a_{i_2}^{(2^{n_1-1}s)}\}\} \\ \text{for } n_1 = 1,2,\ldots; s = 2l_1+1; j_1 \in \{1,2,\ldots,(2m)^n\} \cup \{\varnothing\} \end{array}\right\} \\ \text{for } n \neq 2^{n_2}, \\ \left.\begin{array}{l} x_k^* = x_{k+n}^* = a_{j_1}^{(n)} \notin \cup_{i_2=1}^{(2m+1)^{2^{n_1-1}s}}\{a_{i_2}^{(2^{n_1-1}s)}\} \\ \text{for } n_1 = 1,2,\ldots; s = 1; j_1 \in \{1,2,\ldots,(2m+1)^n\} \cup \{\varnothing\} \end{array}\right\} \\ \text{for } n = 2^{n_2}. \tag{5.287}$$

Such a forward period- n fixed-point is

- *monotonically unstable if* $dx_{k+n}/dx_k|_{x_k^*=a_{i_1}^{(n)}} \in (1, \infty)$;
- *monotonically invariant if* $dx_{k+n}/dx_k|_{x_k^*=a_{i_1}^{(n)}} = 1$, *which is*
  - *a monotonic upper-saddle of the* $(2l_1)^{\text{th}}$ *order for* $d^{2l_1}x_{k+n}/dx_k^{2l_1}|_{x_k^*} > 0$;
  - *a monotonic lower-saddle the* $(2l_1)^{\text{th}}$ *order for* $d^{2l_1}x_{k+n}/dx_k^{2l_1}|_{x_k^*} < 0$;
  - *a monotonic source of the* $(2l_1+1)^{\text{th}}$ *order for* $d^{2l_1+1}x_{k+n}/dx_k^{2l_1+1}|_{x_k^*} > 0$;
  - *a monotonic sink the* $(2l_1+1)^{\text{th}}$ *order for* $d^{2l_1+1}x_{k+n}/dx_k^{2l_1+1}|_{x_k^*} < 0$;
- *monotonically unstable if* $dx_{k+n}/dx_k|_{x_k^*=a_{i_1}^{(n)}} \in (0, 1)$;
- *invariantly zero-stable if* $dx_{k+n}/dx_k|_{x_k^*=a_{i_1}^{(n)}} = 0$;
- *oscillatorilly stable if* $dx_{k+n}/dx_k|_{x_k^*=a_{i_1}^{(n)}} \in (-1, 0)$;
- *flipped if* $dx_{k+n}/dx_k|_{x_k^*=a_{i_1}^{(n)}} = -1$, *which is*
  - *an oscillatory upper-saddle of the* $(2l_1)^{\text{th}}$ *order for* $d^{2l_1}x_{k+n}/dx_k^{2l_1}|_{x_k^*} > 0$;
  - *an oscillatory lower-saddle the* $(2l_1)^{\text{th}}$ *order for* $d^{2l_1}x_{k+n}/dx_k^{2l_1}|_{x_k^*} < 0$;
  - *an oscillatory source of the* $(2l_1+1)^{\text{th}}$ *order for* $d^{2l_1+1}x_{k+n}/dx_k^{2l_1+1}|_{x_k^*} < 0$;
  - *an oscillatory sink the* $(2l_1+1)^{\text{th}}$ *order for* $d^{2l_1+1}x_{k+n}/dx_k^{2l_1+1}|_{x_k^*} > 0$;
- *oscillatorilly unstable if* $dx_{k+n}/dx_k|_{x_k^*=a_{i_1}^{(n)}} \in (-\infty, -1)$.

For a fixed-point of $x_{k+n}^* = x_k^* = a_{i_1}^{(n)}$ ($i_1 \in I_q^{(n)}$, $q \in \{1,2,\ldots,N\}$), *there is a period-doubling of the* q^{th}-set of period-*n fixed-points if*

$$\begin{array}{l} \dfrac{dx_{k+n}}{dx_k}|_{x_k^*=a_{i_1}^{(n)}} = 1 + a_0^{(n)} \prod_{j_2=1, j_2 \neq i_1}^{(2m)^n} (a_{i_1}^{(n)} - a_{j_2}^{(n)}) = -1, \\ \dfrac{d^s x_{k+n}}{dx_k^s}|_{x_k^*=a_{i_1}^{(n)}} = 0, \text{ for } s = 2, \cdots, r-1; \\ \dfrac{d^r x_{k+n}}{dx_k^r}|_{x_k^*=a_{i_1}^{(n)}} \neq 0 \text{ for } 1 < r \leq (2m)^n \end{array} \tag{5.288}$$

with

- *a* r^{th}*-order oscillatory sink for* $d^r x_{k+n}/dx_k^r|_{x_k^*=a_{i_1}^{(n)}} > 0$ *and* $r = 2l_1 + 1$;
- *a* r^{th}*-order oscillatory source for* $d^r x_{k+n}/dx_k^r|_{x_k^*=a_{i_1}^{(n)}} < 0$ *and* $r = 2l_1 + 1$;
- *a* r^{th}*-order oscillatory upper-saddle for* $d^r x_{k+n}/dx_k^r|_{x_k^*=a_{i_1}^{(n)}} > 0$ *and* $r = 2l_1$;
- *a* r^{th}*-order oscillatory lower-saddle for* $d^r x_{k+n}/dx_k^r|_{x_k^*=a_{i_1}^{(n)}} < 0$ *and* $r = 2l_1$.

The corresponding period- $2 \times n$ *discrete system of the* $(2m+1)^{th}$*-degree polynomial discrete system in Eq. (5.276) is*

$$x_{k+2\times n} = x_k + a_0^{(2\times n)} \prod_{i_1 \in I_q^{(n)}} (x_k - a_{i_1}^{(n)})^3 \prod_{j_2=1}^{(2m+1)^{2\times n}} (x_k - a_{j_2}^{(2\times n)})^{(1-\delta(i_1,j_2))} \tag{5.289}$$

with

$$\begin{aligned} &\frac{dx_{k+2\times n}}{dx_k}\Big|_{x_k^*=a_{i_1}^{(n)}} = 1, \; \frac{d^2 x_{k+2\times n}}{dx_k^2}\Big|_{x_k^*=a_{i_1}^{(n)}} = 0; \\ &\frac{d^3 x_{k+2\times n}}{dx_k^3}\Big|_{x_k^*=a_{i_1}^{(n)}} = 6a_0^{(2\times n)} \prod_{i_1 \in I_q^{(n)}, i_2 \neq i_1} (a_{i_1}^{(n)} - a_{i_2}^{(n)})^3 \\ &\qquad \times \prod_{j_2=1}^{(2m+1)^{2\times n}} (a_{i_1}^{(n)} - a_{j_2}^{(2\times n)})^{(1-\delta(i_1,j_2))}. \end{aligned} \tag{5.290}$$

Thus, $x_{k+2\times n}$ *at* $x_k^* = a_{i_1}^{(n)}$ *for* $i_1 \in I_q^{(n)}$, $q \in \{1, 2, \ldots, N\}$ *is*

- *a monotonic sink of the third-order if* $d^3 x_{k+2\times n}/dx_k^3|_{x_k^*=a_{i_1}^{(n)}} < 0$,
- *a monotonic source of the third-order if* $d^3 x_{k+2\times n}/dx_k^3|_{x_k^*=a_{i_1}^{(n)}} > 0$.

(iv) *After l-times period-doubling bifurcations of period-n fixed points, a period-*$2^l \times n$ *discrete system of the* $(2m+1)^{th}$ *degree polynomial discrete system in Eq. (5.276) is*

$$\begin{aligned} x_{k+2^l\times n} &= x_k + [a_0^{(2^{l-1}\times n)} \prod_{i_1=1}^{(2m+1)^{2^{l-1}\times n}} (x_k - a_{i_1}^{(2^{l-1}\times n)})] \\ &\quad \times \{1 + \prod_{i_1=1}^{(2m+1)^{2^{l-1}\times n}} [1 + a_0^{(2^{l-1})} \prod_{i_2=1, i_2\neq i_1}^{(2m+1)^{2^{l-1}\times n}} (x_k - a_{i_2}^{(2^{l-1}\times n)})]\} \\ &= x_k + [a_0^{(2^{l-1}\times n)} \prod_{i_1=1}^{(2m+1)^{2^{l-1}\times n}} (x_k - a_{i_1}^{(2^{l-1}\times n)})] \\ &\quad \times [(a_0^{(2^{l-1}\times n)})^{(2m+1)^{(2^{l-1}\times n)}} \prod_{j_1=1}^{((2m+1)^{2^l\times n} - (2m+1)^{2^{l-1}\times n})/2} (x_k^2 + B_{j_2}^{(2^l\times n)} x_k + C_{j_2}^{(2^l\times n)})] \end{aligned}$$

$$
\begin{aligned}
&= x_k + [a_0^{(2^{l-1}\times n)} \prod_{i_1=1}^{(2m+1)^{2^{l-1}\times n}} (x_k - a_{i_1}^{(2^{l-1}\times n)})] \\
&\quad \times [(a_0^{(2^{l-1}\times n)})^{(2m+1)^{2^{l-1}\times n}} \prod_{j_2=1}^{((2m+1)^{2^l\times n}-(2m+1)^{2^{l-1}\times n})/2} (x_k - b_{j_2,1}^{(2^l\times n)})(x_k - b_{j_2,2}^{(2^l\times n)})] \\
&= x_k + (a_0^{(2^{l-1}\times n)})^{(2m+1)^{2^{l-1}\times n}} \prod_{i=1}^{(2m+1)^{2^l\times n}} (x_k - a_i^{(2^l\times n)}) \\
&= x_k + a_0^{(2^l\times n)} \prod_{i=1}^{(2m+1)^{2^l\times n}} (x_k - a_i^{(2^l\times n)})
\end{aligned}
\tag{5.291}
$$

with

$$
\begin{aligned}
&\frac{dx_{k+2^l\times n}}{dx_k} = 1 + a_0^{(2^l\times n)} \sum_{i_1=1}^{(2m+1)^{2^l\times n}} \prod_{i_2=1,i_2\neq i_1}^{(2m+1)^{2^l\times n}} (x_k - a_{i_2}^{(2^l\times n)}), \\
&\frac{d^2x_{k+2^l\times n}}{dx_k^2} = a_0^{(2^l\times n)} \sum_{i_1=1}^{(2m+1)^{2^l\times n}} \sum_{i_2=1,i_2\neq i_1}^{(2m+1)^{2^l\times n}} \prod_{i_3=1,i_3\neq i_1,i_2}^{(2m+1)^{2^l\times n}} (x_k - a_{i_3}^{(2^l\times n)}), \\
&\vdots \\
&\frac{d^rx_{k+2^l\times n}}{dx_k^r} = a_0^{(2^l\times n)} \sum_{i_1=1}^{(2m+1)^{2^l\times n}} \cdots \sum_{i_r=1,i_r\neq i_1,i_2\cdots,i_{r-1}}^{(2m+1)^{2^l\times n}} \prod_{i_{r+1}=1,i_{r+1}\neq i_1,i_2\cdots,i_r}^{(2m+1)^{2^l\times n}} (x_k - a_{i_{r+1}}^{(2^l\times n)}) \\
&\text{for } r \le (2m+1)^{2^l\times n};
\end{aligned}
\tag{5.292}
$$

where

$$
\begin{aligned}
&a_0^{(2\times n)} = (a_0^{(n)})^{1+(2m+1)^{2\times n}}, a_0^{(2^l\times n)} = (a_0^{(2^{l-1}\times n)})^{1+(2m+1)^{2^{l-1}\times n}}, l = 1,2,3,\ldots; \\
&\cup_{i=1}^{(2m+1)^{2^l\times n}} \{a_i^{(2^l\times n)}\} = \text{sort}\{\cup_{i_1=1}^{(2m+1)^{2^{l-1}\times n}} \{a_{i_1}^{(2^{l-1}\times n)}\}, \cup_{i_2=1}^{M_2} \{b_{i_2,1}^{(2^l\times n)}, b_{i_2,2}^{(2^l\times n)}\}\}; \\
&b_{i,1}^{(2^l\times n)} = -\frac{1}{2}(B_i^{(2^l\times n)} + \sqrt{\Delta_i^{(2^l\times n)}}), \\
&b_{i,2}^{(2^l\times n)} = -\frac{1}{2}(B_i^{(2^l\times n)} - \sqrt{\Delta_i^{(2^l\times n)}}), \\
&\Delta_i^{(2^l\times n)} = (B_i^{(2^l\times n)})^2 - 4C_i^{(2^l\times n)} \ge 0, \text{ for } i \in \cup_{q_2=1}^{N_2} I_{q_2}^{(2^l\times n)} \\
&I_{q_1}^{(2^{l-1}\times n)} = \{l_{(q_1-1)\times(2^{l-1}\times n)+1}, l_{(q_1-1)\times(2^{l-1}\times n)+2}, \cdots, l_{q_1\times(2^{l-1}\times n)}\} \\
&\qquad \subseteq \{1,2,\cdots,M_1\} \cup \{\varnothing\},
\end{aligned}
$$

$$
\begin{aligned}
&\text{for } q_1 \in \{1,2,\cdots,N_1\}, M_1 = N_1 \times (2^{l-1} \times n);\\
&I_{q_2}^{(2^l \times n)} = \{l_{(q_2-1)\times(2^l\times n)+1}, l_{(q_2-1)\times(2^l\times n)+2}, \cdots, l_{q_2\times(2^{l-1}\times n)}\}\\
&\qquad \subseteq \{M_1+1, M_1+2, \cdots, M_2\} \cup \{\varnothing\},\\
&\text{for } q_2 \in \{1,2,\cdots,N_2\}, M_2 = ((2m+1)^{2^l \times n} - (2m+1)^{2^{l-1}\times n})/2;\\
&b_{i,1}^{(2^l\times n)} = -\frac{1}{2}(B_i^{(2^l\times n)} + \mathbf{i}\sqrt{|\Delta_i^{(2^l\times n)}|}),\\
&b_{i,2}^{(2^l\times n)} = -\frac{1}{2}(B_i^{(2^l\times n)} - \mathbf{i}\sqrt{|\Delta_i^{(2^l\times n)}|}),\\
&\Delta_i^{(2^l\times n)} = (B_i^{(2^l\times n)})^2 - 4C_i^{(2^l\times n)} < 0, \ \mathbf{i} = \sqrt{-1},\\
&i \in \{l_{N_2\times(2^l\times n)+1}, l_{N_2\times(2^l\times n)+2}, \cdots, l_{M_2}\}\\
&\quad \subset \{1,2,\cdots,M_2\} \cup \{\varnothing\}
\end{aligned} \tag{5.293}
$$

with fixed-points

$$
\begin{aligned}
&x_{k+2^l\times n}^* = x_k^* = a_i^{(2^l\times n)}, t(i = 1,2,\ldots,(2m+1)^{2^l\times n})\\
&\cup_{i=1}^{(2m+1)^{2^l}\times n}\{a_i^{(2^l\times n)}\} = \text{sort}\{\cup_{i_1=1}^{(2m+1)^{2^{l-1}}\times n}\{a_{i_1}^{(2^{l-1}\times n)}\}, \cup_{i_2=1}^{M_2}\{b_{i_2,1}^{(2^l\times n)}, b_{i_2,2}^{(2^l\times n)}\}\}\\
&\textit{with } a_i^{(2^i\times n)} < a_{i+1}^{(2^l\times n)}
\end{aligned} \tag{5.294}
$$

(v) *For a fixed-point of* $x^*_{k+(2^l\times n)} = x_k^* = a_{i_1}^{(2^{l-1}\times n)}$ $(i_1 \in I_q^{(2^{l-1}\times n)}, q \in \{1,2,\ldots,N_1\})$, *there is a period-* $2^{l-1}\times n$ *discrete system if*

$$
\begin{aligned}
&\frac{dx_{k+2^{l-1}\times n}}{dx_k}\Big|_{x_k^*=a_{i_1}^{(2^{l-1}\times n)}} = 1 + a_0^{(2^{l-1}\times n)} \prod_{i_2=1, i_2\neq i_1}^{(2m+1)^{2^{l-1}\times n}} (a_{i_1}^{(2^{l-1}\times n)} - a_{i_2}^{(2^{l-1}\times n)}) = -1,\\
&\frac{d^s x_{k+2^{l-1}\times n}}{dx_k^s}\Big|_{x_k^*=a_{i_1}^{(2^{l-1}\times n)}} = 0, \ \text{ for } s = 2,\cdots,r-1;\\
&\frac{d^r x_{k+2^{l-1}\times n}}{dx_k^r}\Big|_{x_k^*=a_{i_1}^{(2^{l-1}\times n)}} \neq 0 \ \text{ for } 1 < r \le (2m+1)^{2^{l-1}\times n}
\end{aligned} \tag{5.295}
$$

with

- *a* r^{th}*-order oscillatory sink for* $d^r x_{k+2^l\times n}/dx_k^r|_{x_k^*=a_{i_1}^{(2^l\times n)}} > 0$ *and* $r = 2l_1+1$;
- *a* r^{th}*-order oscillatory source for* $d^r x_{k+2^l\times n}/dx_k^r|_{x_k^*=a_{i_1}^{(2^l\times n)}} < 0$ *and* $r = 2l_1+1$;
- a rth*-order oscillatory upper-saddle for* $d^r x_{k+2^l\times n}/dx_k^r|_{x_k^*=a_{i_1}^{(2^l\times n)}} > 0$ *and* $r = 2l_1$;

- *a* r^{th}*-order oscillatory lower-saddle for* $d^r x_{k+2^l\times n}/dx_k^r|_{x_k^*=a_{i_1}^{(2^l\times n)}}<0$ *and* $r=2l_1$.

The corresponding period- $2^l \times n$ *discrete system is*

$$
\begin{aligned}
x_{k+2^l\times n} = x_k + a_0^{(2^l\times n)} & \prod\nolimits_{i_1\in I_q^{(2^{l-1}\times n)}} (x_k - a_{i_1}^{(2^{l-1}\times n)})^3 \\
& \times \prod\nolimits_{j_2=1}^{(2m+1)^{2^l\times n}} (x_k - a_{j_2}^{(2^l\times n)})^{(1-\delta(i_1,j_2))}
\end{aligned} \tag{5.296}
$$

where

$$
\delta(i_1,j_2)=1 \text{ if } a_{j_2}^{(2^l\times n)} = a_{i_1}^{(2^{l-1}\times n)},\ \delta(i_1,j_2)=0 \text{ if } a_{j_2}^{(2^l\times n)} \neq a_{i_1}^{(2^{l-1}\times n)} \tag{5.297}
$$

with

$$
\begin{aligned}
& \frac{dx_{k+2^l\times n}}{dx_k}|_{x_k^*=a_{i_1}^{(2^{l-1}\times n)}} = 1,\ \frac{d^2x_{k+2^l\times n}}{dx_k^2}|_{x_k^*=a_{i_1}^{(2^{l-1}\times n)}} = 0; \\
& \frac{d^3x_{k+2^l\times n}}{dx_k^3}|_{x_k^*=a_{i_1}^{(2^{l-1})}} = 6a_0^{(2^l\times n)} \prod\nolimits_{i_2\in I_q^{(2^{l-1}\times n)},i_2\neq i_1} (a_{i_1}^{(2^{l-1}\times n)} - a_{i_2}^{(2^{l-1}\times n)})^3 \\
& \qquad \times \prod\nolimits_{j_2=1}^{(2m+1)^{(2^l\times n)}} (a_{i_1}^{(2^{l-1}\times n)} - a_{j_2}^{(2^l\times n)})^{(1-\delta(i_2,j_2))} \neq 0 \\
& (i_1 \in I_q^{(2^{l-1}\times n)}, q\in\{1,2,\ldots,N_1\})
\end{aligned} \tag{5.298}
$$

Thus, $x_{k+2^l\times n}$ *at* $x_k^* = a_{i_1}^{(2^{l-1}\times n)}$ is

- *a monotonic sink of the third-order if* $d^3x_{k+2^l\times n}/dx_k^3|_{x_k^*=a_{i_1}^{(2^{l-1})}}<0$;
- *a monotonic source of the third-order if* $d^3x_{k+2^l\times n}/dx_k^3|_{x_k^*=a_{i_1}^{(2^{l-1})}}>0$.

(v_1) *The period-* $2^l \times n$ *fixed-points are trivial if*

$$
\left.
\begin{aligned}
& x_{k+2^l\times n}^* = x_k^* = a_j^{(2^l\times n)} \in \{\cup_{i_l=1}^{2m+1}\{a_{i_1}^{(1)}\}, \cup_{i_2=1}^{(2m+1)^{2^{l-1}\times n}}\{a_{i_2}^{(2^{l-1}\times n)}\}\} \\
& \text{for } j=1,2,\cdots,(2m+1)^{(2^l\times n)} \\
& \text{for } n\neq 2^{n_1}
\end{aligned}
\right\}
$$

$$
\left.
\begin{aligned}
& x_{k+2^l\times n}^* = x_k^* = a_j^{(2^l\times n)} \in \{\cup_{i_2=1}^{(2m+1)^{2^{l-1}\times n}}\{a_{i_2}^{(2^{l-1}\times n)}\}\} \\
& \text{for } j=1,2,\cdots,(2m+1)^{2^l\times n} \\
& \text{for } n = 2^{n_1}.
\end{aligned}
\right\} \tag{5.299}
$$

(v_2) *The period-* $2^l \times n$ *fixed-points are non-trivial if*

$$\left.\begin{array}{l} x^*_{k+2^l\times n} = x^*_k = a_j^{(2^l\times n)} \notin \{\cup_{i_1=1}^{2m+1}\{a_{i_1}^{(1)}\}, \cup_{i_2=1}^{(2m+1)^{2^{l-1}\times n}}\{a_{i_2}^{(2^{l-1}\times n)}\}\} \\ \text{for } j=1,2,\cdots,(2m+1)^{2^l\times n} \end{array}\right\}$$
$$\text{for } n \neq 2^{n_1}$$
$$\left.\begin{array}{l} x^*_{k+2^l\times n} = x^*_k = a_j^{(2^l\times n)} \notin \{\cup_{i_2=1}^{(2m+1)^{2^{l-1}\times n}}\{a_{i_2}^{(2^{l-1}\times n)}\}\} \\ \text{for } j=1,2,\cdots,(2m+1)^{2^l\times n} \end{array}\right\}$$
$$\text{for } n = 2^{n_1}. \tag{5.300}$$

Such a period- $2^l \times n$ *fixed-point is*

- *monotonically unstable if* $dx_{k+2^l\times n}/dx_k|_{x_k^*=a_{i_1}^{(2^l\times n)}} \in (1,\infty)$;
- *monotonically invariant if* $dx_{k+2^l\times n}/dx_k|_{x_k^*=a_{i_1}^{(2^l\times n)}} = 1$, *which is*
  - *a monotonic upper-saddle of the* $(2l_1)^{\text{th}}$ *order for* $d^{2l_1}x_{k+2^l\times n}/dx_k^{2l_1}|_{x_k^*} > 0$ *(independent* $(2l_1)$*-branch appearance)*;
  - *a monotonic lower-saddle the* $(2l_1)^{\text{th}}$ *order for* $d^{2l_1}x_{k+2^l\times n}/dx_k^{2l_1}|_{x_k^*} < 0$ *(independent* $(2l_1)$*-branch appearance)*;
  - *a monotonic source of the* $(2l_1+1)^{\text{th}}$ *order for* $d^{2l_1+1}x_{k+2^l\times n}/dx_k^{2l_1+1}|_{x_k^*} > 0$ *(dependent* $(2l_1+1)$*-branch appearance from one branch)*;
  - *a monotonic sink the* $(2l_1+1)^{\text{th}}$ *order for* $d^{2l_1+1}x_{k+2^l\times n}/dx_k^{2l_1+1}|_{x_k^*} < 0$ *(dependent* $(2l_1+1)$*-branch appearance from one branch)*;
- *monotonically stable if* $dx_{k+2^l\times n}/dx_k|_{x_k^*=a_{i_1}^{(2^l\times n)}} \in (0,1)$;
- *invariantly zero-stable if* $dx_{k+2^l\times n}/dx_k|_{x_k^*=a_{i_1}^{(2^l\times n)}} = 0$;
- *oscillatorilly stable if* $dx_{k+2^l\times n}/dx_k|_{x_k^*=a_{i_1}^{(2^l\times n)}} \in (-1,0)$;
- *flipped if* $dx_{k+2^l\times n}/dx_k|_{x_k^*=a_{i_1}^{(2^l\times n)}} = -1$, *which is*
  - *an oscillatory upper-saddle of the* $(2l_1)^{\text{th}}$ *order for* $d^{2l_1}x_{k+2^l\times n}/dx_k^{2l_1}|_{x_k^*} > 0$;
  - *an oscillatory lower-saddle the* $(2l_1)^{\text{th}}$ *order for* $d^{2l_1}x_{k+2^l\times n}/dx_k^{2l_1}|_{x_k^*} < 0$;
  - *an oscillatory source of the* $(2l_1+1)^{\text{th}}$ *order for* $d^{2l_1+1}x_{k+2^l\times n}/dx_k^{2l_1+1}|_{x_k^*} < 0$;
  - *an oscillatory sink the* $(2l_1+1)^{\text{th}}$ *order for* $d^{2l_1+1}x_{k+2^l\times n}/dx_k^{2l_1+1}|_{x_i^*} > 0$;
- *oscillatorilly unstable if* $dx_{k+2^l\times n}/dx_k|_{x_k^*=a_{i_1}^{(2^l\times n)}} \in (-\infty,-1)$.

Proof Through the nonlinear renormalization, the proof of this theorem is similar to the proof of Theorem 1.11. This theorem can be easily proved. ■

References

Luo ACJ (2020a) The stability and bifurcation of the $(2m + 1)^{\text{th}}$-degree polynomial systems. J Vibr Test Syst Dynam 4(2):93–144

Luo ACJ (2020b) Bifurcation and stability in nonlinear dynamical system. Springer, New York

Index

A
Antenna switching bifurcation, 284, 307, 369, 401
Appearing bifurcation, 9, 11, 276, 294, 346, 352, 354

B
Backward bifurcation tree, 75
Backward cubic nonlinear discrete system, 150
Backward period-1 appearing bifurcation, 28
Backward period-1 switching bifurcation, 39
Backward period-2 quadratic discrete system, 75
Backward period-2 quartic discrete system, 240
Backward period-3 cubic discrete system, 148
Backward period-doubling renormalization, 79, 153, 243
Backward period-*n* appearing, 82, 157, 247
Backward period-*n* bifurcation tree, 92
Backward quadratic discrete system, 28
Backward quartic discrete system, 239
Broom appearing bifurcation, 355, 357, 380, 381
Broom-sprinkle-spraying appearing bifurcation, 358–360, 386, 388, 391

C
Constant adding discrete system, 2, 5
Cubic nonlinear discrete system, 93

D
$(2m)^{\text{th}}$ -degree polynomial discrete system, 258

F
Flower-bundle switching bifurcation, 290–292, 316, 376, 377, 409
Forward bifurcation tree, 44, 318
Forward cubic discrete system, 121
Forward quadratic discrere system, 7
Forward quartic discrete system, 223

I
Instant fixed-point, 2
Invariant sink, 2

L
Linear backward discrete system, 5
Linear discrete system, 1
l_{p}th mXX appearing bifurcation, 261, 268, 274, 339
l_{p}th mXX switching bifurcation, 260, 261, 263, 266, 268, 270, 273–275, 338, 339, 341, 345, 348, 351–354

M
Monotonically stable node, 2, 6
Monotonically unstable node, 3, 6
Monotonic backward saddle discrete flow, 29
Monotonic lower-saddle, 9, 18, 25, 30, 40, 100, 101, 103
Monotonic lower-saddle discrete flow, 119, 169, 170, 192, 212, 217, 219, 222
Monotonic lower-saddle-node appearing bifurcation, 9, 20, 103–105, 169, 171, 173, 194, 196, 206

A. C. J Luo, *Bifurcation Dynamics in Polynomial Discrete Systems*, Nonlinear Physical Science, https://doi.org/10.1007/978-981-15-5208-3

Monotonic lower-saddle-node bundle-switching bifurcation, 205, 214, 219, 221
Monotonic lower-saddle-node flower-bundle-switching bifurcation, 208
Monotonic lower-saddle-node switching bifurcation, 7, 18, 26, 27, 43, 100, 101, 170, 173, 195, 212, 217
Monotonic saddle, 22, 40
Monotonic saddle discrete flow, 9, 95–97
Monotonic saddle-node appearing bifurcation, 9, 18
Monotonic saddle-node switching bifurcation, 22, 25, 43
Monotonic saddle switching, 3, 6
Monotonic sink, 2, 6, 10
Monotonic sink bundle-switching bifurcation of the third-order, 213
Monotonic sink discrete flow, 96, 97, 171, 213, 218
Monotonic sink switching bifurcation of the third-order, 171, 193, 194, 207, 218
Monotonic source, 3, 6, 10
Monotonic source bundle-switching bifurcation of the third-order, 213
Monotonic source discrete flow, 96, 97, 171, 213, 218
Monotonic source switching bifurcation of the third-order, 171, 193, 207, 215
Monotonic upper-saddle, 9, 18, 24, 29, 100–102, 106
Monotonic upper-saddle discrete flow, 119, 169, 170, 173, 192, 212, 217, 219
Monotonic upper-saddle-node appearing bifurcation, 9, 20, 30, 38, 102, 106, 107, 169, 171, 173, 194, 196, 206
Monotonic upper-saddle-node bundle-switching bifurcation, 205, 214, 215
Monotonic upper-saddle-node flower-bundle-switching bifurcation, 194, 208
Monotonic upper-saddle-node switching bifurcation, 18, 26, 27, 101, 170, 173, 194

N
Negative backward discrete flow, 28
Negative discrete flow, 8, 9, 18, 168, 195, 258

Scan this code for color figures

O
Oscillatorilly stable node, 2, 6
Oscillatorilly unstable node, 3, 6
Oscillatory lower-saddle, 10, 19, 22
Oscillatory saddle switching, 3, 6
Oscillatory sink, 2, 6
Oscillatory source, 3, 6
Oscillatory upper-saddle, 10, 19

P
Period-1 appearing bifurcation, 7, 167
Period-1 cubic discrete system, 93
Period-1 quartic discrete system, 224
Period-1 switching bifurcation, 21
Period-2 appearing bifurcation, 44
Period-doubled cubic discrete system, 121
Period-doubling renormalization, 53, 128, 227
Period-doublization, 62, 82, 138, 157, 231, 326
Period-*n* appearing, 62, 138, 231, 326, 419
Period-*n* appearing bifurcation, 147
Period-*n* bifurcation tree, 71
Permanent invariant discrete system, 2, 5
Positive backward discrete flow, 28
Positive discrete flow, 8, 9, 18, 168, 195, 258

Q
Quadratic nonlinear discrete system, 1
Quartic nonlinear discrete system, 167

S
Sink, 2, 6
Source, 2
Spraying appearing bifurcation, 278, 279, 281, 296, 299, 301
Sprinkler-spraying appearing bifurcation, 278, 279, 281, 299, 301
Stable node, 2, 6
Straw-bundle switching bifurcation, 286, 310, 315, 371, 404, 408
Switching-appearing bifurcation, 289, 311, 374, 407
Switching bifurcation, 21, 24, 39, 95, 114, 282, 305, 369, 398

T
Teethcomb appearing bifurcation, 277, 296

U
Unstable node, 2

NONLINEAR PHYSICAL SCIENCE

(Series Editors: Albert C.J. Luo, Dimitri Volchenkov)

	49 Nonlinear Dynamics, Chaos, and Complexity (2021) by Dimitri Volchenkov (Editor)
ISBN 978-7-04-055802-9 9 787040 558029 >	48 Slowly Varying Oscillations and Waves (2021) by Lev Ostrovsky
	47 不连续动力系统 (2021) 罗朝俊著，闵富红、李欣业译
ISBN 978-7-04-054753-5 9 787040 547535 >	46 连续动力系统 (2021) 罗朝俊著，王跃方、黄金、李欣业译
ISBN 978-7-04-055783-1 9 787040 557831 >	45 Bifurcation Dynamics in Polynomial Discrete Systems (2021) by Albert C. J. Luo
ISBN 978-7-04-055228-7 9 787040 552287 >	44 Bifurcation and Stability in Nonlinear Discrete Systems (2020) by Albert C. J. Luo
ISBN 978-7-04-050615-0 9 787040 506150 >	43 Theory of Hybrid Systems: Deterministic and Stochastic (2019) by Mohamad S. Alwan, Xinzhi Liu
ISBN 978-7-04-050235-0 9 787040 502350 >	42 Rigid Body Dynamics: Hamiltonian Methods, Integrability, Chaos (2018) by A. V. Borisov, I. S. Mamaev
ISBN 978-7-04-048458-8 9 787040 484588 >	41 Galloping Instability to Chaos of Cables (2018) by Albert C. J. Luo, Bo Yu
ISBN 978-7-04-048004-7 9 787040 480047 >	40 Resonance and Bifurcation to Chaos in Pendulum (2017) by Albert C. J. Luo
ISBN 978-7-04-047940-9 9 787040 479409 >	39 Grammar of Complexity: From Mathematics to a Sustainable World (2017) by Dimitri Volchenkov
ISBN 978-7-04-047809-9 9 787040 478099 >	38 Type-2 Fuzzy Logic: Uncertain Systems’ Modeling and Control (2017) by Rómulo Martins Antão, Alexandre Mota, R. Escadas Martins, J. Tenreiro Machado
ISBN 978-7-04-047450-3 9 787040 474503 >	37 Bifurcation in Autonomous and Nonautonomous Differential Equations with Discontinuities (2017) by Marat Akhmet, Ardak Kashkynbayev
ISBN 978-7-04-043231-2 9 787040 432312 >	36 离散和切换动力系统（中文版）(2015) 罗朝俊

ISBN 978-7-04-043102-5 9 787040 431025	35 Replication of Chaos in Neural Networks, Economics and Physics (2015) by Marat Akhmet, Mehmet Onur Fen
ISBN 978-7-04-042835-3 9 787040 428353	34 Discretization and Implicit Mapping Dynamics (2015) by Albert C.J.Luo
ISBN 978-7-04-042385-3 9 787040 423853	33 Tensors and Riemannian Geometry with Applications to Differential Equations (2015) by Nail Ibragimov
ISBN 978-7-04-042131-6 9 787040 421316	32 Introduction to Nonlinear Oscillations (2015) by Vladimir I. Nekorkin
ISBN 978-7-04-038891-6 9 787040 388916	31 Keller-Box Method and Its Application (2014) by K. Vajravelu, K.V. Prasad
ISBN 978-7-04-039179-4 9 787040 391794	30 Chaotic Signal Processing (2014) by Henry Leung
ISBN 978-7-04-037357-8 9 787040 373578	29 Advances in Analysis and Control of Time-Delayed Dynamical Systems (2013) by Jianqiao Sun, Qian Ding (Editors)
ISBN 978-7-04-036944-1 9 787040 369441	28 Lectures on the Theory of Group Properties of Differential Equations (2013) by L.V. Ovsyannikov (Author), Nail Ibragimov (Editor)
ISBN 978-7-04-036741-6 9 787040 367416	27 Transformation Groups and Lie Algebras (2013) by Nail Ibragimov
ISBN 978-7-04-030734-4 9 787040 307344	26 Fractional Derivatives for Physicists and Engineers Volume II. Applications (2013) by Vladimir V. Uchaikin
ISBN 978-7-04-032235-4 9 787040 322354	25 Fractional Derivatives for Physicists and Engineers Volume I. Background and Theory (2013) by Vladimir V. Uchaikin
ISBN 978-7-04-035449-2 9 787040 354492	24 Nonlinear Flow Phenomena and Homotopy Analysis: Fluid Flow and Heat Transfer (2012) by Kuppalapalle Vajravelu, Robert A.Van Gorder
ISBN 978-7-04-034819-4 9 787040 348194	23 Continuous Dynamical Systems (2012) by Albert C.J. Luo
ISBN 978-7-04-034821-7 9 787040 348217	22 Discrete and Switching Dynamical Systems (2012) by Albert C.J. Luo
ISBN 978-7-04-032279-8 9 787040 322798	21 Pseudo chaotic Kicked Oscillators: Renormalization, Symbolic Dynamics, and Transport (2012) by J.H. Lowenstein

ISBN 978-7-04-032298-9 9 787040 322989 >	20 Homotopy Analysis Method in Nonlinear Differential Equations (2011) by Shijun Liao
ISBN 978-7-04-031964-4 9 787040 319644 >	19 Hyperbolic Chaos: A Physicist's View (2011) by Sergey P. Kuznetsov
ISBN 978-7-04-032186-9 9 787040 321869 >	18 非线性变形体动力学（中文版）(2011) 罗朝俊著，郭羽、黄健哲、闵富红译
ISBN 978-7-04-031954-5 9 787040 319545 >	17 Applications of Lie Group Analysis in Geophysical Fluid Dynamics (2011) by Ranis Ibragimov, Nail Ibragimov
ISBN 978-7-04-031957-6 9 787040 319576 >	16 Discontinuous Dynamical Systems (2011) by Albert C.J. Luo
ISBN 978-7-04-031694-0 9 787040 316940 >	15 Linear and Nonlinear Integral Equations: Methods and Applications (2011) by Abdul-Majid Wazwaz
ISBN 978-7-04-029710-2 9 787040 297102 >	14 Complex Systems: Fractionality, Time-delay and Synchronization (2011) by Albert C.J. Luo , Jianqiao Sun (Editors)
ISBN 978-7-04-031534-9 9 787040 315349 >	13 Fractional-order Nonlinear Systems: Modeling, Analysis and Simulation (2011) by Ivo Petráš
ISBN 978-7-04-031533-2 9 787040 315332 >	12 Bifurcation and Chaos in Discontinuous and Continuous Systems (2011) by Michal Fečkan
ISBN 978-7-04-031695-7 9 787040 316957 >	11 Waves and Structures in Nonlinear Nondispersive Media: General Theory and Applications to Nonlinear Acoustics (2011) by S.N. Gurbatov, O.V. Rudenko, A.I. Saichev
ISBN 978-7-04-032187-6 9 787040 321876 >	10动态域上的不连续动力系统（中文版）(2011) 罗朝俊著，闵富红，黄健哲，郭羽译
ISBN 978-7-04-029474-3 9 787040 294743 >	9 Self-organization and Pattern-formation in Neuronal Systems Under Conditions of Variable Gravity (2011) by Meike Wiedemann, Florian P.M. Kohn, Harald Rosner, Wolfgang R.L. Hanke
ISBN 978-7-04-029473-6 9 787040 294736 >	8 Fractional Dynamics (2010) by Vasily E. Tarasov
ISBN 978-7-04-029187-2 9 787040 291872 >	7 Hamiltonian Chaos beyond the KAM Theory (2010) by Albert C.J. Luo, Valentin Afraimovich (Editors)
ISBN 978-7-04-029188-9 9 787040 291889 >	6 Long-range Interactions, Stochasticity and Fractional Dynamics (2010) by Albert C.J. Luo, Valentin Afraimovich (Editors)

ISBN 978-7-04-028882-7 9 787040 288827 >	5 Nonlinear Deformable-body Dynamics (2010) by Albert C.J. Luo
ISBN 978-7-04-018292-7 9 787040 182927 >	4 Mathematical Theory of Dispersion-Managed Optical Solitons (2010) by A. Biswas, D. Milovic, E. Matthew
ISBN 978-7-04-025480-8 9 787040 254808 >	3 Partial Differential Equations and Solitary Waves Theory (2009) by Abdul-Majid Wazwaz
ISBN 978-7-04-025759-5 9 787040 257595 >	2 Discontinuous Dynamical Systems on Time-varying Domains (2009) by Albert C.J. Luo
ISBN 978-7-04-025159-3 9 787040 251593 >	1 Approximate and Renormgroup Symmetries (2009) by Nail H. Ibargimov

郑重声明